Handbook of Enology

Volume 1

Handbook of Enology

VOLUME 1

THE MICROBIOLOGY OF WINE AND VINIFICATIONS

Third Edition

Pascal Ribéreau-Gayon, Denis Dubourdieu,
Bernard B. Donèche and Aline A. Lonvaud
Research Unit in Enology, Institute of Vine
and Wine Sciences, University of Bordeaux,
Villenave d'Ornon, France

Coordinator: Philippe Darriet
Research Unit in Enology, Institute of Vine and
Wine Sciences, University of Bordeaux, Villenave
d'Ornon, France

Translator: John Towey
With assistance from Juliet Towey
Vista Translations LLC, Portland, OR, USA

This edition first published 2021
© 2021 John Wiley & Sons Ltd

Edition History
1e Wiley 2000
2e Wiley 2005

All rights reserved. No part of this publication may be reproduced, stored in a retrieval system, or transmitted, in any form or by any means, electronic, mechanical, photocopying, recording or otherwise, except as permitted by law. Advice on how to obtain permission to reuse material from this title is available at http://www.wiley.com/go/permissions.

The right of Pascal Ribéreau-Gayon, Denis Dubourdieu, Bernard B. Donèche and Aline A. Lonvaud to be identified as the authors of this work has been asserted in accordance with law.

Originally published in France as
Traité d'oenologie
Volume 1 – Microbiologie du Vin, Vinifications
By Pascal Ribéreau-Gayon, Denis Dubourdieu, Bernard B. Donèche, Aline A. Lonvaud
© Dunod Editeur, 2017, Malakoff

Registered Offices
John Wiley & Sons, Inc., 111 River Street, Hoboken, NJ 07030, USA
John Wiley & Sons Ltd, The Atrium, Southern Gate, Chichester, West Sussex, PO19 8SQ, UK

Editorial Office
The Atrium, Southern Gate, Chichester, West Sussex, PO19 8SQ, UK

For details of our global editorial offices, customer services, and more information about Wiley products visit us at www.wiley.com.

Wiley also publishes its books in a variety of electronic formats and by print-on-demand. Some content that appears in standard print versions of this book may not be available in other formats.

Limit of Liability/Disclaimer of Warranty
In view of ongoing research, equipment modifications, changes in governmental regulations, and the constant flow of information relating to the use of experimental reagents, equipment, and devices, the reader is urged to review and evaluate the information provided in the package insert or instructions for each chemical, piece of equipment, reagent, or device for, among other things, any changes in the instructions or indication of usage and for added warnings and precautions. While the publisher and authors have used their best efforts in preparing this work, they make no representations or warranties with respect to the accuracy or completeness of the contents of this work and specifically disclaim all warranties, including without limitation any implied warranties of merchantability or fitness for a particular purpose. No warranty may be created or extended by sales representatives, written sales materials or promotional statements for this work. The fact that an organization, website, or product is referred to in this work as a citation and/or potential source of further information does not mean that the publisher and authors endorse the information or services the organization, website, or product may provide or recommendations it may make. This work is sold with the understanding that the publisher is not engaged in rendering professional services. The advice and strategies contained herein may not be suitable for your situation. You should consult with a specialist where appropriate. Further, readers should be aware that websites listed in this work may have changed or disappeared between when this work was written and when it is read. Neither the publisher nor authors shall be liable for any loss of profit or any other commercial damages, including but not limited to special, incidental, consequential, or other damages.

Library of Congress Cataloging-in-Publication Data
Names: Ribéreau-Gayon, Pascal, author. | Dubourdieu, Denis, author. |
　Donèche, Bernard B., author. | Lonvaud, Aline A., author. | Glories,
　Yves, author. | Maujean, Alain, author. | Towey, John, translator.
Title: Handbook of enology.
Other titles: Traite' d'oenologie. English
Description: Third edition / coordinator, Philippe Darriet ; translator
　John Towey. | Hoboken, NJ : Wiley, 2020-2021. | Includes bibliographical
　references and indexes. | Contents: volume 1. The microbiology of wine
　and vinifications / Pascal Ribéreau-Gayon, Denis Dubourdieu, Bernard B.
　Donèche and Aline A. Lonvaud – volume 2. The chemistry of wine
　stabilization and treatments / Pascal Ribéreau-Gayon, Denis Dubourdieu,
　Yves Glories and Alain Maujean.
Identifiers: LCCN 2020033294 (print) | LCCN 2020033295 (ebook) | ISBN
　9781119584681 (v. 1 ; hardback) | ISBN 9781119587767 (v. 2 ; hardback) |
　ISBN 9781119584629 (v. 1 ; epub) | ISBN 9781119588498 (v. 2 ; epub) |
　ISBN 9781119584698 (v. 1 ; adobe pdf) | ISBN 9781119588443 (v. 2 ; adobe pdf)
Subjects: LCSH: Wine and wine making–Handbooks, manuals, etc. | Wine and
　wine making–Microbiology–Handbooks, manuals, etc. | Wine and wine
　making–Chemistry–Handbooks, manuals, etc.
Classification: LCC TP548 .T7613 2020 (print) | LCC TP548 (ebook) | DDC
　663/.2–dc23
LC record available at https://lccn.loc.gov/2020033294
LC ebook record available at https://lccn.loc.gov/2020033295

Cover Design: Wiley
Cover Image: © Digoarpi/Shutterstock

Set in 10/12pt StixTwo Text by SPi Global, Pondicherry, India

CONTENTS

FOREWORD — XI
PREFACE TO THE SECOND EDITION — XIII
PREFACE TO THE FIRST EDITION — XV
REMARKS CONCERNING THE EXPRESSION OF CERTAIN PARAMETERS OF MUST AND WINE COMPOSITION — XIX

Part I Microbiology of Wine 1

1 Yeasts 3

- 1.1. Introduction 3
- 1.2. The Cell Wall 5
- 1.3. The Plasma Membrane 9
- 1.4. The Cytoplasm and Its Organelles 15
- 1.5. The Nucleus 18
- 1.6. Reproduction and the Yeast Biological Cycle 19
- 1.7. The Killer Phenomenon 23
- 1.8. Classification of Yeast Species 27
- 1.9. Identification of Wine Yeast Strains 49
- 1.10. Ecology of Grape and Wine Yeasts 57
- References 67

2 Yeast Metabolism 73

- 2.1. Introduction 73
- 2.2. Sugar Degradation Pathways 74
- 2.3. Regulation of Sugar-Utilizing Metabolic Pathways 82
- 2.4. Metabolism of Nitrogen Compounds 93
- References 101

3 Conditions of Yeast Development 103

- **3.1.** Introduction 103
- **3.2.** Monitoring and Controlling Fermentations 104
- **3.3.** Yeast Growth Cycle and Fermentation Kinetics 108
- **3.4.** Nutrition Requirements 110
- **3.5.** Fermentation Activators 115
- **3.6.** Inhibition of Fermentation 122
- **3.7.** Physicochemical Factors Affecting Yeast Growth and Fermentation Kinetics 126
- **3.8.** Stuck Fermentations 135
- References 142

4 Lactic Acid Bacteria 145

- **4.1.** The Different Components of the Bacteria Cell 145
- **4.2.** Taxonomy of Lactic Acid Bacteria 154
- **4.3.** Identification of Lactic Acid Bacteria 156
- **4.4.** The *Oenococcus oeni* Species 170
- References 171

5 Metabolism of Lactic Acid Bacteria 175

- **5.1.** Generalities: A Review 175
- **5.2.** Metabolism of Sugars: Lactic Acid Fermentation 177
- **5.3.** Metabolism of the Principal Organic Acids of Wine 182
- **5.4.** Other Transformations Likely to Occur in Winemaking 189
- **5.5.** Effect of the Metabolism of Lactic Acid Bacteria on Wine Composition and Quality 195
- References 197

6 Lactic Acid Bacteria Development in Wine 201

- **6.1.** Lactic Acid Bacteria Nutrition in Wine 201
- **6.2.** Physicochemical Factors of Bacterial Growth 204
- **6.3.** Evolution of Lactic Acid Bacteria Microflora: Influence on Wine Composition 210
- **6.4.** Microbial Interactions During Winemaking 218
- **6.5.** The Importance of Bacteriophages 224
- References 226

7 Acetic Acid Bacteria   229

- 7.1. Principal Characteristics and Cytology   229
- 7.2. Classification and Identification   230
- 7.3. Principal Physiological Characteristics   232
- 7.4. Metabolism of Acetic Acid Bacteria   233
- 7.5. Acetic Acid Bacteria Development in Grape Musts   237
- 7.6. Evolution of Acetic Acid Bacteria During Winemaking and Wine Aging and the Impact on Wine Quality   238

References   240

8 The Use of Sulfur Dioxide in Must and Wine Treatment   243

- 8.1. Introduction   243
- 8.2. Physiological Effects   246
- 8.3. Chemistry of Sulfur Dioxide   248
- 8.4. Molecules Binding Sulfur Dioxide   252
- 8.5. Practical Consequences: The State of Sulfur Dioxide in Wines   261
- 8.6. Antimicrobial Properties of Sulfur Dioxide   262
- 8.7. The Role of Sulfur Dioxide in Winemaking   265
- 8.8. The Use of Sulfur Dioxide in the Winery   270

References   277

9 Products and Methods Complementing the Effect of Sulfur Dioxide   279

- 9.1. Introduction   279
- 9.2. Sorbic Acid   280
- 9.3. Octanoic and Decanoic Acids (Saturated Short-Chain Fatty Acids)   283
- 9.4. Dimethyl Dicarbonate (DMDC)   285
- 9.5. Lysozyme   286
- 9.6. Destruction of Yeasts by Heat (Pasteurization)   289
- 9.7. Ascorbic Acid   293
- 9.8. The Use of Inert Gases   297

References   301

viii Contents

Part II Vinification 303

Reflections on Global Taste and Typicity of Wines 305

10 The Grape and Its Maturation 309

- 10.1. Introduction 309
- 10.2. Description and Composition of the Mature Grape 310
- 10.3. Changes in the Grape During Maturation 322
- 10.4. Definition of Ripeness—Concept of Vintage 343
- 10.5. Impact of Various Other Factors on Maturation and Grape Composition at Ripeness 366
- 10.6. *Botrytis cinerea* 375
- References 391

11 Harvest and Pre-fermentation Treatments 399

- 11.1. Introduction 399
- 11.2. Improving Grape Quality by Overripening 400
- 11.3. Harvest Date and Operations 401
- 11.4. Acidity Adjustments of the Harvested Grapes 408
- 11.5. Increasing Sugar Concentrations 412
- 11.6. Enzymatic Transformations of the Grape After Harvest 418
- 11.7. Use of Commercial Enzymes in Winemaking 427
- References 429

12 Red Winemaking 431

- 12.1. Generalities 431
- 12.2. Mechanical Processing of the Harvested Grapes 433
- 12.3. Tank Filling 440
- 12.4. Controlling Alcoholic Fermentation 446
- 12.5. Maceration 452
- 12.6. Draining Off the Skins and Pressing 469
- 12.7. Malolactic Fermentation 480
- 12.8. Automated Red Winemaking Methods 494
- 12.9. Carbonic Maceration 499
- References 511

13 White Winemaking 513

    13.1. Distinctive Characteristics of White Winemaking 513
    13.2. White Grape Quality and Picking Criteria 517
    13.3. Juice Extraction 526
    13.4. Protecting Juice from Oxidation 538
    13.5. Clarification 543
    13.6. Juice Treatments and the Advisability of Bentonite Treatments 552
    13.7. Fermentation Operations 552
    13.8. Making Dry White Wines in Barrels 557
    13.9. Controlling Reduction Off-Aromas During White Wine Aging 564
    References 568

14 Other Winemaking Methods 571

    14.1. Rosé Wines 571
    14.2. Botrytized Sweet Wines (Sauternes and Tokaji) 577
    14.3. Champagne and Sparkling Wines 588
    14.4. Fortified Wines 602
    14.5. Flor Wines 608
    References 612

INDEX 615

FOREWORD

Over 70 years ago, in 1949, under the direction of Professor Jean Ribéreau-Gayon, the first edition of *Traité d'oenologie* (*Handbook of Enology*) appeared. That publication aimed to gather scientific and technical knowledge in the main fields of a science still in its early stages, originating from the works of Louis Pasteur. Later on, the book was regularly republished by integrating the most recent experimental results, especially those conducted at the Bordeaux School of Enology. In 1997, the fourth edition, published under the direction of Professor Pascal Ribéreau-Gayon and Professors Denis Dubourdieu, Yves Glories, Aline A. Lonvaud, Bernard B. Donèche, and Alain Maujean, was completely reshaped into two volumes.

The current book, which is the seventh French edition (third in English) of the *Handbook of Enology*, updates and enriches the knowledge found in the two-volume edition that first appeared 20 years ago. Subsequent editions were printed in 2004 and then in 2012. As Émile Peynaud said in *Le Vin et les Jours*, "...enology is at the service of wine, it is a science in movement, advancing both in its research and applications..." As an applied science, it is irrigated by knowledge from the fundamental sciences (chemistry, biochemistry, microbiology, bioengineering, psychophysics, cognitive psychology, etc.) and nourished by empirical observations. The approach used in the *Handbook of Enology* is thus still the same. It is about providing practitioners, winemakers, technicians, and enology students with solid knowledge and the most recent research results. This knowledge can be used to contribute to a better definition of the quality of grapes and wine, a greater understanding of chemical and microbiological parameters, with the aim of ensuring satisfactory fermentations and predicting the evolution of wines, and better mastery of wine stabilization processes. As a result, the purpose of this publication is to guide readers in their thought processes with a view to preserving and optimizing the identity and taste of wine and its aging potential. In other words, this publication was meant, as the late Professor Denis Dubourdieu wrote, to help "...obtain original products that are sufficiently complex, fine, and appreciated by modern consumers...," in a context where grape ripeness characteristics are evolving and greater attention is being placed on chemical additives.

The outline of this edition of the *Handbook of Enology* does not differ from the previous ones. Some of the chapters have been significantly reworked in light of recent research, while others have been changed only slightly or not at all. This is a collective work; it is the fruit of efforts by the researchers and professors of the Enology Research Unit at the *Institut des Sciences de la Vigne et du Vin* (ISVV), who provided their expertise to update this Volume 1:

- Patricia Ballestra, Associate Professor, IUT Périgueux, Université de Bordeaux, researcher in the Enology Research Unit, ISVV, Université de Bordeaux (Volume 1, Chapter 5), regarding spoilage microorganisms related to mousiness.
- Philippe Darriet, Director of the Enology Research Unit, Université de Bordeaux (Volume 1, Chapters 10 and 13), regarding the biochemistry of aromas

xi

during grape ripening and overripening in the presence of *Botrytis cinerea*, special winemaking types.
- Marguerite Dols-Laffargue, Professor, ENSCBP, researcher in the Enology Research Unit, ISVV, Université de Bordeaux (Volume 1, Chapter 5), regarding sugar metabolism in lactic acid bacteria.
- Laurence Geny, Professor, ISVV, Université de Bordeaux, researcher in the Enology Research Unit (Volume 1, Chapter 10), regarding the biochemistry of grapes and their maturation.
- Aline A. Lonvaud, Professor Emeritus, ISVV, Université de Bordeaux, researcher in the Enology Research Unit (Volume 1, Chapters 1, 4–7), regarding the microbiology of yeasts and bacteria in wine.
- Patrick Lucas, Professor, ISVV, Deputy Director of the Enology Research Unit, Université de Bordeaux, researcher in the Enology Research Unit (Volume 1, Chapter 4), regarding the microbiology of bacteria in wine.
- Axel Marchal, Professor, ISVV, Université de Bordeaux, researcher in the Enology Research Unit (Volume 1, Chapter 12), regarding red winemaking and the evolution of sensory aspects during the aging of wines.
- Isabelle Masneuf-Pomarede, Professor, Bordeaux Sciences Agro, researcher in the Enology Research Unit, ISVV, Université de Bordeaux (Volume 1, Chapter 1), regarding yeast cytology, ecology, and taxonomy.
- Cécile Thibon, research engineer, INRAE, Bordeaux Sciences Agro, researcher in the Enology Research Unit, ISVV, Université de Bordeaux (Volume 1, Chapter 10), regarding the biochemistry of aromas during grape ripening and overripening in the presence of *B. cinerea*, varietal aromas.

The following persons also participated in writing the sixth French edition of Volume 1:
- Marina Bely, Associate Professor HDR, ISVV, Université de Bordeaux, researcher in the Enology Research Unit (Volume 1, Chapters 1–3), regarding the microbiology of yeasts.
- Bernard B. Donèche, Professor, ISVV, Université de Bordeaux, former director of the training department at ISVCV (Volume 1, Chapter 10), regarding the biochemistry of grapes and their maturation.
- Pierre-Louis Teissedre, Professor, ISVV, Université de Bordeaux (Volume 1, Chapters 8, 9, and 12), regarding sulfur dioxide and complementary products, red winemaking.

PREFACE TO THE SECOND EDITION

The two-volume *Enology Handbook* was published simultaneously in Spanish, French, and Italian in 1999 and has been reprinted several times. The Handbook has apparently been popular with students as an educational reference book, as well as with winemakers, as a source of practical solutions to their specific technical problems and scientific explanations of the phenomena involved.

It was felt appropriate at this stage to prepare an updated, reviewed, corrected version, including the latest enological knowledge, to reflect the many new research findings in this very active field. The outline and design of both volumes remain the same. Some chapters have changed relatively little as the authors decided there had not been any significant new developments, while others have been modified much more extensively, either to clarify and improve the text or, more usually, to include new research findings and their practical applications. Entirely new sections have been inserted in some chapters.

We have made every effort to maintain the same approach as we did in the first edition, reflecting the ethos of enology research in Bordeaux. We use indisputable scientific evidence in microbiology, biochemistry, and chemistry to explain the details of mechanisms involved in grape ripening, fermentations and other winemaking operations, aging, and stabilization. The aim is to help winemakers achieve greater control over the various stages in winemaking and choose the solution best suited to each situation. Quite remarkably, this scientific approach, most intensively applied in making the finest wines, has resulted in an enhanced capacity to bring out the full quality and character of individual terroirs. Scientific winemaking has not resulted in standardization or leveling of quality. On the contrary, by making it possible to correct defects and eliminate technical imperfections, it has revealed the specific qualities of the grapes harvested in different vineyards, directly related to the variety and terroir, more than ever before.

Interest in wine in recent decades has gone beyond considerations of mere quality and taken on a truly cultural dimension. This has led some people to promote the use of a variety of techniques that do not necessarily represent significant progress in winemaking. Some of these are simply modified forms of processes that have been known for many years. Others do not have a sufficiently reliable scientific interpretation, nor are their applications clearly defined. In this Handbook, we have only included rigorously tested techniques, clearly specifying the optimum conditions for their utilization.

As in the previous edition, we deliberately omitted three significant aspects of enology: wine analysis, tasting, and winery engineering. In view of their importance, these topics will each be covered in separate publications.

The authors would like to take the opportunity of the publication of this new edition of Volume 1 to thank all those who have contributed to updating this work:

- Marina Bely for her work on fermentation kinetics and the production of

- volatile acidity (Sections 2.3.4 and 14.2.5).
- Isabelle Masneuf for her investigation of the yeasts' nitrogen supply (Section 3.4.2).
- Gilles de Revel for elucidating the chemistry of SO_2, particularly details of combination reactions (Section 8.4).
- Gilles Masson for the section on rosé wines (Section 14.1).
- Cornelis Van Leeuwen for the impact of vineyard water supply on grape ripening (Section 10.4.6).
- André Brugirard for the section on French fortified wines—*vins doux naturels* (Section 14.4.2).
- Paulo Barros and João Nicolau de Almeida for their work on Port (Section 14.4.3).
- Justo F. Casas Lucas for the section on Sherry (Section 14.5.2).
- Alain Maujean for his in-depth revision of the section on Champagne (Section 14.3).

Bordeaux, May 17, 2004
Professor Pascal Ribéreau-Gayon
Corresponding Member of the
Institut de France
Member of the French Academy of Agriculture

PREFACE TO THE FIRST EDITION

Wine has probably inspired more research and publications than any other beverage or food. In fact, through their passion for wine, great scientists have not only contributed to the development of practical enology but have also made discoveries in the general field of science.

A forerunner of modern enology, Louis Pasteur developed simplified contagious infection models for humans and animals based on his observations of wine spoilage. The following quote clearly expresses his theory in his own words: "when profound alterations of beer and wine are observed because these liquids have given refuge to microscopic organisms, introduced invisibly and accidentally into the medium where they then proliferate, how can one not be obsessed by the thought that a similar phenomenon can and must sometimes occur in humans and animals."

Since the 19th century, our understanding of wine, wine composition, and wine transformations has greatly evolved in function of advances in relevant scientific fields, i.e. chemistry, biochemistry, and microbiology. Each applied development has led to better control of winemaking and aging conditions and of course wine quality. In order to continue this approach, researchers and winemakers must strive to remain up to date with the latest scientific and technical developments in enology.

For a long time, the Bordeaux School of Enology was largely responsible for the communication of progress in enology through the publication of numerous works (Béranger Publications and later Dunod Publications):

Wine Analysis, U. Gayon and J. Laborde (1912); *Treatise on Enology*, J. Ribéreau-Gayon (1949); *Wine Analysis*, J. Ribéreau-Gayon and E. Peynaud (1947 and 1958); *Treatise on Enology* (two volumes), J. Ribéreau-Gayon and E. Peynaud (1960 and 1961); *Wine and Winemaking*, E. Peynaud (1971 and 1981); *Wine Science and Technology* (four volumes), J. Ribéreau-Gayon, E. Peynaud, P. Ribéreau-Gayon, and P. Sudraud (1975–1982).

For an understanding of current advances in enology, the authors propose this book *Handbook of Enology Volume 1: The Microbiology of Wine and Vinifications* and the second volume of the *Handbook of Enology Volume 2: The Chemistry of Wine: Stabilization and Treatments*.

Although written by researchers, the two volumes are not specifically addressed to this group. Young researchers may, however, find these books useful to help situate their research within a particular field of enology. Today, the complexity of modern enology does not permit a sole researcher to explore the entire field.

These volumes are also of use to students and professionals. Theoretical interpretations as well as solutions are presented to resolve the problems encountered most often at wineries. The authors have adapted these solutions to many different situations and winemaking methods. In order to make the best use of the information contained in these works, enologists should have a broad understanding of general scientific knowledge. For example, the understanding and application of molecular biology and genetic engineering have become indispensable in the field of wine microbiology. Similarly, structural and quantitative physiochemical analysis methods such as chromatography, NMR, and mass spectrometry must now be mastered in order to explore wine chemistry.

The goal of these two works was not to create an exhaustive bibliography of each subject. The authors strove to choose only the most relevant and significant publications to their particular field of research. A large number of references to French enological research have been included in these works in order to make this information available to a larger non-French-speaking audience.

In addition, the authors have tried to convey a French and more particularly a Bordeaux perspective of enology and the art of winemaking. The objective of this perspective is to maximize the potential quality of grape crops based on the specific natural conditions that constitute their "terroir." The role of enology is to express the characteristics of the grape specific not only to variety and vineyard practices but also maturation conditions, which are dictated by soil and climate.

It would, however, be an error to think that the world's greatest wines are exclusively a result of tradition, established by exceptional natural conditions, and that only the most ordinary wines, produced in giant processing facilities, can benefit from scientific and technological progress. Certainly, these facilities do benefit the most from high-performance installations and automation of operations. Yet history has unequivocally shown that the most important enological developments in wine quality (for example, malolactic fermentation) have been discovered in ultra-premium wines. The corresponding techniques were then applied to less prestigious products.

High-performance technology is indispensable for the production of great wines, since a lack of control of winemaking parameters can easily compromise their quality, which would be less of a problem with lower quality wines.

The word "vinification" has been used in this work and is part of the technical language of the French tradition of winemaking. Vinification describes the first phase of winemaking. It comprises all technical aspects from grape maturity and harvest to the end of alcoholic and sometimes malolactic fermentation. The second phase of winemaking "wine maturation, stabilization, and treatments" is completed when the wine is bottled. Aging specifically refers to the transformation of bottled wine.

This distinction of two phases is certainly the result of commercial practices. Traditionally, in France, a vine grower farmed the vineyard and transformed grapes into an unfinished wine. The wine merchant transferred the bulk wine to his cellars, finished the wine, and marketed the product, preferentially before bottling. Even though most wines are now bottled at the winery, these long-standing practices have maintained a distinction between "wine grower enology" and "wine merchant enology." In countries with a more recent viticultural history, generally English speaking, the vine grower is responsible for winemaking and wine sales. For this reason, the Anglo-Saxon tradition speaks of winemaking, which covers all operations from harvest reception to bottling.

In these works, the distinction between "vinification" and "stabilization and treatments" has been maintained, since the first phase primarily concerns microbiology and the second, chemistry. In this manner, the individual operations could be linked to their particular sciences. There are of course limits to this approach. Chemical phenomena occur during vinification; the stabilization of wines during storage includes the prevention of microbial contamination.

Consequently, the description of the different steps of enology does not always obey logic as precise as the titles of these works may lead to believe. For example, microbial contamination during aging and storage is covered in Volume 1. The antiseptic properties of SO_2 incited the description of its use in the same volume. This line of reasoning led to the description of the antioxidant-related chemical properties of this compound in the same chapter as well as an explanation of adjuvants to sulfur dioxide: sorbic acid (antiseptic) and ascorbic acid (antioxidant). In addition, the on lees aging of white wines and the resulting chemical transformations cannot be separated from vinification and are therefore also covered in Volume 1. Finally, our understanding of phenolic compounds in red wine is based on complex chemistry. All aspects related to the nature of the corresponding substances, their properties, and their evolution during grape maturation, vinification, and aging are therefore covered in Volume 2.

These works only discuss the principles of equipment used for various enological operations and their effect on product quality. For example, temperature control systems, destemmers, crushers, and presses, as well as filters, inverse osmosis, machines, and ion exchangers are not described in detail. Bottling is not addressed at all. An in-depth description of enological equipment would merit a detailed work dedicated to the subject.

Wine tasting, another essential role of the winemaker, is not addressed in these works. Many related publications are, however, readily available. Finally, wine analysis is an essential tool that a winemaker should master. It is, however, not covered in these works except in a few particular cases, i.e. phenolic compounds, whose different families are often defined by analytical criteria.

The authors thank the following people who have contributed to the creation of this work:

J.F. Casas Lucas, Chapter 14, Sherry; A. Brugirard, Chapter 14, Sweet Wines; J.N. de Almeida, Chapter 14, Port Wines; A. Maujean, Chapter 14, Champagne; C. Poupot for the preparation of material in Chapters 1, 2, and 13; Miss F. Luye-Tanet for her help with typing.

They also thank Madame B. Masclef in particular for her important part in the typing, preparation, and revision of the final manuscript.

Pascal Ribéreau-Gayon

Bordeaux

REMARKS CONCERNING THE EXPRESSION OF CERTAIN PARAMETERS OF MUST AND WINE COMPOSITION

Units

Metric system units of length (m), volume (l), and weight (g) are exclusively used. The conversion of metric units into Imperial units (inches, feet, gallons, pounds, etc.) can be found in the following enological work: *Principles and Practices of Winemaking*, R.B. Boulton, V.L. Singleton, L.F. Bisson, and R.E. Kunkee, 1995, The Chapman & Hall Enology Library, New York.

Expression of Total Acidity and Volatile Acidity

Although EU regulations recommend the expression of total acidity in the equivalent weight of tartaric acid, the French custom is to give this expression in the equivalent weight of sulfuric acid. The more correct expression in milliequivalents per liter has not come into widespread use. In this work, we have generally expressed acid levels as sulfuric acid, but in certain cases, the corresponding weight in tartaric acid, often used in other countries, has been given.

Using the weight of the milliequivalent of various acids, the table below can be used to convert from one expression to another.

Coefficients of conversion from one expression of total or volatile acidity to another

Known expression	Desired expression			
	mEq/l	g/l H_2SO_4	g/l tartaric acid	g/l acetic acid
mEq/l	1.00	0.049	0.075	0.060
g/l H_2SO_4	20.40	1.00	1.53	1.22
g/l tartaric acid	13.33	0.65	1.00	
g/l acetic acid	16.67	0.82		1.00

More particularly, to convert from total acidity expressed in H_2SO_4 to its expression in tartaric acid, add half of the value to the original value:

$$4 \text{ g/l } H_2SO_4 \rightarrow 6 \text{ g/l tartaric acid.}$$

In the other direction, a third of the value must be subtracted.

The French also continue to express volatile acidity in equivalent weight of sulfuric acid. More generally, in other countries, volatile acidity is expressed as acetic acid. It is rarely expressed in milliequivalents per liter. The table above also allows simple conversion from one expression to another.

The expression of volatile acidity as acetic acid is approximately 20% higher than as sulfuric acid.

Evaluating the Sugar Concentration of Musts

This measurement is important for tracking grape ripening, monitoring fermentation kinetics, and, if necessary, determining the need for chaptalization.

This measurement is always determined by physical, densimetric, or refractometric analysis. The expression of the results can be given according to several scales: some are rarely used today (Baumé scale and Oechsle scale). At present, two systems exist (Volume 1, Section 10.4.3):

1. The potential ethanol content (PEC) of musts can be read directly on equipment that is graduated using a scale corresponding to 17.5 or 17 g/l of sugar for 1% volume of alcohol. Today, the EU recommends using 16.83 g/l as the conversion factor. The mustimeter is a hydrometer containing two graduated scales: one shows specific gravity and the other gives a direct reading of the PEC. Different methods varying in precision exist to calculate the PEC from a specific gravity reading. These methods take various elements of must composition into account (Boulton *et al*., 1995).
2. Degrees Brix express the percentage of sugar by weight. By multiplying degrees Brix by 10, the weight of sugar in 1 kg (or slightly less than 1 l) of must is obtained. A conversion table between degrees Brix and PEC is shown in Volume 1, Section 10.4.3. Seventeen (17) degrees Brix corresponds to an approximate PEC of 10%, and 20° Brix corresponds to a PEC of about 12%. Within the alcohol range most relevant to enology, degree Brix can be multiplied by 10 and then divided by 17 to obtain a fairly good approximation of the PEC.

In any case, the determination of the Brix or PEC of a must is approximate. First of all, it is not always possible to obtain a representative grape or must sample for analysis. Secondly, although physical, densimetric, or refractometric measurements are extremely precise and rigorously express the sugar concentration of a sugar and water mixture, these measurements are affected by other substances released into the sample from the grape and other sources. Furthermore, the concentrations of these substances are different for every grape or must sample. Finally, the conversion rate of sugar into alcohol (approximately 17–18 g/l per 1% alcohol by vol.) varies and depends on fermentation conditions and yeast properties. The widespread use of selected yeast strains has lowered the sugar conversion rate.

Measurements Using Visible and Ultraviolet Spectrometry

The measurement of optical density (or absorbance) is widely used to determine wine color (Volume 2, Section 6.4.5) and total phenolic compounds (Volume 2, Section 6.4.1). In these works, optical density is noted as OD: OD 420 (yellow), OD 520 (red), OD 620 (blue), or OD 280 (absorption in ultraviolet spectrum) to indicate the optical density at the indicated wavelengths.

- Wine color intensity is expressed as: CI = OD 420 + OD 520 + OD 620, or is sometimes expressed in a more simplified form: CI = OD 420 + OD 520.
- Hue is expressed as:

$$H = \frac{OD420}{OD520}$$

The total phenolics concentration is expressed by OD 280. The analytical methods are described in Chapter 6 of Volume 2.

PART I

Microbiology of Wine

CHAPTER 1

Yeasts

1.1 Introduction
1.2 The Cell Wall
1.3 The Plasma Membrane
1.4 The Cytoplasm and Its Organelles
1.5 The Nucleus
1.6 Reproduction and the Yeast Biological Cycle
1.7 The Killer Phenomenon
1.8 Classification of Yeast Species
1.9 Identification of Wine Yeast Strains
1.10 Ecology of Grape and Wine Yeasts

1.1 Introduction

Man has been making bread and fermented beverages since the beginning of recorded history. Yet the role of yeasts in alcoholic fermentation, particularly in the transformation of grapes into wine, was only clearly established in the middle of the 19th century. The ancients explained the boiling during fermentation (from the Latin *fervere*, to boil) as a reaction between substances that come into contact with each other during crushing to produce effervescence. In 1680, a Dutch cloth merchant, Antonie van Leeuwenhoek, first observed yeasts in beer wort using a microscope that he designed and produced. He did not, however, establish a relationship between these corpuscles and alcoholic fermentation. It was not until the end of the 18th century that Antoine Lavoisier began the chemical study of alcoholic fermentation. Joseph Louis Gay-Lussac continued Lavoisier's research into the next century. As early as 1785, Adam Fabroni, an Italian scientist, was the first to provide an interpretation of the chemical composition of the ferment responsible for alcoholic fermentation, which he described

---

Handbook of Enology, Volume 1: The Microbiology of Wine and Vinifications, Third Edition.
Pascal Ribéreau-Gayon, Denis Dubourdieu, Bernard B. Donèche and Aline A. Lonvaud.
© 2021 John Wiley & Sons Ltd. Published 2021 by John Wiley & Sons Ltd.

as a plant–animal substance. According to Fabroni, this material, comparable to the gluten in flour, was located in special utricles, particularly on grapes and wheat, and alcoholic fermentation occurred when it came into contact with sugar in the must. In 1837, a French physicist named Charles Cagnard de La Tour proved for the first time that yeast was a living organism. According to his findings, it was capable of multiplying and belonged to the plant kingdom; its vital activities were the basis for the fermentation of sugar-containing liquids. The German naturalist Theodor Schwann confirmed his theory and demonstrated that heat and certain chemical products were capable of stopping alcoholic fermentation. He named the beer yeast *zuckerpilz*, which means sugar fungus—*Saccharomyces* in Latin. In 1838, Franz Meyen used this nomenclature for the first time.

This vitalist or biological conception of the role of yeasts in alcoholic fermentation, obvious to us today, was not readily embraced. Justus von Liebig and certain other organic chemists were convinced that chemical reactions, not living cellular activity, were responsible for the fermentation of sugar. In his famous works, *Studies on Wine* (1866) and *Studies on Beer* (1876), Louis Pasteur gave definitive credibility to the vitalist view of alcoholic fermentation. He demonstrated that the yeasts responsible for spontaneous fermentation of grape must or crushed grapes came from the surface of the grape; he isolated several races and species. He even conceived the notion that the nature of the yeast carrying out the alcoholic fermentation could influence the taste characteristics of wine. He also demonstrated the effect of oxygen on the assimilation of sugar by yeasts and proved that the yeast produced secondary products such as glycerol in addition to alcohol and carbon dioxide.

Since Pasteur, yeasts and alcoholic fermentation have incited a considerable amount of research, making use of progress in microbiology, biochemistry, and now genetics and molecular biology.

In taxonomy, scientists define yeasts as single-celled fungi that reproduce by budding and binary fission. Certain multicellular fungi have a single-celled stage and are also grouped with yeasts. Yeasts form a complex and heterogeneous group found in three classes of fungi, characterized by their reproduction mode: Ascomycetes, Basidiomycetes, and the imperfect fungi (Deuteromycetes). However, yeasts found on the surface of the grape and in wine belong only to Ascomycetes and the imperfect fungi. The haploid spores or ascospores of the Ascomycetes class are contained in the ascus, a type of sac made from vegetative cells. Asporogenous yeasts, incapable of sexual reproduction, are classified with the imperfect fungi.

In this chapter, the morphology, reproduction, taxonomy, and ecology of grape and wine yeasts will be discussed. Cytology is the morphological and functional study of the structural components of the cell (Rose and Harrison, 1991).

Yeasts are the most simple of the eukaryotes. The yeast cell contains cell envelopes, a cytoplasm with various organelles, and a nucleus surrounded by a membrane and enclosing the chromosomes (Figure 1.1). Like all plant cells, the yeast cell has two cell envelopes: the cell wall and the plasma membrane. The periplasmic space is the space between the cell wall and the membrane. The cytoplasm and the membrane make up the protoplasm. The terms protoplast and spheroplast designate a cell whose cell wall has been "artificially" removed in full or in part, respectively. Yeast cell envelopes play an essential role: they contribute to a successful alcoholic fermentation of must and release certain constituents that add to the resulting wine's composition. To take advantage of these properties, the winemaker or enologist must have a profound knowledge of these organelles.

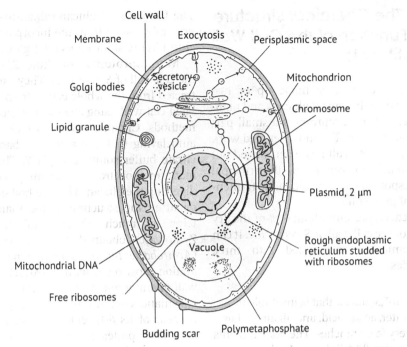

FIGURE 1.1 A yeast cell (Gaillardin and Heslot, 1987).

1.2 The Cell Wall

1.2.1 The General Role of the Cell Wall

The works of various authors (Fleet, 1991; Klis, 1994; Stratford, 1994; Klis *et al.*, 2002) have greatly expanded our knowledge of the yeast cell wall, which represents 15–25% of the dry weight of the cell. The yeast cell wall essentially consists of polysaccharides. It is a rigid envelope, yet endowed with a certain elasticity.

Its first function is to protect the cell. Without its wall, the cell would burst under the internal osmotic pressure, which is determined by the composition of the cell's environment. Protoplasts placed in pure water are immediately lysed in this manner. Cell wall elasticity can be demonstrated by placing whole yeasts, sampled during their log growth phase, in a hypertonic (NaCl) solution. Their cell volume decreases by approximately 50%. The cell wall appears thicker, while the cytoplasmic membrane practically does not detach from it at all. The cells regain their initial form after being placed back into an isotonic medium.

Yet the cell wall cannot be considered an inert, semirigid "armor." On the contrary, it is a dynamic and multifunctional organelle. Its composition and functions evolve during the life of the cell, in response to environmental factors. In addition to its protective role, the cell wall gives the cell its particular shape through its macromolecular organization. It is also the site of molecules that determine certain cellular interactions such as sexual union, flocculation, and the killer factor, which will be examined in detail in Section 1.7. Lastly, a number of enzymes, generally hydrolases, are connected to the cell wall or situated in the periplasmic space. Their substrates are nutritive substances of the environment and the macromolecules of the cell wall itself, which are constantly reshaped during cell morphogenesis.

1.2.2 The Chemical Structure and Function of the Cell Wall Constituents

The yeast cell wall is made up of two principal constituents: β-glucans and mannoproteins. Chitin represents a small part of its composition. The most detailed work on the yeast cell wall has been carried out on *Saccharomyces cerevisiae*—the principal yeast responsible for the alcoholic fermentation of grape must.

Glucan represents about 60% of the dry weight of the cell wall of *S. cerevisiae*. It can be chemically fractionated into three categories:

1. A β-1,3-glucan that is insoluble in water, acetic acid, and alkalis. It has very few branches. The branch points involve β-1,6 linkages. Its degree of polymerization is 1,500. Under an electron microscope, this glucan appears fibrous. It ensures the shape and the rigidity of the cell wall. It is always associated with chitin.
2. A β-1,3-glucan, with about 1,500 glucose units, that is insoluble in water but soluble in alkalis. It has very few branches, like the preceding glucan. In addition to these few branches, it has a small number of β-1,6 glycosidic linkages. It has an amorphous appearance under the electron microscope. It gives the cell wall its elasticity and acts as an anchor for mannoproteins. It can also constitute an extraprotoplasmic reserve substance.
3. A β-1,6-glucan that is obtained from alkali-insoluble glucans by extraction in acetic acid. The resulting product is amorphous, water-soluble, and extensively branched with β-1,3-glycosidic linkages. Its degree of polymerization is 140. It links the different constituents of the cell wall together. It is also a receptor site for the killer factor (Section 1.7).

The fibrous β-1,3-glucan (alkali-insoluble) probably results from the incorporation of chitin in the amorphous β-1,3-glucan.

Mannoproteins constitute 25–50% of the cell wall of *S. cerevisiae*. They can be extracted from the whole cell or from the isolated cell wall using chemical or enzymatic methods. Chemical methods make use of autoclaving in the presence of bases or a citrate buffer solution at pH 7. The enzymatic method frees the mannoproteins by digesting the glucan. This method does not denature the structure of the mannoproteins as much as chemical methods. Zymolyase, obtained from a bacterium (*Arthrobacter luteus*), is the enzymatic preparation most often used to extract the cell wall mannoproteins of *S. cerevisiae*. This enzymatic complex is effective primarily because of its β-1,3-glucanase activity. The action of protease contaminants in the zymolyase also combines with the aforementioned activity to liberate the mannoproteins. Glucanex, another industrial preparation of β-glucanase, produced by a fungus (*Trichoderma harzianum*), has been demonstrated to possess endo- and exo-β-1,3- and endo-β-1,6-glucanase activities (Dubourdieu and Moine, 1995). These activities also facilitate the extraction of the cell wall mannoproteins from the *S. cerevisiae* cell.

The mannoproteins of *S. cerevisiae* have a molecular weight between 20 and 450 kDa. Their degree of glycosylation varies. However, some of them, containing about 90% mannose and 10% peptides, are hypermannosylated.

Four forms of glycosylation are described (Figure 1.2) but do not necessarily exist at the same time in all mannoproteins.

The mannose in mannoproteins may form short, linear chains with one to five residues. They are linked to the peptide chain by O-glycosyl linkages on serine and threonine residues. These glycosidic side-chain linkages are α-1,2 and α-1,3.

The carbohydrate part of the mannoprotein can also be a polysaccharide. It is linked to an asparagine residue of the peptide chain by an N-glycosyl linkage. This

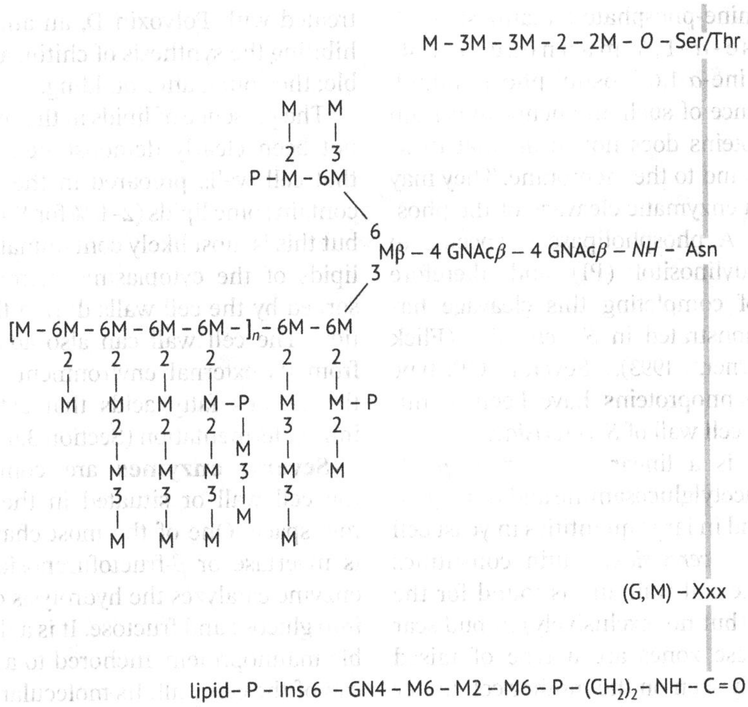

1.2 | The Cell Wall

```
                              M - 3M - 3M - 2 - 2M - O - Ser/Thr

                   M   M
                   |   |
                   2   3
               P - M - 6M
                        \
                         6
                          Mβ - 4 GNAcβ - 4 GNAcβ - NH - Asn
                         3
                        /
   [M - 6M - 6M - 6M - 6M -]ₙ - 6M - 6M
      2    2    2    2      2    2
      |    |    |    |      |    |
      M    M    M    M - P  M    M - P
      2    2    2    3      2
      |    |    |    |      |
      M    M    M    M      M
      3    3    3    |      3
      |    |    |    M      |
      M    M    M           M

                                          (G, M) - Xxx


    lipid - P - Ins 6 - GN4 - M6 - M2 - M6 - P - (CH₂)₂ - NH - C = O
```

FIGURE 1.2 The four types of glucosylation of cell wall yeast mannoproteins (Klis, 1994). M, mannose; G, glucose; GN, glucosamine; GNAc, N-acetylglucosamine; Ins, inositol; Ser, Serine; Thr, threonine; Asn, asparagine; P, Phosphate; Xxx, the nature of the bond is not known.

linkage consists of a double unit of β-1,4-linked N-acetylglucosamine (chitobiose). The mannan linked in this manner to the asparagine includes an attachment region made up of a dozen mannose residues and a highly branched outer chain consisting of 150–250 mannose units. The attachment region beyond the chitobiose residue consists of an α-1,6-linked mannose skeleton with side branches possessing one, two, or three mannose residues with α-1,2 and/or α-1,3 bonds. The outer chain is also made up of a skeleton of α-1,6-linked mannose units. This chain bears short side chains composed of α-1,2-linked mannose residues and an α-1,3-linked terminal mannose. Some of these side chains possess a branch attached by a phosphodiester bond.

A third type of glycosylation was also described. It can occur in mannoproteins, which make up the cell wall of the yeast. It consists of a glucomannan chain containing essentially α-1,6-linked mannose residues and α-1,6-linked glucose residues. The nature of the glycan–peptide point of attachment is not yet clear, but it may be an asparaginyl–glucose bond. Moreover, this type of glycosylation characterizes the proteins freed from the cell wall by the action of a β-1,3-glucanase. Therefore, *in vivo*, the glucomannan chain may also comprise β-1,3-linked glucose residues.

The fourth type of glycosylation of yeast mannoproteins is the glycosylphosphatidylinositol (GPI) anchor. This attachment between the terminal carboxylic group of the peptide chain and a membrane phospholipid permits certain mannoproteins, which cross the cell wall, to anchor themselves in the plasma membrane. The region of attachment is characterized by the following sequence (Figure 1.2):

ethanolamine-phosphate-6-mannose-α-1,2-mannose-α-1,6-mannose-α-1,4-glucosamine-α-1,6-inositol-phospholipid. The presence of such an anchor in certain mannoproteins does not mean that these remain bound to the membrane. They may detach via enzymatic cleavage of the phospholipids. A phospholipase C specific to phosphatidylinositol (PI) and therefore capable of completing this cleavage has been demonstrated in *S. cerevisiae* (Flick and Thorner, 1993). Several GPI-type anchor mannoproteins have been identified in the cell wall of *S. cerevisiae*.

Chitin is a linear polymer of β-1,4-linked *N*-acetylglucosamine and is not generally found in large quantities in yeast cell walls. In *S. cerevisiae*, chitin constitutes 1–2% of the cell wall and is found for the most part (but not exclusively) in bud scar zones. These zones are a type of raised crater easily seen on the mother cell under an electron microscope (Figure 1.3). This chitin scar is formed essentially to ensure cell wall integrity and cell survival. Yeasts treated with Polyoxin D, an antibiotic inhibiting the synthesis of chitin, are not viable; they burst after budding.

The presence of lipids in the cell wall has not been clearly demonstrated. It is true that cell walls prepared in the laboratory contain some lipids (2–15% for *S. cerevisiae*), but this is most likely contamination by the lipids of the cytoplasmic membrane, adsorbed by the cell walls during their isolation. The cell wall can also adsorb lipids from its external environment, especially the various fatty acids that activate and inhibit fermentation (Section 3.6.2).

Several enzymes are connected to the cell wall or situated in the periplasmic space. One of the most characteristic is invertase or β-fructofuranosidase. This enzyme catalyzes the hydrolysis of sucrose into glucose and fructose. It is a thermostable mannoprotein anchored to a β-1,6-glucan of the cell wall. Its molecular weight is 270,000 Da. It contains approximately 50% mannose and 50% protein. Periplasmic acid phosphatase is also a mannoprotein.

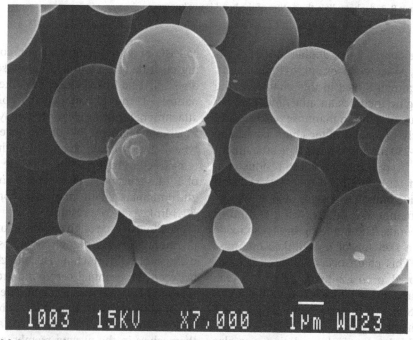

FIGURE 1.3 Scanning electron microscope photograph of proliferating *S. cerevisiae* cells. The budding scars on the mother cells can be observed. (***Source***: Photograph from M. Mercier, Department of Electron Microscopy, Université de Bordeaux I.)

Other periplasmic enzymes that have been noted are β-glucosidase, α-galactosidase, melibiase, trehalase, aminopeptidase, and esterase. Yeast cell walls also contain endo- and exo-β-glucanases (β-1,3 and β-1,6). These enzymes are involved in the reshaping of the cell wall during the growth and budding of cells. Their activity is at a maximum during the exponential growth phase of the population and diminishes notably afterward. However, cells in the stationary phase and even dead yeasts contained in the lees still retain β-glucanase activity in their cell walls several months after the completion of fermentation. These endogenous enzymes are involved in the autolysis of the cell wall during the lees aging of wines. This aging method will be covered in the Chapter 13.

1.2.3 General Organization of the Cell Wall and Factors Affecting Its Composition

The cell wall of *S. cerevisiae* is made up of an outer layer of mannoproteins. These mannoproteins are connected to a matrix of amorphous β-1,3-glucan, which covers an inner layer of fibrous β-1,3-glucan. The inner layer is connected to a small quantity of chitin (Figure 1.4). β-1,6-glucan probably acts as a cement between the two layers. The rigidity and the shape of the cell wall are due to the internal framework of the fibrous β-1,3-glucan. Its elasticity is due to the outer amorphous layer. The intermolecular structure of the mannoproteins of the outer layer (hydrophobic bonds and disulfide bonds) equally determines cell wall porosity for micromolecules (molecular weights less than 4,500) and impermeability to macromolecules. This impermeability can be affected by treating the cell wall with certain chemical agents, such as β-mercaptoethanol. This substance breaks the disulfide bonds, thus destroying the intermolecular network between the mannoprotein chains.

The composition of the cell wall is strongly influenced by nutritive conditions and cell age. The proportion of glucan in the cell wall increases with respect to the amount of sugar in the culture medium. Certain deficiencies (for example, of mesoinositol) also result in an increase in the proportion of glucan compared with mannoproteins. The cell walls of older cells are richer in glucans and in chitin and less rich in mannoproteins than younger ones. For this reason, they are more resistant to physical and enzymatic agents used to break them down. Finally, the composition of cell walls is profoundly modified by morphogenetic alterations (conjugation and sporulation).

1.3 The Plasma Membrane

1.3.1 Chemical Composition and Organization

The plasma membrane is a highly selective barrier controlling exchanges between the living cell and its external environment. This organelle is essential to the life of the yeast.

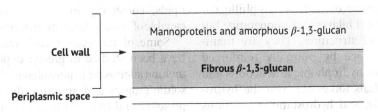

FIGURE 1.4 Cellular organization of the cell wall of *S. cerevisiae*.

Like all biological membranes, the yeast plasma membrane is principally made up of lipids and proteins. The plasma membrane of *S. cerevisiae* contains about 40% lipids and 50% proteins. Glucans and mannans are only present in small quantities (a few percent).

The lipids of the membrane are essentially phospholipids and sterols. They are amphiphilic molecules, i.e. possessing a hydrophilic and a hydrophobic part.

The three principal phospholipids (Figure 1.5) of the plasma membrane of yeast are phosphatidylethanolamine (PE), phosphatidylcholine (PC), and PI, which represent 70–85% of the total. Phosphatidylserine (PS) and diphosphatidylglycerol or cardiolipin (PG) are less prevalent. Free fatty acids and phosphatidic acid are frequently reported in plasma membrane analysis. They are probably extraction artifacts caused by the activity of certain lipid degradation enzymes.

The fatty acids of the membrane phospholipids contain an even number (14–24) of carbon atoms. The most abundant are C16 and C18 acids. They can be saturated, such as palmitic acid (C16) and stearic acid (C18), or unsaturated, as with oleic acid (C18, double bond in position 9), linoleic acid (C18, two double bonds in positions 9 and 12), and linolenic acid (C18, three double bonds in positions 9, 12, and 15). All membrane phospholipids share a common characteristic: they possess a polar or hydrophilic part made up of a phosphorylated alcohol and a nonpolar or hydrophobic part comprising two more-or-less parallel fatty acid chains. Their symbolic representation is shown in Figure 1.6. In an aqueous medium, the phospholipids spontaneously form bimolecular films or a lipid bilayer because of their amphiphilic nature. The lipid bilayers are cooperative but non-covalent structures. They are maintained in place by mutually reinforced interactions: hydrophobic interactions and van der Waals forces between the hydrocarbon tails, and hydrostatic interactions and hydrogen bonds between the polar heads and water molecules. The examination of cross-sections of yeast plasma membranes under an electron microscope reveals a classic lipid bilayer structure with a thickness of about 7.5 nm. The membrane surface appears sculpted with creases, especially during the stationary phase. However, the physiological meaning of this anatomical characteristic remains unknown. The plasma membrane also has a depression under the bud scar.

Ergosterol is the primary sterol of the yeast plasma membrane. In addition, 24(28)-dehydroergosterol and lesser amounts of zymosterol are present (Figure 1.7). Sterols are exclusively produced in the mitochondria under aerobic conditions during the yeast log phase. As with phospholipids, membrane sterols are amphipathic. The hydrophilic part is composed of the hydroxyl group in the C3 position, while the rest of the molecule is hydrophobic, especially the flexible hydrocarbon tail.

The plasma membrane also contains numerous proteins or glycoproteins presenting a wide range of molecular weights (from 10,000 to 120,000). The available information indicates that the organization of the plasma membrane of a yeast cell resembles the fluid mosaic model. This model, proposed for biological membranes by Singer and Nicolson (1972), consists of two-dimensional solutions of proteins and oriented lipids. Certain proteins penetrate the membrane; they are called integral proteins (Figure 1.6). They interact strongly with the nonpolar part of the lipid bilayer. The peripheral proteins are linked to the integral ones by hydrogen bonds. Their location is asymmetrical, at either the inner or the outer side of the plasma membrane. The molecules of proteins and membrane lipids, constantly in lateral motion, are capable of rapidly diffusing in the membrane.

Some of the yeast membrane proteins have been studied in greater depth. These include adenosine triphosphatase (ATPase), solute (sugars and amino acids) transport proteins, and enzymes involved in the production of glucans and chitin of the cell wall.

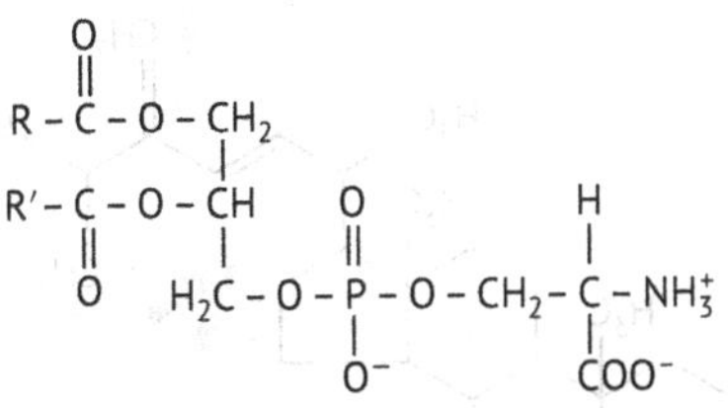

1.3 | The Plasma Membrane

$$\begin{array}{c}
\text{O}\\
\parallel\\
\text{R} - \text{C} - \text{O} - \text{CH}_2\\
\text{R}' - \text{C} - \text{O} - \text{CH} \\
\parallel \quad | \\
\text{O} \quad \text{H}_2\text{C} - \text{O} - \text{P} - \text{O} - \text{CH}_2 - \text{C} - \text{NH}_3^+\\
| \qquad | \\
\text{O}^- \qquad \text{COO}^-
\end{array}$$

Phosphatidylserine

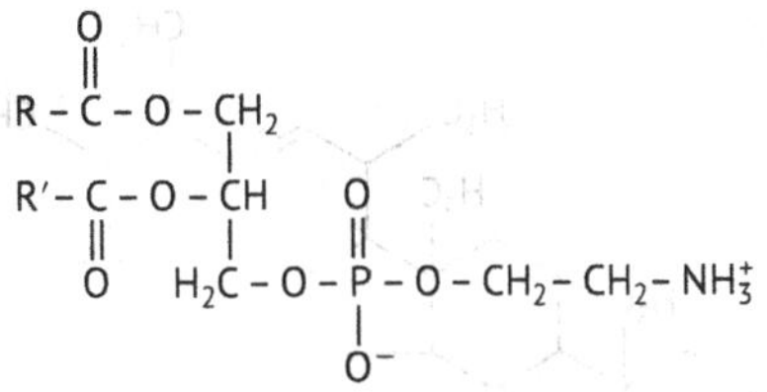

$$\text{R} - \text{C} - \text{O} - \text{CH}_2$$
$$\text{R}' - \text{C} - \text{O} - \text{CH}$$
$$\text{H}_2\text{C} - \text{O} - \text{P} - \text{O} - \text{CH}_2 - \text{CH}_2 - \text{NH}_3^+$$

Phosphatidylethanolamine

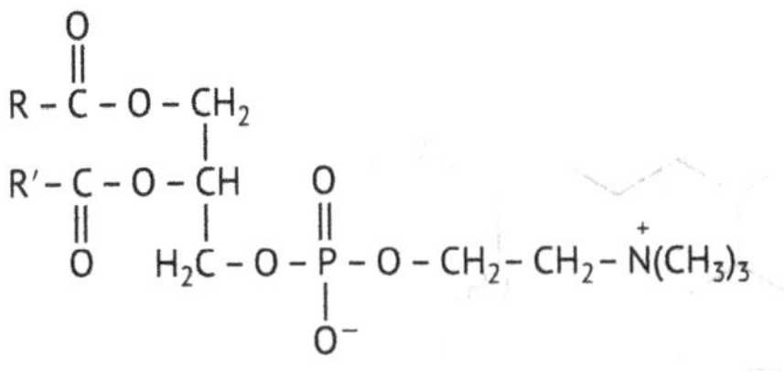

$$\text{R} - \text{C} - \text{O} - \text{CH}_2$$
$$\text{R}' - \text{C} - \text{O} - \text{CH}$$
$$\text{H}_2\text{C} - \text{O} - \text{P} - \text{O} - \text{CH}_2 - \text{CH}_2 - \overset{+}{\text{N}}(\text{CH}_3)_3$$

Phosphatidylcholine

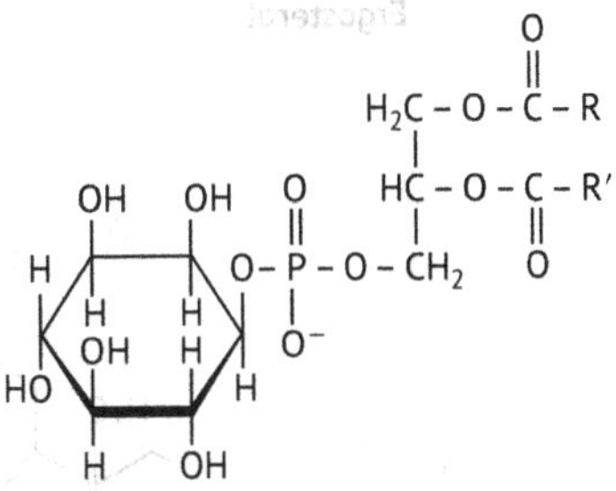

Phosphatidylinositol

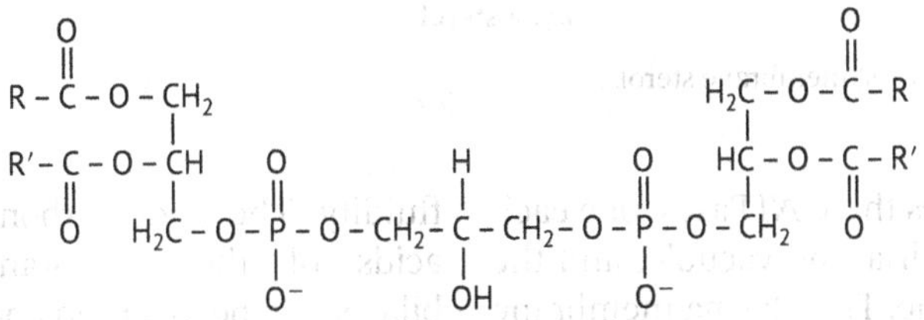

Diphosphatdiylglycerol (cardiolipin)

FIGURE 1.5 Yeast membrane phospholipids.

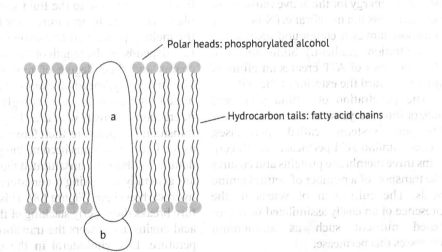

FIGURE 1.6 Diagram of a membrane lipid bilayer. The integral proteins (a) are strongly associated to the hydrocarbon region of the bilayer. The peripheral proteins (b) are linked to the integral proteins.

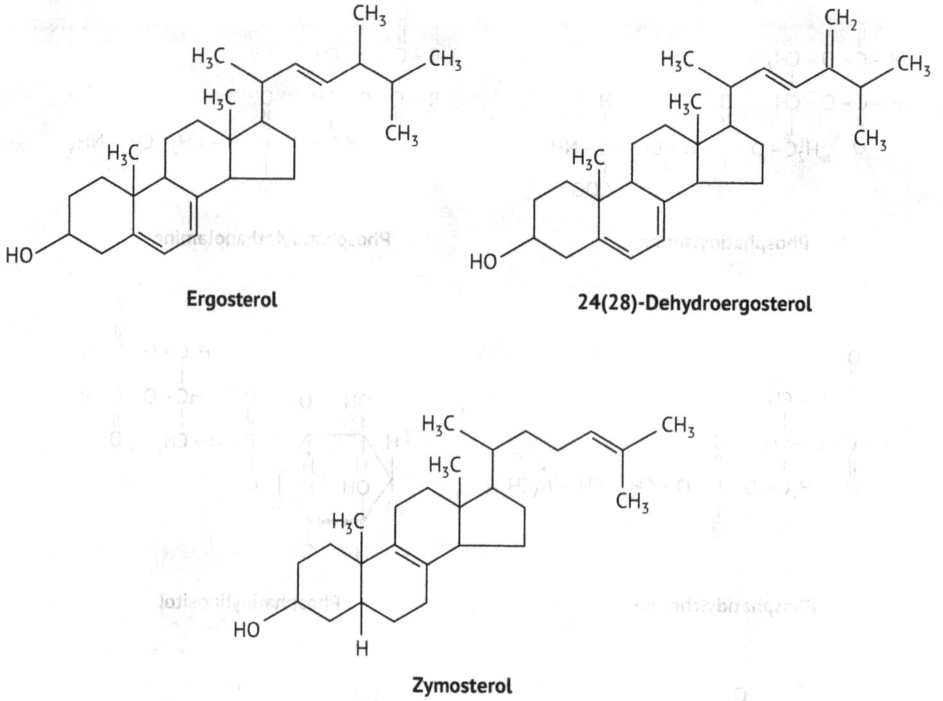

FIGURE 1.7 Principal yeast membrane sterols.

Yeast possesses three ATPases: one each in the mitochondria, the vacuole, and the plasma membrane. The plasma membrane ATPase is an integral protein with a molecular weight of around 100,000 Da. It catalyzes the hydrolysis of adenosine triphosphate (ATP), which furnishes the necessary energy for the active transport of solutes across the membrane. (Note: active transport moves a compound against the concentration gradient.) Simultaneously, the hydrolysis of ATP creates an efflux of protons toward the exterior of the cell.

The penetration of amino acids and sugars into the yeast activates membrane transport systems called permeases. General amino acid permease (GAP) contains three membrane proteins and ensures the transport of a number of neutral amino acids. The cultivation of yeasts in the presence of an easily assimilated nitrogen-based nutrient such as ammonium represses this permease.

The fatty acid composition of the membrane and its sterol content control its fluidity. The hydrocarbon chains of fatty acids of the membrane phospholipid bilayer can be in a rigid and orderly state or in a relatively disorderly and fluid state. In the rigid state, all of the carbon bonds of the fatty acids are *trans*. In the fluid state, some of the bonds become *cis*. The transition from the rigid state to the fluid state takes place when the temperature rises beyond the melting point. This transition temperature depends on the length of the fatty acid chains and their degree of unsaturation. The straight hydrocarbon chains of the saturated fatty acids interact strongly. These interactions intensify with their length. The transition temperature therefore increases as the fatty acid chains become longer. The double bonds of the unsaturated fatty acids are generally *cis*, giving a curvature to the hydrocarbon chain (Figure 1.8). This curvature breaks the orderly stacking of the fatty acid chains and lowers the transition temperature. Like cholesterol in the cells of mammals, ergosterol is also a fundamental regulator of membrane fluidity in yeasts.

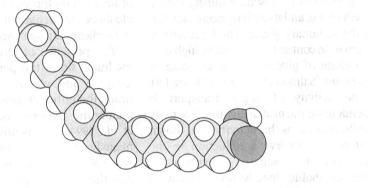

FIGURE 1.8 Molecular models representing the three-dimensional structure of stearic and oleic acids. The *cis* configuration of the double bond of oleic acid produces a curvature of the carbon chain.

Ergosterol is inserted in the bilayer perpendicularly to the plane of the membrane. Its hydroxyl group is bound by hydrogen bonds with the polar head of the phospholipid, and its hydrocarbon tail is inserted in the hydrophobic region of the bilayer. Thus, the membrane sterols insert themselves between the phospholipids. In this manner, they inhibit the crystallization of the fatty acid chains at low temperatures. Conversely, in reducing the movement of these same chains by steric encumbrance, they regulate any excessive membrane fluidity when the temperature rises.

1.3.2 Functions of the Plasma Membrane

The plasma membrane constitutes a stable, hydrophobic barrier between the cytoplasm and the environment outside the cell, owing to its phospholipids and sterols. This barrier presents a certain impermeability to solutes as a function of osmotic properties.

Furthermore, through its system of permeases, the plasma membrane also controls the exchanges between the cell and the medium. The functioning of these transport proteins is greatly influenced by its lipid composition, which affects membrane fluidity. In a defined environmental model, the supplementing of membrane phospholipids with unsaturated fatty acids (oleic and linoleic) promoted the penetration and accumulation of certain amino acids as well as the expression of GAP (Henschke and Rose, 1991). On the other hand, membrane sterols seem to have less influence on the transport of amino acids than the degree of unsaturation of the phospholipids. The synthesis of unsaturated fatty acids is an oxidative process and requires the aeration of the culture medium at the beginning of alcoholic fermentation. Under semi-anaerobic winemaking conditions, the amount of unsaturated fatty acids in the grape, or in the grape must, probably favors the membrane transport mechanisms of amino acids.

The transport systems of sugars across the membrane are far from being completely elucidated. There exists, however, at least two kinds of transport systems: a high affinity one for glucose and a low affinity system with one-tenth of the affinity of the former (Bisson, 1991). The low-affinity transport system is essential during the log growth phase, and its activity decreases during the stationary phase. The high-affinity system is, in contrast, repressed by high concentrations of glucose, as in the case of grape must (Salmon et al., 1993) (Figure 1.9).

The activity of sugar transport is dependent on the protein synthesis activity of cells. As soon as this activity slows down or stops, we observe a reduction in the sugar transport activity (phenomenon called catabolite inactivation) (Busturia and Lagunas, 1986; Salmon et al., 1993). Under winemaking conditions, the protein synthesis rate starts to decrease very early during alcoholic fermentation at the end of cell growth. Thus, the catabolite inactivation process of sugar transport systems intervenes and is accelerated in case of yeast-assimilable nitrogen deficiencies (Salmon, 1989).

The amount of sterols in the membrane, especially ergosterol, as well as the degree of unsaturation of the membrane phospholipids favor the penetration of glucose in the cell. This is especially true during the stationary and decline phases. This phenomenon explains the decisive influence of aeration on the successful completion of alcoholic fermentation during the yeast multiplication phase (Section 3.7.2).

The presence of ethanol, in a culture medium, slows the penetration speed of arginine and glucose into the cell and limits the efflux of protons resulting from membrane ATPase activity (Alexandre et al., 1994; Charpentier, 1994). Simultaneously, the presence of ethanol increases the synthesis of membrane phospholipids and their percentage of unsaturated fatty acids (especially oleic). Temperature and ethanol act in synergy to affect membrane ATPase activity. The concentration of ethanol required to cancel the proton efflux decreases as temperature rises. However, this modification of membrane ATPase activity by ethanol may not be the source of the decrease in plasma membrane permeability in an alcohol medium. The role of

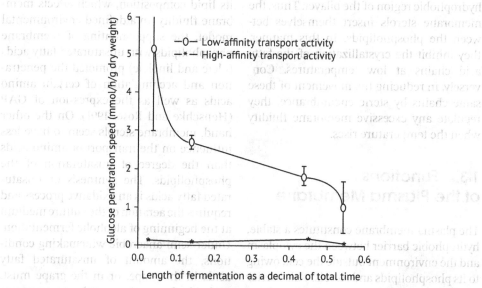

FIGURE 1.9 Evolution of glucose transport system activity of *S. cerevisiae* fermenting a model medium (Salmon et al., 1993).

membrane ATPase in yeast resistance to ethanol has not been clearly demonstrated.

The plasma membrane also produces cell wall glucan and chitin. Two membrane enzymes are involved: β-1,3-glucan synthase and chitin synthase. These two enzymes catalyze the polymerization of glucose and N-acetyl-glucosamine, derived from their activated forms (uridine diphosphate or UDP). Mannoproteins are essentially produced in the endoplasmic reticulum (ER; Section 1.4.2). They are then transported by vesicles that fuse with the plasma membrane and deposit their contents at the exterior of the membrane.

Finally, certain membrane proteins act as specific cell receptors. Therefore, the yeast can react to various external stimuli such as sexual hormones or changes in the concentration of external nutrients. The activation of these membrane proteins triggers the release of compounds such as cyclic adenosine monophosphate (cAMP) into the cytoplasm. These compounds serve as secondary messengers that set off other intercellular reactions. The consequences of these cell mechanisms in the alcoholic fermentation process merit further study.

1.4 The Cytoplasm and Its Organelles

Bounded by the plasma membrane and the nuclear membrane, the cytoplasm contains a basic cytoplasmic substance or cytosol. The organelles (ER, Golgi apparatus, vacuole, and mitochondria) are isolated from the cytosol by membranes.

1.4.1 Cytosol

Cytosol is a buffered solution, with a pH between 5 and 6, containing soluble enzymes, glycogen, and ribosomes.

Glycolysis and alcoholic fermentation enzymes (Sections 2.2.1 and 2.2.2) as well as trehalase (an enzyme catalyzing the hydrolysis of trehalose) are present. Trehalose is a reserve disaccharide, also cytoplasmic, that ensures yeast viability during the dehydration and rehydration phases by maintaining membrane integrity.

The lag phase, preceding the log growth phase in a sugar-containing medium, is marked by a rapid breakdown of trehalose, linked to an increase in trehalase activity. This activity is itself closely related to an increase in the amount of cAMP in the cytoplasm. This compound is produced by a membrane enzyme, adenylate cyclase, in response to the stimulation of a membrane receptor by an environmental factor.

Glycogen is the principal yeast carbohydrate reserve substance. Similar in structure to animal glycogen, it accumulates during the stationary phase in the form of spherical granules of about 40 µm in diameter.

When observed under the electron microscope, the yeast cytoplasm appears rich in ribosomes. These tiny granulations, made up of ribonucleic acids and proteins, are the center of protein synthesis. Associated with polysomes, several ribosomes migrate along the length of the messenger RNA. They translate it simultaneously so that each one produces a complete polypeptide chain.

1.4.2 The ER, the Golgi Apparatus, and the Vacuoles

The ER is a double membrane system partitioning the cytoplasm. It is linked to the cytoplasmic membrane and nuclear membrane. It is, in a way, an extension of the latter. Although less developed in yeasts than in exocrine gland cells of higher eukaryotes, the ER has the same function. It ensures the routing of the proteins synthesized by the attached ribosomes. As a matter of fact, ribosomes can be either free in the cytosol or bound to the ER. The proteins synthesized by free ribosomes remain in the cytosol, as do the enzymes involved in glycolysis. Those produced in the ribosomes bound to the ER have three possible destinations: the vacuole, the plasma membrane, and the external environment (secretion). The presence of a signal sequence (a particular chain of amino acids) at the N-terminal extremity of the newly formed protein determines the association of the initially free ribosomes in the cytosol with the ER. The synthesized protein crosses the ER membrane by an active transport process called translocation. This process requires the hydrolysis of an ATP molecule. Having reached the inner space of the ER, the proteins undergo certain modifications including the necessary excising of the signal peptide by signal peptidase. In many cases, they also undergo a glycosylation. The yeast glycoproteins, in particular the structural or enzymatic cell wall mannoproteins, contain carbohydrate side chains (Section 1.2.2). Some of these are linked to asparagine by N-glycosidic bonds. This oligosaccharidic linkage is constructed in the interior of the ER by the sequential addition of activated sugars (in the form of UDP derivatives) to a hydrophobic lipid transporter called dolichol phosphate. The entire unit is transferred in one piece to an asparagine residue of the polypeptide chain. Thus, the dolichol phosphate is regenerated.

The Golgi apparatus consists of a stack of membrane sacs and associated vesicles. It is an extension of the ER. Transfer vesicles transport the proteins issued from the ER to the sacs of the Golgi apparatus. The Golgi apparatus has a dual function. It is responsible for completing the glycosylation of proteins and then it sorts them so as to direct them via specialized vesicles either into the vacuole or into the plasma membrane. An N-terminal peptide (propeptide) sequence determines the routing of proteins toward the vacuole. This sequence is revealed in the precursors of two cytoplasm-to-vacuole targeting enzymes in the yeast: carboxypeptidase Y and proteinase A. However, the vesicles that transport the proteins of the plasma membrane or the secretion granules, like those that transport periplasmic invertase, are still the default destinations.

The vacuole is a spherical organelle, 0.3–3 μm in diameter, surrounded by a single membrane. Depending on the stage of the cellular cycle, yeasts have one or several vacuoles. Before budding, a large vacuole splits into small vesicles. Some penetrate into the bud, while others gather at the opposite end of the cell and fuse to form one or two large vacuoles. The vacuolar membrane, or tonoplast, has the same general structure (fluid mosaic) as the plasma membrane, but it is more elastic and its chemical composition is somewhat different. It is less rich in sterols and contains less protein and glycoprotein but more phospholipids with a higher degree of unsaturation. The vacuole stores some of the cell hydrolases, in particular carboxypeptidase Y, proteases A and B, aminopeptidase I, X-prolyl-dipeptidylaminopeptidase, and alkaline phosphatase. In this respect, the yeast vacuole can be compared to an animal cell lysosome. Vacuolar proteases play an essential role in the turnover of cellular proteins. In addition, protease A is indispensable in the maturation of other vacuolar hydrolases. It excises a small peptide sequence and thus converts precursor forms (proenzymes) into active enzymes. The vacuolar proteases are also responsible for autolysis, after cell death, when aging white wine on its lees.

Vacuoles also have a second principal function: they store metabolites before their use. In fact, they contain a quarter of the pool of amino acids of the cell, including a lot of arginine as well as S-adenosyl methionine. In this organelle, there is also potassium, adenine, isoguanine, uric acid, and polyphosphate crystals. These are involved in the fixation of basic amino acids. Specific permeases ensure the transport of these metabolites across the vacuolar membrane. An ATPase linked to the tonoplast furnishes the necessary energy for the movement of stored compounds against the concentration gradient. It is different from the plasma membrane ATPase, but also produces a proton efflux.

The ER, Golgi apparatus, and vacuoles must therefore be considered as different components of an internal system of membranes, called the vacuome, participating in the flux of glycoproteins to be excreted or stored.

1.4.3 The Mitochondria

Distributed on the periphery of the cytoplasm, the mitochondria (mt) are spherical or rod-shaped organelles surrounded by two membranes. The inner membrane is highly folded to form cristae. The general organization of mitochondria is the same as in higher plants and animal cells. The membranes delimit two compartments: the intermembrane space and the matrix. Mitochondria are true respiratory organelles for yeasts. Under aerobic conditions, the *S. cerevisiae* cell contains about 50 mitochondria. Under anaerobic conditions, these organelles degenerate, their inner surface decreases, and the cristae disappear. Supplementing the culture medium with ergosterol and unsaturated fatty acids limits the degeneration of mitochondria under anaerobic conditions. In any case, when cells formed under anaerobic conditions are placed under aerobic conditions, the mitochondria regain their normal appearance. Even in aerated grape must, the high sugar concentration represses the synthesis of respiratory enzymes. As a result, the mitochondria no longer function. This phenomenon is called catabolite repression by glucose (Section 2.3.1).

The mitochondrial membranes are rich in phospholipids—principally PC, PI, and PE (Figure 1.5). PG, a minority component in the plasma membrane, is predominant in the inner mitochondrial membrane. The fatty acids of the mitochondrial phospholipids are C16:0, C16:1, C18:0, and C18:1. Under aerobic conditions, the unsaturated residues predominate. When the cells are grown under anaerobic conditions, without lipid supplements, the short-chain saturated residues become predominant; cardiolipin and PE diminish, whereas the proportion of PI increases. Under aerobic conditions, the temperature during the log growth phase influences the degree of unsaturation of the phospholipids—which becomes higher as the temperature decreases.

The mitochondrial membranes also contain sterols, as well as numerous proteins and enzymes (Guerin, 1991).

The two membranes, inner and outer, contain enzymes involved in the synthesis of phospholipids and sterols. The ability to synthesize significant amounts of lipids, characteristic of yeast mitochondria, is not limited by respiratory-deficient mutations or glucose catabolite repression.

The outer membrane is permeable to most small metabolites coming from the cytosol, since it contains porin, a 29 kDa transmembrane protein possessing a large pore. Porin is present in the mitochondria of all eukaryotes as well as in the outer membrane of bacteria.

The intermembrane space contains adenylate kinase, which ensures interconversion of ATP, adenosine diphosphate (ADP), and adenosine monophosphate (AMP).

Oxidative phosphorylation takes place in the inner mitochondrial membrane, while the matrix, on the other hand, is the center of the reactions of the citric acid cycle and of the oxidation of fatty acids.

The majority of mitochondrial proteins are coded by the genes of the nucleus and are synthesized by the free polysomes of the cytoplasm. The mitochondria, however, also have their own machinery for protein synthesis. In fact, each mitochondrion possesses a circular 75 kb (kilobase pairs) molecule of double-stranded DNA and ribosomes. The mtDNA is extremely rich in A (adenine) and T (thymine) bases. It contains a few dozen genes, which code in particular for the synthesis of certain pigments and respiratory enzymes, such as cytochrome *b*, and several subunits of cytochrome oxidase and of the ATP synthase complex. Some mutations affecting these genes can result in the yeast becoming resistant to certain specific mitochondrial inhibitors such as oligomycin. This property has been applied in the genetic marking of wine yeast strains. Some mitochondrial mutants are respiratory deficient and form small colonies on solid agar media. These *"petite"* mutants are not used in winemaking because it is impossible to produce them industrially by respiration.

1.5 The Nucleus

The yeast nucleus is spherical. It has a diameter of 1–2 μm and is barely visible using a phase-contrast optical microscope. It is located near the principal vacuole in non-proliferating cells. The nuclear envelope is made up of a double membrane attached to the ER. It contains many ephemeral pores, whose locations are continually changing. These pores enable the exchange of small proteins between the nucleus and the cytoplasm. Contrary to what happens in higher eukaryotes, the yeast nuclear envelope is not dispersed during mitosis. In the basophilic part of the nucleus, the crescent-shaped nucleolus can be seen by using a nuclear-specific staining method. As in other eukaryotes, it is responsible for the synthesis of ribosomal RNA. During cell division, the yeast nucleus (Williamson, 1991) also contains rudimentary spindle threads composed of microtubules of tubulin, some discontinuous and others continuous (Figure 1.10). The continuous microtubules are stretched between the two spindle pole bodies (SPB). These corpuscles are permanently included in the nuclear membrane and correspond to the centrioles of higher organisms. The cytoplasmic microtubules depart from the SPB toward the cytoplasm.

The size of the yeast genome, about 12,800 kb, is low compared to that of higher eukaryotes (about 10 times lower than that of *Arabidopsis*). It has a genome almost three times larger than in *Escherichia coli*, but its genetic material is organized into true chromosomes. Each one contains a single molecule of linear double-stranded DNA associated with basic proteins known as histones, to form chromatin that contains repetitive units called nucleosomes. Because of their small size and their weak condensation, yeast chromosomes cannot be observed under the microscope.

Pulsed-field electrophoresis (Carle and Olson, 1984; Schwartz and Cantor, 1984) enables the separation of the 16 chromosomes in *S. cerevisiae*, whose sizes range from 200 to 2,000 kb. This species has a very large degree of chromosomic polymorphism. This characteristic has made karyotype analysis one of the main criteria for the identification of *S. cerevisiae* strains (Section 1.9.3).

The full chromosomal DNA sequence of *S. cerevisiae* (S288C) was established in 1996. It has 6,275 genes, including 23% in common with humans (Goffeau et al., 1996). In 2009, the EC1118 diploid genome of a yeast strain in wine was fully sequenced. It reveals the gene transfer mechanisms between *Saccharomyces* and non-*Saccharomyces*. These works show that the genome of winemaking yeast may be constantly remodeled by the addition of exogenous genes (Novo et al., 2009). This detailed knowledge of the yeast genome will constitute a powerful tool, both for the molecular understanding of its physiology and for the selection and improvement of winemaking strains. Current research done in the field of synthetic biology, as well as in enology, aims to create and assemble a full artificial genome of *S. cerevisiae* yeast with a number of potential medical and industrial applications, including in the field of enology (Richardson et al., 2017).

The yeast chromosomes contain relatively few repeated sequences. Most genes are only present in a single copy in the haploid genome, but the ribosomal RNA genes are highly repeated (about 100 copies).

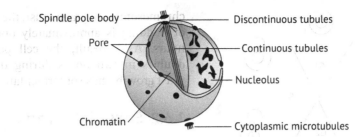

FIGURE 1.10 The yeast nucleus (Williamson, 1991).

The genome of *S. cerevisiae* also contains transposable elements or transposons—specifically, transposon yeast (Ty) elements. These comprise a central ε region (5.6 kb) framed by a direct repeated sequence called the δ sequence (0.25 kb). The δ sequences have a tendency to recombine, resulting in the loss of the central region and a δ sequence. As a result, there are about 100 copies of the δ sequence in the yeast genome. The Ty elements code for noninfectious retrovirus particles. This retrovirus contains Ty messenger RNA as well as a reverse transcriptase capable of copying the RNA into complementary DNA. The latter can reinsert itself into any site of the chromosome. The random excision and insertion of Ty elements in the yeast genome can modify the genes and play an important role in strain evolution.

A single plasmid, called the 2 μm plasmid, has been identified in the yeast nucleus. It is a circular molecule of DNA, containing 6 kb, and there are 50–100 copies per cell. Its biological function is not known. However, it is a very useful tool, used by molecular biologists to construct artificial plasmids and genetically transform yeast strains.

1.6 Reproduction and the Yeast Biological Cycle

Like other spore-forming yeasts belonging to the Ascomycetes class, *S. cerevisiae* can multiply either asexually by vegetative reproduction or sexually by forming ascospores. By definition, yeasts belonging to the imperfect fungi class can only reproduce by vegetative reproduction.

1.6.1 Vegetative Reproduction

Most yeasts undergo vegetative reproduction by a process called budding. Some yeasts, such as species belonging to the genus *Schizosaccharomyces*, reproduce by binary fission.

Figure 1.11 (Tuite and Oliver (1991)) represents the life cycle of *S. cerevisiae* divided into four phases: M, G1, S, and G2. M corresponds to mitosis; G1 is the period preceding S, which is the synthesis of DNA; and G2 is the period before mitosis. As soon as the bud emerges, at the beginning of S, the splitting of the SPB can be observed in the nuclear membrane by electron microscopy, while the cytoplasmic microtubules orient themselves toward the emerging bud. These microtubules seem to guide numerous vesicles that appear in the budding zone and are involved in the reshaping of the cell wall. As the bud grows larger, discontinuous nuclear microtubules begin to appear. The longest microtubules form the mitotic spindle between the two SPBs. At the end of G2, the nucleus begins to push and pull apart in order to penetrate the bud. Some of the mitochondria also pass with some small vacuoles into the bud, whereas a large vacuole is formed at the other pole of the cell. The expansion of the latter seems to push the nucleus into the

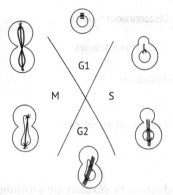

FIGURE 1.11 *Saccharomyces cerevisiae* cell cycle (vegetative reproduction) (Tuite and Oliver, 1991). M, mitosis; G1, period preceding DNA synthesis; S, DNA synthesis; G2, period preceding mitosis.

bud. During mitosis itself, the nucleus stretches to its maximum, and the mother cell separates from the daughter cell. This separation takes place only after the construction of the separation cell wall and the depositing of a ring of chitin on the bud scar of the mother cell. The movement of chromosomes during mitosis is difficult to observe in yeasts, but a microtubule–centromere link almost certainly guides the chromosomes. In grape must, the duration of budding is approximately one to two hours. As a result, the cell population doubles in two hours during the active yeast growth phase of fermentation.

1.6.2 Sexual Reproduction

When the diploid cells of spore-forming yeast are in a hostile nutrient medium (for example, depleted of fermentable sugar, poor in nitrogen, and very aerated), they stop multiplying. Some transform into a kind of sac with a thick cell wall. These sacs are called asci. Each one contains four haploid ascospores arising from meiotic division of the nucleus. Grape must and wine are not propitious to yeast sporulation and, in principle, it never occurs in this medium. Yet Mortimer *et al.* (1994) observed the sporulation of certain wine yeast strains, even in sugar-rich media. Our researchers have often observed asci in "old" agar culture media stored for several weeks in the refrigerator or at ambient temperatures (Figure 1.12). The natural

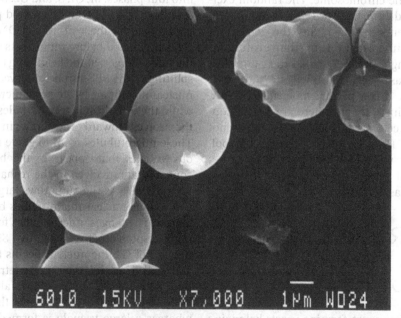

FIGURE 1.12 Scanning electron microscope photograph of *S. cerevisiae* cells kept on a sugar-agar medium for several weeks. Asci containing ascospores can be observed. (***Source***: Photograph from M. Mercier, Department of Electron Microscopy, Université de Bordeaux 1.)

conditions under which wild wine yeasts sporulate and the frequency of this phenomenon are not known. In the laboratory, the agar or liquid media conventionally used to provoke sporulation have a sodium acetate base (1%). In *S. cerevisiae*, aptitude for sporulation varies greatly from strain to strain. Wine yeasts, both wild and selected, do not sporulate easily, and when they do, they often produce nonviable spores.

Meiosis in yeasts and in higher eukaryotes (Figure 1.13) has some similarities. Several hours after the transfer of diploid vegetative cells to a sporulation medium, the SPB splits during the S phase of DNA replication. A dense body (DB) appears simultaneously in the nucleus near the nucleolus. The DB evolves into synaptonemal complexes—structures permitting the coupling and recombination of homologous chromosomes. After eight to nine hours in the sporulation medium, the two SPBs separate and the spindle begins to form. This stage is called metaphase I of meiosis. At this stage, the chromosomes are not yet visible. Then, while the nuclear membrane remains intact, the SPB divides. At metaphase II, a second mitotic spindle stretches itself while the ascospore cell wall begins to form. Spindle stretching and cell wall development of ascospores go hand in hand. Nuclear buds, cytoplasm, and organelles migrate into the ascospores. At this point, edification of the cell wall is completed. The spindle then disappears when the division is completed.

Placed under favorable conditions, i.e. sugar-enriched nutrient media, the ascospores germinate, breaking the cell wall of the ascus, and begin to multiply. In *S. cerevisiae*, the haploid cells have two mating types: **a** and *α*. The ascus contains two **a** ascospores and two *α* ascospores (Figure 1.14). Type **a** (*MAT***a**) cells produce a mating pheromone **a**. This peptide made up of 12 amino acids is called mating factor *α*. In the same manner, type *α* cells produce mating factor *α*, a peptide made up of 13 amino acids. The **a** factor, emitted by the *MAT***a** cells, stops the reproduction of *MATα* cells in G1, and reciprocally, the *α* factor produced by *α* cells stops the biological cycle of type **a** cells. Mating occurs between two cells of the opposite mating type. Their agglutination enables cellular and nuclear fusion and makes use of cell wall glycoproteins, called **a** and *α* agglutinins. The vegetative diploid cell that is formed (**a**/*α*) can no longer produce mating pheromones and is insensitive to their action; it reproduces by budding.

Some strains, from a monosporic culture, can be maintained in a stable haploid state. Their mating type remains constant for many generations. They are heterothallic. Others change mating type during cell division, causing diploid cells to appear in the descendants of an ascospore. They are homothallic and have an *HO* gene that inverses mating type at a high frequency during vegetative division. This interconversion (Figure 1.15) occurs in the mother cell at the G1 stage of the biological cycle, after the first budding but before the DNA replication phase. In this manner, a type *α* M ascospore divides to produce two *α* cells (M and the first daughter cell, D1). During the following cell division, M produces two cells (M and D2) that have become **a** cells. In the same

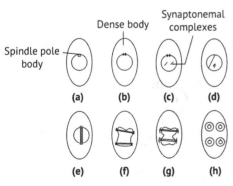

FIGURE 1.13 Meiosis in *S. cerevisiae* (Tuite and Oliver, 1991). (a) Cell before meiosis; (b) dividing of SPB; (c) synaptonemal complexes appear; (d) separation of SPBs; (e) constitution of spindle (metaphase I of meiosis); (f) dividing of the SPBs; (g) metaphase II of meiosis; (h) end of meiosis; formation of ascospores.

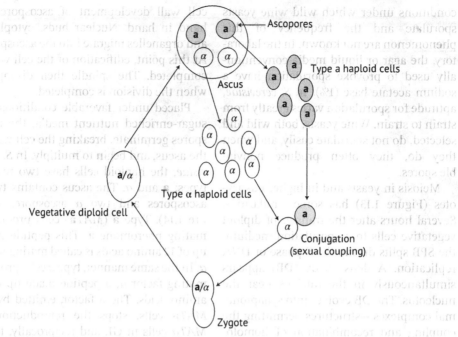

FIGURE 1.14 Reproduction cycle of a heterothallic yeast strain. **a**, α: spore mating types.

manner, the D1 cell produces two α cells after the first division and two **a** cells during its second budding. Laboratory strains that are deficient or missing the *HO* gene have a stable mating type. Heterothallism can therefore be considered the result of a mutation of the *HO* gene or of genes that control its functioning (Herskowitz et al., 1992).

Most wild and selected winemaking strains that belong to the *S. cerevisiae* species are diploid and homothallic. This is also true of almost all of the strains that have been isolated in vineyards of the Bordeaux region. Moreover, recent studies carried out by Mortimer et al. (1994) in Californian and Italian vineyards have shown that the majority of strains (80%) are homozy-

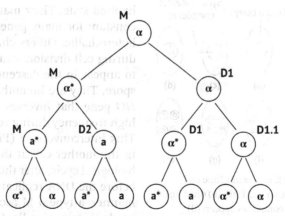

FIGURE 1.15 Mating type interconversion model of haploid yeast cells in a homothallic strain (Herskowitz et al., 1992). * designates cells capable of changing mating type at the next cell division or cells already having undergone budding. M, initial cell carrying the *HO* gene; D1, D2, daughter cells of M; D1.1, daughter cell of D1.

gous for the *HO* character (*HO/HO*); heterozygosis (*HO/ho*) is in the minority. Heterothallic strains (*ho/ho*) are rare (less than 10%). We have made the same observations for yeast strains isolated in the Bordeaux region; for example, the F10 strain, which is fairly prevalent in spontaneous fermentations in certain Bordeaux wines, is *HO/HO*. In other words, the four spores arising from an ascus give monoparental diploids, capable of forming asci when placed in a pure culture. This generalized homozygosis for the *HO* character of wild winemaking strains is probably an important factor in their evolution, according to the genome renewal phenomenon proposed by Mortimer *et al.* (1994) (Figure 1.16). According to this author, the continuous reproduction of a yeast strain in its natural environment is accompanied by the accumulation of heterozygotic damage to the DNA. Certain slow-growth or functional loss mutations of certain genes decrease strain vigor in the heterozygous state. Sporulation, however, produces haploid cells containing various combinations of these heterozygotic characters. All of these spores will become homozygous diploid cells with a series of genotypes because of the homozygosity of the *HO* character. Certain diploids that prove to be more vigorous than others will in time supplant the parents and less vigorous collaterals. This very persuasive model is supported by the characteristics of the wild winemaking strains analyzed. In these, the spore viability rate is an inverse function of the heterozygosis rate for a certain number of mutations. The completely homozygous strains present the highest spore viability and vigor.

In conclusion, we can question whether sporulation of strains under natural conditions is indispensable to ensure their growth and fermentation performance. We can also raise the question of conserving selected strains of active dry yeasts (ADYs) for use as yeast starters. It may be necessary to regenerate them periodically to eliminate possible mutations from their genome, which could diminish their vigor.

1.7 The Killer Phenomenon

1.7.1 Introduction

Certain yeast strains, known as killer strains (K), secrete protein toxins into their environment that are capable of killing other, sensitive strains (S). The killer strains are not sensitive to their own toxin but can be killed by a toxin that they do not

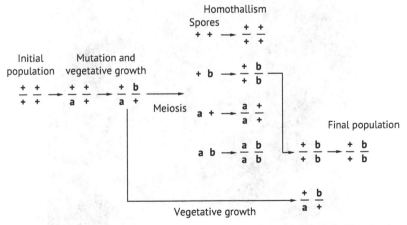

FIGURE 1.16 Genome renewal of a homozygote yeast strain for the *HO* gene of homothallism, having accumulated recessive mutations during vegetative reproduction (Mortimer et al., 1994). (a and b), mutation of certain genes.

produce. Additionally, neutral strains (N) do not produce a toxin but are resistant. The action of a killer strain on a sensitive strain is easy to demonstrate in the laboratory on an agar culture medium at pH 4.2–4.7 at 20°C. The sensitive strain is inoculated into the mass of agar before it solidifies; then the strain to be tested is inoculated in streaks on the solidified medium. If it is a killer strain, a clear zone in which the sensitive strain cannot grow encircles the inoculum streaks (Figure 1.17).

This phenomenon, known as the killer factor, was discovered in *S. cerevisiae*, but killer strains also exist in other yeast genera such as *Hansenula*, *Candida*, *Kloeckera*, *Hanseniaspora*, *Pichia*, *Torulopsis*, *Kluyveromyces*, and *Debaryomyces*. Killer yeasts have been classified into 11 groups according to the sensitivity reaction between strains as well as the nature and properties of the toxins involved. The killer factor is a cell interaction model mediated by the excreted protein toxin. It has given rise to much fundamental research (Tipper and Bostian, 1984; Young, 1987). Barre (1984, 1992), Radler (1988), and Van Vuuren and Jacobs (1992) have described the technological implications of this phenomenon for wine yeasts and the fermentation process.

1.7.2 Physiology and Genetics of the Killer Phenomenon

The determinants of the killer factor are both cytoplasmic and nuclear. In *S. cerevisiae*, the killer phenomenon is associated with the presence of double-stranded RNA particles, called virus-like particles (VLP), in the cytoplasm. They are in the same category as noninfectious mycoviruses. There are two kinds of VLP: M and L. The M genome (1.3–1.9 kb) codes for the toxin (K) and for the immunity factor (R). The L genome (4.5 kb) codes for an RNA polymerase and the protein capsid that encapsulates the two genomes. Killer strains (K^+R^+) secrete the toxin and are immune to it. The sensitive cells (K^-R^-) do not possess

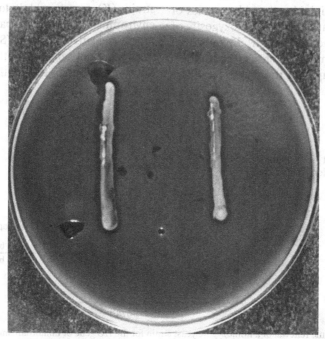

FIGURE 1.17 Identification of the K2 killer phenotype in *S. cerevisiae*. The presence of a halo around the two streaks of the killer strain is due to the death of the sensitive strain cultivated in the medium.

M VLP, but most of them have L VLP. The two types of viral particles are necessary for the yeast cell to express the killer phenotype (K^+R^+), since the L mycovirus is necessary for the maintenance of the M type.

There are four types of killer activity in *S. cerevisiae* strains, corresponding to K1, K2, K28, and Klus toxins. They are, respectively, coded by M1, M2, M28, and Mlus VLPs with sizes of 1.8, 1.7, 2.1, and 2.3 kb. The K2 activity group is by far the most widespread in the *S. cerevisiae* strains encountered in wine (Rodríguez-Cousiño *et al.*, 2011). Neutral strains (K^-R^+) are insensitive to a given toxin without being capable of producing it. They possess M VLPs of normal size that code only for the immunity factor, but they either do not produce toxins or are inactive because of mutations affecting the M-type RNA.

Many chromosomic genes are involved in the maintenance and replication of L and M RNA particles as well as in the maturation and transport of the toxin produced.

The K1 toxin is a small protein made up of two subunits (9 and 9.5 kDa); it is active and stable in a very narrow pH range (4.2–4.6). It is therefore inactive in grape must. The K2 toxin, a 16 kDa glycoprotein produced by homothallic strains of *S. cerevisiae* encountered in wine, is active between pH 2.8 and 4.8 with a maximum activity between 4.2 and 4.4. It is therefore active at the pH of grape must and wine.

K1 and K2 toxins attack sensitive cells by attaching themselves to a receptor located in the cell wall—a β-1,6-glucan. Two chromosomal genes, *KRE*1 and *KRE*2 (*k*iller *re*sistant), determine the possibility of this linkage. The *KRE*1 gene produces a cell wall glycoprotein that has β-1,6-glucan synthase activity. *KRE*1 mutants are resistant to K1 and K2 toxins because they are deficient in this enzyme and devoid of a β-1,6-glucan receptor. The *KRE*2 gene is also involved in the fixation of toxins to the cell wall receptor; *KRE*2 mutants are also resistant. The toxin linked to a glucan receptor is then transferred to a membrane receptor site by a mechanism requiring energy. Cells in the log phase are therefore more sensitive to the killer effect than cells in the stationary phase. When the plasma membrane of the sensitive cell is exposed to the toxin, it manifests serious functional alterations after a lag phase of about 40 minutes. Serious functional alterations include the interruption of the coupled transport of amino acids and protons, the acidification of the cell contents, and potassium and ATP leakage. The cell dies in two to three hours after contact with the toxin because of the above damage, due to the formation of pores in the plasma membrane.

The killer effect exerts itself exclusively on yeasts, and the pharmacological tests reported on killer toxins are negative.

1.7.3 The Role of the Killer Phenomenon in Winemaking

Depending on the authors and viticultural regions studied, the frequency of the killer character varies a lot among wild winemaking strains isolated on grapes or in fermenting grape must. In a work by Barre (1978) studying 908 wild strains, 504 manifested the K2 killer character, 299 were sensitive, and 95 neutral. Investigating the killer phenomenon among winemaking yeasts from French vineyards, Cuinier and Gros (1983) reported a high frequency (65–90%) of K2 strains in Mediterranean and Beaujolais vineyards, whereas none of the strains analyzed in the Touraine region manifested the killer effect. In the Bordeaux region, the K2 killer character is extremely widespread. Thus, in a study carried out in 1989 and 1990 on the ecology of native strains of *S. cerevisiae* in several tanks of red must in a Pessac-Léognan vineyard, all of the isolated strains manifested K2 killer activity, i.e. about 30, differentiated by their karyotype (Frezier, 1992). Rossini *et al.* (1982) reported an extremely varied frequency (12–80%) of killer strains in spontaneous fermentations in Italian wineries. Some K2 killer strains were also isolated in the Southern Hemisphere (Australia, South Africa, and Brazil). On the other hand, most of the killer strains isolated in Japan presented the K1 characteristic, and K2 strains are the minority. The Klus characteristic was highlighted in wine yeasts isolated in Spain (Rodríguez-Cousiño et al., 2011). Most research on the killer character of wine yeasts concerns the species *S. cerevisiae*. Recent works have reported data on the killer effect of non-*Saccharomyces* species, which make up the great proportion of grape microflora. Killer strains of *Hanseniaspora uvarum* and *Pichia kluyveri* have been identified by Zorg *et al.* (1988) and those of *Torulaspora delbrueckii* by Velázquez *et al.* (2015). Two toxins (PMKT and PMKT 2) have also been described in *Pichia membranifaciens* (review by Belda *et al.*, 2017). Certain non-*Saccharomyces* species are described for their capacity to produce an active killer toxin that affects the spoilage yeast *Brettanomyces bruxellensis*. For example, *Candida pyralidae* (Mehlomakulu *et al.*, 2017) may constitute a potential biocontrol agent to fight against *B. bruxellensis*.

In *S. cerevisiae*, Barre (1992) studied the activity and stability of the K2 killer toxin under winemaking conditions (Figure 1.18). The killer toxin only manifests pronounced activity on cells in the log phase. Cells in the stationary phase are relatively insensitive. The amount of ethanol or SO_2 in the wine has practically no effect on killer toxin activity. On the other hand, it is quickly destroyed by heat, since its half-life is around 30 minutes at 32°C. It is also quickly inactivated by the presence of phenolic compounds. Lastly, it is easily adsorbed by bentonite.

The role of the killer factor in the competition between species during alcoholic fermentation merits further study. However, its role in the competition between *S. cerevisiae* strains during grape must fermentation has been diversely reported in the literature. In an example given by Barre (1992), killer cells inoculated at 2% can completely supplant the sensitive strain during the alcoholic fermentation of must (Heard and Fleet, 1987). In other works, the killer yeast/sensitive yeast ratio able to affect the implantation of sensitive yeasts during winemaking varies between 1/1,000 and 100/1, depending on the author. This considerable discrepancy can probably be attributed to the implantation and fermentation speeds of the strains present. The killer phenomenon seems more important to interstrain competition when the killer strain implants itself quickly and the sensitive strain slowly. In the opposite situation, a high percentage of killer yeasts would be necessary to eliminate the sensitive population. Some authors have observed spontaneous fermentations dominated by sensitive strains despite a non-negligible proportion of killer strains (2–25%). In Bordeaux, we have always observed that certain sensitive strains establish themselves well in red wine fermentation, despite a strong presence of killer yeasts in the wild microflora (for example, 522M, an active dry yeast [ADY] starter). In white winemaking, the neutral yeast VL1 and sensitive strains such as EG8, a slow-growth strain, also successfully establish themselves. The wild killer population does not appear to compete with a sensitive yeast starter, and therefore is not a significant cause of fermentation difficulties in real-life applications.

As such, in light of the high frequency of killer strains among native yeasts in many viticultural regions, there is little advantage conferred on a selected killer strain in terms of implantation capacity. In other words, this character is not sufficient to guarantee the implantation of a certain strain during winemaking over a wild strain that also has killer activity. On the other hand, under certain conditions, inoculating with a sensitive strain will fail because of the killer effect of a wild population. Therefore, resistance to the K2 toxin (killer or neutral phenotype) should be included among the selection criteria

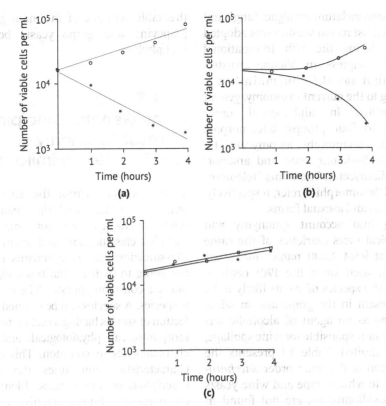

FIGURE 1.18 Yeast growth and survival curves in a grape juice medium containing killer toxin (Barre, 1992). (a) White juice at pH 3.4; cells in exponential growth phase introduced at t_0. (b) Same juice, cells in stationary phase introduced at t_0. (c) Red juice extracted by hot maceration, pH 3.4; cells in exponential growth phase introduced at t_0. *, 10% of an active supernatant of a K2 strain culture; o, 10% of supernatant inactivated by heating.

for winemaking strains. The high frequency of the K2 killer character in native wine yeasts greatly facilitates this strategy.

1.8 Classification of Yeast Species

1.8.1 General Remarks

As mentioned in the introduction to this chapter, yeasts constitute a vast group of single-celled fungi that are taxonomically heterogeneous and very complex. Hansen's first classification at the beginning of the 20th century only distinguished between sporogenous and asporogenous yeasts. Since then, yeast taxonomy has been the subject of considerable research. This research has been gathered in successive works, thus progressively creating the classification known today. The previous enological textbook from the University of Bordeaux (Ribéreau-Gayon et al., 1975) was based on Lodder's (1970) classification. Between the last edition of that book and the previous classification (Lodder and Kregger-Van Rij, 1952), the designation and classification of yeasts had already changed profoundly. In this book, we use the latest classification, given by Kurtzman et al. (2011), relying on the recent methods of molecular biology and genome analysis for the demarcation of species and genera.

Rules concerning the taxonomy of yeasts and other fungi fall under the International

Code of Nomenclature for algae, fungi, and plants. Its most recent version was adopted in 2005, during the 17th International Botanical Congress in Vienna, Austria (http://ibot.sav.sk/icbn/main.htm). According to the current taxonomy, genera are classified in alphabetical order according to four groups: teleomorphic ascomycetes, anamorphic ascomycetes, teleomorphic basidiomycetes, and anamorphic basidiomycetes. The terms "teleomorphic" and "anamorphic" refer, respectively, to the sexual and asexual forms.

Taking into account synonymy and physiological races (varieties of the same species), at least 4,000 names for yeasts have been used since the 19th century. However, the species of yeasts likely to be highly present in the grape and in wine, intervening as an agent of alcoholic fermentation or responsible for wine spoilage, are more limited. Table 1.1 presents the classification of the main order *Saccharomycetales*, to which grape and wine yeasts belong. Basidiomycetes are not found in this table, as none of the main genera of grapevine and grape yeasts belong to that phylum.

1.8.2 Evolution of the General Principles of Yeast Taxonomy and Species Delimitation

Yeast taxonomy (from the Greek *taxis*: putting in order), and the taxonomy of other microorganisms for that matter, includes classification and identification. Classification groups organisms into *taxa* according to their similarities and/or their ties to a common ancestor. The basic taxon is species. A species can be defined as a collection of strains having a certain number of morphological, physiological, and genetic characteristics in common. This group of characteristics constitutes the standard description of the species. Identification compares an unknown organism to individ-

TABLE 1.1

Classification of Grape and Wine Yeast Genera (Kurtzman et al., 1994)

Phylum	Subphylum	Order	Families	Genera
Ascomycota	Saccharomycotina	Saccharomycetales	Saccharomycetaceae	Saccharomyces
				Kluyveromyces
				Torulaspora
				Zygosaccharomyces
				Kazachstania
				Lachancea
			Pichiaceae	Pichia
				Brettanomyces
				Dekkera
			Metschnikowiaceae	Metschnikowia
			Debaryomycetaceae	Debaryomyces
				Lodderomyces
			Saccharomycodaceae	Hanseniaspora
				Kloeckera
				Saccharomycodes
		Saccharomycetales Incertae sedis	Wickerhamomycetaceae	Candida

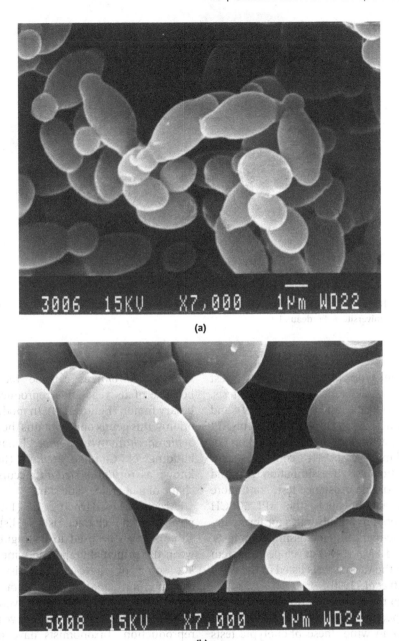

FIGURE 1.19 Observation of two winemaking yeast species having an apiculate form (*Source*: Photographs from M. Mercier, Department of Electron Microscopy, Université de Bordeaux 1.) (a) *H. uvarum*. (b) *S. ludwigii*.

uals already classified and named that have similar characteristics.

Taxonomists first delimited yeast species using morphological and physiological criteria. The first classifications were based on the phenotypic differences between yeasts: cell shape and size, spore formation, cultural characteristics, fermentation and assimilation of various sugars, assimilation of nitrates, growth-factor needs, and resistance to cycloheximide. Ribéreau-Gayon *et al.* (1975) described the application of these methods to wine yeasts in detail. Since then, many rapid, ready-to-use diagnostic kits

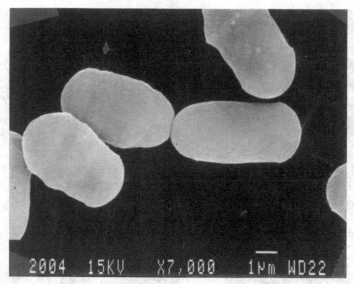

FIGURE 1.20 Binary fission of *S. pombe*. (*Source*: Photographs from M. Mercier, Department of Electron Microscopy, Université de Bordeaux 1.)

have been proposed to determine yeast response to different physiological tests. Lafon-Lafourcade and Joyeux (1979) and Cuinier and Leveau (1979) designed the API 20 C system for the identification of winemaking yeasts. It contains eight fermentation tests, 10 assimilation tests, and a cycloheximide resistance test. For a more complete identification, the API 50 CH system contains 50 substrates for fermentation (under paraffin) and assimilation tests. Heard and Fleet (1990) developed a system that uses the various tests listed in the work of Barnett *et al.* (1990).

Due to the relatively limited number of yeast species significantly present on grapes or in wine, these phenotypic tests identify winemaking yeast species in certain genera without difficulty. Certain species can be identified by observing growing cells under the microscope alone. Small apiculate cells, with small lemon-like shapes, are indicative of the species *H. uvarum* and its imperfect form *Kloeckera apiculata* (Figure 1.19). *Saccharomycodes ludwigii* is characterized by much larger (10–20 µm) apiculate cells. Since most yeast multiply by budding, the genus *Schizosaccharomyces* can be recognized because of its vegetative reproduction by binary fission (Figure 1.20). In modern taxonomy, this genus only contains the species *Schizosaccharomyces pombe*. Finally, the budding of *Candida stellata* (formerly known as *Torulopsis stellata*) occurs in the shape of a characteristic star.

According to Barnett *et al.* (1990), the physiological characteristics listed in Table 1.2 can be used to distinguish between the principal grape and wine yeasts. Yet some of these characteristics (for example, fermentation profiles of sugars) vary within the species and are even unstable for a given strain during vegetative reproduction. Taxonomists have realized that they cannot differentiate species based solely on phenotypic discontinuity criteria. They have progressively established a delimitation founded on the biological and genetic definition of a species.

In theory, a species can be defined as a collection of interfertile strains whose hybrids are themselves fertile—capable of producing viable spores. This biological definition runs into several problems when applied to yeasts. First of all, many

TABLE 1.2

Physiological Characteristics of the Principal Grape and Wine Yeasts (Barnett et al., 1990)

	F1 D-Glucose fermentation	F2 D-Galactose fermentation	F3 Maltose fermentation	F4 α-D-glucoside fermentation	F5 Sucrose fermentation	F6 α,α-Trehalose fermentation	F7 Melibiose fermentation
Candida stellata	+	-	-	-	+	-	-
Candida vini	v	-	-	-	-	-	-
Candida famata	v	v	-	-	v	v	v
Dekkera anomala	+	v	v	-	+	v	-
Dekkera bruxellensis	v	v	v	+	+	+	-
Hanseniaspora uvarum	+	-	-	-	-	-	-
Melschnikowia pulcherrima	+	v	-	-	-	v	-
Pichia anomala	+	v	v	v	+	v	-
Pichia fermentans	+	-	-	-	-	-	-
Pichia membranifaciens	v	-	-	-	-	-	-
Saccharomyces cerevisiae	+	v	v	v	v	v	v
Saccharomycodes ludwigii	+	-	-	v	+	-	-
Kluyveromyces thermolerens	+	v	v	v	+	v	-
Schizosaccharomyces pombe	+	v	v	v	+	v	v
Zygosaccharomyces bailii	+	-	-	-	v	v	-

	C30 Galactitol growth	C31 myo-Inositol growth	C32 D-Glucono-1,5-lactone growth	C33 2-Keto-D-gluconate growth	C34 5-Keto-D-gluconate growth	C35 D-gluconate growth	C38 DL-Lactate growth
Candida stellata	-	-	-	-	-	-	-
Candida vini	-	-	-	v	-	v	v
Candida famata	v	-	v	+	-	v	v
Dekkera anomala	-	-	v	-	-	v	v
Dekkera bruxellensis	-	-	v	v	-	v	v
Hanseniaspora uvarum	-	-	+	+	-	-	-
Melschnikowia pulcherrima	-	-	+	+	-	v	v
Pichia anomala	-	-	+	-	-	v	+
Pichia fermentans	-	-	v	-	-	-	+
Pichia membranifaciens	-	-	v	-	-	-	v
Saccharomyces cerevisiae	-	-	v	-	-	-	v
Saccharomycodes ludwigii	-	-	v	-	-	-	v
Kluyveromyces thermolerens	-	-	v	v	-	v	-
Schizosaccharomyces pombe	-	-	v	v	-	v	-
Zygosaccharomyces bailii	-	-	v	-	-	-	-

(Continued)

TABLE 1.2 (CONTINUED)

	F8 Lactose fermentation	F9 Cellobiose fermentation	F10 Melezitose fermentation	F11 Raffinose fermentation	F12 Inulin fermentation	F13 Starch fermentation	C1 D-Glucose growth
Candida stellata	-	-	-	v	-	-	+
Candida vini	-	-	-	-	-	-	+
Candida famata	-	v	v	v	-	-	+
Dekkera anomala	v	v	v	v	-	-	v
Dekkera bruxellensis	-	v	v	v	-	-	v
Hanseniaspora uvarum	-	v	-	-	-	-	+
Melschnikowia pulcherrima	-	-	-	-	-	-	+
Pichia anomala	-	v	-	v	-	v	+
Pichia fermentans	-	-	-	-	-	-	+
Pichia membranifaciens	-	-	-	-	-	-	+
Saccharomyces cerevisiae	-	-	-	v	-	-	+
Saccharomycodes ludwigii	-	-	v	v	-	v	+
Kluyveromyces thermolerens	-	-	-	+	v	-	+
Schizosaccharomyces pombe	-	-	v	v	v	v	+
Zygosaccharomyces bailii	-	-	-	v	-	-	+

	C39 Succinate growth	C40 Citrate growth	C42 Ethanol growth	N1 Nitrate growth	N2 Nitrite growth	N3 Ethylamine growth	N4 L-Lysine growth
Candida stellata	-	-	-	-	-	-	+
Candida vini	+	-	+	-	-	+	+
Candida famata	+	+	+	-	v	+	+
Dekkera anomala	v	-	v	v	+	+	+
Dekkera bruxellensis	-	-	v	v	+	+	+
Hanseniaspora uvarum	-	-	-	-	-	+	+
Melschnikowia pulcherrima	+	+	+	-	-	+	+
Pichia anomala	+	+	-	+	+	+	+
Pichia fermentans	+	+	+	-	-	+	+
Pichia membranifaciens	v	-	+	-	-	+	+
Saccharomyces cerevisiae	v	-	-	-	-	-	-
Saccharomycodes ludwigii	-	-	v	-	-	+	+
Kluyveromyces thermolerens	v	-	v	-	-	+	+
Schizosaccharomyces pombe	-	-	-	-	-	v	v
Zygosaccharomyces bailii	-	-	+	-	-	+	+

	C2 D-Galactose growth	C3 D-Sorbose growth	C4 D-Glucosamine growth	C5 D-Ribose growth	C6 D-Xylose growth	C10 Sucrose growth	C11 Maltose growth
Candida stellata	-	v	-	-	-	+	-
Candida vini	-	-	-	-	-	-	-
Candida famata	+	v	v	v	-	+	+
Dekkera anomala	v	-	v	v	+	v	v
Dekkera bruxellensis	v	-	v	v	-	v	v
Hanseniaspora uvarum	-	-	-	-	-	-	-
Melschnikowia pulcherrima	+	v	v	v	v	+	+
Pichia anomala	v	-	-	v	v	+	+
Pichia fermentans	-	-	v	-	+	-	-
Pichia membranifaciens	-	v	v	-	v	-	v
Saccharomyces cerevisiae	v	-	-	-	-	v	-
Saccharomycodes ludwigii	-	-	-	-	-	+	-
Kluyveromyces thermolerens	v	+	-	-	-	+	+
Schizosaccharomyces pombe	v	-	-	-	-	+	+
Zygosaccharomyces bailii	v	v	-	-	-	v	-

	N5 Cadaverine growth	N6 Creatine growth	N7 Creatinine growth	N8 Glucosamine growth	V1 Growth W/O vitamins	V2 Growth W/O myo-Inositol	V3 Growth W/O Pantothenate
Candida stellata	-	-	-	-	-	-	v
Candida vini	+	-	-	-	+	+	+
Candida famata	+	v	-	-	v	+	+
Dekkera anomala	+	-	-	-	-	v	-
Dekkera bruxellensis	+	-	-	-	-	+	+
Hanseniaspora uvarum	+	-	-	-	-	-	-
Melschnikowia pulcherrima	+	-	-	-	-	+	+
Pichia anomala	+	-	-	-	+	+	+
Pichia fermentans	+	-	-	+	-	+	+
Pichia membranifaciens	+	-	-	v	v	+	+
Saccharomyces cerevisiae	-	-	-	-	v	v	v
Saccharomycodes ludwigii	+	-	-	+	-	+	-
Kluyveromyces thermolerens	+	-	-	-	-	-	v
Schizosaccharomyces pombe	v	-	-	-	-	-	v
Zygosaccharomyces bailii	+	-	-	-	-	+	+

TABLE 1.2 (CONTINUED)

	C12 α,α-Trehalose growth	C13 Me α-D-Glucoside growth	C14 Cellobiose growth	C15 Salicin growth	C16 Arbutin growth	C17 Melibiose growth	C18 Lactose growth
Candida stellata	–	–	–	–	–	–	–
Candida vini	–	–	–	–	–	–	–
Candida famata	+	+	–	–	–	–	–
Dekkera anomala	v	v	v	v	v	v	v
Dekkera bruxellensis	v	v	v	v	v	–	v
Hanseniaspora uvarum	v	v	–	–	–	–	–
Melschnikowia pulcherrima	+	+	+	+	+	–	–
Pichia anomala	v	+	+	+	+	–	–
Pichia fermentans	–	–	v	–	–	–	–
Pichia membranifaciens	–	–	–	–	–	–	–
Saccharomyces cerevisiae	v	v	–	–	–	v	–
Saccharomycodes ludwigii	–	–	+	+	+	–	–
Kluyveromyces thermolerens	+	+	–	–	–	–	–
Schizosaccharomyces pombe	v	v	–	–	–	v	–
Zygosaccharomyces bailii	v	–	–	–	–	–	–

	V4 Growth W/O Biotin	V5 Growth W/O Thiamin	V6 Growth W/O Biotin and Thiamin	V7 Growth W/O Pyridoxine	V8 Growth W/O Pyridoxine and Thiamin	V9 Growth W/O Niacin	T2 Growth at 30°C
Candida stellata	–	–	–	+	–	v	+
Candida vini	v	v	v	+	–	+	v
Candida famata	v	+	v	+	–	+	+
Dekkera anomala	–	–	–	+	–	+	+
Dekkera bruxellensis	–	–	–	+	–	+	+
Hanseniaspora uvarum	v	v	–	–	–	–	+
Melschnikowia pulcherrima	–	+	+	+	–	+	+
Pichia anomala	+	+	+	+	+	+	v
Pichia fermentans	+	–	–	+	–	+	+
Pichia membranifaciens	v	v	v	+	v	+	+
Saccharomyces cerevisiae	v	v	v	v	v	+	+
Saccharomycodes ludwigii	–	–	–	–	–	–	+
Kluyveromyces thermolerens	v	v	v	–	–	v	+
Schizosaccharomyces pombe	v	v	v	v	–	–	+
Zygosaccharomyces bailii	–	+	–	+	–	+	+

	C19 Raffinose growth	C20 Melezitose growth	C21 Inulin growth	C22 Starch growth	C23 Glycerol growth	C24 Erythritol growth	C25 Ribitol growth
Candida stellata	+	-	-	-	-	-	-
Candida vini	-	-	-	-	v	-	v
Candida famata	+	+	v	v	+	v	+
Dekkera anomala	v	v	-	-	v	-	-
Dekkera bruxellensis	v	v	v	-	v	-	-
Hanseniaspora uvarum	-	-	-	-	-	-	-
Melschnikowia pulcherrima	-	+	-	+	+	+	v
Pichia anomala	v	v	-	-	+	-	v
Pichia fermentans	-	-	-	-	+	-	-
Pichia membranifaciens	-	-	-	-	v	-	-
Saccharomyces cerevisiae	v	v	-	v	v	-	-
Saccharomycodes ludwigii	v	-	-	-	+	-	-
Kluyveromyces thermolerens	+	+	v	-	+	-	v
Schizosaccharomyces pombe	+	-	v	v	-	-	-
Zygosaccharomyces bailii	v	-	-	-	v	-	v

	T3 Growth at 35°C	T4 Growth at 37°C	T5 Growth at 40°C	O1 0.01% Cyclo-heximide growth	O3 1% Acetic acid growth	O4 50% D-Glucose growth	M2 Acetic acid production
Candida stellata	v	-	-	-	-	v	-
Candida vini	-	-	-	-	-	+	-
Candida famata	v	v	-	v	-	-	+
Dekkera anomala	+	v	-	+	-	-	+
Dekkera bruxellensis	v	v	-	+	v	-	-
Hanseniaspora uvarum	-	-	-	-	-	v	-
Melschnikowia pulcherrima	v	-	-	-	-	v	-
Pichia anomala	v	v	-	-	-	v	-
Pichia fermentans	+	+	-	-	-	v	-
Pichia membranifaciens	v	v	-	-	-	v	-
Saccharomyces cerevisiae	v	v	v	-	-	-	-
Saccharomycodes ludwigii	+	v	-	-	-	-	-
Kluyveromyces thermolerens	+	-	-	-	-	+	-
Schizosaccharomyces pombe	+	+	-	v	+	+	-
Zygosaccharomyces bailii	v	v	-	-	-	+	-

TABLE 1.2 (CONTINUED)

	C26 Xylitol growth	C28 D-Glucitol growth	C29 D-Mannitol growth
Candida stellata	–	–	–
Candida vini	v	+	+
Candida famata	+	+	+
Dekkera anomala	–	–	–
Dekkera bruxellensis	–	–	–
Hanseniaspora uvarum	–	v	–
Melschnikowia pulcherrima	+	+	+
Pichia anomala	v	+	+
Pichia fermentans	–	–	–
Pichia membranifaciens	v	–	–
Saccharomyces cerevisiae	–	v	v
Saccharomycodes ludwigii	–	v	–
Kluyveromyces thermolerens	+	v	v
Schizosaccharomyces pombe	–	–	–
Zygosaccharomyces bailii	v	v	v

	M3 Urea hydrolysis
Candida stellata	–
Candida vini	–
Candida famata	–
Dekkera anomala	–
Dekkera bruxellensis	–
Hanseniaspora uvarum	–
Melschnikowia pulcherrima	–
Pichia anomala	–
Pichia fermentans	–
Pichia membranifaciens	–
Saccharomyces cerevisiae	–
Saccharomycodes ludwigii	–
Kluyveromyces thermolerens	–
Schizosaccharomyces pombe	+
Zygosaccharomyces bailii	–

+, test positive; –, test negative; v, variable result.
[a] Cannot be differentiated with these tests from S. bayanus, S. paradoxus, and S. pastorianus.

Ascomycetes yeasts are homothallic; hybridization tests are thus especially tedious and difficult for routine identification. Finally, certain wine yeast strains have little or no sporulation aptitude, which makes the use of strain infertility criteria even more difficult.

To overcome these difficulties, researchers have developed a molecular taxonomy based on the following tests: DNA–DNA reassociation, the similarity of DNA base composition, the similarity of enzymes, ultrastructure characteristics, and cell wall composition. The development of DNA sequence analysis has profoundly changed the way in which yeasts are identified. More specifically, the comparison of gene sequences coding for ribosome RNA now resolves the taxonomic position of closely related species, as well as the most distantly related taxa. The sequence analysis of variable domains 1 and 2 (D1/D2) has been successfully applied to the identification of ascomycetous yeasts (Kurtzman and Robnett, 1998). Then, Fell et al. (2000) developed the method by bringing in an additional database to identify basidiomycetous yeasts. Moreover, analysis of internal transcribed spacer (ITS) sequences is often used to identify species (Granchi et al., 1999). Currently, the sequencing of all genomes is used to refine the taxonomy of yeasts in phylogenomic studies.

1.8.3 Successive Classifications of the Genus *Saccharomyces* and the Position of Wine Yeasts in the Current Classification

Due to many changes in yeast classification and nomenclature since the beginning of taxonomic studies, wine-related yeast names and their positions in the classification have often changed. This has inevitably resulted in some confusion for winemakers. Even the most recent enological textbooks (Fleet, 1993; Delfini, 1995; Boulton et al., 1995) use a number of species names (*cerevisiae, bayanus, uvarum*, etc.) attached to the genus name *Saccharomyces* to designate yeasts responsible for alcoholic fermentation. Although still in use, this enological terminology is no longer accurate to designate the species currently delimited by taxonomists.

The evolution of species classification for the genus *Saccharomyces* since the early 1950s (Table 1.3) has created this difference between the naming of wine yeasts and current taxonomy. By taking a closer look at this evolution, the origin of the differences may be understood.

In Lodder and Kregger-Van Rij (1952), the names *cerevisiae, oviformis, bayanus, uvarum*, etc., referred to a number of the 30 species of the genus *Saccharomyces*. Ribéreau-Gayon and Peynaud (1960) considered that two principal fermentation species were found in wine: *S. cerevisiae* (formerly called *ellipsoideus*) and *Saccharomyces oviformis*, which was encountered especially toward the end of fermentation and was considered more ethanol resistant. The difference in their ability to ferment galactose distinguished the two species: *S. cerevisiae* (Gal$^+$) fermented galactose, whereas *S. oviformis* (Gal$^-$) did not. According to the same authors, the species *Saccharomyces bayanus* was rarely found in wines. Although it possessed the same physiological fermentation and sugar assimilation characters as *S. oviformis*, its cells were more elongated, its fermentation was slower, and it displayed specific behavior toward growth factors. As for the species *Saccharomyces uvarum*, identified in wine by many authors, it differed from *S. cerevisiae, S. oviformis*, and *S. bayanus* because it could ferment melibiose.

In Lodder's following edition (1970), the number of species of the genus *Saccharomyces* increased from 30 to 41. Some species formerly grouped with other genera were integrated into the genus *Saccharomyces*. Moreover, several species names were considered to be synonyms and disappeared altogether. Such was the case of *S. oviformis*, which was moved to the species *S. bayanus*.

TABLE 1.3 Evolution of the Nomenclature for the Saccharomyces Genus, 1952–2011 (Libkind et al. 2011)

1952	1970	1984	2011
Saccharomyces cerevisiae ⇒	Saccharomyces cerevisiae	Saccharomyces cerevisiae	Saccharomyces arboricolus[a]
Saccharomyces pastorianus	Saccharomyces aceti		Saccharomyces bayanus[a]
Saccharomyces bayanus ⇒	Saccharomyces bayanus		Saccharomyces cariocanus
Saccharomyces oviformis	Saccharomyces capensis		Saccharomyces cerevisiae
Saccharomyces logos	Saccharomyces prostoserdovii		Saccharomyces eubayanus
Saccharomyces chevalieri ⇒	Saccharomyces chevalieri		Saccharomyces kudriavzevii
Saccharomyces fructuum	Saccharomyces coreanus		Saccharomyces mikatae
Saccharomyces lactis	Saccharomyces diastaticus		Saccharomyces paradoxus
Saccharomyces elegans	Saccharomyces globosus		Saccharomyces pastorianus[a]
Saccharomyces heterogenicus	Saccharomyces heterogenicus		Saccharomyces uvarum
Saccharomyces fermentati	Saccharomyces hienipiensis		
Saccharomyces mellis	Saccharomyces inusitatus		
Saccharomyces italicus ⇒	Saccharomyces italicus		
Saccharomyces steineri	Saccharomyces norbensis		
Saccharomyces pastori	Saccharomyces oleaceus		
Saccharomyces carlsbergensis ⇒	Saccharomyces oleaginosusss		
Saccharomyces uvarum ⇒	Saccharomyces uvarum		

1952	1970	1984	2011
Saccharomyces acidifaciens	Saccharomyces incompspicuus		
Saccharomyces bailii ⇒	Saccharomyces amurcae		
Saccharomyces fragilis	Saccharomyces bailii		
	Saccharomyces cidri		
Saccharomyces delbrueckii ⇒	Saccharomyces dairensis⇒	Saccharomyces dairensis	
Saccharomyces marxianus	Saccharomyces delbrueckii		
Saccharomyces exiguus ⇒	Saccharomyces eupagycus		
Saccharomyces veronae	Saccharomyces exiguus ⇒	Saccharomyces exiguus	
Saccharomyces florentinus ⇒	Saccharomyces fermentati		
Saccharomyces bisporus ⇒	Saccharomyces florentinus		
Saccharomyces willianus	Saccharomyces bisporus		
	Saccharomyces kloeckerianus		
	Saccharomyces kluyveri ⇒	Saccharomyces kluyveri	
Saccharomyces microellipsodes ⇒	Saccharomyces microellipsodes		
	Saccharomyces montanus		
	Saccharomyces mrakii		
	Saccharomyces pretoriensis		

TABLE 1.3 (CONTINUED)

1952	1970	1984	2011
Saccharomyces rosei ⇒	Saccharomyces rosei		
Saccharomyces rouxii ⇒	Saccharomyces rouxii		
	Saccharomyces saitoanus		
	Saccharomyces telluris ⇒	Saccharomyces telluris	
	Saccharomyces transvaalensis		
	Saccharomyces unisporus ⇒	Saccharomyces unisporus	
	Saccharomyces vafer	Saccharomyces servazzii	

Source: From Barnett et al. (2000), Lodder and Kregger-Van Rij (1952), Lodder (1970), Kregger-Van Rij (1984), and Kurtzman et al. (2011).
[a] Interspecific hybrids.

Ribéreau-Gayon et al. (1975) considered, however, that the distinction between *S. oviformis* and *S. bayanus* was of enological interest because of the different technological characteristics of these two yeasts. Nevertheless, by the early 1980s, most enology texts had abandoned the name *S. oviformis* and replaced it with *S. bayanus*.

The new classification by Kregger-Van Rij (1984), based on Yarrow's work on percentages of guanine and cytosine bases in yeast DNA, brought forth another important change in the designation of *Saccharomyces* species. Only seven species continued to exist, while 17 names became synonyms of *S. cerevisiae*. As with the preceding yeast classification, these yeast were differentiated by their sugar utilization profile (Table 1.4), and some authors considered them to be races or physiological varieties of *S. cerevisiae*. However, this was nothing more than an artificial taxonomy without biological significance. Enologists took the habit of adding the varietal name to *S. cerevisiae* to designate wine yeasts: *S. cerevisiae* var. *cerevisiae*, var. *bayanus*, var. *uvarum*, var. *chevalieri*, etc. In addition, two species, *bailii* and *rosei*, were removed from the genus *Saccharomyces* and integrated into two other genera to become *Zygosaccharomyces bailii* and *T. delbrueckii*, respectively.

Based on recent advances in genetics and molecular taxonomy, the latest yeast classification (Kurtzman et al., 2011) and recent studies (Libkind et al., 2011) have again thrown the species delimitation of the genus *Saccharomyces* into confusion. There are now 10 species (Table 1.3). The species *Saccharomyces paradoxus*, *Saccharomyces arboricolus*, *Saccharomyces mikatae*, and *Saccharomyces eubayanus* include strains initially isolated from tree exudates, insects, and soil (Naumov et al., 1998, 2000a; Wang and Bai, 2008; Libkind et al., 2011). Using genome analysis, Redzepovic has nevertheless identified a high percentage of *S. paradoxus* in Croatian grape microflora in Redzepovic et al. (2002).

TABLE 1.4

Physiological Races of *S. cerevisiae* Grouped Under a Single Species *S. cerevisiae* by Yarrow and Nakase (1975)

	Fermentation					
	Ga	Su	Ma	Ra	Me	St
Saccharomyces						
aceti	−	−	−	−	−	−
bayanus	−	+	+	+	−	−
capensis	−	+	−	+	−	−
cerevisiae	+	+	+	+	−	−
chevalieri	+	+	−	+	−	−
coreanus	+	+	−	+	+	−
diastaticus	+	+	+	+	−	+
globosus	+	−	−	−	−	−
heterogenicus	−	+	+	−	−	−
hienipiensis	−	−	+	−	+	−
inusitatus	−	+	+	+	+	−
norbensis	−	−	−	−	+	−
oleaceus	+	−	−	+	+	−
oleaginosus	+	−	+	+	+	−
prostoserdovii	−	−	+	−	−	−
steineri	+	+	+	−	−	−
uvarum	+	+	+	+	+	−

Ga, D-galactose; Su, sucrose; Ma, maltose; Ra, raffinose; Me, melibiose; St, soluble starch.

TABLE 1.5

DNA/DNA Reassociation Percentages Between the Four Species Belonging to the *Saccharomyces* Genus in the Strict Sense (Vaughan Martini and Martini, 1987)

	Saccharomyces cerevisiae	Saccharomyces bayanus	Saccharomyces pastorianus	Saccharomyces paradoxus
Saccharomyces cerevisiae	100			
Saccharomyces bayanus	20	100		
Saccharomyces pastorianus	58	70	100	
Saccharomyces paradoxus	53	24	24	100

Saccharomyces pastorianus replaces the former name *Saccharomyces carlsbergensis*, given to brewery yeast strains, used for low-temperature fermentations (lager) and until then included in the *cerevisiae* species. The placement of *S. bayanus*, *S. uvarum*, and *S. pastorianus* within the *Saccharomyces* genus was the subject of controversies among yeast scientists for years. The latest works, based on DNA sequencing and the recent discovery of the *S. eubayanus* species, have definitively determined the taxonomic positions of *S. bayanus* and *S. pastorianus* (Tables 1.4 and 1.5; Nguyen and Gaillardin, 2005; Libkind *et al.*, 2011; Nguyen and Boekhout, 2017). It has been clearly established that *S. bayanus* and *S. pastorianus* refer to hybrid individuals, composed of *S. eubayanus*, *S. uvarum*, and *S. cerevisiae* genomes. On the other hand, *S. uvarum* and *S. eubayanus* are considered as being of genetically pure lineage.

Thus, *S. bayanus* is now considered a distinct hybrid species of *S. cerevisiae* and *S. uvarum* according to taxonomists. Nevertheless, enologists and winemakers use the name of *bayanus (ex oviformis)* to designate a physiological race of *S. cerevisiae* that does not ferment galactose. It possesses a greater resistance to ethanol than *S. cerevisiae*. The implementation of molecular biology methods, based on DNA analysis, has helped establish the position of the winemaking yeasts formerly designated as var. *bayanus* and var. *uvarum* in agreement with the current taxonomy.

For some 30 years, fragment amplification of the genome by the polymerase chain reaction (PCR) has provided an excellent discrimination tool for winemaking yeast species.

Since its discovery by Saiki *et al.* (1985), PCR has often been used to identify different plant and bacteria species. This technique consists in enzymatically amplifying one or several gene fragments *in vitro*. The reaction is based on the hybridization of two oligonucleotides that frame a target region on a double strand of DNA or template. These oligonucleotides have different sequences and are complementary to the DNA sequences that frame the strand being amplified. Figure 1.21 illustrates the various stages of the amplification process. The DNA is first denatured at a high temperature (95°C). The reaction mixture is then cooled to a temperature between 37 and 55°C, enabling the hybridization of these oligonucleotides on the denatured strands. The strands serve as primers from which a DNA polymerase enables the step-by-step addition of deoxyribonucleotide units in the 5′–3′ direction. The DNA polymerase (Figure 1.22) requires four deoxyribonucleoside-5Π-triphosphates (dATP, dGTP, dTTP, and dCTP). A

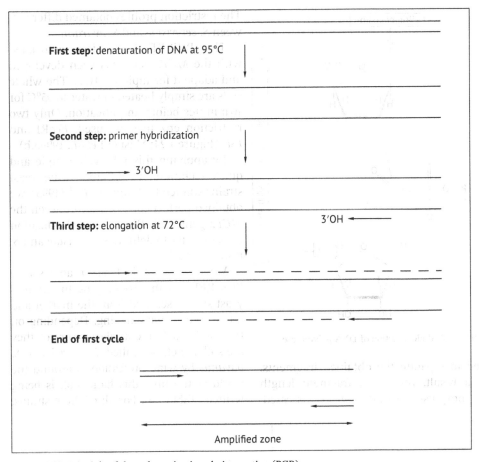

FIGURE 1.21 Principle of the polymerization chain reaction (PCR).

phosphodiester bond is formed between the 3П–OH end of the primer and the innermost phosphorus of the activated deoxyribonucleoside. Pyrophosphate is thus released. The newly synthesized strand is elongated on the template. A heat-resistant enzyme, TAQ DNA polymerase, comes from the heat-resistant bacteria *Thermus aquaticus*. It is used to conduct a large number of amplification cycles (25–40) *in vitro* without having to add DNA polymerase after each denaturation. In this manner, the DNA fragment amplified during the first cycle serves as the template for the following cycles. In consequence, each successive cycle doubles the target DNA fragment, which is thus amplified by a factor of 10^5 to 10^6 during 25–30 amplification cycles.

Hansen and Kielland-Brandt (1994) proposed *MET*2 gene PCR amplification to differentiate between *S. cerevisiae* and *S. uvarum* (formerly *S. bayanus*), when working on strain types of these two species. This gene, which codes for the synthesis of homoserine acetyltransferase, has different sequences in the two species. Part of the gene is initially amplified by using two complementary oligonucleotides of the sequences bordering the fragment to be amplified. The fragment obtained (about 600 bp) is the same size for typical strains of the *S. cerevisiae* and *S. uvarum* species. Different restriction endonucleases, which recognize certain specific DNA sequences, then digest the amplified fragment. Figure 1.23 gives an example of the mode of action of the *Eco*R1 restriction endonuclease. This enzyme recognizes the base sequence GAATTC and cuts at the location indicated by the arrows. Electrophoresis is

44 Chapter 1 Yeasts

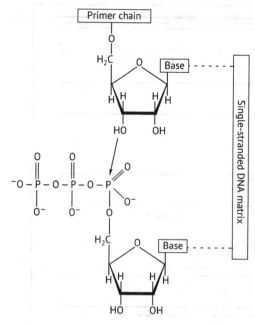

FIGURE 1.22 Mode of action of DNA polymerase.

used to separate the obtained fragments. As a result, restriction fragment length polymorphism (RFLP) can be assessed.

The restriction profiles obtained differ between *S. cerevisiae* and *S. uvarum*.

This PCR-RFLP technique associated with the *MET2* gene has been developed and adapted for rapid analysis. The whole cells are simply heated in water to 95°C for 10 minutes before amplification. Only two restriction enzymes are used: *Eco*R1 and *Pst*1 (Figure 1.24; Masneuf *et al.*, 1996a,b).

By applying this relatively simple and quick technique to different wine yeast strains studied by Naumov *et al.* (1993), we obtained perfect concordance between the *MET2* gene PCR-RFLP and hybridization tests in order to delimit *S. cerevisiae* and *S. uvarum* species.

We extended this type of analysis by PCR-RFLP of the *MET2* gene to different yeast strains selected from the market and often used in winemaking. Depending on their ability to ferment galactose or not, they are still sometimes called *S. cerevisiae* or *S. bayanus* by wine professionals around the world at the time this handbook is being written (Table 1.6). For all of these strains,

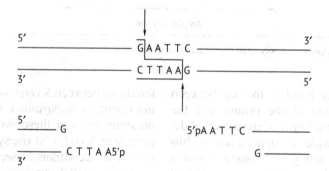

FIGURE 1.23 Recognition site and cutting mode of an *Eco*R1 restriction endonuclease.

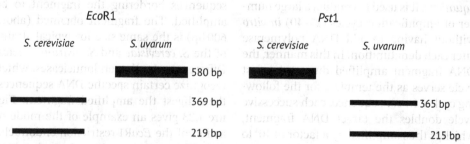

FIGURE 1.24 Identification principles for the *S. cerevisiae* and *S. uvarum* species by the *MET2* gene PCR-RFLP technique, after cutting the amplified fragment with *Eco*R1 and *Pst*1 restriction enzymes.

TABLE 1.6
Characterization by PCR-RFLP of the *MET2* Gene of Various Commercial Strains of the *S. cerevisiae* Species Used in Winemaking (Masneuf, 1996)

Strains	Commercial brand	Origin	Enological designation	Species
VL1	Zymaflore VL1	FŒB	*Saccharomyces cerevisiae*	*Saccharomyces cerevisiae*
VL3c	Zymaflore VL3	FŒB	*Saccharomyces cerevisiae*	*Saccharomyces cerevisiae*
WET 136	Siha levactif 3	Dormstadt	*Saccharomyces cerevisiae*	*Saccharomyces cerevisiae*
71B	Actiflore primeur	INRA Narbonne	*Saccharomyces cerevisiae*	*Saccharomyces cerevisiae*
F10	Zymaflore F10	FŒB	*Saccharomyces cerevisiae*	*Saccharomyces cerevisiae*
R2	Vitlevure KD	NA	*Saccharomyces bayanus*	*Saccharomyces cerevisiae*
BO213	Actiflore bayanus	Institut Pasteur	*Saccharomyces bayanus*	*Saccharomyces cerevisiae*
CH158	Siha levactif 4	NA	*Saccharomyces bayanus*	*Saccharomyces cerevisiae*
QA23	Lalvin QA23	UTM	*Saccharomyces bayanus*	*Saccharomyces cerevisiae*
IOC182007	IOC 182007	IŒC	*Saccharomyces bayanus*	*Saccharomyces cerevisiae*
DV10	Vitlevure DV10	CIVC	*Saccharomyces bayanus*	*Saccharomyces cerevisiae*
O16	Lalvin O16	UB	*Saccharomyces bayanus*	*Saccharomyces cerevisiae*
Epernay2	Uvaferm CEG	NA	*Saccharomyces bayanus*	*Saccharomyces cerevisiae*

FŒB: Faculté d'Œnologie de l'Université de Bordeaux II, Talence, France. UTM: Université de Tras os Montes, Portugal. ŒC: Institut Œnologique de Champagne, France. CIVC: Comité Interprofessionnel des Vins de Champagne (Interprofessional Champagne Committee), Epernay, France. UB: Université de Bourgogne, Dijon, France. NA: not available.

TABLE 1.7

Characterization by PCR-RFLP of the *MET2* Gene of Various Species of Native *Saccharomyces* Isolated on the Grape and in Wine (Masneuf, 1996)

Number of different strains analyzed	Origin	Collection	Enological designation	Species
8	Sauternes wines	FŒB	*Saccharomyces cerevisiae*	*Saccharomyces cerevisiae*
2	Dry white Bordeaux wines	FŒB	*Saccharomyces bayanus*	*Saccharomyces cerevisiae*
9	Sauternes wines	FŒB	*Saccharomyces bayanus*	*Saccharomyces cerevisiae*
2	Dry white Bordeaux wines	FŒB	*Saccharomyces chevalieri*	*Saccharomyces cerevisiae*
1	Sauternes wines	FŒB	*Saccharomyces capensis*	*Saccharomyces cerevisiae*
36	Unknown	Lallemand	*Saccharomyces bayanus*	*Saccharomyces cerevisiae*
11	Sauternes wines	FŒB	*Saccharomyces uvarum*	*Saccharomyces uvarum*
1	Sauternes wines	FŒB	*Hanseniaspora uvarum*	*Saccharomyces cerevisiae*
10	Sancerre and Pouilly Fumé (Loire) wines	FŒB	*Hanseniaspora uvarum*	*Saccharomyces uvarum*
1	Sancerre grapes	FŒB	*Hanseniaspora uvarum*	*Hanseniaspora uvarum*
2	Unknown	Lallemand	*Hanseniaspora uvarum*	*Saccharomyces cerevisiae*

FŒB: Faculté dOEnologie de l'Université de Bordeaux II, Talence, France. Lallemand: Lallemand Inc, Montreal, Quebec, Canada.

we have obtained the same characteristic restriction profiles of the species *S. cerevisiae*.

Moreover, we have determined the species of 82 strains of native *Saccharomyces* isolated from fermenting wine or from grapes (Table 1.7). For the eight Gal⁺ Mel⁻ strains analyzed, as for the 47 Gal⁻ Mel⁻, respectively, called *S. cerevisiae* and *S. bayanus* by enologists, restriction profiles of the amplified fragment of the *MET2* gene are characteristic of *S. cerevisiae*. The same goes for the two *S. chevalieri* strains, which ferment galactose but not maltose (Ma⁻), as well as for the *S. capensis* strain (Gal⁻ Ma⁻). As for Mel⁺ strains, called *S. uvarum* until now, most of them (11 out of 12 for Sauternes isolates and 11 out of 11 for Sancerre isolates) belong to the *S. uvarum* species. However, some Mel⁺ strains are *S. cerevisiae* (one strain from Sauternes and two strains from the Lallemand collection). In short, regarding the classification of the main winemaking yeasts (Section 1.8.3), we can distinguish between three stages. First, we considered them to be several species: *S. cerevisiae*, *S. bayanus*, and/or *S. oviformis*, *S. uvarum*. Then, we decided they were different races of the *S. cerevisiae* species. The current taxonomy, based on molecular biology results, has made substantial changes. It has defined three species: *S. cerevisiae*, *S. uvarum*, and *S. paradoxus*. The involvement of *S. paradoxus* in the fermentation microflora of grapes remains to be confirmed.

All of the results of molecular taxonomy presented above show that the former phenotypic classifications, based on physiological identification criteria, are not suitable even for delimiting the small number of fermentative species of the *Saccharomyces* genus found in winemaking. Moreover, specialists have long known about the instability of the physiological properties of *Saccharomyces* strains. Rossini *et al.* (1982) reclassified a thousand strains from the yeast collection of the Microbiology Institute of Agriculture at the University of Perugia. They observed that 23 out of 591 *S. cerevisiae* strains conserved on malt agar had lost the ability to ferment galactose. The use of genetic methods is thus essential to identify winemaking yeasts.

1.8.4 Interspecific Hybrids

PCR-RFLP associated with the *MET2* gene can be used to demonstrate the existence of hybrids between the species *S. cerevisiae* and *S. uvarum*. This method has been used to prove the existence (Masneuf *et al.*, 1998) of one such natural hybrid (strain S6U) among commercial dry yeasts sold by Lallemand Inc. (Montreal, Canada). Ciolfi (1992, 1994) isolated this yeast in an Italian winery, and it was selected for certain enological properties, in particular its aptitude to ferment at low temperatures, its low production of acetic acid, and its ability to preserve must acidity. The *MET2* gene restriction profiles of this strain by *Eco*R1 and *Pst*1, composed of three bands, are identical (Figure 1.25). In addition to the amplified fragment, two bands characteristic of *S. cerevisiae* with *Eco*R1 and two bands characteristic of the species *S. bayanus* with *Pst*1 are obtained. The bands are not artifacts due to an impurity in the strain, because the amplification of the *MET2* gene carried out on subclones (obtained from the multiplication of unique cells isolated by a micromanipulator) produces identical results. Hansen from the Carlsberg laboratory (Denmark) sequenced two of the *MET2* gene alleles from this strain. The sequence of one of the alleles is identical to that of the *S. cerevisiae* *MET2* gene, with the exception of one nucleotide. The sequence of the other allele is 98.5% similar to that of *S. uvarum*. The presence of this allele is thus probably due to an interspecific cross.

Subsequently, more recent research (Naumov *et al.*, 2000b) has shown that the S6U strain is, in fact, a tetraploid hybrid. Indeed, the percentage germination of spores from 24 tetrads, isolated using a micromanipulator, was very high (94%), whereas it would have been very low for a

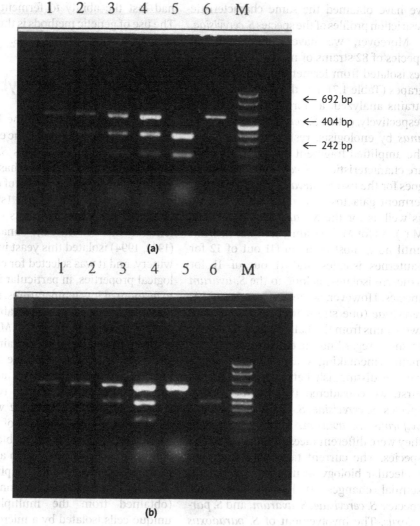

FIGURE 1.25 Electrophoresis in agarose gel (1.8%) of (a) *Eco*R1 and (b) *Pst*1 digestions of the amplified fragments of the *MET2* gene of the hybrid strain. Bands 1, 2, 3, subclones of the hybrid strain; band 4, hybrid strain; band 5, *S. cerevisiae* control; band 6, *S. uvarum* control; M, molecular weight marker.

"normal" diploid interspecific hybrid. The monospore clones in this first generation (D1) were all capable of sporulating. However, none of the ascospores of the second-generation tetrads was viable. The hybrid nature of the monospore clones produced by D1 was confirmed by the presence of the *S. cerevisiae* and *S. uvarum MET2* gene, identified by PCR/RFLP. Finally, measuring the DNA content per cell using flux cytometry estimation confirmed that the descendants of S6U were interspecific diploids and that S6U itself was an allotetraploid.

The molecular characterization of wine yeasts isolated on grapes and in must undergoing spontaneous fermentation revealed the existence of many natural hybrids between *S. cerevisiae* and *S. uvarum* (Le Jeune et al., 2007). This is also true between *S. cerevisiae* and *Saccharomyces kudriavzevii* (Sipiczki, 2008; Arroyo-López et al., 2009; Erny et al., 2012). Interspecific hybridization leads to new gene combinations in a given cell and may confer a selective advantage with respect to parental strains. Interspecific hybrids possess interesting technological properties for winemaking. For example, *S. uvarum* and *S. kudriavzevii* are better adapted to low-temperature growth, and *S. cerevisiae* presents high tolerance to ethanol. Natural hybrids between these species are more adapted to growth in a large range of temperatures and at high ethanol concentrations, thanks to the genetic heritage of one or both parents, thus yielding properties of interest (Da Silva *et al.*, 2015). This mechanism for acquiring superior phenotypic characteristics in hybrid descendants during cross-breeding of parental strains is well described in plants. This is called a heterosis effect.

1.9 Identification of Wine Yeast Strains

1.9.1 General Principles

The principal yeast species involved in grape must fermentation, particularly *S. cerevisiae* and *S. uvarum*, comprise a very large number of strains with extremely varied technological properties. The yeast strains involved during winemaking influence fermentation speed, the nature and quantity of secondary products formed during alcoholic fermentation, and the aroma characteristics of the wine. The ability to differentiate between the different strains of *S. cerevisiae* is required for the following fields: the ecological study of wild yeasts responsible for the spontaneous fermentation of grape must, the selection of strains presenting the best enological qualities, production and marketing controls, the verification of the implantation of selected yeasts used as yeast starter, and the constitution and maintenance of wild or selected yeast collections.

The initial research on infraspecific differentiation within *S. cerevisiae* attempted to distinguish strains by electrophoretic analysis of their exocellular (Bouix *et al.*, 1981) or intracellular (Van Vuuren and Van Der Meer, 1987) proteins or glycoproteins. Other teams proposed identifying the strains by the analysis of long-chain fatty acids using gas chromatography (Tredoux *et al.*, 1987; Augustyn *et al.*, 1991; Bendova *et al.*, 1991; Rozes *et al.*, 1992). Although these different techniques differentiate between certain strains, they are irrefutably less discriminating than genetic differentiation methods. They also present the major drawback of depending on the physiological state of the strains and on cultural conditions, which must always be identical.

In the late 1980s, owing to the development of genetics, certain techniques from molecular biology were successfully applied to characterize wine yeast strains. They are based on the clonal polymorphism of the mitochondrial and genomic DNA of *S. cerevisiae* and *S. uvarum*. These genetic methods are independent of the physiological state of the yeast, unlike the previous techniques based on the analysis of metabolism by-products.

1.9.2 Mitochondrial DNA Analysis

The mtDNA of *S. cerevisiae* has two remarkable properties: it is extremely polymorphic, depending on the strain; and stable (it mutates very little) during vegetative reproduction. Restriction endonucleases (such as *Eco*R5) cut this DNA at specific sites. This process generates fragments of variable size that are few in number and can be separated by electrophoresis on agarose gel.

Aigle *et al.* (1984) first applied this technique to brewer's yeasts. Since 1987, it has been used for the characterization of enological strains of *S. cerevisiae* (Dubourdieu *et al.*, 1987; Hallet *et al.*, 1988).

The extraction of mtDNA comprises several stages. The protoplasts obtained by enzymatic digestion of the cell walls are lysed in a hypotonic buffer. The mtDNA is then separated from the chromosomal

DNA by ultracentrifugation in a cesium chloride gradient, in the presence of bisbenzimide, which acts as a fluorescent intercalating stain. This agent amplifies the difference in density between chromosomal DNA and mtDNA. The mtDNA has a high number of adenine and thymine base pairs, for which bisbenzimide has a strong affinity. Finally, the mtDNA is purified by a phenol–chloroform extraction and an ethanol precipitation.

Defontaine et al. (1991) and Querol et al. (1992) simplified this protocol by separating the mitochondria from the other cell constituents before extracting the DNA. In this manner, they avoided the ultracentrifugation step. The coarse cellular debris is eliminated from the yeast lysate by centrifuging at 1,000 g. The supernatant is then recentrifuged at 15,000 g to obtain the mitochondria. The mitochondria are then lysed in a suitable buffer to liberate the DNA.

Unlike the industrial brewer's yeast strains analyzed by Aigle et al. (1984), which have the same mtDNA restriction profile, implying that they are of common origin, the winemaking yeast strains have a large degree of mtDNA diversity. This method easily differentiates between most of the selected yeasts used in winemaking as well as wild strains of S. cerevisiae found in spontaneous fermentations (Figure 1.26). This method may also help differentiate S. uvarum strains (Naumova et al., 2010).

This technique is very discriminating and not too expensive, but it is long and requires several complex manipulations. It is useful for the subtle characterization of a small number of strains. Inoculation effectiveness can also be verified by this method. In the laboratory, the lees, sampled during or toward the end of alcoholic fermentation, are cultured in a liquid medium. The mtDNA restriction profiles of this total biomass and of the yeast starter strain are compared. The absence of any extra bands, with respect to the yeast starter strain restriction profile, demonstrates that the yeast starter has been properly implanted, with an accuracy of 90%. In fact, in the case of a binary mixture, the minority strain must represent around 10% of the total population to be detected (Hallet et al., 1989).

1.9.3 Karyotype Analysis

Saccharomyces cerevisiae has 16 chromosomes with a size range between 250 and 2,500 kb. Its genomic DNA is very polymorphic; thus, it is possible to differentiate

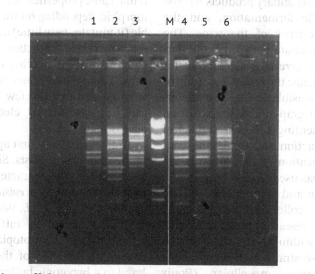

FIGURE 1.26 Restriction profile by *Eco*R5 of mtDNA of different strains of *S. cerevisiae*. Band 1, FIO; band 2, BO213; band 3, VLI; M, marker; band 4, 522; band 5, Sita 3; band 6, VL3c.

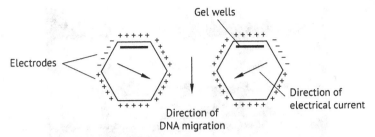

FIGURE 1.27 CHEF pulsed-field electrophoresis device (*contour clamped electrophoresis field*).

strains of the species according to the size distribution of their chromosomes. Pulsed-field electrophoresis is used to separate *S. cerevisiae* chromosomes and to compare karyotypes of the strains. This technique uses two electric fields oriented differently (90–120°). The electrodes placed on the sides of the apparatus apply the fields alternately (Figure 1.27).

The user can define the duration of the electric current that will be applied in each direction (pulse). With each change in the direction of the electric field, the DNA molecules reorient themselves; the smaller chromosomes reorient themselves more quickly than the larger ones (Figure 1.28).

Blondin and Vezhinet (1988), Petering *et al.* (1988), and Dubourdieu and Frezier (1990) applied this technique to identify wine yeast strains. Sample preparation is relatively easy. The yeasts are cultivated in a liquid medium, collected during the log phase, and then placed in suspension in a warm agarose solution that is poured into a partitioned mold to form small plugs.

Figure 1.29 gives an example of the identification of *S. cerevisiae* strains isolated from a grape must undergoing spontaneous fermentation. Vezhinet *et al.* (1990) have shown that karyotype analysis can distinguish between strains of *S. cerevisiae* as well or better than the use of mtDNA restriction profiles. Furthermore, karyotype analysis is much quicker and easier to use than mtDNA analysis. In the case of ecological studies of spontaneous fermentation microflora, pulsed-field electrophoresis of chromosomes is extensively used today to characterize strains of *S. cerevisiae* (Frezier and Dubourdieu, 1992; Versavaud *et al.*, 1993, 1995).

Very little research on the chromosomal polymorphism in other species of grape and wine yeasts is currently available. Naumov *et al.* (1993) suggested that *S. uvarum* and *S. cerevisiae* karyotypes can be easily distinguished. Other authors (Vaughan Martini and Martini, 1993; Masneuf, 1996) have confirmed their results. In fact, a specific chromosomal band systematically appears in *S. uvarum*. Furthermore, there are only two chromosomes whose sizes are less than 400 kb in *S. uvarum* but generally more in *S. cerevisiae*, in all of the strains that we have analyzed.

Non-*Saccharomyces* species, in particular apiculate yeasts (*H. uvarum* and

Position of fragments

Instant of first pulse   After first pulse   Instant of second pulse   After second pulse

FIGURE 1.28 Mechanism of DNA molecule separation by pulsed-field electrophoresis.

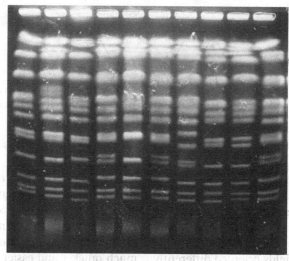

FIGURE 1.29 Example of electrophoretic (pulsed field) profile of S. cerevisiae strain karyotypes.

K. apiculata), are present on grapes and are sometimes found at the beginning of fermentations. These species have fewer polymorphic karyotypes and fewer bands than in *Saccharomyces*. Versavaud *et al.* (1993) differentiated between strains of apiculate yeast species and *Candida famata* by using restriction endonucleases at rare sites (*Not*1 and *Sfi*1). The endonucleases cut the chromosomes into a limited number of fragments, which were then separated by pulse-field electrophoresis.

1.9.4 Genomic DNA Restriction Profile Analysis Associated with DNA Hybridization by Specific Probes (Fingerprinting)

The yeast genome contains DNA sequences that repeat from dozens to hundreds of times, such as the δ sequences or Y1 elements of the chromosome telomeres. The distribution, or more specifically, the number and location of these elements, has a certain intraspecific variability. This genetic fingerprint is used to identify strains (Pedersen, 1986; Degre *et al.*, 1989).

The yeast strains are cultivated in a liquid medium and are sampled during the log phase, as in the preceding techniques. The entire DNA is isolated and digested by restriction endonucleases. The generated fragments are separated by electrophoresis on agarose gel and then transferred to a nylon membrane (Southern, 1975). Complementary radioactive probes (nucleotide sequences taken from δ and Y1 elements) are used to hybridize with fragments having homologous sequences. The result gives a hybridization profile containing several bands.

Genetic fingerprinting after hybridization is a more complicated and involved method than mtDNA or karyotype analysis. It is, however, without doubt the most discriminating strain identification method and may even discriminate too well. It has correctly indicated minor differences between very closely related strains. In fact, in the Bordeaux region (Frezier, 1992), *S. cerevisiae* clones isolated from spontaneous fermentations in different wineries have been encountered, which have the same karyotype and the same mtDNA restriction profile. Yet their hybridization profiles differ depending on sample origin. These strains, probably descendants of the same mother strain, have therefore undergone minor random modifications, maintained during vegetative reproduction.

1.9.5 PCR Associated with δ Sequences

This method consists of using PCR to amplify certain sequences of the yeast genome (Section 1.8.4), occurring between the repeated δ elements, whose separation distance does not exceed a certain value (1 kb). This method was developed (Ness *et al.*, 1992; Masneuf and Dubourdieu, 1994; Legras and Karst, 2003) to characterize *S. cerevisiae* strains. The amplification is carried out directly on whole cells. They are simply heated to make the cell envelopes permeable. The resulting amplification fragments are separated according to their size by agarose gel electrophoresis and viewed using ultraviolet fluorescence (Figure 1.30).

This analysis can distinguish between most *S. cerevisiae* ADY strains used in winemaking (Figure 1.31). Out of the 26 selected commercial yeast strains analyzed, 25 can be differentiated by their PCR profile associated with δ sequences. Lavallée *et al.* (1994) also observed excellent discriminating power with this method while analyzing industrially produced commercial strains from Lallemand Inc. (Montreal, Canada). In addition, this

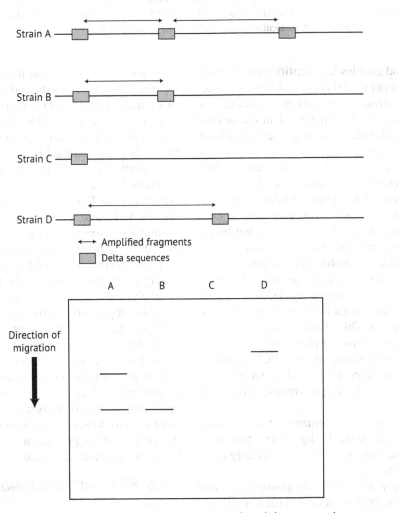

FIGURE 1.30 Principle of identification of *S. cerevisiae* strains by PCR associated with δ elements.

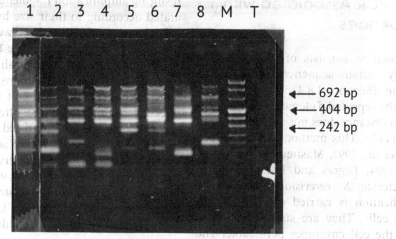

FIGURE 1.31 Electrophoresis in agarose gel (at 1.8%) of amplified fragments obtained from various commercial yeast strains. Band 1, F10; band 2, BO213; band 3, VL3c; band 4, UP30Y5; band 5, 522 D; band 6, EG8; band 7, L-1597; band 8, WET 136; M, molecular weight marker; T, negative control.

method enables the identification of 25–50 strains per day; it is the quickest of the different strain identification techniques currently available. When used for native yeast strain identification in a given viticultural region, however, it seems to be less discriminating than karyotype analysis. PCR profiles of wild yeasts isolated in a given location often appear similar. They have several constant bands and only a small number of variable discriminating bands. Certain strains have the same PCR amplification profile while having different karyotypes. In a given location, the polymorphism witnessed by PCR associated with δ sequences is lower than that of the karyotypes. This method is therefore complementary to other methods for characterizing winemaking strains. PCR enables a rapid primary sorting of a native yeast population. Karyotype analysis refines this discrimination.

Saccharomyces uvarum strains cannot be distinguished by this technique because their genome contains only a few Ty elements.

Lastly, because of its convenience and rapidity, PCR associated with δ sequences facilitates verification of the implantation of yeast starters used in winemaking. The analyses are conducted on the entire biomass derived from lees, placed beforehand in a liquid medium in a laboratory culture. The amplification profiles obtained are compared with inoculated yeast strain profiles. They are identical with a successful implantation, and different if the inoculation fails. Figure 1.32 gives examples of successful (yeasts B and C) and unsuccessful (yeasts A, D, and E) implantations. Contaminating strains have a different amplification profile than the yeast starter.

The detection threshold of a contaminating strain was studied in the laboratory by analyzing a mixture of two strains in variable proportions. In the example given in Figure 1.33, the contaminating strain is easily detected at 1%. In winery fermentations, however, several native strains can coexist with the inoculated strain in small proportions. When must undergoing fermentation or lees is analyzed by PCR, the yeast implantation rate is at least 90% when the amplification profiles of the lees and the yeast starter are identical.

1.9.6 PCR with Microsatellites

Microsatellites are tandem repeat units of short DNA sequences (1–10 nucleotides),

1.9 | Identification of Wine Yeast Strains

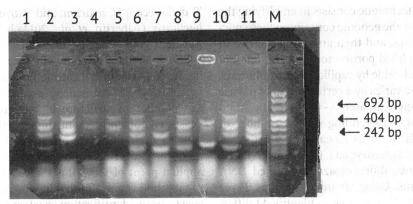

FIGURE 1.32 Electrophoresis in agarose gel (1.8%) of amplified fragments illustrating examples of yeast implantation tests (successful: yeasts B and C; unsuccessful: yeasts A, D, and E). Band 1, negative control; band 2, lees A; band 3, ADY A; band 4, lees B; band 5, ADY B; band 6, lees C; band 7, ADY C; band 8, lees D; band 9, ADY D; band 10, lees E; band 11, ADY E; M, molecular weight marker.

i.e. in the same direction and dispersed throughout the eukaryote genome. The number of motif repetitions is extremely variable from one individual to another, making these sequences highly polymorphic in size. These regions are easily identified, thanks to knowledge of the full sequence of the *S. cerevisiae* genome. Approximately 275 sequences have been listed, mainly AT dinucleotides and AAT and AAC trinucleotides (Field and Wills, 1998; Hennequin *et al.*, 2001; Perez *et al.*, 2001). Furthermore, these sequences are allelic markers, transmitted to the offspring in a Mendelian fashion. Consequently, these are ideal genetic markers for identifying specific yeast strains, making it possible not only to distinguish between strains but also to arrange them in related groups. This technique has many applications in humans: paternity tests, forensic medicine, etc. In viticulture, this molecular identification method has already been applied to *Vitis vinifera* grape varieties (Bowers *et al.*, 1999).

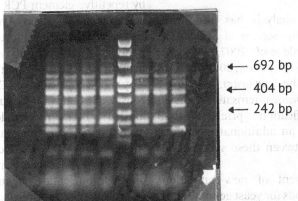

FIGURE 1.33 Determination of the detection threshold of a contaminating strain. T, negative control; band 1, strain A 70%, strain B 30%; band 2, strain A 80%, strain B 20%; band 3, strain A 90%, strain B 10%; band 4, strain A 99%, strain B 1%; M, molecular weight marker; band 5, strain A 99.9%, strain B 0.1%; band 6, strain A; band 7, strain B.

The technique consists in amplifying the region of the genome containing these microsatellites, and then analyzing the size of the amplified portion to a level of detail of one nucleotide by capillary electrophoresis. This size varies by a certain number of base pairs (approximately 8–40) from one strain to another, depending on the number of times the motif is repeated. A yeast strain may be heterozygous for a given locus, giving two different-sized amplified DNA fragments. Using six microsatellites, Perez et al. (2001) were able to identify 44 different genotypes within a population of 51 strains of *S. cerevisiae* used in winemaking. Other authors (Gonzalez Techera et al., 2001; Hennequin et al., 2001, Klis et al., 2002) have shown that the strains of *S. cerevisiae* used in winemaking are weakly heterozygous for the loci studied. However, interstrain variability of the microsatellites is very high. The results are expressed in numerical values for the size of the microsatellite in base pairs or the number of repetitions of the motifs on each allele. These digital data are easy to interpret, unlike the karyotype images on agarose gel, which are not really comparable from one laboratory to another. Based on 41 microsatellites, Legras et al. (2005) selected six very discriminating and reproducible loci. They highlight the relationships between strains from different geographical zones or industrial environments.

Microsatellite analysis has also been used to identify the strains of *S. uvarum* (Masneuf-Pomarède et al., 2007, 2016) and of *S. kudriavzevii* (Erny et al., 2012) used in winemaking. As the *S. uvarum*, *S. kudriavzevii*, and *S. cerevisiae* microsatellites have different amplification primers, this method provides an additional means of distinguishing between these species and their hybrids.

The development of new-generation sequencing methods for yeast genomes has made sequences available for non-*Saccharomyces* species. A typing method of yeast strains by microsatellite marker analysis is now offered for *B. bruxellensis*, *T. delbrueckii*, *H. uvarum*, and *Starmerella bacillaris* (Albertin et al., 2014a,b, 2016; Masneuf-Pomarede et al., 2015, 2016). When applied to the study of a great number of yeast isolates, these methods help to better describe the genetic diversity and the population structure of winemaking yeasts. Factors influencing this structure as well as their life cycle and reproduction mode have also been described. From an applied point of view, this molecular typing method is a useful tool in winemaking yeast strain identification, ecological surveys, and quality control of industrial production batches.

1.9.7 Genome Sequencing

With the development of new-generation sequencing methods, new approaches to yeast characterization have been suggested. The multilocus sequence typing (MLST) method is a standardized approach to full or partial sequence analysis of certain gene expressions in yeast. These genes are characterized by a slow accumulation of mutations, which help differentiate between individuals, as well as deduce phylogenetic relationships between strains. Applied to *S. cerevisiae*, the results obtained do not indicate a superior ability to discriminate among yeast strains when compared with analysis by repetitive-element PCR or microsatellite marker polymorphism (Ayoub et al., 2006). However, studies have revealed the specific population structure of wine yeasts, confirming the domestication of these yeasts (Fay and Benavides, 2005). Other approaches consist in establishing sequences of regions located between randomly selected restriction sites in the genome (restriction site-associated DNA sequencing or RAD-seq). Many positions of variation of a base, or single nucleotide polymorphism (SNP), can thus be used for phylogenetic analyses. The RAD-seq method has established the diversity and genetic structure of *S. cerevisiae* strains from a variety of ecological niches (Cromie et al., 2013; Hyma and Fay, 2013).

1.10 Ecology of Grape and Wine Yeasts

1.10.1 Succession of Grape and Wine Yeast Species

A large amount of research was focused on the description and ecology of wine yeasts. It concerned the distribution and succession of species found on the grape and then in wine during fermentation and conservation (Ribéreau-Gayon et al., 1975; Lafon-Lafourcade, 1983).

The ecological study of grape and wine yeast species represents a considerable amount of research. De Rossi began his research in the 1930s (De Rossi, 1935). Castelli (1955, 1967) pursued the study of yeast ecology in Italian vineyards. Peynaud and Domercq (1953) and Domercq (1956) published the first results on the ecology of wine yeasts in France. They described not only the species found on the grape and during alcoholic fermentation but also contaminating and spoilage yeasts. Among the many publications on this topic since the 1960s in viticultural regions around the world, the following works are worth noting: Brechot et al. (1962), Sapis-Domercq (1970), Barnett et al. (1972), Minarik (1971), Cuinier and Guerineau (1976), Park (1975), Soufleros (1978), Belin (1979, 1981), Poulard et al. (1980), Poulard and Lecocq (1981), Bureau et al. (1982), Rossini et al. (1982), Fleet et al. (1984), Mills et al. (2002), Baleiras Couto et al. (2005), Hierro et al. (2006), and Nisiotou and Nychas (2007).

Yeasts are widespread in nature and are found in soils, on the surface of plants, and in the digestive tract of animals. Wind, insects, and birds disseminate them. Yeasts are distributed irregularly on the surface of the grapevine; found in small quantities on leaves, the stems, and unripe grapes, they colonize the grape skin during maturation. During the winter season, the main natural reservoir of yeasts is the trunk, while during the vegetative growth phase of the vine, it is found in the ground and in berries (Cordero-Bueso et al., 2011). Observations under a scanning electron microscope have identified the location of yeasts on the grape. They are rarely found on the bloom, but multiply preferentially on exudates released from microlesions in zones situated around the stomatal apparatus. *Botrytis cinerea* spores and lactic acid and acetic acid bacteria also develop in the proximity of these peristomatic fractures (Figure 1.34).

The number of yeasts on the grape berry, just before harvest, is between 10^3 and 10^5, depending on the geographical location of the vineyard, weather conditions during maturation, the healthiness of the harvested grapes, and pesticide treatments applied to the vine. The most abundant yeast populations are obtained under warm weather conditions (lower latitudes and higher temperatures). Insecticide treatments and certain fungicidal treatments can make the native grape microflora less populous. Quantitative results available on this subject, however, are few. After the harvest, transport, and crushing of the crop, the number of cells capable of forming colonies on an agar medium generally reaches 10^6 cells/ml of must.

The number of yeast species significantly present on the grape is limited. From fruit set to maturity, *Aureobasidium pullulans*, *Cryotococcus*, and *Rhodotorula* are the main genera and species encountered (Prakitchaiwattana et al., 2004; Martins et al., 2014). Their proportion then decreases in favor of Ascomycetes. Among the latter, the apiculate species (*K. apiculata* and its sporogenous form *H. uvarum*) are the most common. They comprise up to 99% of the yeasts isolated in certain grape samples. The following are generally found but in lesser proportions: *Metschnikowia pulcherrima*, *C. famata*, *C. stellata*, *P. membranifaciens*, *Pichia fermentans*, and *Hansenula anomala*. The increase of grape exudate sugar content

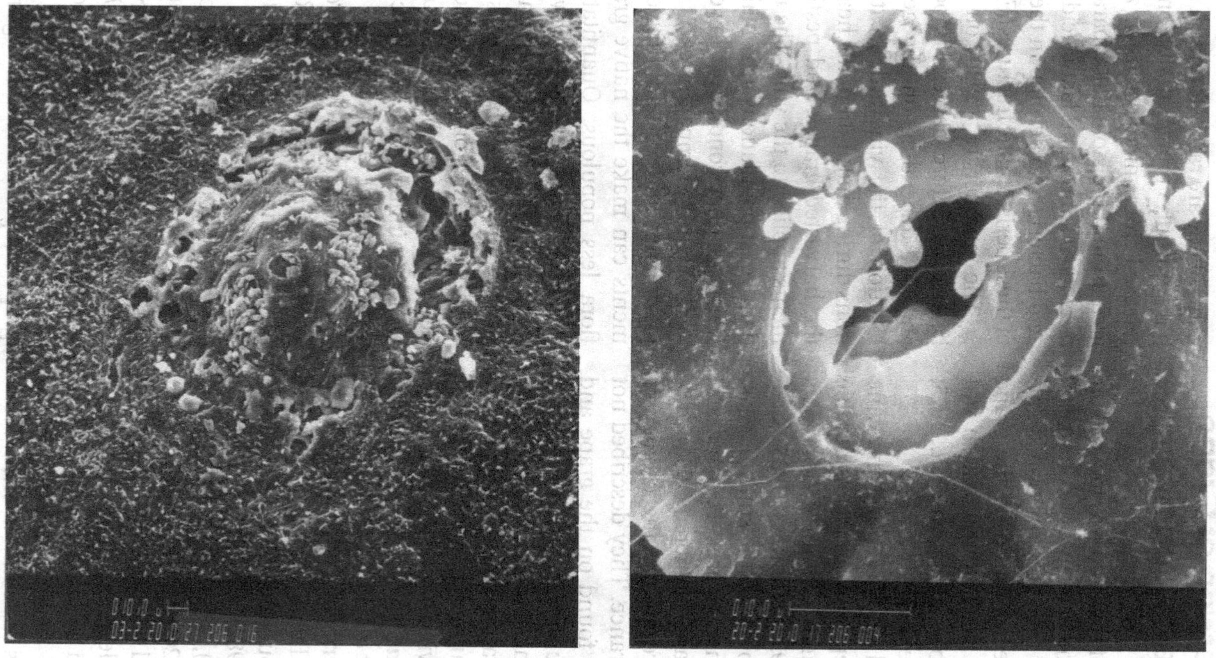

FIGURE 1.34 Grape surface under scanning electron microscope, with detail of yeast peristomatic zones. (*Source:* Photographs from B. Pucheu-Plante and M. Mercier, Department of Electron Microscopy, Université de Bordeaux 1.)

may partly explain these changes in yeast populations (Martins, 2012).

All research confirms the extreme rarity of *S. cerevisiae* on grapes (Mortimer and Polsinelli, 1999). However, these yeasts are not totally absent. Their existence cannot be proven by streaking diluted must on a solid medium prepared under aseptic conditions, but their presence on grapes can be proven by analyzing the spontaneous fermentative microflora of grape samples placed in sterile bags, then aseptically crushed and vinified in the laboratory in the absence of all contamination. Red and white grapes from the Bordeaux region were treated in this manner. At mid-fermentation in the majority of cases, *S. cerevisiae* represented almost all of the yeasts isolated. In some rare cases, no yeast of this species developed and non-*Saccharomyces* yeasts began the fermentation. Nevertheless, growth of *S. cerevisiae* under these conditions from healthy grapes is not guaranteed. Thus, based on 134 samples of grapes collected in the Bordeaux region, 31% of samples were positive for *S. cerevisiae* at the two-thirds mark of alcoholic fermentation (Börlin, 2015). Under the same operating conditions, 28–38% of positive fermentation was obtained for grape samples from the Douro region in Portugal (Schuller *et al.*, 2012). The frequency of isolation of *S. cerevisiae* may approach 100% on damaged grapes, which are a very favorable environment for the development of fermentation yeasts (Mortimer and Polsinelli, 1999).

Ecological surveys carried out at the Bordeaux Faculty of Enology from 1992 to 1999 (Naumov *et al.*, 2000a) demonstrated the presence of *S. uvarum* yeasts on grapes and in spontaneously fermenting white musts from the Loire Valley, Jurançon, and Sauternes. The frequency of the presence of this species alongside *S. cerevisiae* varies from 4 to 100%. On one estate in Alsace, strains of *S. uvarum* were identified on grapes, in the press, and in tanks, where they represented up to 90% of the yeasts involved throughout fermentation in three consecutive years (Demuyter *et al.*, 2004). More recently, other authors (Naumov *et al.*, 2002; Zhang *et al.*, 2015) have shown that *S. uvarum*, identified on grapes and in fermenting must, is involved in making Tokaji and New Zealand wines.

The adaptation of *S. uvarum* to relatively low temperatures (6–10°C) certainly explains its presence in certain ecological niches: northerly vineyards, late harvests, and spontaneous "cold" fermentation of white wines. In contrast, this strain is sensitive to high temperatures and has not been found in spontaneous fermentations of red Bordeaux wines.

Between two harvests, winery walls, floors, equipment, and sometimes even the winery building itself are colonized mostly by the various non-fermentation species previously cited. Winemakers believe, however, that spontaneous fermentations are more difficult to initiate in new tanks than in tanks that have already been used. This empirical observation leads to the supposition that *S. cerevisiae* can also survive in the winery between two harvests. Moreover, this species was found in non-negligible proportions in the wooden fermentors of some of the best vineyards in Bordeaux during the harvest, just before they were filled.

Recent studies relying on global approaches to sequencing of DNA extracted from a biological sample demonstrate the presence of the main yeast species in fermenting must (*H. uvarum*, *S. bacillaris*, and *S. cerevisiae*) in the various zones of winemaking cellars (pressing, fermentation, and storage) before, during, and after the harvest period (Bokulich *et al.*, 2013).

In the first hours of spontaneous fermentations, the first tanks filled have a very similar microflora to that of the grapes, with a large proportion of *H. uvarum*, *S. bacillaris* (formerly known as *Candida zemplinina*), and *M. pulcherrima* species. After about 20 hours, *S. cerevisiae* develops and coexists with the grape yeasts (Zott *et al.*, 2008). Non-*Saccharomyces* yeasts quickly disappear at the start of

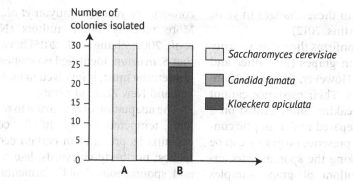

FIGURE 1.35 Comparison of yeast species present at the start of alcoholic fermentation ($d = 1.06$). A, in a tank of sulfited red grapes in Bordeaux (Frezier, 1992); B, in a tank of unsulfited white must, for the production of Cognac (Versavaud, 1994).

spontaneous fermentation (Figure 1.35). The microflora of tanks inoculated with yeast is very similar to that of tanks undergoing spontaneous fermentation (Figure 1.36). The major difference resides in the nature of *S. cerevisiae* strains that conduct fermentation: there is a dominant clone in the case of adding yeast. Meanwhile, in the case of spontaneous fermentation, several clones coexist (Section 1.10.2). Thus, the practice of adding yeast does not eliminate populations of non-*Saccharomyces* yeasts at the start of alcoholic fermentation. In red winemaking in the Bordeaux region, as soon as must specific gravity drops below 1.070–1.060, the colony samples obtained by plating diluted must on a solid medium generally isolate *S. cerevisiae* (10^7–10^8 cells/ml) exclusively. This species plays an essential role in the alcoholic fermentation process. Environmental conditions influence its selection. This selection pressure is exhibited by four main parameters: anaerobic conditions, must or grape sulfiting, sugar concentration, and the increasing presence of ethanol. The increase in temperature, especially in the case of red winemaking, also favors the development of *S. cerevisiae* to the detriment of non-*Saccharomyces* yeasts (Goddard, 2008). In winemaking, where no sulfur dioxide is used, such as white wines for the production of spirits, the dominant grape microflora can still be found. It is largely present at the beginning of alcoholic fermentation (Figure 1.35). Even in this type of winemaking, the presence of apiculate yeasts is limited at the midpoint of alcoholic fermentation.

During dry white winemaking, the separation of the pomace after pressing, combined with clarification by racking, greatly reduces yeast populations, at least in the first few days of the harvest. The yeast population of a severely racked must rarely exceeds 10^4–10^5 cells/ml.

A few days into the harvest, the *S. cerevisiae* yeasts colonize the harvest equipment, grape transport machinery, and especially the grape receiving equipment, the crusher, stemmer, the wine press, and cellar atmosphere (Grangeteau *et al.*, 2015). For this reason, *S. cerevisiae* is already widely present at the time of filling the tanks (around 50% of yeasts isolated during the first homogenization pump-over of a red grape tank). Fermentations are initiated more rapidly as harvest goes on. In fact, the last tanks filled often complete their fermentations before the first ones. Similarly, static racking in dry white winemaking becomes more and more difficult to achieve, even at low temperatures, from the second week of the harvest onward, especially in hot years. The entire facility inoculates the must with a sizeable fermentation yeast population. General weekly disinfection of the pumps, piping,

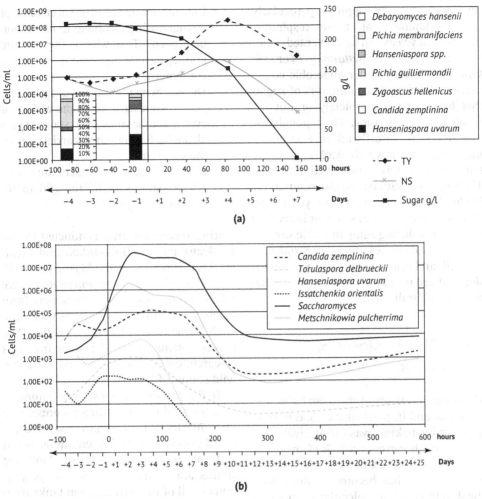

FIGURE 1.36 Dynamics of total yeasts and non-*Saccharomyces* yeasts during red winemaking monitored by (a) culture/RFLP-ITS-PCR and (b) specific quantitative PCR applied on a DNA pellet extracted directly from fresh must. Hour 0/day 0, time of inoculation with commercial yeast; − hours/days, cold soaking; + hours/days, alcoholic fermentation (Zott *et al.*, 2010).

wine presses, settling tanks, etc., is therefore strongly recommended.

During the final part of alcoholic fermentation (the yeast decline phase), the population of *S. cerevisiae* progressively decreases while still remaining greater than 10^6 cells/ml. Under favorable winemaking conditions, characterized by a rapid and complete exhaustion of sugars, no other yeast species significantly appears at the end of fermentation. Under poor conditions, spoilage yeasts can contaminate the wine. One of the most frequent and most dangerous contaminations is due to the development of *B. bruxellensis*, which is responsible for serious off-odors (Volume 2, Section 8.4.5).

In the weeks that follow the completion of alcoholic fermentation, the viable populations of *S. cerevisiae* drop rapidly, falling below a few hundred cells/ml. In many cases, other yeast species (spoilage yeasts) can develop in wines during bulk or bottle aging. Some yeasts have an oxidative metabolism of ethanol and form a veil on the surface of the wine, such as *Pichia* or *Candida*, or even certain strains of *S. cerevisiae*—sought after in the production of specialty wines. By topping up regularly, the development of these respiratory metabolism yeasts can be prevented. Some other yeasts, such as *Brettanomyces* or *Dekkera*, can develop under anaerobic conditions, consuming trace amounts of sugars that have been incompletely or not fermented by *S. cerevisiae*. Their population can attain 10^4–10^5 cells/ml in a contaminated red wine in which alcoholic fermentation is otherwise completed normally. These contaminations can also occur in the bottle. Lastly, refermentation yeasts can develop significantly in sweet or botrytized sweet wines during aging or bottle storage. The principal species found are *S. ludwigii*, *Z. bailii*, and also some strains of *S. cerevisiae* that are particularly resistant to ethanol and sulfur dioxide.

1.10.2 The Ecology of *S. cerevisiae* Strains

The ecological study of the clonal diversity of yeasts, and in particular of *S. cerevisiae* during winemaking, was inconceivable for a long time because of a lack of means to distinguish yeast strains from one another. Such research has become possible with the development of molecular yeast strain identification methods (Section 1.9). This section focuses on recent advances in this field.

The alcoholic fermentation of grape must or grapes is essentially carried out by a single yeast species, *S. cerevisiae*. Therefore, an understanding of the clonal diversity within this species is much more important for the winemaker than investigations on the partially or non-fermentative grape microflora.

The analysis of *S. cerevisiae* strains under practical winemaking conditions in particular is intended to answer the following questions:

- Is spontaneous fermentation carried out by a dominant strain, a small number or a very large number of strains?
- Can the existence of a succession of strains during alcoholic fermentation be proven? If so, what is their origin: grapes, harvest equipment, or winery equipment?
- During winemaking and from one year to another in the same winery or even the same vineyard, is spontaneous alcoholic fermentation carried out by the same strains?
- Can the practice of inoculating with selected strains modify the wild microflora of a vineyard?

During recent research conducted in the Bordeaux region (Dubourdieu and Frezier, 1990; Frezier, 1992; Masneuf, 1996), many samples of yeast microflora were taken in the vineyard and the winery from batches of white and red wines spontaneously fermenting or inoculated with ADYs. Several conclusions can be drawn from this research, carried out on several thousand wild strains of *S. cerevisiae*.

In the majority of cases, a small number of major strains (one to three) representing up to 70–80% of the colonies isolated, carry out the spontaneous fermentations of red and dry white wines. These dominant strains are found in comparable proportions in all of the fermentation tanks from the same winery from the start to end of alcoholic fermentation. This phenomenon is illustrated by the example given in Figure 1.37, describing the native microflora of several tanks of red must from a Pessac-Léognan vineyard in 1989. The strains of *S. cerevisiae*, possessing different karyotypes, are identified by an alphanumeric code comprising the initial of the vineyard, the tank number, the time of the sampling, the isolated colony number, and the year of the sample. Two strains, Fzlb1 (1989) and Fzlb2 (1989), are encountered in all of the tanks throughout the entire alcoholic fermentation process.

The spontaneous fermentation of dry white wines from the same vineyard is also carried out by the same dominant yeast strains in all of the barrels.

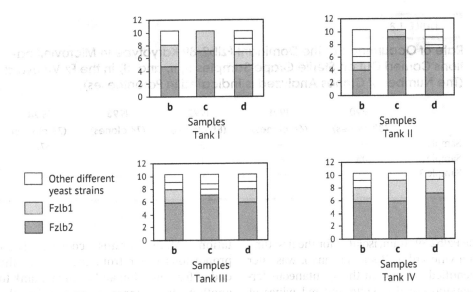

FIGURE 1.37 Breakdown of *S. cerevisiae* karyotypes during alcoholic fermentation in red grape tanks in the Fz vineyard (Pessac-Léognan, France) in 1989 (Frezier, 1992). b, c, and d designate the start, middle, and end of alcoholic fermentation, respectively. Tanks I and II (Merlot) and III and IV (Cabernet Sauvignon) were filled on the 1st, 3rd, 17th, and 23rd day of the harvest, respectively.

The tank filling order and the grape variety have little effect on the clonal composition of the populations of *S. cerevisiae* spontaneously found in the winery. The daily practice of pumping over the red grape must, with pumping equipment used for all of the tanks, and the contact of must with the atmosphere in the cellar enriched with *S. cerevisiae* populations probably ensure the dissemination of the same strains in the winery. In white winemaking, the wine press facility plays the same role as an inoculator.

The same major strain is frequently encountered for several consecutive vintages in the same vineyard in spontaneously fermenting red must tanks. In 1990, one of the major strains was the same as the previous year in the red must tanks of the Fz vineyard. Other strains appeared, however, which had not been isolated in 1989.

When sterile grape samples are taken, pressed sterilely, sulfited at winemaking levels, and fermented in the laboratory in sterile containers, one or several dominant strains responsible for spontaneous fermentations in the winery exist in some samples. These strains are therefore present in the vineyard. In practice, they probably begin to multiply as soon as the grapes arrive at the winery. A few days into the harvest, they infest the winery equipment, which in turn ensures a systematic inoculation of the fresh grape crop.

The presence each year of the same dominant strain in the vineyard is not systematic (Table 1.8). In the Fz vineyard, the Fzlb2-89 strain could not be isolated in 1991, although it was present in certain vineyard samples in 1990, 1992, and 1994. In 1993, another strain proved to be dominant in spontaneous fermentations of sterile grape samples.

The spontaneous microflora of *S. cerevisiae* seems to fluctuate. At present, the factors involved in this fluctuation have not been identified. In a given vineyard, spontaneous fermentation is not systematically carried out by the same strains each year, and there is thus no specific strain that is one of the vineyard's characteristics. Ecological observations do not confirm the notion of a vineyard-specific yeast. Furthermore, some native strains, dominant in a given vineyard, have been found in other nearby or distant vineyards. For example,

TABLE 1.8

Rate of Occurrence of the Dominant Fzlb2-89 Karyotype in Microvinifications Carried Out on Sterile Grape Samples (I, II, and III) in the Fz Vineyard (The Number of Clones Analyzed is Indicated in Parentheses)

	1990 (30 clones)	1991 (60 clones)	1992 (85 clones)	1993 (74 clones)	1994 (79 clones)
Sample 1	—	—	25%	—	87%
Sample 2	70%	—	31%	—	—
Sample 3	—	—	3%	—	40%

the Fzlb2-89 strain, isolated for the first time in a vineyard in Pessac-Léognan, was later identified not only in the spontaneous fermentation of dry white and red wines of other vineyards in the same appellation but also in relatively distant wineries as far away as the Médoc. This strain has since been selected and commercialized under the name Zymaflore F10.

In some cases (Figure 1.38), *S. cerevisiae* populations with a large clonal diversity carry out spontaneous must fermentation. Many strains coexist. Their proportions differ from the start to the end of fermentation and from one tank to another. Is it more favorable for the sensory characteristics of wine to have a fermentation with low or large clonal diversity? Few scientific works have addressed this question until now. Howell *et al.* (2006) showed that analytical profiles of wines obtained from mixed fermentations are different from those obtained with yeast monocultures. These cannot be reproduced by blending different monoculture wines. Evidently, metabolic interactions explain the differences observed between mixed fermentation and blending of monocultures. The contribution of a yeast clone to the aroma profile of a wine resulting from polyclonal fermentation cannot simply be predicted by this clone's population. Thus, the direct relationship between the diversity of yeast strains and wine complexity has not received until now any scientific demonstration and must be considered with care. In the Bordeaux region, the coexistence of a large number of yeast strains is often associated with slow fermentations, and sometimes even stuck fermentations before full depletion of sugars. No single strain seems to be capable of asserting itself. On the other hand, the presence of a small number of dominant strains generally characterizes complete and rapid spontaneous fermentations. These dominant strains are found from the start to the end of fermentation.

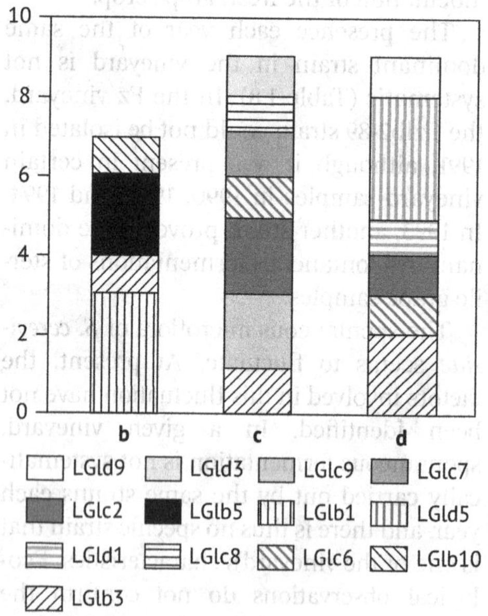

FIGURE 1.38 Breakdown of *S. cerevisiae* karyotypes in tank I of red grapes from the LG vineyard (Pomerol, France) in 1989 (Frezier, 1992). b, c, and d designate the start, middle, and end of alcoholic fermentation, respectively.

1.10 | Ecology of Grape and Wine Yeasts

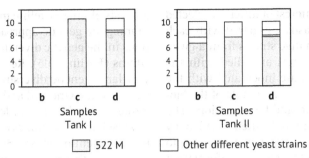

FIGURE 1.39 Breakdown of karyotypes for 10 strains analyzed in tank I and tank II from vineyard (P) in 1990. Tank I was inoculated with 522M dry yeast and tank II underwent spontaneous fermentation. b, c, d: start, middle, and end of alcoholic fermentation, respectively.

Under normal red winemaking conditions, the inoculation of the first tanks in a winery influences the wild microflora of non-inoculated tanks. The strain(s) used for inoculating the first tanks are frequently found as the predominant strain(s) in the latter. Figure 1.39 provides an example comparing the microflora of a tank of Merlot from Pomerol, inoculated with an ADY strain (522M) on the first day of the harvest, with a non-inoculated tank filled later. From the start of alcoholic fermentation, the selected strain is successfully established in the inoculated tank. Even in the non-inoculated tank, the same strain is equally established throughout the fermentation. It is therefore difficult to select dominant wild strains in red winemaking tanks when some of the tanks have been inoculated. An early and massive inoculation of the must, however, enables the successful establishment of different selected yeasts in several tanks at the same winery (Figure 1.40).

In white winemaking, inoculating rarely influences the microflora of spontaneous fermentations in wineries. For the most part, dominant native strains in non-inoculated barrels of fermenting dry white wine are observed, even though in the same wine cellar, other batches have been inoculated with different selected yeasts. The absence of pump-overs probably hinders the dissemination of the same yeasts in all of the fermenting barrels. This situation means the fermentative behavior

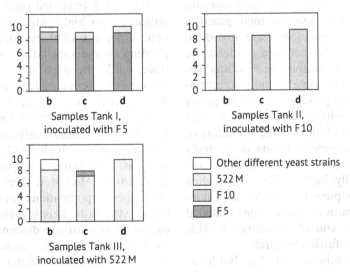

FIGURE 1.40 Breakdown of karyotypes of 10 strains analyzed in tanks I, II, and III from vineyard (F) in 1990, with massive early inoculation with F5, F10, and 522M, respectively (Frezier, 1992). b, c, and d: see Figure 1.39.

and enological interest of different selected strains can be easily compared with each other and with native strains from a given vineyard. The barrels are filled with the same must; some are inoculated with the yeast to be compared. A sample of the biomass is taken at mid-fermentation. The desired implantation is then verified by PCR combined with δ sequences. Due to the ease of use of this method, information on the characteristics of selected strains and their influence on wine quality can be gathered at the winery.

Vezhinet et al. (1992) and Versavaud et al. (1995) have also studied the clonal diversity of yeast microflora in other wine regions. Their results confirm the polyclonal character of fermentative populations of S. cerevisiae. The notion of dominant strains (one to two per fermentation) is obvious in the work carried out in the Charentes region. As in Champagne and the Loire Valley, some Charentes strains are found for several years in a row in the same winery. The presence of these dominant strains on the grape has been confirmed before any contact with winery equipment during several harvests.

Why do some S. cerevisiae strains coming from a very heterogeneous population become dominant during spontaneous fermentation? Why can they be found two to three years in a row in the same vineyard and wine cellar? Despite their practical interest, these questions have not often been studied, and there are no definitive responses. It seems that these strains rapidly start and complete alcoholic fermentation and have good resistance to sulfur dioxide (up to 10 g/hl). Furthermore, during mixed inoculations in the laboratory of either non-fermented musts or partially fermented musts (8% ethanol by vol.), these strains rapidly become dominant when placed in the presence of other wild non-dominant strains of S. cerevisiae isolated at the start and end of fermentation. This subject merits further research.

Studies conducted in the Bordeaux region were recently revisited by using the analysis of microsatellite markers, in order to integrate genetic approaches to populations in the genetic diversity results for the strains (Börlin, 2015). Results confirm the very large genetic diversity of S. cerevisiae strains in the vineyard and in spontaneous fermentation in the Bordeaux region. Of 1,374 isolates from grapes and 1,078 isolates from fermenting must, 75% have different genotypes. During spontaneous fermentation, the presence of yeasts related to commercial yeasts, i.e. that share at least 75% of genetic markers, is low (7%). This result, which mainly concerns organic wine cellars, merits confirmation in conventional cellars. However, a higher percentage (25%) of strains related to industrial yeast is obtained for the population of S. cerevisiae isolated from grapes in the vineyard. Therefore, this result highlights the importance, underestimated until now, of cellar yeasts being returned to the vineyard environment. Prior work showed that the detection of strains in the vineyard was possible in a radius of 10–200 m around the cellar where they were used (Valero et al., 2007; Schuller et al., 2007). Our data show a dissemination of winery yeasts in the vineyard over a maximum distance of 400 m from the cellar.

In Bordeaux vineyards, structural analyses of S. cerevisiae yeast populations on grapes have highlighted a significant genetic difference of these populations depending on the appellation. At the estate scale, while the populations involved in spontaneous fermentations are not very distinct over two to three consecutive years, they are clearly different over a longer period of time (>20 years). This confirms that in a given vineyard, populations of yeast fluctuate over time (Börlin et al., 2016). Moreover, there are clearly exchanges of populations between neighboring vineyards. This is probably due to the fact that yeast can disseminate in the environment through various vectors (insects, birds, and humans) to arrive in neighboring parcels of land.

References

Aigle M., Erbs D. and Moll M. (1984) Am. Soc. Brew. Chem., 42, 1, 1.

Albertin W., Chasseriaud L., Comte G., Panfili A., Delcamp A., Salin F., Marullo P. and Bely M. (2014a) PLoS One, 9, 4, e94246.

Albertin W., Panfili A., Miot-Sertier C., Goulielmakis A., Delcamp A., Salin F., Lonvaud-Funel A., Curtin C. and Masneuf-Pomarede I. (2014b) Food Microbiol., 42, 188.

Albertin W., Setati M.E., Miot-Sertier C., Mostert T.T., Colonna-Ceccaldi B., Coulon J., Girard P., Moine V., Pillet M., Salin F., Bely M., Divol B. and Masneuf-Pomarede I. (2016) Front. Microbiol., 20, 6, 1569.

Alexandre E.H., Rousseaux I. and Charpentier C. (1994) Biotechnol. Appl. Biochem., 19.

Arroyo-López F.N., Orlié S., Querol A. and Barrio E. (2009) Int. J. Food Microbiol., 31, 663.

Augustyn O.P.H., Kock J.L.F. and Ferreira D. (1991) Syst. Appl. Microbiol., 15, 105–115.

Ayoub M.J., Legras J.L., Saliba R. and Gaillardin C. (2006) J. Appl. Microbiol., 100, 699.

Baleiras Couto M.M., Reizinho R.G. and Duarte F.L. (2005) Int. J. Microbiol., 102, 49.

Barnett J.A., Delaney M.A., Jones E., Magson A.B. and Winch B. (1972) Arch. Microbiol., 83, 52–55.

Barnett J.A., Payne R.W. and Yarrow D. (1990) *Yeast: Characteristics and Identification*, 2nd edition. Cambridge University Press, Cambridge.

Barnett J.A., Payne R.W. and Yarrow D. (2000) *Yeasts: Characteristics and Identification*, 3rd edition. Cambridge University Press, Cambridge.

Barre P. (1978) *Killer factor activity under vinification conditions*, Vth International Symposium on Yeasts, Montpellier.

Barre P. (1984) Bull. OIV, 57, 635–643.

Barre P. (1992) Le facteur killer. In *Les acquisitions récentes de la microbiologie du vin*, pp. 63–69 (Ed B. Donèche). Tec & Doc, Lavoisier, Paris.

Belda I., Ruiz J., Alonso A., Marquina D. and Santos A. (2017) Toxins, 23, 9.

Belin J.M. (1979) Mycopathologia, 67, 2, 67–81.

Belin J.M. (1981) *Biologie des levures liées à la vigne et au vin*, Thèse Docteur ès Sciences Naturelles, Université de Dijon.

Bendova O., Richter V., Janderova B. and Haüsler J. (1991) Appl. Microbiol. Biotechnol., 35, 810–812.

Bisson L.F. (1991) Yeasts–metabolism of sugars. In *Wine Microbiology and Biotechnology*, pp. 55–75 (Ed G.H. Fleet). Harwood Academic Publishers, Chur.

Blondin B. and Vezhinet F. (1988) Rev. Fr. Oenol., 115, 7.

Bokulich N.A., Ohta M., Richardson P.M. and Mills DA. (2013) PLoS One, 8, 6, e66437.

Börlin M. (2015) *Diversité et structure de population des levures Saccharomyces cerevisiae à l'échelle du vignoble bordelais: Impact de différents facteurs sur la diversité*, Thèse de Doctorat, Université de Bordeaux 2.

Börlin M., Venet P., Claisse O., Salin F., Legras J.L. and Masneuf-Pomarede I. (2016) Appl. Environ. Microbiol., 82, 2909–2918.

Bouix M., Leveau J.Y. and Cuinier C. (1981) Œno One, 15, 41–52.

Boulton R.B., Singleton V.L., Bisson L.F. and Kunkee R. (1995) *Principles and Practices of Winemaking*. Chapman & Hall Enology Library, New York.

Bowers J., Boursiquot J.M., This P., Chu K., Johansson H. and Meredith C. (1999) Science, 285, 1562–1565.

Brechot P., Chauvet J. and Girard H. (1962) Ann. Technol. Agric., 11, 3, 235–244.

Bureau G., Brun O., Vigues A., Maujean A., Vesselle G. and Feuillat A. (1982) Œno One, 16, 1, 15.

Busturia A. and Lagunas R. (1986) J. Gen. Microbiol., 132, 379.

Carle G.F. and Olson M.V. (1984) Nucleic Acids Res., 12, 5647–5664.

Castelli T. (1955) Am. J. Enol. Vitic., 6, 18–20.

Castelli T. (1967) *Ecologie et systématique des levures du vin*, 2ème Symposium International d'Enologie, Bordeaux-Cognac, INRA, Paris.

Charpentier C. (1994) Rev. Oenol., 73S, 25–28.

Ciolfi G. (1992) L'enotecnico, 11, 87.

Ciolfi G. (1994) L'enotecnico, 71.

Cordero-Bueso G., Arroyo T., Serrano A. and Valero E. (2011) FEMS Microbiol. Ecol., 77, 429.

Cromie G.A., Hyma K.E., Ludlow C.L., Garmendia-Torres C., Gilbert T.L., May P., Huang A.A., Dudley A.M. and Fay J.C. (2013) G3 (Bethesda), 3, 12, 2163–2171.

Cuinier C. and Gros C. (1983) Vigne Vin, 318, 25–27.

Cuinier C. and Guerineau L. (1976) Vigne Vin, 269, 29–41.

Cuinier C. and Leveau J.Y. (1979) Vigne Vin, 283, 44–76.

Da Silva T., Albertin W., Dillmann C., Bely M., la Guerche S., Giraud C., Huet S., Sicard D., Masneuf-Pomarede I., de Vienne D. and Marullo P. (2015) PLoS One, 10, 5, e0123834.

De Rossi G. (1935) *Il lieviti della fermentazione nella regione umbra, IVème Congrès International de la Vigne et du Vin*, Lausanne.

Defontaine A., Lecocq F.M. and Hallet J.N. (1991) Nucleic Acids Res., 19, 1, 185.

Degre R., Thomas D.Y., Frenette J. and Mailhot K. (1989) Rev. Fr. Oenol., 119, 23–26.

Delfini C. (1995) *Scienza e technica di microbiologia enologica*. Il Lievito, Asti.

Demuyter C., Lollier M., Legras J.L. and Le Jeune C. (2004) J. Appl. Microbiol., 97, 6, 1140–1148.

Domercq S. (1956) *Etude et classification des levures de vin de la Gironde*, Thèse de Docteur-Ingénieur de l'Université de Bordeaux.

Dubourdieu D. and Frezier V. (1990) Rev. Fr. Oenol., 30, 37–40.

Dubourdieu D. and Moine V. (1995) *Actualités oenologiques 95, Compte-rendus du 5ème Symposium International d'Enologie de Bordeaux*, Bordeaux.

Dubourdieu D., Sokol A., Zucca J., Thalouarn P., Datee A. and Aigle M. (1987) Œno One, 4, 267.

Erny C., Raoult P., Alais A., Butterlin G., Delobel P., Matei-Radoi F., Casaregola S. and Legras J.L. (2012) Appl. Environ. Microbiol., 78, 3256.

Fay J.C. and Benavides J.A. (2005) PLoS Genet., 1, 1.

Fell J.W., Boekhout T., Fonseca A., Scorzetti G. and Statzell-Tallman A. (2000) Int. J. Syst. Evol. Microbiol., 50, 3, 1351–1371.

Field D. and Wills C. (1998) Proc. Natl. Acad. Sci., 95, 1647.

Fleet G.H. (1991) Cell walls. In *The Yeasts, Vol. 4: Yeast Organelles*, p. 199 (Eds A.H. Rose and J.S. Harrison). Academic Press, London.

Fleet G.H. (1993) *Wine Microbiology and Biotechnology*. Harwood Academic Publishers, Chur.

Fleet G.H., Lafon-Lafourcade S. and Ribereau-Gayon P. (1984) Appl. Environ. Microbiol., 48, 1034–1652.

Flick J.S. and Thorner J. (1993) Mol. Cell. Biol. 13, 5861–5876.

Frezier V. (1992) *Recherche sur l'écologie des souches de Saccharomyces cerevisiae au cours des vinifications bordelaises*. Thèse de Doctorat de l'Université de Bordeaux II.

Frezier V. and Dubourdieu D. (1992) Am. J. Enol. Vitic. 43, 375–380.

Gaillardin C. and Heslot H. (1987) La Recherche, 188, 18, 586.

Goddard M.R. (2008) Ecology, 89, 2077.

Goffeau A., Barrell B.G., Bussey H., Davis R.W., Dujon B., Feldmann H., Galibert F., Hoheisel J.D., Jacq C., Johnston M., Louis E.J., Mewes H.W., Murakami Y., Philippsen P. and Tettelin H. (1996) Science, 274, 563–567.

Gonzalez Techera A., Jubany S., Carrau F.M. and Gaggero C. (2001) Lett. Appl. Microbiol., 33, 71.

Granchi L., Bosco M., Messini A. and Vincenzini M. (1999) J. Appl. Microbiol., 87, 949.

Grangeteau C., Gerhards D., Rousseaux S., von Wallbrunn C., Alexandre H. and Guilloux-Benatier M. (2015) Food Microbiol., 50, 70.

Guerin B. (1991) Mitochondria. In *The Yeasts, Vol. 4: Yeast Organelles*, p. 541 (Eds A.H. Rose and J.S. Harrison). Academic Press, London.

Hallet J.N., Craneguy B., Zucca J. and Poulard A. (1988) Prog. Agric. Vitic., 105, 328.

Hallet J.N., Craneguy B., Daniel P. and Poulard A. (1989) *Actualités oenologiques 89, Comptes rendus du 4ème Symposium d'oenologie de Bordeaux*, Dunod.

Hansen J. and Kielland-Brandt M.C. (1994) Gene, 140, 33–40.

Heard G.M. and Fleet G.H. (1987) Appl. Environ. Microbiol. 51, 539–45.

Heard G.M. and Fleet G. (1990) J. Appl. Bacteriol., 68, 445.

Hennequin C.A., Thierry G.F., Richard G., Lecointre H., Nguyen V., Gaillardin C. and Dujon B. (2001) J. Clin. Microbiol., 39, 551.

Henschke P.A., and Rose A.H. (1991) Plasma membrane. In *The Yeasts, Vol. 4: Yeast Organelles*, p. 297 (Eds A.H. Rose and J.S. Harrison). Academic Press, London.

Herskowitz I., Rine J. and Strathern J.N. (1992) Mating-type determination and mating-type interconversion in *Saccharomyces cerevisiae*. In *The Molecular and Cellular Biology of the Yeast Saccharomyces: Gene Expression*, p. 583 (Eds E.W. Jones, J.R. Pringle, and J.R. Broach). Cold Spring Harbor Laboratory Press, New York.

Hierro N., Esteve-zarzozo B., Mas A. and Guillamon J.M. (2006) FEMS Yeast Res., 6, 102.

Howell K.S., Cozzolino D., Bartowsky E., Fleet G.H. and Henschke P.A. (2006) FEMS Yeast Res., 6, 91.

Hyma K.E. and Fay J.C. (2013) Mol. Ecol., 22, 2917.

Klis F.M. (1994) Yeast, 10, 851.

Klis F.M., Moll P., Hellingwerf K. and Brul S. (2002) FEMS Microbiol. Rev., 26, 3, 239–256.

Kregger-Van Rij N.J.W. (1984) *The Yeasts, a Taxonomic Study*. Elsevier, Amsterdam.

Kurtzman C.P. and Robnett C.J. (1998) Antoine Van Leeuwenhoek, 73, 4, 331.

Kurtzman C., Fell J.W. and Boekhout T. (Eds) (1994) *The Yeasts: A Taxonomic Study*, Vol. 1, 5th edition, 2354 pp. Elsevier, Amsterdam.

Kurtzman C.P., Fell J.W. and Boekhout T. (2011) *The Yeasts, a Taxonomic Study*, 5th edition, Elsevier: Amsterdam.

Lafon-Lafourcade S. (1983) Wine and brandy. In *Biotechnology Vol. V: Food and Feed Production with Microorganisms*, pp. 100–109 (Eds H.J. Rehm and G. Reed). Verlag Chemie, Weinheim.

Lafon-Lafourcade S. and Joyeux A. (1979) Œno One, 4, 295.

Lavallée F., Salves S., Lamy S., Thomas D.Y., Degre R. and Dulau L. (1994) Am. J. Enol. Vitic., 45, 86.

Le Jeune C., Lollier M., Demuyter C., Erny C., Legras J.L., Aigle M. and Masneuf-Pomarède I. (2007) FEMS Yeast Res., 7, 4, 540–549.

Legras J.L. and Karst F. (2003) FEMS Microbiol. Lett., 221, 2, 249–255.

Legras J.L., Ruh O., Merdinoglu D. and Karst F. (2005) Int. J. Food Microbiol., 102, 1, 73–83.

Libkind D., Hittinger C.T., Valério E., Gonçalves C., Dover J., Johnston M., Gonçalves P. and Sampaio J.P. (2011) Proc. Natl. Acad. Sci. U. S. A., 108, 35, 14539–14544.

Lodder J. (1970) *The Yeasts, a Taxonomic Study*. Elsevier, Amsterdam.

Lodder J. and Kregger-Van Rij N.J.W. (1952) *The Yeasts, a Taxonomic Study*. Elsevier, Amsterdam.

Martins, G.M. (2012) *Communautés microbiennes de la baie de raisin: Incidence des facteurs biotiques et abiotiques*, Thèse de Doctorat, Université de Bordeaux 2.

Martins G., Vallance J., Mercier A., Albertin W., Stamatopoulos P., Rey P., Lonvaud A. and Masneuf-Pomarède I. (2014) Int. J. Food Microbiol., 2, 177, 21–28.

Masneuf I. (1996) *Recherches sur l'identification génétique des levures de vinification. Applications oenologiques*, Thèse de Doctorat de l'Université de Bordeaux II.

Masneuf I. and Dubourdieu D. (1994) Œno One, 28, 2, 153.

Masneuf I., Aigle M. and Dubourdieu D. (1996a) FEMS Microbiol. Lett., 138, pp. 239–244.

Masneuf I., Aigle M. and Dubourdieu D. (1996b) Œno One, 30, 1, 15.

Masneuf I., Hansen J., Groth C., Piskpur J. and Dubourdieu D. (1998) Applied. Environ. Microbiol., 64, 3887.

Masneuf-Pomarède I., Le Jeune C., Durrens P., Lollier M., Aigle M. and Dubourdieu D. (2007) Syst. Appl. Microbiol., 30, 1, 75–82.

Masneuf-Pomarede I., Juquin E., Miot-Sertier C., Renault P., Laizet Y., Salin F., Alexandre H., Capozzi V., Cocolin L., Colonna-Ceccaldi B., Englezos V., Girard P., Gonzalez B., Lucas P., Mas A., Nisiotou A., Sipiczki M., Spano G., Tassou C., Bely M. and Albertin W. (2015) FEMS Yeast Res., 15, 5.

Masneuf-Pomarede I., Salin F., Börlin M., Coton E., Jeune C.L. and Legras J.L. (2016) FEMS Yeast Res., 16, 2, fow002.

Mehlomakulu N.N., Prior K.J., Setati M.E. and Divol B. (2017) J. Appl. Microbiol., 122, 747–758.

Mills D.A., Johansen E.A. and Cocolin L. (2002) Applied Env. Microbiol., 68, 4884.

Minarik E. (1971) Œno One, 2, 185–198.

Mortimer R.K. and Polsinelli M. (1999) Res. Microbiol., 150, 199.

Mortimer R.K., Romano P., Suzzi G. and Polsinelli M. (1994) Yeast, 10, 1543.

Naumov G., Naumova E. and Gaillardin C. (1993) Syst. Appl. Microbiol., 16, 274.

Naumov G.I., Naumova E.S. and Sniegowski P.D. (1998) Can. J. Microbiol., 44, 1045.

Naumov G.I., Masneuf I., Naumova E.S., Aigle M. and Dubourdieu D. (2000a) Res. Microbiol., 151, 683.

Naumov G.I., Naumova E.S., Masneuf I., Aigle M., Kondratieva V.I. and Dubourdieu D. (2000b) Syst. Appl. Microbiol., 23, 442.

Naumov G.I., Naumova E.S., Antunovics Z. and Sipiczki M. (2002) Appl. Microbiol. Biotechnol., 59, 6, 727.

Naumova E.S., Naumov G.I., Barrio E. and Querol A. (2010) Mikrobioligiia, 79, 543.

Ness F., Lavallee F., Dubourdieu D., Aigle M. and Dulau L. (1992) J. Sci. Food Agric., 62, 89.

Nguyen H.V. and Boekhout T. (2017) FEMS Yeast Res., 17, 2. Doi: 10.1093/femsyr/fox014.

Nguyen H.V. and Gaillardin C. (2005) FEMS Yeast Res., 5, 471.

Nguyen H.V., Lepingle A. and Gaillardin C.A. (2000) Syst. Appl. Microbiol., 23, 1, 71.

Nisiotou A.A. and Nychas G.J.E. (2007) Applied Env. Microbiol., 73, 6705.

Novo M., Bigey F., Beyne E., Galeote V., Gavory F., Mallet S., Cambon B., Legras J.L., Wincker P., Casaregola S. and Dequin S. (2009) Proc. Natl. Acad. Sci. U. S. A., 106, 38, 16333–16338.

Park Y.H. (1975) Œno One, 3, 253.

Pasteur L. (1866) *Etudes sur le vin*. Imprimerie Impériale, Paris.

Pasteur L. (1876) *Etudes sur la bière*. Gauthier-Villars, Paris.

Pedersen M.B. (1986) Carlsb. Res. Commun., 51, 163.

Perez M.A., Gallego F.J., Martinez I. and Hidalgo P. (2001) Lett. Appl. Microbiol., 33, 461.

Petering J., Langridge P. and Henschke P. (1988) Aust. N. Z. Wine Ind. J., 3, 48.

Peynaud E. and Domercq S. (1953) Ann. Technol. Agric., 4, 265.

Poulard A. and Lecocq M. (1981) Rev. Fr. Oenol., 82, 31.

Poulard A., Simon L. and Cuinier C. (1980) Œno One, 14, 219.

Prakitchaiwattana C.J., Fleet G.H. and Heard G.M. (2004) FEMS Yeast Res., 4, 8, 865.

Querol A., Barrio E. and Ramon D. (1992) Syst. Appl. Microbiol., 15, 439.

Radler F. (1988) *Application à l'oenologie des progrès récents en microbiologie et en fermentation*, p. 273. Office International de la Vigne et du Vin, Paris.

Redzepovic S., Orlic S., Sikora S., Madjak A. and Pretorius I.S. (2002) Lett. Appl. Microbiol., 35, 305.

Ribéreau-Gayon J. and Peynaud E. (1960) *Traité d'oenologie*, 2nd edition. Librairie Polytechnique Ch. Béranger, Paris.

Ribéreau-Gayon J., Peynaud E., Ribéreau-Gayon P. and Sudraud P. (1975) *Traité d'oenologie. Sciences et techniques du vin*, Vol. 2. Dunod, Paris.

Richardson S.M., Mitchell L.A., Stracquadanio G., Yang K., Dymond J.S., DiCarlo J.E., Lee D., Cheng L., Huang C.L.V., Chandrasegaran S., Cai Y., Boeke J.D., Joel S. and Bader J.S. (2017) Science, 355, 1040.

Rodríguez-Cousiño N., Maqueda M., Ambrona J., Zamora E., Esteban R. and Ramírez M. (2011) Appl. Environ. Microbiol., 77, 1822.

Rose A.H. and Harrison J.S. (1991) *The Yeasts, Vol. 4: Yeasts Organelles*. Academic Press, London.

Rossini G., Federici F. and Martini A. (1982) Microb. Ecol., 8, 83.

Rozes N., Garcia-Jares C., Larue F. and Lonvaud-Funel A. (1992) J. Sci. Food Agric., 59, 351.

Saiki R., Sharf S., Falcona F., Mullis K., Horn G., Erlich H. and Arnheim N. (1985) Science, 230, 1350.

Salmon J.M. (1989) Appl. Environ. Microbiol., 55, 953.

Salmon J.M., Vincent O., Mauricio J.C. and Bely M. (1993) Am. J. Enol. Vitic., 44, 1, 56.

Sapis-Domercq S. (1970) Œno One, 4, 45.

Schuller D., Pereira L., Alves H., Cambon B., Dequin S. and Casal M. (2007) Yeast, 24, 625.

Schuller D., Cardoso F., Sousa S., Gomes P., Gomes A.C., Santos M.A. and Casal M. (2012) PLoS One, 7, 2, e32507.

Schwartz D.C. and Cantor C.R. (1984) Cell, 37, 67.

Singer S.J. and Nicolson G.L. (1972) Science, 175, 720–731.

Sipiczki M. (2008) FEMS Yeast Res., 8, 996.

Soufleros E. (1978) *Les levures de la région viticole de Naoussa, Grèce, Identification et classification. Etude des produits volatils formés au cours de la fermentation*, Thèse Docteur-Ingénieur Université de Bordeaux II.

Southern E. (1975) J. Biol. Chem., 98, 503.

Stratford M. (1994) Yeast, 10, 1741.

Tipper D.J. and Bostian K.A. (1984) Microbiol. Rev. 48, 2, 125.

Tredoux H.G., Kock J.L.F., Lategan P.L. and Muller H.B. (1987) Am. J. Enol. Vitic., 38, 161.

Tuite M.F. and Oliver S.G. (1991) *Saccharomyces Biotechnology Handbooks*. Plenum Press, New York, London.

Valero E., Cambon B., Schuller D., Casal M. and Dequin S. (2007) FEMS Yeast Res., 7, 317.

Van Vuuren H.J.J. and Jacobs C.J. (1992) Am. J. Enol. Vitic., 43, 2, 119.

Van Vuuren H.J.J. and Van Der Meer L. (1987) Am. J. Enol. Vitic., 38, 49.

Vaughan Martini A. and Martini A. (1987) Antonie Van Leeuwenhoek, 53, 77.

Vaughan Martini A. and Martini A. (1993) Syst. Appl. Microbiol., 16, 113.

Velázquez R., Zamora E., Alvarez M.L., Hernández L.M. and Ramírez M. (2015) Front. Microbiol., 3, 1222.

Versavaud A. (1994) *Analyse de la diversité génétique de la microflore levurienne de la région des Charentes: application à la sélection de souches œnologiques*, Thèse de Doctorat, Université de Nantes.

Versavaud A., Dulau L. and Hallet J.N. (1993) Rev. Fr. Oenol., 142, 20.

Versavaud A., Courcoux P., Roulland C., Dulau L. and Hallet J.N. (1995) Appl. Environ. Microbiol., 61, 10, 3521.

Vezhinet F., Blondin B. and Hallet J.N. (1990) Appl. Microbiol. Biotechechnol., 32, 568.

Vezhinet F., Hallet J.N., Valade M. and Poulard A. (1992) Am. J. Enol. Vitic., 43, 1, 83.

Wang S.A. and Bai F.Y. (2008) Int. J. Syst. Evol. Microbiol., 58, Pt 2, 510.

Williamson D.H. (1991) Nucleus, chromosomes and plasmids. In *The Yeasts, Vol. 4: Yeast organelles*, p. 433 (Eds A.H. Rose and J.S. Harrison). Academic Press, London.

Yarrow D. and Nakase T. (1975) Antonie Van Leeuwenhoek, 41, 81.

Young T.W. (1987) Killer yeasts. In *The Yeasts, Vol. 2, Yeasts and the Environment*, p. 131 (Eds A.H. Rose and J.S. Harrison). Academic Press, New York.

Zhang H., Richards K.D., Wilson S., Lee S.A., Sheehan H., Roncoroni M. and Gardner R.C. (2015) Front. Microbiol., 46, 92.

Zorg J., Kilian J. and Radler F. (1988) Arch. Microbiol., 149, 261.

Zott K., Miot-Sertier C., Claisses O., Lonvaud-Funel A. and Masneuf-Pomarede I. (2008) Int. J. Food Microbiol., 125, 197.

Zott K., Claisse O., Lucas P., Coulon J., Lonvaud-Funel A. and Masneuf-Pomarede I. (2010) Food Microbiol., 27, 559.

CHAPTER 2

Yeast Metabolism

2.1 Introduction
2.2 Sugar Degradation Pathways
2.3 Regulation of Sugar-Utilizing Metabolic Pathways
2.4 Metabolism of Nitrogen Compounds

2.1 Introduction

The synthesis of living material is endergonic, requiring the consumption of energy. Chlorophyllous plants, called phototrophs, collect solar energy. Some bacteria obtain energy from the oxidation of minerals: they are chemolithotrophs. Like animals and most bacteria, fungi, including yeast, are chemoorganotrophs: they draw their necessary energy from the degradation of organic nutrients, which are their true "fuel."

In a growing organism, energy produced by degradation reactions (catabolism) is transferred to the chain of synthesis reactions (anabolism). Conforming to the laws of thermodynamics, energy furnished by the degradation of a substrate is only partially converted into work; this is called free energy (the rest is dissipated in the form of heat). Part of this free energy can be used for transport, movement, or synthesis. In most cases, the free energy transporter particular to biological systems is adenosine triphosphate (ATP). This molecule is rich in energy because its triphosphate unit contains two phosphoanhydride bonds (Figure 2.1). The hydrolysis of ATP into adenosine diphosphate (ADP) results in the liberation of a large quantity of free energy (7.3 kcal/mol). Biosynthesis and the active transport of metabolites make use of this free energy.

$\Delta G^{o\prime} = -7.3$ kcal/mol

In this reaction, $\Delta G^{o\prime}$ is the change in free energy. ATP is thus considered to be

---

Handbook of Enology, Volume 1: The Microbiology of Wine and Vinifications, Third Edition.
Pascal Ribéreau-Gayon, Denis Dubourdieu, Bernard B. Donèche and Aline A. Lonvaud.
© 2021 John Wiley & Sons Ltd. Published 2021 by John Wiley & Sons Ltd.

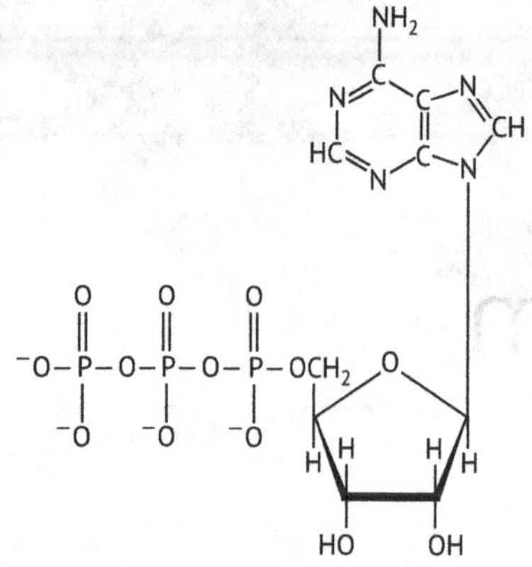

FIGURE 2.1 Structure of adenosine triphosphate (ATP).

"the universal currency of free energy in biological systems" (Stryer, 1992). In reality, microorganism growth or, in this case, yeast growth is directly related to the quantity of ATP furnished by the metabolic pathways used for degrading a substrate. It is very indirectly related to the quantity of substrate degraded.

In the living cell, there are two processes that produce ATP: substrate-level phosphorylation and oxidative phosphorylation. Both of these pathways exist in wine yeasts.

Substrate-level phosphorylation can be either aerobic or anaerobic. During oxidation by electron loss, a phosphate ester bond is formed. It is a high-energy bond between the oxidized carbon of the substrate and a molecule of inorganic phosphate. This bond is then transferred to the ADP by transphosphorylation, thus forming ATP. This process takes place during glycolysis (Section 2.2.1).

Oxidative phosphorylation is an aerobic process. The production of ATP is linked to the transport of electrons to an oxygen molecule by the cytochrome respiratory chain. This oxygen molecule is the final acceptor of the electrons. These reactions occur in the mitochondria.

This chapter describes the principal biochemical reactions occurring during grape must fermentation by wine yeasts. It covers sugar metabolism, i.e. the biochemistry of alcoholic fermentation and nitrogen metabolism.

2.2 Sugar Degradation Pathways

Depending on aerobic conditions, yeast can degrade sugars using two metabolic pathways: alcoholic fermentation and respiration. These two processes begin in the same way, sharing the common trunk of glycolysis.

2.2.1 Glycolysis

This series of reactions, transforming glucose into pyruvate with the formation of ATP, constitutes a quasi-universal pathway in biological systems. The elucidation of the different steps of glycolysis is intimately associated with the birth of modern biochemistry. The starting point was the fortuitous discovery by Hans and Eduard Buchner (1897), of the fermentation of sucrose by an acellular yeast extract. Studying possible therapeutic applications for their yeast extracts, the Buchners discovered that the sugar used to preserve their yeast extract was rapidly fermented into alcohol. Several years later, Harden and Young demonstrated that inorganic phosphate must be added to acellular yeast extract to ensure a constant glucose fermentation rate. The depletion of inorganic phosphate during *in vitro* fermentation led them to believe that it was incorporated into a sugar phosphate. They also observed that the yeast extract activity was due to a nondialyzable component, denaturable by heat, and a thermostable dialyzable component. They named these two components "zymase" and "cozymase." Today, zymase is known to be a series of enzymes, and cozymase is composed of their cofactors as well as metal ions and ATP. The complete description of glycolysis dates back to the 1940s, due in particular to the contributions of Embden, Meyerhof, and Neuberg. For that reason, glycolysis is often called the Embden–Meyerhof–Parnas pathway.

The transport of must hexoses (glucose and fructose) across the plasma membrane activates a complex system of protein transporters that is not fully explained (Section 1.3.2). This mechanism facilitates the diffusion of must hexoses in the cytoplasm, where they are rapidly metabolized. Since solute moves in the direction of the concentration gradient, from the concentrated outer medium to the diluted inner medium, it is not an active transport system requiring energy. This is favorable from an energy standpoint.

Next, glycolysis (Figure 2.2) is carried out entirely in the cytosol of the cell. It includes a first stage, in which glucose is converted into fructose 1,6-bisphosphate, requiring two ATP molecules. This transformation itself comprises three steps: an initial phosphorylation of glucose into glucose 6-phosphate, the isomerization of the latter into fructose 6-phosphate, and a second phosphorylation forming fructose 1,6-bisphosphate. These three reactions are catalyzed by hexokinase, phosphoglucoisomerase, and phosphofructokinase, respectively.

In fact, *Saccharomyces cerevisiae* has two hexokinases (PI and PII) capable of phosphorylating glucose and fructose. Hexokinase PII is essential and is active predominantly during the yeast log phase in a sugar-rich medium. Hexokinase PI, partially repressed by glucose, is not active until the stationary phase (Bisson, 1993).

Mutant strains devoid of phosphoglucoisomerase have been isolated. Their inability to develop on glucose suggests that glycolysis is the only catabolic pathway of glucose in *S. cerevisiae* (Caubet et al., 1988). The pentose phosphate pathway, by which some organisms utilize sugars, serves only as a means of synthesizing ribose 5-phosphate, incorporated in nucleic acids and in reduced nicotinamide adenine dinucleotide phosphate (NADPH) in *Saccharomyces*.

The second stage of glycolysis forms glyceraldehyde 3-phosphate. Under the catalytic action of aldolase, fructose 1,6-bisphosphate is cleaved, thus forming two triose phosphate isomers: dihydroxyacetone phosphate and glyceraldehyde 3-phosphate. Triose phosphate isomerase catalyzes the isomerization of these two compounds. Although at equilibrium, the ketose form is more abundant than the aldose form, the transformation of dihydroxyacetone phosphate into glyceraldehyde 3-phosphate is rapid, since this compound is continually eliminated by the ensuing glycolysis reactions. In other words, a molecule of glucose leads to the formation of two molecules of glyceraldehyde 3-phosphate.

The third phase of glycolysis is composed of two steps that recover part of the energy from glyceraldehyde 3-phosphate. Initially, it is converted into 1,3-bisphosphoglycerate (1,3-BPG). This reaction is catalyzed by glyceraldehyde 3-phosphate dehydrogenase. It is an oxidation coupled with a substrate-level phosphorylation. Nicotinamide adenine dinucleotide (NAD^+) is the cofactor of the dehydrogenation. At this stage, it is in its oxidized form; nicotinamide is the reactive part of the molecule (Figure 2.3). Simultaneously, an energy-rich phosphate ester bond is established between the oxidized carbon of the substrate and the inorganic phosphate. NAD^+ accepts two electrons and a hydrogen atom lost by the oxidized substrate. Next, phosphoglycerate kinase catalyzes the transfer of the phosphoryl group of the acylphosphate from 1,3-BPG to ADP; and 3-phosphoglycerate and ATP are formed.

The last phase of glycolysis transforms 3-phosphoglycerate into pyruvate. Phosphoglyceromutase catalyzes the conversion of 3-phosphoglycerate into 2-phosphoglycerate. Enolase catalyzes the dehydration of the latter, forming phosphoenolpyruvate. This compound has a high phosphoryl group transfer potential. By phosphorylation of ADP, pyruvic acid and ATP are formed; pyruvate kinase catalyzes this reaction. In this manner, glycolysis creates four ATP molecules. Two are immediately

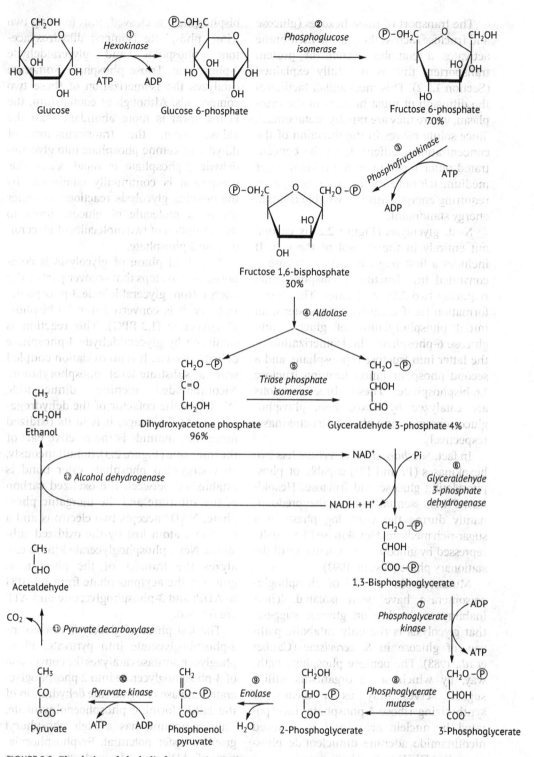

FIGURE 2.2 Glycolysis and alcoholic fermentation pathway.

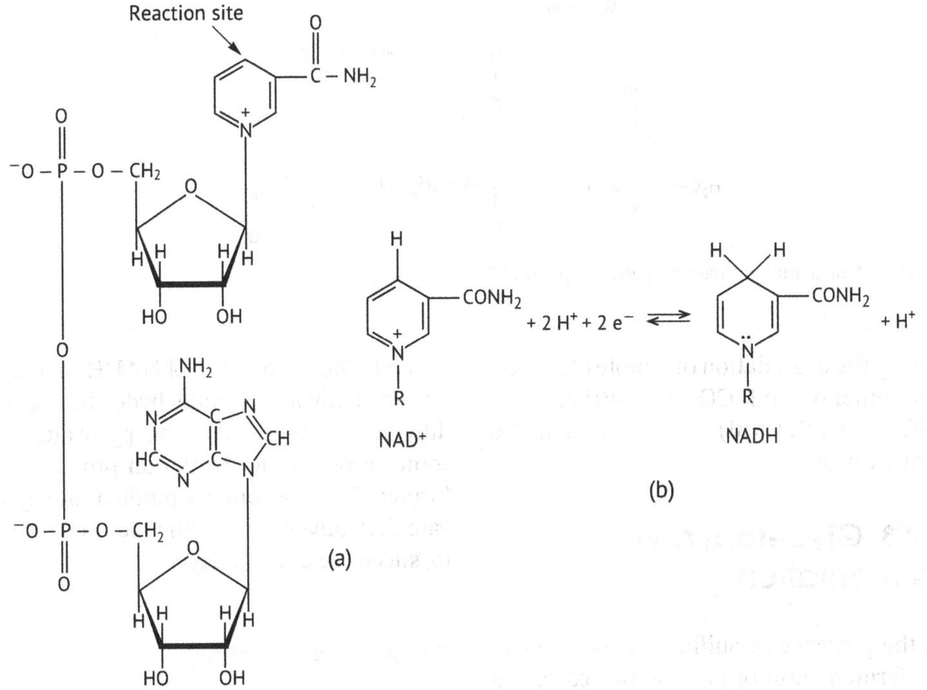

FIGURE 2.3 (a) Structure of nicotinamide adenine dinucleotide in the oxidized form (NAD⁺). (b) Equilibrium reaction between the oxidized (NAD⁺) and reduced (NADH) forms.

used to activate a new hexose molecule and the net gain of glycolysis is therefore two ATP molecules per molecule of hexose metabolized. This stage marks the end of the common trunk of glycolysis, which differentiates between alcoholic fermentation, glyceropyruvic fermentation, and respiration.

2.2.2 Alcoholic Fermentation

The reducing power of NADH produced by glycolysis must be transferred to an electron acceptor in order to regenerate NAD^+. In alcoholic fermentation, it is not pyruvate but rather acetaldehyde, its decarboxylation product, that serves as the terminal electron acceptor. With respect to glycolysis, alcoholic fermentation contains two additional enzymatic reactions.

The first decarboxylates pyruvic acid, catalyzed by pyruvate decarboxylase (PDC). The cofactor is thiamine pyrophosphate (TPP) (Figure 2.4). TPP and pyruvate form an intermediate compound. More precisely, the carbon atom located between the nitrogen and the sulfur of the thiazole ring of TPP is ionized. It forms a carbanion, which readily combines with the pyruvate carbonyl group. The second step reduces acetaldehyde into ethanol by NADH. This reaction is catalyzed by alcohol dehydrogenase, whose active site contains a Zn^{2+} ion.

Saccharomyces cerevisiae PDC comprises two isoenzymes: a major form, PDC1, representing 80% of the decarboxylase activity, and a minor form, PDC5, whose function remains uncertain.

From an energy viewpoint, glycolysis followed by alcoholic fermentation therefore supplies the yeast with two molecules of ATP per molecule of glucose degraded or 14.6 biologically usable kcal per mole of glucose fermented. From a thermodynamic viewpoint, the change in free energy

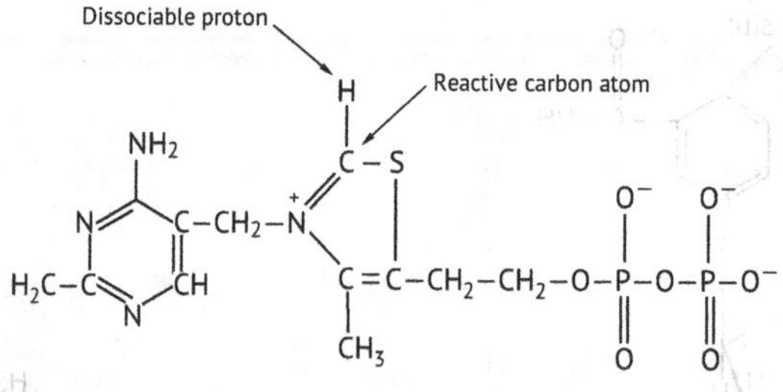

FIGURE 2.4 Structure of thiamine pyrophosphate (TPP).

during the degradation of a mole of glucose into ethanol and CO_2 is −40 kcal. The difference (25.4 kcal) is dissipated in the form of heat.

2.2.3 Glyceropyruvic Fermentation

In the presence of sulfite (Neuberg, 1946), the fermentation of glucose by yeasts produces equivalent quantities of glycerol, carbon dioxide, and acetaldehyde in its bisulfite form. This glyceropyruvic fermentation takes place in the following manner. Since the sulfite-bound acetaldehyde cannot be reduced into ethanol, dihydroxyacetone phosphate becomes the terminal acceptor of electrons from the oxidation of glyceraldehyde 3-phosphate, which is then reduced to glycerol 3-phosphate. The latter is dephosphorylated into glycerol. This mechanism was used for the industrial production of glycerol. In this fermentation, only two molecules of ATP are produced for every molecule of hexose degraded. ATP is required to activate the glucose in the first step of glycolysis (Figure 2.5). Glyceropyruvic fermentation, whose net gain in ATP is nil, does not furnish biologically assimilable energy for yeasts.

Glyceropyruvic fermentation does not occur solely in a highly sulfited environment. At the beginning of alcoholic fermentation of grape must, the inoculum consists of yeasts initially grown in the presence of oxygen. Their PDC and alcohol dehydrogenase are weakly expressed. As a result, acetaldehyde accumulation is limited. The reoxidation of NADH therefore does not involve acetaldehyde, but rather dihydroxyacetone. Glycerol, pyruvate, and some secondary fermentation products are formed. These secondary products are pyruvate derivatives—including, but not limited to, succinate and diacetyl.

2.2.4 Respiration

When sugar is used by the respiratory pathway, pyruvic acid (originating from glycolysis) undergoes an oxidative decarboxylation in the presence of coenzyme A (CoA) (Figure 2.6) and NAD^+. This process generates carbon dioxide, NADH and, acetyl-CoA:

$$Pyruvate + CoA + NAD^+ \rightarrow acetyl-CoA + CO_2 + NADH + H^+$$

The enzymatic complex of pyruvate dehydrogenase catalyzes this reaction. It takes place inside the mitochondria. TPP, lipoamide, and flavin adenine dinucleotide (FAD) participate in this reaction and serve as catalytic cofactors.

The acetyl unit coming from pyruvate is activated in the form of acetyl-CoA. The reactions of the citric acid cycle, also called the tricarboxylic acid cycle or Krebs cycle (Figure 2.7), completely oxidize the acetyl-CoA into CO_2. These reactions also occur in the mitochondria.

This cycle begins with the condensation of a two-carbon acetyl unit with a four-carbon compound, oxaloacetate, to produce a

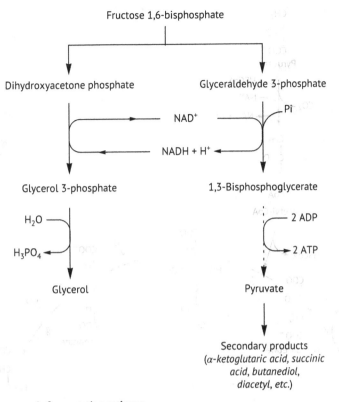

FIGURE 2.5 Glyceropyruvic fermentation pathway.

tricarboxylic acid with six carbon atoms: citric acid. Four oxidation–reduction reactions regenerate the oxaloacetate. The oxidative pathway involves decarboxylation of isocitrate, an isomer of citrate, into α-ketoglutarate. Isocitrate dehydrogenase catalyzes this reaction. A five-carbon compound, α-ketoglutarate, undergoes an oxidative decarboxylation to become succinate, catalyzed by α-ketoglutarate

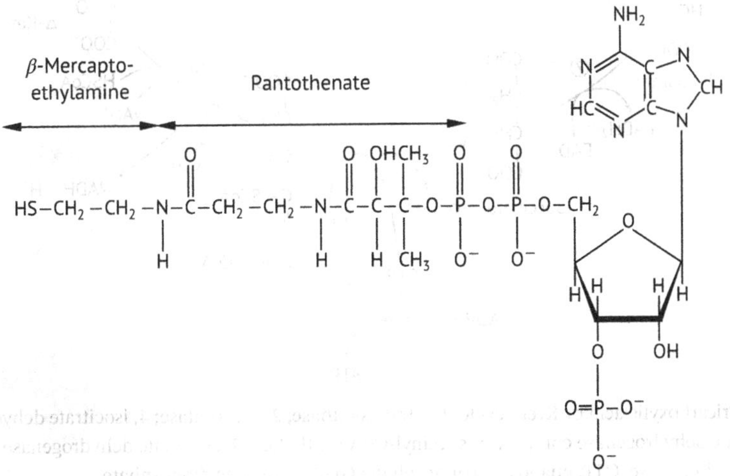

FIGURE 2.6 Structure of coenzyme A. The reaction site is the terminal thiol group.

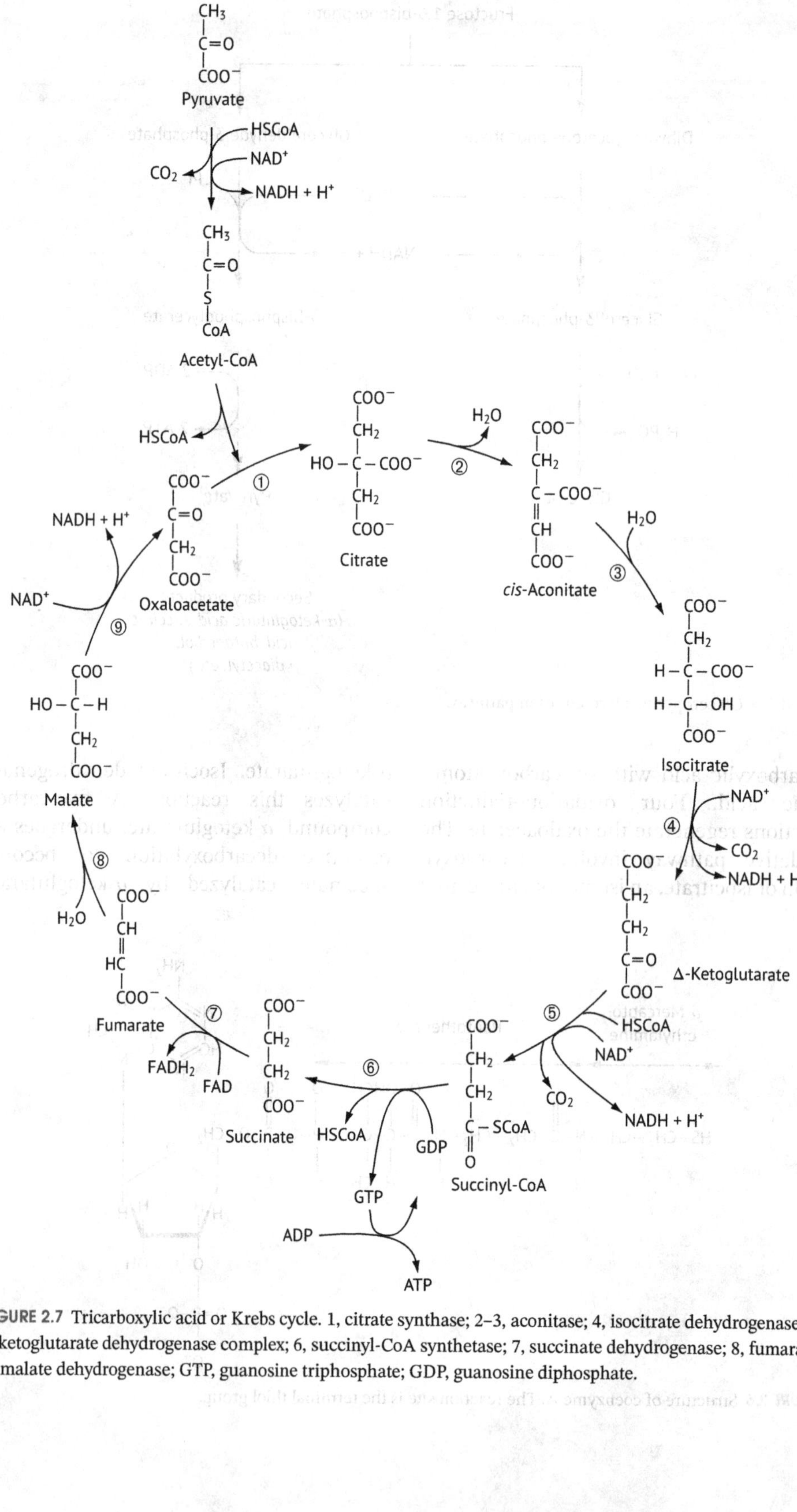

FIGURE 2.7 Tricarboxylic acid or Krebs cycle. 1, citrate synthase; 2–3, aconitase; 4, isocitrate dehydrogenase; 5, α-ketoglutarate dehydrogenase complex; 6, succinyl-CoA synthetase; 7, succinate dehydrogenase; 8, fumarase; 9, malate dehydrogenase; GTP, guanosine triphosphate; GDP, guanosine diphosphate.

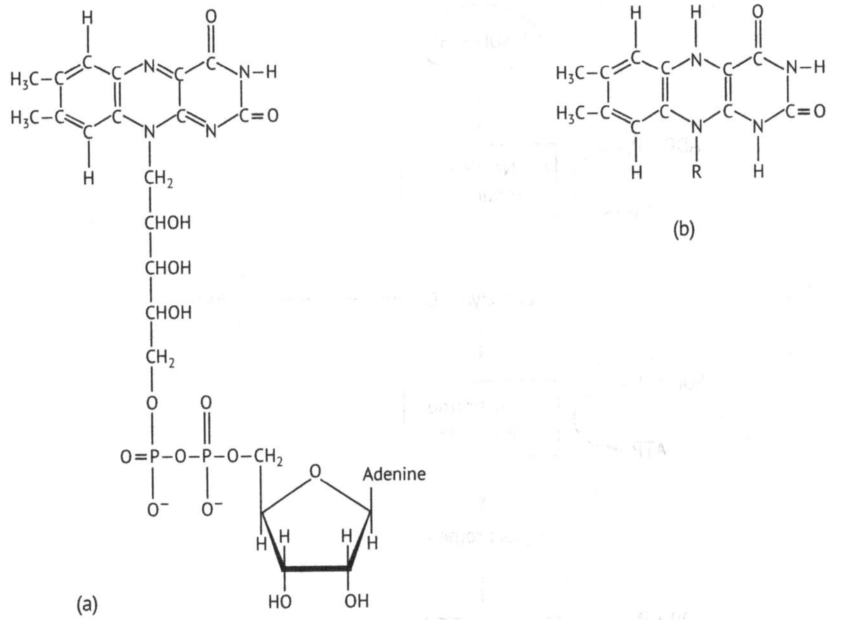

FIGURE 2.8 Structure of flavin adenine dinucleotide (FAD): (a) oxidized form (FAD); (b) reduced form ($FADH_2$).

dehydrogenase. In these two reactions, NAD^+ is the hydrogen acceptor. Fumarate dehydrogenase is responsible for the reduction of succinate into fumarate. FAD is the hydrogen acceptor (Figure 2.8). Finally, fumarate is hydrated into L-malate. The latter is reduced into oxaloacetate by malate dehydrogenase. In this case, the NAD^+ is the electron acceptor once again.

From acetate, each complete cycle produces two CO_2 molecules, three hydride ions transferred to three NAD^+ molecules (six electrons), and a pair of hydrogen atoms (two electrons) transferred to one FAD molecule. The cytochrome chain transports these electrons toward oxygen. ATP is formed during this process. This oxidative phosphorylation (Figure 2.9) takes place in the mitochondria. This process makes use of three enzymatic complexes (NADH-Q reductase, cytochrome reductase, and cytochrome oxidase). Two electron transport complexes (ubiquinone, or coenzyme Q, and cytochrome *c*) link these enzymatic complexes.

Oxidative phosphorylation yields three ATP molecules per pair of electrons transported between NADH and oxygen—or two ATP molecules with $FADH_2$. In the Krebs cycle, substrate-level phosphorylation also forms one ATP molecule during the transformation of succinyl-CoA into succinate.

The respiration of a glucose molecule (Table 2.1) produces 36 or 38 ATP molecules. Two originate from glycolysis, 28 from the oxidative phosphorylation of NADH and $FADH_2$ generated by the Krebs cycle, and two from substrate-level phosphorylation during the formation of succinate. Four to six ATP molecules result from the oxidative phosphorylation of two NADH molecules produced in glycolysis. The precise number depends on the transport system used to move the electrons from the cytosolic NADH to the respiratory chain in the mitochondria. The respiration of the same amount of sugar produces 18–19 times more biologically usable energy available to yeasts than fermentation. Respiration is used for industrial yeast production.

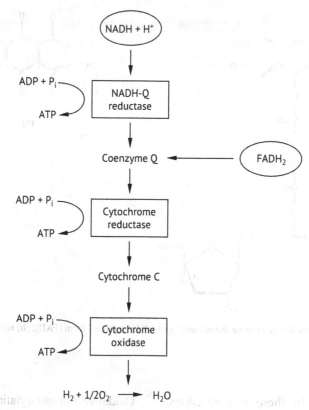

FIGURE 2.9 Oxidative phosphorylation during electron transport in the respiratory chain.

2.3 Regulation of Sugar-Utilizing Metabolic Pathways

2.3.1 Regulation Between Fermentation and Respiration: Pasteur Effect and Crabtree Effect

Pasteur was the first to compare yeast growth under aerobic and anaerobic conditions and to observe an inhibition of fermentation by respiration. At low concentrations of glucose on culture media, yeasts utilize sugars through either respiration or fermentation. Aeration induces an increase in biomass formed (total and per unit of sugar degraded) and a decrease in alcohol production and sugar consumption. Pasteur therefore deduced that respiration inhibits fermentation.

The "Pasteur effect" has been interpreted in several ways. Two enzymes compete to catalyze either the respiration or fermentation of pyruvate. This competition explains the respiratory inhibition of fermentation. PDC is involved in the fermentative pathway. It has a much lower affinity toward pyruvate than does pyruvate dehydrogenase. Furthermore, oxidative phosphorylation consumes a lot of ADP and inorganic phosphate, which migrate to the mitochondria. A lack of ADP and inorganic phosphate in the cytoplasm ensues. This deficit can limit the phosphorylation and thus slow the glycolytic flux. The inhibition of glycolysis enzymes by ATP explains the Pasteur effect for the most part. The ATP from oxidative phosphorylation inhibits phosphofructokinase in particular. Phosphorylated hexoses accumulate as a result. The transmembrane

TABLE 2.1
Energy Balance of Oxidation of Glucose in Respiration

Stage	Reduction coenzyme	Number of molecules of ATP formed
Glycolysis	2NADH	4 or 6
Net gain of ATP from glycolysis		2
Pyruvate → acetyl-CoA	NADH	6
Isocitrate → α-ketoglutarate	NADH	6
α-Ketoglutarate → succinyl-CoA	NADH	6
Succinyl-CoA → succinate		2
Succinate → fumarate	FADH$_2$	4
Malate → oxaloacetate	NADH	6
Net yield from glucose		36–38

transport of sugars and thus glycolysis is slowed down.

For high glucose concentrations—for example, in grape must—*S. cerevisiae* only metabolizes sugars by the fermentative pathway. Even in the presence of oxygen, respiration is impossible. Discovered by Crabtree (1929) on tumor cells, this phenomenon is known by several names: catabolite repression of respiration by glucose, the inverted Pasteur effect, and the Crabtree effect. Yeasts manifest the following signs during this effect: a degeneration of the mitochondria, a decrease in the proportion of cellular sterols and fatty acids, and a repression of the synthesis of Krebs cycle mitochondrial enzymes and constituents of the respiratory chain. With *S. cerevisiae*, there must be at least 2 g of glucose per liter for the Crabtree effect to occur. The catabolite repression exerted by glucose on wine yeasts is very strong. In grape must, at any level of aeration, yeasts are forced to ferment because of the high glucose and fructose concentrations. From a technological viewpoint, yeasts consume sugars via the respiratory pathway during the industrial production of dry yeast, but not in winemaking. If must aeration helps the alcoholic fermentation process (Section 3.7.2), the fatty acids and sterols synthesized by yeasts, which proliferate in the presence of oxygen, are responsible, not respiration.

Saccharomyces cerevisiae can metabolize ethanol via the respiratory pathway in the presence of small quantities of glucose. After alcoholic fermentation, oxidative yeasts develop in a similar manner on the surface of wine (Sections 14.5.2 and 14.5.3) as part of the process of making certain specialty wines (Sherry and *vin jaune* from Jura in France).

2.3.2 Regulation Between Alcoholic Fermentation and Glyceropyruvic Fermentation: Glycerol Accumulation

Wines contain about 8 g of glycerol per 100 g of ethanol. During grape must fermentation, about 8% of the sugar molecules undergo glyceropyruvic fermentation and 92% undergo alcoholic fermentation. The fermentation of the first 100 g of sugar forms the majority of the glycerol, after which glycerol production slows but is never nil. Glyceropyruvic fermentation is therefore more than just an inductive fermentation that regenerates NAD$^+$ when acetaldehyde, normally reduced into ethanol, is not yet present. Alcoholic fermentation and glyceropyruvic fermentation overlap slightly throughout the fermentation process.

Pyruvic acid is derived from glycolysis. When this molecule is not used by alcoholic fermentation, it participates in the formation of secondary products. In this case, a molecule of glycerol is formed by the reduction of dihydroxyacetone.

Glycerol production therefore equilibrates the yeast endocellular oxidation–reduction potential or $NAD^+/NADH$ balance. This "relief valve" eliminates surplus NADH, which appears at the end of the synthesis of amino acids, proteins, and the oxidations that generate secondary products.

Some winemakers place too much importance on the sensory role of glycerol. This compound has a sugary taste similar to glucose. In the presence of other constituents of wine, however, the sweetness of glycerol is practically imperceptible. For the majority of tasters, even well trained, the addition of 3–6 g of glycerol per liter to a red wine is not discernible. Therefore, the pursuit of winemaking conditions that are more conducive to glyceropyruvic fermentation has no enological interest. On the contrary, the winemaker should favor a pure alcoholic fermentation and should strive to minimize glyceropyruvic fermentation. The production of glycerol is accompanied by the formation of other secondary products, derived from pyruvic acid, whose increased presence (such as carbonyl function compounds and acetic acid) decreases wine quality.

2.3.3 Secondary Products Formed from Pyruvate by Glyceropyruvic Fermentation

When a molecule of glycerol is formed, a molecule of pyruvate is also formed. The latter cannot be transformed into ethanol following its decarboxylation into acetaldehyde. Under anaerobic conditions, oxaloacetate is the means of entry of pyruvate into the cytosolic citric acid cycle. Although the mitochondria are no longer functional, the enzymes of the citric acid cycle are present in the cytoplasm. Pyruvate carboxylase (PC) catalyzes the carboxylation of pyruvate into oxaloacetate. The prosthetic group of this enzyme is biotin; it serves as a CO_2 transporter. The reaction makes use of an ATP molecule:

$$Biotin - PC + ATP + CO_2 \rightarrow CO_2 - biotin - PC + ADP + P_i$$

$$CO_2 - biotin - PC + pyruvate \rightarrow biotin - PC + oxaloacetate$$

Under these anaerobic conditions, the citric acid cycle cannot be completed since the succinate dehydrogenase activity requires the presence of FAD, a strictly respiratory coenzyme. The chain of reactions is therefore interrupted at succinate, which accumulates (Figure 2.7) up to levels of 0.5–1.5 g/l. The NADH generated by this portion of the citric acid cycle (from oxaloacetate to succinate) is reoxidized by the formation of glycerol from dihydroxyacetone.

Under anaerobic conditions, α-ketoglutarate dehydrogenase has a very low activity; some authors therefore believe that the oxidative reactions of the citric acid cycle are interrupted at α-ketoglutarate. In their opinion, a reductive pathway of the citric acid cycle forms succinic acid under anaerobic conditions:

oxaloacetate $\rightarrow$ malate $\rightarrow$ fumarate $\rightarrow$ succinate.

Bacteria have a similar mechanism. Camarasa *et al.* (2003) demonstrated that this is the main pathway found in *S. cerevisiae* yeast under anaerobic conditions. Furthermore, additional succinate is formed during alcoholic fermentation in a glutamate-enriched medium. Glutamate is deaminated to form α-ketoglutarate, which is oxidized into succinate.

Among secondary products, compounds with a ketone function (pyruvic acid and α-ketoglutaric acid) and acetaldehyde predominantly bind with sulfur dioxide in

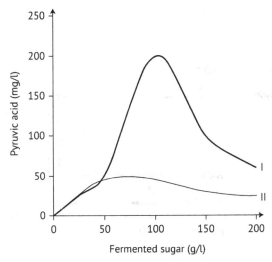

FIGURE 2.10 Effect of thiamine addition on pyruvic acid production during alcoholic fermentation (Lafon-Lafourcade, 1983). I, control must; II, thiamine-supplemented must.

wines made from healthy grapes. Their excretion is significant during the yeast proliferation phase and decreases toward the end of fermentation. Additional acetaldehyde is released in the presence of excessive quantities of sulfur dioxide in must. A high pH and high fermentation temperature, anaerobic conditions, and a deficiency in thiamine and pantothenic acid increase the production of ketoacids. Adding thiamine to must limits the accumulation of ketone compounds in wine (Figure 2.10).

Other secondary products of fermentation are also derived from pyruvic acid: acetic acid, lactic acid, butanediol, diacetyl, and acetoin. Their formation mechanisms are described in the following paragraphs.

2.3.4 Formation and Accumulation of Acetic Acid by Yeasts

Acetic acid is the principal volatile acid in wine. It is produced in particular during bacterial spoilage (acetic acid spoilage and lactic acid spoilage) but is always formed by yeasts during fermentation. Beyond a certain limit, which varies depending on the wine, acetic acid has a detrimental sensory effect on wine quality. In healthy grape must with a moderate sugar concentration (less than 220 g/l), *S. cerevisiae* produces relatively small quantities (100–300 mg/l), varying according to the strain. However, under certain winemaking conditions, even without bacterial contamination, yeast acetic acid production can be abnormally high and becomes a problem for the winemaker.

The biochemical pathway for the formation of acetic acid in wine yeasts has not yet been clearly identified. The hydrolysis of acetyl-CoA can produce acetic acid. Pyruvate dehydrogenase produces acetyl-CoA beforehand by the oxidative decarboxylation of pyruvic acid. This reaction takes place in the mitochondrial matrix but is limited under anaerobic conditions. Aldehyde dehydrogenase can also form acetic acid by the oxidation of acetaldehyde (Figure 2.11). This enzyme, whose cofactor is $NADP^+$, is active during alcoholic fermentation. The NADPH thus formed can be used to synthesize lipids. When pyruvate dehydrogenase is repressed, this pathway forms acetyl-CoA through the action of acetyl-CoA synthetase. Under

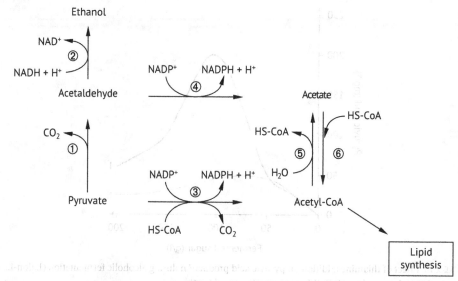

FIGURE 2.11 Acetic acid formation pathways in yeasts. 1, pyruvate decarboxylase; 2, alcohol dehydrogenase; 3, pyruvate dehydrogenase; 4, aldehyde dehydrogenase; 5, acetyl-CoA hydrolase; 6, acetyl-CoA synthetase.

anaerobic conditions in a model medium, yeast strains producing the least amount of acetic acid have the highest acetyl-CoA synthetase activity (Verduyn et al., 1990).

The acetaldehyde dehydrogenase in S. cerevisiae has five isoforms, three located in the cytosol (Section 1.4.1) (Ald6p, Ald2p, and Ald3p) and the remaining two (Ald4p and Ald5p) in the mitochondria (Section 1.4.3). These enzymes differ by their specific use of the NAD⁺ or NADP⁺ cofactor (Table 2.2).

Remize et al. (2000) and Blondin et al. (2002) studied the impact of the deletion of each gene and demonstrated that the NADP-dependent cytoplasmic isoform coded by *ALD6* played a major role in the formation of acetic acid during the fermentation of dry wines, while the *ALD5* mitochondrial isoform was also involved, but to a lesser extent (Figure 2.12).

Practical winemaking conditions likely to lead to abnormally high acetic acid production by *S. cerevisiae* are well known. As is the case with glycerol formation, acetic acid production is closely dependent on the initial sugar level of the must, independent of the quantity of sugars fermented (Table 2.3). The higher the sugar content of the must, the more acetic acid (and glycerol) the yeast produces during fermentation. This is due to the yeast's

TABLE 2.2

Isoforms of Acetaldehyde Dehydrogenase in *S. cerevisiae* (Navarro Avino et al., 1999)

Chromosome	Gene	Location	Cofactor
XIII	ALD2	Cytosol	NAD+
XIII	ALD3	Cytosol	NAD+
XV	ALD4	Mitochondria	NAD+ and NADP+
V	ALD5	Mitochondria	NADP+
XVI	ALD6	Cytosol	NADP+

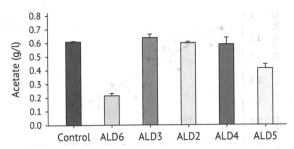

FIGURE 2.12 Acetate production by strains of *S. cerevisiae* (V5) following deletion of different gene coding for isoforms of acetaldehyde dehydrogenase (Blondin *et al.*, 2002).

mechanism for adapting to a medium with a high sugar concentration: *S. cerevisiae* increases its intracellular accumulation of glycerol to counterbalance the osmotic pressure of the medium (Blomberg and Alder, 1992). This regulation mechanism is controlled by a cascade of signal transmissions leading to an increase in the transcription level of genes involved in the production of glycerol (*GPD1*), but also of acetate (*ALD2*, *ALD3*, and *ALD4*) (Attfield *et al.*, 2000; Erasmus *et al.*, 2003; Pigeau and Inglis, 2005). Acetate formation plays an important physiological role in the intracellular redox equilibrium by regenerating reduced equivalents of NADH. Thus, it is clear that an increase in acetate production is inherent to the fermentation of high-sugar musts. However, Bely *et al.* (2003) demonstrated that it was possible to reduce acetate production by supplying more NADH to the redox equilibrium process. This may be done indirectly by stimulating biomass formation, which generates an excess of NADH during amino acid synthesis (Bakker *et al.*, 2001). Assimilable nitrogen in the must plays a key role in this stimulation process. Thus, in high-sugar musts, acetate production is inversely correlated with the maximum cell population (Figure 2.13), which is, in turn, related to the assimilable nitrogen content of the must. It is strongly recommended to monitor the assimilable nitrogen content of botrytized musts and to supplement them with ammonium sulfate, if necessary. The optimum assimilable nitrogen concentration in this type of must to minimize acetic acid production is approximately 190 mg/l (Figure 2.14). The best time for adding nitrogen supplements is at the very beginning of fermentation, as later additions are less effective and may even increase acetate production. Indeed, in

TABLE 2.3

Effect of Initial Sugar Concentration of the Must on the Formation of Secondary Products of the Fermentation (Lafon-Lafourcade, 1983)

Initial sugar (g/l)	Fermented sugar (g/l)	Secondary products		
		Acetic acid (g/l)	Glycerol (g/l)	Succinic acid (g/l)
224	211	0.26	4.77	0.26
268	226	0.45	5.33	0.25
318	211	0.62	5.70	0.26
324	179	0.84	5.95	0.26
348	152	1.12	7.09	0.28

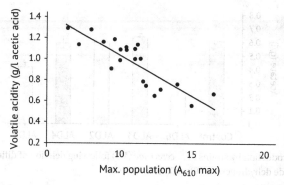

FIGURE 2.13 Correlation between volatile acidity production and the maximum cell population in high-sugar botrytized musts.

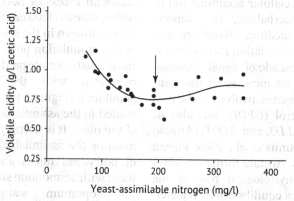

FIGURE 2.14 Effect of the yeast-assimilable nitrogen content in must (with or without ammonium supplements) on the production of volatile acidity (initial sugar content: 350 g/l).

view of the unpredictable increase in acetic acid production that sometimes occurred in botrytized musts supplemented with ammonium sulfate, many enologists had given up the practice entirely. It is now known that, provided the supplement is added at the very beginning of fermentation, adjusting the assimilable nitrogen content to the optimum level (190 mg/l) always minimizes acetic acid production in botrytized wines. Furthermore, Bely et al. (2005) demonstrated that direct inoculation with industrial preparations of active dry yeast leads to greater acetate production than does a yeast starter with a preculture period of 24 hours in a botrytized grape must diluted by half. This preculture period enables S. cerevisiae to adapt to the osmotic stress. Bely et al. (2008) showed that the species *Torulaspora delbrueckii*, which is a highly osmotolerant yeast, can be used in combination with S. cerevisiae to minimize acetic acid production during the making of wines from botrytized grapes.

In wines made from botrytized grapes, certain substances in the must inhibit yeast growth (Volume 2, Section 3.7.2) and increase the production of acetic acid and glycerol during fermentation. *Botrytis cinerea* secretes these "botryticine" substances (Ribéreau-Gayon et al., 1952, 1979). Fractional precipitation with ethanol partially purifies these compounds from must and culture media of B. cinerea. These heat-stable glycoproteins have molecular weights between 10,000 and 50,000. They are composed of a peptide (10%) and a carbohydrate part containing mostly mannose

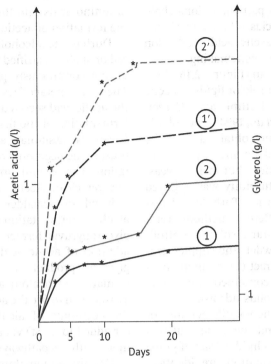

FIGURE 2.15 Effect of an alcohol-induced precipitate of a botrytized grape must on glycerol and acetic acid formation during the alcoholic fermentation of healthy grape must (Dubourdieu, 1982). 1, Evolution of acetic acid concentration in the control must; 2, evolution of acetic acid concentration in the must supplemented with the freeze-dried precipitate; 1', evolution of glycerol concentration in the control must; 2', evolution of glycerol concentration in the must supplemented with the freeze-dried precipitate.

and galactose as well as some rhamnose and glucose (Dubourdieu, 1982). When added to healthy grape must, these compounds provoke an increase in glyceropyruvic fermentation and a more-or-less significant excretion of acetic acid at the end of fermentation (Figure 2.15). The mode of action of these glycoproteins on yeasts has not yet been identified. The physiological state of yeast populations at the time of inoculation seems to play an important role in the fermentation of botrytized grape must. Commercial dry yeast preparations are much more sensitive to alcoholic fermentation inhibitors than yeast starters obtained by preculture in healthy grape must.

Other winemaking factors favor the production of acetic acid by *S. cerevisiae*: anaerobic conditions, very low pH (<3.1) or very high pH (>4), certain amino acid or vitamin deficiencies in the must, and excessively high temperature (25–30°C) during the yeast growth phase. In red winemaking, temperature is the most important factor, especially when the must has a high sugar concentration. In hot climates, the grapes should be cooled when filling the tanks. The temperature should not exceed 20°C at the beginning of fermentation. The same procedure should be followed during thermovinification immediately following the heating of the grapes.

In dry white and rosé winemaking, excessive must clarification can also lead to the excessive production of volatile acidity by yeast. This phenomenon can be particularly pronounced with certain yeast strains. Therefore, must turbidity should be adjusted to the lowest possible level that enables a complete and rapid fermentation (Sections 3.7.3 and 13.5.3). The input of lipids made available to yeasts via solid sediments (grape solids), in particular long-chain unsaturated fatty acids (C18:1 and C18:2), greatly influences acetic acid production during white and rosé winemaking.

The experiment in Figure 2.16 illustrates the important role of lipids in acetic acid metabolism (Delfini and Cervetti, 1992; Alexandre et al., 1994). The volatile acidity of three wines obtained from the same Sauvignon Blanc must was compared. After filtration but before yeast inoculation, must turbidity was adjusted to 250 Nephelometric Turbidity Units (NTU) by three different methods: reincorporating fresh grape solids (control), adding cellulose powder, and supplementing with a lipid extract (using methanol–chloroform) that contained the same quantity of grape solids adsorbed on the cellulose powder. The volatile acidities of the control wine and the wine that was supplemented with a lipid extract of grape solids before fermentation are identical and perfectly normal. Although the fermentation was normal, the volatile acidity of the wine made from the must supplemented with cellulose (therefore devoid of lipids) was practically twice as high (Lavigne, 1996). Supplementing the medium with lipids appears to favor the penetration of amino acids into the cell, which limits the formation of acetic acid.

During the alcoholic fermentation of red or slightly clarified white wines, yeasts do not continuously produce acetic acid. The yeast metabolizes a large portion of the acetic acid secreted in must during the fermentation of the first 50–100 g of sugar. It can also assimilate acetic acid added to must at the beginning of alcoholic fermentation. The assimilation mechanisms are not yet clear. Acetic acid appears to be reduced to acetaldehyde, which favors alcoholic fermentation to the detriment of glyceropyruvic fermentation. In fact, the addition of acetic acid to a must lowers glycerol production but increases the formation of acetoin and 2,3-butanediol. Yeasts seem to use the acetic acid formed at the beginning of alcoholic fermentation (or added to must) via acetyl-CoA in their lipid synthesis pathways.

Certain winemaking conditions produce abnormally high amounts of acetic acid. Since this acid is not used during the second half of the fermentation, it accumulates until the end of fermentation. When refermenting a tainted wine, yeasts can lower volatile acidity by metabolizing acetic acid. The wine is incorporated into

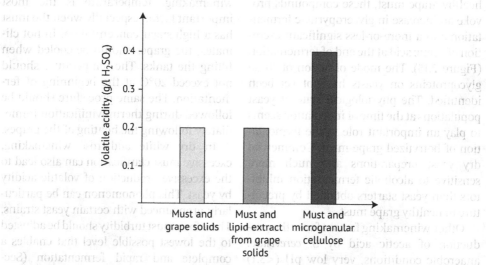

FIGURE 2.16 Effect of the lipid fraction of grape solids on acetic acid production by yeasts during alcoholic fermentation (Lavigne, 1996).

freshly crushed grapes at a proportion of no more than 20–30%. The wine should be sulfited or filtered before incorporation to eliminate bacteria. The volatile acidity of this mixture should not exceed 0.6 g/l expressed as H_2SO_4. The volatile acidity of this newly made wine rarely exceeds 0.3 g/l expressed as H_2SO_4. The concentration of ethyl acetate decreases simultaneously.

2.3.5 Other Secondary Products of the Fermentation of Sugars

Lactic acid is another secondary product of fermentation. It is also derived from pyruvic acid, which is directly reduced by yeast L(+)- and D(−)-lactate dehydrogenases. Under anaerobic conditions (the case in alcoholic fermentation), the yeast synthesizes predominantly D(−)-lactate dehydrogenase. Yeasts form 200–300 mg of D(−)-lactic acid per liter and only a few dozen milligrams of L(+)-lactic acid. The latter is formed essentially at the start of fermentation. By determining the D(−)-lactic acid concentration in a wine, it can be ascertained whether the origin of acetic acid is yeast or lactic acid bacteria (Section 14.2.3). Wines that have undergone malolactic fermentation can contain several grams per liter of exclusively L(+)-lactic acid. On the other hand, the lactic acid fermentation of sugars (lactic spoilage) forms D(−)-lactic acid. When D(−)-lactic acid concentrations exceed 200–300 mg/l, it is clear that lactic acid bacteria have transformed substrates other than malic acid.

Yeasts also make use of pyruvic acid to form acetoin, diacetyl, and 2,3-butanediol (Figure 2.17). This process begins with the condensation of a pyruvate molecule and a molecule of active acetaldehyde bound to TPP, leading to the formation of α-acetolactic acid. The oxidative decarboxylation of α-acetolactic acid produces diacetyl. Acetoin is produced by either the nonoxidative decarboxylation of α-acetolactic acid or the reduction of diacetyl. The reduction of acetoin leads to the formation of 2,3-butanediol; this last reaction is reversible.

From the start of alcoholic fermentation, yeasts produce diacetyl, which is rapidly reduced to acetoin and 2,3-butanediol. This reduction takes place in the days that follow the end of alcoholic fermentation, when wines are conserved on the yeast

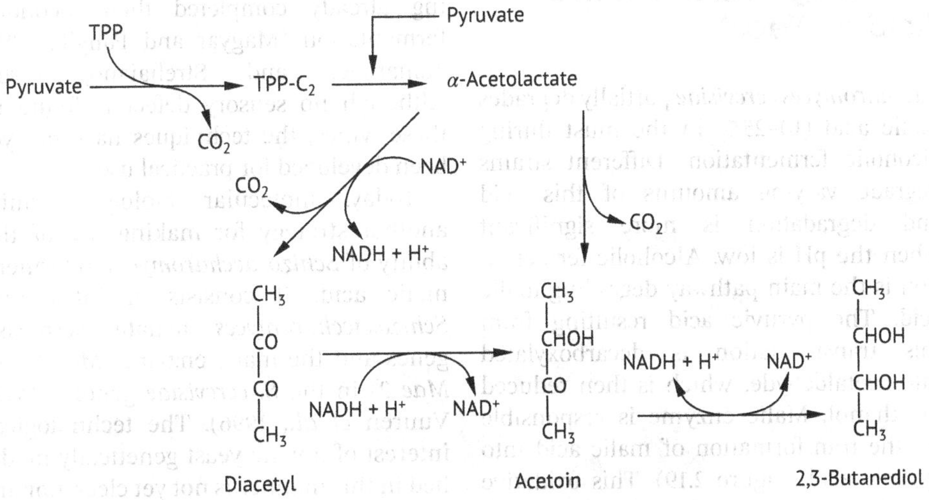

FIGURE 2.17 Acetoin, diacetyl, and 2,3-butanediol formation by yeasts under anaerobic conditions. TPP, thiamine pyrophosphate; TPP-C_2, active acetaldehyde.

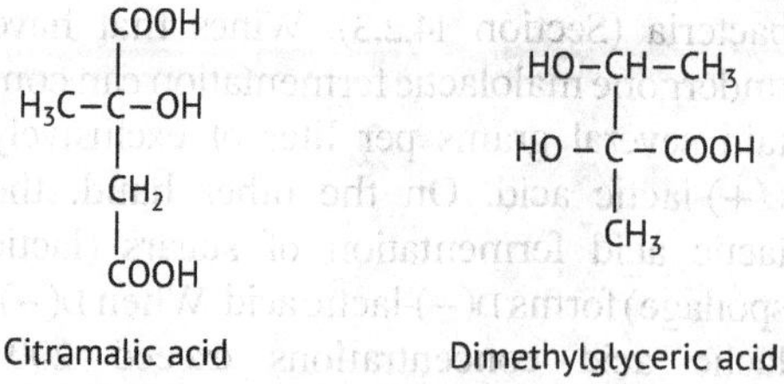

```
      COOH                    HO—CH—CH₃
       |                          |
H₃C—C—OH                      HO—C—COOH
       |                          |
      CH₂                        CH₃
       |
      COOH

Citramalic acid            Dimethylglyceric acid
```

FIGURE 2.18 Citramalic acid and dimethylglyceric acid.

biomass (de Revel et al., 1996). Acetoin and especially diacetyl are strong-smelling compounds that evoke a buttery aroma. Above a certain concentration, they have a negative effect on wine aroma. However, in wines that have not undergone malolactic fermentation, their concentration is too low (a few milligrams per liter for diacetyl) to have a sensory impact. On the other hand, lactic acid bacteria can degrade citric acid to produce much higher quantities of these carbonyl compounds than yeasts (Section 5.3.2).

Finally, yeasts condense acetic acid (in the form of acetyl-CoA) and pyruvic acid to produce citramalic acid (0–300 mg/l) and dimethylglyceric acid (0–600 mg/l) (Figure 2.18). These compounds have little sensory impact.

2.3.6 Degradation of Malic Acid by Yeast

Saccharomyces cerevisiae partially degrades malic acid (10–25%) in the must during alcoholic fermentation. Different strains degrade varying amounts of this acid and degradation is more significant when the pH is low. Alcoholic fermentation is the main pathway degrading malic acid. The pyruvic acid resulting from this transformation is decarboxylated into acetaldehyde, which is then reduced to ethanol. Malic enzyme is responsible for the transformation of malic acid into pyruvic acid (Figure 2.19). This oxidative decarboxylation requires NAD^+ (Fuck and Radler, 1972). This malo-alcoholic fermentation lowers wine acidity significantly more than malolactic fermentation does.

Schizosaccharomyces differs from wine yeasts. The alcoholic fermentation of malic acid is complete in yeasts of this genus, which possess an active malate transport system. In *S. cerevisiae*, malic acid penetrates the cell by simple diffusion. Yet at present no attempts to use *Schizosaccharomyces* in winemaking to break down the malic acid in musts have been successful (Peynaud et al., 1964; Carre et al., 1983). First of all, the implantation of these yeasts in the presence of *S. cerevisiae* is difficult in a non-sterilized must. Secondly, their optimum growth temperature (30°C), higher than for *S. cerevisiae*, imposes warmer fermentation conditions. Sometimes, the higher temperature adversely affects the sensory quality of wine. Finally, some grape varieties fermented by *Schizosaccharomyces* do not express their varietal aromas. The acidic Gros Manseng variety produces a very fruity wine when correctly vinified with *S. cerevisiae*, but has no varietal aroma when fermented by *Schizosaccharomyces*. To resolve these problems, some researchers have used non-proliferating populations of *Schizosaccharomyces* enclosed in alginate balls. These populations degrade the malic acid in wines having already completed their alcoholic fermentation (Magyar and Panyik, 1989; Taillandier and Strehaiano, 1990). Although no sensory defect is found in these wines, the techniques have not yet been developed for practical use.

Today, molecular biology permits another strategy for making use of the ability of *Schizosaccharomyces* to ferment malic acid. It consists of integrating *Schizosaccharomyces* malate permease genes and the malic enzyme (*Mae* 1 and *Mae* 2) in the *S. cerevisiae* genome (Van Vuuren et al., 1996). The technological interest of a wine yeast genetically modified in this manner is not yet clear, nor are the risks of its proliferation in wineries and nature.

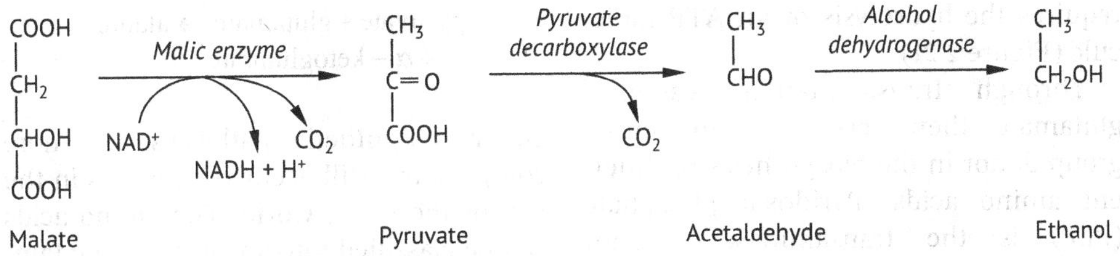

FIGURE 2.19 Decomposition of malic acid by yeasts during alcoholic fermentation.

2.4 Metabolism of Nitrogen Compounds

The nitrogen requirements of wine yeasts and the nitrogen supply in grape musts are discussed later (Section 3.4.2). The following section covers the general mechanisms of assimilation, biosynthesis, and degradation of amino acids in yeasts. The consequences of these metabolisms, which occur during alcoholic fermentation and affect the production of higher alcohols and their associated esters in wine, are also discussed.

2.4.1 Amino Acid Synthesis Pathways

The ammonium ion and amino acids found in grape must supply the yeast with nitrogen. The yeast can also synthesize most of the amino acids necessary for constructing its proteins. It fixes an ammonium ion on a carbon skeleton derived from the metabolism of sugars. The yeast uses the same reaction pathways as all organisms. Glutamate and glutamine play an important role in this process (Cooper, 1982; Magasanik, 1992).

$NADP^+$ glutamate dehydrogenase ($NADP^+$-GDH), a product of the GDH1 gene, produces glutamate (Figure 2.20) from an ammonium ion and an α-ketoglutarate molecule. The latter is an intermediate product of the citric acid cycle. The yeast also possesses an NAD^+ glutamate dehydrogenase (NAD^+-GDH), produced by the GDH2 gene. This dehydrogenase is involved in the oxidative catabolism of glutamate. It produces the inverse reaction of the previous one, liberating the ammonium ion used in the synthesis of glutamine. NADP-GDH activity is at its maximum when the yeast is cultivated on a medium containing exclusively ammonium as its source of nitrogen. The NAD-GDH activity, however, is at its highest level when the principal source of nitrogen is glutamate. Glutamine synthetase (GS) produces glutamine from glutamate and ammonium. This amidation

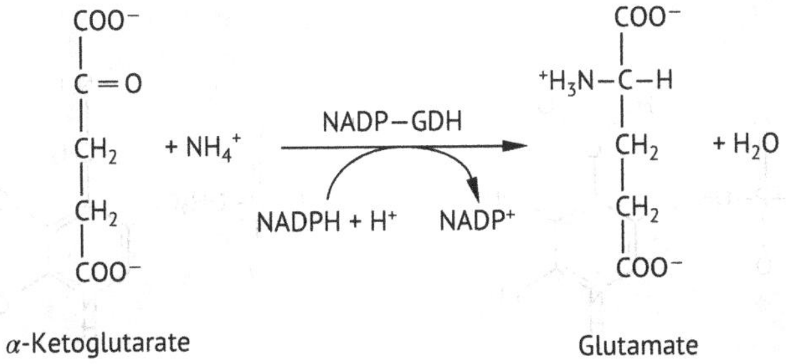

FIGURE 2.20 Incorporation of the ammonium ion in α-ketoglutarate catalyzed by NADP glutamate dehydrogenase (NADP-GDH).

requires the hydrolysis of an ATP molecule (Figure 2.21).

Through transamination reactions, glutamate then serves as an amino group donor in the biosynthesis of different amino acids. Pyridoxal phosphate (PLP) is the transaminase cofactor (Figure 2.22); it is derived from pyridoxine (vitamin B_6).

The carbon skeleton of amino acids originates from intermediates of glycolysis (pyruvate, 3-phosphoglycerate, and phosphoenolpyruvate), the citric acid cycle (α-ketoglutarate and oxaloacetate), or the pentose phosphate cycle (ribose 5-phosphate and erythrose 4-phosphate). Some of these reactions are very simple, such as the formation of aspartate or alanine by transamination of glutamate into oxaloacetate or pyruvate:

$$\text{oxaloacetate} + \text{glutamate} \rightarrow \text{aspartate} + \alpha\text{-ketoglutarate}$$

$$\text{pyruvate} + \text{glutamate} \rightarrow \text{alanine} + \alpha\text{-ketoglutarate}$$

Other biosynthetic pathways are more complex, but still occur in yeasts as in the rest of the living world. The amino acids can be classified into six biosynthetic families, depending on their nature and their carbon precursor (Figure 2.23):

1. In addition to glutamate and glutamine, proline and arginine are formed from α-ketoglutarate.
2. Asparagine, methionine, lysine, threonine, and isoleucine are derived from aspartate, which comes from oxaloacetate. ATP can activate methionine to form S-adenosylmethionine, which can be demethylated to form S-adenosylhomocysteine, the hydrolysis of which liberates adenine to produce homocysteine.

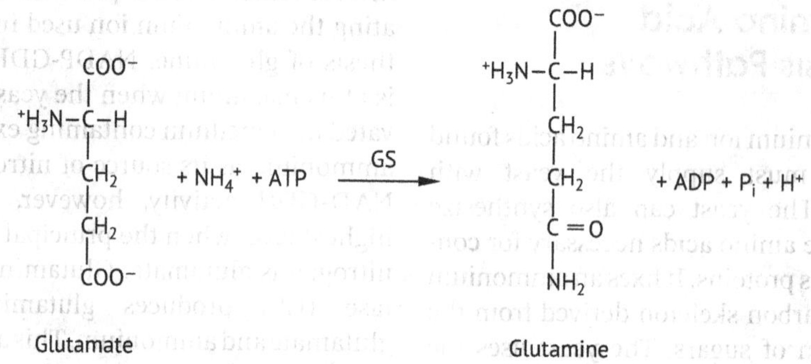

FIGURE 2.21 Amidation of glutamate into glutamine by glutamine synthetase (GS).

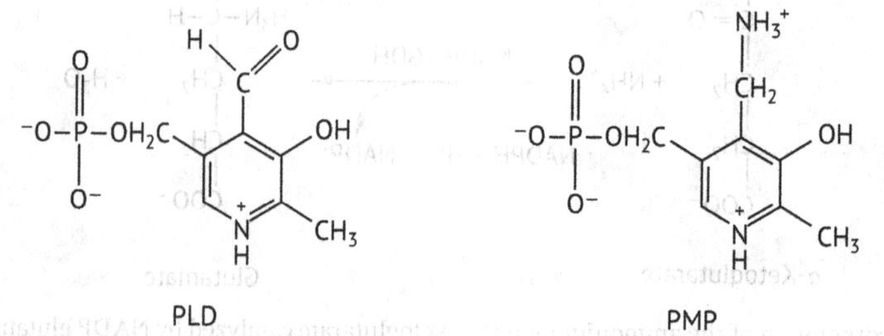

FIGURE 2.22 Pyridoxal phosphate (PLD) and pyridoxamine phosphate (PMP).

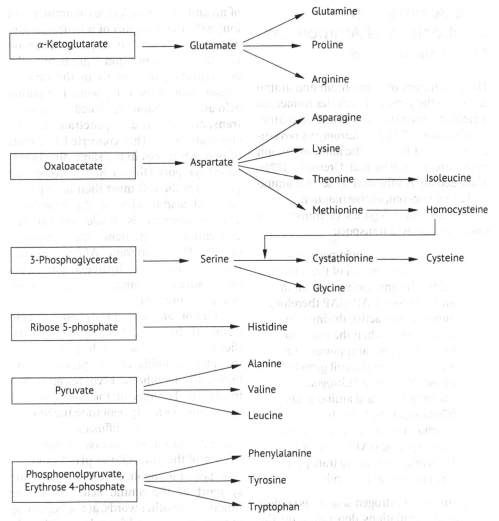

FIGURE 2.23 General biosynthesis pathways of amino acids.

3. Pyruvate is the starting point for the synthesis of alanine, valine, and leucine.
4. 3-Phosphoglycerate leads to the formation of serine and glycine. The condensation of homocysteine and serine produces cystathionine, a precursor of cysteine.
5. The imidazole ring of histidine is formed from ribose 5-phosphate and adenine coming from ATP.
6. The amino acids possessing an aromatic ring (tyrosine, phenylalanine, and tryptophan) are derived from erythrose 4-phosphate and phosphoenolpyruvate. These two compounds are intermediates of the pentose cycle and glycolysis, respectively. Their condensation forms shikimate. The condensation of this compound with another molecule of phosphoenolpyruvate produces chorismate, a precursor of aromatic amino acids.

2.4.2 Assimilation Mechanisms of Ammonium and Amino Acids

The penetration of ammonium and amino acids into the yeast cell activates numerous membrane protein transporters or permeases (Section 1.3.2). *Saccharomyces cerevisiae* has at least two specific ammonium ion transporters (Dubois and Grenson, 1979). Their activity is inhibited by several amino acids, in a noncompetitive manner.

Two distinct categories of transporters ensure amino acid transport:

1. A general amino acid permease (GAP) transports all of the amino acids. The ammonium ion inhibits and represses GAP. GAP therefore appears to be active during winemaking only when the must no longer contains ammonium, i.e. after the end of the cell growth phase. It acts as a "nitrogen scavenger" toward amino acids (Cartwright *et al.*, 1989).
2. *Saccharomyces cerevisiae* also has many specific GAPs (at least 11). Each one ensures the transport of one or more amino acids.

The transport of nitrogen sources is subject to complex regulations depending on the nitrogen content of the medium, by means of a system called nitrogen catabolite repression (NCR) (Beltran *et al.*, 2004).

Toward the end of fermentation, yeasts excrete significant but variable amounts of different amino acids. During fermentation, yeasts assimilate between 1 and 2g/l of amino acids. Finally, at the end of alcoholic fermentation, a few hundred milligrams of amino acids per liter remain; proline generally represents half of this amount.

Contrary to must hexoses that penetrate the cell by facilitated diffusion, ammonium and amino acids require active transport. Their concentration in the cell is generally higher than in the external medium. The permeases involved couple the transport of an amino acid molecule (or ammonium ion) with the transport of a hydrogen ion. The hydrogen ion moves in the direction of the concentration gradient: the concentration of protons in the must is higher than in the cytoplasm. The amino acid and the proton are linked to the same transport protein and penetrate the cell simultaneously. This concerted transport of two substances in the same direction is called symport. Obviously, the proton that penetrates the cell must then be exported to avoid acidification of the cytoplasm. This movement is made against the concentration gradient and requires energy. The membrane ATPase ensures the excretion of the hydrogen ion across the plasma membrane, acting as a proton pump (Figure 2.24).

Ethanol strongly limits amino acid transport. In an alcohol medium, it modifies the composition and the properties of the phospholipids of the plasma membrane. The membrane becomes more permeable to H^+ ions in the medium, and these ions massively penetrate the interior of the cell by simple diffusion. The membrane ATPase must increase its operation to control the intracellular pH. As soon as this task monopolizes the ATPase, the symport of the amino acids no longer functions. In other words, at the beginning of fermentation, and for as long as the ethanol concentration in the must is low, yeasts can rapidly assimilate amino acids and concentrate them in the vacuoles for later use, according to their biosynthesis needs.

2.4.3 Catabolism of Amino Acids

The ammonium ion is essential for the synthesis of amino acids necessary for building proteins, but yeasts cannot always find sufficient quantities in their environment. Fortunately, they can obtain ammonium from available amino acids through various reactions.

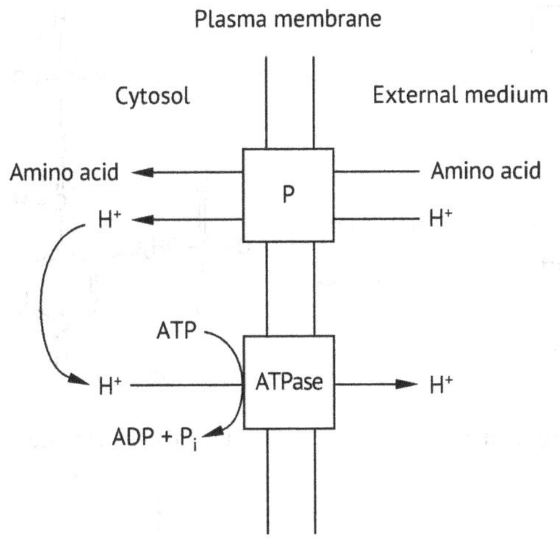

2.4 | Metabolism of Nitrogen Compounds

FIGURE 2.24 Active amino acid transfer mechanisms in the yeast plasma membrane. P, protein playing the role of an amino acid "symporter."

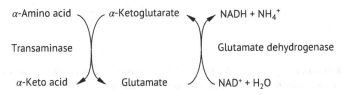

FIGURE 2.25 Oxidative deamination of an amino acid, catalyzed by a transaminase and glutamate dehydrogenase.

The most common pathway is the transfer of an α-amino group, originating from one of many different amino acids, onto α-ketoglutaric acid to form glutamate. Aminotransferases or transaminases catalyze this reaction, whose prosthetic group is PLP. Glutamate is then deaminated by an oxidative pathway to form NH_4^+ (Figure 2.25). These two reactions can be summarized as follows:

$$\alpha-\text{amino acid} + NAD^+ + H_2O \rightarrow \alpha-\text{keto acid} + NH_4^+ + NADH + H^+$$

During transamination, PLP is temporarily transformed into pyridoxamine phosphate (PMP). The PLP aldehyde group is bound to a lysine residue ε-amino group on the active site of the aminotransferase to form an intermediate (E-PLP) (Figure 2.26). The α-amino group of the amino acid that is the transamination substrate displaces the lysine ε-amino group bound to PLP. The cleavage of this intermediate liberates PMP and the keto acid corresponding to the amino acid substrate. PMP can in turn react with another keto acid to furnish a second amino acid and regenerate PLP.

The partial reactions can be written in the following manner:

$$\text{amino acid} 1 + E - PLP \rightarrow \text{keto acid} 1 + E - PMP$$

$$\text{keto acid} 2 + E - PMP \rightarrow \text{amino acid} 2 + E - PLP$$

the balance for which is:

$$\text{amino acid} 1 + \text{keto acid} 2 \rightarrow \text{keto acid} 1 + \text{amino acid} 2$$

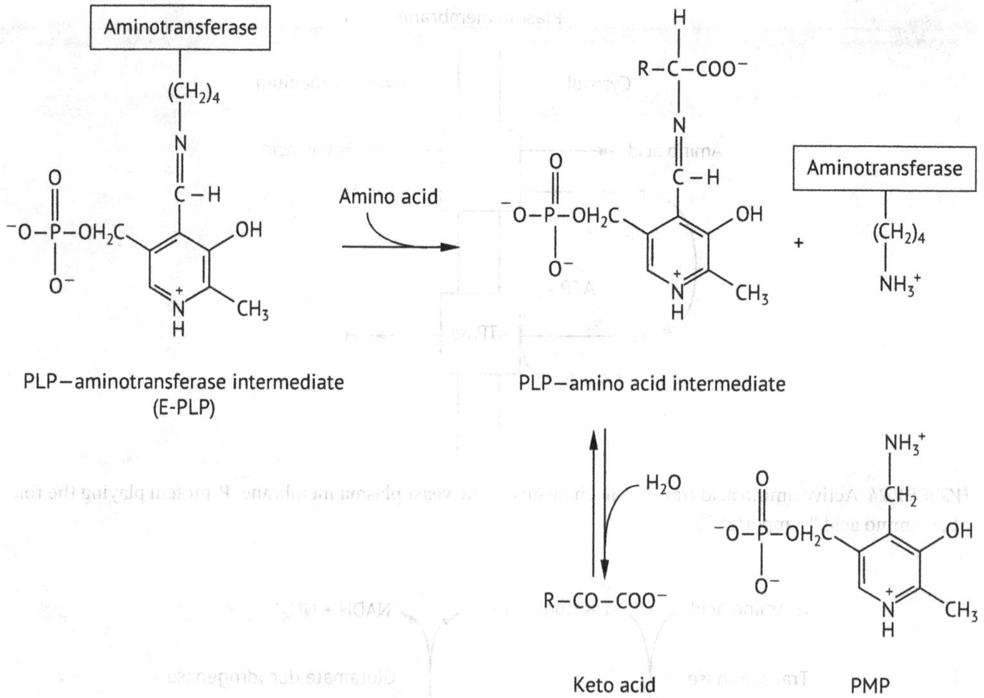

FIGURE 2.26 Mode of action of pyridoxal phosphate (PLP) in transamination reactions. Formation of intermediates between PLP and aminotransferase or the amino acid substrate.

Some amino acids, such as serine and threonine, possess a hydroxyl group on their β carbon. They can be directly deaminated by dehydration. A dehydratase catalyzes this reaction, producing the corresponding keto acid and ammonium (Figure 2.27).

2.4.4 Formation of Higher Alcohols and Esters

Yeasts can excrete keto acids originating from the deamination of amino acids only after their decarboxylation into aldehyde and reduction into alcohol (Figure 2.28). This mechanism, known as the Ehrlich reaction, explains in part the formation of higher alcohols in wine. Table 2.4 lists the principal higher alcohols and their corresponding amino acids, possible precursors of these alcohols.

Several experiments clearly indicate, however, that the degradation of amino acids is not the only pathway for forming higher alcohols in wine. In fact, certain ones, such as propan-1-ol and butan-1-ol, do not have amino acid precursors. Moreover, certain mutants deficient in the synthesis of specific amino acids do not produce the corresponding higher alcohol, even if the amino acid is present in the culture medium. There is no relationship between the amount of amino acids in must and the amount of corresponding higher alcohols in wine.

The production of higher alcohols by yeasts appears to be linked not only to the catabolism of amino acids but also to their synthesis via the corresponding keto acids. These acids are derived from the metabolism of sugars. For example, propan-1-ol has no corresponding amino acid. It is derived from α-ketobutyrate that can be formed from pyruvate and acetyl-CoA. α-Ketoisocaproate is a precursor of isoamyl alcohol and an intermediate in the synthesis of leucine. It too can be produced from α-acetolactate,

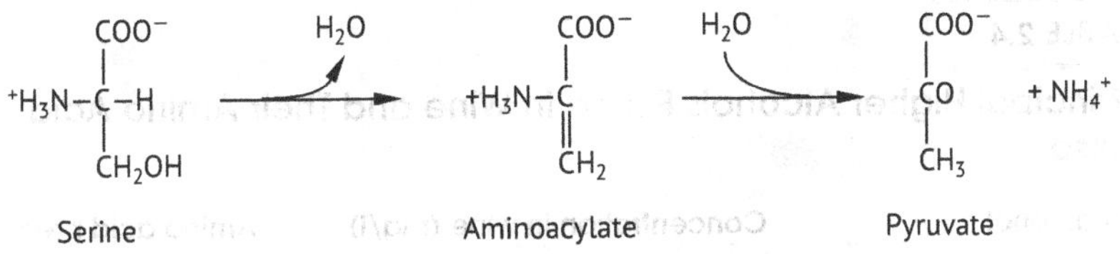

$$\underset{\text{Serine}}{\overset{COO^-}{\underset{CH_2OH}{\overset{|}{\underset{|}{{}^+H_3N-C-H}}}}} \xrightarrow{H_2O} \underset{\text{Aminoacylate}}{\overset{COO^-}{\underset{CH_2}{\overset{|}{\underset{||}{{}^+H_3N-C}}}}} \xrightarrow{H_2O} \underset{\text{Pyruvate}}{\overset{COO^-}{\underset{CH_3}{\overset{|}{\underset{|}{CO}}}}} + NH_4^+$$

FIGURE 2.27 Deamination of serine by a dehydratase.

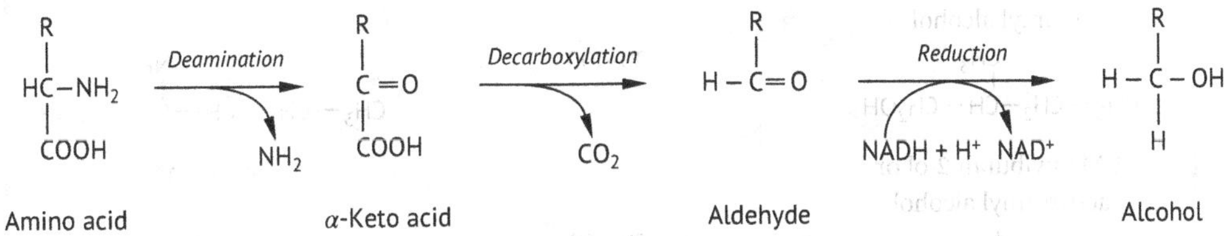

$$\underset{\text{Amino acid}}{\overset{R}{\underset{COOH}{\overset{|}{\underset{|}{HC-NH_2}}}}} \xrightarrow[NH_2]{\text{Deamination}} \underset{\alpha\text{-Keto acid}}{\overset{R}{\underset{COOH}{\overset{|}{\underset{|}{C=O}}}}} \xrightarrow[CO_2]{\text{Decarboxylation}} \underset{\text{Aldehyde}}{\overset{R}{\overset{|}{H-C=O}}} \xrightarrow[NADH+H^+\quad NAD^+]{\text{Reduction}} \underset{\text{Alcohol}}{\overset{R}{\underset{H}{\overset{|}{\underset{|}{H-C-OH}}}}}$$

FIGURE 2.28 Formation of higher alcohols from amino acids (Ehrlich reactions).

which is derived from pyruvate. Most higher alcohols in wine can also be formed by the metabolism of glucose, without the involvement of amino acids.

The physiological function of higher alcohol production by yeasts is not clear. It may be a simple waste of sugars, a detoxification process of the intracellular medium, or a means of regulating the metabolism of amino acids.

With the exception of phenylethanol, which has a rose-like fragrance, higher alcohols smell bad. Most, such as isoamyl alcohol, have heavy solvent-like odors. Methionol is a peculiar alcohol because it contains a sulfur atom. Its cooked-cabbage odor has the lowest perception threshold (1.2 mg/l). It can be responsible for the most persistent and unpleasant off-odors of reduction, especially in white wines (Volume 2, Section 8.6.2). In general, the winemaker should avoid excessive higher alcohol odors. Fortunately, their sensory impact is limited at their usual concentrations in wine, but it depends on the overall aroma intensity of the wine. Excessive yields and rain at the end of ripening can dilute the must, in which case the wine will have a low aroma intensity and the heavy, common character of higher alcohols can be pronounced.

The winemaking parameters that increase higher alcohol production by yeasts are well known: high pH, high fermentation temperature, and aeration. In red winemaking, the extraction of pomace constituents and the concern for rapid and complete fermentations impose both aeration and high temperatures. In this case, the production of higher alcohols by yeast cannot be limited. In white winemaking, a fermentation temperature between 20 and 22°C limits the formation of higher alcohols.

Ammonium and amino acid deficiencies in must lead to an increased formation of higher alcohols. Under these conditions, the yeast appears to recuperate all of the available amino nitrogen by transamination. It releases the unused carbon skeleton in the form of higher alcohols. Settling of white must for clarification purposes also limits the production of higher alcohols (Section 13.5.2).

The nature of the yeast (species and strain) responsible for fermentation also affects the production of higher alcohols. Certain species, such as *Hansenula anomala*, have long been known to produce a lot of these compounds, especially under aerobic conditions (Guymon et al., 1961). However, production by wine yeasts is limited, even in spontaneous fermentation. More recently, various researchers (Kishimoto, 1994; Masneuf, 1996; Masneuf-Pomarède et al., 2010) have shown that

TABLE 2.4

The Principal Higher Alcohols Found in Wine and Their Amino Acid Precursors

Higher alcohol	Concentration in wine (mg/l)	Amino acid precursor
$CH_3-CH(CH_3)-CH_2-CH_2OH$ 3-Methylbutan-1-ol or isoamyl alcohol	80–300	$CH_3-CH(CH_3)-CH_2-CH(NH_2)-COOH$ Leucine
$CH_3-CH_2-CH(CH_3)-CH_2OH$ 2-Methylbutan-2-ol or active amyl alcohol	30–100	$CH_3-CH_2-CH(CH_3)-CH(NH_2)-COOH$ Isoleucine
$CH_3-CH(CH_3)-CH_2OH$ 2-Methylpropan-1-ol or isobutyl alcohol	50–150	$CH_3-CH(CH_3)-CH(NH_2)-COOH$ Valine
$C_6H_5-CH_2-CH_2OH$ Phenylethanol	10–100	$C_6H_5-CH_2-CH(NH_2)-COOH$ Phenylalanine
$HO-C_6H_4-CH_2-CH_2OH$ Tyrosol	20–50	$HO-C_6H_4-CH_2-CH(NH_2)-COOH$ Tyrosine
$CH_3-CH_2-CH_2OH$ Propan-1-ol	10–50	?
$CH_3-CH_2-CH_2-CH_2OH$ Butan-1-ol	1–10	?
Indole-CH_2-CH_2OH Tryptophol	0–1	Indole-$CH_2-CH(NH_2)-COOH$ Tryptophan
γ-Butyrolactone	0–5	$COOH-CH_2-CH_2-CH(NH_2)-COOH$ Glutamic acid
$CH_3-S-CH_2-CH_2-CH_2OH$ Methionol	0–5	$CH_3-S-CH_2-CH_2-CH(NH_2)-COOH$ Methionine

most *Saccharomyces bayanus* strains (formerly called *uvarum*) produce considerably more phenylethanol than does *S. cerevisiae*. Lastly, the production of higher alcohols in *S. cerevisiae* depends on the strain. Limited production of higher alcohols (with the exception of phenylethanol) should be among the selection criteria for wine yeasts.

Due to their esterase activities, yeasts form various esters (a few milligrams per liter). The most important acetates of higher alcohols are isoamyl acetate (banana aroma) and phenylethyl acetate (rose aroma). Although they are not linked to nitrogen metabolism, ethyl esters of medium-chain fatty acids are also involved. They are formed by the condensation of acetyl-CoA. These esters have more interesting aromas than the others. Hexanoate has a flowery and fruity aroma reminiscent of green apples. Ethyl decanoate has a soap-like odor. In white winemaking, the production of these esters can be increased by lowering the fermentation temperature and increasing must clarification. Certain yeast strains (71B) produce large quantities of these compounds, which contribute to the fermentation aroma of young wines. They are rapidly hydrolyzed during their first year in bottle and have no long-term influence on the aroma of white wines.

References

Alexandre H., Nguyen Van Long T., Feuillat M. and Charpentier C. (1994) Rev. Fr. Œnol., cahiers scientifiques, 146, 11.

Attfield P.V., Kletsas S., Veal D.A., Van Rooijen R. and Bell P.J.L. (2000) J. Appl. Microbiol., 89, 207–214.

Bakker B.M., Overkamp K.M., van Maris A.J., Kötter P., Luttik M.A., van Dijken S.P. and Pronk J.T. (2001) FEMS Microbiol. Rev., 25, 1, 15.

Beltran G., Novo M., Rozès N., Mas A. and Guillamon J.M. (2004) FEMS Yeast Res., 4, 625.

Bely M., Rinaldi A. and Dubourdieu D. (2003) J. Biosci. Bioeng., 96, 6, 507.

Bely M., Masneuf-Pomarède I. and Dubourdieu D. (2005) Œno One, 39, 191.

Bely M., Stoeckle P., Masneuf-Pomarède I. and Dubourdieu D. (2008) Int. J. Food Microbiol., 122, 312.

Bisson L.F. (1993) Yeasts-metabolism of sugars. In *Wine Microbiology and Biotechnology*, 55–75 (Ed G.H. Fleet). Harwood Academic Publishers, Chur.

Blomberg A. and Alder L. (1992) Adv. Microbiol. Physiol., 33, 145.

Blondin B., Dequin S., Saint Prix F. and Sablayrolles J.M. (2002) *La formation d'acides volatils par les levures*, dans *13ième Symposium international d'œnologie 9–10 Juin 2002*, INSA/INRA, Montpellier, France.

Buchner E. (1897) Über die alkoholische Gärung ohne Hefezellen. In *Berichte der deutschen chemischen Gesellschaft*, January–April 1897, p. 1110.

Camarasa C., Grivet J.P. and Dequin S. (2003) Microbiol., 149, 2669.

Carre E., Lafon-Lafourcade S. and Bertrand A. (1983) Œno One, 17, 43.

Cartwright C.P., Rose A.H., Calderbank J. and Keenan M.H.J. (1989) Solute transport. In *The Yeasts*, Vol. 3, p. 5 (Eds A.H. Rose and J.S. Harrison). Academic Press, London.

Caubet R., Guérin B. and Guérin M. (1988) Arch. Microbiol., 149, 324.

Cooper T.G. (1982) Nitrogen metabolism in *Saccharomyces cerevisiae*. In *The Molecular Biology of the Yeast Saccharomyces: Metabolism and Gene Expression*, pp. 39–99 (Eds J.N. Strathern, E.W. Jones and J.R. Broach). Cold Spring Harbor Laboratory Press, New York.

Crabtree H.G. (1929) Biochem. J., 23, 536.

De Revel G., Lonvaud-Funel A. and Bertrand A. (1996) Étude des composés dicarbonylés au cours des fermentations alcoolique et malolactique. In *Œnologie 95, 5e Symposium international d'œnologie*, p. 321 (Eds A. Lonvaud). Tec & Doc, Lavoisier, Pairs.

Delfini C. and Cervetti F. (1992) Vitic. Enol. Sci., 46, 142.

Dubois E. and Grenson M. (1979) Mol. Gen. Genet. 175, 67.

Dubourdieu D. (1982) *Recherches sur les polysaccharides sécrétés par Botrytis cinerea dans la baie de raisin*, Thèse Doctorat és Sciences, Université de Bordeaux II.

Erasmus D.J, Van der Merwe G.K. and Van Vuuren H.J.J. (2003) FEMS Yeast Res., 3, 375.

Fuck E. and Radler F. (1972) Archiv. für Mikrobiologie, 87, 149.

Guymon J.F., Ingraham J.L. and Crowell E.A. (1961) Arch. Biochem. Biophys., 95, 163.

Kishimoto M. (1994) J. Ferment. Bioeng., 77, 4, 432.

Lafon-Lafourcade S. (1983) Wine and brandy. In *Biotechnology*, Vol. 5, pp. 100–109 (Eds H.J. Rehm and G. Reed). Verlag Chemie, Weinheim.

Lavigne V. (1996) *Recherche sur les composés volatils formés par la levure au cours de la vinification et de l'élevage des vins blancs secs*, Thèse de Doctorat, Université de Bordeaux II.

Magasanik B. (1992) Regulation of nitrogen utilization. In *The Molecular Biology of the Yeast Saccharomyces. Gene Expression*, pp. 283–317 (Eds E.W. Jones, J.R. Pringle and J.R. Broach. Cold Spring Harbor Laboratory Press, New York.

Magyar I. and Panyik I., 1989, Am. J. Oenol. Vitic. 40, 233.

Masneuf I. (1996) *Recherches sur l'identification génétique des levures de vinification. Applications oenologiques*, Thèse de Doctorat de l'Université de Bordeaux II.

Masneuf-Pomarède I., Bely M., Marullo P., Lonvaud-Funel A. and Dubourdieu D. (2010) Int. J. Food Microbiol., 139, 79.

Navarro-Avino J.P., Prasd R., Miralles V.J., Benito R.M. and Serrano R. (1999) Yeast, 15, 929.

Neuberg C. (1946) Ann. Rev. Biochem., 15, 435.

Peynaud E., Domercq S., Boidron A., Lafon-Lafourcade S. and Guimberteau G. (1964) Archiv. für Mikrobiologie, 48, 150.

Pigeau G.M. and Inglis D.L. (2005) J. Appl. Microbiol., 99, 112.

Remize F., Andrieu E. and Dequin S. (2000) Appl. Environ. Microbiol., 66, 3151.

Ribéreau-Gayon J., Peynaud E. and Lafourcade S. (1952) C. R. Acad. Sci. 234, 478.

Ribéreau-Gayon P., Lafon-Lafourcade S., Dubourdieu D., Lucmaret V. and Larue F. (1979) C. R. Acad. Sci., 289 D, 441.

Stryer L. (1992) Métabolisme: concepts et vue d'ensemble. In *La biochimie de Lubert Stryer*, p. 315. Flammarion, Paris.

Taillandier P. and Strehaiano P. (1990) Dégradation de l'acide malique par *Schizosaccharomyces*. In *Actualités œnologiques 89, C. R. du 4e Symposium international d'œnologie, Bordeaux 1989* (Eds P. Ribéreau- Gayon and A. Lonvaud. Dunod-Bordas, Paris.

Van Vuuren H.J.J., Viljoen M., Grobler J., Volschenk H., Bauer F. and Subden R.E. (1996) Genetic analysis of the *Schizosaccharomyces pombe* malate permeases, *Mae I* and malic enzyme, *Mae II*, genes and their expression in *Saccharomyces cerevisiae*. In *Œnologie 95, 5e Symposium international d'œnologie*, p. 195 (Ed. A. Lonvaud). Tec & Doc, Lavoisier, Paris.

Verduyn C., Postma E., Scheffers W.A. and Van Dijken J.P. (1990) J. Gen. Micro., 136, 359.

CHAPTER 3

Conditions of Yeast Development

3.1 Introduction
3.2 Monitoring and Controlling Fermentations
3.3 Yeast Growth Cycle and Fermentation Kinetics
3.4 Nutrition Requirements
3.5 Fermentation Activators
3.6 Inhibition of Fermentation
3.7 Physicochemical Factors Affecting Yeast Growth and Fermentation Kinetics
3.8 Stuck Fermentations

3.1 Introduction

Grape must is a highly fermentable medium in which yeasts find the necessary substances to ensure their vital functions. Carbohydrates (glucose and fructose) are used as carbon and energy sources—the yeasts make ethanol from them. Ethanol gives wines their principal characteristic. Organic acids (tartaric and malic acids) and mineral salts (phosphate, sulfate, chloride, potassium, calcium, and magnesium) ensure a suitable pH. Nitrogen compounds exist in several forms: ammonia, amino acids, polypeptides, and proteins. Grape must also contains substances serving as growth (vitamins) and survival factors. Other grape constituents (phenolic compounds and aroma compounds) contribute to wine character, but do not play an essential role in fermentation phenomena.

In general, with an adequate inoculation (10^6 cells/ml), fermentation is easily initiated in grape must. It is complete, if the initial carbohydrate concentration is not excessive, but a number of factors can disrupt yeast growth and fermentation kinetics. Some factors are chemical in nature and correspond with either nutritional deficiencies or the presence of inhibitors

---

Handbook of Enology, Volume 1: The Microbiology of Wine and Vinifications, Third Edition.
Pascal Ribéreau-Gayon, Denis Dubourdieu, Bernard B. Donèche and Aline A. Lonvaud.
© 2021 John Wiley & Sons Ltd. Published 2021 by John Wiley & Sons Ltd.

formed during fermentation (ethanol, fatty acids, etc.). Others are physicochemical in nature—for example, oxygenation, temperature, and must clarification. Finally, difficult fermentations can lead to the development of undesirable microorganisms. They can be antagonistic toward the desired winemaking yeast strains.

The successful completion of fermentation depends on all of these factors. A perfect mastery of fermentation is one of the primary responsibilities of an enologist, who must use the necessary means to avoid microbial deviations and bring the fermentation to completion—the complete depletion of sugars in dry wines. Stuck fermentations are a serious problem. They are not only often difficult to restart but can also lead to bacterial spoilage such as volatile acidity in sugar-containing media. These issues are discussed here in connection with practical winemaking applications (Ribéreau-Gayon et al., 1975a, 1976). Theoretical information concerning yeast physiology can be found in more fundamental works (Fleet, 1992).

3.2 Monitoring and Controlling Fermentations

3.2.1 Counting Yeasts

Various methods (described elsewhere, Section 1.9) enable the industrial characterization of yeast strains involved in fermentation. Tracking the yeast population as fermentation progresses can be useful.

Lafon-Lafourcade and Joyeux (1979) described counting and identification techniques for the various microorganisms in must and wine (yeast, acetic acid, and lactic acid bacteria) (Volume 2, Section 11.3.4).

After the appropriate dilution of fermenting must, the total number of yeast cells can be estimated under the microscope, using a Malassez counting chamber.

After calibration, this determination can also be made by measuring the optical density of the fermentation medium at 620 nm. This measurement enables a yeast cell count by interpretation of the medium's haziness, caused by yeasts.

However, the total cells counted in this manner include both "dead" yeast and "live" yeast. The two must be differentiated in enology. In fact, counting "viable microorganisms" is preferable. When placed in a suitable, solid nutrient medium, viable cells are capable of developing and forming a microscopic cluster, visible to the naked eye, called a colony. The number of viable yeast cells can be determined by counting the colonies formed on this medium after two to three days of incubation at 30°C.

Viable yeast populations can also be estimated directly by counting under the microscope using specific dye or epifluorescence techniques. Viable populations can also be determined with ATP measurements using bioluminescence. Bouix et al. (1997) proposed the use of immunofluorescence to detect bacterial contaminations during winemaking.

Flow cytometry is a rapidly growing technique in which high-speed particles pass through a laser beam. In the future, it should provide a precious tool for the precise characterization of yeast populations (counts and physiological state).

Microbiological testing is useful for research work, but these methods are relatively long and difficult. For this reason, fermentations are generally monitored by more simple chemical methods in the winery.

3.2.2 Monitoring Fermentation Kinetics

Winemakers must closely monitor wine fermentation in each tank of the winery. This close supervision allows them to observe transformations, anticipate their evolution, and act quickly if necessary.

They should conduct both fermentation and temperature checks daily.

Fermentations can be monitored by measuring the amount of sugar consumed, alcohol formed, or carbon dioxide released, but the measurement of the mass per unit volume (density) is the simplest method. Mass per unit volume constitutes an approximate measure of the amount of sugar contained in the grape must. Since a relationship exists between the amount of alcohol produced during fermentation and the initial concentration of sugar in the must, must density can directly give an approximate potential alcohol. The density and potential alcohol are generally indicated on the stem of the hydrometer: approximately 17 g of sugar produces 1% alcohol by volume (Section 10.4.3, Table 10.5). Expressing potential alcohol is without doubt the best solution.

During fermentation, sugar depletion and ethanol formation result in decreased density. A hydrometer is used to monitor must density. Samples are taken from the middle of the tank by using the sampling valve. Before taking a sample, the valve should be cleared by letting a few centiliters flow out. The density is then corrected for must temperature. No other conversions or interpretations are necessary. Plotting the points in the form of a graph (Figure 3.1) helps the winemaker evaluate fermentation progress and kinetics. More importantly, sluggish fermentations, which lead to stuck fermentations, can be identified at an early stage. Due to the heterogeneity of fermentation kinetics in a red winemaking tank (fermentation is most active under the pomace cap), homogenizing the tank is recommended before taking samples.

3.2.3 Measuring Temperature

The daily monitoring of tank temperature during fermentation is indispensable, but this measurement must be taken properly. In red winemaking, in particular, tank temperature is never homogeneous. The temperature is highest in the pomace cap and lowest at the bottom of the tank. In the first hours of fermentation, sudden temperature increases occur in the pomace and are sometimes very localized. As a result, the must temperature against the tank lining is always less than in the center of the tank. The temperature taken under these conditions, even after properly clearing the sampling valve, is not representative of the entire tank. The temperature should be taken after a pump-over, which homogenizes tank temperature. In this manner, an average temperature can be obtained, but the maximum temperature, also important, remains unknown.

The must temperature can be taken with a dial thermometer having a 1.5 m probe. This effective method can measure must temperature directly in different areas, especially just below the pomace cap in the hottest part of the tank. This zone has the most significant fermentation activity. The temperature can also be taken by thermoelectric probes judiciously placed in each tank. The probes are linked to a measurement system in the winery laboratory. With this system, the winemaker can verify the temperature of the tanks at any moment. Certain temperature control systems automatically regulate tank temperature when the temperature reaches a certain value.

3.2.4 Fermentation Control Systems

Various automated systems simplify the monitoring of fermentation and make it more rigorous. These systems can also automatically heat and cool the must in order to control its temperature.

Some of these systems can be very sophisticated. For example, when the temperature exceeds the set limit, the apparatus initially homogenizes the must in the tank by pumping. If the temperature is still too

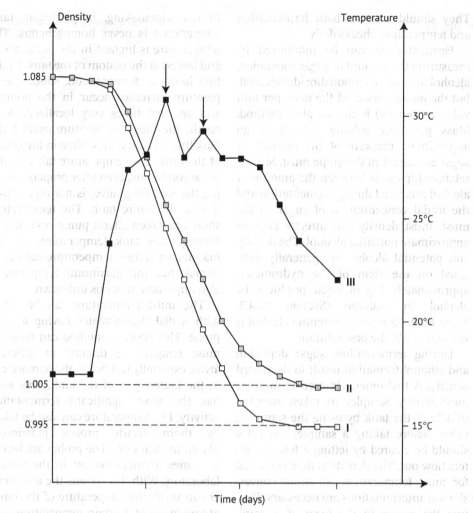

FIGURE 3.1 Example of daily fermentation monitoring in two tanks (initial must density = 1.085). I, Normal fermentation curve: after a latency period, fermentation begins, accelerates, and then slows before stopping on the 10th day at a density below 0.995, when all sugar is fermented. II, Fermentation curve leading to a stuck fermentation: fermentation stops on the 11th day at a density of 1.005; unfermented sugar remains. Fermentation slows early enough for the winemaker to take preventive action. III, Change in temperature (red wine fermentation): arrows indicate must cooling operations.

high, the refrigeration unit of the system cools the must. Owing to the seasonal character of winemaking, some winemakers must make do with manual control systems. Automation can, however, be fully justified, when it enables greater precision in winemaking. For example, a temperature gradient can be produced in this manner during red winemaking. At the beginning, a moderate temperature (18–20°C) favors cell growth and vitality; at the end, a higher temperature (>30°C) facilitates the extraction of pomace constituents.

The concept of automated process control goes further (Flanzy, 1998). For the moment at least, these systems control temperature, modulating it according to fermentation kinetics. In addition to online temperature control, a system should be developed to monitor fermentation kinetics. Various methods have been tested: measurement of carbon dioxide given off, weight loss, and decrease in density (measured as the difference in pressure between the top and the bottom of the tank). Measurement of gas released seems to be the most reliable method (by weighing in the laboratory or by using a domestic gas meter for large-capacity tanks; El Halaoui *et al.*, 1987). This process assumes that the tanks are completely airtight, which is not particularly recommended, especially in red winemaking where pumping-over is indispensable. In the future, new devices may well be developed, based on other physical principles. They could enable continuous monitoring of fermentation kinetics and temperature. Data would be recorded as curves, which could be used to schedule the many operations needed for full control of the fermentation process.

The first examples of control systems concern temperature control (Sablayrolles and Barre, 1989; Sablayrolles *et al.*, 1990). The aim of these systems is to regulate temperature in order to control the fermentation better and minimize the use of cooling systems. The law of control introduces the concept of intervention based on fermentation rate and thus on yeast activity. This latter parameter is definitely as important as temperature itself. When the rate of CO_2 production exceeds a previously established threshold, the temperature should be adjusted to the agreed level. As soon as the rate decreases, the apparatus should let the temperature rise in order to revive the fermentation. This enables the fermentation to be completed more quickly. Temperature modulation depends on must fermentability. Letting the temperature increase results in a decrease in total energy demand (Table 3.1). Of course, the apparatus should maintain the temperature within a range that is compatible with good winemaking practice.

Another example involves the detection of nitrogen deficiencies in musts, by measuring the maximum rate of CO_2 production at the start of fermentation and then adding nitrogen only if necessary and at the optimal moment (Section 3.4.2) (Bely *et al.*, 1990).

Another approach consists in creating a model of the alcoholic fermentation process. Various calculations concerning time, temperature, and alcohol and sugar concentrations are used to predict fermentation behavior—especially the risk of a stuck fermentation (Bovée *et al.*, 1990). In this way, the need and moment of a certain operation (especially temperature control) could be anticipated.

Lastly, these control laws must be adapted to the particular needs of each tank using quantifiable data and the enologist's know-how and experience. A highly advanced automated system optimizing alcoholic fermentations in winemaking could make use of artificial intel-

TABLE 3.1

Comparison of a Temperature-Controlled (17–22°C) Fermentation and Isothermal Fermentations at 17 and 22°C (Sablayrolles *et al.*, 1990)

Temperature	Isothermal 17°C	Isothermal 22°C	Temperature-controlled 17–22°C
Duration of fermentation (h)	263	174	183
Maximum rate (g CO_2/l/h)	0.69	1.12	0.72
Total refrigeration units needed (kcal/l)	18.2	16.1	10
Maximum refrigeration unit demand (kcal/l/h)	0.174	0.257	0.179

ligence to take into account the enologist's expertise (Grenier *et al.*, 1990).

3.2.5 Avoiding Foam Formation

During fermentation, foam may be formed as carbon dioxide is released. This can result in the tank overflowing. To avoid the problem, tanks are sometimes filled to only half their capacity. This space constraint can be unacceptable in many wineries. Factors that influence foam formation (Pavlenko, 1989; Guinand, 1989) include must nitrogen composition (especially protein concentration), fermentation temperature, and the nature of the yeast strain. Attempts to create fermentation conditions (for example, removal of proteins with bentonite) capable of limiting this phenomenon have not led to satisfactory results.

For this reason, some American wineries have adopted the use of products that increase surface tension and thus reduce foam formation and stability. Two antifoaming agents are gaining popularity: dimethyl polysiloxane and a mixture of oleic acid mono- and diglyceride. They are used at a concentration of less than 10 mg/l and do not leave a residue in wine, especially after filtration. Due to their effectiveness, red wine tanks can be filled to 75–80% capacity and white wine tanks to 85–90%. These products are not toxic. The International Organisation of Vine and Wine (OIV) recommends the exclusive use of mixtures of oleic acid mono- and diglyceride. This mixture is authorized for sale in the United States only.

3.3 Yeast Growth Cycle and Fermentation Kinetics

In an unsulfited and non-inoculated must, contamination yeasts can begin to develop within a few hours of filling the tank. Apiculate yeasts (*Kloechera* and *Hanseniaspora*) are the most frequently encountered. Aerobic yeasts may also develop (*Candida*, *Pichia*, and *Hansenula*), producing acetic acid and ethyl acetate. *Brettanomyces* and its characteristic barnyard odors are rare in must. Although such yeasts can be relatively resistant to sulfur dioxide (Fleet, 1992), sulfiting followed by inoculating with a selected strain of *Saccharomyces cerevisiae* constitute, in practice, an effective means of avoiding contamination (Section 3.5.4).

In general, *S. cerevisiae* inoculated at 10^6 cells/ml, either naturally or by a selected strain inoculation, induces grape must fermentation in large-scale applications.

The yeast growth cycle and grape must fermentation kinetics are depicted in Figure 3.2 (Lafon-Lafourcade, 1983). To accentuate certain phenomena, the figure concerns a must containing particularly high sugar concentrations, which cannot be completely fermented. Analysis of this figure prompts the following remarks:

1. The growth cycle has three principal phases: a limited growth phase (two to five days) increases the population to between 10^7 and 10^8 cells/ml. A quasi-stationary phase follows and lasts about eight days. Lastly, the death phase progressively reduces the viable population to 10^5 cells/ml. The final phase can last for several weeks.
2. During this particularly long cycle, growth is limited to four or five generations.
3. The stoppage of growth is not the result of depletion of energy-yielding nutrients.
4. The durations of these different phases are not equal. The death phase, in particular, is three to four times longer than the growth phase.
5. Fermentation kinetics are directly linked to the growth cycle. The fermentation rate is at its maximum

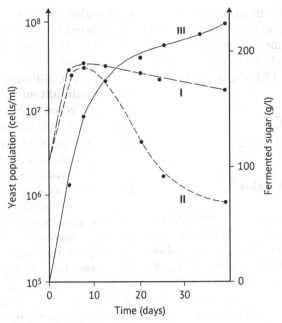

FIGURE 3.2 Yeast growth cycle and fermentation kinetics of grape must containing high sugar concentrations (320 g/l) (Lafon-Lafourcade, 1983). I, Total yeast population; II, viable yeast population; III, fermented sugar.

during the growth cycle. The fermentation rate then progressively slows, but fermentation still lasts several weeks more. At this stage, the yeast population is in the survival phase. Lastly, the end of fermentation is not simply the result of insufficient yeast growth. Inhibition of the metabolic activity of non-proliferating cells is also a factor.

These phenomena have several technological consequences. For a limited sugar concentration (less than 200 g/l), fermentation occurs during the first two phases of the cycle. It takes place rapidly and most often without a problem. On the other hand, if there is a higher amount of sugar, a population in its death phase carries out the last part of fermentation. In this case, its metabolic activity continues to decrease throughout the process. The total transformation of sugar into alcohol (i.e. fermentation to dryness) depends on the survival capacity of this population; this is as important as the initial growth phase.

Excessive temperatures and sugar concentrations can provoke sluggish or stuck fermentations. Nutritional deficiencies and inhibition phenomena can also be involved. All of them have either chemical or physicochemical origins. Fermentation kinetics can be improved by different methods that influence these phenomena. Early action appears to increase their effectiveness; yeasts in their growth phase and in a medium containing little ethanol are more receptive to external stimuli. The winemaker should anticipate fermentation difficulties: the possible treatments are much less effective after these difficulties occur.

Moreover, certain operations, intended to activate fermentation, affect yeast growth and improve fermentation kinetics at its beginning but do not always affect yeast survival or the final stages of fermentation, at least in high-sugar musts (the most difficult to ferment). Increasing must temperature is a classic example of an operation that increases the fermentation rate at the beginning but leads to stuck fermentations (see Section 3.7.1, Figure 3.8).

3.4 Nutrition Requirements

3.4.1 Carbon Supply

In grape must, yeasts find glucose and fructose—sources of carbon and energy. The total sugar concentration in must is between 190 and 255 g/l, corresponding to wines with between 11 and 15% vol. ethanol after fermentation. The amount of dissolved sugar can be even higher in grapes for the production of sweet wines—up to 350 g/l in Sauternes musts. The must sugar concentration can influence the selection of yeast strains for specific fermentations (Fleet, 1992).

Fermentation is slow in a medium containing a few grams of sugar per liter. Its speed increases in musts that have 15–20 g/l and remains stable until about 200 g/l. Above this concentration, fermentation slows. In fact, alcohol production can be lower in a must containing 300 g/l than in another containing only 200 g/l. At 600–650 g of sugar per liter, the concentrated grape must becomes practically unfermentable. The presence of sugar, as well as alcohol, contributes to the stability of fortified wine.

Thus, an elevated amount of sugar hinders yeast growth and decreases the maximum population. Consequently, fermentation slows even before the production of a significant quantity of ethanol—which normally has an antiseptic effect (Section 3.6.1).

Fermentation slows down in the same way when the high sugar concentration is due to the addition of sugar (chaptalization) or concentrated must, or the elimination of water by reverse osmosis or vacuum evaporation (Section 11.5.1). The effect is exacerbated if sugar is added when fermentation is already well advanced and alcohol has started to inhibit yeast development, although intervening at this stage has the advantage of not overheating the must. When sugar is added, it is advisable to wait until the second day after fermentation starts, i.e. the end of the yeast growth phase. Under these conditions, the population reaches a higher value, because it grows in a medium with a relatively low sugar concentration. Next, the addition of sugar in a medium containing yeast in full activity increases the fermentation capacity of the cells and therefore the transformation of sugar. Of course, a refrigeration system is necessary to compensate for the corresponding temperature increase of the must.

3.4.2 Nitrogen Supply

Henschke and Jiranek (1992) analyzed many theoretical works on this subject in detail. The research for these different works was carried out under a large range of conditions and so the results are not always applicable to the winemaking conditions analyzed by Ribéreau-Gayon et al. (1975a).

Grape must contains a relatively high concentration of nitrogen compounds (0.1–1 g of soluble nitrogen per liter), although only representing about a quarter of total berry nitrogen. These constituents include the ammonium cation (3–10% of total nitrogen), amino acids (25–30%), polypeptides (25–40%), and proteins (5–10%). The grape nitrogen concentration depends on variety, rootstock, environment, and growing conditions—especially nitrogen fertilization. It decreases when mold develops on the grapes or the vines suffer from drought conditions. Water stress, however, is generally a positive factor in quality red wine production. Planting cover crops in the vineyard to control yields also reduces the grapes' nitrogen levels. The effects are variable, e.g. 118 mg/l nitrogen in grapes from the control plot and 46 mg/l nitrogen for the plot with cover crop in one case and 354 and 210 mg/l nitrogen, respectively, in another. The nitrogen content of overripe grapes may increase due to concentration of the juice.

In dry white winemaking, juice extraction methods influence the amino nitrogen and protein concentration in must. Slow pressing and skin maceration (Section 13.3.5), which favor the extraction of skin constituents, increase their concentration (Dubourdieu et al., 1986).

Yeasts find the nitrogen supply necessary for their growth in grape must. The ammonium cation is easily assimilated and can satisfy yeast nitrogen needs, in particular, for the synthesis of amino acids. Polypeptides and proteins do not participate in *S. cerevisiae* growth, since this yeast cannot hydrolyze these substances. *Saccharomyces cerevisiae* does not need amino acids as part of its nitrogen supply, since it is capable of synthesizing them individually, but their addition stimulates yeasts more than ammonia nitrogen does. A mixture of amino acids and ammonia nitrogen is an even more effective stimulant. Yeasts use amino acids according to three mechanisms (Henschke and Jiranek, 1992):

1. Direct integration into proteins, without transformation.
2. Decomposition of the amino group, which is used for the biosynthesis of different amino constituents. The corresponding carbon-containing molecule is excreted. Such a reaction is one of the pathways of higher alcohol formation in wine:

$$R-CHNH_2-COOH + H_2O \rightarrow R-CH_2OH + CO_2 + NH_3$$

3. Yeasts are probably capable of obtaining ammonia nitrogen from amino acids through other pathways.
4. The amino acid molecule can be used as a source of carbon in metabolic reactions. The yeast simultaneously retrieves the corresponding ammonia nitrogen.

The assimilation of different amino acids depends on the functioning of transport systems and the regulation of metabolic systems (nitrogen catabolite repression, NCR) (Beltran et al., 2004).

The assimilation rate of amino acids and the ammonium ion is variable. According to Jiranek et al. (1995), who monitored the order of incorporation of these nitrogen sources under winemaking conditions, they can be classified in four groups:

Group A, which corresponds to the immediately yeast-assimilable nitrogen sources, is composed of the ammonium ion, arginine, asparagine, aspartic acid, glutamine, iso-leucine, leucine, lysine, methionine, serine, and threonine.

Group B contains the amino acids absorbed after a latency period and at a lower rate: alanine, glutamate, histidine, phenylalanine, and valine.

Group C is composed of amino acids that disappear after a long latency period, which coincides with the time for groups A and B to disappear: glycine, tryptophan, and tyrosine.

Lastly, group D is composed solely of proline, which is one of the main amino acids in must that is not assimilable or very poorly so by yeast under anaerobic conditions.

Although complex mixtures of ammonium salts and amino acids are more effective for promoting yeast growth and fermentation speed, ammonium salts are used almost exclusively to increase nitrogen concentrations in must, for reasons of simplicity. The addition of yeast-assimilable nitrogen is not always sufficient for resolving difficult final stages of fermentation, although it accelerates fermentation in the early stages.

Diammonium phosphate and diammonium sulfate are the two forms of ammonium salts in use. EU regulations authorize the addition of 100 g of one of these two salts per hectoliter, corresponding to 210 mg/l nitrogen. The addition of this form of nitrogen to must increases acidity, due to the contribution of the anion. For 10 g of diammonium salt per hectoliter, the must acidity can increase by 0.35 g/l (expressed as H_2SO_4) or 0.52 g/l (as tartaric acid).

The initial concentration of ammonium cations and amino acids in the must is one of the most important elements in determining the need for supplements. When the NH_4^+ concentration is less than 25 mg/l, nitrogen addition is necessary. It is useful for concentrations between 25 and 50 mg/l. Above this concentration, supplementing has no adverse effects. It is, however, unlikely that an addition will activate the fermentation. If the values are expressed in free amino nitrogen (FAN), a concentration between 70 and 140 mg/l is necessary to have a complete fermentation of musts containing between 160 and 250 g of sugar per liter (Henschke and Jiranek, 1992).

Bely et al. (1990) determined that adding nitrogen was effective if the available nitrogen content (NH_4^+ + FAN) in the must was below 140 mg/l, but was unnecessary and could even be harmful at initial concentrations above 300 mg/l. From a technological standpoint, these authors claim that the optimal time for nitrogen addition is when fermentation is 40% complete. Aerny's (1996) formol number provides a simple estimate of the free amino acid and ammonium cation content. A formol number of one corresponds to 14 mg of amino nitrogen per liter. According to Lorenzini (1996), the addition of nitrogen to Swiss varietal musts is indispensable if the index is less than 10, and is recommended if the index is between 10 and 14. The formol method provides a quick, simple assessment of available nitrogen deficiency by assaying NH_4^+ and free amino acids, except proline. It should be more widely used to monitor ripeness and fermentation.

For example, in a study of musts from Bordeaux vineyards in the 1996–1999 vintages (Table 3.2), Masneuf et al. (2000) found nitrogen levels of 36–270 mg/l in white musts, with deficiencies in 22% of the samples (nitrogen concentrations under 140 mg/l). In reds, levels ranged from 46 to 354 mg/l, with deficiencies in 49%; in rosés, levels ranged from 42 to 294 mg/l, with deficiencies in 60%, while 89% of botrytized musts were nitrogen-deficient.

Choné (2000) analyzed Cabernet Sauvignon musts in 1997 and found significant variations in nitrogen levels, from 95 to 218 mg/l.

Finally, different nitrogen concentrations have also been found in individual

TABLE 3.2

Available Nitrogen Content (NH_4^+ and Free Amino Acids Expressed in milligram per liter) in Musts from Bordeaux Vineyards (1996–1999 Vintages) Determined by the Formol Method (Masneuf et al., 2000)

	White	Red	Rosé	Botrytized
Number of samples	32	55	48	9
Minimum value	36	46	42	22
Maximum value	270	354	294	157
Mean	181.9	157	119	82.8
Standard deviation	32	55	48	9
Deficient musts (%) <140 mg/l	22	49	60	89

plots within the same vineyard, e.g. 25–45 mg/l in vines with less vegetative growth and 152–294 mg/l in grapes from more vigorous vines.

All these analytical findings show the extent and frequency of nitrogen deficiencies, which are much more common than was generally thought in the past, perhaps due to changes in vineyard management techniques. It was generally accepted previously that nitrogen concentrations in musts from northerly vineyards in the Northern Hemisphere (temperate, oceanic climate) were sufficiently high.

Adding nitrogen to musts containing insufficient levels is extremely useful in achieving good fermentation kinetics. Some observations suggest that an excessive increase in nitrogen content, especially if it is done at the start of fermentation, may have a negative effect on fermentation kinetics; so it is advisable to modulate nitrogen additions according to the natural level in the must and ensure that the total never exceeds 200 mg/l.

Adding excessive amounts of nitrogen may also result in the presence of non-assimilated residual nitrogen at the end of fermentation. Although there are no specific data on this issue, residual nitrogen may have a negative impact on a wine's microbiological stability. An excess of ammonia nitrogen can also lead to a modification of the aroma characters of wine. Since the yeast no longer needs to deaminate amino acids, it forms fewer secondary products (higher alcohols and their esters). This modifies wine aroma, especially in white wines. Finally, it is well known that the nitrogen supply affects ethyl carbamate production. This undesirable constituent has carcinogenic properties and is controlled by legislation. Bell and Henschke (2005) stress the importance of controlling the total yeast-assimilable nitrogen content (initial and added) for the aroma profile of wines. A very low nitrogen content increases the production of undesirable sulfur compounds and higher alcohols and leads to a decrease in ester production.

The must sugar concentration also affects the impact of nitrogen addition on fermentation kinetics, especially the successful completion of fermentation. For moderate concentrations of sugar (less than 200 g/l), the addition of nitrogen increases the biomass of yeast formed and in consequence the fermentation speed; the fermentation is completed a few days in advance. For high concentrations of sugar, the fermentation is accelerated at the beginning with respect to the control sample but, as the fermentation continues, the gap between the control sample and the supplemented sample decreases. Eventually, their fermentations spontaneously stop with similar quantities of residual sugar remaining. The ineffectiveness of this nitrogen addition at the start of fermentation in high-sugar musts can be explained by a drop in sugar transport activity. Nitrogen affects not only yeast growth but also sugar transport kinetics throughout the fermentation (Section 1.3.2). In the case of high-sugar musts, a split dose of nitrogen (once at the start of fermentation and then again during the stationary phase) is recommended. This nitrogen management mode promotes yeast growth and maintains a sufficient level of yeast activity at the end of reaction in order to prevent stuck fermentation (Bely *et al.*, 1994).

Other factors affect the assimilation of nitrogen during fermentation. Yeasts have strain-specific capabilities. Henschke and Jiranek (1992) reported that different *S. cerevisiae* strains fermenting grape must assimilated quantities of nitrogen varying from 329 to 451 mg/l at 15.5°C and from 392 to 473 mg/l at 20°C. These last figures also show, among other things, that temperature increases nitrogen assimilation. Julien *et al.* (2000) compared the nitrogen requirements of several *S. cerevisiae* yeast strains as estimated during the stationary phase of yeast growth. This test measures the quantity of nitrogen required to maintain a constant fermentation rate during this phase. Nitrogen requirements vary by a factor of two depending on the strain (26 winemaking strains tested). The evaluation of the nitrogen requirements of yeasts is certainly an important criterion in strain selection for nitrogen-deficient musts.

Oxygen, however, has the most effect on the assimilation of nitrogen (Section 3.7.2). Yeasts have long been known to use considerably more nitrogen in the presence of oxygen (Ribéreau-Gayon et al., 1975a), probably via the formation of sterols. During white winemaking, grape solids may have the same effect (Section 3.7.3). It has been observed that yeasts fermenting in the complete absence of oxygen assimilate 200 mg of nitrogen per liter. When they develop in the presence of oxygen, their assimilation increases to 300 mg/l. Under aerobic conditions, they can assimilate up to 735 mg/l without a proportional increase in cell multiplication (Ribéreau-Gayon et al., 1975a).

The impact of oxygen on fermentation kinetics, irrespective of any addition of NH_4^+, is apparently complex and dependent on several factors (Sablayrolles et al., 1996a,b), and probably on the type of must (sugar content and possible nitrogen deficiency). It is accepted that adding nitrogen accelerates fermentation, resulting in faster completion, at least for musts with moderate sugar contents. It is, however, more difficult to identify the conditions under which adding nitrogen can prolong sugar conversion by the yeasts and prevent fermentation from becoming stuck, at least in musts rich in sugar.

In an experiment carried out by Rozès et al. (1988), using a must containing 222 g/l of sugar with a normal nitrogen content (35 mg/l NH_4^+ corresponding to approximately 200 mg/l nitrogen), fermentation stopped prematurely in the absence of air. Adding NH_4^+ (0.15 or 0.50 g/l $(NH_4)_2SO_4$) initially accelerated fermentation but did not increase the amount of sugar fermented. With aeration on the third day, fermentation was faster and all the sugar was fermented. Adding NH_4^+ did not improve fermentation kinetics, and, on the contrary, after an initial acceleration, yeast activity stopped when 9 g/l of sugar was still unfermented. Of course, these experimental results must be interpreted in the light of the specific conditions (sugar and nitrogen content of the must). The results would not necessarily have been the same under different conditions, particularly if there had been a significant nitrogen deficiency in the must. In any case, this experiment shows quite clearly that adding nitrogen does not necessarily eliminate problems at the end of fermentation. Further experiments using nitrogen-deficient must are required to identify a possible improvement.

Sablayrolles et al. (1996a,b) reported that nitrogen supplements were most effective in mid-fermentation, together with aeration. This combined operation had more impact on fermentation kinetics than aeration alone and provided the best solution for avoiding stuck fermentation (Sablayrolles and Blateyron, 2001).

In conclusion, adding ammonium salts to musts with naturally low nitrogen levels (YAN ≤ 140 mg/l) is likely to improve fermentation kinetics, with varying effects on yeast growth and sugar conversion. For maximum effectiveness, total nitrogen after supplementation should not exceed 200 mg/l. Some experimental findings indicate that fermentation may slow down following the addition of excessive amounts of nitrogen. If the must already has a sufficiently high nitrogen content, further supplementation is likely to cause an initial acceleration in fermentation, but the effect wears off gradually. Adding nitrogen cannot be expected to remedy problems in the final stages of fermentation (high-sugar musts or strictly anaerobic conditions). It is, however, true that nitrogen deficiencies (such as in musts from old vines or vineyards with cover crop) have not been given sufficient consideration in the past, and that completion of fermentation is facilitated in these cases by adding ammonium salts. Total nitrogen in must should be analyzed in tanks before the start of fermentation as a matter of course, together with sugar and acidity levels.

Adding oxygen at the start of fermentation (Section 3.7.2) when the yeast population is in the growth phase is still the most effective way of accelerating fermentation and preventing premature stoppages. Opinions diverge on the correct time to add ammonium salts, varying from the beginning of fermentation to halfway through. In any case, nitrogen supplements are more effective at accelerating fermentation than preventing it from becoming stuck with unwanted residual sugar.

3.4.3 Mineral Requirements

The yeasts that Pasteur cultivated in the following medium proliferated well: water, 1,000 ml; sugar, 100 g; ammonium tartrate, 1 g; yeast ashes, 10 g. Yeast ashes supply the yeast with all of its required minerals. Dry yeast contains 5–10% mineral matter, whose average composition (in percentage weight of ashes) is as follows:

K_2O	23–48
Na_2O	0.06–2.2
CaO	1.0–4.5
MgO	3.7–8.5
Fe_2O_3	0.06–7.3
P_2O_5	45–59
SO_3	0.4–6.3
SiO_2	0–1.8
Cl	0.03–1.0

Other minerals not listed above are present in trace amounts: Al, Br, Cr, Cu, Pb, Mn, Ag, Sr, Ti, Sn, Zn, etc. These are called trace elements. Not all of them are indispensable, but some are essential at very small doses as components of enzymatic systems.

The precise function of only a few minerals is known. Grape must contains, both qualitatively and quantitatively, a sufficient mineral supply to ensure yeast development.

3.5 Fermentation Activators

3.5.1 Growth Factors

Growth factors affect cell multiplication and activity, even in small concentrations. They are indispensable to microorganisms and a deficiency in these substances disturbs metabolism. Microorganisms behave differently in relation to growth factors. Some can totally or partially synthesize them; others cannot and must find them in their environment.

The substances that are growth factors for microorganisms are also necessary vitamins for higher organisms (Figure 3.3). They are essential components of coenzymes and are involved in metabolic reactions. Grape must has an ample supply of growth factors (Table 3.3), but alcoholic fermentation alters its vitamin composition. For example, thiamine disappears almost entirely: yeasts are capable of consuming greater amounts of thiamine (600–800 µg/l) than the must contains. In contrast, yeasts form riboflavin. The concentration of nicotinamide remains constant in red musts and wines, but only 60% remains in white wines. Pantothenic acid, pyridoxine, and biotin are used by yeasts and then released; their concentrations are nearly identical in musts and wines. Mesoinositol is practically untouched.

Although musts contain sufficient amounts of growth factors to ensure yeast development and alcoholic fermentation, natural concentrations do not necessarily correspond with optimal concentrations. For this reason, supplementing must with certain growth factors is recommended.

A deficiency in pantothenic acid causes the yeast to accumulate acetic acid, but it has not been proven that the (unauthorized) addition of pantothenic acid to a fermenting must lowers the wine's volatile acidity originating from yeast (Section 2.3.4). The production by yeasts of

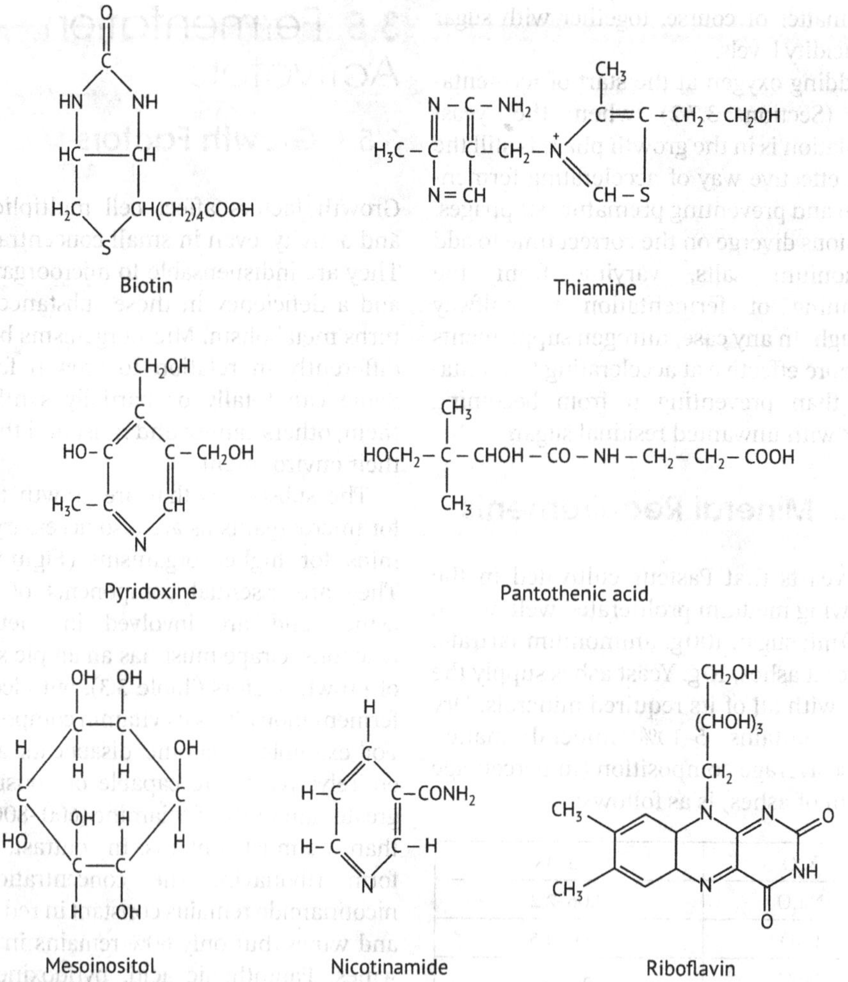

FIGURE 3.3 Yeast growth factors.

TABLE 3.3

Maximum and Minimum Growth Factor Concentrations (μg/l) in Musts and Wines (Ribéreau-Gayon et al., 1975a)

Vitamins	Grape musts	Wines	
		Whites	Reds
Thiamine	160–450	2–58	103–245
Riboflavin	3–60	8–133	0.47–1.9
Pantothenic acid	0.5–1.4	0.55–1.2	0.13–0.68
Pyridoxine	0.16–0.5	0.12–0.67	0.13–0.68
Nicotinamide	0.68–2.6	0.44–1.3	0.79–1.7
Biotin	1.5–4.2	1–3.6	0.6–4.6
Mesoinositol	380–710	220–730	290–334
Cobalamin (vitamin B_{12})	0	0–0.16	0.04–0.10
Choline	19–39	19–27	20–43

abnormally high levels of volatile acidity is probably due to the must's deficiencies in certain lipids. These deficiencies are most likely linked to deficiencies in pantothenic acid, which is involved in the formation of acetyl coenzyme A, which is responsible for fatty acid and lipid synthesis.

The addition of biotin and especially thiamine has been shown to improve the must fermentation kinetics in numerous experiments. An addition of 0.5 mg of thiamine per liter can increase the viable population by 30%; the fermentation of sugar is also quicker. According to Bataillon *et al.* (1996), the presence of about 0.25 mg/l is needed, so that thiamine does not limit fermentation rate, this requirement being lesser in musts containing low amounts of yeast-assimilable nitrogen.

These results, although regularly observed in laboratory experiments, are not always obtained under practical conditions. The natural concentration of thiamine may or may not be a limiting factor of fermentation kinetics, depending on the nature of the grape, the ripening conditions, and pre-fermentation treatments.

The addition of thiamine is legal in several countries (EU, at a dose of 60 mg/hl), but it is rarely used to accelerate fermentation in winemaking. However, it effectively decreases high keto acid concentrations (pyruvic and α-ketoglutaric acids) via decarboxylation. Large quantities of these acids bind to sulfur dioxide in botrytized sweet wines (Section 8.4.2).

3.5.2 Survival Factors

The idea of survival factors is derived from the interpretation of how sterols and certain long-chain fatty acids affect yeast activity and fermentation kinetics. The first works on this subject (Andreasen and Stier, 1953; Bréchot *et al.*, 1971) were analyzed by Ribéreau-Gayon *et al.* (1975a). The growth factor activity of ergosterol under complete anaerobic conditions is optimal at a concentration of 7 mg/l; it is solubilized with Tween 80. For example, in a must with a high sugar concentration (260 g/l), *S. cerevisiae* ferments, under complete anaerobic conditions, 175 g of sugar per liter in 10 days in the control sample, and 258 g/l in the presence of 5 mg/l of ergosterol. Under aerobic conditions, on the other hand, a slight inhibition of fermentation is observed when ergosterol is added. The authors concluded that these sterols are indispensable to yeasts under complete anaerobic conditions, because they cannot be synthesized under these conditions. Sterols are necessary for ensuring cell membrane permeability. In the presence of oxygen, yeasts are capable of producing sterols themselves. Under anaerobic conditions, ergosterol is in some ways an oxygen substitute for yeasts.

Other sterols and long-chain fatty acids share most of the properties of ergosterol. Some are constituents of grape bloom and cuticular wax, such as oleanolic acid—especially when associated with oleic acid (Figure 3.4).

Later works (Larue *et al.*, 1980; Lafon-Lafourcade, 1983) showed that the action mechanism of sterols is in fact more complex. These authors confirmed the growth factor effect in a strictly anaerobic fermentation: the maximum population increases. They also witnessed the inhibitory effect of sterols on a fermentation with permanent aeration. Neither of these two conditions corresponds exactly to winemaking conditions.

In the winery, large-volume fermentations are certainly anaerobic, but the must is aerated during crushing and inoculated with a yeast starter that was precultivated under aerobic conditions; the yeasts are therefore well equipped with sterols. Both commercial active dry yeasts (ADYs) and native yeasts, which develop on the surfaces of equipment in contact with the harvested grapes, initially develop under aerobic conditions. Under these conditions, the addition of ergosterol or oleanolic acid does not increase the maximum population (Table 3.4). The fermentation speed is also

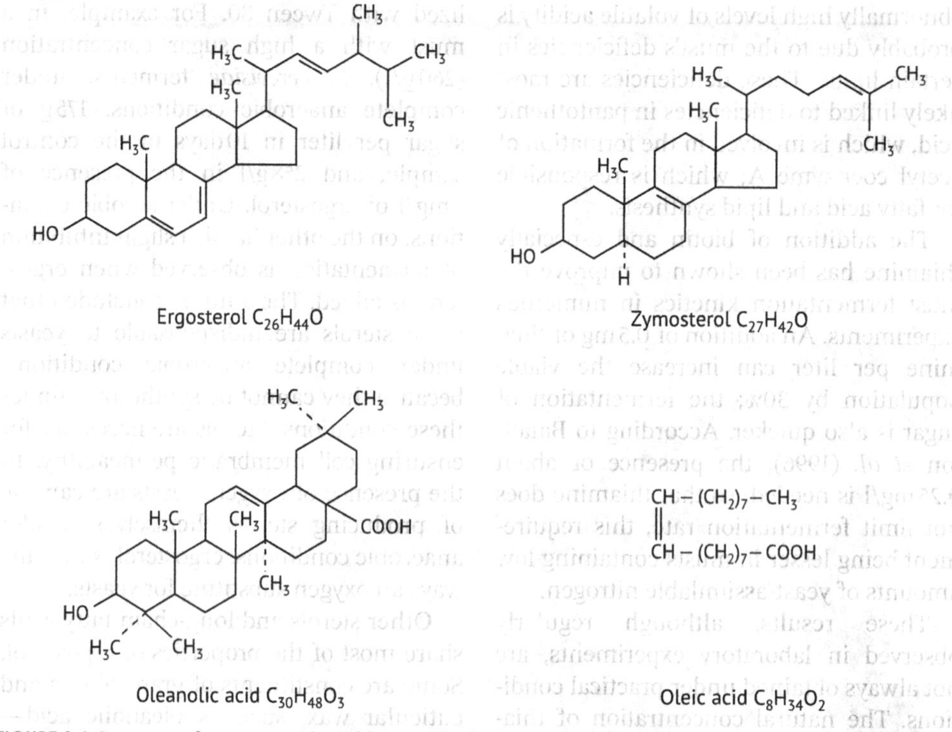

FIGURE 3.4 Structure of some steroids and fatty acids playing a role in yeast growth.

Ergosterol $C_{26}H_{44}O$

Zymosterol $C_{27}H_{42}O$

Oleanolic acid $C_{30}H_{48}O_3$

Oleic acid $C_8H_{34}O_2$

TABLE 3.4

Sterol Concentrations in Yeasts During Alcoholic Fermentation of Grape Must (Larue et al., 1980)

Conditions	Constant aeration			Anaerobic conditions		
	C	+E	+OA	C	+E	+OA
Day 2						
Fermented sugar (g/l)	30	27	24	37	36	23
Sterols (% of dry weight)	2.70	2.80	2.30	1.60	1.40	1.70
Viable cells (10⁶/ml)				22	20	17
Day 5						
Fermented sugar (g/l)	116	101	95	113	111	105
Sterols (% of dry weight)	1.90	1.90		0.60	1.10	0.40
Viable cells (10⁶/ml)				13	10	12
Day 9						
Fermented sugar (g/l)	187	175	155	164	169	154
Sterols (% of dry weight)	1.20	1.10	0.7	0.40	1.00	0.30
Viable cells (10⁶/ml)				5	7	5
End of fermentation						
Fermented sugar (g/l)	256	234	211	170	199	185
Sterols (% of dry weight)	1.00	0.80	0.40	0.30	0.60	0.20
Viable cells (10⁶/ml)				0.05	0.5	0.1

C, control must; +E, ergosterol (25 mg/l); +OA, oleanolic acid (50 mg/l).
Must sugar concentration after addition: 250 g/l; active dry yeast: *S. cerevisiae*: initial sterol concentration: 1.5%; initial population: 2.2×10^6 cells/ml.

not affected during the first 10 days. However, since the yeast cells are rich in sterols, they maintain their fermentation activity for a longer time. At the end of fermentation, they will also have broken down a larger amount of sugar than non-supplemented cells. The term "survival factor" has been proposed for this action that does not correspond to an increase in growth. The evolution of yeast populations during fermentation in the presence of sterols is represented in Figure 3.5; the effect of temperature is also indicated.

The notion of survival factors complements the notion of growth factors. They are especially interesting in the case of difficult fermentations—for example, high-sugar musts. Of course, the direct addition of sterols to tanks during fermentation should not be considered. Winemaking can, however, be orientated toward processes that promote sterol synthesis. Moreover, their existence in the solid parts of the grape should be taken into account: crushed red grapes ferment better than white grape musts because of solids contact during fermentation. In addition, the elimination of sterols during the excessive clarification of white grape must can result in extremely difficult fermentations (Section 3.7.3). This concept can also explain past experiments that show increased fermentation speeds with the addition of ground grape skins and seeds.

3.5.3 Other Fermentation Activators

Ribéreau-Gayon et al. (1975a) examined other fermentation activators. These activators generally help the yeast to make better use of must nitrogen. Incidentally, the same phenomenon is observed each time that the fermentation is accelerated by the presence of air or by the addition of yeast extract nutrients or survival factors (Table 3.5).

Hydrolyzed yeast extracts are rich in yeast-assimilable nitrogen, survival factors, and mineral salts. They have often been

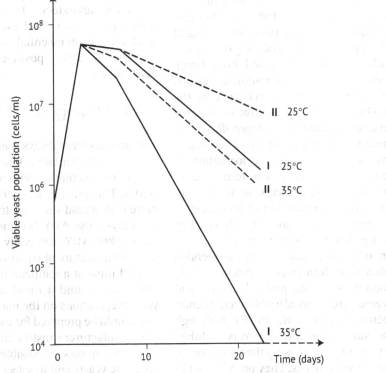

FIGURE 3.5 Influence of sterols on yeast survival during the death phase at different temperatures (Lafon-Lafourcade, 1983). I, Control; II, added ergosterol.

TABLE 3.5
Effect of Air and Survival Factors on Grape Must Nitrogen Assimilation

Atmosphere	Additions	Nitrogen utilization (mg of FAN/l)	Number of cells ($\times 10^7$/l)	Fermentation speed (g of sugar per 100 ml/24 h)
Air	0	87	35.0	1.80
Nitrogen	0	65	1.3	0.50
Air	YE	314	36.0	4.00
Nitrogen	YE	107	3.0	1.22
Nitrogen	YE + SF	293	32.0	3.60

FAN, free amino nitrogen; YE, yeast extract; SF, survival factors (Tween 80, ergosterol). Musts without and with addition of YE contain 99 and 386 mg FAN/l, respectively.

Source: Adapted from Ingledew and Kunkee (1985), by Cantarelli (1989), cited by Henschke and Jiranek (1992).

used (at high concentrations, up to 4 g/l) to accelerate fermentation in the food industry. Some can impart foreign odors and tastes.

Among the activator formulas proposed (Ribéreau-Gayon *et al.*, 1975a), 200 mg/l of the following mixture may facilitate fermentation in musts with vitamin and nitrogen deficiencies: 100 g of diammonium sulfate, 250 mg of thiamine, 250 mg of calcium pantothenate, and 2 mg of biotin.

Other products, extracted from fungi, are also alcoholic fermentation activators. Some have been commercialized in the past. One of the most effective is prepared from a culture medium of *Aspergillus niger*. It modifies the concentration of secondary products by promoting the formation of glycerol and pyruvate via fermentation from sugar. A yeast activator is also obtained from the mycelium of *Botrytis cinerea*. These activators are not authorized by wine legislation, at least in the EU.

In dry white winemaking, suspended solids activate fermentation (Section 3.7.3). Certain constituents, probably sterols and fatty acids, are involved in this phenomenon (Ribéreau-Gayon *et al.*, 1975b). Although these substances are not very soluble, yeasts are capable of using them to improve fermentation kinetics. They probably act in conjunction with other factors, such as oxygenation and possibly nitrogen additions. Yeast hulls have a similar effect, independent of their ability to eliminate inhibition (Section 3.6.2).

As with nitrogen or growth factor supplementation, fermentation is not always activated to the same degree in the winery as in the laboratory. Oxygenation, yeast starter preparation, and the fermentation medium play an essential role, in addition to the must's possible nutritional deficiencies.

3.5.4 Adding Yeast Starter

Winemakers have always been interested in improving fermentation kinetics and wine quality by inoculating with activated yeast starter. This practice has certainly become more widespread since relatively economical, easy-to-use ADY became available on the market. ADYs are easily reactivated in water or a mixture of equal volumes of water and must at a temperature of 35–40°C. There are around a hundred commercial yeast preparations on the market, and each one should be prepared for use according to the manufacturer's instructions. The use of ADY also makes it possible to eliminate apiculate yeasts and to select strains with a high sugar-to-alcohol conversion rate. Together with other winegrowing practices, the use of ADY has contributed to the general increase in average alcohol content in wines and to the corresponding decrease in the need for chaptalization (adding sugar) in certain regions. In contrast, in some situations, there is an interest in using yeast strains with a lower conversion rate. This mainly concerns hot areas where there may be a high sugar level in the grape pulp, although the other ripeness indicators (skin) have not reached optimum levels. In this case, it is necessary to delay the harvest, with the risk of obtaining very high-sugar musts that are difficult to ferment and that produce high-alcohol wines.

An initial inoculum of 10^6 cells/ml is generally considered necessary to obtain good fermentation kinetics. In view of the current constraints on white winemaking, this initial level is rarely achieved, and, as such, the use of yeast starter has become practically compulsory. In red winemaking, there may be insufficient inoculum in the first few tanks filled, but grape and must handling operations in the winery rapidly result in proliferation of the yeasts. The use of ADY is mainly justified in the first tanks. Later tanks either require no starter at all or ADY can be replaced with 2–5% must from a tank where fermentation is going well.

There has, however, been interest for some time in the possibility of improving a wine's quality by selecting the appropriate strain for fermentation. It is certainly true that the composition of grapes and other natural factors (e.g. *terroir*) are the main elements of the specific characteristics ("typicity") recognized as forming the basis of quality, especially in the concept of *appellation d'origine contrôlée*. It is true that positive results have been obtained, especially with white wines (Section 13.7). Several strains capable of fermenting musts with low turbidity and without producing excessive volatile acidity have been isolated. Strains have also been identified that do not produce an excessive amount of vinylphenols, with their unpleasant chemical odor, and other undesirable characteristics linked to high levels of fermentation esters. The latter mask varietal aromas and are acceptable only in wines made from neutral varieties. It is clear that using yeast starter is a good way of avoiding these types of defects and, consequently, making the most of the grapes' intrinsic quality. Another example is the growth of aerobic yeasts likely to produce ethyl acetate as soon as tanks are filled with non-sulfited grapes or must. Reports in the literature indicate that it is possible to avoid these defects by using appropriate yeast strains after the grapes/must have been sulfited (Section 3.3).

Today, there is increasing interest in selecting yeast strains capable of enhancing the varietal aromas of various grape varieties by releasing variable quantities of aroma molecules from their odorless precursors. Research has focused on Muscat varieties and, above all, Sauvignon Blanc (Section 13.7.2). Although different yeast strains have varying impacts on wine aromas, it is impossible to say that the use of yeast starter leads to the development of a uniform character, since this depends mainly on the grapes' composition in terms of aroma precursors.

In the case of red wines, it has been reported that the use of specific yeast starters has an impact on color intensity and the aroma character of some grape varieties. These effects may occur during fermentation itself or result from the autolysis of dead yeast cells, which justifies the practice of aging wine on the lees. These observations, however, require a more detailed theoretical investigation.

A dose of 10^6 cells/ml yeast starter is generally recommended, which corresponds to 10–20 g/hl of ADY. As the yeast population in strongly fermenting must is on the order of 10^8 cells/ml, a 1% inoculum is theoretically sufficient, but 2% is more commonly used, or even 5%, to offset any potential difficulties. Experiments carried out with higher doses of yeast starter (20–25 g/hl of ADY) indicated that there was a lower risk of fermentation becoming sluggish toward the end, but some off-aromas could be produced. In any case, the most important selection criteria for winemaking yeast starters are temperature resistance and the ability to complete fermentation in high-sugar musts.

When the yeast starter is added, it is important to avoid antagonism with other strains naturally present in the must. Antagonistic reactions may reduce the fermentation rate and contribute to stuck fermentation (Section 3.8.1). For inoculation with ADY to be successful, the yeast starter must be more abundant and more active than the native yeasts, which must be inhibited by proper hygiene, sufficiently low temperatures, and appropriate use of sulfites.

3.6 Inhibition of Fermentation

This section covers the phenomenon of inhibition during grape must fermentation. A large number of substances exist that may hinder yeast cell growth: chemical antiseptics and fungicidal antibiotics, which prevent yeast growth (Ribéreau-Gayon et al., 1975a). Inhibitors used for the conservation of wines (in particular sulfur dioxide) are described in Chapters 8 and 9.

3.6.1 Inhibition by Ethanol

Ethanol produced by fermentation slows the assimilation of nitrogen and paralyzes the yeast. Ethanol acts by modifying active transport systems across the cell membrane (Henschke and Jiranek, 1992). The quantity of alcohol necessary to block fermentation depends on many factors, including yeast strain, temperature, and aeration.

The presence of ethanol at the time of inoculation prolongs the latent phase and reduces cell growth. High temperature increases this inhibitory action. This effect of ethanol on yeast growth and fermentation speed occurs even at low concentrations, from the start of fermentation onward. The difficulty of restarting a stuck fermentation is, therefore, understandable.

The experiment in Table 3.6 shows the effect of the addition of alcohol to grape must. It delays the initiation of fermentation and limits the assimilation of nitrogen and the formation of alcohol. Yet the yeasts can continue their activity up to a higher alcohol content, as long as the inhibitory action of ethanol is not excessive. In this experiment, the variation of the glycerol concentration represents significant metabolic modification. As seen in Section 3.4.1, ethanol intensifies the inhibitory effect of high sugar concentration in must.

TABLE 3.6

Effect of Ethanol Addition to Must on Fermentation (Limited Aerobic Conditions at 25°C) (Ribéreau-Gayon et al., 1975a)

Alcohol addition (% vol.)	Delay in initiation of fermentation (days)	Yeast population (10^6/ml)	Alcohol content attained (% vol.)	Alcohol formed (% vol.)	Residual sugar (g/l)	Nitrogen assimilated (mg/l)	Glycerol (mmol/l)
+0	1	80	14.0	14.0	2	252	57
+2	2	67	15.6	13.6	6	233	65
+6	4	62	18.2	12.2	15	194	72
+10	12	30	16.0	6.0	125	81	80

3.6.2 Inhibition by Fermentation By-Products: Use of Yeast Hulls

Past observations have indicated that substances other than ethanol, formed during fermentation, can have an inhibitory action on yeast. Geneix *et al.* (1983) confirmed this hypothesis in a significant experiment. Synthetic fermentation media containing variable concentrations of ethanol were inoculated with *S. cerevisiae*. A first series consisted of non-fermented and entirely synthetic media. A second series consisted of pre-fermented media. The ethanol in the second series came from a fermentation stopped by double centrifugation in the growth, stationary, and death phases of the population growth cycle.

The composition of the non-fermented and pre-fermented media was as follows:

A: alcohol content = 1.7% vol.; sugar = 160 g/l
B: alcohol content = 7.0% vol.; sugar = 65 g/l
C: alcohol content = 9.5% vol.; sugar = 23 g/l

Figure 3.6 shows that yeasts grow in all the non-pre-fermented media. In the pre-fermented media, on the other hand, growth

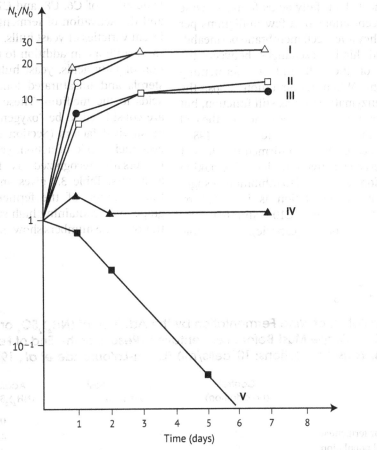

FIGURE 3.6 Evolution of *S. cerevisiae* population in fermenting media containing different alcohol concentrations (A = 1.7% vol.; B = 7.0% vol.; C = 9.5% vol., obtained by fermentation or alcohol additions) (Geneix *et al.*, 1983). N_t = cell count at time t; N_0 = cell count at start (approximately 10^7 cells/ml). I, Non-fermented media A and B; II, non-fermented medium C; III, pre-fermented medium A; IV, pre-fermented medium B; V, pre-fermented medium C.

is only possible in medium A, which has a low alcohol concentration. Population decline is significant in pre-fermented medium C, which has a high alcohol content. According to this experiment, fermentation creates other substances besides ethanol, which inhibit yeast growth and alcoholic fermentation. A complementary experiment indicated that these substances are eliminated by charcoal, confirming past observations. For a stuck fermentation, charcoal helps to restart yeast activity by removing yeast metabolism products from wine.

Research into the impact of various fermentation by-products on yeast demonstrated the inhibiting effect of C6, C8, and C10 short-chain fatty acids found in wine at concentrations of a few milligrams per liter. They affect cell membrane permeability and hinder exchanges between the inside of the cell and the fermenting medium. When fermentation stops, the yeast enzymatic systems still function, but the sugars can no longer penetrate the cell to be metabolized (Larue et al., 1982). Salmon et al. (1993) confirmed that loss of *S. cerevisiae* activity under winemaking conditions was linked to inhibition of sugar transport. Fermentation is inhibited by these C6, C8, and C10 saturated fatty acids—hexanoic (caproic), octanoic (caprylic), and decanoic (capric) acids. Other unsaturated long-chain fatty acids (C18), however, are activators under certain conditions: oleic acid, one double bond; and linoleic acid, two double bonds (Section 3.5.2). The term "fatty acid" used in both cases can lead to confusion.

The preceding facts explain the use of yeast hulls in winemaking. They are currently the most effective fermentation activators known for winemaking (Lafon-Lafourcade et al., 1984). Yeast hulls eliminate the inhibition of the fermentation by fixing the toxic fatty acids (Lafon-Lafourcade et al., 1984). The permeability of the cell membranes is reestablished in this manner. Munoz and Ingledew (1989) confirmed the toxic effect of C6, C8, and C10 fatty acids and the activation of fermentation by different varieties of yeast hulls. According to these authors, in addition to their adsorption of fatty acids, yeast hulls contribute sterols and unsaturated long-chain fatty acids to the medium. These constituents are considered to be "oxygen substitutes" or survival factors (Section 3.5.2). Whatever their mode of action, yeast hulls are universally recognized as fermentation activators. Table 3.7 gives an example of the activation of the fermentation of a grape must containing high sugar concentrations. The numbers show the superiority

TABLE 3.7

Stimulation of Wine Fermentation by the Addition of $(NH_4)_2SO_4$ or Yeast Hulls to Grape Must Before Fermentation (Results at the End of Fermentation; Yeast Populations: 10^7 cells/ml) (Lafon-Lafourcade et al., 1984)

	Control (no addition)	Addition of yeast hulls (g/l)		Addition of $(NH_4)_2SO_4$ (g/l)
		0.2	1	0.2
Sugar fermented (g/l)	206	247	257	212
Total population	9	11	14	10
Viable population	3.5	10	26	1.7

Initial sugar concentration: 260 g/l; initial viable yeast population: 10^6 cells/ml; dry yeast: *S. cerevisiae*; fermentation temperature: 19°C.

of yeast hulls with respect to ammonium salts for the activation of the fermentation. During the final stage of fermentation, the total cell population has not greatly increased, but the viable yeast population is clearly more significant in the presence of yeast hulls. This characteristic survival factor effect does not exist when only ammonium salts are added. The addition of yeast hulls on the fifth day following the onset of fermentation, i.e. after the growth phase, has a more pronounced effect on population survival than an addition before fermentation.

Yeast hulls have proven to be effective in musts that are difficult to ferment; for example, those containing high sugar concentrations or pesticide residues. Yeasts are also more temperature-resistant in their presence (Table 3.8). They may be used, although less effectively, in cases of stuck fermentation (Section 3.8.3).

Yeast hulls must be perfectly purified to avoid a sensory impact on wines. The industrial preparation of yeast extract results in pronounced odors or tastes in the product, and these must be removed from the hulls before use in wine. Moreover, if the hulls are not sufficiently purified, spoilage (due to the presence of residual lipids) may occur during storage under certain conditions. This spoilage leads to an unfavorable development in the wine's organoleptic characters. Such circumstances have incited excessive criticism concerning the use of yeast hulls.

The involvement of yeast hulls in fermentation processes is also accompanied by changes in the concentration of secondary products (higher alcohols, fatty acids, and their esters). As a result, wine aromas and tastes can be modified. All operations that affect fermentation kinetics affect the wine—temperature, oxygenation, addition of ammonium salts, etc.—and yeast hulls have no more of an impact on the fermentation than these other factors, and certainly less than temperature, for example. Whatever the case, yeast hulls should be used with prudence for the fermentation of lighter-bodied wines, such as certain dry white wines. In this case, their effect can be more significant.

3.6.3 Inhibition from Different Origins

Some vine treatment spray residues are well known to inhibit fermentation. Sulfur- and chloride-based compounds are the most harmful to yeasts. Inoculation with active yeast once the inhibiting residue has broken down is generally sufficient to reactivate fermentation in the must (Hatzidimitriou et al., 1997). However, certain difficult final stages of fermentation can be attributed to the presence of these residues. The minimum time between the last application of a product and the harvest date indicated by the manufacturer is not always sufficient.

TABLE 3.8

Stimulation of Red Wine Fermentation by the Addition of Yeast Hulls (Maximum Fermentation Temperature of 34°C Attained on Day 7) (Lafon-Lafourcade et al., 1984)

	Residual sugars during fermentation (g/l)			
	Day 4	Day 8	Day 9	Day 16
Control	194	22	15	15
Addition of yeast hulls (0.3 g/l)	206	15	2	

Elevated concentrations of tannins and coloring matter found in certain varieties of red wines can hinder yeast activity. They bind to the cell wall by a kind of tanning process. The effect of these substances is not clear: some activate fermentation, while others inhibit it.

Carbon dioxide produced by fermentation is known to have an inhibitory effect. This occurs during fermentations under pressure (sparkling wines). A slight internal pressure in the tank is sufficient to slow the fermentation, and above 7 bar, fermentation becomes impossible. Under normal winemaking conditions, carbon dioxide is released freely and exercises no inhibition on the fermentation.

The difference in fermentability of various grapes and musts is linked with many poorly understood factors. In the same way that specific activators must exist, the presence of natural inhibitors in grapes has been considered. Their interaction would affect must fermentability.

More specifically, the preliminary development of yeasts, lactic acid bacteria, or *B. cinerea* can hinder the alcoholic fermentation process. In the case of alcoholic fermentation difficulties, bacterial growth and malolactic fermentation exacerbate these difficulties. Stuck fermentations often result (Sections 3.8.1 and 6.4.1).

Among these causes of fermentation difficulties, the involvement of *B. cinerea* has been the most studied. Must derived from moldy grapes (noble or gray rot) is more difficult to ferment than must originating from healthy grapes. Past works, summarized by Ribéreau-Gayon *et al.* (1975a), identified a substance with antibiotic properties; the authors named it botryticin. Sulfiting and prolonged heating at 120°C destroy this substance. Ethanol at 80% can precipitate it. Subsequent work showed that this fungistatic substance is either partially or completely a mannose-based neutral polysaccharide (Ribéreau-Gayon *et al.*, 1979). The phytotoxic properties of such substances are known, and this polysaccharide affects fermentation kinetics. It is also the cause of certain metabolic deviations induced by *B. cinerea*—in particular an increase in the production of glycerol and acetic acid (Section 2.3.4; Volume 2, Section 3.7.2).

3.7 Physicochemical Factors Affecting Yeast Growth and Fermentation Kinetics

3.7.1 Effect of Temperature

Alcoholic fermentation, depicted by the following chemical equation:

$$C_6H_{12}O_6 \rightarrow 2C_2H_5OH + 2CO_2,$$

releases 40 kcal of free energy per molecule. Yeasts use part of this energy to ensure their vital functions, in particular their growth and multiplication, and to form two ATP molecules from one sugar molecule. These ATP molecules have a high energy potential:

$$2ADP + 2H_3PO_4 \rightarrow 2ATP + 2H_2O.$$

It takes 7.3 kcal of energy to form one ATP molecule. The difference:

$$40 - 14.6 = 25.4 \text{ kcal}$$

is non-utilized energy that is dissipated in the form of heat, causing the fermentation tanks to heat.

This estimate of unutilized energy is open to discussion. In situations where part of the ATP formed is not needed by the yeast, it is hydrolyzed by the corresponding enzymes. Yet 25 kcal corresponds fairly well with past thermodynamic measures of dissipated heat by the fermentation of one molecule of sugar.

The fermentation of must containing 180 g (one molecule) of sugar per liter therefore liberates 25 kcal in the form of heat. This release of heat theoretically could raise the temperature from 20 to 45°C. Such an increase in temperature would kill the yeast. Fortunately, this increase is the hypothetical case of an explosive, instantaneous fermentation or a fermentation in a fully insulated tank. In reality, fermentation takes place over several days. During this time, the calories produced are dissipated by several phenomena: by being entrained with the large quantity of carbon dioxide released during fermentation, by cooling resulting from the evaporation of water and alcohol, and by exchanges across the tank wall.

Temperature increases in fermentors depend on several factors.

1. The must sugar concentration, which determines the amount of calories liberated.
2. The initial must temperature.
3. The fermentation speed, which depends on must composition (nitrogen-based substances) and yeast inoculation conditions. Operations such as aeration, chaptalization, and inoculation will increase the fermentation speed, limit the dissipation of calories, and increase the maximum temperature. Reciprocally, not crushing red grapes, as is the case in carbonic maceration (Section 12.9.1), will slow fermentation kinetics and lower the maximum temperature.
4. Tank dimensions. When the volume increases, the relative surface area of the walls and their heat exchange capacity decrease when the same proportions are maintained.
5. Tank material. The overall heat exchange coefficient (K, expressed in cal/h/m² for each degree difference in temperature) is from 0.7 to 0.78 for a concrete wall, 10 cm thick; from 1.46 to 1.49 for a 5 cm wooden wall; and from 5.34 to 32.0 for a 3 mm stainless steel wall. The coefficient varies the most in the case of stainless steel. A stainless steel tank is sensitive to the external conditions (temperature and aeration) in which it is placed.
6. Aeration and cellar temperature. The ventilation of the winery limits fermenting tank temperatures by dissipating heat.

The maximum tank temperature is related to all of these factors by complex laws and is difficult to predict. Depending on the circumstances, the maximum temperature may be compatible with red winemaking. In this case, a maximum temperature between 25 and 30°C ensures sufficient extraction of phenolic compounds from the solid parts during maceration. In other cases, refrigeration is necessary to avoid exceeding the maximum temperature limit. Refrigeration is always necessary for white wines: their fermentation must be carried out at around 20°C to retain their aromas.

Current refrigeration methods include circulating cold water or another cold fluid through the double lining of metal tanks or through a heat exchanger submerged in the tank. In certain cases, spraying the exterior of a metal tank can be sufficient. The must can also be sent through tubular exchangers cooled by circulating water, itself refrigerated by an air exchanger.

Temperature has an impact on yeast development and fermentation kinetics. According to Fleet and Heard (1992), temperature can affect native yeast ecology. The authors suggest that different strains are more or less adapted to different temperatures, ranging from 10 to 30°C. More precisely, the growth rate varies for each strain according to temperature—for instance, with *Kloeckera apiculata* and *S. cerevisiae*. The possible enological consequences of this phenomenon merit further research.

Numerous overlapping factors make it difficult to anticipate the impact of temperature on fermentation kinetics. Ough (1964, 1966) developed equations for the estimation of the impact of temperature on fermentation kinetics as a function of numerous parameters.

Bordeaux enologists dedicated much research to this subject, summarized by Ribéreau-Gayon et al. (1975a). Temperature profoundly affects yeast respiratory and fermentation intensity (Table 3.9): fermentation intensity doubles for every 10°C temperature increase. It reaches its maximum at 35°C and begins to decrease at 40°C. These numbers show the importance of fermentation temperature. The fermentation of sugar is twice as fast at 30°C as at 20°C, and for each temperature increase of 1°C, the yeast transforms 10% more sugar in the same elapsed time. The optimal fermentation temperature varies according to the yeast species.

Temperature influences fermentation kinetics. The alcohol yield is generally lower at high temperatures, in which case some of the alcohol may be entrained with the intense release of carbon dioxide. Additionally, most of the secondary products of glyceropyruvic fermentation (coupled glycerol/pyruvate formation) are found in greater concentrations. Fatty acids, higher alcohols, and their esters are the most affected: their formation is at its maximum at about 20°C and then progressively diminishes. Low fermentation temperatures are justified when these products are desired in white winemaking.

In addition to its influence on yeast activity, temperature affects fermentation speed and limits. Between 15 and 35°C, the duration of the latent phase and the delay before the onset of fermentation become shorter as the temperature increases. Simultaneously, yeast consumption of nitrogen increases (Section 3.4.2).

For example, a grape must with a limited sugar concentration (less than 200 g/l) takes several weeks to ferment at 10°C, 15 days at 20°C, and three to four days at 30°C. For musts with higher sugar concentrations, fermentation becomes more limited as the temperature increases; in fact, fermentation can stop, leaving high levels of non-fermented sugar.

Table 3.10 concerns a must from Sauternes containing more than 300 g of sugar per liter. The same phenomenon occurred in Müller-Thurgau's experiment in 1884 (cited in Ribéreau-Gayon et al., 1975a). He fermented the same must with increasing concentrations of sugar (Table 3.11).

The initial sugar concentration of must and excessive temperatures limit ethanol production. Other factors, such as the amount of oxygen present, can limit or stop the fermentation at relatively low alcohol concentrations (11–12% vol.) and

TABLE 3.9

Average Fermentation and Respiratory Intensities (mm³ of O_2 Consumed or of CO_2 Released/g of Dry Yeasts/Hour) of Various S. cerevisiae Species According to Temperature (Ribéreau-Gayon et al., 1975a)

Temperature (°C)	Respiratory intensity (Q_{O_2})	Fermentation intensity (Q_{CO_2})
15	4.2	118
20	6.7	168
25	9.6	229
30	11.4	321
35	6.2	440
40	3.0	376

TABLE 3.10

Fermentation Onset Speed and Limits According to Temperature (Initial Sugar Concentration: Approximately 300 g/l) (Ribéreau-Gayon et al., 1975a)

Temperature (°C)	Onset of fermentation	Final alcohol content (% vol.)
10	8 days	16.2
15	6 days	15.8
20	4 days	15.2
25	3 days	14.5
30	36 hours	10.2
35	24 hours	6.0

temperatures (less than 30°C). An excessively high temperature (25–30°C) during the yeast multiplication phase affects yeast viability and favors stuck fermentations. The impact of temperature on fermentation kinetics also varies from one yeast strain to another.

These facts are important for winery practices and show the difficulty of determining a maximum acceptable temperature limit.

The impact of temperature on fermentation is depicted in Figure 3.7. The latency time decreases and the initial fermentation speed increases as the temperature rises. The risks and severity of a stuck fermentation also increase with temperature. Of course, if the initial sugar concentration had been lower in this example, the fermentation would have been complete at all temperatures. On the other hand, a higher sugar concentration would have resulted in a stuck fermentation even at 25°C. This shows that fermentation speed increases as the temperature rises, but that the fermentation is also increasingly limited.

An abrupt temperature change during fermentation can lead to thermal shock. This phenomenon is different from the notion of shock or stress that is used in microbiology (Section 6.2.5). In certain wineries, white wine fermentation tanks are situated outdoors due to space limitations. These wines, fermenting at moderate temperatures (20°C), have difficulty withstanding the abrupt temperature variations between day and night when autumn cold first arrives. The fermentation progressively slows and finally stops. A second inoculation is not effective. If these wines are transferred to a tank at a constant temperature before the fermentation has com-

TABLE 3.11

Alcohol Formation (% vol.) According to Fermentation Temperature (Müller-Thurgau, 1884)

Sugar concentration (g/l)	127	217	303
Potential alcohol (% vol.)	7.2	12.4	17.3
Alcohol produced at 9°C (% vol.)	7	11.8	9.9
Alcohol produced at 18°C (% vol.)	6.9	11.0	9.1
Alcohol produced at 27°C (% vol.)	6.9	9.4	7.7
Alcohol produced at 36°C (% vol.)	4.2	4.8	5.1

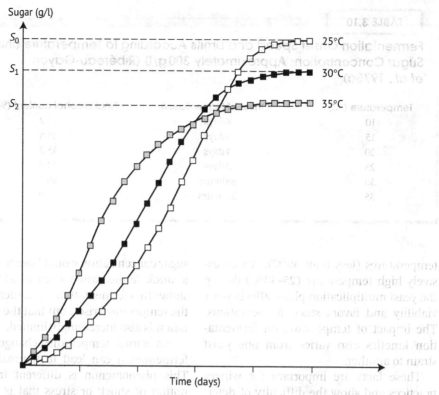

FIGURE 3.7 Influence of temperature on fermentation speed and limit (S_0 = initial sugar concentration). At 25°C, fermentation is slower, but complete. At 30°C, and especially at 35°C, it is more rapid, but stops at fermented sugar concentrations S_1 and S_2, respectively, below S_0.

pletely stopped, fermentation will go to completion, possibly after reinoculation. Laboratory tests confirm that an abrupt temperature change, in one direction or the other, affects yeast activity. This phenomenon merits more in-depth study.

The data in Table 3.12 are taken from a laboratory experiment. At 12°C, the fermentation is slow but complete. At 19°C, the fermentation is quicker and sugar transformation is also complete. If the fermentation begins at 19°C and is abruptly lowered to 12°C, it stops, leaving unfermented sugar. The same is true for a fermentation whose temperature abruptly increases from 12 to 19°C. If thermal shock occurs during the final stage of fermentation (after the fermentation of 120 g of sugar per liter, for example), its effect is less significant.

To recapitulate, an increase in temperature accelerates the fermentation but also adversely affects its end point. A stuck fermentation can also occur if other limiting factors add their effects (high sugar content and anaerobic conditions). For this reason, there is no fixed temperature limit above which fermentation is no longer possible. In red winemaking, it is generally recommended for fermentation temperature not to exceed 30°C: above this temperature, the risks are certain. Of course, this does not mean that a fermentation cannot be complete at temperatures at or above 35°C.

Yeast is most sensitive to temperature at the beginning of its development; it is more resistant during the final stage of fermentation. For this reason, in red winemaking, the fermentation should begin at 18–20°C

3.7 | Physicochemical Factors Affecting Yeast Growth and Fermentation Kinetics

TABLE 3.12

Effect of Temperature Variations (Thermal Shocks) on Grape Must Fermentation (Initial Sugar Concentration: 220 g/l) (Larue et al., 1987)

	Duration of fermentation (days)	Residual sugar concentration at end of fermentation (g/l)
Fermentation at 12°C	93	<2
Fermentation at 12°C, changed to 19°C after transformation of 40 g sugar/l	56	27
Fermentation at 19°C	50	<2
Fermentation at 19°C, changed to 12°C after transformation of 40 g sugar/l	21	108

and be allowed to increase progressively to 32°C or even a little higher. The higher final temperature favors maceration phenomena. An excessively high temperature at the onset of fermentation (around 30°C) can result in a difficult end of fermentation.

3.7.2 Influence of Oxygen: Effect of Must Aeration

Yeasts use energy derived from the breakdown of sugars. This breakdown is carried out by either the respiratory or the fermentation pathway. In grape must, due to the catabolite repression of respiration exerted by must glucose in *S. cerevisiae* (Section 2.3.1), sugar breakdown is carried out exclusively by alcoholic fermentation.

Yeast respiratory capacity is put to good use in enology for the production of flor wines, such as Sherry (Section 14.5.2). In this case, yeasts oxidize ethanol into aldehyde in dry wines (i.e. containing no residual sugar). Flor yeasts can also develop during winemaking: they oxidize ethanol into carbon dioxide and are considered to be spoilage yeasts.

Pasteur spoke of must fermentation as a type of "life without air." Yeast development and fermentation have long been known to be impossible in the complete absence of oxygen (Ribéreau-Gayon et al., 1951). The complete absence of oxygen supposes the fermentation of a must devoid of oxygen under strict anaerobic conditions. The must would also have to be inoculated by a yeast starter cultivated in the absence of oxygen. Complete anaerobic experimental conditions are difficult to maintain. For this reason, some authors might have thought that oxygen was not absolutely necessary for fermentation.

On the contrary, oxygen has a considerable impact on the fermentation kinetics of wine. The addition of oxygen is probably the most effective method available to the winemaker for controlling must fermentation. For this reason, the terms "complete anaerobic, "semi-anaerobic," or "limited aerobic" conditions are sometimes used to explain the amount of oxygen added during fermentation. In laboratory experiments, samples are sealed with a sterile wad that permits the controlled introduction of oxygen and ensures an exchange in both directions between the interior and exterior environment. Complete anaerobic conditions are obtained by obturating the opening of the samples with a fermentation lock.

In the winery, open tanks leaving the wine in contact with air enable permanent aeration. They are not recommended because of the risk of bacterial development—closed tanks are preferable, but a controlled amount of oxygen should be added to the fermenting wine in these tanks during pumping-over, for example. This technique has been widely used in red winemaking for many years. The fermenting must is easily saturated with oxygen (6–8 mg/l). Pumping-over, however, is less used in white winemaking because of fears of oxidizing the must and modifying the aromas. In fact, this fear has not been confirmed in practice, since the yeast can absorb a large amount of oxygen during fermentation. The aromas of must before fermentation and those of white wine separated from its lees after fermentation are susceptible to oxidation, but the aromas of must during fermentation are probably less affected by aeration.

Aeration accelerates fermentation and as a result increases the demand for nitrogen-containing nutrients. Oxygen promotes the synthesis of sterols and unsaturated fatty acids, improving cell membrane permeability and thus the penetration of carbohydrates into the cell. The addition of oxygen has an effect similar to the addition of sterols, which are considered to be oxygen substitutes.

The data in Table 3.13, taken from an experiment carried out many years ago in Ribéreau-Gayon's laboratory (Ribéreau-Gayon et al., 1951), show the effects of controlled aeration on the fermentation kinetics of a relatively high-sugar must. Under strict anaerobic conditions, fermentation stops on the 14th day, leaving 75 g of sugar per liter. The yeast population is 5×10^7 cells/ml. Under limited aerobic conditions, the fermentation is complete, with a yeast population twice as great. Moreover, the onset of fermentation and its initial speed are greater in the presence of oxygen. Of course, if the initial sugar concentration had been lower in the experiment in Table 3.13, the fermentation would have been complete in both cases, although slower under anaerobic conditions. A higher sugar concentration would have led to a stuck fermentation in both cases before the complete depletion of sugar.

In winemaking, permanent aeration is rarely possible, but occasional aerations are a suitable replacement. Table 3.14 is also taken from a past experiment (Ribéreau-Gayon et al., 1951). It confirms the difference in fermentability between musts in the presence and in the absence of oxygen. The effect of oxygen on fermentation kinetics increases with the quantity of oxygen introduced. The timing of the addition of oxygen appears to be especially important. The acceleration of fermentation is most significant when oxygen is added on the second day following the onset of fermentation, during the growth phase of the yeast population. In this case, the fermentation is nearly as rapid as a fermentation with limited permanent aeration; the same amount of sugar is also transformed. In fact, it is not the must that needs to be aerated but rather the yeasts

TABLE 3.13

Evolution of the Fermentation of a Grape Must Containing a High Sugar Concentration (270 g/l) as a Function of Aeration Conditions (Initial Inoculation: 10^6 cells/ml) (Ribéreau-Gayon et al., 1951)

Time (days)	Limited aerobic conditions (cotton stoppered flasks)		Anaerobic conditions (flasks stoppered with a fermentation lock)	
	Residual sugars (g/l)	Total cells (10^7/ml)	Residual sugars (g/l)	Total cells (10^7/ml)
7	86		140	
14	2		75	
21	2	10.7	75	5

TABLE 3.14

Effect of Oxygen Addition at Different Stages of Grape Must Fermentation Containing 228 Sugar/gl (Measurements Made on Day 14 of Fermentation) (Ribéreau-Gayon et al., 1951)

	Oxygen added[a] (ml/l)	Total yeast cells (10^7/ml)	Fermented sugars (g/l)
Fermentation in contact with air			
Limited aeration		9.3	225
Fermentation without air contact			
Without aeration	0	5.1	164
Brief aeration before fermentation	6.0	6.2	164
Brief aeration on day 2	0.15	5.8	190
	0.75	6.1	196
	1.5	6.3	205
	6.0	7.5	223
Brief aeration on day 4	0.75	5.3	184
	1.5	6.0	202
	6.0	6.0	202
Brief aeration on day 8	6.0	5.2	173

[a] Oxygen can be expressed in mg/l by multiplying the values by 1.43 (density of oxygen with respect to air).

fermenting the must—especially the population in the growth phase. Other experiments confirm that yeasts use oxygen primarily during the first stages of fermentation. They do not benefit from aeration when the fermentation is in its advanced stages and the alcohol content is excessive.

These observations are significant for practical purposes and they should be taken into account during winemaking—especially red winemaking (Section 12.4.2). The risks of winemaking in open tanks, or permanent aerobic conditions, are known (Section 12.3.1). Anaerobic conditions in tanks can be recommended if oxygen is added at the right moment. The acceleration of the fermentation induced by brief aeration must nevertheless be anticipated. This acceleration results in a greater temperature increase.

The experiments cited in this section date back to Ribéreau-Gayon et al. (1951). Even though no longer cited in the literature, the results have been confirmed in more recent works (Fleet, 1992). In particular, Sablayrolles and Barre (1986) retained slightly higher values for yeast oxygen needs (10 mg/l). They also confirmed the significant influence of oxygen and the moment of its addition on fermentation kinetics. Furthermore, Julien et al. (2000) compared the respective oxygen needs of various strains of winemaking yeasts.

Other research has shown that aeration, combined with the addition of nitrogen at mid-fermentation, is more effective than aeration alone (Sablayrolles et al., 1996a,b) (Section 3.4.2). This effect is apparently more marked in certain media under certain fermentation conditions. Also a study conducted on about 100 musts with sluggish or stuck fermentations shows that combined addition of nitrogen and oxygen at mid-fermentation systematically leads to a shorter fermentation time and complete breakdown of sugar (Blateyron and Sablayrolles, 2001).

3.7.3 Effect of Must Clarification on White Grapes

Must clarification before the onset of fermentation has long been known to affect the quality of white wine (Section 13.5.2). Yeasts fermenting clear must form more higher alcohols, fatty acids, and corresponding esters. In addition, suspended solids in the must can impart unpleasant vegetal odors. The settling and racking of must is therefore essential.

Must clarification also affects fermentation phenomena (Section 13.5.3). It eliminates some of the wild yeasts, along with the natural plant sediment. However, inoculation compensates for this loss, and is also often recommended to compensate for the small population present at the time of filling the tank. In this manner, fermentation is carried out by selected strains that best express must quality, without the development of sensory flaws.

During static juice settling, certain conditions useful for sedimentation (such as low temperature and sulfiting) can promote the development of certain strains resistant to these conditions. Of course, this growth must not lead to actual fermentation; otherwise, it would put the sediments back in suspension. Nevertheless, these strains can develop preferentially during fermentation even after an active yeast inoculation (Fleet, 1992).

Clarification essentially modifies must fermentability. Clear must is known to ferment with more difficulty than cloudy must (though the elimination of yeasts is not the only reason for this fermentation difficulty, as was once thought). However, clarification simultaneously favors the aroma finesse of the wine. White wine quality is thought to be enhanced by a somewhat difficult and slow fermentation. In general, all operations that accelerate fermentation will lower wine quality, and vice versa. A compromise permitting complete fermentation and satisfactory wine quality must be sought.

Several researchers have studied the effect of grape solids and other solid materials on fermentation kinetics (Ough and Groat, 1978; Delfini and Costa, 1993). Ribéreau-Gayon et al. (1975b) evaluated the effect of several related factors on must fermentability: the elimination of yeast, the elimination of nutritive elements released by grape solids, and a possible support effect that would permit greater yeast activity, possibly by the fixation of toxic compounds (short-chain fatty acids). A number of fermentations were carried out in the laboratory under different conditions. For this experiment, the medium was heated at 100°C for five minutes to destroy the yeasts and ensure the solubilization of the nutritive elements likely to be involved in fermentation. In view of the destruction of the yeasts under certain conditions, the medium was systematically inoculated using a strain of *S. cerevisiae* with good fermentation potential at a relatively low concentration (10^5/ml) to assess any possible effect of natural elimination of the yeast. In this manner, the subsequent effect of the natural elimination of yeasts can be appreciated. By comparing the different samples, the contribution of each of the three parameters to the loss of fermentability could be evaluated.

The results of two tests with different concentrations of sugar are given in Table 3.15. Fermentability loss due to juice clarification appeared to be significant. The measurements were taken on the second and fifth days of fermentation, when the slowing of the fermentation due to juice settling and clarification is most obvious. The differences tend to become less apparent over time. In test A, the fermentation is complete even after juice settling. In test B, juice settling results in a stuck fermentation. In spite of the necessary approximations made in carrying out this experiment, it shows the multitude of effects of juice settling. Furthermore, the sum of the individual fermentability losses corresponds fairly well with the total loss.

TABLE 3.15

Analysis of the Effect of Different Must Clarification Operations on Fermentability Loss[a] (in %) (Ribéreau-Gayon et al., 1975b)

Cause of fermentability loss	Trial A (initial sugar concentration 220 g/l) measured on day 2 (%)	Trial B (initial sugar concentration 285 g/l) measured on day 5 (%)
Total fermentability loss due to clarification	62	37
Elimination of yeasts in grape solids	27	10
Elimination of support effect of grape solids	9	13
Elimination of nutritive elements released by grape solids	23	21
Sum of individual fermentability losses	59	44

[a] Fermentability loss is defined as the percentage of sugar remaining unfermented, in comparison to a non-clarified control, all other conditions being the same.

The nature of the sediment and its effect on must fermentability are related to grape origin, grape mold status, and must extraction conditions. Lafon-Lafourcade et al. (1980) demonstrated the essential role of the crushing and pressing conditions of white grapes. Under identical clarification conditions, musts extracted after the energetic crushing of grapes ferment less well than musts originating from pressing without crushing. This difference in fermentation can result in a stuck fermentation when associated with other unfavorable conditions (high sugar concentrations, strict anaerobic conditions, etc.). Pressing without crushing maintains the juice in contact with the skins for a certain period, and this contact seems to permit the diffusion of grape skin steroids. In the same manner, pre-fermentation skin contact generally results in a must with good fermentability, even after careful juice settling. Delfini et al. (1992) noticed that must clarification eliminates long-chain fatty acids, which are known to have an effect on yeast. Their elimination has been linked to the increased production of acetic acid frequently reported in white musts with fermentation problems, particularly those that have been highly clarified.

Grape solids and even carbohydrate macromolecules, which contribute to the colloidal turbidity of musts, can, like yeast hulls, adsorb short-chain fatty acids (C8 and C10) (Section 3.6.2) (Ollivier et al., 1987).

Consequently, the level of must clarification should be controlled for each type of white winemaking by measuring must haziness or turbidity, expressed in NTU (Sections 13.5.2 and 13.5.3). For vineyards in the Bordeaux region, a turbidity of less than 60 NTU can lead to serious fermentation difficulties. Above 200 NTU, there are definite risks of sensory deviations due to the presence of grape solids.

3.8 Stuck Fermentations

3.8.1 Causes of Stuck Fermentations

Stuck fermentations have always been a major problem in winemaking. French enological literature has mentioned them since the beginning of the 20th century.

The production of fortified wines was definitely a response to difficult final stages of fermentation and to the ensuing microbial spoilage, especially in countries with warm climates. These wines were rapidly stabilized by the addition of pure alcohol. Around the world, many of these wines disappeared as progress in microbiology permitted the elaboration of dry wines.

Stuck fermentation continues to be a much-discussed subject. In some cases, a modification of grapevine varieties has produced grapes that have high sugar concentrations. These grapes are more difficult to ferment than past varieties. In other cases, winemakers have realized that sluggish fermentations, spread out over several months, are not ideal for making wine.

As red winemaking techniques used in Bordeaux (grapes with a relatively high sugar content and long maceration requiring closed tanks) are particularly conducive to this problem, it has been studied here for many years. In the 1950s, in-depth research in Bordeaux resulted in important discoveries concerning temperature regulation and, above all, aeration. The spread of stuck fermentations in other winemaking regions has led to new research confirming past work.

The slowing of fermentation can be monitored by tracking must density (mass per unit volume). If a density below 1.005 decreases by only 0.001 or 0.002/day, a stuck fermentation can be anticipated before the complete depletion of sugar. From the past experience, the consequences of a stuck fermentation are not too serious if it occurs with at least 15 g of sugar per liter and a moderate alcohol content (<12% vol.). In this case, the restarting of fermentation does not pose any major problems. On the other hand, with a sugar concentration of less than 10 g/l, it is often very difficult to restart stuck fermentations.

A stuck fermentation can result from a particular cause. For example, excessive must sugar concentration makes complete fermentation impossible; in this case, only a sweet wine can be made. Stuck fermentations can be expected in the case of excessive temperatures, and this type of stuck fermentation is generally the result of several causes. The effects are cumulative, although sometimes individually they are without consequence. Winemakers do not always understand possible cumulative risks and the precautions that must be taken.

To summarize, the following factors can be involved in stuck fermentations:

1. The must sugar concentration has an inhibitory effect that compounds the toxicity of the alcohol formed. The addition of sugar to must (chaptalization), when it is too late, requires yeasts to pursue their metabolic activity when already hindered by the alcohol formed.

2. An excessive temperature may result from the initial temperature, the quantity of sugar fermented, and/or the type of tank used (dimensions and material). All operations that accelerate the sugar transformation rate increase the maximum temperature. The temperature becomes a limiting factor at about 30°C. The effect is more pronounced when the temperature is high in the early stages of fermentation. Normally, fermentation should begin at a moderate temperature (20°C).

3. Conversely, too low of an initial temperature can limit yeast growth and lead to an insufficient yeast population. At moderate temperatures, yeasts have difficulty supporting extreme temperature changes (thermal shocks).

4. Strict anaerobic conditions do not enable satisfactory yeast activity (growth and survival). Aeration increases fermentation speed. It must take place in the early stages of fermentation, during the

population growth phase. Oxygen substitutes, such as steroids and long-chain fatty acids, can also improve fermentation kinetics.

5. Yeast activity can be affected by nutritional deficiencies: nitrogen compounds, growth factors, and possibly minerals. Combined additions of oxygen and ammonia nitrogen appear to be particularly effective.

   These nitrogen deficiencies probably occur in specific situations that we are now capable of predicting. The effectiveness of the addition of nutrients, observed in laboratory work, should be interpreted with respect to other fermentation conditions (sugar concentration and aeration). Certain grape growing conditions, such as water stress, old vines, and cover cropping to decrease vine vigor, can lead to less fermentable musts. Under these conditions, stuck fermentation is probably due to nutritional deficiencies (nitrogen), which probably require further investigation.

6. Metabolic by-products (C6, C8, and C10 saturated fatty acids) inhibit yeast growth, intensifying alcohol toxicity.

7. Antifungal substances can be present in must—pesticide residues used to protect the vine or compounds produced by *B. cinerea* in moldy grapes.

8. In white winemaking, must extraction conditions have a significant influence: grape crushing, juice runoff, pressing of the crushed grapes, and especially the level of must clarification (juice settling). These operations may result in the excessive elimination of steroids, which act as survival factors for the yeast.

In the acidity range of must, high acidity does not seem to favor fermentation, but an elevated pH can make the consequences of a stuck fermentation much more serious. A low pH combines with the effect of sulfiting to inhibit bacterial growth. In this case, antagonistic phenomena between bacteria and yeasts diminish and the fermentation is steadier. Another unfortunate consequence of low pH is that it promotes the formation of volatile acidity by the yeast.

In addition to chemical and physicochemical causes of stuck fermentations, microbial phenomena also are involved. First of all, the quantity of the initial yeast inoculum can be insufficient (Section 3.5.4) if, for example, must temperature is much too low. Antagonistic phenomena between different yeast strains can also occur, and the killer factor (Section 1.7) explains this fairly widespread antagonism. Fermentation can be rapid in some tanks while being slower in others, and strain identification techniques have shown that fermentation is carried out almost exclusively by one strain in the first case, whereas several strains ferment the must in the second case (Section 1.10.2). These antagonistic phenomena can affect an inoculation. In certain conditions (for example, a significant natural inoculum in full activity), inoculating with dry commercial yeast leads to a slower fermentation than not inoculating. Yeast strains must therefore be selected according to the type of wine being made, ensuring that they are more active and numerous than the native yeasts. The necessary conditions for controlling fermentation include: cleanliness; inhibiting native yeasts sufficiently early by maintaining low temperatures; sulfiting appropriately; and inoculating with an active yeast starter as soon as the tank is filled in order to ensure its rapid establishment.

Antagonistic phenomena between yeasts and lactic acid bacteria can also cause fermentation difficulties (Section 6.4.1), especially in red winemaking (Section 12.4.3). The initial sulfiting of the grapes has to inhibit the bacteria temporarily while at the same time permitting yeast development and sugar fermentation. Bacteria do not develop as long as yeast activity is sufficient, but if alcoholic fermentation slows for some reason, bacteria can begin to grow—especially if the initial sulfiting was insufficient. This bacterial development aggravates yeast difficulties and increases the risks of a prematurely stuck alcoholic fermentation (Section 6.4.1). The bacterial risk is an additional justification for red grape sulfiting (5 g/hl) before fermentation (Section 8.7.4). The addition of lysozyme (200–300 mg/l), extracted from egg whites, has been suggested to reinforce the inhibitory effect of sulfiting on bacteria in difficult fermentations (Section 9.5.2). In addition, the inoculation of lactic acid bacteria (*Oenococcus oeni*) before alcoholic fermentation to activate a subsequent malolactic fermentation is not recommended; in the case of difficult alcoholic fermentations, this operation increases the risk of bacterial spoilage (Section 3.8.2). Yet this is standard practice in some winemaking regions. The relationship between this practice and an increase in volatile acidity should be considered.

For a long time, difficult final stages of fermentation and stuck fermentations were a real problem during red winemaking. Temperature control systems and the general practice of pumping-over with aeration have helped to limit these incidents. White musts, however, have become increasingly difficult to ferment because of excessive clarification and mechanized must extraction conditions. This excessive clarification removes must constituents essential for a complete fermentation. In white winemaking, the must is often not aerated in order to avoid oxidation, yet a lack of aeration during fermentation also contributes to fermentation difficulties. Today, controlled aeration of white musts is recommended during fermentation as the CO_2 being released protects them from oxidation.

Human error is another factor that certainly has an impact on stuck fermentations, although it is difficult to prove. It is not unusual to find wineries where stuck fermentations occur with some regularity, as though there was a specific, technical cause that could be identified and corrected, but then these disappear completely following a change in winemaker.

3.8.2 Consequences of Stuck Fermentations

Residual sugar is not acceptable in dry white and most red wines. A stuck fermentation therefore requires the restarting of yeast activity in a hostile medium. Obviously, if the alcohol content is already high (13% vol.), the chances of restarting the fermentation are slim.

The risk of bacterial spoilage is the principal danger of a stuck fermentation.

Figure 3.8 schematizes the involvement of different microbial phenomena (red winemaking is specifically represented since malolactic fermentation is taken into account). The various stages of fermentation are understood. Yeasts ferment sugars, and yeast activity should stop only when all of the sugar molecules have been consumed. The lactic acid bacteria then assert themselves, exclusively decomposing malic acid molecules in a process called malolactic fermentation. If yeast activity stops before the complete depletion of must sugar, bacteria can develop in their place. Bacterial development depends on several factors, including the initial sulfiting of the grapes and the possible addition of lysozyme (Section 9.5.2). Acetic acid is formed when lactic acid bacteria, mainly heterofermentative *Oenococcus*, are present in a medium containing sugar. In these situations, volatile acidity can rapidly increase to unacceptable levels, even if the residual sugar concentration is relatively low. In fact, bacteria form acetic acid from sugar after their growth phase, during which malic

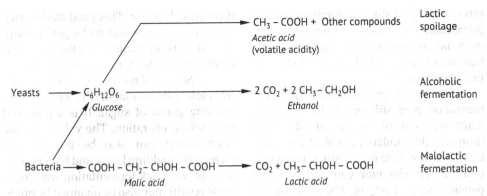

FIGURE 3.8 Effect of different microbial phenomena during primary and secondary fermentations (Ribéreau-Gayon, 1999).

acid is fully consumed. In consequence, in the case of a stuck alcoholic fermentation, the winemaker can let the malolactic fermentation continue until its completion before inhibiting the bacteria.

However, we can also observe that a stuck alcoholic fermentation is much harder to restart when lactic acid bacteria have intervened and malolactic fermentation is complete. Antagonistic phenomena between yeast and bacteria are undoubtedly responsible for these difficulties. This observation confirms the need for proper sulfiting of the freshly picked grapes, in order to delay bacterial growth until alcoholic fermentation is complete.

The understanding of the processes represented in Figure 3.8 has constituted an unquestionable progress in wine microbiology. As a result, certain operations have been initiated to control the alcoholic fermentation and avoid stuck fermentations. The volatile acidity of top-ranked red wines has decreased substantially, with a corresponding improvement in quality. In the 1930s in the Bordeaux region, the volatile acidity of these wines was often 1.0 g/l (expressed as H_2SO_4) or 1.2 g/l (expressed as acetic acid). Average content has been decreased to half this value, and today's higher figures are due to various problems during fermentation and storage, which can and must be avoided.

3.8.3 Action in Case of a Stuck Fermentation

Many stuck fermentations result from winemaking errors. More often than not, the necessary operations are known but not carried out properly. Of course, a higher sugar concentration in the must increases the risks, but in certain situations there is no satisfactory explanation for stuck fermentations.

A must density that remains stable for 24 or 48 hours confirms a stuck fermentation. In this case, various procedures exist to restart the fermentation while avoiding bacterial spoilage. However, the success of such operations is by no means guaranteed.

First of all, this medium is rich in alcohol and poor in sugar—it is not conducive to the development of a second fermentation. In addition, the yeasts are exhausted from the first fermentation, and they react poorly to the different stimuli employed. They benefit more from operations such as nitrogen supplementation and oxygenation at the beginning of their development, when the medium contains high sugar concentrations and does not contain ethanol. Therefore, an operation after a stuck fermentation cannot compensate for a winemaking error. All operations beneficial to the fermentation should be employed from the start of the winemaking process to avoid stuck fermentations. The prevention of stuck fermentations is essential to winemaking, and it should take into account all of the recommendations previously stated.

In spite of all precautions, a stuck fermentation may still occur. In this case, white wines must be treated differently from reds that undergo malolactic fermentation. At the time of the stuck fermentation, the red wine tank contains juice and pomace rich in bacteria. The wine should be drained rapidly, even if the skin and seed maceration is not complete. Draining eliminates part of the bacterial contamination and introduces oxygen, which favors the restarting of fermentation and decreases the temperature. The wine can be sulfited at the same time to inhibit bacterial development. In some cases, the fermentation restarts spontaneously.

Even if the stuck fermentation results from the combination of several basic causes, each has its own effect on the ease of restarting the fermentation. Excessive temperatures destroy yeasts but do not make the medium unfermentable, as does fermentation under strict anaerobic conditions.

If the fermentation does not restart on its own, an inoculation with active yeast is required. At present, commercial dry yeasts are inactive in media containing more that 8–9% vol. of alcohol, due to manufacturing conditions. In the future, industrially prepared yeast capable of developing in a high-alcohol medium would be desirable. Bacteria with this property have now been developed for malolactic fermentation.

An active yeast starter must be prepared using the stuck fermentation medium adjusted to 9% vol. alcohol and 15 g of sugar per liter; 3 g of SO_2 per hectoliter is also added. The ADYs are added at a concentration of 20 g/hl. Their growth at 20°C requires several days and is monitored by measuring the density or sugar content. When all of the sugar has been consumed, the yeasts are at the peak of their growth phase. This yeast starter, rich in activated yeasts and no longer containing sugar, is inoculated into the stuck fermentation medium at a concentration of 5–10%. Several days are required for the complete exhaustion of the last few remaining grams of sugar. It is a long and painstaking operation. The volume of the yeast starter can also be progressively increased by adding larger and larger quantities of the stuck fermentation wine to it. Good results may also be obtained by pitching pomace and juice from a tank that has just completed fermentation into the stuck tank.

In choosing a yeast strain, the yeasts should certainly be resistant to ethanol. Commercially available *S. cerevisiae* yeasts (formerly *Saccharomyces bayanus*), known for their resistance to ethanol and low probability of producing volatile acidity, are recommended for this purpose.

The volatile acidity of the wine tends to increase during the restarting of a stuck fermentation. This generally occurs when the yeasts encounter unfavorable conditions. Certain yeast strains are more predisposed to forming it than others. The addition of 50 mg of pantothenic acid per hectoliter (not authorized by EU legislation) not only can limit its formation but also can contribute to the disappearance of excess acetic acid.

The temperature for restarting the fermentation must also be considered. A slightly elevated temperature favors cell multiplication, but the antiseptic properties of ethanol increase with temperature. The risk of an increase in volatile acidity also seems to be in function of temperature. For these reasons, the restarting of the fermentation should be carried out at a temperature between 20 and 25°C maximum.

Existing activated yeasts in the winery can also be used to restart a stuck fermentation. If there is a large volume of freshly harvested grapes available at the right moment, they can be added to the tank with the stuck fermentation. This operation, however, is in conflict with the legitimate desire to select and separate cuvées. The practice of adding 5–20% of a medium in full fermentation to a stopped tank should be carried out with prudence. Active yeasts are certainly added but sugar is too. In this situation, the fermentation has sometimes been observed to restart and then stop again, leaving about the same amount of sugar that existed before the operation.

The lees of a tank that has normally completed its fermentation can also be used as a yeast starter to restart a stuck fermentation. Supplemental sugar is not introduced into the medium, but yeast in their death phase are no longer very active. The correct restarting of a fermentation requires the introduction of active yeasts in this alcoholic medium without introducing supplemental sugar. The preparation of a yeast starter using dry yeast gives the most satisfactory results.

In white winemaking, at least when malolactic fermentation is not sought, the wine with a stuck fermentation should be lightly sulfited to protect against bacterial development. The fermentation can then be restarted using a yeast starter prepared according to the preceding instructions.

Many possible aids helping to restart a stuck fermentation have been proposed. The addition of ammonium salts does not raise any concerns, but no appreciable improvement of the second fermentation has been observed. The addition of ammonium sulfate should be limited to 5 g/hl due to the limited use of nitrogen by yeasts.

Active charcoal has also been used for a long time to reactivate fermentations (10–20 g/hl). Such an addition is hardly conceivable in red wines (owing to color loss), but its effectiveness for stuck fermentations in white wines is recognized. It works by eliminating yeast inhibitors (fatty acids) (Section 3.6.2).

The addition of yeast hulls is certainly the most effective way of helping to restart a stuck fermentation, although less so than in preventing fermentation from stopping in the first place (Section 3.6.2). They can be added to the yeast starter preparation or directly to the medium with the stuck fermentation (20–30 g/hl).

In results of an experiment given in Table 3.16, the first fermentation of a must initially containing 250 g of sugar per liter stops at 67 g of non-fermented sugar per liter. The second fermentation is conducted after an inoculation at 10^6 cells of *S. cerevisiae* per milliliter, without the addition of yeast hulls in the control sample and with an addition of 0.5 g/l in the test sample 24 hours later. This addition permits a complete fermentation in 36 days, which is not possible in the control sample.

A massive addition of yeast hulls combined with an inoculation of active yeast can result in sensory changes for light wines such as whites and rosés. Doses between 20 and 30 g/hl (maximum) are therefore recommended.

In white as well as red winemaking, the restarting of a stuck fermentation should be closely monitored, especially by measuring volatile acidity to ensure that the alcoholic fermentation is pure. The smallest increase in volatile acidity represents a bacterial contamination, which should absolutely be avoided. A judicious sulfiting should prevent contamination, even though it definitely slows the fermentation. Yet if the doses are adapted to the situation (3–5 g/hl), the fermentation will not be definitively compromised; it will restart after inoculating. However, it must be enough to prevent all bacterial development before the complete depletion of sugars, even though its addition can make malolactic fermentation more difficult.

Tanks with stuck fermentations must be restarted as soon as possible. In the middle of winter, this operation can become impossible and in these situations, it is preferable to wait until the following spring, when fermentation may restart spontaneously.

References

Aerny J. (1996) Rev. Suisse Arboric. Hortic., 28, 3, 161.

Andreasen A.A. and Stier T.J.B. (1953) J. Cell. Comp. Physiol., 41, 23.

Bataillon M., Rico A., Sablayrolles J.M., Salmon J.M. and Barre P. (1996) J. Ferm. Bioeng., 82, 101.

Bell S. and Henschke P.A. (2005) Aust. J. Grape Wine Res., 11, 242.

Beltran G., Novo M., Rozès N., Mas A. and Guillamon J.M. (2004) FEMS Yeast Res., 4, 625.

Bely M., Sablayrolles J.M. and Barre P. (1990) J. Ferment. Bioeng., 70, 246.

Bely M., Salmon J.M. and Barre P. (1994) J. Inst. Brew., 100, 279.

Blateyron L. and Sablayrolles J.M. (2001) J. Biosc. Bioeng., 91, 184.

Bouix M., Busson C., Charpentier M., Leveau J.Y. and Dutertre B. (1997) Œno One, 31, 1, 11.

Bovée J.P., Blouin J., Marion J.M. and Strehaiano P. (1990) In *Actualités Oenologiques*, Vol. 89 (Eds P. Ribéreau-Gayon and A. Lonvaud. Dunod, Paris.

Bréchot P., Chauvet J., Dupuy P., Crosson M. and Rabatu U. (1971) C.R. Acad. Sci., 272, 890.

Cantarelli C. (1989) Factors affecting the behaviour of yeast in wine fermentation. In *Biotechnology Application in Beverage Production*, pp. 127–151 (Eds C. Cantarelli and G. Lanzarem). Elsevier Applied Sciences, London and New York.

Choné X. (2000) *Contribution à l'étude des terroirs bordelais: étude de l'influence des déficits hydriques modérés et de l'alimentation en azote sur le potentiel aromatique des raisins de Vitis vinifera, cv. sauvignon blanc*, Thèse de doctorat, Université Victor Segalen Bordeaux 2.

Delfini C. and Costa A. (1993) Am. J. Enol. Vitic., 44, 1, 86.

Delfini C., Conterno L., Giacosa D., Cocito C., Ravaglia S. and Bardi L. (1992) Vitic. Enol. Sci., 47, 69.

Dubourdieu D., Ollivier C. and Boidron J.N. (1986) Œno One, 20, 1, 53.

El Halaoui N., Ricque D. and Corrieu, G. (1987) Sci. Aliments, 7, 241.

Flanzy C. (1998) *OEnologie. Fondements Scientifiques et Technologiques*. Tec & Doc, Lavoisier, Paris.

Fleet G.H. (1992) *Wine Microbiology and Biotechnology*. Harwood Academic Publishers, Chur.

Fleet G.H. and Heard G.M. (1992) Yeasts: growth during fermentation. In *Wine Microbiology and Biotechnology*, pp. 27–54 (Ed. G.H. Fleet). Harwood Academic Publishers, Chur.

Geneix C., Lafon-Lafourcade S. and Ribéreau-Gayon P. (1983) C.R. Acad. Sci., Série III, 296, 943.

Grenier P., Feuilloley P. and Sablayrolles J.M. (1990) In *Actualités Oenologiques*, Vol. 89 (Eds P. Ribéreau-Gayon and A. Lonvaud. Dunod, Paris.

Guinand G. (1989) *Groupe d'experts Technologie du vin*. Vigne et Vin Publications Internationales, Paris.

Hatzidimitriou E., Darriet P., Bertrand A. and Dubourdieu D. (1997) Œno One, 31, 1, 51.

Henschke P.A. and Jiranek V. (1992) Yeast: metabolism of nitrogen compounds. In *Wine Microbiology and Biotechnology*, pp. 77–164 (Ed. G.H. Fleet). Harwood Academic Publishers, Chur.

Ingledew W.M. and Kunkee R. (1985) Am. J. Enol. Vit., 36, 65.

Jiranek V., Langridge P. and Henschke P.A. (1995) Am. J. Enol. Vitic., 46, 75.

Julien A., Roustan J.L., Dulau L. and Sablayrolles J.M. (2000) Am. J. Viti. Oenol., 51, 3, 215.

Lafon-Lafourcade S. (1983) Wine and brandy. In *Biotechnology*, Vol. 5 (Eds H.J. Rehm and G. Read). Verlag Chemie, Weinheim.

Lafon-Lafourcade S. and Joyeux A. (1979) Œno One, 13, 4, 295.

Lafon-Lafourcade S., Dubourdieu D., Hadjinicolaou D. and Ribéreau-Gayon P. (1980) Œno One, 14, 2, 127.

Lafon-Lafourcade S., Geneix C. and Ribéreau-Gayon P. (1984) Appl. Environ. Microbiol., 47, 6, 1246.

Larue F., Lafon-Lafourcade S. and Ribéreau-Gayon P. (1980) Appl. Environ. Microbiol., 39, 4, 808.

Larue, F., Lafon-Lafourcade, S. and Ribéreau-Gayon, P. (1982) C.R. Acad. Sci., 294, 587–590.

Larue F., Couralet M. and Ribéreau-Gayon P. (1987) *Compte-rendu des activités de recherches (1984–1986) de l'Institut d'Enologie, Université de Bordeaux II*. Université de Bordeaux, Bordeaux.

Lorenzini F. (1996) Rev. Suisse Arboric. Hortic., 28, 3, 169.

Masneuf I., Murat M.L., Choné X. and Dubourdieu D. (2000) Vitis, 249, 41.

Müller-Thurgau M. (1884), *Der Generalversamm. des Deutschen Weinbau-Ver. Ber., 50–57, Geisenheim*.

Munoz E. and Ingledew W.M. (1989) Am. J. Enol. Vitic., 40, 1, 61.

Ollivier C., Stonestreet T., Larue F. and Dubourdieu D. (1987) Œno One, 21, 59.

Ough C.S. (1964) Am. J. Enol. Vitic., 15, 4, 167.

Ough C.S. (1966) Am. J. Enol. Vitic., 17, 1, 20–74.

Ough C.S. and Groat M.L. (1978) Appl. Environ. Microbiol., 35, 881.

Pavlenko N.M. (1989) *Groupe d'experts Technologie du vin*. Office International de la Vigne et du Vin, Paris.

Ribéreau-Gayon P. (1999) Œno One, 33, 1, 39.

Ribéreau-Gayon J., Peynaud E. and Lafourcade S. (1951) Ind. Agri. Alim., 68, 141.

Ribéreau-Gayon J., Peynaud E., Ribéreau-Gayon P. and Sudraud P. (1975a) *Sciences et Techniques du Vin. Vol 2: Caractères des Vins, Maturation du raisin, Levures et bactéries*. Dunod, Paris.

Ribéreau-Gayon P., Lafon-Lafourcade S. and Bertrand A. (1975b) Œno One, 2, 117.

Ribéreau-Gayon J., Peynaud E., Ribéreau-Gayon P. and Sudraud P. (1976) *Sciences et Techniques du Vin, Vol. 3: Vinifications. Transformations du vin*. Dunod, Paris.

Ribéreau-Gayon P., Lafon-Lafourcade S., Dubourdieu D., Lucmaret V. and Larue F. (1979) C.R. Acad. Sci., Série D, 289, 491.

Rozès N., Cuzange B., Larue F. and Ribéreau-Gayon P. (1988) Œno One, 22, 2, 163.

Sablayrolles J.M. and Barre P. (1986) Sci. Alim., 6, 177 and 373.

Sablayrolles J.M. and Barre P. (1989) Sci. Alim., 9, 239.

Sablayrolles J.M. and Blateyron L. (2001) Bull. OIV, 845–846, 464.

Sablayrolles J.M., Grenier P. and Corrieu G. (1990) In *Actualités Oenologiques*, Vol. 89 (Eds P. Ribéreau-Gayon and A. Lonvaud. Dunod, Paris.

Sablayrolles J.M., Dubois C., Manginot C., Roustan J.L. and Barre P. (1996a) J. Ferm. Bioeng., 82, 4, 377.

Sablayrolles J.M., Salmon J.M. and Barre P. (1996b) Progrès Agric. Vitic., 113, 339.

Salmon J.M., Vincent O., Mauricio J.C., Bely M. and Barre P. (1993) Am. J. Enol. Vitic., 44, 1, 58.

CHAPTER 4

Lactic Acid Bacteria

4.1 The Different Components of the Bacteria Cell
4.2 Taxonomy of Lactic Acid Bacteria
4.3 Identification of Lactic Acid Bacteria
4.4 The *Oenococcus oeni* Species

Lactic acid bacteria are present in all grape musts, in almost all wines during the winemaking process, and sometimes in wines after several years of bottle aging. Depending on the wine's characteristics and on the stage of the winemaking process, the species and strains encountered are variable and more or less abundant. Environmental conditions determine their ability to multiply. When they develop, they metabolize numerous substrates. Lactic acid bacteria therefore play an important role in the transformation of grape must into wine. All lactic acid bacteria have a relatively similar cellular organization, yet their metabolic characteristics are diverse. Their impact on wine quality thus depends on the species and strains present, on the substrates they transform, and on the stage at which they conduct these transformations. Many analytical methods based on their different phenotypic or genotypic differences have been developed to enable their taxonomic classification and their identification in musts and wines.

4.1 The Different Components of the Bacteria Cell

Bacteria are prokaryotic cells with a relatively simple organization. They can be distinguished from eukaryotes (to which yeast belong) by their small size and a lack of a nuclear membrane delimiting a true nucleus.

Handbook of Enology, Volume 1: The Microbiology of Wine and Vinifications, Third Edition.
Pascal Ribéreau-Gayon, Denis Dubourdieu, Bernard B. Donèche and Aline A. Lonvaud.
© 2021 John Wiley & Sons Ltd. Published 2021 by John Wiley & Sons Ltd.

The cellular organization of all bacteria is very similar. It can be divided into three principal elements (Figure 4.1):

- Cell envelopes, including the plasma membrane, which delimits the inside of the cell and separates it from the extracellular medium, and the wall, which is found around the outside of the cell. In certain cases, an additional layer of polysaccharides and/or proteins may be observed on the wall surface.
- The cytoplasm, which includes all the cellular content delimited by the plasma membrane.
- The nucleoid, which is not a true nucleus, but rather a disseminated zone of the cytoplasm. The genetic material is concentrated there.

4.1.1 The Cell Wall

There are two very different types of walls in bacteria: those of Gram-positive bacteria (Section 4.3.2), such as lactic acid bacteria, and those of Gram-negative bacteria, such as acetic acid bacteria (Chapter 7). The cell wall of Gram-positive bacteria is a macromolecule called peptidoglycan, or murein, that is only found in prokaryotes (Figure 4.2). The peptidoglycan is formed of polysaccharide chains aligned parallel with each other and linked together by peptide bonds. These give it great elasticity. This network surrounds the bacterial cell and piles up in several layers, reaching a total thickness of several dozen nanometers. The polysaccharide chains are composed of two glucose derivatives: N-acetylmuramic acid and N-acetylglucosamine. They alternate along the entire length of the chain, linked by β-(1–4) glycoside bonds (Figure 4.2). Note that these bonds can be hydrolyzed by enzymes such as lysozyme or mutanolysin, which causes bacterial cell lysis.

Peptide bonds linking polysaccharide chains are grafted on N-acetylmuramic acid. In general, these are chains of four or five amino acids, often including L- and/or D-alanine, D-glutamic acid, L-lysine, and diaminopimelic acid. These are directly bound together or through other amino acids (Figure 4.3). The peptide chains vary depending on the species of the bacteria. The sequence of their amino acids can be used in taxonomy.

The cell walls of lactic acid bacteria, like those of nearly all Gram-positive bacteria, also contain teichoic and lipoteichoic acids. They represent up to 50% of the wall's dry weight. These are polymers of ribitol phosphate or glycerol phosphate connected by phosphodiester bonds, on which amino acids and sugars can be fixed. They cross through the peptidoglycan layers, with which they create covalent bonds, and thus contribute to stabilizing the wall structure. In addition, lipoteichoic acids contain a glycolipid by which they attach themselves to the external layer of the plasma membrane. This ensures cohesion between the wall and the plasma membrane. They pass through the peptidoglycan and are at the surface of the cell wall, acting as the antigen sites of bacteria. The proportion of peptidoglycans, teichoic acids, and lipoteichoic acids varies depending on the species and also the phase of the cell development cycle.

The cell wall is rigid and gives the cell its form: round for cocci and elongated for bacilli. It permits the cell to resist very high internal osmotic pressures (up to 20 bar). The culture of cells in the presence of penicillin, inhibiting the synthesis of the cell wall, leads to the formation of protoplasts (cells without walls): the latter are viable only in isotonic media. Similarly, lysozyme hydrolyzes the glycoside bonds of peptidoglycan, provoking the bursting of the cell in a hypotonic medium.

The wall is permeable to water, mineral ions, substrates, and metabolic products, which diffuse freely across the peptidoglycan layers. Substrates present in the environment may thus diffuse in this structure up to the plasma membrane surface.

The wall also contains numerous enzymes, participating in proper cell

4.1 | The Different Components of the Bacteria Cell

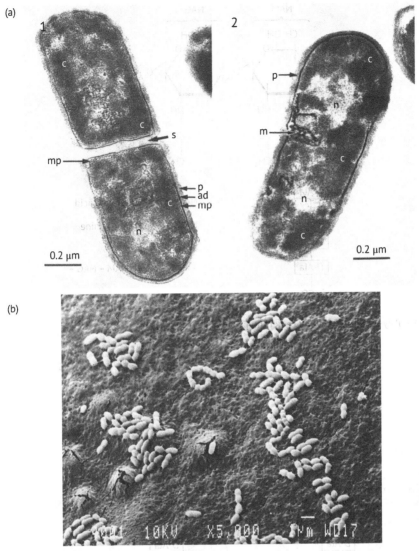

FIGURE 4.1 Lactic acid bacteria isolated in wine under a scanning electron microscope (Department of Electron Microscopy, Université de Bordeaux I). (a) Photograph of *L. plantarum* cells (transmission electron microscope, Lonvaud, 1975). ad, dense base of wall; c, cytoplasm; mp, plasma membrane; p, cell wall; s, septum; m, mesosome; n, nucleus; 1, cell at end of division; 2, growing cell. (b) Photograph of *O. oeni* (scanning electron microscope).

operation. For example, these are proteases and glycosidases, which hydrolyze complex substrates present in the environment and release small peptides, amino acids, or sugars. Those compounds may then be transported into the cell.

The cell wall surface is sometimes covered with a protein layer (S-layer). Its presence depends on species, strains, and conditions of the medium. The study of this S-layer in wine bacteria has not yet been attempted.

Finally, polysaccharides are often present on the cell wall surface. Their nature varies, and they are more or less abundant, according to environmental conditions, species, and strains. In enology, *Pediococcus damnosus* or *Pediococcus parvulus* gives the best example of this phenomenon. Under certain conditions, strains of this

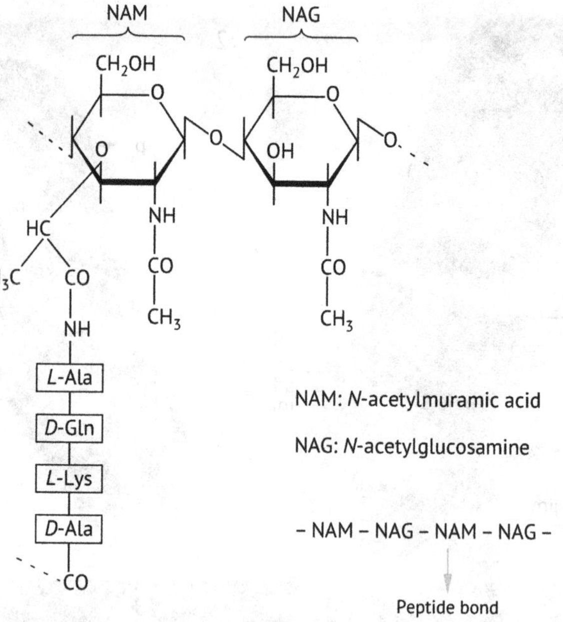

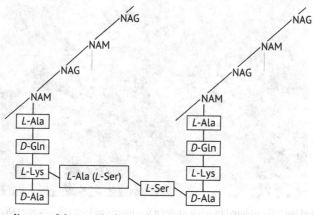

FIGURE 4.2 Polysaccharide chain of bacterium peptidoglycan.

NAM: *N*-acetylmuramic acid

NAG: *N*-acetylglucosamine

FIGURE 4.3 Structure diagram of the peptidoglycan of *O. oeni* bacteria.

species synthesize significant quantities of polysaccharides that make the wine viscous (ropiness) (Section 5.4.4). These cells are easily recognized under an optical microscope by the refringent halo that surrounds them. Nevertheless, most wine bacteria produce more or less significant quantities of other types of surface polysaccharides, which have no significant impact on wine quality. They may give bacteria a certain resistance with regard to the environment or surface adhesion capabilities (Bastard *et al.*, 2016).

4.1.2 The Plasma Membrane

The plasma membrane of lactic acid bacteria has the classic structure of all biological membranes: a lipid bilayer creating a central hydrophobic zone (Section 1.3.1, Figure 1.6). The proteins are more or less tightly attached to it. Among them, the water-soluble proteins are only attached to the surface by ionic or hydrogen bonds (peripheral proteins, 30% of the proteins). The others are lodged in the membrane by hydrophobic bonds (integral proteins). The

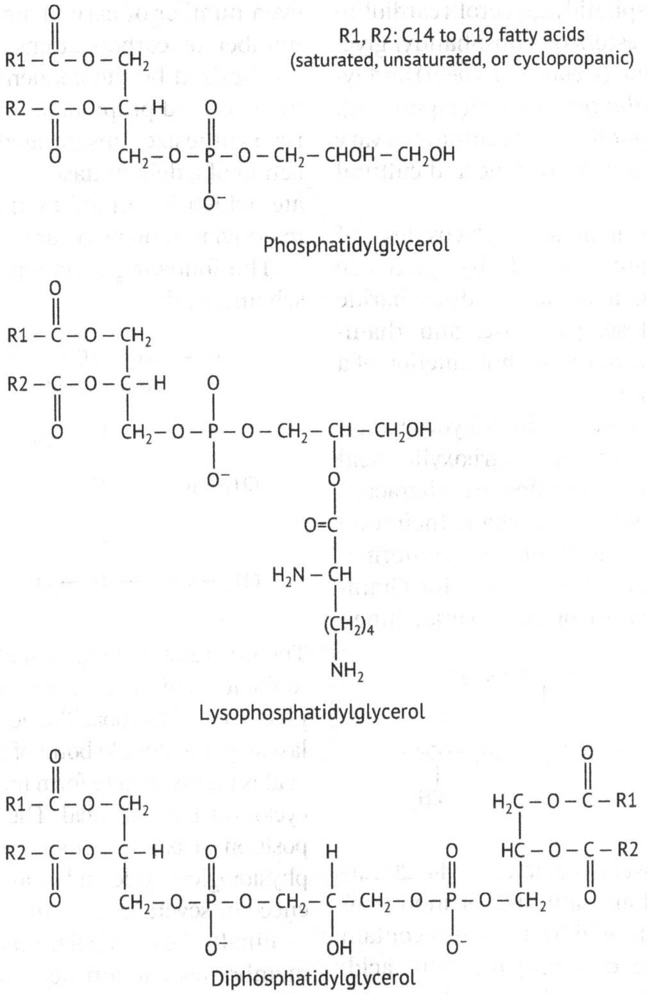

Phosphatidylglycerol

R1, R2: C14 to C19 fatty acids
(saturated, unsaturated, or cyclopropanic)

$R1-C(=O)-O-CH_2$
$R2-C(=O)-O-CH-H$
$CH_2-O-P(=O)(O^-)-O-CH_2-CHOH-CH_2OH$

Lysophosphatidylglycerol

$R1-C(=O)-O-CH_2$
$R2-C(=O)-O-CH-H$
$CH_2-O-P(=O)(O^-)-O-CH_2-CH(O-C(=O)-CH(NH_2)-(CH_2)_4-NH_2)-CH_2OH$

Diphosphatidylglycerol

$R1-C(=O)-O-CH_2$
$R2-C(=O)-O-CH-H$
$CH_2-O-P(=O)(O^-)-O-CH_2-CH(OH)-CH_2-O-P(=O)(O^-)-O-CH_2-CH(O-C(=O)-R2)-CH_2-O-C(=O)-R1$

FIGURE 4.4 Chemical formulae of some membrane phospholipids.

peripheral proteins have a certain mobility in the periplasmic space between the peptidoglycan and the membrane, whereas the integral proteins are almost immobile. Some protrude from the membrane, while others only appear on the surface. Hydrophobic bonds between aliphatic lipid chains and protein chains create the framework of the membrane. The high number of these bonds ensures the solidity of this structure, but there are no covalent bonds and so the framework created remains fluid. The biochemical functions ensured by the membrane depend on this fluidity, i.e. lipid–protein interactions. The structure can be destroyed by organic solvents and detergents. It is also disturbed by wine components. Finally, on the surface, the hydrophilic parts of the lipids and the ionized groups of the proteins establish ionic bonds between themselves.

Membrane lipids represent nearly all (95–99%) lipids in the bacteria cell. They essentially include phospholipids and glycolipids. Phospholipids are most abundant; they consist of a glycerol molecule, which has a primary alcohol function, and a secondary alcohol function esterified by fatty acids. The other primary function is esterified by phosphoric acid, which is in turn esterified by glycerol, forming phosphatidyl glycerol. Lactic acid bacteria also contain diphosphatidylglycerol (cardiolipin) and amino esters of phosphatidyl glycerol with alanine (*Oenococcus oeni*) and lysine (*Lactobacillus plantarum*) (Figure 4.4). Bacterial phospholipid concentrations vary according to the growth stage and cultural conditions.

Glycolipids—generally glycosides of diglycerides—are formed by glycoside bonds between a mono- or disaccharide (glucose, fructose, galactose, and rhamnose) and the primary alcohol function of a diglyceride (Figure 4.5).

Fatty acids possess a long hydrocarbon chain and a terminal carboxylic acid function. These molecules are characterized by the length of their chain, their level of unsaturation, the *cis* or *trans* conformation of the double bonds, and (for Gram-positive bacteria) *iso* or *anteiso* branching.

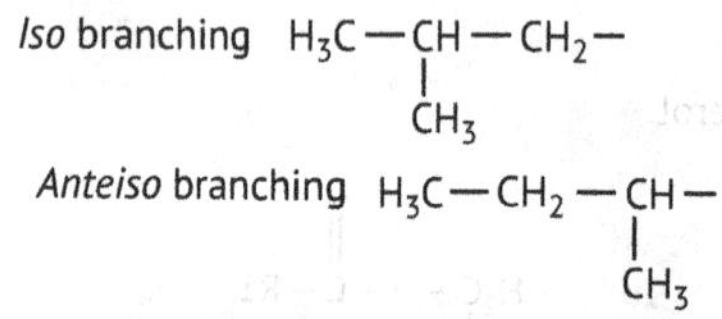

$$\text{Iso branching} \quad H_3C-CH-CH_2-$$
$$\qquad\qquad\qquad\quad |$$
$$\qquad\qquad\qquad\quad CH_3$$

$$\text{Anteiso branching} \quad H_3C-CH_2-CH-$$
$$\qquad\qquad\qquad\qquad\qquad\quad |$$
$$\qquad\qquad\qquad\qquad\qquad\quad CH_3$$

In bacteria, most fatty acids have 14–20 carbon atoms and are saturated or mono-unsaturated. Lactic acid bacteria also contain a characteristic cyclopropane fatty acid: lactobacillic acid (*cis*-11,12-methylene-octodecanoic acid). Table 4.1 lists the principal fatty acids of lactic acid bacteria found in wine—notably *O. oeni* (Lonvaud-Funel and Desens, 1990). Malonyl-CoA and acetate condense to form fatty acids with an even number of carbon atoms. For an odd number of carbon atoms, fatty acids are synthesized by the condensation of malonyl-CoA and propionate. Anaerobic bacteria synthesize unsaturated acids by the action of a dehydratase on hydroxydecanoate, which is formed by the addition of a malonyl unit on an octanoic acid molecule.

The following reactions are given very schematically:

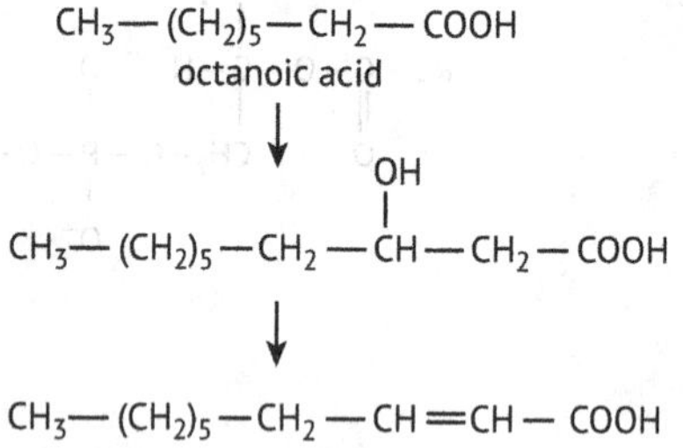

$$CH_3-(CH_2)_5-CH_2-COOH$$
$$\text{octanoic acid}$$
$$\downarrow$$
$$\qquad\qquad\qquad OH$$
$$\qquad\qquad\qquad |$$
$$CH_3-(CH_2)_5-CH_2-CH-CH_2-COOH$$
$$\downarrow$$
$$CH_3-(CH_2)_5-CH_2-CH=CH-COOH$$

The progressive elongation of this acid leads to the formation of *cis*-vaccenic acid (C18), a precursor of lactobacillic acid (C19). In this last step, the double bond of the unsaturated acid is methylated to form the corresponding cyclopropane fatty acid. The fatty acid composition of bacteria lipids varies during the physiological cycle and is also strongly influenced by several environmental factors.

Finally, besides polar lipids, the bacteria membranes contain neutral lipids, analogous to sterols in eukaryotes. These triterpene and pentacyclic molecules are called hopanoids. They are formed by the cyclization of squalene in an anaerobic process. They have not been clearly identified in lactic acid bacteria.

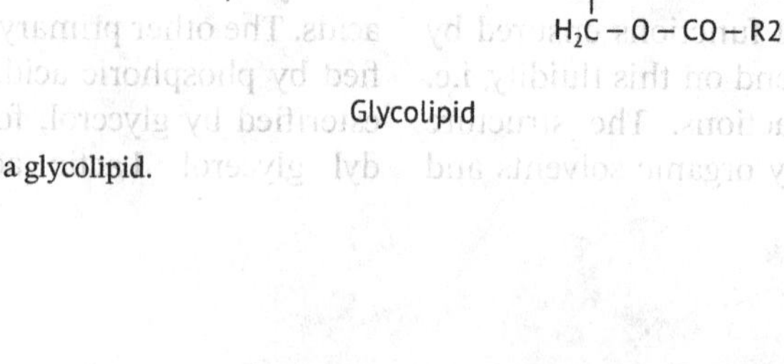

FIGURE 4.5 Formula of a glycolipid.

TABLE 4.1
Principal Fatty Acids of Lactic Acid Bacteria

		Chain
Myristic acid	Tetradecanoic	C14:0
Palmitic acid	Hexadecanoic	C16:0
Palmitoleic acid	cis-9-Hexadecanoic	C16:1 Δ 9
Stearic acid	Octadecanoic	C18:0
Oleic acid	cis-9-Octadecanoic	C18:1 Δ 9
cis-Vaccenic acid	cis-11-Octadecanoic	C18:1 Δ 11
Dehydrosterculic acid[a]	cis-9-10-Methylene octadecanoic	C19 cyc-9
Lactobacillic acid[a]	cis-11-12-Methylene octadecanoic	C19 cyc-11

[a] These two acids contain a cyclopropane ring

$$-CH-CH-$$
$$\backslash \ / $$
$$CH_2$$

The plasma membrane is even more vital to bacteria than the cell wall. Numerous proteins in the membrane ensure essential enzymatic functions. For example, these include membrane transporters, which ensure the transport of substrates and metabolites, or ATP synthase, which ensures the production or synthesis of ATP. It should be noted, however, that the plasma membrane of lactic acid bacteria does not contain all the components of the respiratory chain. This explains why these bacteria do not have a respiratory metabolism, except in special cases.

The plasma membrane acts as a selective barrier. This selective permeability controls the entry and exit of ions and compounds participating in cell metabolism. The lipid bilayer is permeable to ions and to most compounds, while membrane transporters, ion channels, and other membrane proteins ensure the selection in and out of the cell. Selective permeability is necessary for multiple cell functions. For example, it controls the entry of protons, which is essential for the cell to maintain an optimum intracellular pH (close to neutral) for correct operation of numerous reactions of cell metabolism. It also helps maintain an electrochemical potential difference and a proton gradient on either side of the membrane. The two components of the proton-motor force are used by ATP synthase to produce ATP.

The plasma membrane constitutes a barrier whose optimal functioning is guaranteed by fluidity. This fluidity determines the specific activity of the proteins according to the lipid environment, but it must be under control for the membrane to remain an effective barrier between the cytoplasm and the outside environment. During the cell growth cycle and in response to multiple external parameters such as temperature, pH, and the presence of toxic substances (ethanol), the cell manages to modify membrane composition to adapt to and resist environmental effects. The physical properties of the membrane are maintained at least as long as the stress factor remains within certain limits. The mechanisms put into play act together on the same properties. They affect the average length and the unsaturation, branching, and cyclization level of fatty acid chains; the proportion of neutral and polar lipids; and the quantity of proteins. In this way, from the growth phase until the stationary phase, cis-vaccenic acid diminishes greatly to the point where it represents less than 10% of the total fatty acids, whereas lactobacillic acid attains a proportion of 55% in *O. oeni*, *L. plantarum*, and *P. damnosus* (Lonvaud-Funel and Desens, 1990).

The effect of temperature on membrane composition is one of the most understood effects. At low temperatures, to compensate for branching, the fatty acid unsaturation rate increases, as does the proportion of acids with branched chains. At the same time, the length of the chains decreases. In this manner, palmitic acid (C16) increases in concentration, and *cis*-vaccenic and lactobacillic acids decrease in *O. oeni* and *L. plantarum* when the temperature of the culture increases from 25 to 30°C. The introduction of a methyl group, the formation of a cyclopropane ring, has the same effect on the physical properties of bacteria as a double bond. The inverse phenomena occur when the culture temperature is higher. The unsaturated fatty acids are less abundant. Neutral lipids also participate in cell adaptation to the medium by increasing membrane viscosity.

The presence of ethanol in the medium provokes significant modifications in membrane structure. It exerts a detergent effect by intercalating in the hydrophobic zone of the membrane, whose polarity increases as a result. The fluidity is increased, and the proteins are denatured. In general, an increase in the unsaturated/saturated fatty acid ratio is observed. In *O. oeni*, this ratio increases from 0.4 to 2.1, when bacteria are cultivated in the presence of 9% ethanol. The results are the same for the species *Lactobacillus hilgardii*, whose strains, like those of *O. oeni*, are capable of growing better than other species in an alcoholic medium (Desens, 1989).

The membrane proteins also participate in cell response to an environmental change. The stress proteins in microorganisms are becoming better known. Their synthesis is increased, for example, by temperature, acidity, or ethanol concentration. Certain proteins also change when the cell enters the stationary phase. Several families of these proteins have been defined, and the specific functions of some of them have been identified. Their overexpression in the cell is related to a better resistance to stress factors. Their induction by heat shock protects the cell not only against the toxic effect of heat but also against the effect of other factors, such as ethanol and acidity. In certain wine lactic acid bacteria, especially *O. oeni*, these proteins exist, but their precise role is not known. Their synthesis is increased when wine is added to their culture medium or when the cells are directly inoculated into wine. The concentrations found with *O. oeni* are up to five times higher than with other species (Garbay, 1994). Among these, two proteins have been identified and coded by the *Omr* A and *Fts* H genes (Bourdineaud et al., 2003, 2004). The stress protein Lo18 may play a role in maintaining the *O. oeni* membrane intact in an alcohol environment (Delmas et al., 2001; Maitre et al., 2014).

4.1.3 The Cytoplasm

The cytoplasm contains the main elements for cell operation: enzymes, nuclear material, and sometimes reserve substances. The entire metabolism—both degradation reactions (catabolism) and synthesis reactions (anabolism)—is carried out in a programmed manner according to exchanges with the external environment, to produce the energy necessary for cell growth.

Coded by the genome, the cytoplasmic proteins may vary from one strain to another within a given species. This translates into phenotypic differences. For example, certain strains have enzymes that metabolize a sugar, organic acid, or another specific substrate. Other strains are incapable of using these if they do not have the suitable enzymes.

Moreover, the synthesis level of most proteins varies depending on environmental conditions. Such is the case of stress proteins, whose synthesis is induced or increased by changes in environmental conditions. This is also the case of numerous metabolism enzymes, whose synthesis is regulated by the presence of specific substrates.

Cytoplasmic granulations can be revealed by specific staining techniques. They are insoluble reserve substances of an organic nature: polymers of glucose or of the polyester of β-hydroxybutyric acid. These reserve substances accumulate in the event of a nitrogen deficiency, when a source of carbon is still present. Inclusions of volutin (a polymer of insoluble, inorganic phosphate) are characteristic in lactic acid bacteria, especially certain strictly homofermentative species of the *Lactobacillus* genus. Volutin comprises a phosphate reserve available for the synthesis of phosphorylated molecules such as nucleic acids.

Under the transmission electron microscope, the interior of the bacterial cell appears granular. This is due to the ribosomes, which are essential players in protein synthesis. They ensure, along with the tRNA, the translation of the genetic code. The ribosomes consist of two parts characterized by their sedimentation speed, expressed in Svedberg values (S). These two subunits are different in size: 30S and 50S in prokaryotes. The assembled ribosome has a sedimentation constant of 70S. The 30S subunit contains a 16S ribosomal RNA molecule (1,500 nucleotides) and 21 different protein molecules; its molecular mass is 900 kDa. The larger 50S subunit contains two ribosomal RNA molecules, a 23S and a 5S molecule (2,900 and 120 nucleotides, respectively). It also contains 35 proteins and has a molecular mass of 1,600 kDa. The genes encoding proteins and ribosomal RNA are known in many bacterial species. They are organized in operons, ensuring the control of the synthesis of ribosomal components. The operon of genes encoding the three rRNA molecules has the following structure:

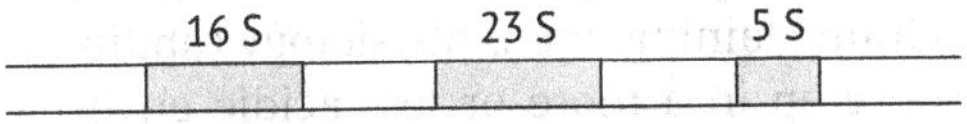

```
    16 S        23 S         5 S
──────────────────────────────────
    ▓▓▓        ▓▓▓▓▓         ▓
──────────────────────────────────
```

The nucleotide sequences of these genes, especially those of 16S rRNA, are known in all currently described species and are used to identify and differentiate bacterial species.

4.1.4 The Genetic Material

The genetic material of lactic acid bacteria consists of a single circular chromosome of double-stranded DNA. It forms a more or less diffuse region of the cytoplasm, called the nucleoid. Its size varies depending on the species. In *L. plantarum*, its length is about 3,300 kilobase pairs (kb). It is much smaller in *O. oeni* (about 1,800 kb) and *P. damnosus* (2,500 kb). The chromosome carries the essential genetic information of a cell. The average size of genes is about 1 kb, and there are few non-coding regions in bacterial chromosomes. Thus, a chromosome such as that of *O. oeni* has about 1,800 genes, most of which code for proteins. The others code for rRNA or tRNA. All genes essential to the cell's survival are carried by the chromosome. Nevertheless, bacteria often have other genetic components: plasmids. They are smaller than chromosomes: about 3 kb for the smaller ones and over 200 kb for the larger ones (equivalent of three to over 200 genes). Unlike the chromosome, plasmids contain genes with functions not vital to the cell. However, these genes may improve its survival and growth. For example, plasmids determine functions such as the fermentation of certain sugars, the hydrolysis of proteins, and resistance to phages, antibiotics, heavy metals, etc. Plasmids vary in size and number depending on the species and strain of bacteria. In *O. oeni*, plasmids of 3–20 kb are often identified. Most of the genes identified code for the replication and mobilization of the plasmid (for example, pLo13; Fremaux *et al.*, 1993). This does not grant any special properties to the bacteria hosting them. However, the pOENI-1 plasmid and its variant pOENI-1v2 seem to be linked to a good technological aptitude of strains (Favier *et al.*, 2012).

In wine lactic acid bacteria of the genus *Pediococcus*, a plasmid has been identified as a determinant of exopolysaccharide synthesis. Strains that contain it are responsible for ropiness in wines (Section 5.4.4) (Lonvaud-Funel *et al.*, 1993). This plasmid contains glucosyl transferase, which is responsible for synthesizing the expolysaccharide (Walling, 2003; Dols-Lafargue *et al.*, 2008). Characteristically, plasmids are relatively unstable during cell division. Unlike the chromosome, they cannot be transmitted to one of the daughter cells. Certain conjugative plasmids can naturally transfer from one bacterium to another, through a protein channel, forming between them.

4.1.5 Reproduction of Bacteria

All bacteria reproduce by binary fission (Figure 4.1). A mother cell yields two completely identical daughter cells. Reproduction supposes, on the one hand, division of nuclear material, and on the other hand, synthesis for the construction of new cell envelopes and cytoplasmic elements, in particular ribosomes and enzymes.

The genetic material is transmitted after the duplication of the chromosomal DNA and any plasmids. DNA replication, according to the semiconservative mechanism, leads to the formation of two molecules that are identical to the parent chromosome or plasmid. The replication occurs during virtually the entire cellular cycle in the mesosomes. When it is finished, the fission of the cytoplasm begins.

A septum is formed in the middle of the cell as a result of the synthesis of portions of the membrane and the cell wall. It separates the mother cell little by little into two daughter cells. The genetic material and the other cell components are simultaneously distributed between them. Finally, when the septum is completely formed, the two daughter cells separate. Cell division and genetic material division are not synchronous; replication is quicker. Moreover, a replication cycle can start before cell division is completed. For this reason, bacteria cells in their active growth phase contain more than one chromosome per cell. Bacteria develop numerous strategies to coordinate chromosome replication and cell division. During division, plasmids (much smaller than the chromosome) are not always correctly distributed between the daughter cells after their replication, hence their instability over generations.

4.2 Taxonomy of Lactic Acid Bacteria

4.2.1 Bacterial Species

The objective of taxonomy is to describe and group living organisms, in order to name, classify, and identify them. Classification is done according to species. By convention, species are named in Latin and written in italics with two terms: the genus (starting with an uppercase letter) and the species (all lowercase). For example: *Oenococcus oeni* or *O. oeni* (Stackebrandt et al., 2002).

In general, a biological species is defined as a community of living organisms, recognized by their characters and able to reproduce by creating fertile offspring. However, this definition does not apply to bacteria, which have few morphological differences and no sexual reproduction. Thus, a bacterial species is constituted by its typical strain, and all strains are considered sufficiently close to it. The typical strain is generally one of the first strains of a species to be studied and is one of the best characterized strains.

The criteria used to measure the proximity of strains have evolved as new analytical methods for bacteria appear. The first were phenotypic methods analyzing cell morphology (shape, arrangement, Gram staining, etc.), physiology (ability to develop in a more or less acidic environment, in the presence of alcohol, in the presence of certain substrates, etc.), or their biochemical and immunological characters and the composition of certain cell components. Then, genotypic methods appeared based on DNA composition, properties, or sequence (G+C percentage, DNA–DNA hybridization, polymerase chain reaction [PCR] and derived methods, gene sequencing, etc.). The latter not only differentiate between species with great precision, but also make it possible to classify them according to their phylogenetic proximity.

Nowadays, the method commonly used to determine whether two bacteria belong to the same species is to measure the hybridization percentage of their genomic DNA. The closer the sequences are, the higher this percentage is. Two bacterial strains belong to the same species if the hybridization percentage is greater than 70%. Nevertheless, the subcommittee of the International Code of Nomenclature of Bacteria recommends including many other genotypic, phenotypic, and ecological characteristics for the description of lactic acid bacteria species (Mattarelli *et al.*, 2014).

A bacterial species thus groups together a set of strains that share many characteristics. A strain is a set of bacteria that are impossible to differentiate. The genomic DNA sequence of these clones is strictly identical. For example, these bacteria form a colony in a Petri dish, because they all come from a single mother cell. If one of these bacteria loses a plasmid or is subject to a DNA mutation, it no longer belongs to the same strain.

4.2.2 Classification of Lactic Acid Bacteria in Wine

Lactic acid bacteria belong to a phylum named Firmicutes, which includes Gram-positive bacteria (Section 4.3.2). They are in the Lactobacillales order. The major product of glucose metabolism in these bacteria is lactic acid. The lactic acid bacteria isolated from grape must and wine belong to the Lactobacillaceae family for the *Lactobacillus* and *Pediococcus* genera, and Leuconostocaceae for *Leuconostoc* and *Oenococcus*.

Among the characteristics easily observed during cultures, besides their morphology in coccal or rod-like forms, the homofermentative or heterofermentative character is a decisive factor in their classification. Homofermentative bacteria produce more than 85% lactic acid from glucose. Heterofermentative bacteria produce carbon dioxide, ethanol, and acetic acid in addition to lactic acid.

Among the cocci, the bacteria from the genus *Pediococcus* are homofermenters and those from the genus *Leuconostoc* are heterofermentative. In the genus *Lactobacillus*, bacteria can display both behaviors. They are divided into three groups:

- Strict homofermenters (Group I)
- Facultative heterofermenters (Group II)
- Strict heterofermenters (Group III)

The strictly homofermentative lactobacilli do not ferment pentose and form two molecules of lactic acid from one molecule of glucose (from one and/or the other of the two isomers) via the Embden–Meyerhof pathway (Section 5.2.1).

In facultative heterofermenters (Group II), one glucose molecule, as in the case of Group I, leads to two molecules of lactic acid. However, pentoses are fermented into lactic and acetic acids via the heterofermentative pentose phosphate pathway (Section 5.2.2). The strictly heterofermentative bacteria in Group III do not possess fructose 1,6-bisphosphate aldolase, which is characteristic of the Embden–Meyerhof pathway. They ferment glucose into CO_2, lactic and acetic acids, and ethanol via the pentose phosphate pathway, and pentose into lactic and acetic acids in the same manner as bacteria from Group II.

Table 4.2 lists the lactic acid bacteria most often encountered in grape must and wine. The most frequent genera are therefore divided into facultative and strict heterofermenters for lactobacilli and into homofermenters (*Pediococcus*) and heterofermenters (*Leuconostoc* and *Oenococcus*) for cocci. It is likely that this classification will be modified—on the one hand, due to anticipated progress in the identification of new species in wine, and on the other hand, due to potential reclassifications of lactobacilli in the groups described above.

Bacteria do not have complete respiratory chains. However, certain species have cytochromes and may have a respiratory metabolism when they are cultivated in the presence of heme.

Catalase activity is generally assumed not to exist. Yet several species of bacteria (*Lactobacillus*, *Pediococcus*, and *Oenococcus*) can synthesize a manganese-dependent, non-hemic pseudocatalase. A hemic catalase activity has also been identified in many strains.

The following is a general description of the four genera of lactic acid bacteria in wine:

Leuconostoc and *Oenococcus* genera:

- Spherical or slightly elongated cells, assembled in pairs or small chains; diameter 0.5–0.7 µm, length 0.7–1.2 µm, nonmobile, and non-sporulating.
- Facultative anaerobes.
- Chemoorganotrophs: require a rich medium and fermentable sugars.
- Mesophilic: optimum growth temperature of 20–30°C.
- Metabolites of glucose: CO_2, D-lactic acid, ethanol, and acetic acid.
- Arginine is hydrolyzed by certain strains of *O. oeni*, whereas all *Leuconostoc* species respond negatively to this test.
- G + C percentage from 38 to 44%.
- No teichoic acid.

Pediococcus genus:

- Spherical (never elongated) cells; diameter 1–2 µm; division in two orthogonal planes that leads to the formation of tetrads. No chains, sometimes isolated, nonmobile, and non-sporulating.
- Facultative anaerobes.
- Chemoorganotrophs: require a rich medium for growth and fermentable sugars.
- Metabolites of glucose: DL- or L-lactic acid, no CO_2 gas.
- G + C percentage from 34 to 42%.
- No teichoic acid.

Lactobacillus genus:

- Regular elongated cells. 0.5–1.2 µm by 1.0–10 µm, often long rod-like forms. Some are very small (nearly the same dimensions as *Leuconostoc*). Isolated, assembled in pairs or in variably sized chains, nonmobile, and non-sporulating.
- Facultative anaerobes.
- Chemoorganotrophs: require a rich medium for growth and fermentable sugars.
- Fermentative metabolism: lactic acid represents at least half of the products of the metabolism of glucose. Homofermentative metabolism leads to this molecule alone. Heterofermentative metabolism also produces acetic acid, ethanol, and CO_2.
- G + C percentage from 36 to 47%.
- Many species contain teichoic acids in the cell wall. There are few species frequently isolated from grape must and wine. Others are described, especially among lactobacilli, but they are not very characteristic of the lactic acid bacteria flora of wines because relatively few are present (Table 4.2).

4.3 Identification of Lactic Acid Bacteria

4.3.1 General Principles

The identification of bacteria was first based on their phenotypic traits (Section 4.3.2). In addition to its morphology, which gives little information, a strain was identified essentially by physiological and metabolic characteristics (the substrates and products formed). When more discriminating analytical methods appeared, the chemical composition of microorganisms (fatty acids and proteins) also participated in their identification.

Identification by phenotypic analysis of clones isolated in wine often poses two kinds of problems. First, these clones are difficult to multiply under laboratory conditions. Numerous subcultures are needed

TABLE 4.2

List of Lactic Acid Bacteria Species Isolated in Grape Must and Wine

Lactobacilli	Homofermenters (Group I)	*Lactobacillus delbrueckii, L. mali, Lactobacillus nagelii, Lactobacillus oeni, Lactobacillus vini, Lactobacillus uvarum*
	Facultative heterofermenters (Group II)	*Lactobacillus casei, Lactobacillus curvatus, L. plantarum*
	Strict heterofermenters (Group III)	*Lactobacillus brevis, Lactobacillus buchneri, L. collinoides, Lactobacillus diolivorans, L. fructivorans, L. hilgardii, L. kunkeei*
Cocci	Homofermenters	*Pediococcus damnosus, P. parvulus, P. pentosaceus*
	Heterofermenters	*Oenococcus oeni, L. mesenteroides* subsp. *mesenteroides*

to obtain a sufficient biomass to carry out all of the physiological and metabolic tests. Second, a phenotypic trait, such as the assimilation of a substrate or the formation of a particular product, represents the result of a metabolic chain that depends on the entirety of cell enzymatic activity. For a phenotype to be positive, all of the genes encoding the enzymes of the particular chain must be expressed; the enzymes must also be functional. The induction or repression of enzyme synthesis as well as inhibitions due to certain medium conditions can therefore modify a phenotype.

Since the 1990s, molecular biology tools have become much more prevalent. These have made it possible to measure genotypic characteristics. Identification in terms of the genus, species, and strain has become quicker, more reliable, and more precise (Section 4.3.3).

The DNA composition (composition and sequence) is strictly specific to a strain and invariable. Unlike the phenotype, it is not influenced by culture conditions. It can, however, undergo occasional mutations or potential insertions/eliminations over generations. At the laboratory culture scale, these mutations do not significantly affect the genomic DNA characteristics. Strain identification by genomic analysis therefore appears to be the most reliable approach. Several types of analysis exist that permit different levels of identification: strain, species, and genus.

The general principle consists in looking for similarities between the DNA of the unidentified strain and the DNA of the reference strain. There are several methods based on the properties and structure of the DNA molecule. For example, the study of restriction polymorphism is based on the specific action of restriction enzymes, which hydrolyze the chain at very precise points, determined by the nucleotide sequence. Hybridization is based on the ability of single-strand DNA chains with identical or nearly identical sequences to be reassembled because of the complementarity of their bases. Lastly, it is possible *in vitro* to synthesize a DNA strand that is complementary with another one taken as a template, via nucleotide polymerization. That is the principle of the PCR, which enables the amplification of portions of the genome delimited by markers. These markers are primers (oligonucleotides) that must hybridize with the template DNA, at specific sites of the genome sequence, for amplification to start. PCR thus calls on DNA properties: the ability to be denatured, hybridized, and synthesized by nucleotide polymerization, while complying with base sequencing. It is used routinely in accordance with different protocols to differentiate, identify, or quantify species or strains with great precision. DNA sequencing has become a method of choice to differentiate species and strains. Initially limited to one gene, in general 16S RNA, whose sequence differentiates between species, it has been extended to entire genomes with the arrival of new-generation sequencing methods. In a few hours, these methods produce millions of sequences covering entire genomes. Back in the early 2000s, it would take months or years to obtain the same result using the traditional sequencing method (Sanger method). Simultaneously, spectacular progress in bioinformatics has provided the necessary tools for processing masses of data. Genome sequences now help differentiate and compare species and strains with unrivalled precision.

4.3.2 Phenotypic Analysis

A bacterium's phenotype is all of its observable characteristics. They directly result from the expression of genes, but also depend on the environment in which the bacterium is. Phenotypic analysis encompasses morphology, the assimilation of diverse substrates, the nature of metabolites, and the composition of some cell components.

Microscopic observation Morphological observations of lactic acid bacteria can be made with fresh cells, i.e. by placing a liquid sample containing bacteria on a microscope slide. This often helps differentiate between cocci (*Leuconostoc*, *Pediococcus*, and *Oenococcus* genera) and bacilli (*Lactobacillus*) as well as determine cell arrangement (pairs, tetrads, and small chains). However, this never makes it possible to identify species. Moreover, this observation is more difficult when the sample contains a blend of various lactic acid bacteria, as is the case for wines.

Microscopic observation can be coupled with the Gram stain test. After the bacteria are placed and fixed on the slide, the preparation is dipped first in a crystal violet solution, in an iodine solution, then in an alcohol–acetone solution, and finally in a safranin solution. The cell wall of Gram-positive lactic acid bacteria is not altered by the organic solvent: these bacteria retain the violet color. Conversely, Gram-negative bacteria are rose-colored.

Analysis of metabolism Lactic acid bacteria have metabolic characteristics that may vary depending on genera and species. The homofermentative or heterofermentative character is first determined. The strain to be identified is cultivated in a medium with glucose as the energy source. After cell growth, the metabolites are characterized or measured. A release of CO_2 manifests the heterofermentative character of the strain. This result is regularly confirmed by measuring acetic acid and ethanol concentrations. Their presence is also proof of a heterofermentative pathway for glucose. Conversely, the exclusive formation of lactic acid attests to a homofermentative character. Under these culture conditions, facultative heterofermentative bacilli (for example, *L. plantarum*) have a homofermentative metabolism with respect to glucose.

During the same test, it is interesting to determine the stereochemical nature of lactic acid formed from glucose. This analysis makes use of an enzymatic process. The two stereoisomers of lactic acid (L and D) are analyzed separately. This result is particularly adapted to the identification of heterofermentative cocci (*O. oeni* and *Leuconostoc mesenteroides*), which only form the D isomer, and of *Lactobacillus casei* and *Lactobacillus kunkeei*, which only form L-lactic acid. Often, other species form a blend of L and D isomers.

These initial investigations enable bacterial identification at the genus level: *Lactobacillus* by morphology, and *Pediococcus*, *Oenococcus*, and *Leuconostoc* by morphology and determination of their

TABLE 4.3

Determination of the Genus of Lactic Acid Bacteria Isolated in Wine

Morphology	Round cells (cocci)		Elongated cells (bacilli)
Cell arrangement	Pairs Small chains	Tetrads	Pairs or small chains
Glucose fermentation	Heterofermentative	Homofermentative	Heterofermentative or homofermentative
Lactic acid stereoisomer	D	L or D, L	L, D or D, L
Genus	Leuconostoc, Oenococcus	Pediococcus	Lactobacillus

homofermentative or heterofermentative character (Table 4.3). Classification at the species level makes use of the analysis of the fermentation profiles of a large number of sugars.

For this type of analysis, biochemical identification galleries are commonly used. In this system, the classic tests that were once performed in test tubes are miniaturized. The unidentified strain is inoculated into a nutritive medium that contains all of the nitrogen-based nutrients, vitamins, and salts necessary for its growth. There is no carbohydrate energy source. A colored indicator in the culture medium changes if the medium acidifies. Each microtube in the gallery contains a different sugar: hexoses, pentoses, disaccharides, etc.

To carry out this test, 0.1 ml of bacterial suspension is deposited in each of the microtubes. The tubes are sealed with a drop of paraffin to ensure anaerobic conditions. Generally, the system is incubated at 25°C for 24 hours. Tubes in which the color changes from blue to yellow indicate positive characters because sugars were fermented and lactic acid has been produced. In this manner, the fermentation profile of the examined strain can be established (Figure 4.6). These characters as a whole make the determination of a species possible without too much ambiguity, by referring to *Bergey's Manual of Systematic Bacteriology* (Vos et al., 2009). This method is well adapted for the identification of numerous lactic acid bacteria, but it must be carried out very carefully with bacteria isolated in wine. Experience has shown that the strain should undergo several successive subcultures in the standard laboratory medium beforehand. This preparation is essential for obtaining profile stability, which is indispensable before referring to the identification key. In any case, a strain cannot be identified solely on these results. The method's discriminating powers are not sufficient, and the similarity in profiles of *L. plantarum* and *L. mesenteroides* demonstrates this point (Figure 4.6). All of the other phenotypes previously described should also be taken into consideration at the same time.

Tables 4.3 and 4.4 summarize the phenotypic traits used to determine genus and species. Most of the strains for each species have fermentative profiles that correspond with those listed in Table 4.4. Nevertheless, besides differences that can be introduced by the pre-culture of the strain before the test, the more pronounced intraspecific variability of certain traits must be taken into account. For example, *O. oeni* was long thought not to possess the arginine hydrolysis trait, but recent studies have proven that numerous strains of *O. oeni* possess all of the

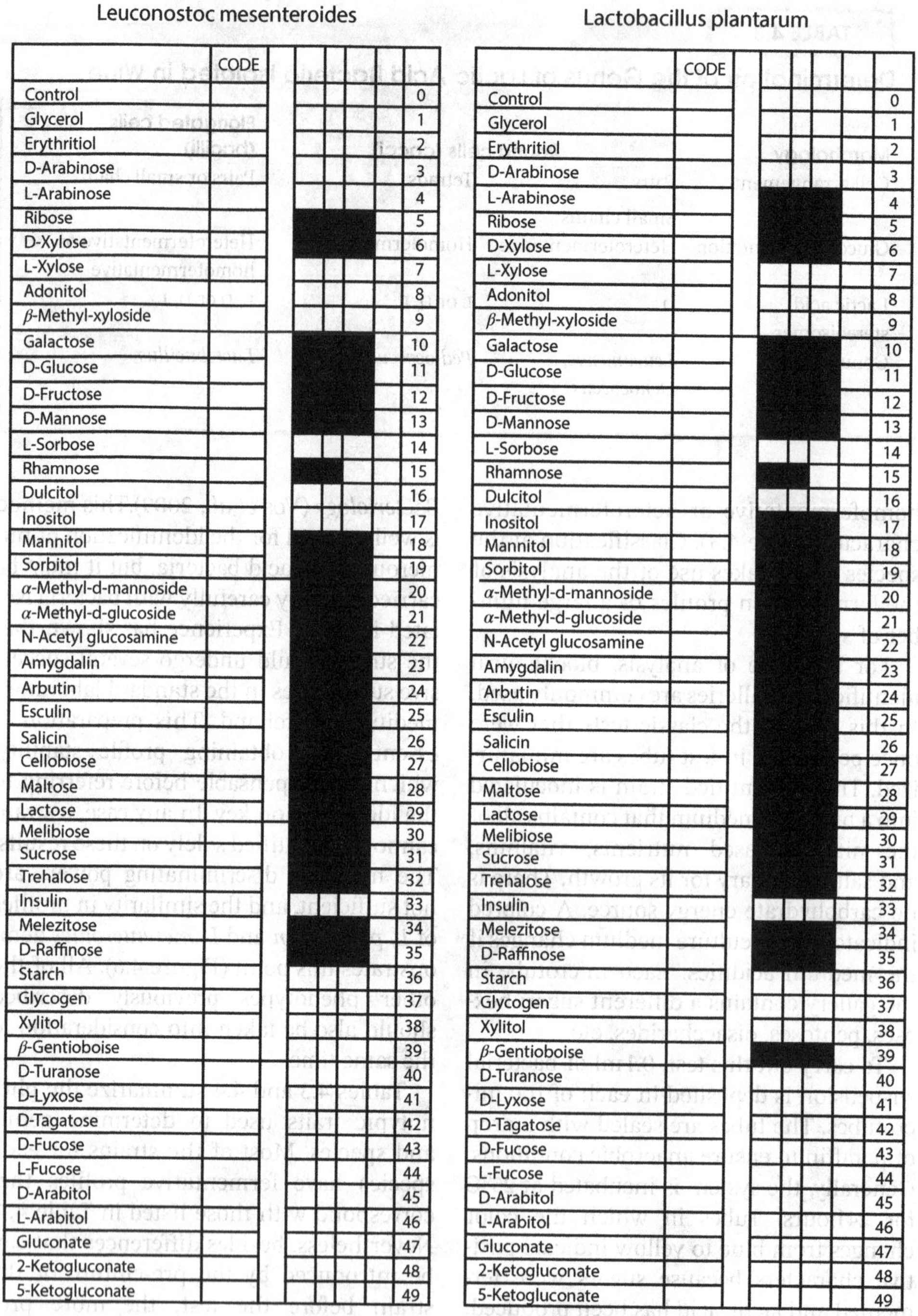

Leuconostoc mesenteroides	CODE		Lactobacillus plantarum	CODE
Control	0		Control	0
Glycerol	1		Glycerol	1
Erythritol	2		Erythritol	2
D-Arabinose	3		D-Arabinose	3
L-Arabinose	4		L-Arabinose	4
Ribose ■	5		Ribose	5
D-Xylose ■	6		D-Xylose	6
L-Xylose	7		L-Xylose	7
Adonitol	8		Adonitol	8
β-Methyl-xyloside	9		β-Methyl-xyloside	9
Galactose ■	10		Galactose ■	10
D-Glucose ■	11		D-Glucose ■	11
D-Fructose ■	12		D-Fructose ■	12
D-Mannose ■	13		D-Mannose ■	13
L-Sorbose	14		L-Sorbose	14
Rhamnose ▨	15		Rhamnose ▨	15
Dulcitol	16		Dulcitol	16
Inositol	17		Inositol	17
Mannitol	18		Mannitol ■	18
Sorbitol	19		Sorbitol ■	19
α-Methyl-d-mannoside ■	20		α-Methyl-d-mannoside	20
α-Methyl-d-glucoside	21		α-Methyl-d-glucoside	21
N-Acetyl glucosamine	22		N-Acetyl glucosamine ■	22
Amygdalin	23		Amygdalin ■	23
Arbutin	24		Arbutin ■	24
Esculin	25		Esculin ■	25
Salicin	26		Salicin ■	26
Cellobiose	27		Cellobiose ■	27
Maltose	28		Maltose ■	28
Lactose	29		Lactose ■	29
Melibiose ■	30		Melibiose ■	30
Sucrose ■	31		Sucrose ■	31
Trehalose ■	32		Trehalose ■	32
Insulin ▨	33		Insulin ▨	33
Melezitose ▨	34		Melezitose ■	34
D-Raffinose ■	35		D-Raffinose ■	35
Starch	36		Starch	36
Glycogen	37		Glycogen	37
Xylitol	38		Xylitol	38
β-Gentioboise	39		β-Gentioboise ▨	39
D-Turanose	40		D-Turanose	40
D-Lyxose	41		D-Lyxose	41
D-Tagatose	42		D-Tagatose	42
D-Fucose	43		D-Fucose	43
L-Fucose	44		L-Fucose	44
D-Arabitol	45		D-Arabitol	45
L-Arabitol	46		L-Arabitol	46
Gluconate	47		Gluconate	47
2-Ketogluconate	48		2-Ketogluconate	48
5-Ketogluconate	49		5-Ketogluconate	49

FIGURE 4.6 Biochemical substrate assimilation profiles (API 50 CHL gallery) of two lactic acid bacteria species.

necessary enzymatic equipment to hydrolyze arginine, leading to the production of citrulline, ornithine, and carbamoyl phosphate. The hydrolysis activity depends on environmental conditions, which determine not only the enzyme's activity but also its synthesis (Liu et al., 1994; Makaga, 1994). Furthermore, the "arc" operon,

TABLE 4.4
Determination of Wine Lactic Acid Bacteria Species and Similar Species (Sneath et al., 1986)

Species	Fructose	Glucose	Galactose	Glycerol	Arabinose	Ribose	Xylose	Esculin hydrolysis	Lactic acid isomer	Arginine dihydrolase
Lactobacillus casei	+	+	+	−	−	+	−	+	L	−
Lactobacillus homohiochii	+	+	−	−	−	+	−	−	L	−
Lactobacillus pentosus	+	+	+	+	+	+	v	+	DL	−
Lactobacillus plantarum	+	+	+	−	v	+	v	+	DL	−
Lactobacillus rhamnosus	+	+	+	(+)	v	+	−	+	L	−
Lactobacillus sake	+	+	+	−	+	+	−	+	DL	−
Lactobacillus brevis	+	+	v	−	+	+	+	v	DL	+
Lactobacillus fructivorans	+	+	−	−	−	−	−	−	DL	+
Lactobacillus fructosus	+	+	+	−	−	−	+	−	D or L	+
Lactobacillus hilgardii	+	+	v	−	−	−	+	−	DL	+
Lactobacillus kefir	+	+	−	−	v	+	−	−	DL	−
Lactobacillus sanfrancisco	−	+	+	−	−	−	+	−	DL	+
Pediococcus acidilactici	+	+	+	−	v	+	−	−	L or DL	+
Pediococcus damnosus	+	+	+	−	−	−	−	−	DL	−
Pediococcus parvulus	+	+	+	−	v	−	−	−	DL	v
Pediococcus pentosaceus	+	+	+	−	+	+	v	+	DL	+
Leuconostoc mesenteroides Subsp. mesenteroides	+	+	+	−	+	+	v	+	D	−
Oenococcus oeni	+	v	v	−	v	−	v	+	D	v

+, positive; −, negative; v, variable.

containing genes coding for the various enzymes in the metabolic pathway, has been identified and sequenced in strains of *O. oeni* (Tonon et al., 2001).

The use of fermentative profiles as an identification method should therefore be standardized. Bacterial traits can then be expressed in the most reproducible manner possible. Finally, these tests place bacteria under optimal growth and metabolic conditions—they thus give no indication of their true metabolism in wine. The fermentation of a carbohydrate found in the biochemical identification gallery may be totally impossible in wine: its nutritional conditions are far from those in the synthetic medium. Conversely, a substrate that cannot be metabolized under optimal conditions may be metabolized in wine because of the totally different metabolic regulations in play.

Fatty acid and protein composition

Like metabolic traits, the protein and fatty acid composition of bacteria is also determined by the mass of information in the genome and can therefore be used for identification purposes. In both cases, these components result from a succession of genetically determined syntheses. Differences in fatty acid and protein composition therefore reflect differences between strains. They can possibly even lead to identification of genus and species.

The total fatty acids are dosed in the form of esters after saponification of the lipids that contain them. The analysis makes use of gas chromatography (Rozes et al., 1993). Even though this analysis is reliable, it must be used with caution for identification. In fact, several studies have clearly proven that, for a given lactic acid bacterium, the same fatty acids are always represented, but their proportion varies significantly according to the cell cycle phase and even more so the physicochemical conditions of growth. Modifications essentially concern the level of saturation and the length of the carbon chains. Moreover, for a given species, bacteria are capable, with greater or lesser efficiency depending on strains, of synthesizing very long-chain fatty acids (more than 20 carbon atoms) to adapt to growth under conditions of stress (Desens, 1989; Kalmar, 1995).

Bacteria can therefore be identified by their composition of total fatty acids only when the culture of the cells to be analyzed is standardized. In the genus *Pediococcus*, this has been used to characterize three groups in which six species are classified. *Pediococcus damnosus* and *Pediococcus pentosaceus*, encountered in enology, belong to two of these groups. The authors of this work observed that culture age and environmental conditions modify the proportions of fatty acids without affecting the separation of the groups (Uchida and Mogi, 1972).

Bacterial cell proteins constitute another level of genome expression. The amino acid sequence of proteins is the result of a direct translation of genes. It is therefore normal to distinguish bacteria from one another by the proteins that they contain. Proteins are separated according to size in an electrophoretic gel under denaturing conditions. This identification method therefore involves subjecting the total bacterial cell contents to electrophoresis. After staining, the protein profiles are compared. According to Kersters (1985), protein profiles of strains are identical when genomes present a level of homology greater than 90%; they are quite similar up to 70%. Bacteria can therefore be identified in this manner at the species level. Nevertheless, as with the method using fatty acids, all of the following conditions must be rigorously standardized: bacterial culture conditions, the moment of sampling, and extraction and electrophoretic protocols. Recently, this method has been used to discover lactic acid spoilage bacteria in fortified wines: *O. oeni* (Dicks et al., 1995b), and various species of lactobacilli (*L. hilgardii*, *Lactobacillus fructivorans*, *Lactobacillus collinoides*, and *Lactobacillus mali*—the last three being rare) (Couto and Hogg, 1994).

4.3.3 Methods of Genotypic Analysis

The genotype of a bacterium is all the genetic information contained in its genome (DNA). Unlike the phenotype, the genotype does not change depending on the medium in which the bacterium is found. The genomic DNA of a bacterium is characterized both by its composition and its sequence. The first analytical methods developed were based on DNA composition or on that molecule's properties. Then, the PCR method appeared and revolutionized analytical methods. Finally, sequencing methods (with or without prior PCR) brought a new level of precision in identifying lactic acid bacteria.

Determining G+C percentage

Determining the guanine and cytosine base percentage of a genome was one of the first typing methods ever used. The two DNA strands are linked by hydrogen bonds created between complementary bases: three between G and C bases and only two between A and T bases. Consequently, the higher the G+C percentage in DNA, the more stable the interaction between two strands. It is easy to determine this experimentally by measuring the DNA's melting point. This consists in progressively heating it and following the dissociation between the two strands by measuring OD at 260 nm. It is also possible to calculate this percentage when the full genome sequence is known. Lactic acid bacteria have a G+C percentage under 50%. Two strains of the same species have an identical G+C percentage. However, two bacteria with the same G+C percentage do not necessarily belong to the same species. This is because two genomes may have the same composition but different sequences. Therefore, this method is not very precise in identifying a bacterium; yet, it is still an important criterion in the taxonomy of species.

DNA–DNA hybridization: species identification

The generally accepted reference method to determine whether two bacteria belong to the same species is DNA–DNA hybridization. It relies on the ability of two complementary single DNA strands to reconnect specifically. The closer the genomic DNA sequences are, the more the DNA strands can hybridize together. Technically, genomic DNA of two bacteria are fragmented, denatured by heat to form single strands, blended, and cooled to enable the reconnection of complementary strands. This also facilitates the formation of hybrids between two species. Two bacterial strains belong to the same species if their hybridization percentage is over 70%. For a lower value, they may be part of the same genus if hybridization can be measured. This method is very efficient, but difficult to implement. Its use is thus reserved for taxonomy purposes, such as the description of a new species.

DNA–DNA hybridization using probes: detecting species or strains carrying specific genes

A variant of DNA–DNA hybridization is based on the use of DNA probes. It is often used in molecular genetics to identify species or strains. A probe is a single-stranded DNA, labeled by radioactivity or chemically (for example, a fluorescent compound) in order to visualize it. Most often, this is a known DNA sequence corresponding to a specific gene or region of the genomic DNA that is characteristic of a species or strain. The method consists in determining whether a probe can hybridize on the DNA of the test bacterium. The DNA of the test bacterium is denatured to form single-stranded DNA and then is fixed on a support such as a nylon membrane. The single-stranded DNA probe is then put in contact with the membrane to enable its hybridization with bacterial DNA. If they are complementary, the probe may hybridize, and its marking makes it possible to visualize this. These operations are diagrammed in Figure 4.7.

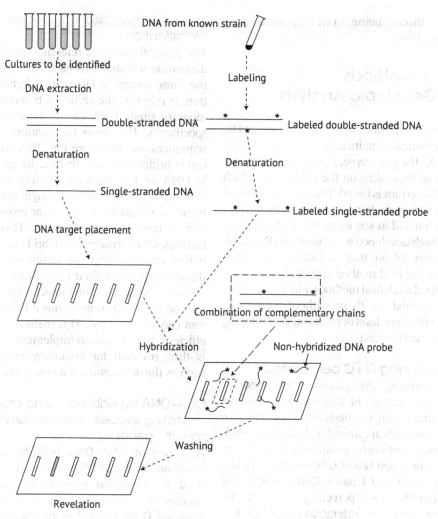

FIGURE 4.7 Schematic diagram of the general principle for the identification of lactic acid bacteria using a specific DNA/DNA probe.

The nature of the probe—complete genome, gene, or portion of a gene—determines the specificity of the analysis. Most lactic acid bacteria of wine can be identified at the species level in this manner. The first working probe was developed for the species *O. oeni*. It is so far from other species that its genomic DNA can be used as a probe (Lonvaud-Funel et al., 1989). Bacteria of this species can be identified even in a mixture containing other bacteria. The same is true for other species found in grape must and wine (Lonvaud-Funel et al., 1991b). However, the similarity of the *L. hilgardii* and *Lactobacillus brevis* species required the development of a more specific probe targeting *L. hilgardii* (Sohier et al., 1999). Hybridization on the membrane is particularly interesting because it can be done directly by taking colonies on an agar medium. It is possible to easily study a mixed population of lactic acid bacteria in consecutive hybridization cycles with the probes we have. Each clone is identified. At least five different species can successively be detected in a mixture with this protocol (Figure 4.8) (Lonvaud-Funel et al., 1991a). Nevertheless, this method is rarely used these days, because it is difficult and long to implement, while

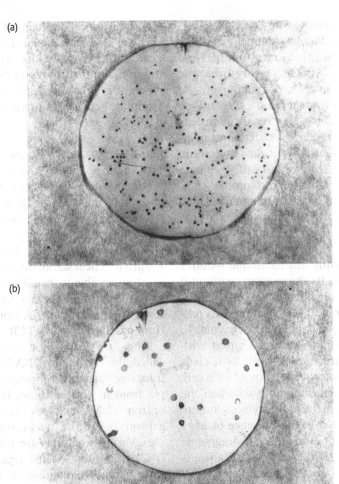

FIGURE 4.8 Specific lactic acid bacteria population counts by hybridization on colonies with reference DNA probes. (a) Hybridization on colonies of cultured *O. oeni*. (b) Detection of *L. hilgardii* colonies in a mix of five species (*O. oeni*, *L. mesenteroides*, *P. damnosus*, *L. plantarum*, and *L. hilgardii*). Results obtained after four successive hybridizations and dehybridizations with probes from four other species.

results of equivalent quality are obtained by PCR, which is simpler and quicker to conduct.

REA-PFGE: identification of strains

Restriction enzyme analysis by pulsed-field gel electrophoresis (REA-PFGE) is one of the rare genotyping methods that can differentiate strains without using PCR. The genome of any bacteria is composed of successive A, T, G, and C bases that sometimes form repeating patterns. For example, statistically, the "AT" pattern is found every 16 base pairs (bp) because with each genome position, the probability of finding them is a one-in-four chance of having an A, multiplied by a one-in-four chance that the following base is a T. Restriction enzymes have the particularity of recognizing and cutting special patterns of 4–8 bp or more. They are thus able to cleave the DNA of a bacterium into fragments. An enzyme such as *Eco*R1 recognizes the "GAATTC" pattern, which statistically repeats every 4,096 bp. In a bacterial genome with 2×10^6 bp, it will thus produce about 500 fragments, with an average size of 4,096 bp. The NotI enzyme, which recognizes the site of 8 bp "GCGGCCGC," only produces about 30 fragments of 65 kb in average size. The exact number and the size of each fragment vary from one bacterium to another, depending on the distribution of restriction sites along the genome. Consequently, all fragments produced constitute "a restriction profile," which is a characteristic of each bacterium. There are no techniques to easily separate and visualize a restriction profile composed of several hundred fragments. However, it is possible to separate several dozen large fragments by pulsed-field electrophoresis (Section 1.9.3, Figure 1.29). Analyzing a bacterium using this technique requires the digestion of its chromosome without it mechanically breaking apart, thus avoiding the appearance of DNA fragments other than those generated by the enzyme. To do so, cells of a culture are mixed with lukewarm agarose, which gels as it cools and forms a protective gel around each cell. Their walls then undergo lysis when the gel is incubated in a lysozyme solution. Then, incubation in the presence of a restriction enzyme, such as NotI, fragments their genomic DNA. A piece of agarose is simply placed in an agarose gel positioned in a pulsed-field gel electrophoresis device. The DNA fragments then migrate and separate. The result obtained is a restriction profile comparable to yeast karyotypes obtained by the same type of electrophoresis (Section 1.9.3, Figure 1.30).

This method is very effective for the differentiation of bacteria strains of the same species because most have different restriction profiles. First, the species to which bacteria belong must be established because it is impossible to connect a restriction profile to one species in particular. This method is used to monitor the development of *O. oeni* strains selected and used as malolactic starters during winemaking. The restriction profiles of the biomass collected in wine during malolactic fermentation are compared with that of the selected strain that was added (Gindreau et al., 1997, 2003). However, this is time-consuming and cannot be used routinely to check the establishment of malolactic starters in wine. Its current scope of interest is limited to research laboratories, where it is still an efficient method to differentiate between strains of any lactic acid bacteria species.

Polymerase chain reaction

PCR produces a very large number of copies of almost any DNA fragment a user has chosen. Its main advantage is that it is very easy to visualize a large number of identical DNA molecules, while it is extremely complicated to observe a limited number. From a practical standpoint, it would be almost impossible to visualize the DNA of a single bacterium. It is easy, however, to observe a fragment of DNA from this bacterium once amplified by PCR.

The DNA used for amplification is called the "template DNA." It can come from one single microorganism (for example, from an isolated colony in a Petri dish) or from a DNA mixture of several microorganisms (e.g. those in a must or wine sample). Various methods are used to extract DNA, depending on the type of microorganism or PCR conducted. In all cases, cells must undergo lysis to release their contents. Then, DNA is purified to eliminate debris and other cell components that may inhibit PCR.

The template DNA fragment, amplified by PCR, is chosen by selecting two oligonucleotides, or PCR primers, which frame it. These primers are generally 100–3,000 bp apart, which can easily be amplified. They are more or less 20 nucleotides long, are each complementary to either one of the two DNA strands, and are placed to face each other.

PCR is conducted in a tube containing a reaction medium and the template DNA, primers, thermoresistant DNA polymerase (Taq polymerase), and deoxynucleotide triphosphates (dATP, dCTP, dGTP, and dTTP), which will be used to synthesize DNA fragments. During the first PCR cycle, the template DNA is denatured at 95°C to separate the two DNA strands, then cooled to about 50–60°C for the primers to hybridize on each of the two strands. It is finally reheated to 72°C, which is the optimal temperature for Taq polymerase activity. The latter recognizes hybrids formed by the template DNA and primers, and enables the elongation, i.e. the synthesis, of the chosen DNA fragment. This starts from the primer's 3' end and successively adds the complementary dNTP molecules to the template DNA strand on which each primer is attached. When this elongation stage is completed, the reaction medium contains twice as much of the DNA fragment framed by primers (Section 1.8.4, Figure 1.21). These new fragments may then be used as template DNA for a new PCR cycle. A second cycle, identical to the first, is thus conducted, then a third and so forth until about 30 or 40 cycles in total. At the end of each cycle, the quantity of DNA fragment obtained is doubled. By starting from a single template DNA molecule, theoretically, up to 2^{30} or 2^{40} copies of the selected DNA fragment are obtained.

Electrophoresis in agarose gel is the most popular and simplest technique for analyzing DNA fragments. At an alkaline pH, the phosphate groups of DNA are ionized. The molecules placed in an electric field therefore migrate toward the anode. In the gel, electrophoretic mobility depends on the size of the molecules. The smaller the molecules are, the faster they migrate. If the DNA obtained from PCR is placed in an agarose gel, it is thus possible to make it migrate in the gel. It is thus customary to make DNA markers, of known sizes and used as migration references, migrate simultaneously with the unknown fragments. Molecules separated by electrophoresis are revealed by staining with ethidium bromide or other similar molecules. This compound is an analog of an aromatic base; it intercalates in the DNA and fluoresces orange under ultraviolet light. As previously indicated, PCR can be used to obtain a significant quantity of same-sized DNA fragments. These migrate to the same position in a gel and fix enough ethidium bromide to generate a visible signal under UV light. As described in the following paragraphs, this simple technique has also been used for various applications to identify with great precision the species and strains of lactic acid bacteria present in any type of sample.

Specific PCR: detection of species or strains possessing specific genes

PCR is especially well adapted to analysis without any prior culturing of the sample, i.e. the analysis of non-isolated microorganisms.

It is possible to detect the presence either of strains of a species or of strains carrying specific genes in an unknown mixture. To do so, it is sufficient to have the primers specific to the species or gene. A positive amplification signal, revealed either by a electrophoresis gel band or by an increase in real-time PCR fluorescence, proves the existence of at least one strain of the target species in a mixture. There are specific primers to identify and detect strains of O. oeni via classical PCR (Zapparoli et al., 1998), but also in real-time PCR (Pinzani et al., 2004). Primers have also been developed for other species of wine lactic acid bacteria. Petri et al. (2013) suggested a "multiplex" PCR protocol, where several DNA strands are amplified in the same reaction tube, each with their specific probes: 13 species are detected in two reactions containing six and seven primer couples in mixture.

According to the same principle, specific primers, determined from the gene sequence we are looking for in a population, are used to assess the presence of undesirable strains. This method is especially powerful since only the gene is amplified, whatever the species or strain within a species.

In enology, this method is used to detect strains responsible for ropiness and strains producing biogenic amines. Primers have been determined from sequences of the *gtf* gene (coding for the production of the glucan found in ropy wines) (Section 5.4.4) (Gindreau *et al.*, 2001). This is also true of genes coding for enzymes involved in the production of biogenic amines, such as histamine, putrescine, and tyramine (Nanelli *et al.*, 2008; Coton *et al.*, 2010). It is now also possible to detect lactic acid bacteria that break down glycerol to produce acrolein, by targeting the glycerol dehydratase gene (Claisse and Lonvaud-Funel, 2001).

Quantitative PCR

A variant of PCR is real-time PCR or quantitative PCR, which helps determine the population level of a microorganism in a sample. The principle is identical to PCR, but the DNA amplification is monitored in "real time," at the end of each PCR cycle. It is measured by fluorescence, which is proportional to the quantity of synthesized DNA. Reactions are conducted in a thermal cycler comparable to the one used for PCR, but it also possesses a sensor for the measurement of fluorescence and is managed by a microcomputer. Various detection types are possible: molecules that are only fluorescent between the two synthesized strands; or precise sequence probes that are only fluorescent when fixed to the synthesized DNA. The more template DNA molecules initially present in the reaction, the faster a fluorescence threshold is reached. The PCR cycle, in which this fluorescence threshold appears, is called the threshold cycle. It is inversely proportional to the number of DNA molecules (in log base 2). Insofar as this number of template DNA molecules is directly linked to the number of cells that were present in the analyzed sample, the measurement of the threshold cycle helps to calculate the cell population in the sample.

Quantitative PCR may be used like specific PCR, but with the considerable advantage that it indicates the population level of the targeted species or strain. This information is especially important in enology. For example, it is common for a spoilage microorganism, such as a bacterium producing biogenic amines, to start having a harmful effect on wine quality once it reaches a population over 10^3 cells/mm. Without information about the population level, it is impossible to assess the risk to which the wine is exposed. Lastly, in addition to its use in determining microorganism populations, this method also presents the advantage of being more sensitive than classical PCR.

PCR and sequencing of the 16S RNA gene: species identification

While DNA-DNA hybridization is officially recognized for identifying a bacterial species, it is rarely used in practice because it is difficult to implement. Amplification by PCR followed by 16S RNA gene sequencing is the method used routinely and is just as reliable to identify lactic acid bacteria species. All bacteria possess this gene, which codes for RNA entering the composition of the ribosome's small subunit. This gene accumulates mutations throughout speciation; its sequence is thus different from one species to another. More accurately, certain regions of the gene are variable from one species to another. To form a functional ribosome, it cannot undergo mutations in "conserved" regions, which have a functional role on the RNA. It is thus possible to synthesize "universal" PCR primers, which hybridize in these conserved regions, whatever the bacterium is. To identify a bacterium using this method, its genomic DNA is purified and used as a template in PCR with universal primers specific to this gene. The DNA fragment amplified by PCR has the same size, whatever the species. However, its sequence is variable from one species to another because it encompasses variable regions of the gene. Sequencing this DNA is easily conducted in a specialized laboratory. Then, one simply compares the sequence obtained with sequences of bacterial 16S RNA genes present in public databases to precisely identify the bacterium. The BLAST algorithm helps conduct this search (Altschul *et al.*, 1990). It generates an alignment of the sequence introduced with all available sequences and lists the best matches. It is generally accepted that two bacteria belong to the same species if the sequence of their 16S RNA genes is over 97% identical.

PCR RAPD and PCR VNTR: strain identification

To answer certain practical enology questions, it is necessary to be able to identify strains of a species. That is especially the case for *O. oeni*, whether one is assessing the number of strains in a sample from the cellar, or in the commercial starter industry to ensure the strain's identity. PCR is once again at the source of these methods. Historically, randomly amplified polymorphic DNA (RAPD) was the first method used (Zapparoli et al., 2000). Primers are short-sequence oligonucleotides (6–10 nucleotides), which statistically can hybridize in very many regions of the chromosome. When two primers hybridize near each other (less than 2,000 bp apart) and are facing each other, they can amplify the DNA fragment separating them. This situation occurs several times for a given bacterial DNA and leads to the amplification of several DNA fragments of various sizes. From one bacterial strain to another, primers hybridize in different regions, producing batches of various DNA fragments. After separation in an electrophoresis gel, each bacterial strain is thus identifiable by a specific "band profile." This method is quick and simple to use, which makes it a good routine method. It is sometimes implemented by analytical laboratories to check the establishment of malolactic starters in wine. Nevertheless, the same principle of random binding of primers makes it only slightly discriminating and not very reproducible. In fact, even if there are only a few variations in template DNA quality or in PCR conditions, it is possible to obtain different profiles for a given strain. Moreover, results obtained are rarely comparable from one laboratory to another.

The most discriminating and most reliable of the current typing methods is another use of PCR: analysis by variable number of tandem repeats (VNTR), also called multiple loci VNTR analysis (MLVA). It is based on the presence, in the genome, of regions containing sequences repeated in tandem. These repetitions range from three to over 100 bp, depending on the regions. Also, their number is variable from one region to another and one strain to another (from one to over 100 repetitions). Prior knowledge of the complete genomes of the species is necessary to locate polymorphic regions. Chromosome regions located immediately upstream and downstream from repetition zones are conserved in all species and are used to choose PCR primers. Since there are several VNTR regions on the genome, several primer pairs are chosen. During PCR, VNTR sites are amplified, and due to the varying number of repetitions, the sizes of amplifiers are different. They form a profile in electrophoresis gel or in capillary electrophoresis. This method was developed with *O. oeni*; five VNTR markers, called TR1 through TR5, were found to discriminate between most strains. Sequences repeated in tandem for each strain are quite variable in size, ranging from 6 to 147 nucleotides. This method gives a perfectly reproducible result. Each strain is characterized by a "formula," giving the number of pattern repetitions for each TR (Claisse and Lonvaud-Funel, 2012, 2014).

PCR-DGGE: simultaneous detection of several species

PCR followed by denaturing gradient gel electrophoresis (PCR-DGGE) simultaneously detects several species present in a sample. To do this, the "total" DNA of all microorganisms in the sample is extracted and used as the template. "Universal" PCR primers are chosen; they hybridize on a gene present in all bacteria, such as the 16S RNA gene or *rpo*B gene, which codes for a subunit of RNA polymerase. These genes have regions whose sequence is conserved in all species and others that vary from one species to another. By using primers that fix on the conserved regions, it is possible to amplify the gene of each species present in the sample via PCR. All PCR products have the same size, yet their sequences are different. It would be impossible to separate them by electrophoresis in a classical gel. However, by introducing an agent able to denature the DNA in the gel, for example, urea, PCR products are more or less quickly denatured depending on their sequence. As a result, they migrate more or less quickly in the gel. Electrophoresis with a denaturing gradient thus enables the separation of DNA fragments, depending on their composition, and leads to a profile where each band corresponds to a species. By comparison with known distances of species migration, it is possible to identify species in the mixture (Renouf et al., 2006a,b). This method enables complex microflora studies. It does not require prior cultures or the isolation of colonies. However, it is not quantitative, and it is possible for species found in low numbers in the mixture to escape amplification.

Genome sequencing: strain description and comparison

Genome sequencing is without a doubt the most robust method to describe a strain and to compare it to other strains. The first complete genome of a lactic acid bacterium was reported in 2001 and the first from an *O. oeni* strain was obtained from wine in 2005 (Mills et al., 2005). Sequencing was then conducted by the Sanger method, which required months or years of work to produce a single genome. Since then, "new-generation" high-speed sequencing methods have been developed and enable the production of a complete genome in several hours. These methods are generally offered by specialized institutions because they require specific, constantly evolving equipment. For a laboratory, sequencing a genome often consists in purifying the genomic DNA of the bacterium and transmitting it to a specialized institution for sequencing. Then, the bioinformatics processing of millions of sequences is conducted. In fact, new-generation sequencing methods produce a very large number of sequences (average length varies between 150 and 15,000 bases, depending on the methods) that must be "cleaned" and assembled to reconstruct the full genome. They are then annotated to identify the positions of all genes and to determine the functions of all proteins coded by these genes. Today, genome sequencing is not yet of any interest for the routine analysis of wines. Nevertheless, it helps identify characteristics of strains, especially those of commercial strains used to conduct malolactic fermentation (Borneman et al., 2012; Sternes and Borneman, 2016). Sequencing also provides precious information on the diversity and specificity of strains encountered in various regions and types of wines (Campbell-Sills et al., 2015).

4.4 The *Oenococcus oeni* Species

Oenococcus oeni is the main lactic acid bacterium of interest to enology because it conducts malolactic fermentation in the great majority of spontaneous fermentations. This species was first described in 1967 based on morphological, physiological, and biochemical criteria and was classified in the *Leuconostoc* genus under the species name *Leuconostoc oenos* (Garvie, 1967). However, once it was possible to analyze its genotypic characteristics, especially the sequence of its 16S RNA gene, it became evident that it is actually very far removed from *Leuconostoc*. As a result, it was reclassified in a new genus, *Oenococcus* (Dicks et al., 1995a). For a long time, *O. oeni* remained the only species known to belong to this genus. However, two new species have been described recently. *Oenococcus kitaharae* was isolated in Japan in Shochu distillation residue (Endo and Okada, 2006). *Oenococcus alcoholitolerans* was discovered in Brazil in bioethanol and Cachaça production factories (Badotti *et al.*, 2014). Neither of these two species has been detected in wine at present.

Since the 1970s, it has been widely understood that there is a large diversity of *O. oeni* strains, more or less able to survive and develop in wine and to conduct malolactic fermentation. After inoculation in wine, various strains not only perform fermentation more or less quickly and completely but also create differences between wines at the organoleptic level. Phenotypic analyses confirm that strains differ in their ability to metabolize sugars, citric acid, and many other compounds, which have an impact on quality.

From the 1980s–1990s onward, the use of genotyping methods generating band profiles, such as restriction profiles, RAPD, or the VNTR method, have enabled the differentiation between strains to be much quicker and more precise. However, these methods do not give information about the genetic proximity of strains. To understand the genetic diversity of the *O. oeni* species in a cellar, production region, or in the species in general, it is necessary to use methods based on gene sequencing. Those used at present are multilocus sequence typing (MLST) and the single nucleotide polymorphism (SNP) method. MLST consists in sequencing several genes in each strain and comparing all sequences obtained. The SNP method is based on the same principle but is limited to sequencing several dozen pairs of selected bases and genome sequencing.

Applied to collections of several dozens or hundreds of strains, the MLST method has revealed that the *O. oeni* species is structured into several large genetic groups of strains: A, B, and C. These are then divided into subgroups (Bilhère *et al.*, 2009; Bridier *et al.*, 2010). These groups are not linked to specific regions, but some of them are related to certain products. For example, it is clear that *O. oeni* strains developing in cider are genetically different from those in wine. Moreover, certain strain groups come from certain wines, such as white wines from Burgundy or Champagne. A study conducted using the SNP method confirms this structuring into genetic groups and suggests that strains are genetically adapted to specific wines, but not to wineries or regions, because the same strains are sometimes found in very distant regions (El Khoury *et al.*, 2017). Also, dominant strains do not seem to exist in the environment; on the contrary, there is a large diversity of strains, each represented in low numbers.

Today, over 200 genomes are described in public databases (Sternes and Borneman, 2016). Genome sequencing also confirms that the *O. oeni* species is structured in genetic groups and subgroups; some of them are related to products or specific types of wines. Additionally, it shows that strains have been domesticated by humans. When humans make wines, new environments are created, and certain strains have evolved to adapt to them (Campbell-Sills *et al.*, 2015). Nevertheless, the link between the phenotypes and genotypes of strains remains to be established, in order to understand the characteristics leading to their adaptation to a particular type of wine.

References

Altschul S.F., Gish W., Miller W., Myers E.W. and Lipman D.J. (1990) J. Mol. Biol., 215, 403.

Badotti F., Moreira A.P., Tonon L.A., de Lucena B.T., Gomes Fátima de, C., Kruger R., Thompson C.C., de Morais M.A. Jr., Rosa C.A. and Thompson F.L. (2014) Ant. Van Leeuw., 106, 1259.

Bastard A., Coelho C., Briandet R., Canette A., Gougeon R., Alexandre H., Guzzo J. and Weidmann S. (2016) Front. Microbiol., 7, 613.

Bilhère E., Lucas P., Claisse O. and Lonvaud-Funel A. (2009) Appl. Environ. Microbiol., 75, 1291.

Borneman A.R., McCarthy J.M., Chambers P.J. and Bartowsky E.J. (2012) BMC Genomics, 13, 373.

Bourdineaud J.P., Nehmé B., Tessé S. and Lonvaud-Funel A. (2003) Appl. Environ. Microbiol., 69, 2512.

Bourdineaud J.P., Nehmé B., Tessé S. and Lonvaud-Funel A. (2004) Int. J. Food Microbiol., 92, 1.

Bridier J., Claisse O., Coton M., Coton E. and Lonvaud-Funel A. (2010) Appl. Environ. Microbiol., 76, 7754.

Campbell-Sills H., El Khoury M., Favier M., Romano A., Biasioli F., Spano G., Sherman D.J., Bouchez O., Coton E., Coton M., Okada S., Tanaka N., Dols-Lafargue M. and Lucas P.M. (2015) Genome Biol. Evol., 7, 1506.

Claisse O. and Lonvaud-Funel A. (2001) J. Food Prot., 64, 833.

Claisse O. and Lonvaud-Funel A. (2012) Food Microbiol., 30, 340.

Claisse O. and Lonvaud-Funel A. (2014) Food Microbiol., 38, 80.

Coton M., Romano A., Spano G., Ziegler K., Vetrana C., Desmarais C., Lonvaud-Funel A., Lucas P. and Coton E. (2010) Food Microbiol., 27, 1078.

Couto J.A. and Hogg T. (1994) J. Appl. Bacteriol., 76, 487.

Delmas F., Pierre F., Coucheney F., Divies C. and Guzzo J. (2001) J. Mol. Microbiol. Biotechnol., 3, 601.

Desens C. (1989) *Les phénomènes d'inhibition des bactéries lactiques du vin. Relations avec la constitution lipidique des membranes*. Thèse de Doctorat, Université de Bordeaux II.

Dicks L.M.T., Dellaglio F. and Collins M.D. (1995a) Int. J. Syst. Bacteriol., 45, 395.

Dicks L.M.T., Loubser P.A. and Augustyn O.P.H. (1995b) J. Appl. Bacteriol., 79, 43.

Dols-Lafargue M., Lee H.Y., Le Marrec C., Heyraud A., Chambat G. and Lonvaud-Funel A. (2008) Appl. Environ. Microbiol., 74, 4079–4090.

El Khoury M., Campbell-Sills H., Salin F., Guichoux E., Claisse O. and Lucas P.M. (2017) Appl. Environ. Microbiol., 83, 1.

Endo A. and Okada S. (2006) Int. J. Syst. Evol. Microbiol., 56, 2345.

Favier M., Bilhère E., Lonvaud-Funel A., Moine V. and Lucas P.M. (2012) PLoS One, 7, e49082.

Frémaux C., Aigle M. and Lonvaud-Funel A. (1993) Plasmids, 30, 212.

Garbay S. (1994) *Recherche du mode d'adaptation de Leuconostoc oenos au vin: Relation avec la constitution lipidique et protéique de la membrane plasmique*. Thèse de Doctorat, Université de Bordeaux II.

Garvie E.I. (1967) J. Gen. Microbiol., 48, 431.

Gindreau E., Joyeux A., de Revel G., Claisse O. and Lonvaud-Funel A. (1997) Œno One, 31, 197.

Gindreau E., Walling E. and Lonvaud-Funel A. (2001) J. Appl. Microbiol., 90, 535.

Gindreau E., Kheim H., de Revel G., Bertrand A. and Lonvaud-Funel A. (2003) Œno One, 51, 37.

Kalmar (1995) *Recherches sur la tolérance à l'éthanol des bactéries lactiques des vins: Relation avec la constitution lipidique membranaire*. Thèse de Doctorat, Université de Bordeaux II.

Kersters K. (1985) Numerical methods in the classification of bacteria by protein electrophoresis. In *Computer-Assisted Bacterial Systematics* (Eds M. Goodfellow and D.E. Minnkin). Academic Press, London.

Liu S.Q., Pritchard G.G., Hardman M.J. and Pilone G.J. (1994) Am. J. Enol. Vitic., 45, 235.

Lonvaud M. (1975) *Recherches sur l'enzyme des bactéries lactiques assurant la transformation du malate en lactate*. Thèse de Doctorat 3e cycle, Université de Bordeaux II.

Lonvaud-Funel A. and Desens C. (1990) Sci. Aliment., 10, 817.

Lonvaud-Funel A., Biteau N. and Frémaux C. (1989) Sci. Aliment., 9, 533.

Lonvaud-Funel A., Joyeux A. and Ledoux O. (1991a) J. Appl. Bacteriol., 71, 501.

Lonvaud-Funel A., Fremaux C., Biteau N. and Joyeux A. (1991b) Food Microbiol., 8, 215.

Lonvaud-Funel A., Guilloux Y. and Joyeux A. (1993) J. Appl. Bacteriol., 74, 41.

Maitre M., Weidmann S., Dubois-Brissonnet F., David V., Covès J. and Guzzo J. (2014) Appl. Environ. Microbiol. 80, 2973.

Makaga E. (1994) *Etude du métabolisme de l'arginine chez l'espèce Leuconostoc oenos*. Thèse de Doctorat, Université de Reims.

Mattarelli P., Holzapfel W., Franz C., Endo A., Felis G.E., Hammes W., Pot B., Dicks L. and Dellaglio F. (2014) Int. J. Syst. Evol. Microbiol., 64, 1434.

Mills D.A., Rawsthorne H., Parker C., Tamir D. and Makarova K. (2005) FEMS Microbiol. Rev., 29, 465.

Nanelli F., Claisse O., Gindreau E., de Revel G., Lonvaud-Funel A. and Lucas P. (2008) Lett. Appl. Microbiol., 47, 594.

Petri A., Pfannebecker J., Frohlich J. and Konig H. (2013) Food Microbiol., 33, 48.

Pinzani P., Bonciani L., Pazzagli M., Orlando C., Guerrini S. and Granchi L. (2004) Lett. Appl. Microbiol., 38, 118.

Renouf V., Claisse O. and Lonvaud-Funel A. (2006a) J. Microbiol. Methods, 67, 162.

Renouf V., Claisse O., Miot-Sertier C. and Lonvaud-Funel A. (2006b) Food Microbiol., 23, 136.

Rozès N., Garbay S., Denayrolles M. and Lonvaud-Funel A. (1993) Lett. Appl. Microbiol., 17, 126.

Sneath P.H.A., Mair N.S., Sharpe M.E. and Holt J.G. (Eds) (1986) *Bergey's Manual of Systematic Bacteriology*, 1st edition, Vol. 2. Williams & Wilkins, Baltimore.

Sohier D., Coulon J. and Lonvaud-Funel A. (1999) Int. J. Syst. Bacteriol., 49, 1075.

Stackebrandt E., Frederiksen W., Garrity G.M.M., Grimont P.A.D., Kampfer P., Maiden M.C.J., Nesme X., Rossello-Mora R., Swings J., Truper H.G., Vauterin L., Ward A.C. and Whitman W.B. (2002) Int. J. Syst. Evol. Microbiol., 52, 1043.

Sternes P.R. and Borneman A.R. (2016) BMC Genomics, 17, 308.

Tonon T., Bourdineaud J.P. and Lonvaud-Funel A. (2001) Res. Microbiol., 152, 653.

Uchida K. and Mogi K. (1972) J. Gen. Appl. Microbiol., 18, 109.

Vos P., Garrity G., Jones D., Krieg N.R., Ludwig W., Rainey F.A., Schleifer K.-H. and Whitman W.B. (2009) [1984 (Williams & Wilkins)]. *The Firmicutes*, 2nd edition, Vol. 3, p. 1450. Bergey's Manual of Systematic Bacteriology. Springer, New York.

Walling E. (2003) *La biosynthèse d'exopolysaccharides par les bactéries lactiques du vin: approche génétique, enzymatique et physiologique de la production de glucane par Pediococcus damnosus*. Thèse Doctorat, Université de Bordeaux II.

Zapparoli G., Torriani S., Pesente P. and Dellaglio F. (1998) Lett. Appl. Microbiol., 27, 243.

Zapparoli G., Reguant C., Bordons A., Torriani S. and Dellaglio F. (2000) Curr. Microbiol., 40, 351.

CHAPTER 5

Metabolism of Lactic Acid Bacteria

5.1 Generalities: A Review
5.2 Metabolism of Sugars: Lactic Acid Fermentation
5.3 Metabolism of the Principal Organic Acids of Wine
5.4 Other Transformations Likely to Occur in Winemaking
5.5 Effect of the Metabolism of Lactic Acid Bacteria on Wine Composition and Quality

5.1 Generalities: A Review

Metabolism represents all of the biochemical reactions of degradation and synthesis carried out by the bacteria cell during its reproduction. Catabolic reactions provide energy, transforming assimilable substrates from the environment or reserve substances of the cell; anabolic reactions guarantee cell synthesis from environmental substrates and intermediate catabolites.

Lactic acid bacteria are chemoorganotrophic: they find the energy required for their entire metabolism from the oxidation of organic compounds. The oxidation of substrates represents a loss of electrons and protons, which must be accepted by another molecule, which is reduced. The transport of these two species, H^+ and e^-, to the final acceptor involves a chain of successive oxidation–reductions of various molecules.

Thus, the biological oxidation of a substrate is always coupled with the reduction of another molecule. In the diagram below, the oxidized substance is noted as DH_2 and the final electron and proton acceptor as A.

$$DH_2 \rightarrow D + 2H^+ + 2e^- + energy$$

$$A + 2H^+ + 2e^- \rightarrow AH_2$$

Handbook of Enology, Volume 1: The Microbiology of Wine and Vinifications, Third Edition.
Pascal Ribéreau-Gayon, Denis Dubourdieu, Bernard B. Donèche and Aline A. Lonvaud.
© 2021 John Wiley & Sons Ltd. Published 2021 by John Wiley & Sons Ltd.

The overall reaction is:

$$DH_2 + A \rightarrow AH_2 + D + \text{energy}.$$

The nature of the final electron acceptor A determines the type of metabolism: fermentative or respiratory. The presence of oxygen also distinguishes aerobic and anaerobic microorganisms. Relationships of microorganisms with oxygen are diverse. Some microorganisms cannot use oxygen as the final acceptor and do not survive in its presence. Other microorganisms are also anaerobic, but oxygen has no effect on them. Some operate under aerobic conditions or anaerobic conditions but with different metabolic pathways depending on conditions. The latter microorganisms strictly depend on whether oxygen is present or not.

Under aerobic conditions, the electrons and protons are transported to oxygen, which is reduced to water. This process is called aerobic respiration. The transport system consists of a group of cytochromes. The proton flux creates a proton motive force, which enables the synthesis of ATP molecules. The conservation of the oxidation energy is ensured by the synthesis of the pyrophosphate bond of ATP. This bond generates energy when it is hydrolyzed. This system does not exist in lactic acid bacteria, although some species can synthesize cytochromes from precursors. It characterizes acetic acid bacteria (Chapter 7).

Some lactic acid bacteria reduce oxygen from the environment by forming hydrogen peroxide according to the following reaction:

$$O_2 + 2e^- + 2H^+ \rightarrow H_2O_2$$

Hydrogen peroxide must be eliminated since it is a toxic molecule. Cells that are not capable of eliminating it cannot develop in the presence of oxygen: they are strict anaerobes. Depending on their behavior with respect to oxygen, lactic acid bacteria are classified as strict anaerobes, facultative anaerobes, microaerophiles, or aerotolerant anaerobes. The distinction between these different categories is often difficult to establish for a given strain.

Most lactic acid bacteria tolerate the presence of oxygen but do not use it in energy-producing mechanisms. Depending on the species, they use different pathways to eliminate the toxic peroxide, activating peroxidases that use NADH as a reducer, a superoxide dismutase, a pseudocatalase, and sometimes Mn^{2+} ions (Desmazeaud and de Roissart, 1994). To date, this subject has not been specifically studied for the species isolated in wine.

If the final electron and proton acceptor is a mineral ion (sulfate, nitrate), the microorganism functions under anaerobic conditions, but a respiratory mechanism is still involved. This process is called anaerobic respiration.

Under anaerobic conditions, the reduced molecule can also be an endogenous substance, i.e. one of the products of metabolism. This is the case with fermentation. In lactic acid bacteria, this molecule is pyruvate. It is reduced into lactate in the reaction that characterizes lactic acid fermentation:

$$\text{Pyruvate} + NADH + H^+ \rightarrow \text{lactate} + NAD^+$$

Contrary to the reoxidation of the coenzyme by the respiratory chain, this reaction is not energy producing.

Other kinds of reactions can lead to ATP synthesis. They occur under aerobic or anaerobic conditions. During these reactions, the oxidation of the substrate is accompanied by the creation of an energy-rich bond between the oxidized carbon and a phosphate molecule (phosphorylation at the substrate level).

$$XH_2 + P_i \rightarrow X \sim P + 2H^+ + 2e^-$$

The energy of the phosphate bond is then stored in a pyrophosphate bond of the ATP:

$$X \sim P + ADP \rightarrow X + ATP$$

In this manner, phosphoenolpyruvate and acetyl phosphate can transfer their phosphate group to the ADP in the same type of reaction. The two intermediate molecules in the catabolism of sugar are therefore very important from an energy standpoint.

5.2 Metabolism of Sugars: Lactic Acid Fermentation

The oxidation of sugars constitutes the principal pathway for producing the energy essential for bacterial growth. In lactic acid bacteria, fermentation is the pathway for the assimilation of sugars. For a given species, the type of sugar fermented and environmental conditions (the presence of electron acceptors, pH, etc.) modify the energy yield and the nature of the final products. In must and wine, lactic acid bacteria metabolize hexoses, pentoses, and possibly the sugars (especially glucose) released by glycosidases from various glycosylated derivatives (Section 5.5). Among the latter, nonvolatile glycosides are precursors of wine aroma compounds.

The cytoplasmic membrane is an effective barrier separating the external environment from the cellular cytoplasm. Although permeable to water, salts, and low-molecular-weight molecules, it is impermeable to many organic substances. Various works describe the different active sugar transport systems in lactic acid bacteria. They are for the most part ATP-dependent ABC transporters and major facilitator superfamily (MFS) permeases and phosphotransferase system (PTS) permeases using phosphoenolpyruvate for the translocation and simultaneous phosphorylation of the substrate. The phosphorylated sugar then enters the metabolic pathway. These transport mechanisms are not very well characterized in wine bacteria. In *Oenococcus oeni*, depending on the sugars, transport is supposed to use various systems; yet hypotheses are only based on genomic data and must be verified (Cibrario *et al.*, 2016).

Lactic acid bacteria of the *Lactobacillus*, *Leuconostoc*, *Oenococcus*, and *Pediococcus* genera assimilate sugars by either a homofermentative or heterofermentative pathway. Among the cocci, *Pediococcus* bacteria are homofermentative, while *Leuconostoc* and *Oenococcus* are heterofermentative. Among lactobacilli, heterofermenters and homofermenters are distinguished according to the pathway used for hexose degradation. Pentoses, when degraded, are metabolized by heterofermentation.

5.2.1 Homofermentative Metabolism of Hexoses

Homofermentative bacteria transform nearly all of the hexoses that they use, especially glucose, into lactic acid. Depending on the species, either the L- or D-lactate isomer is formed (Section 4.2.2). The homofermentative pathway, or Embden–Meyerhof pathway, includes a first phase containing all of the reactions of glycolysis that lead from hexose to pyruvate (Section 2.2.1). During this stage, the oxidation reaction (from glyceraldehyde 3-phosphate to 3-phosphoglyceric acid) takes place, generating the reduced coenzyme $NADH + H^+$. This pathway is used by numerous organisms. For aerobic organisms, this pathway is followed by the citric acid, or Krebs, cycle.

In lactic acid bacteria, the second phase involves the characteristic lactic acid fermentation reaction. The reduced coenzyme ($NADH + H^+$) is oxidized into NAD^+ during the reduction of pyruvate into lactate.

The reactions of glycolysis are listed in Figure 5.1. In the first stage, glucokinase catalyzes the phosphorylation of glucose into glucose 6-phosphate (glucose 6-P). This molecule then undergoes an isomerization to become fructose 6-P. Another

178 Chapter 5 Metabolism of Lactic Acid Bacteria

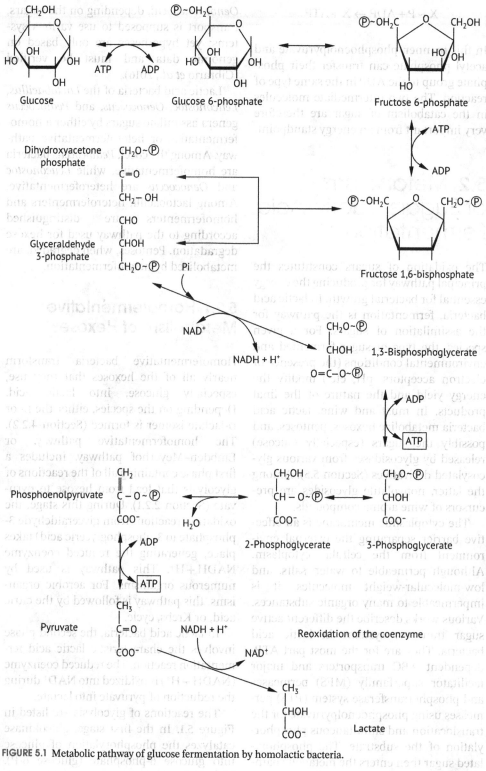

FIGURE 5.1 Metabolic pathway of glucose fermentation by homolactic bacteria.

phosphorylation leads to the formation of fructose 1,6-bisphosphate. At this stage, two important reactions have already occurred. They involve kinases, which require bivalent ions (Mg^{2+}, Mn^{2+}) and use an ATP molecule each time. One of them, phosphofructokinase, an allosteric enzyme controlled by ATP, determines the speed of glycolysis.

Fructose 1,6-bisphosphate is then split into two molecules of triose phosphate. This reaction is catalyzed by aldolase, a key enzyme of the glycolytic pathway. Homofermentative bacteria have a high fructose-1,6-bisphosphate aldolase activity. The products of this reaction are glyceraldehyde 3-P and dihydroxyacetone-P.

Only glyceraldehyde 3-P pursues the transformation pathway. Dihydroxyacetone-P is rapidly isomerized into aldehyde. In reality, the equilibrium between these two molecules favors dihydroxyacetone-P, but it is continually reversed, since glyceraldehyde 3-P is eliminated by the reaction that follows. In the next stage, energy production processes begin. The aldehyde is oxidized into 1,3-bisphosphoglycerate. A phosphorylation from inorganic phosphate accompanies the oxidation. The NAD^+ coenzyme is reduced to $NADH + H^+$. These reactions permit the synthesis of an acyl-phosphate bond—a high potential energy bond. During the hydrolysis of this bond, the reaction immediately following recuperates the energy by the synthesis of an ATP molecule. The 1,3-bisphosphoglycerate is transformed into glycerate 3-P. This molecule undergoes a rearrangement: its phosphate group passes from position 3 to position 2, esterifying in this manner the secondary alcohol function of the glycerate. An internal dehydration of the molecule then occurs. The important reaction that follows generates an enolphosphate, a high potential energy molecule, called phosphoenolpyruvate. Lastly, this energy is used for the synthesis of ATP from ADP in a reaction that forms pyruvate.

From the moment when the triose molecules are utilized, the second part of glycolysis comprises the most important energy-producing phases. Two reactions ensure the synthesis of ATP for each of the glyceraldehyde 3-P molecules coming from hexose. The total reaction energy from the transformation of a glucose molecule is therefore the synthesis of two ATP molecules and incidentally the reduction of NAD^+.

For each hexose molecule assimilated, the cell requires an NAD^+ molecule. The cell must therefore make use of a system that maintains an acceptable NAD^+ level. Lactic acid bacteria use the pyruvate formed by glycolysis as an electron acceptor to oxidize NADH. This defines lactic acid fermentation.

Bacteria therefore transform one hexose molecule almost exclusively into two lactate molecules by the homolactic pathway. In an anaerobic environment with a limited glucose concentration, homofermentative bacteria such as *Lactobacillus casei* form less lactic acid; the primary products can become acetic acid, formic acid, and ethanol. The change is linked to the regulation of the L-LDH enzyme by fructose 1,6-bisphosphate. The change is less obvious when the homofermentative species possess both LDH types, L and D. FBP does not regulate D-LDH.

5.2.2 Heterofermentative Metabolism of Hexoses

Bacteria using the heterofermentative pathway transform hexoses principally but not exclusively into lactate. The other molecules produced by this metabolism are essentially CO_2, acetate, and ethanol; this is the pentose phosphate pathway. After being transported into the cell, a glucokinase catalyzes the phosphorylation of glucose into glucose 6-P. Its fate is completely different from that of glucose 6-P in the homofermentative pathway. Two oxidation reactions occur successively: the first leads to gluconate 6-P; the second, accompanied by a decarboxylation, forms ribulose 5-P

(Figure 5.2). In each of these reactions, a molecule of the coenzyme NAD$^+$ or NADP$^+$ is reduced. Ribulose 5-P is then epimerized into xylulose-5.

Xylulose 5-P phosphoketolase is the key enzyme of this pathway: it catalyzes the cleavage of the pentulose 5-P molecule into acetyl-P and glyceraldehyde 3-P. This reaction requires inorganic phosphate. The glyceraldehyde 3-P is finally metabolized into lactic acid by following the same pathway as in the homofermentative pathway.

The acetyl-P has two possible destinations, depending on environmental conditions. This molecule can be successively reduced into acetaldehyde and then ethanol. In that case, the molecules of the coenzyme NADH + H$^+$ or NADPH + H$^+$, formed during the two oxidation reactions of hexose at the beginning of the heterofermentative pathway, are reoxidized. This reoxidation is essential for regenerating the coenzymes necessary for the assimilation of sugar.

Under certain conditions, when the cell makes use of other coenzyme reoxidation systems, acetyl-P leads to acetate. This reaction is coupled with recovery of the bond energy of the ~P group for the synthesis of an ATP molecule, catalyzed by acetate kinase. In this case, the coenzyme reoxidation systems use NADH or NADPH oxidases, when the cells are under aerobic conditions, or else they are reoxidized in substrate reductions. Fructose is thus reduced into mannitol. This very active mechanism is well known in the heterofermentative bacteria in wine. Other reactions are likely, but not yet demonstrated for wine bacteria. Such is the case for the reduction of hydroxycinnamic acid or vinylphenols, which are the result of decarboxylation (Filannino *et al.*, 2014). When acetyl-P follows this pathway, there is a definite energy advantage. An additional ATP molecule is formed for each hexose molecule transformed.

The final quantity of glucose metabolism products (ethanol and acetic acid) from heterofermentative bacteria zdemonstrates that acetyl-P is nearly always dephosphorylated, but the extent depends on the degree of aeration and the presence of other proton and electron acceptors. Under anaerobic conditions, NADH oxidase cannot regenerate NAD. In this case, glucose preferentially leads to the formation of lactic acid and ethanol. When NADH can be reoxidized by another process, the amount of ethanol formed decreases, resulting in an increase in acetic acid. This occurs under aerobic conditions or in the presence of another electron acceptor substrate.

In addition to products formed, conditions and mechanisms of coenzyme reoxidation change the energy yield, which influences growth. In the laboratory, the growth of heterofermentative bacteria is quicker in environments containing a glucose and fructose mixture compared with growth in glucose alone. Additionally, these are not strict anaerobic bacteria, and thus they reproduce better from glucose in contact with air for good functioning of the oxidases.

5.2.3 Metabolism of Pentoses

Certain strains of *Lactobacillus*, *Pediococcus*, *Oenococcus*, or *Leuconostoc* ferment pentoses such as ribose, arabinose, and xylose, whether they are homofermenters or heterofermenters, according to the same schema (Figure 5.3). The pentoses are phosphorylated by reactions involving kinases and using ATP. Specific isomerases then lead to the formation of the xylulose 5-P molecule. The subsequent reactions are described in the heterofermentative pathway for glucose assimilation. While glyceraldehyde 3-P has the same fate in this case, acetyl-P exclusively leads to the formation of the acetate molecule, generating one ATP molecule in this manner. In fact, a reduced coenzyme molecule is not available to reduce acetyl-P into ethanol. Pentoses are always at the origin of acetic acid in addition to lactic acid production.

5.2 | Metabolism of Sugars: Lactic Acid Fermentation

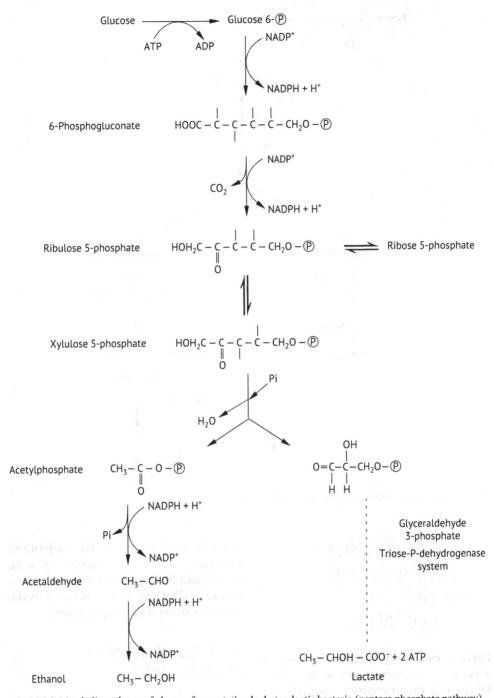

FIGURE 5.2 Metabolic pathway of glucose fermentation by heterolactic bacteria (pentose phosphate pathway).

The pathway furnishes two ATP molecules for each pentose molecule fermented. This pathway therefore has a greater yield than the fermentation of a hexose by the pentose phosphate pathway.

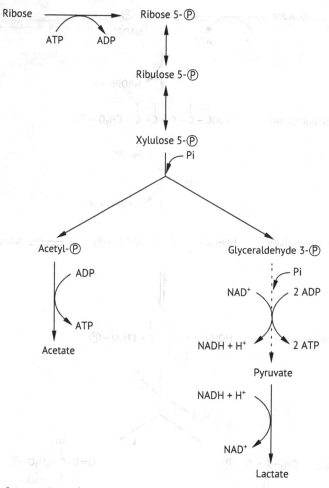

FIGURE 5.3 Pentose fermentation pathway by lactic acid bacteria.

5.3 Metabolism of the Principal Organic Acids of Wine

Bacteria essentially degrade two organic acids in wine: malic acid and citric acid. Other acids can of course be degraded but are of less interest in enology—with the exception of tartaric acid, which has rarely been studied. Since the initial research on lactic acid bacteria and their role in winemaking, malic acid has been the focus of a large number of studies. However, the degradation of citric acid also plays an important role in winemaking. The majority of bacterial species preponderant in wine after alcoholic fermentation degrade these two acids. This degradation is evidently the source of many organoleptic changes.

5.3.1 Transformation of Malic Acid

In the case of non-proliferating cells in a laboratory medium and during winemaking, lactic acid bacteria of wine transform malic acid exclusively into lactic acid. Seifert (1901) established the reaction of the malolactic transformation according to the following equation:

$$\text{Malic acid} \rightarrow \text{lactic acid} + CO_2.$$

This equation was confirmed and specified when the stereoisomers could be separately determined for each of the two acids. Each molecule of L-malic acid (the only metabolizable isomer) leads to an L-lactic acid molecule.

This reaction therefore involves a decarboxylation without an intermediate product capable of following another metabolic pathway. Several authors have reported that certain bacterial strains form other molecules from malic acid, suggesting in this manner the existence of other reactions. Even if their existence cannot be ruled out, malolactic transformation is the only reaction that exists in the lactic acid bacteria involved in winemaking.

Alizade and Simon (1973) studied the stereochemistry of this transformation. Enzymatic methods were used to determine the specific quantities of the stereoisomers. In addition, the fermentation of radioactively labeled glucose and malic acid enabled the study of their products. The heterofermentative cocci (*O. oeni*), which are abundant or exclusive during winemaking, were found to present several properties. They form exclusively D-lactic acid from glucose (Section 4.2.2) and exclusively L-lactic acid from L-malic acid (Figure 5.4).

This observation suggests that the transformation of malic acid does not pass via the intermediary of pyruvic acid. Peynaud (1968) concluded that the substrate was decarboxylated directly. A lot of research has been carried out to elucidate this mechanism. It naturally leads to the examination of the enzymatic aspect of this transformation.

Malate dehydrogenase (MDH) and malic enzyme fix and catalyze a reaction whose substrate is L-malic acid. These two enzymes have been described in numerous plant and animal cells and in diverse microorganisms. They catalyze the following reactions:

$$\text{MDH: L-malate} \underset{+ NAD^+}{\overset{Mn^{2+}}{\rightleftarrows}} \text{oxaloacetate} + NADH + H^+$$

$$\text{malic enzyme: L - malate} + NAD^+ \rightleftarrows \text{pyruvate} + CO_2 + NADH + H^+$$

Since oxaloacetate (metabolic intermediate) is easily decarboxylated into pyruvate and CO_2, these two reactions lead to the formation of pyruvate from L-malate. Since the final product of the malolactic transformation in wine is L-lactic acid, MDH or the malic enzyme would appear to be associated with an L-LDH catalyzing the reduction of pyruvate into L-lactate via this metabolic pathway. At least for wine bacteria (*O. oeni*), this concept is not acceptable, since these bacteria only possess a D-LDH, as demonstrated by the exclusive formation of D-lactate from glucose. Malic acid would only lead to the formation of D-lactic acid.

Therefore, the hypothesis of the existence of an enzyme catalyzing the direct decarboxylation of L-malic acid into L-lactic acid was proposed. The enzyme, called the malolactic enzyme, was purified for the first time in *Lactobacillus plantarum* (Schütz and Radler, 1974; Lonvaud, 1975). From bacterial extracts, enzymatic fractions respond to the functional

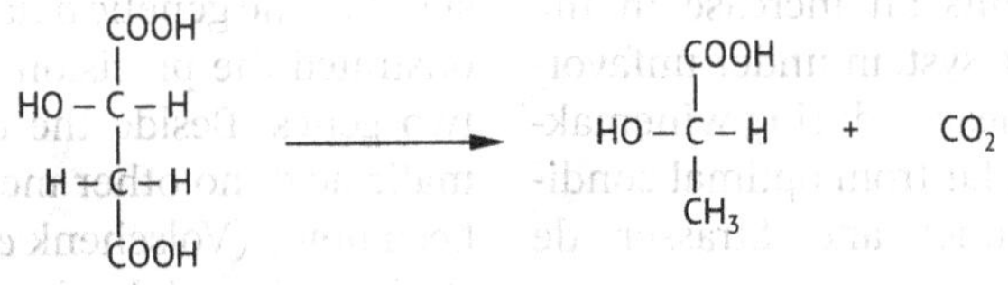

$$\begin{array}{c} COOH \\ | \\ HO-C-H \\ | \\ H-C-H \\ | \\ COOH \end{array} \longrightarrow \begin{array}{c} COOH \\ | \\ HO-C-H \\ | \\ CH_3 \end{array} + CO_2$$

L-Malic acid $\qquad\qquad\qquad$ L-Lactic acid

FIGURE 5.4 Equation of the malolactic reaction.

characteristics of the malolactic enzyme. L-Malic acid is transformed stoichiometrically into L-lactic acid. These fractions do not have any LDH activity.

At least in *Leuconostoc mesenteroides* and *O. oeni*, the malolactic enzyme is inducible. Cultivated without malic acid for numerous generations, the cells conserve a very small residual activity. They regain their maximum activity as soon as malic acid is added (1g/l or more). The presence of fermentable sugars (hexose or pentose) also favors its activity.

The same enzyme was purified in other strains and species of lactic acid bacteria, notably in strains of *L. plantarum*, *Lactobacillus murinus*, *L. mesenteroides*, *O. oeni*, and *Lactococcus lactis*. Overall, the physical characteristics and kinetics of all of the described malolactic enzymes are the same. The enzyme is a dimeric or tetrameric protein formed by the association of two or four 60 kDa monomers. The isoelectric pH of the enzyme is 4.35. It functions only in the presence of the NAD^+ cofactor and bivalent ions, Mn^{2+} being the most effective. It has the kinetics of a sequential ordered mechanism, in which Mn^{2+} and NAD^+ fix themselves on the protein before the L-malate. At the optimum pH, the Michaelis constants are 2×10^{-3} M for malate and 4×10^{-5} M for NAD^+. The optimum pH of the enzymatic reaction is 5.9. At this pH, the kinetics are Michaelian. At a pH far from the optimum pH, it is sigmoidal—demonstrating a positive cooperative mechanism that signifies a growing affinity for malate. Homopolymeric enzymes share this characteristic: the binding of the first substrate molecule on the first promoter transmits a deformation, increasing the affinity of the others. This cooperativity permits an increase in the effectiveness of the system under unfavorable conditions. Clearly, during winemaking, bacteria are in far from optimal conditions (Lonvaud-Funel and Strasser de Saad, 1982).

The carboxylic acids of wine—succinic, citric, and L-tartaric acids—are competitive inhibitors with the following inhibition constants: 8×10^{-2} M, 1×10^{-2} M, and 0.1 M, respectively. L-Lactic acid, a product of the reaction, is an inefficient, noncompetitive inhibitor whose inhibition constant of 0.3 M indicates a weak affinity.

Although this enzyme is becoming better known, a question still remains unanswered: what is the real role of NAD^+ in the oxidation–reduction exchange? This indispensable coenzyme of the reaction is not involved in the exchange, at least in a conventional manner.

The malolactic enzyme purified from *L. lactis*, a lactic acid bacterium of milk origin, has exactly the same characteristics as the enzyme found in wine strains. It was used to study the structure of the protein, which strongly resembles that of the malic enzyme (Denayrolles *et al.*, 1994; Ansanay *et al.*, 1993). Coenzyme attachment sites on the protein have also been located. This gene was transferred to laboratory *Saccharomyces cerevisiae* yeasts where it expresses itself (Denayrolles *et al.*, 1995; Ansanay *et al.*, 1993).

In the late 1980s, a program aimed at developing a "malolactic yeast" capable of carrying out the malolactic transformation during alcoholic fermentation was supported by winemakers in France and abroad. The first stage consisted of cloning the gene of the malolactic enzyme and expressing it in yeast. Unfortunately, it very rapidly became obvious that the system was limited by the uptake of malic acid into the yeast. To overcome this problem, the solution consisted in cloning in the malate permease gene from *Schizosaccharomyces pombe*, another yeast found in wine (Grobler *et al.*, 1995). These two genes were inserted into a yeast chromosome to stabilize the genetic data. Sequencing demonstrated the precision of insertion of the two genes. Beside the decarboxylation of malic acid, no other metabolic impact has been noted (Volschenk *et al.*, 1997). A yeast strain with malolactic activity now exists, can be produced on an industrial scale, and has shown a certain level of performance

(Husnik *et al.*, 2007). However, this yeast is a genetically modified microorganism and, as such, is far from gaining authorization in certain countries. Furthermore, its use cannot be compared with conventional malolactic fermentation, in which other bacterial activities are involved, thus enhancing the sensory qualities of wine. It may only be useful as an organic agent for deacidifying wines when the winemaker wishes to preserve the primary aromas and avoid effects of bacteria.

The production of the malolactic enzyme for direct use in wines is of no use, since this protein is rapidly inhibited by diverse substances in wine—acids, alcohol, and polyphenols.

The malolactic reaction takes place inside the bacterium in a medium protected from inhibitors by the bacterial membrane. Although the optimal pH for enzyme activity is around 6.0, it is around 3.0–3.5 for whole cells of *O. oeni*. At this pH, malic acid penetrates more easily into the bacterium than at higher pHs.

The inhibitory action of tartaric and succinic acids is even stronger on whole cells than on proteins. Citric acid at a concentration of 0.5 g/l, normally not reached in wine, only slows cellular activity by around 5% (Lonvaud-Funel and Strasser de Saad, 1982).

Lastly, among the questions raised from the very onset of research on malolactic fermentation, the physiological role of malic acid remains to be interpreted. The addition of malic acid to a culture medium of lactic acid bacteria simultaneously increases the yield and the growth rate. The malolactic reaction itself is not very exergonic, yet it indirectly constitutes a real energy source for the cell. Poolman (1993) demonstrated that, following the decarboxylation reaction, the increase of the internal pH (which imposes an influx of protons), the uptake of malic acid, and the efflux of lactic acid combine to create a proton motive force, permitting the conservation of energy via the membrane ATPase.

5.3.2 Metabolism of Citric Acid

Certain lactic acid bacteria (heterofermentative cocci and homofermentative bacilli) degrade citric acid. Among the species found in wine, *L. plantarum*, *L. casei*, *O. oeni*, and *L. mesenteroides* rapidly use citric acid. Strains of the genus *Pediococcus* and of the species *Lactobacillus hilgardii* and *Lactobacillus brevis* cannot.

In certain dairy industry bacteria, the lack of utilization of citric acid is linked to the loss of a plasmid that codes for citrate permease, which is essential for the uptake of this acid. In bacteria isolated in wine, especially *O. oeni*, citrate permease has not been studied as much, but its role is inconsequential; since at the pH of wine, the undissociated substrate diffuses across the membrane without needing the permease. The species and the strains that do not degrade citric acid are therefore at least deficient in the first enzyme of the metabolic pathway: citrate lyase. This enzyme was studied in wine lactic acid bacteria and more particularly in a strain of *L. mesenteroides* (Weinzorn, 1985).

Within bacteria, citric acid is split into an oxaloacetate molecule and an acetate molecule by the lyase (Figure 5.5). The largest quantities of this enzyme are synthesized in low sugar media containing citric acid. Glucose acts as a repressor. The protein is active in an acetylated form. The inactive deacetylated form can be reacetylated *in vivo* by citrate lyase ligase with acetyl-CoA or acetate and ATP. This first degradation stage leads to the formation of an acetate molecule for each molecule of the substrate.

Oxaloacetate is then decarboxylated into pyruvate in *O. oeni*, the most important bacteria in enology. In certain *Lactobacillus* species, it can also lead in part to formation of succinate and formate. Pyruvate is the source of acetoinic compounds: diacetyl, acetoin, and 2,3-butanediol. The first is particularly

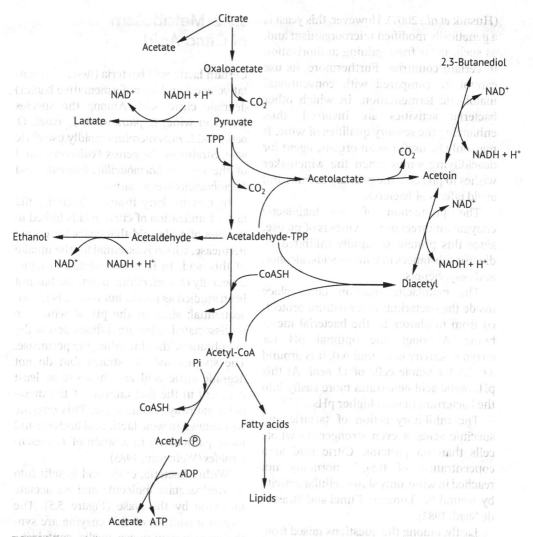

FIGURE 5.5 Metabolic pathway for citric acid degradation by lactic acid bacteria.

important organoleptically. It is the very aromatic molecule that gives butter its smell. The aroma intensity of acetoin and butanediol, which are derived from diacetyl by reduction of the ketone functions, is much lower (Figure 5.5).

Two diacetyl synthesis pathways have been proposed. In one, diacetyl results from the reaction of acetyl-CoA with acetaldehyde-TPP (active acetaldehyde), catalyzed by a diacetyl synthetase, which has never been isolated. The other pathway supposes that from two pyruvate molecules, α-acetolactate synthetase produces α-acetolactate that is then decarboxylated into acetoin. The diacetyl is derived from it by oxidation. This is an aerobic pathway.

In addition to acetoinic substances, the pyruvate molecules coming from citrate have other fates. First of all, if the coenzyme NADH, produced by other pathways, is available, it leads to the formation of lactate. Next comes the decarboxylation of pyruvate and then a reduction produces ethanol. Finally, the pyruvate derived from citrate participates in the synthesis of fatty acids and lipids via acetyl-CoA. The radioactivity of labeled citrate supplied to the bacteria is incorporated into the cellular material. In this pathway, part of the acetyl-CoA can also generate acetate molecules (Figure 5.5).

The products of citric acid metabolism are therefore very diverse. Whatever the conditions, more than one molecule of acetic acid is surely formed from a substrate molecule. The production of others is largely determined by factors influenced by growth conditions. Under conditions of limited glucose concentration, low pH, and the presence of growth inhibitors, citric acid preferentially leads to the formation of acetoinic substances. In fact, owing to these conditions, pyruvate is orientated neither toward the synthesis of cellular material, since growth is difficult, nor toward lactate and ethanol, because of a lack of reduced coenzymes. The acetoinic substance synthesis pathway is considered to be a detoxification process of the cell. In order to maintain its intracellular pH, it must eliminate pyruvate. Conversely, when growth is easy, pyruvate is utilized by fatty acid synthesis pathways; acetic acid is produced in larger quantities. In a laboratory medium, *O. oeni* cells form more than two acetic acid molecules from one citric acid molecule at a pH of 4.8 and only 1.2 molecules at a pH of 4.1. Conversely, the production of acetoinic molecules is four times higher (Lonvaud-Funel *et al.*, 1984).

It is therefore not surprising that some wines contain more than 10 mg of diacetyl per liter. Yet it has been determined that a few milligrams of diacetyl per liter (2–3 mg/l for white wines and about 5 mg/l for red wines) contribute favorably to wine bouquet. Above these concentrations, the clearly perceptible buttery aroma diminishes wine quality. Malolactic fermentation conditions, the quantity of citric acid degraded (from 0.2 to 0.3 g/l), and also without doubt the strains of *O. oeni* involved determine the quantity of diacetyl produced. Several other reactions contribute to its final concentration. First of all, yeasts also synthesize diacetyl during alcoholic fermentation by a completely different pathway linked to the metabolism of amino acids. Diacetyl is then reduced into acetoin by diacetyl reductase, an enzyme present in yeasts and lactic acid bacteria. The diacetyl concentration attains two maxima in this manner: one during alcoholic fermentation, the other during the degradation of citric acid by bacteria. It diminishes between the two fermentations and at the final stage of bacterial activity. Maintaining wine on yeast and bacteria lees at the end of fermentation ensures this reduction and also determines the final diacetyl level (De Revel *et al.*, 1989). Finally, sulfur dioxide further diminishes its concentration by combining with the ketone functions.

Citric acid is always metabolized during fermentation because in nearly every case the *O. oeni* species is involved. Its degradation begins at the same time as malic acid degradation, but it is much slower—so much so that at the end of malolactic fermentation, citric acid often remains (up to 0.15 g/l or sometimes even more). It represents an additional energy source. In fact, ATP is formed from acetyl-P, derived from pyruvate, which is directed toward the production of cell components. In the presence of residual sugars (glucose or fructose) degraded by the heterofermentative pathway, part of the pyruvate derived from citric acid acts as an electron and proton acceptor. Part of the acetyl-P originating from sugars leads to the formation of acetate by producing ATP. In this manner the presence of citric acid in a wine favors bacterial growth and survival, notably in the presence of residual sugars. This metabolism therefore participates, along with the malic acid metabolism, in the microbiological stabilization of wine by eliminating energy sources (Lonvaud-Funel, 1986).

5.3.3 Metabolism of Tartaric Acid

Wine lactic acid bacteria can degrade tartaric acid, but this metabolism differs from malic and citric acid metabolisms. It is a true bacterial spoilage. Pasteur described it in the last century and named it *tourne* disease (tartaric acid degradation in English). It is dangerous since the disappearance of tartaric acid, an essential acid in wine, lowers the fixed acidity and is accompanied by an increase in the volatile acidity. The degradation can be total or partial, depending on the level of bacterial growth, but it always lowers wine quality.

This spoilage is rare since the strains capable of degrading tartaric acid seem to be relatively few in number. Studies carried out on this subject in the 1960s and 1970s showed that this property is not linked to a particular bacterial species. Strains belonging to different species have been isolated by various authors, but they are most often lactobacilli. Radler and Yannissis (1972) found four strains of *L. plantarum*, and one strain of *L. brevis* having this trait out of the 78 strains examined. Peynaud (1967) discovered 30 or so strains capable of partially or totally using tartaric acid in a study carried out on more than 700 strains. The scarcity of this phenotypic trait is thus the first protection against this disease.

Since a high pH is always propitious to the development of a larger number of

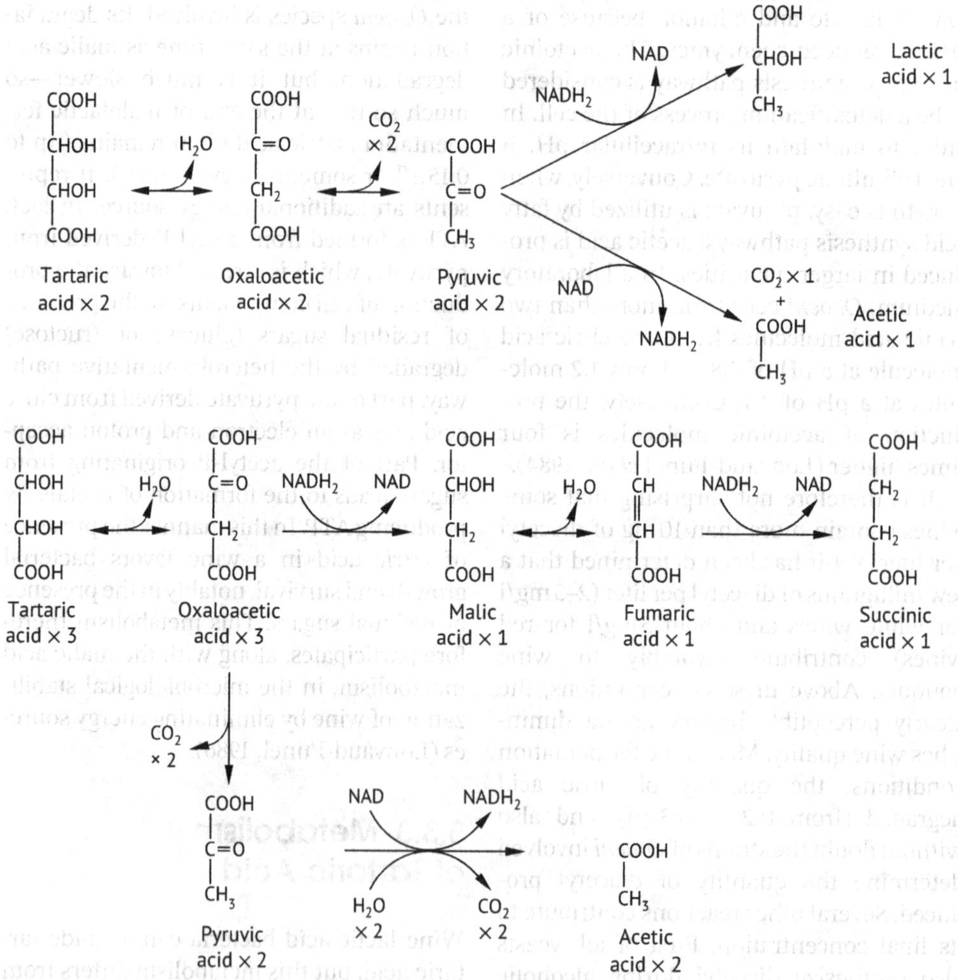

FIGURE 5.6 Tartaric acid metabolism by lactic acid bacteria (Radler and Yannissis, 1972).

bacteria, higher acidity wines are less affected. Moreover, these bacteria are sensitive to SO_2. Therefore, respecting the current rules of hygiene in the wine cellar and in wine should be sufficient to avoid this problem.

Few studies exist examining the metabolic pathways of the transformation of tartaric acid. The only existing results describe different pathways for *L. plantarum* and *L. brevis* (Radler and Yannissis, 1972). From one molecule of tartaric acid, *L. plantarum* produces 0.5 molecules of acetic acid, 0.5 of lactic acid, and 1.5 of CO_2. *Lactobacillus brevis* forms 0.7 molecules of acetic acid, 0.3 of succinic acid, and 1.3 of CO_2 (Figure 5.6).

5.4 Other Transformations Likely to Occur in Winemaking

5.4.1 Degradation of Glycerol

Glycerol is one of the principal components of wine, both in its concentration (5–8 g/l) and in its contribution to taste. Yeasts form glycerol by means of the glyceropyruvic fermentation at the beginning of vinification. The degradation of glycerol harms wine quality, partly because of the decrease in its concentration and partly because of the resulting products of the metabolism. Certain bacterial strains produce bitterness in wine—a fact known since the time of Pasteur. Lactic acid bacteria make use of a glycerol dehydratase to transform glycerol into β-hydroxypropionaldehyde (Figure 5.7). This molecule is the precursor of acrolein, which is formed in wine by heating or slowly during aging. The combination of wine tannins and acrolein, or its precursor, gives a bitter taste.

Like tartaric acid degradation, this spoilage is not widespread, due to the rarity of strains capable of degrading glycerol by this pathway. No single species of bacteria is responsible for degrading glycerol in wine. Little research has been devoted to this problem, but strains of at least two species of bacteria, *L. hilgardii* and *Lactobacillus diolivorans*, have been isolated from wine following degradation of glycerol (Claisse, 2002). A key metabolic enzyme, glycerol dehydratase, has been studied in several strains. In strains of *L. brevis* and *Lactobacillus buchneri*, Schutz and Radler (1984) demonstrated the degradation of glycerol by glycerol dehydratase to 1,3-propanediol via β-hydroxypropionaldehyde when the medium also contained glucose or fructose. NADH (or NADPH), produced by the fermentation of sugar, reduces β-hydroxypropionaldehyde to 1,3-propanediol. As described earlier, this cometabolism leads to a deviation of acetyl-P derived from sugar toward the production of additional ATP and acetate. It therefore facilitates bacterial growth. Other bacteria also use the glycerol 3-P dehydrogenase pathway.

The genes coding the enzymes for the first pathway have been studied. They are organized in a set including a total of 13 genes, probably all necessary for the functioning of glycerol dehydratase, which consists of three protein subunits and 1,3-propanediol dehydrogenase, leading, finally, to 1,3-propanediol. *Lactobacillus hilgardii* and *L. diolivorans* are organized in the same way, as are strains of *Lactobacillus collinoides* isolated from cider affected by acrolein spoilage (Gorga *et al.*, 2002). Oligonucleotide primers for detecting these bacteria have been selected from the gene sequence coding for one of the glycerol dehydratase subunits (Claisse and Lonvaud-Funel, 2001). Degradation of glycerol results not only in the production of 3-hydroxypropionaldehyde, the precursor of acrolein, but also, by metabolic coupling, to an increase in volatile acidity produced from the L-lactic acid in the wine.

The other pathway consists of glycerol kinase phosphorylating the glycerol and glycerol-3-P dehydrogenase resulting in

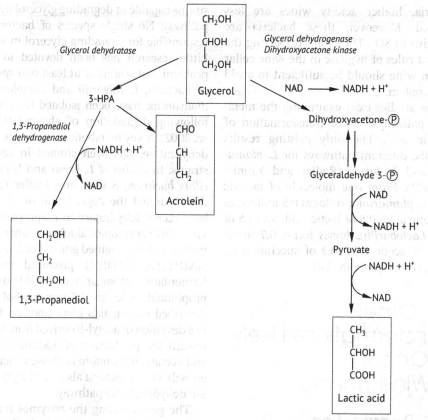

FIGURE 5.7 Glycerol degradation pathways by lactic acid bacteria.

dihydroxyacetone phosphate. This molecule enters into glycolysis reactions that result in the formation of pyruvate. The final products of this pathway are those previously described from pyruvate degradation. The quantity of products varies depending on environmental conditions, in particular the amount of fermentable sugar and aeration.

In particular, large amounts of lactic acid are formed, increasing the wine's total and fixed acidity and causing its pH to drop. Acidification can be as high as 0.8 g/l (expressed in tartaric acid) in wines where the pH has dropped by 0.25–0.30.

Acetoinic molecules may also be formed from pyruvate. The fate of the pyruvate molecules is probably, as usual, determined by the availability of NADH/NAD coenzymes, i.e. the intracellular redox condition. If NADH is available, lactate is produced. If not, the pyruvate is eliminated in the form of acetoinic compounds. Interactions between the metabolic pathways and the bacteria's environment are the decisive factor. In any case, the presence of bacteria capable of degrading glycerol is a risk to the extent that, even if it is not totally metabolized, its metabolic byproducts may spoil the wine to varying degrees.

5.4.2 Formation of Biogenic Amines

Histidine, an amino acid in wine, is decarboxylated into histamine (Figure 5.8), whose toxicity, although low, is additive to the toxicity of other biogenic amines (tyramine, phenylethylamine, putrescine, and cadaverine) (Volume 2, Section 5.4.3).

Tyrosine is decarboxylated to form tyramine by a similar reaction.

In general, biogenic amines, histamine in particular, are more abundant in wines after malolactic fermentation. This explains the results presented in various works: red wines appear to be richer in amines than white wines. Some researchers (Aerny, 1985) also proved that histamine is formed mainly at the end of malolactic fermentation, and even later. For a long time, many attempts were made to isolate bacteria capable of decarboxylating histidine, and they led to the conclusion that only contamination strains belonging to the *Pediococcus* genus had this property. According to some authors, the presence of histamine indicates a lack of winemaking skill and hygiene in the winery.

Yet this phenomenon occurs even during the malolactic fermentations of wines whose microflora is almost exclusively *O. oeni*. This fact contradicts the above results. An in-depth study of the microflora of wines rich in histamine finally led to the isolation of *O. oeni* strains that produce histamine from histidine (Lonvaud-Funel and Joyeux, 1994). In laboratory media, the formation of histamine by these strains increases as the growth conditions become less favorable, the absence of other substrates (in particular, sugar) at a low pH and in the presence of ethanol. The histamine concentration also, of course, depends on the histidine concentration. The addition of yeast lees, which progressively liberate amino acids and peptides, increases the concentration in the medium. Bacteria that also possess peptidases find a significant histidine source there. It is therefore not surprising that the histamine concentration increases toward the end and even after fermentation, as long as lactic acid bacteria are present. A precise study of the histidine carboxylase activity of a strain of *O. oeni* shows that even non-viable cells conserve their activity for a very long time in wine.

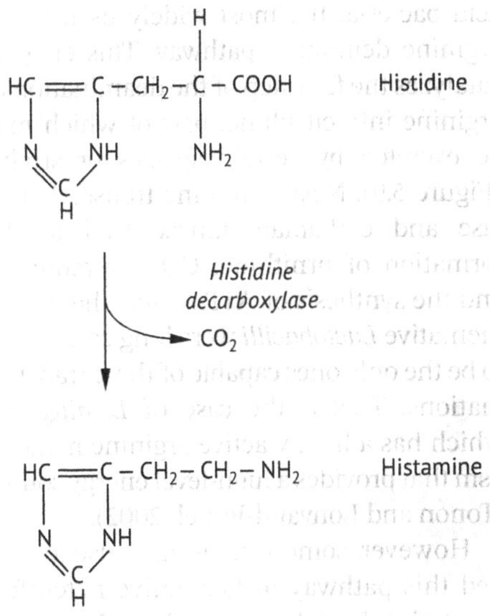

This enzyme has been fully purified (Coton, 1996). Its characteristics and properties are similar or identical to those of the enzyme purified from a *Lactobacillus* of non-enological origin. The protein comprises two different subunits grouped in an $[\alpha\beta]_6$ hexamer. The activity is optimal at a pH of 4.5. Optimum pH is the same for whole cells. Like the malolactic reaction, the decarboxylation of histidine does not directly generate cellular energy, but the strains of *O. oeni* that are capable of using it benefit from its presence in the medium and grow faster than strains that do not have this capability. The energy advantage is explained (as in *L. buchneri*) according to the process described for malolactic fermentation by Poolman (1993). The exchange between histidine and histamine at the membrane level creates a proton gradient and a proton motive force, generating ATP.

The strains of *O. oeni* that use histidine, therefore, have an additional advantage that can be a deciding factor after vinification when the medium is poor in nutrients. Their survival can be facilitated in this manner.

Not all *O. oeni* strains possess histidine decarboxylase; in fact, there are only a few. The detection of these particular strains has been made possible by the use of

FIGURE 5.8 Histidine decarboxylation reaction.

molecular tools. DNA amplification by PCR, by using the appropriate oligonucleotide primers, enables the identification of these particular strains in a mixture and quantification by quantitative PCR (qPCR in real time). The labeling of a gene fragment, or a whole gene, that encodes the enzyme also supplies the specific probe, which can be used in DNA–DNA hybridization. Thus, the detection and counting of strains capable of forming histamine no longer present a problem. For example, the analysis of 250 samples of Bordeaux wines showed that nearly 50% contained such strains (Le Jeune et al., 1995; Coton, 1996).

Until now, the only strains isolated from wine capable of producing tyramine have been identified as L. brevis and L. hilgardii (Moreno-Arribas et al., 2000). The tyrosine decarboxylase enzyme was studied in a particularly active strain of L. brevis, and then the entire operon coding for proteins necessary for this activity was sequenced (Moreno-Arribas and Lonvaud-Funel, 2001; Lucas and Lonvaud-Funel, 2002; Lucas et al., 2003).

Lactic acid bacteria form putrescine by using two distinct metabolic pathways, which are both related to arginine metabolism. The agmatine deiminase pathway is the most common. It has been detected in several species of various origins, notably in wine (Lucas et al., 2007). Agmatine, its substrate, is the product of arginine decarboxylation. Its origin in wine is still uncertain. Agmatine may form in grapes through yeasts or certain bacteria. The agmatine deiminase pathway is multi-enzymatic and similar to the arginine deiminase pathway (Section 5.4.3). Even though bacteria possessing this pathway are detectable in wines, it seems that the presence of high putrescine concentrations is more correlated to bacteria possessing the ornithine decarboxylase pathway (Nannelli et al., 2008). It includes a specific decarboxylase, converting ornithine into putrescine, and a membrane transporter, ensuring the entry of ornithine into the cell and the exit of putrescine. Ornithine, which is present in wine, may come from the metabolism of arginine by the arginine deiminase pathway. At present, the ornithine decarboxylase pathway has been detected in several O. oeni strains, but not in other wine bacteria.

Tools for the rapid, accurate identification and quantification of strains producing biogenic amines are available. A quantitative PCR protocol simultaneously targeting genes coding for essential proteins, leading to histidine, tyramine, and putrescine (via two pathways), is used to quantify bacteria potentially producing amines, during or after vinification (Nannelli et al., 2008). It may be used to carry out studies on the ecology of these strains, particularly their distribution and conditions encouraging their presence in some wineries.

5.4.3 Metabolism of Arginine

Among grape must amino acids, arginine is the most rapidly and completely consumed at the beginning of alcoholic fermentation. It is then secreted by the yeast and released during autolysis. The use of this amino acid by bacteria during fermentation may be the source of ethyl carbamate (Liu et al., 1994; Makaga, 1994).

Microorganisms employ several metabolic pathways for using arginine. In lactic acid bacteria, the most widely used is the arginine deiminase pathway. This enzyme catalyzes the first step of the deamination of arginine into citrulline, part of which may be excreted by certain species or strains (Figure 5.9). Next, ornithine transcarbamylase and carbamate kinase lead to the formation of ornithine, CO_2, ammonium, and the synthesis of ATP. Strictly heterofermentative Lactobacilli were long considered to be the only ones capable of these transformations. This is the case of L. hilgardii, which has a highly active arginine metabolism that provides a high-level energy source (Tonon and Lonvaud-Funel, 2002).

However, some studies have also identified this pathway in facultative heterofermentative Lactobacilli, such as L. plantarum (Arena et al., 1999).

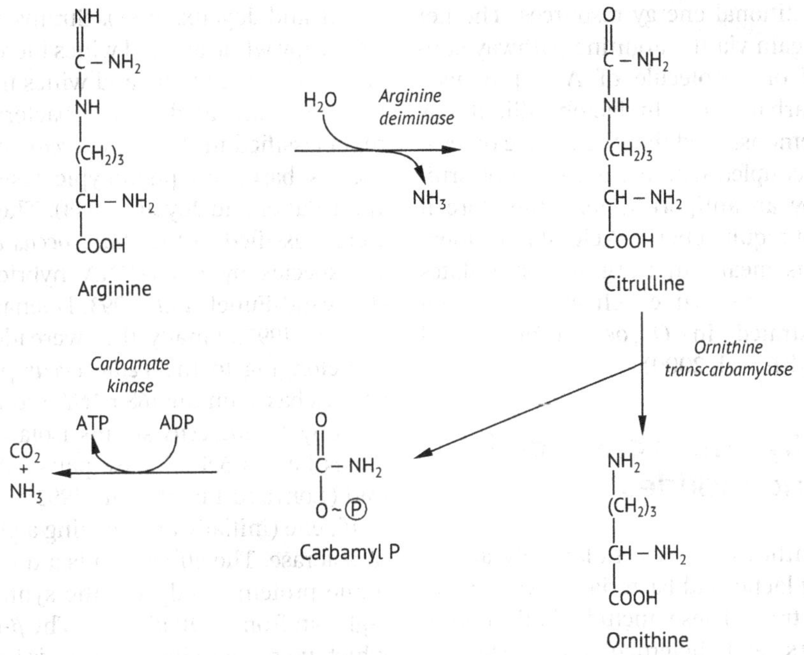

5.4 | Other Transformations Likely to Occur in Winemaking

FIGURE 5.9 Arginine degradation mechanism by *O. oeni*.

Interestingly, *O. oeni* was classified in Bergey's reference manual for the identification of bacteria as being among those that do not hydrolyze arginine. However, in reality, certain strains of *O. oeni* are capable of degrading arginine via the arginine deiminase pathway.

The synthesis of three enzymes—arginine deiminase, ornithine transcarbamylase, and carbamate kinase—is induced by the presence of arginine in the medium. In *O. oeni*, the genes that code for the enzymes of this metabolic pathway are organized in an operon on the chromosome including *arc* A, B, and C genes coding for arginine deiminase, ornithine transcarbamylase, and carbamate kinase; preceded by a regulation gene, *arc* R; and followed by two genes coding for transport proteins, *arc* D1 and *arc* D2 (Tonon et al., 2001; Divol et al., 2003). In a collection of *O. oeni*, the strains do not systematically have this complete region of the chromosome. Of course, only those that have the complete region are capable of assimilating arginine via this pathway.

A strain of *O. oeni* studied by Liu et al. (1994) transforms arginine stoichiometrically into one mole of ornithine and two moles of ammonium. A small quantity of citrulline is still released in the medium (up to 16%); it is not taken up in the catabolism of *O. oeni* as opposed to what occurs in lactobacilli.

The formation of ethyl carbamate corresponds with the degradation of arginine by wine bacteria. However, quantities formed are considerably lower than those originating from the urea released by yeasts during alcoholic fermentation. There is no need to worry about the development of strains of *O. oeni* that degrade arginine during malolactic fermentation. It is, however, advisable to prevent these strains, as well as heterofermentative *lactobacilli*, from proliferating after malolactic fermentation. While the wine no longer contains much free arginine, it does contain residual proteins or peptides released from lees.

As is the case in decarboxylation reactions involving malic acid and amino acids, degrading arginine provides yeasts with additional energy resources. The net energy gain via the arginine pathway consists of one molecule of ATP produced from carbamoyl-P. In lactobacilli, it has been demonstrated that the uptake of arginine is coupled with the excretion of ornithine by an antiport system; therefore it does not require energy. At least in lactobacilli, this means that arginine stimulates growth. The same effect has been demonstrated in *O. oeni* (Tonon and Lonvaud-Funel, 2000).

5.4.4 Synthesis of Exocellular Polysaccharides

The synthesis of exocellular polysaccharides by lactic acid bacteria is a very widespread trait. These include both homopolymers and heteropolymers. Dextran (glucose homopolymer) from *L. mesenteroides* is the best known, as much for its different structures and its biosynthesis as for its various applications.

In wine, the production of exocellular polysaccharides increases viscosity. They give rise to ropiness, studied by Pasteur. The wine is thick and flows like oil.

In the literature, increased viscosity in ciders and beers is attributed to different lactic acid bacteria species (Williamson, 1959; Beech and Carr, 1977; Duenas-Chasco et al., 1998; Walling et al., 2005b; Garai-Ibabe et al., 2010). Luthi (1957) established that the symbiosis between lactic acid bacteria and acetic acid bacteria accelerates the increase in viscosity of the medium.

In *O. oeni*, genetic determinants related to biosynthesis pathways are not identical for all strains (Dimopoulou et al., 2014). Polymers accumulate around cells, forming a more or less dense capsule. Some are released in the medium. These polysaccharides protect cells against stress and phagic attacks and regulate the adhesion properties of strains on surfaces (Caggianiello et al., 2016; Dimopoulou et al., 2016).

Pediococcus bacteria are the most frequently involved in ropiness (Lonvaud-Funel and Joyeux, 1988). Strains isolated from ropy white and red wines increase the viscosity of the media and wines in which they are cultivated. These bacteria were first classified in the *Pediococcus cerevisiae* species based on phenotypic tests (Lonvaud-Funel and Joyeux, 1988). Then, they were classified in the *Pediococcus damnosus* species by DNA–DNA hybridization (Lonvaud-Funel et al., 1993; Duenas-Chasco et al., 1997). Finally, they were identified as belonging to the *Pediococcus parvulus* species based on the 16S rRNA sequence.

Ropy *Pediococcus* strains isolated from wines have a 5.5 kilobase pairs (kb) plasmid (Lonvaud-Funel et al., 1993), carrying a *gtf* gene (initially dps), coding a glucosyltransferase. The *gtf* protein is a transmembrane protein, catalyzing the synthesis of β-glucan from UDP-glucose. The β-glucan, which increases wine viscosity, is a glucose homopolymer. Its structure comprises a β-1,3 chain on which a glucose unit is attached via β-1,2 linkages every two units (Llaubères et al., 1990; Dols-Lafargue et al., 2008).

$$-[glc-glc]-_n$$
$$|$$
$$glc$$

Wine microorganisms other than *P. parvulus* have the *gtf* gene and are able to produce β-glucan. This is the case of some strains of *P. damnosus*, *L. diolivorans*, *Lactobacillus suebicus*, and *O. oeni*. Nevertheless, their ability to produce enough glucan to trigger ropiness in wine has not been demonstrated (Dols-Lafargue et al., 2008).

Compared with non-ropy strains, the ropy strains are distinguished by the existence of a sort of refringent capsule around the cell, clearly visible under the microscope. The colonies formed on a solid medium by ropy strains are also easily identified by the formation of a thick filament when picked with a platinum needle. In a given wine, the increase in viscosity is directly correlated to the production of the polysaccharide. However, it is not possible to diagnose the spoilage of a wine by only measuring viscosity. Many components other than β-glucan contribute to wine viscosity, notably alcohol content.

At the physiological level, the ropy strains demonstrate the occasional ability to adapt to growth in wine. Above all, they develop with the same ease, whatever the alcoholic content, up to 15% or more. Their growth rate is hardly reduced at a pH of 3.0 compared with a pH of 4.5 and is not much affected by sulfur dioxide. Growth remains normal at pH 3.7 with a free SO_2 concentration of 30 mg/l (Lonvaud-Funel and Joyeux, 1988; Dols-Lafargue et al., 2008).

Ropy P. parvulus strains increase wine viscosity when they multiply, and the production yield of glucan is higher in a medium poor in sugars or nitrogen, as is the case of wine. The pH and acidity of spoiled wines are comparable with those of clean, fault-free wine. However, their microbial population is high (10^5–10^7 cell/ml), and the free SO_2 content is often under 10 mg/l. It is generally accepted that the spoilage is clearly visible when the population exceeds 10^7 CFU/ml.

If spoilage occurs in the winery or a warehouse, the first precaution is clearly the rigorous disinfection of all of the tanks and winery equipment in order to avoid future contaminations. Ropiness can occur in the tank at the end of fermentation, but most problems are posed by spoilage in the bottle—mainly a few months after bottling.

In general, a ropy wine does not present any other organoleptic defects, and it can therefore be marketed after the appropriate treatments. The viscosity can generally be lowered by the mechanical action of agitating the wine. Bacteria must be eliminated naturally.

Microbiological analysis enables the detection of spoilage Pediococcus bacteria: colonies grown in viscous count media and cells observed under the microscope surrounded by a halo. To reduce the detection threshold to the necessary level (≈100 cells/ml), methods based on the amplification of the gtf gene by PCR have been developed (Gindreau et al., 2001; Delaherche et al., 2004; Walling et al., 2005b). The analysis is especially recommended in order to assess the presence of bacteria as soon as the spoilage is suspected and, throughout the following years, in a cellar where the contamination has already caused this problem. Since the production of β-glucan only occurs if Pediococcus cells multiply, their development should be stopped, or, better yet, appropriate means should be implemented to ensure their elimination (Walling et al., 2005a). Sulfiting is not very effective, especially if the pH is over 3.5. Progressive filtrations up to a sterile filtration and the heat treatment of wine, just before bottling, are reliable solutions for these fragile wines. Lysozyme and β-glucanase treatments have also been suggested (Coulon et al., 2012).

5.5 Effect of the Metabolism of Lactic Acid Bacteria on Wine Composition and Quality

Unless the appropriate inhibitory treatments are applied, lactic acid bacteria are part of the normal microflora of all white and red wines. From the start to the end of fermentation, and even during aging and storage, they alternate between successive growth and regression periods depending on the species and the strains. All multiplication or survival of the cell involves a metabolism that is perhaps very active or, on the contrary, hardly perceptible and even impossible to detect with current analytical methods. Substrates are transformed, and consequently organoleptic characteristics are modified. Some metabolic activities are favorable, and others are without consequence, while some are totally detrimental to wine quality (Volume 2, Section 8.3).

The main substrates for wine bacteria known to date are simple molecules: sugars and organic acids. Other more complex wine components, such as phenolic compounds, aromatic compounds, or aroma precursors, present in small quantities, are without doubt partially metabolized. Such is the case for highly varied glycosylated molecules, which release fragrant molecules, after bacterial glycosidase action. In *O. oeni*, activities vary depending on strains and on the substrate (Boido et al., 2002; Ugliano et al., 2003, Grimaldi et al., 2005; Gagne et al., 2011). During malolactic fermentation in barrels, they transform wood molecules (Bloem et al., 2008). Similarly, esterases intervene in the synthesis and hydrolysis of esters whose sensory impact has been demonstrated. Their activity depends on strains and substrates (Antalick et al., 2012; Costello et al., 2012). The repercussions of these transformations can have an impact on the characteristics of wines. Yet it remains minor and also quite unpredictable: first because enzymatic activities depend on strains but also for a given strain on its environment and on physicochemical conditions, and secondly, because a bacterium is not alone in the wine. Interactions with other microorganisms (yeasts or bacteria) involve competition for the use of all substrates, whether or not they are the source of aroma and flavor molecules, which all of these microorganisms are able to use with more or less affinity.

The only substrate always metabolized via the same pathway by all species of wine bacteria is L-malic acid. Cellular activity is modulated by the presence of other compounds acting on transport or on enzyme activity. The growth of lactic acid bacteria in wine is sought after because of this activity; indeed, it is the only activity truly desired. It permits the softening of wine provoked by deacidification and by the replacement of malic acid with lactic acid, a compound with a less aggressive flavor.

Bacteria degrade must and wine sugars with a different affinity depending on the species. Hexoses are fermented into L- or D-lactic acid, or a mixture of the two forms, depending on the species. In general, bacterial development occurs after yeast development. Therefore, the lactic acid formed from sugars is in negligible quantities compared with the amount coming from malic acid. Several bacterial species produce D-lactic acid, but it is the exclusive form for heterofermentative cocci and thus for *O. oeni*, the most important bacterium in enology. Among the sugar fermentation products of *O. oeni*, acetic acid is significant because of its contribution to the volatile acidity of wines. Like D-lactic acid, it is produced in small quantities as long as the bacteria do not ferment too much residual sugar. An increase in volatile acidity can therefore be attributed to lactic acid bacteria, if an abnormal amount (>0.3 g/l) of D-lactic acid is simultaneously formed. In this case, *O. oeni* ferments a significant quantity of sugars (a few grams per liter). This situation is called lactic spoilage.

Acetic acid is also one of the unavoidable metabolites of citric acid, produced by homofermentative lactobacilli and especially by heterofermentative cocci. The fermentation of a few hundred milligrams of sugars per liter increases the volatile acidity during malolactic fermentation. Although carried out on a small quantity of the substrate, the degradation of citric acid is certainly important on account of the production of diacetyl.

Diacetyl and other α-dicarbonyl compounds in wine, e.g. glyoxal, methylglyoxal, and pentanedione, produced partly by the metabolism of lactic acid bacteria, are highly reactive. Reactions, in particular those with cysteine in wine, produce heterocycles such as thiazole; described as smelling of popcorn, toast, and hazelnuts; thiophene; and furan, with aromas of coffee and burnt rubber (Marchand et al., 2000).

Methionine and cysteine are metabolized into volatile sulfur compounds. The *O. oeni* species is particularly active in converting cysteine into hydrogen sulfide and 2-sulfanyl ethanol and methionine into dimethyl disulfide, 3-(methylsulfanyl)-propan-1-ol, and 3-(methylsulfanyl)-propionic acid. The most interesting of these compounds from a sensory point of view is 3-(methylsulfanyl)-propionic acid, with its earthy, red-berry fruit nuances (Pripis-Nicolau, 2002).

The other known metabolisms of lactic acid bacteria in wine all participate in one way or another in the spoilage of wine. The degradation of essential wine components such as tartaric acid and glycerol into volatile acidity and bitter-tasting substances, respectively, completely destroys the organoleptic quality of the wine. The metabolism of amino acids (arginine, histidine, etc.) does not actually affect taste, but at a toxicological level, it creates a problem by increasing the concentrations of undesirable biogenic amines and ethyl carbamate precursors in the wine. Lysine and ornithine are key molecules in the production of tetrahydropyridines formed by *Brettanomyces bruxellensis* and, to a lesser extent, by lactic acid bacteria. This defect, called mousiness, is perceived by retronasal olfaction and reminds tasters of the odor emanating from a rodent's cage or from toasted or puffed grain foods. As highlighted by Heresztyn (1986), it is associated with the synthesis of 2-acetyl-tetrahydropyridine (ATHP), 2-ethyl-tetrahydropyridine (ETPH), and 2-acetyl-1-pyrroline (APY). The proposed synthesis mechanism involves ethanol and acetaldehyde, molecules related to the heterofermentative metabolism of sugars. Heterofermentative species, including *L. hilgardii* and *L. brevis*, and *O. oeni* are highly capable of forming these compounds (Costello *et al.*, 2001; Costello and Henschke, 2002). Conditions for the development of this defect, which remains rare with regard to the omnipresence of heterofermentative bacteria in wines, have not been identified. All things considered, ropiness seems to be the most widespread and spectacular disease, but even if it causes economic loss, the damages can be limited since the spoiled wine can be treated and, in the end, released on the market.

In contrast with the metabolisms of malic acid, sugars, and citric acid, these last transformations are carried out only by certain strains. Bacterial spoilage cannot be attributed to a specific bacterial species. Genes determining enzymatic activities involved in most bacterial spoilage have been identified and sequenced: they exist in several species. However, their presence is not enough to predict what will occur in wine, because from the gene until the synthesis of the protein and the enzymatic activity, many regulations intervene. The knowledge of genetic sequences is an essential contribution to the progress in developing tools for the detection and monitoring of undesirable strains. Precautions must be taken to reduce microbiological risks based on results of analyses conducted at appropriate times.

References

Aerny J. (1985) Bull. OIV, 656–657, 1016.

Alizade M.A. and Simon H. (1973) Hoppe-Seyler's Z. Physiol. Chem., 354, 163.

Ansanay V., Dequin S., Blondin B. and Barre P. (1993) FEBS Lett., 332, 74.

Antalick G., Perllo M.C. and De Revel G. (2012) J. Agric. Food Chem., 60, 12371

Arena M.E., Saguir F.M. and Manca de Nadra M.C. (1999) Int. J. Food Microbiol., 47, 203.

Beech F.W. and Carr J.G. (1977) Cider and Perry. In *Economic Microbiology, Vol. 1, Alcoholic Beverages*, pp. 139–313 (Ed A.H. Rose). Academic Press, London.

Bloem A., Lonvaud-Funel A. and De Revel G. (2008) Food Microbiol., 25, 99.

Boido E., Lloret A., Medina K., Carrau F. and Dellacassa F. (2002) J. Agric. Food Chem., 50, 2344.

Caggianiello G., Kleerebezem M. and Spano G. (2016) Appl. Microbiol. Biotechnol., 100, 3877–3886.

Cibrario A., Peanne C., Lailheugue M. and Sols-Lafargue M. (2016) BMC Genomics, 17, 984.

Claisse O. (2002) *Diplôme Expérimentation et Recherche en Oenologie*, Université Victor Segalen, Bordeaux 2.

Claisse O. and Lonvaud-Funel A. (2001) J. Food Prot., 64, 833.

Costello P.J. and Henschke P.A. (2002) J. Agric. Food Chem., 50, 7079.

Costello P.J., Lee T.H. and Henschke P.A. (2001) Aust. J. Grape Wine Res., 7, 160.

Costello P.J., Francis I.L. and Bartowsky E.J. (2012) Aust. J. Grape Wine Res., 18, 287.

Coton E. (1996) *Etude de l'histidine décarboxylase de Leuconostoc oenos (Oenococcus oeni): aspects physiologique, enzymologique et génétique, conséquences oenologiques*. Thèse de Doctorat, Université de Bordeaux II.

Coulon J., Houlès A., Dimopoulou M., Maupeu J. and Dols-Lafargue M. (2012) Int. J. Food Microbiol., 159, 25–29.

De Revel G., Bertrand A. and Lonvaud-Funel A. (1989) Œno One, 23, 39.

Delaherche A., Claisse O. and Lonvaud-Funel A. (2004) J. Appl. Microbiol., 97, 910–915.

Denayrolles M., Aigle M. and Lonvaud-Funel A. (1994) FEMS Microbiol. Lett., 116, 79.

Denayrolles M., Aigle M. and Lonvaud-Funel A. (1995) FEMS Microbiol. Lett., 125, 37.

Desmazeaud M. and de Roissart H. (1994) Métabolisme general des bactéries lactiques. In *Les Bactèries Lactiques*, Vol. 1, p. 194 (Eds H. de Roissart and F.M. Luquet). Lorica, Uriage.

Dimopoulou M., Vuillemin M., Campbell-Sills H., Lucas P.M., Ballestra P., Miot-Sertier C., Favier M., Coulon J., Moine V., Doco T., Roques M., Williams P., Petrel M., Gontier E., Moulis C., Remaud-Simeon M. and Dols-Lafargue M. (2014) PLoS One, 9, 6, e98898.

Dimopoulou M., Bardeau T., Ramonet P.Y., Miot-Sertier C., Claisse O., Doco T., Petrel M., Lucas P. and Dols-Lafargue M. (2016) Food Microbiol., 53, 10–17.

Divol B., Tonon T., Morichon S., Gindreau E. and Lonvaud-Funel A. (2003) J. Appl. Microbiol., 94, 738.

Dols-Lafargue M., Lee H.Y., Le Marrec C., Heyraud A., Chambat G. and Lonvaud-Funel A. (2008) Appl. Environ. Microbiol., 74, 4079–4090.

Duenas-Chasco M.T., Rodriguez-Carjaval M.A., Tejero-Mateo P., Franco-Rodriguez G., Espartero J.L., Irastorza-Iribas A. and Gil-Serrano A.M. (1997) Carbohydr. Res., 303, 453–458.

Duenas-Chasco M.T., Rodriguez-Carjaval M.A., Tejero-Mateo P., Espartero J.L., Irastorza-Iribas A. and Gil-Serrano A.M. (1998) Carbohydr. Res., 307, 125–133.

Filannino P., Cardinali G., Rizzello C.G., Buchin S., De Angelis M, Gobbetti R., and Di Cagno R. (2014) Appl. Environ. Microbiol., 80, 2206–2215.

Gagne S., Lucas P.M., Perello M.C., Claisse O., Lonvaud-Funel A. and de Revel G. (2011) J. Appl. Microbiol., 110, 218.

Garai-Ibabe G., Areizaga J., Aznar R., Elizaquivel P., Prieto A., Irastorza A. and Dueña M.T. (2010) J. Agric. Food Chem., 58, 6149.

Gindreau E., Walling E. and Lonvaud-Funel A. (2001) J. Appl. Microbiol., 90, 535.

Gorga A., Claisse O. and Lonvaud-Funel A. (2002) Sci. Aliment., 22, 113.

Grimaldi A., Bartowsky E.J. and Jiranek V. (2005) Int. J. Food Microbiol., 105, 233.

Grobler J., Bauer F., Subden R.E. and Van Vuuren H.J.J. (1995) Yeast, 11, 613.

Heresztyn T. (1986) Am. J. Enol. Vitic., 37, 127.

Husnik J.I., Delaquis P.J., Cliff M.A. and Van Vuuren H.J.J. (2007) Am. J. Enol. Vitic., 58, 42.

Le Jeune C., Lonvaud-Funel A., Ten Brink B., Hofstra H. and Van der Voosen J.M.B.M. (1995) J. Appl. Bacteriol., 78, 316.

Liu S.Q., Pritchard G.G., Hardman M.J. and Pilone G.J. (1994) Am. J. Enol. Vitic., 45, 235.

Llaubères R.M., Richard B., Lonvaud-Funel A. and Dubourdieu D. (1990) Carbohydr. Res., 203, 103.

Lonvaud M. (1975) *Recherches sur l'enzyme des bactéries lactiques assurant la transformation du malate en lactate*. Thèse Doctorat 3ème cycle, Université de Bordeaux II.

Lonvaud-Funel A. (1986) *Recherches sur les bactéries lactiques du vin: Fonctions métaboliques, croissance, génétique plasmidique*. Thèse de Doctorat ès Sciences, Université de Bordeaux II.

Lonvaud-Funel A. and Joyeux A. (1988) Sci. Aliment., 8, 33–49.

Lonvaud-Funel A. and Joyeux A. (1994) J. Appl. Bacteriol., 77, 401.

Lonvaud-Funel A. and Strasser de Saad A.M. (1982) Appl. Environ. Microbiol., 43, 357.

Lonvaud-Funel A., Zmirou-Bonnamour C. and Weinzorn F. (1984) Sci. Aliment., 4, HS III, 81.

Lonvaud-Funel A., Guilloux Y. and Joyeux A. (1993) J. Appl. Bacteriol., 74, 41.

Lucas P. and Lonvaud-Funel A. (2002) FEMS Microbiol. Lett., 21, 85.

Lucas P., Landete J., Coton M., Coton E. and Lonvaud-Funel A. (2003) FEMS Microbiol. Lett., 229, 65.

Lucas P., Blancato V., Claisse O., Magni C., Lolkema J.S. and Lonvaud-Funel A. (2007) Microbiology, 153, 2221.

Luthi H. (1957) Am. J. Enol. Vitic., 8, 176–181.

Makaga E. (1994) *Etude du métabolisme de l'arginine chez l'espèce Leuconostoc oenos*. Thèse de Doctorat, Universitè de Reims.

Marchand S., de Revel G. and Bertrand A. (2000) J. Agric. Food Chem., 48, 4890.

Moreno-Arribas V. and Lonvaud-Funel A. (2001) FEMS Microbiol., 195, 103.

Moreno-Arribas V., Torlois S., Joyeux A., Bertrand A. and Lonvaud-Funel A. (2000) J. Appl. Microbiol., 88, 584.

Nannelli F., Claisse O., Gindreau E., de Revel G., Lonvaud-Funel A. and Lucas P. (2008) Lett. Appl. Microbiol., 47, 594.

Peynaud E. (1967) *Etudes récentes sur les bactèries lactiques du vin, IIème Symposium International d'Œnologie*, Bordeaux.

Peynaud E. (1968) CR Acad. Sci., 267D, 121.

Poolman B. (1993) FEMS Microbiol. Rev., 12, 125.

Pripis-Nicolau L. (2002) *Composés soufrés volatils du vin et métabolismes des bactéries lactiques*. Thèse de Doctorat, Université Victor Segalen Bordeaux 2.

Radler F. and Yannissis C. (1972) Arch. Mikrobiol., 82, 219.

Schütz M. and Radler F. (1974) Arch. Mikrobiol., 96, 329.

Schütz H. and Radler F. (1984) Arch. Mikrobiol., 139, 366.

Seifert W. (1901) Z. Landwirtsch. Versuchsu. Dent. Oest., 4, 980.

Tonon T. and Lonvaud-Funel A. (2000) J. Appl. Microbiol., 89, 526.

Tonon T. and Lonvaud-Funel A. (2002) Food Microbiol., 19, 451.

Tonon T., Bourdineaud J.P. and Lonvaud-Funel A. (2001) Res. Microbiol., 152, 653.

Ugliano M., Genovese A. and Moio L. (2003) J. Agric. Food Chem., 51, 5073.

Volschenk H., Vilijoen M., Grobler J., Bauer F., Lonvaud-Funel A., Denayrolles M., Subden R.E. and van Vuuren H.J.J. (1997) Am. J. Enol. Vitic., 48, 193.

Walling E., Dols-Lafargue M. and Lonvaud-Funel A. (2005a) Food Microbiol., 22, 71–78.

Walling E., Gindreau E. and Lonvaud-Funel A. (2005b) Food Microbiol., 98, 53–62.

Weinzorn F. (1985) *La dégradation de l'acide citrique du vin par les bactéries lactiques: étude de la citrate lyase*. Thèse de Doctorat 3ème cycle, Universitè de Bordeaux II.

Williamson D.H. (1959) J. Appl. Bacteriol., 22, 392.

CHAPTER 6

Lactic Acid Bacteria Development in Wine

6.1 Lactic Acid Bacteria Nutrition in Wine
6.2 Physicochemical Factors of Bacterial Growth
6.3 Evolution of Lactic Acid Bacteria Microflora: Influence on Wine Composition
6.4 Microbial Interactions During Winemaking
6.5 The Importance of Bacteriophages

6.1 Lactic Acid Bacteria Nutrition in Wine

Like all microorganisms, lactic acid bacteria cells multiply when the conditions are favorable: presence of nutritional factors, absence of toxic factors, and adequate temperature. All of the principal reactions of their metabolism are directed toward the biosynthesis of cell components: nucleic acids for the transmission of genetic heritage, carbohydrates, lipids, structure proteins, and, of course, enzymatic proteins. To ensure these syntheses, the cell must first find the necessary chemical elements in the medium: carbon, nitrogen, and minerals—in usable forms. Since all of these synthesis reactions are endergonic, the medium must also supply molecules capable of liberating the necessary energy. Most of the energy is supplied by the assimilation of various substrates. In addition, the cell receives energy from sophisticated systems that activate electron and proton transport phenomena. Although these systems cannot ensure the totality of cell growth, they contribute to it, very actively in certain cases, particularly when the cells are under nutritionally limited conditions.

Handbook of Enology, Volume 1: The Microbiology of Wine and Vinifications, Third Edition.
Pascal Ribéreau-Gayon, Denis Dubourdieu, Bernard B. Donèche and Aline A. Lonvaud.
© 2021 John Wiley & Sons Ltd. Published 2021 by John Wiley & Sons Ltd.

6.1.1 Energy Sources

Most of the energy comes from the assimilation of numerous organic substrates, sugars, amino acids, and organic acids. Lactic acid bacteria are chemoorganotrophic organisms. The oxidation of these substrates is principally represented by the fermentation of sugars (Section 5.2). Heterofermentative and homofermentative lactic acid bacteria degrade hexoses and pentoses. At different stages of their metabolism, exergonic reactions permit the stocking of energy in ATP molecules (Section 5.2). The oxidation of sugars is always coupled with the reduction of coenzymes. Under anaerobic conditions, the lactic acid fermentation process is responsible for their reoxidation. In the metabolism of other substrates, the liberation of energy by reactions can be accompanied by the synthesis of ATP: this is the case with the degradation of citric acid (Section 5.3.2) and arginine (Section 5.4.3).

The energetic importance of the proton motive force created at the membrane level has been demonstrated in several lactic acid bacteria species. The bacterial membrane in fact has a dual role: on the one hand, it is a barrier opposing the free diffusion of components of the medium and the cytoplasm. On the other hand, it is the site of proton and electron exchange. The proton motive force has two components: a difference in electric potential (negative inside) and a proton gradient (of pH). Maintaining a proton motive force requires an H^+-ATPase system of the membrane, which functions reversibly. An influx of protons leads to the synthesis of ATP; conversely, the efflux of protons consumes energy. In lactic acid bacteria, the efflux of lactate from the metabolism is associated with the efflux of two protons (symport). In this manner, the efflux of protons does not require energy.

During malolactic fermentation, the use of malate produces a sufficient proton motive force for the synthesis of ATP. The influx of negatively charged malate into the cell is coupled with the efflux of neutral lactate; a difference in potential is created. Furthermore, the decarboxylation provokes the alkalization of the cytoplasm and thus increases the pH gradient. All of this leads to the creation of the proton motive force. The energy indirectly furnished by the malolactic transformation is therefore conserved (Section 5.3.1). This same process explains the energy gain from the decarboxylation of histidine and tyrosine (Section 5.4.2). As in the previous case, the histidine/histamine exchange, accompanied by the transfer of a negative charge, and the decarboxylation reaction provoke the alkalization of the internal environment, ensuring the conservation of energy (Poolman, 1993).

6.1.2 Nutrients, Vitamins, and Trace Elements

Apart from water (the most important component), cells draw carbon, nitrogen, and mineral elements, such as phosphorus and sulfur, from their environment. These substances enter into the composition of cell components.

Carbon essentially comes from sugars and sometimes organic acids. Glucose and fructose are the most widely represented sugars in wine after alcoholic fermentation (a few hundred milligrams per liter). Mannose, galactose, pentoses (arabinose, xylose, and ribose), rhamnose, and a few disaccharides are also present in small concentrations (a few dozen milligrams of each per liter). The sugar degradation capacity depends on the bacterial species and (for example, for glucose) on environmental factors.

Oenococcus oeni degrades fructose slightly more easily than glucose. The presence of fructose in a mixture with glucose is very beneficial to growth. Its reduction into mannitol regenerates coenzyme molecules necessary for the oxidation of glucose. Owing to a lack of reduced coenzymes, acetyl phosphate does not lead to the formation of ethanol, but rather to acetic acid and ATP.

The energy obtained by the fermentation of residual sugars largely suffices to ensure the necessary growth for successfully starting and completing malolactic fermentation. According to Radler (1967), less than 1 g of glucose per liter covers the needs of the bacteria to form the biomass necessary for malolactic fermentation. In fact, much less than 1 g/l of glucose suffices, since other sugars in the medium are also used. The available sugars not only come directly from grape must but probably also from the hydrolysis of some of its components, notably polysaccharides.

Amino acids and sometimes peptides supply lactic acid bacteria with their assimilable nitrogen. Amino acid requirements vary as a function of species and even strain. These acids can be strictly indispensable or simply growth activators. According to Ribéreau-Gayon et al. (1975), the following amino acids are necessary as a whole or in part, depending on the strain: Ala, Arg, Cys, Glu, His, Leu, Phe, Ser, Trp, Tyr, and Val. Cocci have stricter demands than bacilli. The results of auxotrophic studies are, however, difficult to obtain and interpret. In a more recent study on strains of O. oeni, Frémaux (1990) demonstrated their auxotrophy for Ile, Leu, and Val. The synthesis pathways for these acids have enzymes in common for the production of aromatic amino acids (Phe, Trp, and Tyr), derived from the same precursor, chorismic acid, and for Arg, His, Ser, and Met. New observations suggest that His is a stimulant and not an essential amino acid.

Although these data remain very imprecise, an amino acid deficiency does not appear to be responsible for growth difficulties of lactic acid bacteria in wine. Temporary deficiencies can be noted at the beginning of alcoholic fermentation during the rapid yeast growth phase, but at the end, this is no longer the case. The metabolism and then the autolysis of yeasts release a large variety and quantity of amino acids into the environment. The culturing of *Leuconostoc* and *Lactobacillus* in a synthetic laboratory medium shows that all of the amino acids of the medium can be consumed during growth. In wine, the balance, conducted before and after malolactic fermentation, demonstrates that certain amino acids diminish while others increase in concentration, probably because of the simultaneous hydrolysis of peptides or proteins. In addition, the ammonia nitrogen concentration increases owing to deamination reactions (Ribéreau-Gayon et al., 1975). Amino acids are essentially used for protein synthesis. Depending on the strain, some can be catabolized and serve as energy sources (arginine, histidine, tyrosine, and ornithine).

Among nitrogen compounds, purine and pyrimidine bases play an important role in activating growth. In this case, the needs for adenine, guanine, uracil, thymine, and thymidine are also dependent on the strain. They are not always essential.

Minerals such as Mg^{2+}, Mn^{2+}, K^+, and Na^+ are necessary. The first two are often used as key enzyme cofactors of the metabolism (kinases and malolactic enzyme). The following trace elements are involved in the nutrition of *Lactococcus* species: Cu^{2+}, Fe^{3+}, Mo^{4+}, and Se^{4+}. Yet the role of these metal ions is not yet established for lactic acid bacteria in wine.

Vitamins are coenzymes or their precursors. Lactic acid bacteria are incapable of synthesizing B-group vitamins, in particular nicotinic acid, thiamin, biotin, and pantothenic acid (Section 3.5.1). A glycosylated derivative of pantothenic acid has been identified in grape juice; it was initially purified from tomato juice (Tomato Juice Factor, Amachi, 1975).

Finally, among the important chemical elements, phosphorus plays an essential role in lactic acid bacteria, as in all cells, in the composition of nucleic acids, phospholipids, and in the storage of energy in the form of ATP.

All of the minerals and vitamins cited, as well as carbon substrates and nitrogen nutrients, are found in sufficient quantities in wine. Only in exceptional cases are developmental difficulties of bacteria after alcoholic fermentation likely to be due to nutritional deficiencies. A simple experiment suffices to prove this statement: a favorable modification of one of the physicochemical factors that will be studied later (temperature and pH) usually enables the growth of the population. Independent of these physicochemical factors, any growth difficulties should be considered as being caused by inhibitors.

6.2 Physicochemical Factors of Bacterial Growth

Four parameters very distinctly determine the growth rate of lactic acid bacteria in wine: pH, temperature, alcohol content, and SO_2 concentration. Other factors are also in play but to a lesser degree and can only be determinant under some conditions.

These four essential factors have been known for a long time. They led to the establishment of "winemaking rules." Progress in winery equipment has made these rules progressively easier to follow (Section 12.7.4). None of these factors can be considered independently of the others; the four act together as a unit. A favorable level of one compensates an unfavorable value of one or several others. It is also rather difficult to give an exact limit for each of them. In this way, bacteria tolerate higher alcohol contents and SO_2 concentrations in wines with favorable pHs than in wines with low pHs.

6.2.1 Influence of pH

The variation in growth rate related to the pH presents an optimum value and extreme limits. Most bacteria develop better at a pH near neutrality. This is not the case with acidogenic bacteria such as lactic acid bacteria: their acidophilic nature permits their active growth in wine at low pHs, around 3.5. At pHs as low as 2.9–3.0, growth remains possible but slow. At the upper pH limits of wine (3.7–3.8), it is much quicker. Stopped growth due to environmental acidity occurs when the intracellular pH attains a certain limit (pH_i). It not only depends on the environmental pH but also on the nature of the acids (McDonald et al., 1990). In fact, the acids that freely penetrate in non-dissociated form are dissociated inside the cell, resulting in a decrease in pH. Consequently, the activity of intracellular enzymes is more or less inhibited, depending on the optimum pH of their activity. The proton motive force and the transports that depend on it are also slowed, interfering with the overall metabolism of the cell and thus reproduction. The lower limit tolerated for pH_i varies depending on the species. It is approximately 4.7 and 5.5, respectively, for Lactobacillus plantarum and Leuconostoc mesenteroides, according to McDonald et al. (1990). At pH 3.5, O. oeni maintains a higher pH_i than L. plantarum (Henick-Kling, 1986). The strains of this species adapt better to acidity than other species. Moreover, when cultivated in an acidic environment, they have a higher pH_i and thus a greater proton motive force—linked to the higher proton gradient.

Acidity adaptation mechanisms are not known but actively participate in the natural selection of O. oeni and more resistant strains of this species. It has been established for a long time that wines with relatively high pHs present not only a more abundant lactic acid bacteria microflora but also a much more varied one than do acidic wines. These wines are more microbiologically fragile, since some of the bacteria are spoilage factors, and also because a broader range of substrates are metabolized. High pH facilitates the growth of bacteria in wine, as well as promotes their survival, not only directly but also by reducing the effectiveness of free sulfur dioxide. Spoilage may develop several months, or even years, after vinification.

The pH also has an impact on the malolactic activity of the entire cell. Although the optimum pH of the purified enzyme is 5.9, it is not the same for cells. The malolactic activity of *O. oeni* strains is optimum at a pH between 3.0 and 3.2 and around 60% of its maximum activity at pH 3.8. The usual pH range of wines, therefore, corresponds well with the maximum malolactic activity of the bacterial cell. Yet the malolactic fermentation rate depends on not only the activity but also the quantity of cells. Finally, at usual wine values, pH affects both in the same way. Consequently, when all other conditions are equal, malolactic fermentation is quicker at higher pHs. For example, malolactic fermentation lasts 164 days for a wine adjusted to pH 3.15 and 14 days for a wine adjusted to 3.83 (Bousbouras and Kunkee, 1971).

According to Ribéreau-Gayon *et al.* (1975), pH also determines the nature of the substrates transformed. The authors defined the threshold pH for malic acid and sugar assimilation. It corresponds to the lowest pH at which the substrate is transformed, and it varies according to strain. The threshold pH for malic acid is lower than the threshold pH for sugars. In the zone between these two pHs, bacteria degrade malic acid without fermenting a large quantity of sugars and thus without producing volatile acidity. The larger the zone, the better adapted for winemaking is the strain. The average threshold pH of 400 heterofermentative cocci strains tested is 3.23 for malic acid and 3.51 for sugars. These values are, respectively, 3.38 and 3.32 for the 250 heterofermentative lactobacilli strains tested. The growth of the latter therefore does not guarantee malolactic fermentation without the risk of volatile acidity production.

The pH is therefore very important and comes into play at several levels: in the selection of the best adapted strains, in growth rate and yield, in malolactic activity, and even in the nature of the substrates transformed.

The role of pH has diverse practical consequences in the control of malolactic fermentation. First of all, malolactic fermentation is initiated more easily and rapidly in press wines than in the corresponding free-run wine. Press wines may thus be incorporated to help in starting malolactic fermentation, taking into consideration their overall impact on sensory quality. A partial chemical deacidification of wine may then be advisable in the most difficult cases, but rarely to treat entire tanks. It is especially recommended in the preparation of a malolactic fermentation starter— used for the inoculation of recalcitrant wine tanks. Finally, particular attention must be paid to musts and wines with high pHs. They sustain more or less anarchic bacterial growth of a large variety of bacteria and are thus subject to spoilage. Sensible sulfiting is the only tool for controlling these microorganisms.

6.2.2 Effect of Sulfur Dioxide

In wine, sulfur dioxide (SO_2) is in equilibrium between its free and bound forms. Its effectiveness as an antimicrobial agent (anti-yeast and antibacterial) and as an antioxidant is directly linked to wine composition and pH (Section 8.3.1). The active form, in fact, is molecular SO_2, which depends on the concentration of free SO_2 and pH. To calculate it, the Sudraud and Chauvet (1985) formula can be used; it gives the percentage of molecular SO_2 as a function of pH.

$$\% \text{molecular } SO_2 = 100 / \left(10^{pH-1.81} + 1\right)$$

For example, at pH 3.2, this percentage is 3.91%. It is, respectively, 2.00 and 1.01% at pH 3.5 and pH 3.8. These numbers demonstrate the influence of pH. Four times more free SO_2 is necessary at pH 3.8 than at pH 3.2 to obtain the same effectiveness.

The mechanism of the action of SO_2 has been studied in yeasts in particular, but it is most likely very similar in bacteria. According to Romano and Suzzi (1992),

SO_2 penetrates into the cell in molecular form, most likely by diffusion. In the cytoplasm, where the pH is highest, it dissociates and reacts with essential biological molecules: enzymatic proteins with their disulfide bonds, coenzymes, and vitamins. The result is cessation of growth and, finally, cell death. The inhibitory action of SO_2 on the malolactic enzyme of *Oenococcus* cells is added to its effect on cell growth.

For the same concentration of total SO_2, bacterial inhibition depends on the binding power of the wine (Section 8.3.2), which in turn determines the free SO_2 remaining, and on the pH, which establishes the amount of active molecular SO_2. It is only possible to give an approximation of the quantity of SO_2 necessary to inhibit bacterial development. As a general rule, lactic acid bacteria have difficulty in developing at concentrations ≥100 mg of total SO_2 per liter and 10 mg of free SO_2 per liter. Evidently, the result is not the same at pH 3.2 and at pH 3.8. Their sensitivity also varies according to the strain. Finally, for a given strain, sensitivity varies according to environmental growth conditions and physiological adaptation possibilities.

Lafon-Lafourcade and Peynaud (1974) found that cocci seem less resistant than lactobacilli. Thus, *O. oeni* growth is hindered more than *Lactobacillus hilgardii* growth, for example. The effect is also strain dependent. *Pediococcus damnosus* is a useful example: the ropy strains are insensitive to SO_2 doses that inhibit or kill other strains. After two months of bottle storage, ropy-type bacteria can maintain populations between 10^4 and 10^6 CFU/ml (colony-forming units, CFU) (Section 6.3.1) in wines containing 50 mg of free SO_2 per liter (Lonvaud-Funel and Joyeux, 1988).

Bound SO_2 also exerts a growth inhibitor effect, demonstrated by Fornachon (1963) (Section 8.6.3). Lactic acid bacteria may be capable of metabolizing the aldehyde fraction of the bound form and releasing SO_2. The SO_2 then exerts its activity on the cell, but it is less effective than free SO_2.

From their tests, Lafon-Lafourcade and Peynaud (1974) concluded that bound SO_2 is 5–10 times less active than free SO_2. Other authors have observed that its concentration in wine can easily be 5–10 times higher.

Technological consequences can be drawn from these results. When malolactic fermentation is desired, it is important to sulfite the grapes with this in mind. The sulfiting must exert a transitory inhibitory effect on the lactic acid bacteria. At the end of alcoholic fermentation, the bound SO_2 persists and can delay bacterial growth. Obviously, sulfiting red wine when draining it off the skins is not recommended, except in very unusual cases (Section 12.6.2).

6.2.3 Influence of Ethanol

Like most microorganisms, lactic acid bacteria are sensitive to ethanol, which affects the cell membrane (Section 6.2.7). Generally, under laboratory conditions, bacteria isolated from wine are inhibited at an alcohol content of around 8–10% vol. Results vary according to the genus, species, and strain. Ribéreau-Gayon et al. (1975) found that cocci are altogether more sensitive to ethanol than are lactobacilli. At an alcohol content of 13% vol., more than 50% of the lactobacilli resist, as opposed to only 14% of the cocci.

The growth of *O. oeni* strains isolated from wine and cultivated in the laboratory is activated at around 5–6% vol. of ethanol; it is inhibited in environments richer in ethanol and difficult at or above 13–14% vol. The ethanol tolerance of laboratory strains is much less than for the same strains cultivated in wine. Bacteria that multiply in wine adapt to the presence of ethanol but also probably to the wine environment as a whole. In addition to the intrinsic strain tolerance of ethanol, their adaptation capacity varies. It is therefore difficult to set a limit above which lactic acid bacteria no longer multiply.

Strains of *Lactobacillus fructivorans*, *Lactobacillus brevis*, and *L. hilgardii* (heterofermentative bacilli) are frequently isolated from fortified wines with alcohol contents from 16 to 20% vol. They seem to be naturally adapted to ethanol but lose this adaptation after isolation. Strains of *L. fructivorans* nevertheless remain very tolerant of ethanol, which has an activator role in their case (Kalmar, 1995). *Pediococcus damnosus* bacteria are not particularly resistant to alcohol, but the ropy strains multiply at the same rate and with the same yield in the presence or absence of 10–12% vol. of alcohol. The adaptation phenomena are definitely dissimilar in nature. In most cases, they are the result of a structural (fatty acid, phospholipid, and protein composition) and functional modification of the membrane. In the case of ropy *P. damnosus* strains, the polysaccharide capsule possibly acts as a supplementary protector.

6.2.4 Effect of Temperature

Temperature influences the growth rate of all microorganisms. As with chemical reactions, it accelerates biochemical reactions. Cell activity (resulting from all of the involved enzyme activities) and consequently growth vary with temperature according to a bell curve. At the optimum temperature, generation time is the quickest. This curve not only varies with the species and strains but also with the environment in which the bacteria grow.

In a laboratory culture medium, lactic acid bacteria strains isolated from wine are mesophilic, i.e. they multiply between 15 and 45°C, but their optimum growth range is 20–37°C. Thus, the optimum growth temperature for *O. oeni* is from 27 to 30°C. However, it is not the same in an alcoholic medium, especially in wine. The optimum temperature range is more limited: from 20 to 23°C. When the alcohol content increases to 13–14% vol., the optimum temperature decreases. Growth slows as the temperature decreases, becoming nearly impossible around 14–15°C.

The ideal temperature for lactic acid bacteria growth (notably *O. oeni*) and for malic acid degradation in wine is around 20°C. An excessive temperature of 25°C or above always slows malolactic fermentation—principally by inhibiting the bacterial biomass. Additionally, an excessive temperature increases the risk of bacterial spoilage and increases volatile acidity. In practice, therefore, maintaining a wine at 20°C is recommended. It should not be allowed to cool too much after alcoholic fermentation. If the temperature of the winery decreases, the wine should be warmed.

When the temperature is less than 18°C, the onset of malolactic fermentation is delayed. However, once it is underway, malolactic fermentation can continue even in a wine with a temperature between 10 and 15°C. In this case, the bacterial biomass was normally constituted when conditions were favorable. The cooling blocks the reproduction of bacteria but does not eliminate them. The cell activity, however, is slower. Malolactic fermentation of a wine therefore continues after its onset even in the case of being cooled, but the duration is much longer. The time frame for degrading all of the malic acid can range from five to six days to several weeks or months.

Along with pH, temperature is certainly the factor that most strongly influences the malolactic fermentation rate of a properly vinified wine that is not excessively sulfited. This factor is also the most easily monitored and controlled.

6.2.5 Effects of Phenolic Compounds

The action of phenolic compounds on lactic acid bacteria growth remains relatively unknown. Past results have shown that polyphenols tested alone or in a mixture have an inhibitory effect. Saraiva (1983) noticed, on the contrary, that gallic acid stimulates yeasts and lactic acid bacteria. Conversely, different phenolic acids (coumaric, protocatechuic acids, etc.) and condensed anthocyanins inhibit them. Enological tannins have been found to have an antibacterial effect (Ribéreau-Gayon et al., 1975). After much work on the subject, the effect of phenolic compounds on lactic acid bacteria growth remains unclear.

Nevertheless, a systematic study of several types of molecules clearly demonstrated the inhibitory effect of vanillic acid, seed procyanidins, and oak ellagitannins and, at the same time, the stimulating effect of gallic acid and free anthocyanins (Vivas et al., 1995). These results pertain to O. oeni growth but may also be valid for other bacterial species. By favoring growth, gallic acid and anthocyanins activate malolactic fermentation. Bacteria degrade these two compounds. The transformation of anthocyanins seems to activate a β-glucosidase—freeing the aglycone fraction and the glucose, which is metabolized by bacteria. The decarboxylation of gallic acid by O. oeni is possible. By analogy to the decarboxylation of other substrates, it may be an additional source of energy. The hydrolysis of glucosyl gallate into gallic acid, however, is a probability, which explains the positive effect of gallotannins on O. oeni (Chasseriaud et al., 2015). The phenolic composition of wine after alcoholic fermentation varies depending on numerous factors: grape ripeness, pre-fermentative operations, and fermentation, as well as the grape variety. Chasseriaud et al. (2015) demonstrated that O. oeni growth is affected differently by total polyphenols extracted from wines of different grape varieties.

Polyphenols, along with wine components as a whole, affect bacteria. Some are favorable and others unfavorable to bacterial growth and activity, but they play a secondary role compared with the other four parameters examined earlier. These elements, among many others (most of them unknown), determine the malolactic fermentability of wines.

6.2.6 Effect of Oxygen

In a laboratory culture medium, growth is activated in an inert gas atmosphere: CO_2 and N_2. In fact, the behavior of bacterial species present in wine can be diverse with respect to oxygen. They can be indifferent to its presence, adapt better in its absence (facultative anaerobic conditions), tolerate oxygen at its partial pressure in air but be incapable of using it (aerotolerant), or finally can require a small oxygen concentration for optimal growth (microaerophiles). Similarly, oxygen, or its repercussions on the redox potential of wine, can influence the growth of lactic acid bacteria in wine, but its effect is not always clear. Routine observations indicate that limited aeration, after draining off the skins or racking the wine, can strongly favor the onset of malolactic fermentation. In short, the growth of O. oeni is favored under seemingly opposing conditions: on the one hand, when the redox potential is artificially reduced by adding cysteine or naturally by adding certain tannins, and on the other, when it momentarily increases by controlled oxygen dissolution. These two are not incompatible, and we must consider that the bacterium benefits in the second case from oxygen or oxidized quinone molecules as electron acceptors to reoxidize NADH. Mechanisms implemented by the cell to use or foil the effects of the presence/absence of oxygen are very diverse and interdependent. Some may dominate depending on the physiological condition of the bacterium and depending on its environment.

It is therefore difficult to specify the possible oxygen needs of lactic acid bacteria in wine. The only studies touching on the subject were conducted with O. oeni, and it is likely that the rare results obtained cannot be transposed to other species.

6.2.7 Adaptation of Lactic Acid Bacteria to Growth in Wine

Wine is an extremely complex environment, and it is not possible to study the effects of all of its components on lactic acid bacteria. In any case, this would not help the enologist, since these individual effects are cumulative—acting in synergy or, on the contrary, offsetting each other. In this medium, lactic acid bacteria, particularly *O. oeni*, develop under extreme conditions. Acidity and ethanol combine with other molecules (known or not) to inhibit their growth, yet the effect may be lessened or cancelled by others.

It has long been known that a strain of *O. oeni* isolated from wine undergoing malolactic fermentation, and thus capable of multiplying, then cultivated in a laboratory medium, loses its viability when reinoculated in wine. This is actually a major problem for industrial producers of malolactic starters. Many observations, both in the laboratory and in the winery, suggest the existence of adaptation phenomena that ensure the survival and growth of bacteria under these extreme conditions. Isolated cells cultured in a laboratory medium with wine added have a generally higher tolerance to low pH, to SO_2, to ethanol, and to wine in general than isolated cells cultured in the absence of wine (Table 6.1).

The plasma membrane probably participates actively in these adaptation phenomena, which have been shown to exist in *O. oeni* and other bacterial species. The first adaptation mechanism to be discovered was the modification of the fatty acid composition of the membrane. All stress from the medium (addition of ethanol or wine, a temperature change, etc.) capable of provoking a modification of membrane fluidity, and thus membrane functions, is offset by an adjustment of the length and degree of unsaturation of the fatty acid chains. For all species studied (*O. oeni*, *P. damnosus*, *L. plantarum*, *L. hilgardii*, and *L. fructivorans*), the presence of ethanol in the medium, for example, greatly increases the proportion of unsaturated fatty acids and cyclopropanoic fatty acids formed from unsaturated fatty acids, especially lactobacillic acid (Desens, 1989; Garbay et al., 1995; Kalmar, 1995). In *O. oeni*, a gene coding for cyclopropane synthase appears to be transcription-regulated under the influence of alcohol and acidity (Granvalet et al., 2008).

A second phenomenon, quantitatively and qualitatively significant, concerns membrane proteins. Their concentration increases following a shock—whether physical (cold or heat) or chemical (acidity, addition of ethanol, fatty acids, or wine) (Table 6.2). In this manner, as with all living cells, *O. oeni* and the other lactic acid bacteria react to a shock by inducing the synthesis of "stress proteins." These proteins participate in the reaction of the cell against environmental stress (Garbay and Lonvaud-Funel, 1996). The Lo18 protein is induced in *O. oeni* by heat, ethanol, and acidity. After induction by stress, it participates in maintaining membrane integrity and fluidity and in protecting membrane proteins (Guzzo et al., 1997; Delmas et al., 2001; Maitre et al., 2014).

Lactic acid bacteria are extremely demanding during their growth in laboratory media. However, these microorganisms develop spontaneously in wine. Their growth is due to their complex set of adaptation phenomena—notably the induction of proteins, including proteins involved in membrane structures and enzymatic activities, as well as metabolic changes, which still mostly remain unknown.

Genome studies were first engaged to attempt to understand what distinguishes strains able to easily readapt to wine from others. The compared genomes demonstrated that in high-performance strains, homologous parts of the genome were

TABLE 6.1

Influence of Culture Medium on the Evolution of Bacterial Populations (CFU/ml) of Four *O. oeni* Strains (A, B, C, and D) After Inoculation in Red Wine (Garbay, 1994)

	Strains							
	A		B		C		D	
Time (days)	M	MW	M	MW	M	MW	M	MW
0	3×10^6	6×10^6	4×10^5	3×10^6	1×10^7	6×10^6	1×10^7	3×10^6
1	$<10^4$	6×10^6	$<10^4$	4×10^6	4×10^5	3×10^6	3×10^5	2×10^6
3	$<10^4$	5×10^6	$<10^4$	6×10^6	$<10^4$	2×10^6	5×10^4	4×10^5
8	$<10^3$	9×10^7	$<10^4$	1×10^7	$<10^3$	6×10^7	1×10^5	1×10^7

M, cells cultured in laboratory medium; MW, cells cultured in laboratory medium with wine added (V:V). Populations expressed in CFU/ml.

TABLE 6.2

Influence of Different Types of Stress on the Protein Concentration of the *O. oeni* Plasma Membrane (Garbay, 1994)

Stress[a]	Proteins (mg per 10^{12} cells)
Control	2.2
Heating to 37°C	2.6
Heating to 42°C	2.8
Heating to 50°C	6.0
Incubation in 10% ethanol	2.7
Incubation in 10% ethanol + fatty acids	4.8

[a] Stress exposure time = 30 minutes.

et al., 2012). Despite their convergence, the available results do not answer the question of the adaptability of wine strains, which is multi-factored and must also take into account the variability of the wine's composition. In particular, this approach has not been successful in the study of strains that are supposedly especially tolerant to ethanol (Bordas et al., 2013).

6.3 Evolution of Lactic Acid Bacteria Microflora: Influence on Wine Composition

6.3.1 Evolution of the Total Lactic Acid Bacteria Population

Usually, bacteria are counted by culturing the sample in an agar nutritional medium, followed by counting the colonies formed. Results are expressed in colony-forming units per milliliter (CFU/ml). This expression corresponds to living cells that are able to form colonies. They do not take into account dead cells and viable but non-culturable (VBNC) cells (Section 6.3.2).

identical, including in noncoding regions. In contrast, they are much more different among strains that are less adapted to wine (Delaherche et al., 2006). Then, by studying vaster collections of strains, the phylogeny of this species demonstrated that group A contains many strains that are well adapted to growth in wine. Such is the case of starters selected by industrial producers (Bridier et al., 2010). Moreover, genes carried by plasmids also seem more present in malolactic starters and dominant strains during malolactic fermentation (Favier

In the production of wines requiring malolactic fermentation, the total bacterial population passes through several phases (Figure 6.1). During the first days of fermentation, as soon as the tanks are filled, bacteria are present in highly variable quantities—most often from 10^2 to 10^4 CFU/ml. The extent of the population depends on weather conditions during the last days of grape maturation. It is generally lower when the conditions are propitious for healthy grapes, and under these situations, it can be impossible to isolate a single bacterial colony on the berry. The physiological condition of these microorganisms in the berry surface biofilm is particular and not very adapted to an immediate culture in a laboratory environment. However, during the diverse operations from harvest to filling the tank, the must is seeded with lactic acid bacteria very rapidly—probably by the winery equipment. During the harvest period, bacteria, like yeasts, progressively colonize the winery. In general, the last tanks filled have the highest populations.

During the first days of alcoholic fermentation, the bacteria and yeasts multiply. The latter, better adapted to grape must, rapidly invade the medium with high populations. During this time, the bacteria multiply, but their growth remains limited, with a maximum population of 10^4–10^5 CFU/ml. To a large extent, their behavior at this time depends on the pH of the medium and the grape sulfiting level. Normally, the doses of SO_2 added (about 5 g/hl) at pH between 3.2 and 3.4 do not prevent their growth, but simply limit it.

Then, from the most active phase of alcoholic fermentation to the depletion of sugars, the bacteria population rapidly regresses to 10^2–10^3 CFU/ml. This level also depends on environmental conditions (pH and SO_2).

Following alcoholic fermentation, the bacterial population remains in a latent phase for a varying period, which can last several months when the pH, ethanol, and temperature parameters are at their lower limits. Usually, this phase lasts only for a few days, and in certain cases, it does not occur at all. In the most frequently encountered situation, the growth phase takes place after the wine has been drained off the skins.

One microorganism follows the other: the yeasts first and then the lactic acid bacteria. These are ideal winemaking conditions, in which all of the fermentable

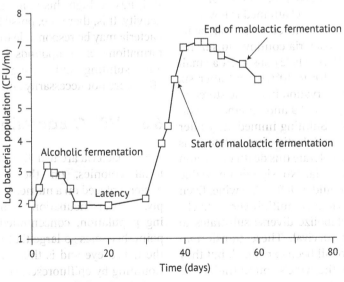

FIGURE 6.1 Evolution of lactic acid bacteria population during winemaking.

sugars are depleted before the bacteria invade the medium. In the opposite case, the bacteria multiply actively toward the end of alcoholic fermentation: they ferment sugars using the heterofermentative pathway and increase the volatile acidity of the wine, causing lactic spoilage (Section 3.8.2).

The growth phase lasts for several days and raises the population to around 10^7 CFU/ml or more. Obviously, its duration also depends on the composition of the medium. The subsequent stationary phase also varies. The bacteria then begin the decline phase.

The malolactic fermentation phase begins during the growth phase, as soon as the total population exceeds 10^6 CFU/ml. It continues and is completed during the stationary phase, or sometimes at the beginning of the death phase. Under very favorable conditions with a limited concentration of malic acid, malolactic fermentations are often completed even before the end of the growth phase. The optimum population in these cases exceeds 10^8 CFU/ml. The malic acid degradation rate is strictly related to biomass quantity and the specific malolactic activity of the cells. Both are highly dependent on physicochemical conditions and on temperature. The transformation of 2 g of malic acid per liter can take more time than 4 g/l if the population level attained is lower.

If wine is not sulfited after malolactic fermentation, bacteria continue to survive for months. Carre (1982) observed a small decrease from 10^7 to 10^5 CFU/ml after six months of conservation in a wine stored at 19°C with a pH of 3.9 and an ethanol volume of 11.25%. Sulfiting immediately after the end of malolactic fermentation is intended to accelerate this death phase and eliminate bacteria. No significant viable population should be left in the wine. Even if they can no longer multiply very actively, cells can metabolize diverse substrates to ensure their survival. These transformations have not all been explained, but they may increase the wine's concentrations of undesirable substances.

Sulfiting at the end of secondary fermentation is done to adjust the free SO_2 concentration to 30–40 mg/l. At this concentration, nearly all of the lactic acid bacteria disappear within a few days ($\approx$1–10 CFU/ml). The results also depend on the composition of the medium (Figure 6.2). Additionally, numerous observations have shown that the lactic acid bacteria population is maintained more easily in barrels than in tanks. During 18 months of barrel aging, a decrease from 10^6 CFU/ml to 10^3 CFU/ml only was noted in spite of a free SO_2 concentration between 20 and 30 mg/l. In red wines, the last fining with egg whites effectively helps to eliminate lactic acid bacteria.

In fact, the drop in the bacterial population assessed by counting the colonies developed on a nutrient medium does not apparently provide an accurate representation of the situation after sulfiting. Counting the bacteria by epifluorescence shows that part of the population retains some metabolic activity, although the cells are incapable of growth in a nutrient medium. This physiological condition is described as "VBNC" (Millet and Lonvaud-Funel, 2000) (Section 6.3.2). According to Coton (1996), histamine may be accumulated in wines by this type of cell, which still has a high histidine decarboxylase activity. It is, therefore, possible that VBNC bacteria may be responsible for other transformations of compounds in wine, even after sulfiting and during aging. These effects are not necessarily entirely negative.

6.3.2 VBNC Bacteria

Viable bacteria are usually counted as bacterial colonies, on the principle that bacteria placed on a nutrient gel will multiply. After an incubation period, the resulting population, concentrated in a single point, becomes so large that it is visible to the naked eye and is thus easily counted. Counting by epifluorescence is based on a completely different principle. Bacteria

6.3 | Evolution of Lactic Acid Bacteria Microflora: Influence on Wine Composition

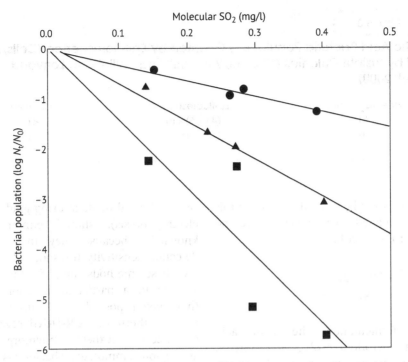

FIGURE 6.2 Influence of molecular SO_2 concentration on lactic acid bacteria survival (Lonvaud-Funel, unpublished results). N_t, viable bacteria population 20 days after sulfiting; N_0, initial bacteria population; ■, wine A; ▲, wine B; ●, wine C.

cells on a microscope slide or filter membrane are placed in contact with a substrate that is transformed by passing through each cell. The most common substrate is a fluorescence ester hydrolyzed by an esterase in the cell, which makes the bacterial content fluoresce under UV light. All cells that fluoresce under these conditions are considered viable, because hydrolysis of the ester indicates the existence of enzyme activity.

Both methods give the same results for bacterial suspensions in the growth phase. Almost all cells multiply at that time. When they move into the stationary, and then the decline phase, the difference between the results increases. While counting by epifluorescence shows a slight decrease in the number of cells, there is a sharp drop in the number of colonies visible. This difference may be explained by the fact that part of the population of fluorescent cells is still biologically active but is incapable of the metabolic and physiological functions necessary for multiplication. They are described as VBNC cells. Table 6.3 shows the lactic acid bacteria count after sulfiting a wine. As expected, sulfiting eliminates the viable (and revivable) population. However, there are still large numbers of VBNC cells (Millet and Lonvaud-Funel, 2000).

This phenomenon is not exclusive to lactic acid bacteria, but rather applies to many other microorganisms. It is easily demonstrated for acetic acid bacteria in winemaking. As soon as they are deprived of oxygen, the difference between the viable and VBNC populations increases rapidly, then disappears completely as soon as the wine is aerated (Millet and Lonvaud-Funel, 2000; Du Toit et al., 2005). The same experiments showed that yeast and bacteria in the VBNC state decrease in size, and some of them may pass through filters intended to retain them (Divol and Lonvaud-Funel, 2005). The energy needs of these cells are low but not nil. Residual

TABLE 6.3

Lactic Acid Bacteria Populations Counted by Epifluorescence (Cells/ml) and by Visible Colonies (CFU/ml) After Sulfiting (Millet and Lonvaud-Funel, 2000)

Free SO_2 (mg/l)	Epifluorescence	Colonies
30	$(44 \pm 6) \times 10^5$	<1
50	$(44 \pm 5) \times 10^4$	<1

activity may not be insignificant when this state is extended over time (Laforgue and Lonvaud-Funel, 2012).

6.3.3 Evolution of Various Bacterial Species

During fermentation, the lactic acid bacteria microflora evolves not only in number but also in variety of species. Strains of the *Lactobacillus* genus on grapes can be isolated, as well as *L. mesenteroides*. The species *O. oeni*, which becomes the most significant one later, is barely present at the beginning of fermentation. Just after its arrival in the tank, grape must contains a very diverse microflora, generally belonging to the eight usual lactobacilli and cocci species: *L. plantarum*, *Lactobacillus casei*, *L. hilgardii*, *L. brevis*, *P. damnosus*, *Pediococcus pentosaceus*, *L. mesenteroides*, and *O. oeni*. However, all species are not always represented, or at least cannot be identified. Then, a spontaneous selection occurs during alcoholic fermentation. The lactic acid bacteria population regresses after reaching its optimum. At the same time, the homofermentative and then heterofermentative lactobacilli disappear to the benefit of *O. oeni*. Afterward, the homofermentative cocci and *L. mesenteroides* also give way to *O. oeni* (Table 6.4).

Certain species may also subsist at very low residual populations—less than 10 or 10^2 CFU/ml, thus explaining why we find them again later. Molecular methods, such as PCR (Section 4.3.3) and PCR-DGGE (PCR followed by denaturing gradient gel electrophoresis), should enhance our knowledge because they increase the detection sensitivity threshold of species. As these methods amplify the signal specific to a particular microorganism, they make it possible to identify minority flora. Furthermore, PCR-DGGE reveals the presence of unexpected microorganisms, as a region common to all bacteria is amplified, then each species is identified individually (Claisse and Lonvaud-Funel, 2003; Renouf et al., 2007). After fermentation, some can multiply if the wine is poorly protected. In fact, species other than *O. oeni* are most often responsible for wine spoilage.

The spontaneous evolution of a mixture of species corresponds with the selection of those best adapted to wine—which is a hostile acidic and alcoholic environment. The composition of the plasma membrane, and the various mechanisms that permit it to react to the aggressiveness of the medium, seem to influence this adaptation. Certain species or strains may also differ in their ability to carry out these transformations. Strains of *L. fructivorans* adapt better to ethanol than *L. plantarum* and *L. hilgardii*, due to a more effective modification of their fatty acids (unsaturation and chain length) (Kalmar, 1995). Unsurprisingly, strains of this species, along with others such as *L. hilgardii*, *L. collinoides*, and *L. mali*, are often identified in fortified wines with an alcohol content of over 15% (Couto and Hogg, 1994).

TABLE 6.4
Population (CFU/ml) of the Different Lactic Acid Bacteria Species During the Alcoholic Fermentation of Cabernet Sauvignon Must (Lonvaud-Funel et al., 1991)

Day	Alcohol content (% vol.)	Oenococcus oeni	Leuconostoc mesenteroides	Pediococcus damnosus	Lactobacillus hilgardii	Lactobacillus brevis	Lactobacillus plantarum	Lactobacillus casei	Total population
0	0	nd	2.9×10^2	6.0×10^2	1.1×10^3	nd	7.5×10^1	7.7×10^1	2.5×10^3
3	7	nd	1.7×10^4	3.8×10^4	8.0×10^4	2.0×10^4	2.0×10^4	2.0×10^4	1.7×10^5
6	9	nd	9.6×10^4	3.7×10^4	4.0×10^4	4.5×10^3	nd	nd	1.5×10^5
10	13	4.2×10^3	3.2×10^3	4.9×10^3	4.4×10^3	nd	nd	nd	1.8×10^4
18	13	3.4×10^6	nd	nd	nd	nd	nd	nd	3.4×10^6

nd: not detected.

6.3.4 Evolution of Wine Composition During the Different Phases of Bacterial Development

As soon as lactic acid bacteria multiply, they inevitably modify wine composition. In fact, their growth requires the assimilation of substrates to supply the cell with energy and with carbon and nitrogen. The division of a bacterium into two daughter cells clearly supposes the neosynthesis of all of the structural components and molecules with biological activity. In wine, bacteria transform sugars, organic acids, and a multitude of other components. Their impact on wine quality varies depending on the type of reactions and the nature and concentration of these substances.

The only bacterial intervention truly sought after in winemaking is the transformation of malic acid into lactic acid (Section 12.7.2). It is the source of the most manifest organoleptic change resulting from malolactic fermentation: the deacidification and the softening of wine. Malic acid, a dicarboxylic acid, is transformed molecule for molecule into lactic acid, which is monocarboxylic. The loss of one acid function per molecule is intensified by the replacement of an acid with a particularly aggressive taste by a much softer acid. This transformation is carried out on 1.5–8 g/l maximum, depending on the variety and grape maturation conditions.

Bacteria do not, however, transform all of the malic acid contained in the grape. Figure 6.3 shows its evolution during winemaking. From the start, during alcoholic fermentation, yeasts metabolize a maximum of 30% of the malic acid. The product, pyruvate, then enters one of many yeast metabolic pathways—notably leading to the formation of ethanol. This "malo-alcoholic" fermentation is catalyzed in the first stage by the malic enzyme. The bacteria must develop a sufficient population before malolactic fermentation can truly start. The production of L-lactic acid is coupled with the decrease in malic acid.

The degradation of citric acid is also very important in enology. First of all, its disappearance from the medium contributes to the natural microbiological stabilization of wine by eliminating a potentially energy-generating substrate. Additionally, the sensory impact of its metabolites, now fairly well known, has been proven (Section 5.3.2) (De Revel et al., 1996). Diacetyl is certainly involved, and at low concentrations, it gives wine an aroma complexity that is much appreciated. In certain kinds of wine, tasters even prefer wine that has a very pronounced odor of this component. The degradation of citric acid also increases volatile acidity—a maximum of approximately 70 mg/l (expressed as H_2SO_4). Sensory deviations due to an excess of volatile acidity or diacetyl coming from the degradation of citric acid, however, are rare; they can have other origins. Recently, enologists and researchers have shown increased interest in this subject, but opinions differ to such a point that some currently advocate avoiding the degradation of citric acid, which represents a real factor in microbial stability. They recommend the use of bacterial strains unable to use citric acid.

In any case, citric acid is always degraded during malolactic fermentation, since O. oeni bacteria have all of the necessary enzymatic equipment. Its transformation is nevertheless slower than that of malic acid, and several dozen milligrams of citric acid per liter often remain at the time of sulfiting. Yet since lactic acid bacteria are not eliminated immediately and remain active for several days (sometimes several weeks), only traces of citric acid remain in wine.

During winemaking, lactic spoilage is a dreaded bacterial spoilage. By definition, it corresponds with the increase of volatile acidity caused by the heterofermentative fermentation of sugars. Normally, lactic acid bacteria multiply only after the completion of alcoholic fermentation. The residual sugars—glucose, fructose, and pentose—are in small but sufficient

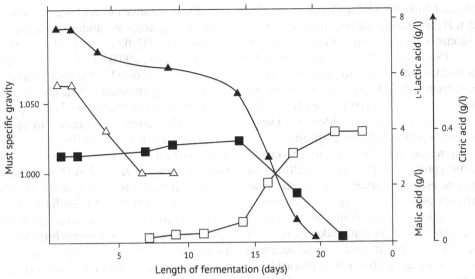

FIGURE 6.3 Evolution of different compound concentrations during fermentation (Lonvaud-Funel, 1986). ▲, malic acid; ■, citric acid; Δ, density; □, L-lactic acid.

concentrations to ensure the essential energy needs of the bacteria. If bacterial growth occurs before the end of alcoholic fermentation, when more than 4–5g of reducing sugar per liter remains in the wine, lactic spoilage can result. In fact, heterofermentative bacteria (*O. oeni*) ferment sugars and produce a significant amount of acetic acid. Moreover, the fermentation of glucose in the presence of fructose, which is the case in wine, preferentially directs the acetyl phosphate molecules toward acetic acid—fructose being reduced into mannitol (Section 5.2.2).

Lactic spoilage, therefore, occurs when environmental conditions are favorable to bacterial growth, even though yeasts have not yet completely fermented the sugars. This category essentially comprises wines derived from particularly ripe harvests. In such cases, the sugar concentration is high, the medium is often low in nitrogen, and the pH is high. A stuck alcoholic fermentation (Section 3.8.2) or simply a sluggish fermentation should be expected. These phenomena may lead to rapid growth of lactic acid bacteria. The proper control over alcoholic fermentation is the best defense against lactic spoilage.

Lactic spoilage is also a widespread form of bacterial spoilage in fortified wines. These wines are made via the addition of alcohol to grape must that has been partially (or not at all) fermented. They are generally stabilized due to their high alcohol content. However, lactic acid bacteria, most often heterofermentative lactobacilli, are particularly resistant to ethanol (Section 6.3.3). They develop easily in this very sugar-rich medium. Not only is there malolactic fermentation (which is not a real problem) but there is also lactic spoilage. The volatile acidity of these wines frequently reaches 1–1.5g/l (expressed as H_2SO_4). This phenomenon often occurs in the bottle, producing carbon dioxide and a cloudy wine.

High volatile acidity in red wines does not necessarily indicate lactic spoilage. Such wines can also have been the site of acetic acid bacteria growth or a metabolic deviation of yeasts. To establish the diagnosis, enologists can use the fact that heterofermentative lactic acid bacteria also form lactic acid from sugars—more precisely, exclusively D-lactic acid for cocci and DL-lactic acid for lactobacilli. Lafon-Lafourcade (1983) demonstrated that yeasts produce little D-lactic acid and concluded that lactic spoilage has occurred when the D-lactic concentration exceeds 0.2 g/l in wine (Section 14.2.3). A simple enzymatic determination of the quantity of D-lactic acid therefore indicates the origin of the wine spoilage. However, a limit of 0.3–0.4 g/l of D-lactic acid per liter would seem to ensure a more reliable diagnosis.

In wines, the first step in preventing lactic spoilage is the proper sulfiting of grapes, especially when they are very ripe. The corresponding musts are more subject to stuck fermentations than others (Section 3.8.1). The winemaker must react accordingly and, if need be, use additives such as nitrogen, vitamins, and yeast hulls whose effectiveness is clearly established. Of course, basic operations, notably aeration and temperature control, must also be scrupulously followed. The catabolic reactions of sugars, malic acid, and citric acid are normal occurrences during fermentation. Many other transformations also occur, and some depend on the nature of the species or even the strain.

Lesser known "secondary" transformations are those involving the phenolic compounds of wine. Malolactic fermentation has been confirmed to cause chromatic changes in wines and a decrease in their color, while stabilizing it. Color molecules, i.e. anthocyanins, as well as tannins are substrates with diverse hydrolysis reactions.

The sulfur-based amino acids, Met and Cys, are converted into volatile aroma compounds. They contribute to the increasing complexity in a wine's aromas and bouquet after malolactic fermentation. *Oenococcus oeni* produces methanethiol, dimethyl sulfide, 3-(methylsulfanyl)-propan-1-ol, and 3-(methylsulfanyl)-propionic acid. Synthetic solutions of 3-(methylsulfanyl)-propionic acid, described as having chocolate and toasty aromas, have a perception threshold of 50 µg/l. Concentrations increase significantly after malolactic fermentation, and interaction with other components of wine produces an aroma reminiscent of red berry fruit (Pripis-Nicolau, 2002). Aroma molecules are released from their precursors under the action of glycosidases (Ugliano and Moio, 2005; Grimaldi *et al.*, 2005) (Section 5.5). Esters that are very involved in the characteristics of wine are synthesized or hydrolyzed (Antalick *et al.*, 2012; Sumby *et al.*, 2013). The role of other esterases remains unknown, since they are not necessarily involved in direct synthesis of aroma molecules. Cinnamyl esterase, identified in *O. oeni*, is one of these. It is able to hydrolyze the tartaric esters of p-coumaric and ferulic acids in wine. Cinnamyl esterase releases cinnamic acids, which are *Brettanomyces bruxellensis* substrates for ethylphenol synthesis. This activity does not exist in all strains (Chescheir *et al.*, 2015).

After the secondary fermentation, sulfiting stabilizes the wine by eliminating viable bacteria and definitively blocking all microbial growth. Even then, profiting from weak sulfite protection or more often due to a natural resistance, spoilage strains sometimes succeed in multiplying. Wine spoilage such as ropiness, *amertume* (acrolein spoilage), and tartaric acid degradation can be triggered (Section 5.4). Nonviable lactic acid bacteria, or at least those that are no longer capable of multiplying, are still active. Some *O. oeni* strains produce histamine in this manner—they decarboxylate histidine from the wine and increase histamine content during aging. The histidine decarboxylase enzyme maintains a high level of activity for several months in nonviable or at least non-culturable cells (Coton, 1996). Consequently, these residual populations can be responsible for other minor, unidentified transformations of the wine during aging.

6.4 Microbial Interactions During Winemaking

When the must arrives in the tank, it contains an extensive variety of microorganisms:

fungi, yeasts, and lactic and acetic acid bacteria. Initially, this microflora comes from the grape and from harvest equipment and then later from equipment that transports whole and crushed grapes to the tank. From this mixture, the microorganisms involved in winemaking are selected naturally—very quickly at first and afterward more progressively. This selection takes place due to changes in environmental conditions (composition and oxidation–reduction potential) and specific antagonistic and synergistic interactions between the different microorganisms. Fungi disappear quickly, giving way to yeasts and bacteria. There are many interactions among them, at the species level but also at the strain level. Only a few of these interactions are well known. Some, on the contrary, are very difficult to identify and study. The yeast/bacteria interactions during fermentation seem to be the most important.

6.4.1 Interactions Between Yeasts and Lactic Acid Bacteria

Yeasts are well adapted to growth in grape must. From the first days of fermentation, their growth is very rapid. Lactic acid bacteria also multiply very easily when inoculated alone in this same environment. Yet under practical conditions, yeasts and bacteria are mixed; yeasts are always observed to dominate bacteria, resulting from numerous inhibitor or activator effects. To study interactions, the system is generally limited to one strain of each microorganism. The objective is to identify the molecules or mechanisms involved. Whether these are active molecules produced by one or the other, they may be specific to the strains present, or, on the contrary, their effects could be more general in nature.

The experimental inoculation of grape must with *Saccharomyces cerevisiae* and a mixture of diverse lactobacilli and *O. oeni* clearly shows a behavioral difference between the bacteria. When yeasts and lactic acid bacteria are inoculated in approximately equal concentrations (7×10^5 CFU/ml), lactobacilli are completely eliminated after eight days. *Oenococcus oeni* disappears more slowly and subsists at a very low concentration. If the same must is inoculated with 10–100 times more bacteria, they remain viable for a longer period but eventually disappear—with the exception of *O. oeni*. This species is better adapted than the others to winemaking conditions.

The interactions between *S. cerevisiae* and *O. oeni* have therefore been studied in greater detail. Grape must (220 g of sugar per liter) was simultaneously inoculated with both microorganisms. Figure 6.4 illustrates their evolution in must at pH 3.4. In an initial phase, corresponding with the explosive growth of yeasts, the bacterial population regresses. After a transitory phase, the inverse phenomenon occurs. The yeast death phase coincides with the rapid growth phase of the bacteria. This evolution can be interpreted as an antagonism exerted by yeasts on the *O. oeni* population. The bacteria not only do not multiply but also are partially eliminated. At this stage, nutritional deficiencies may also be responsible in part. Moreover, during their rapid growth period at the beginning of fermentation, yeasts have been proven to deplete the medium of amino acids. Arginine can be totally consumed. These deficiencies, hindering bacterial growth, are combined with the toxic effects of metabolites released by yeasts. In the first three to four days, the alcohol formed cannot explain this effect. Moreover, at low concentrations (5–6% vol.), it activates bacterial growth. Other substances are involved, including fatty acids released by yeasts, such as hexanoic, octanoic, decanoic, and especially dodecanoic acids (Table 6.5) (Lonvaud-Funel *et al.*, 1988a). These acids target and alter the bacterial membrane. The incubation of whole cells in the presence of these fatty acids

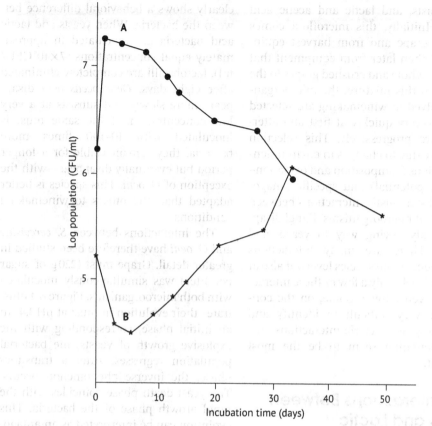

FIGURE 6.4 Evolution of yeast and lactic acid bacteria (*O. oeni*) populations inoculated in a mix in grape must (Lonvaud-Funel et al., 1988b). Grape must pH = 3.4; sugar concentration = 220 g/l; A, yeasts; B, lactic acid bacteria.

results in an ATP leak and a loss of malolactic activity.

As alcoholic fermentation takes place, the alcohol concentration increases in the medium. The negative effects of yeast metabolism are compensated in the end by the positive ones. When the yeast population enters the stationary phase, the viable population count is composed of cells that actively multiply while others are lyzed. The latter cells play an important role with respect to the bacteria—they release vitamins, nitrogen bases, peptides, and amino acids. All of these components act as growth factors for the bacteria.

Therefore, in the final stage of alcoholic fermentation, yeasts stimulate bacterial growth. This effect is also combined with a lesser known phenomenon corresponding to an inhibition of yeasts by bacteria (Section 3.8.1) (Figure 6.5). More precisely,

TABLE 6.5

Influence of the Addition of Fatty Acids (C10 and C12) on the Malolactic Fermentation Rate (Concentration of Malic Acid in gram per liter) (Lonvaud-Funel et al., 1988a)

Lot	Days			
	2	4	6	8
Red wine	2.5	2.1	1.7	1.4
Red wine + C10 (23 µM)	2.8	2.7	2.5	2.5
White wine	3.5	2.4	0.6	0.2
White wine + C12 (2.5 µM)	4.5	4.5	4.5	4.5

the bacteria accelerate the yeast death phase (Lonvaud-Funel *et al.*, 1988b). Bacterial glucosidase and protease activities are responsible for the hydrolysis of the yeast cell wall and lead to the lysis of the entire cell.

At the end of alcoholic fermentation, bacteria therefore accelerate yeast autolysis. Their growth is equally stimulated by the released products. These phenomena amplify each other and finally lead to a rapid decrease in yeast activity and viability. They contribute to slow or even stuck alcoholic fermentations. Furthermore, bacteria probably also produce yeast inhibitors. In fact, grape must that is pre-cultured with bacteria (cocci or bacilli, homofermentative or heterofermentative) is less fermentable by yeast than the control must. The wines obtained still have several dozen grams of non-fermented sugar per liter. Among the species tested, *O. oeni* has the greatest effect. The role of *L. plantarum*, a species very common in musts, nevertheless needs to be emphasized. A strain of this species inhibits not only bacteria but also a large proportion of yeasts from the genera *Saccharomyces*, *Zygosaccharomyces*, and *Schizosaccharomyces*. The inhibitory substance is an extracellular protein that is stable but inactivated by heat (Rammelsberg and Radler, 1990). Another mechanism is suggested by Ramakrishnan *et al.* (2016). Lactic acid bacteria induce in *S. cerevisiae* a prion state, in which hexose transport is altered. These yeasts do not ferment as well and make more nutrients available to bacteria, which is a benefit for bacteria in their competition with yeasts.

Environmental conditions, in particular grape pH and sulfiting, play an important role in the evolution of these mixed cultures (Figure 6.6). A high pH is favorable to bacterial growth. Obviously, the inverse is true of low pHs. But sulfiting considerably limits bacterial survival and growth at the beginning of the primary fermentation. Its role is essential. Yeasts should be allowed to multiply without leaving room for the bacteria. The latter must regress but remain in the medium, all the same, to take advantage of the yeast death phase and

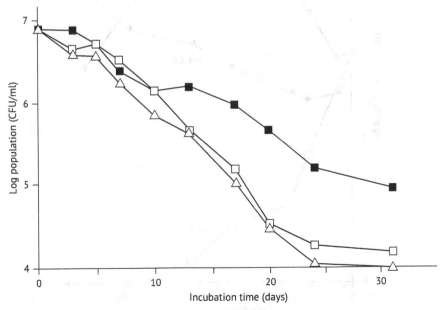

FIGURE 6.5 Effect of lactic acid bacteria on the evolution of the yeast population after alcoholic fermentation (Paraskevopoulos, 1988). ■, Pure yeast culture; □, mixed culture (bacteria 10^6 CFU/ml); △, mixed culture (bacteria 10^7 CFU/ml).

then multiply. These observations illustrate the importance of sulfiting grapes correctly. By taking the pH into account, winemaking incidents caused by the competition between yeasts and bacteria—such as lactic spoilage or, on the contrary, malolactic fermentation difficulties—can be avoided (Section 3.8.1).

The nature and quantity of peptides, polysaccharides, and other macromolecules in wine released by yeasts are different depending on the must, winemaking techniques, and the yeast strain. As a result, the malolactic fermentability of wines obtained from the same must but fermented by different yeast strains varies greatly. The toxic contribution of sulfur dioxide and fatty acids released by yeast, however, remains certain. Yet macromolecules, particularly polysaccharides, are capable of adsorbing these fatty acids and carrying out a veritable detoxification of the medium. Yeast autolysis, the autolysis rate (Lonvaud-Funel et al., 1985; Guilloux-Benatier and Feuillat, 1993), and the nature of the molecules, which vary with the strain, influence their release in the medium.

For example, different wines were obtained from fermentation of the same must by eight commercial yeast strains. The difference in the duration of malolactic fermentation was then compared (Table 6.6). The wines had similar ethanol contents and pHs. After sterile filtration and inoculation by four different *O. oeni* strains, they underwent malolactic fermentation. Their fermentation times varied from 6 to 21 days on average. Determinations of the quantity of sulfur dioxide and dodecanoic acid

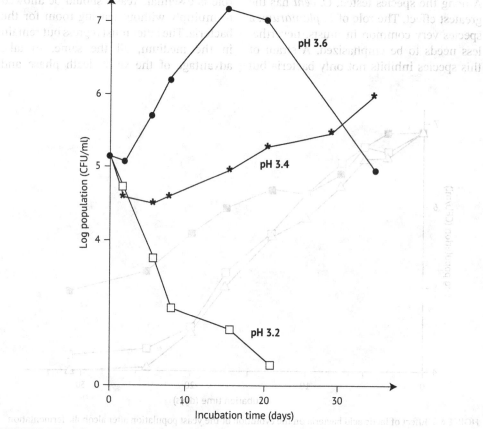

FIGURE 6.6 Effect of must pH on the evolution of lactic acid bacteria populations in the presence of yeasts: grape must with 220 g of sugar per liter (Lonvaud-Funel et al., 1988b).

showed extremely varied concentrations, depending on the wine. A relationship between the duration of the malolactic fermentation and their concentrations appears. The role of lees is very important: when added to a synthetic medium containing only sugars, malic acid, and salts, they permit bacterial activity and growth. Yeast cell compounds (identical to autolysates) have an evident nutritional activity. The prolonged contact with lees favors the onset of malolactic fermentation. It may also create stability problems in the making of lees-aged wines.

6.4.2 Interactions Between Lactic Acid Bacteria

The succession of bacterial species during alcoholic fermentation can be explained by a difference in the sensitivity of bacteria to interactions with yeasts, whether by tolerance to toxicity or the use of favorable components. Interactions between species of lactic acid bacteria must also exist, simultaneously.

Like other microorganisms, they can synthesize and liberate substances with some degree of antimicrobial activity. These substances are simple (hydrogen peroxide, organic acids, etc.) or more complex. Bacteriocins are a class of proteins whose bactericidal activity generally has a narrow range of action. It is sometimes even limited to the same species as the producing strain. A large range of these substances produced by a large variety of lactobacilli and cocci is now known.

Rammelsberg and Radler (1990), Lonvaud-Funel and Joyeux (1993), and Strasser de Saad et al. (1996) tackled this problem for wine lactic acid bacteria. The first of these works reported the discovery of two bacteriocins: brevicin from an *L. brevis* strain and caseicin from an *L. casei* strain. The first has a large range of action and inhibits *O. oeni* and *P. damnosus* strains in addition to *L. brevis*. Caseicin is only active on *L. casei*. Brevicin is a small thermostable protein (3 kDa) and is stable in a large pH range. Caseicin is less stable, with a much higher molecular weight (40–42 kDa). The same authors observed that a strain of *L. plantarum* has an antibacterial activity toward many bacterial species, including lactobacilli and cocci, notably *O. oeni*. The active protein synthesized by this strain has not yet been isolated. In a *P. pentosaceus* strain, Strasser de Saad et al. (1995)

TABLE 6.6

Influence of the Yeast Strain on Malolactic Fermentability of Wines (Lonvaud-Funel, Unpublished Results)

Yeast number	Wine concentrations		Average duration of malolactic fermentation[a] (days)
	SO_2 (mg/l)	Dodecanoic acid (mg/l)	
1	8	0.27	12
2	10	0.42	9.25
3	3	0.22	6.25
4	21	0.32	9.25
5	35	0.28	8.5
6	51	0.44	21
7	16	0.31	10.25
8	2	0.42	7

[a] Calculated for the eight wines inoculated with four strains of *O. oeni*.

demonstrated the production of a bactericidal protein with regard to several strains of *L. hilgardii*, *P. pentosaceus*, and *O. oeni*. This bacteriocin, produced in large quantities in grape juice, is stable under the acidic conditions and ethanol concentrations of wine.

In a strain collection (CRB Oenologie, ISVV Bordeaux), the most obvious interactions were recorded for one strain of *P. pentosaceus* and one strain of *L. plantarum*, both strongly inhibiting the growth of *O. oeni* and *L. mesenteroides*. This inhibition not only exists in mixed cultures but also when a medium cultured with these two strains is added to an *O. oeni* culture medium (Lonvaud-Funel and Joyeux, 1993). Different experiments have permitted the characterization of the possible roles of hydrogen peroxide, pH, and lactic acid. For both strains, the inhibitory molecules that accumulate in the culture medium are small thermostable peptides that are degraded by proteases. Their toxic effect is only temporary; they do not kill the bacteria but merely lower the growth rate and the final population. A more resistant subpopulation may develop in the end or, more simply, these peptides are degraded by the growing population.

In addition to interactions between lactic acid bacteria, we should also look at the fungi and acetic acid bacteria present on infected grapes. The media pre-cultured with these other microorganisms of the grape and wine microbiota have varying effects on lactic acid bacteria multiplication with respect to the control media (San Romao, 1985; Lonvaud-Funel *et al.*, 1987). Organic acids and polysaccharides accumulate in the medium and either impede or activate bacterial growth. In practice, it is difficult to assess their role because it depends on the soundness of the harvested grapes.

The discovery of these active molecules—bacteriocins or simple effectors—gives only an indication of the true situation in wine, with its extremely complex microbiota. The active molecules are specific not only to genera but also to species and especially strains. It is therefore impossible to try to identify them all. We can imagine they work in opposition, in synergy, or in isolation and finally produce the selection of strains observed in all winemaking.

6.5 The Importance of Bacteriophages

Bacteriophages are viruses capable of massively destroying cultures of sensitive bacterial strains. For lactic acid bacteria, bacteriophages were first discovered in the dairy industry: they provided explanations of incidents during cheese production. Phage accidents increased in this industry with the use of single-strain ferments. Considerable research led to the use of mixed fermentation starters and the rotation of cultures in fermentors, which minimized these problems.

The bacteriophage must infect a bacterium in order to multiply. Inside the cell, it uses its own genome as code and the enzyme equipment of the cell to ensure the necessary syntheses. Depending on whether the phage is moderate or virulent, the multiplication cycle does not have the same effect on the development of the culture. With a moderate phage, the genome remains integrated in the bacterial chromosome in the form of a prophage and is replicated and transmitted altogether normally to the daughter cells. With a virulent phage, the virus multiplies into many copies—liberated in the medium after cell lysis. Each one of these copies then infects another cell, and so the destruction of the culture is massive. Under certain conditions, the prophage carried by the lysogen can excise itself from the chromosome and start another lytic cycle.

In enology, the Sozzi team carried out the first research on bacteriophages of lactic acid bacteria from the *O. oeni* species (Sozzi *et al.*, 1976, 1982). The phages were first discovered under an electron microscope after centrifugation of the wine. Subsequently, studies were simplified by isolating sensitive indicator strains. In culture on agar nutritional media, it is possible to obtain lysis plaque formed on indicator strain fields and to isolate and purify these phages. According to the Sozzi team, abrupt stoppages of malolactic fermentation are caused by a phage attack, which entirely destroys the *O. oeni* population. Other authors, such as Davis *et al.* (1985), Henick-Kling *et al.* (1986), and Arendt *et al.* (1990), also demonstrated the existence of bacteriophages, without linking them to winemaking incidents. These phages have a classic morphology, composed of a head and tail (Figure 6.7).

The DNA extracts of all of the *O. oeni* phages hybridize together, and the marking of any of them furnishes a probe. By DNA/DNA hybridization, this probe permits the detection of lysogenic strains in a mixture. In this manner, we have established that nearly 90% of the *O. oeni* strains from our collection, isolated during malolactic fermentation, are lysogenic (Poblet and Lonvaud-Funel, 1996). The restriction profiles of isolated phages are not all identical, which confirms that several *O. oeni* strains coexist in wine during malolactic fermentation. In addition to the interactions described above (Section 6.4), there is also the question of variable sensitivity to phages.

Bacteria and phage counts (phage count is expressed in lysis plaque-forming units on culture medium, PFU/ml), in two tanks during malolactic fermentation, showed that both populations developed in a

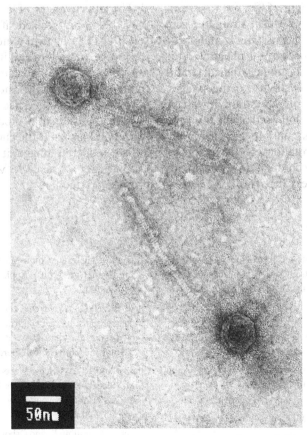

FIGURE 6.7 Electron microscope photograph of *O. oeni* phages. (***Source***: Photograph from Center of Electron Microscopy, Université de Bordeaux I.)

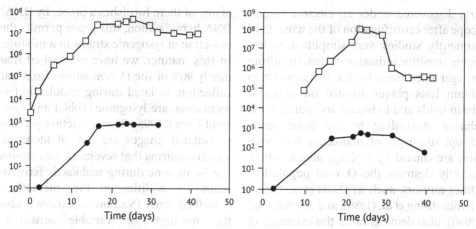

FIGURE 6.8 Evolution of phage and lactic acid bacteria populations during malolactic fermentation (Poblet and Lonvaud-Funel, 1996). □, Bacteria (CFU/ml); ●, phages (PFU/ml).

similar way. Phage populations appeared after a short lag, decreased as the bacteria populations did, and reached a maximum two logarithmic units lower than the viable bacteria count (Figure 6.8).

The diversity of *O. oeni* strains present in wine ensures against stuck malolactic fermentations caused by the phage destruction of bacteria. It is unlikely that all strains have the same sensitivity to the phages. The elimination of one strain by phage attack is probably followed by the multiplication of other strains. In fact, a natural bacterial strain rotation can occur during winemaking. According to Poblet-Icart et al. (1998), the capacity of *O. oeni* to induce the lytic cycle varies depending on strains, including in wine. A stuck malolactic fermentation can be feared only under exceptional circumstances.

The effectiveness of a selected starter may be compromised if the strain is sensitive to a phage in wine resulting from the native microflora and if the infection is not prevented by wine compounds, which remains very unlikely. The problem of *O. oeni* phages should especially be considered in the industrial production of pure starter biomasses. If the chosen strain contains a moderate phage, it may spontaneously trigger a lytic cycle, which will compromise production.

References

Amachi T. (1975) Chemical structure of a growth factor (TJF) and its physical significance for malolactic bacteria. In *Lactic Acid Bacteria in Beverages and Food*, pp. 103–118 (Eds J.G. Carr, C.V. Cutting and G.C. Whitings). Academic Press, London.

Antalick G., Perello M.C. and de Revel G. (2012) J. Agric. Food Chem., 60, 12371.

Arendt E., Neve H. and Hammes W.P. (1990) Appl. Microbiol. Biotechnol., 34, 220.

Bordas M., Araque I., Alegret J., El Khoury M., Lucas P., Rozès N., Reguant C. and Bordons A. (2013) Int. Microbiol., 16, 113.

Bousbouras G.E. and Kunkee R.E. (1971) Am. J. Enol. Vitic., 22, 121.

Bridier J., Claisse O., Coton M., Coton E. and Lonvaud-Funel A. (2010) Appl. Environ. Microbiol., 76, 7754.

Carre E. (1982) *Recherche sur la croissance des bactéries lactiques en vinification. Désacidification biologique des vins*. Thèse de Doctorat, Institut d'Œnologie, Université de Bordeaux II.

Chasseriaud L., Krieger-Weber S., Déléris-Bou M., Sieczkowski N., Jourdes M., Teissedre P.L., Claisse O. and Lonvaud-Funel A. (2015) Food Microb., 52, 131.

Chescheir S., Philbin D. and Osborne J.P. (2015) Am. J. Enol. Vitic., 66, 357.

Claisse O. and Lonvaud-Funel A. (2003) *In Œnologie*. Tec & Doc, Lavoisier, (to be published), Paris.

Coton E. (1996) *Etude de l'histidine décarboxylase de Leuconostoc oenos (Oenococcus oeni): aspects physiologique, enzymologique et génétique, conséquences œnologiques*. Thèse de Doctorat, Faculté d'Œnologie, Université de Bordeaux II.

Couto J.A. and Hogg T.A. (1994) J. Appl. Microbiol., 76, 487.

Davis C.R., Silveira N.F.A. and Fleet G.H. (1985) Appl. Environ. Microbiol., 50, 872–876.

De Revel G., Lonvaud-Funel A. and Bertrand A. (1996) *In Œnologie 95*. Tec & Doc, Lavoisier, Paris.

Delaherche A., Bon E., Dupé A., Lucas M., Arveiler B., De Daruvar A. and Lonvaud-Funel A. (2006) Appl. Microbiol. Biotechnol., 73, 394.

Delmas F., Pierre F., Coucheney F., Divies C. and Guzzo J. (2001) J. Mol. Microbiol. Biotechnol., 3, 601.

Desens C. (1989) *Les phénomènes d'inhibition des bactéries lactiques du vin. Relations avec la constitution lipidique des membranes*. Thèse de Doctorat, Institut d'Œnologie, Université de Bordeaux II.

Divol B. and Lonvaud-Funel A. (2005) J. Appl. Microbiol., 99, 85

Du Toit W., Pretorius I.S. and Lonvaud-Funel A. (2005) J. Appl. Microbiol., 98, 862.

Favier M., Bilhère E., Lonvaud-Funel A., Moine V. and Lucas P.M. (2012) PLoS One, 7, 11, e49082.

Fornachon J.-C.M. (1963) J. Sci. Food Agric., 14, 857.

Frémaux C. (1990) *Application de la biologie moléculaire à la connaissance des bactéries lactiques du vin: identification par sondes nucléiques, études de plasmides de Leuconostoc oenos*. Thèse de Doctorat, Institut d'Œnologie, Université de Bordeaux II.

Garbay S. (1994) *Recherche du mode d'adaptation de Leuconostoc oenos au vin: relation avec la constitution lipidique et protéique de la membrane plasmique*. Thèse de doctorat.Université de Bordeaux 2.

Garbay S. and Lonvaud-Funel A. (1996) J. Appl. Bacteriol., 81, 613.

Garbay S., Rozès N. and Lonvaud-Funel A. (1995) Food Microbiol., 12, 387.

Granvalet C., Assad-García J.S., Chu-Ky S., Tollot M., Guzzo J., Gresti J. and Tourdot-Marechal R. (2008) Microbiology, 154, 2611.

Grimaldi A., Bartowsky M. and Jiranek V. (2005) Int. J. Food Microbiol., 105, 233.

Guilloux-Benatier M. and Feuillat M. (1993) J. Int. Sci. Vigne Vin, 27, 299.

Guzzo J., Delmas F., Pierre F., Jobin M., Samyn B., Van Beeumen J., Carvin J.F. and Divies C. (1997) Lett. Appl. Microbiol., 214, 393.

Henick-Kling T. (1986) *Growth and metabolism of Leuconostoc oenos and Lactobacillus plantarum in wine*. PhD Thesis, University of Adelaide, Australia.

Henick-Kling T., Lee T.H. and Nicholas D.J.D. (1986) J. Appl. Bacteriol., 61, 525.

Kalmar Z.P. (1995) *Recherches sur la tolérance à l'éthanol des bactéries lactiques des vins: relation avec la constitution lipidique membranaire*. Thèse de doctorat.Université de Bordeaux 2.

Lafon-Lafourcade S. (1983) Wine and brandy. In *Biotechnology*, Vol. 5, pp. 81–163 (Eds H.J. Rehm and G. Reed). Verlag-Chemie, Weinheim.

Lafon-Lafourcade S. and Peynaud E. (1974) Conn. Vigne Vin, 8, 187.

Laforgue R. and Lonvaud-Funel A. (2012) Food Microbiol., 32, 230.

Lonvaud-Funel A. (1986) *Recherche sur les bactéries lactiques du vin. Fonctions métaboliques, croissance, génétique plasmidique*. Thèse de Doctorat ès Sciences, Institut d'Œnologie, Université de Bordeaux II.

Lonvaud-Funel A. and Joyeux A. (1988) Sci. Aliments, 8, 33–49.

Lonvaud-Funel A. and Joyeux A. (1993) Food Microbiol., 10, 411.

Lonvaud-Funel A., Desens C. and Joyeux A. (1985) Conn. Vigne Vin, 19, 229.

Lonvaud-Funel A., San Romao M.V., Joyeux A. and Chauvet S. (1987) Sci. Nutriments, 7, 267.

Lonvaud-Funel A., Joyeux A. and Desens C. (1988a) J. Sci. Food Agric., 44, 183.

Lonvaud-Funel A., Masclef J.-Ph., Joyeux A. and Paraskevopoulos Y. (1988b) Conn. Vigne Vin, 22, 11.

Lonvaud-Funel A., Joyeux A. and Ledoux O. (1991) J. Appl. Bacteriol., 71, 501.

Maitre M., Weidmann S., Dubois-Brissonnet F., David V., Covès J. and Guzzo J. (2014) Appl. Environ. Microbiol., 80, 2973.

McDonald L.C., Fleming H.P. and Hassan H.M. (1990) Appl. Environ. Microbiol., 56, 2120.

Millet V. and Lonvaud-Funel A. (2000) Lett. Appl. Microbiol., 30, 136.

Paraskevopoulos Y. (1988) *Utilisation des enveloppes cellulaires de levure pour la stimulation de la fermentation malolactique. Interprétation de leur mode d'action*. Thèse de Docteur-Ingénieur, Institut d'Œnologie, Université de Bordeaux II.

Poblet M. and Lonvaud-Funel A. (1996) In Œnologie 95, pp. 313–316 (Ed A. Lonvaud-Funel). Tec & Doc, Lavoisier, Paris.

Poblet-Icart M., Bordons A. and Lonvaud-Funel A. (1998) Curr. Microbiol., 36, 365.

Poolman B. (1993) FEMS Microbiol. Rev., 12, 125–148.

Pripis-Nicolau L. (2002) Composés soufrés volatils du vin et métabolismes des bactéries lactiques. Thèse Doctorat de l'Université Victor Segalen, Bordeaux 2.

Radler F. (1967) Conn. Vigne Vin, 1, 73.

Ramakrishnan V., Walker G.A., Fan G., Ogawa M., Luo Y., Luong P., Joseph L. and Bisson L.F. (2016) Front. Ecol. Evol., 4, 137.

Rammelsberg M. and Radler F. (1990) J. Appl. Microbiol., 69, 177.

Renouf V., Claisse O. and Lonvaud-Funel A. (2007) Appl. Microbiol. Biotechnol., 75, 149.

Ribéreau-Gayon J., Peynaud E. and Ribéreau-Gayon P. (1975) Traité d'Œnologie. Sciences et Techniques du Vin, Vol. 2. Dunod, Paris.

Romano P. and Suzzi G. (1992) Sulfur dioxide and wine microorganisms. In Wine Microbiology and Biotechnology, pp. 373–393 (Ed. G.H. Fleet). Harwood Academic Publishers, Chur.

San Romao M.V. (1985) Conn. Vigne Vin, 19, 109.

Saraiva R. (1983) Incidence des composés phénoliques à l'égard de l'activité des levures et des bactéries. Thèse Doctorat 3ème cycle, Institut d'Œnologie, Université de Bordeaux II.

Sozzi T., Maret R. and Poulin J.M. (1976) Experientia, 32, 568.

Sozzi T., Gnaegi F., d'Amico N. and Hose H. (1982) Rev. Suisse Vitic. Arboric. Hortic., 14, 2.

Strasser de Saad A.M., Pasteris S. and Manca de Nadra M.C. (1995) J. Appl. Bacteriol., 78, 473.

Strasser de Saad A.M., Pasteris S. and Manca de Nadra M.C. (1996) In Œnologie 95, p. 329. Tec & Doc, Lavoisier, Paris.

Sudraud P. and Chauvet S. (1985) Conn. Vigne Vin, 19, 31.

Sumby K.M., Jiranek V. and Grbin P.R. (2013) Food Chem., 14, 1673.

Ugliano M. and Moio L. (2005) J. Sci. Food. Agri., 86, 2221.

Vivas N., Bellemère L., Lonvaud-Funel A. and Glories Y. (1995) Rev. Fr. Œnol., 151, 39.

CHAPTER 7

Acetic Acid Bacteria

7.1 Principal Characteristics and Cytology
7.2 Classification and Identification
7.3 Principal Physiological Characteristics
7.4 Metabolism of Acetic Acid Bacteria
7.5 Acetic Acid Bacteria Development in Grape Musts
7.6 Evolution of Acetic Acid Bacteria During Winemaking and Wine Aging and the Impact on Wine Quality

7.1 Principal Characteristics and Cytology

Acetic acid bacteria are very prevalent in nature (especially fruits and flowers) and are well adapted to growth in sugar-rich and alcohol-rich environments. Wine, beer, and cider are natural habitats for these bacteria, which reproduce when production and storage conditions are not correctly controlled. Beverage quality is clearly lowered, except in the case of certain very particular beers, where acetic acid bacteria are part of the microbial system of beermaking. These bacteria are also found in environments considered to be rather anaerobic and as human pathogens.

Acetic acid bacteria cells generally have an ellipsoidal or rodlike form, with dimensions of 0.6–0.8 μm by 1–4 μm. They can be either single or organized in pairs or small chains. Some are equipped with cilia, surrounding the cell or at its ends. These locomotive organs give the cells a mobility that is visible under the microscope. These bacteria, like lactic bacteria, do not sporulate. Their metabolism is strictly aerobic. Cellular oxidations of sugars, ethanol, or other substrates are coupled with respiratory chain electron transport

---

Handbook of Enology, Volume 1: The Microbiology of Wine and Vinifications, Third Edition.
Pascal Ribéreau-Gayon, Denis Dubourdieu, Bernard B. Donèche and Aline A. Lonvaud.
© 2021 John Wiley & Sons Ltd. Published 2021 by John Wiley & Sons Ltd.

mechanisms. Oxygen is the ultimate acceptor of electrons and protons (coming from oxidation reactions). They are also characterized by the "oxidative fermentation" pathway: incomplete oxidation of substrates by dehydrogenases linked to the membrane and release of oxidation products in the medium.

The cellular structure of an acetic acid bacterium is similar to that of other bacteria: a cytoplasm containing genetic material (chromosome, plasmids), ribosomes and all of the enzymatic equipment, a plasma membrane, and cell wall. At the structural level, only the cell wall clearly distinguishes it from lactic acid bacteria. Acetic acid bacteria are Gram-negative, whereas lactic acid bacteria are Gram-positive (Section 4.3.2).

The Gram stain reflects a significant structural difference of the cell wall. Peptidoglycan is the principal constituent of Gram-positive cell walls, but it is much less present in Gram-negative walls. In the latter, an essentially lipid-based external membrane is present over the thin layer of peptidoglycan. It is destroyed by ethanol, which acts as a solvent in the Gram test, resulting in the washing away of the purple dye. The external membrane is composed of phospholipids, lipoproteins, and lipopolysaccharides. Like the plasma membrane, it is organized into a lipid bilayer: a hydrophobic zone is contained between the layers. The lipopolysaccharides comprise a lipid zone integrated into the external layer of the membrane, an oligosaccharide, and a polysaccharidic chain at the exterior of the membrane. This chain carries the antigenic specificity of the bacterium. The lipoproteins join the thin peptidoglycan layer to the external membrane. Buried in the lipid layers, crossing the entire membrane, proteins called porins form canals that enable exchanges across the cell wall.

7.2 Classification and Identification

7.2.1 Classification

Acetic acid bacteria belong to the *Acetobacteraceae* family. The (G + C) base composition of their DNA is from 51 to 65%. They are chemoorganotrophic.

The bacteria of this family are separated into two genera: *Acetobacter* and *Gluconobacter*. The key distinguishing features according to *Bergey's Manual* (De Ley et al., 1984) are as follows:

- *Acetobacter* genus: These bacteria oxidize lactic and acetic acid into CO_2; nonmobile or peritrichous flagella.
- *Gluconobacter* genus: These species do not oxidize lactic or acetic acid; nonmobile or polar flagella.

For a long time, the *Acetobacter* and *Gluconobacter* genera were the basis of the classification, and they perfectly correspond to their ecological origin. *Acetobacter* find their niche in alcoholic and acidic environments; *Gluconobacter* find it in more sugary environments, such as fruits. In general, this corresponds to enological reality: *Gluconobacter* in must and *Acetobacter* in wine.

The phylogenic structure of this family, obtained after sequence analysis of DNA coding for 16S rRNA, has completely changed and further complicated the classification. According to Komagata et al. (2014), 32 genera compose the entire family, with three major groups, which include (1) *Acetobacter, Gluconobacter, Neokomagataea,* and *Saccharibacter*; (2) *Asaia, Kozakia, Neoasia,* and *Swaminathania*; and (3) *Gluconacetobacter* and *Komagataeibacter*. Based on phylogenetics, phenotypes, and ecology, the *Gluconacetobacter* genus was split into two groups: *Gluconacetobacter liquefaciens* and *Komagataeibacter xylinus*, which comes from *Gluconacetobacter xylinus*.

TABLE 7.1
Bacteria Genera and Their Standard Species Described in Wine

Genus	Standard species	Number of species
Acetobacter	Acetobacter aceti	25
Gluconobacter	Gluconobacter oxydans	15
Gluconacetobacter	Gluconacetobacter liquefaciens	10
Komagataeibacter	Komagataeibacter xylinus	14

Source: Excerpted from Trcek and Barja (2015).

Depending on the authors, the number of species in each genus is not always the same, but this classification still demonstrates the great diversity of these bacteria. According to Saichana *et al.* (2015), there are 34 species of *Acetobacter*, 16 species of *Gluconobacter*, 24 species of *Gluconacetobacter*, 14 species of *Komagataeibacter*, eight species of *Asaia*, and two species of *Neokomagataea*.

The classification of acetic acid bacteria in grapes and wine has been updated over time in view of new taxonomy data. Many new species have been created, with reference to the 16S rRNA gene sequence. They mostly belong to the genera and species listed in Table 7.1.

However, in accordance with older results (Joyeux *et al.*, 1984), it appears that the species first colonizing grapes and then wine are mainly *Gluconobacter oxydans* and *Acetobacter aceti* (Gonzalez *et al.*, 2005).

The species *G. oxydans*, *A. aceti*, and *Acetobacter pasteurianus* are the ones that are most frequently found in the course of winemaking. The three species succeed each other during winemaking. The *G. oxydans* present on the grape disappears and gives way to *Acetobacter*, which subsists in wine (Lafon-Lafourcade and Joyeux, 1981).

7.2.2 Isolation and Identification

The isolation of acetic acid bacteria from grape must or wine is carried out by culture on a solid nutrient medium. The composition of the medium varies, depending on the researcher. Nevertheless, taking into account their nutritional demands, these bacteria only develop well on rich media containing yeast extract, amino acids, and glucose as the principal energy source. Swings (1992) described diverse media for isolating bacteria from different ecological niches. Culture and isolation conditions play a significant role in the evaluation of species colonizing the media. With certain protocols, results are biased: for example, we can describe a strong dominance of *Gluconobacter* (especially *Gluconobacter cerinus*) on grapes, the presence of other marginal species and the total absence of *Acetobacter*, even on moldy grapes (Mateo *et al.*, 2014). To isolate acetic acid bacteria from wine, the same medium may be used as for lactic acid bacteria (Ribéreau-Gayon *et al.*, 1975): 5g/l of yeast extract, 5g/l of amino acids from casein, 10g/l of glucose, and 10ml/l of tomato juice, with the pH adjusted to 4.5. It is also possible to use grape juice diluted with an equal amount of water plus 5g/l yeast extract. The medium is solidified by adding 20g/l of agar. Other media with similar composition are described in the collection of OIV analytical methods. To ensure that the medium supports only the growth of acetic acid bacteria, 0.2ml of 0.5% pimaricin and 0.1ml of 0.125% penicillin are added per 10ml of culture medium to eliminate yeasts and lactic acid bacteria. The culture must be incubated under aerobic conditions.

After isolation, the species can be identified by a group of tests and identification keys in *Bergey's Manual* (De Ley *et al.*, 1984). The first test is Gram staining. Researchers also depend on the aptitude of the strain for developing on diverse components and on its metabolism in relation to different substrates. It is easy to distinguish between the three main enological species.

Gluconobacter and *Acetobacter* are differentiated by their ability to oxidize lactate: 20 g of lactate per liter is added to the medium, already composed of yeast extract at 5 g/l. *Acetobacter* oxidizes lactate; a cloudy zone is formed by the precipitation of calcium around the colony. Moreover, *A. aceti* oxidizes glycerol, which *A. pasteurianus* is incapable of doing (Joyeux *et al.*, 1984). Ethanol oxidation by the two genera of bacteria is verified by culture in a medium containing 5 g of yeast extract per liter and 2–3% ethanol. The acidification of the medium is demonstrated either by titration or by the addition of a color-changing indicator (bromocresol green).

Molecular methods are the most reliable and the fastest and are now the most used. They implement the most diverse techniques described in Chapter 4 for lactic acid bacteria (Sections 4.3.3). Most often they involve a PCR reaction, more or less specific depending on how the amplifier is then analyzed (simple migration profile in denaturing gel or not, restriction or sequencing) and depending on the desired level of identification. Each of these methods has a certain ability to discriminate between species or strains. These are the methods of PCR-RAPD, RFLP-PCR of the 16S–23S inter-gene regions coding for the two ribosomal subunits, PCR-DGGE of variable regions of 16S DNA, the sequence analysis of "housekeeping genes" (MLST), etc. Phylogenetic groups built from the 16S–23S rRNA ITS inter-gene sequence are identical to those obtained with sequences of the 16S rRNA gene. It follows that the ITS region is suitable for identification by restriction polymorphism. It is easy to implement (Trcek and Barja, 2015). However, like for other wine microorganisms, significant progress is expected with the analysis by MALDI-TOF, which will accelerate routine analysis and enable the study of large numbers of isolates. While the first evaluations of the method are positive, they also show some limits in identifying certain isolates. Perfecting this method will come from both the precise specification of culture protocols to obtain colonies and the quality of spectra databases, which are the key to spectra identification (Andres-Barrao *et al.*, 2013; Trcek and Barja, 2015). This method has already been developed to identify acetic acid bacteria in spoiled beer (Wieme *et al.*, 2014).

7.3 Principal Physiological Characteristics

The bacteria of the *Acetobacter*, *Gluconacetobacter*, and *Gluconobacter* genera are obligate aerobes with an exclusively respiratory metabolism. Their growth, at the expense of substrates that they oxidize, is therefore determined by the presence of dissolved oxygen in the medium. All of these species develop on the surface of liquid media and form a halo or skin, less often a cloudiness and a deposit.

While very dependent on oxygen for their metabolism and their growth, acetic acid bacteria survive well in certain rather anaerobic environments. Such is the case in wines stored away from air, where they are found to be viable, and partially culturable, at the bottom of barrels for prolonged periods of time (Joyeux *et al.*, 1984; Du Toit *et al.*, 2005).

Although possible in all genera, the characteristic metabolism of *Acetobacter* is the oxidation of ethanol into acetic acid with the highest transformation yield. This is not the case for *Gluconobacter* species, which are characterized by a high oxidation activity of sugars into ketone compounds (this activity is low in *Acetobacter*). In sum, bacteria of the *Acetobacter* genus prefer ethanol to glucose for their growth; the inverse is true for bacteria of the *Gluconobacter* genus. In addition, ethanol tolerance goes the same way. Consequently, *Acetobacter* bacteria are more common in alcoholic environments (partially fermented musts and wines) than *Gluconobacter*, which are more present on the grape and in the must.

Some *Acetobacter* strains form cellulose in non-agitated liquid culture media. Certain *Gluconobacter* strains produce other polysaccharides (glucans, levans, etc.), which make the medium viscous.

The vitamin demands are approximately identical for all acetic acid bacteria. Growth is only possible in environments enriched in yeast extract and peptone, which furnish the necessary carbon substrates. In order of preference, the best substrates for *Acetobacter* are ethanol, glycerol, and lactate; for *Gluconobacter* they are mannitol, sorbitol, glycerol, fructose, and glucose. Acetic acid bacteria are not known to require any specific amino acids. Certain *Acetobacter* and *Gluconobacter* species are capable of using ammonium from their environment as a nitrogen source. Acetic acid and lactic acid are oxidized by *Acetobacter* and *Gluconacetobacter*, but not by *Gluconobacter*.

The optimum pH range for growth is from 5 to 6, but the majority of strains can easily multiply in acidic environments as low as pH 3.5.

Although they oxidize ethanol, acetic acid bacteria are not especially resistant to it. On average, *Gluconobacter* species do not tolerate more than 5% vol. ethanol, and few *Acetobacter* species develop above 10% vol. Evidently, adaptation phenomena (probably similar to those described for lactic acid bacteria) occur, ensuring their ethanol tolerance in wine. Acidity and ethanol concentration simultaneously influence the physiology and the resistance of acetic acid bacteria to one of these two factors.

7.4 Metabolism of Acetic Acid Bacteria

The oxidative fermentation pathway is the main originality of the acetic acid bacteria metabolism. This is an incomplete oxidation, in which primary alcohols are oxidized into aldehydes and themselves oxidized into carboxylic acids. Sugars and secondary alcohols are oxidized into mono- or diketone compounds. These products accumulate in the medium. Dehydrogenases are bound to the external face of the plasma membrane in the periplasmic space. The coenzymes are pyrroloquinoline quinine (PQQ) and flavin adenine dinucleotide (FAD). During oxidation, electrons removed from the substrate are transferred through the transporter chain to the terminal oxidase. The *Acetobacter* ubiquinone is type Q9; the *Gluconobacter* and *Gluconacetobacter* ubiquinone is type Q10. This specificity has been used in classification.

Moreover, acetic acid bacteria oxidize products of incomplete oxidation in the cytoplasm using cytosolic dehydrogenases. These enable complete assimilation and biomass production.

7.4.1 Metabolism of Sugars

The direct incomplete oxidation of sugars without phosphorylation leads to the formation of the corresponding ketones. The aldoses are oxidized into aldonic acids in which the aldehyde function of the sugar is transformed into a carboxylic acid function. Glucose is oxidized into gluconic acid by glucose oxidase (PQQ-dependent glucose dehydrogenase).

Bacteria of the *Gluconobacter* genus in particular also have the property of oxidizing gluconic acid, leading to the formation of 5-oxogluconic, 2-oxogluconic, and 2,5-dioxogluconic acids (Figure 7.1). These molecules energetically bind with sulfur dioxide (Section 8.4.3). Wines made from grapes tainted by *Gluconobacter* are therefore very difficult to conserve (Section 7.5).

Certain *Acetobacter* strains also form 2,5-dioxogluconic acid. Similarly, other aldoses, i.e. mannose and galactose, lead to the formation of mannonic and galactonic acid.

Ketoses are less easily oxidized by acetic acid bacteria. The oxidation of fructose can lead to the formation of gluconic acid and

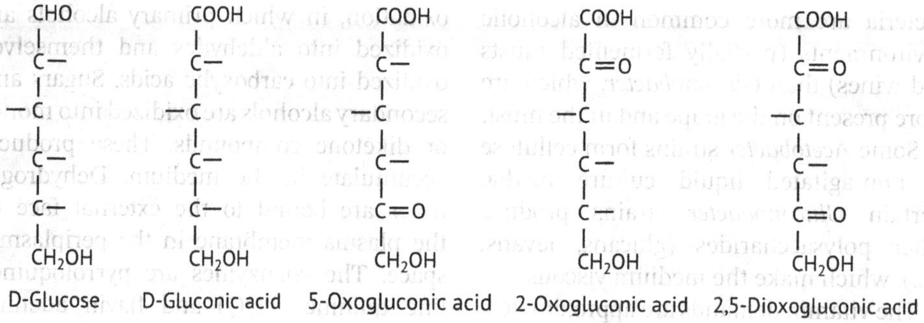

```
    CHO         COOH         COOH         COOH         COOH
     |            |            |            |            |
     C—           C—           C=O          C=O          C=O
     |            |            |            |            |
    —C—          —C—          —C—          —C—          —C—
     |            |            |            |            |
     C—           C—           C—           C—           C=O
     |            |            |            |            |
     C—           C—           C—           C—           C—
     |            |            |            |            |
    CH₂OH        CH₂OH        CH₂OH        CH₂OH        CH₂OH
  D-Glucose  D-Gluconic acid 5-Oxogluconic acid 2-Oxogluconic acid 2,5-Dioxogluconic acid
```

FIGURE 7.1 Products of incomplete oxidation of glucose.

5-oxofructose. The carbon chain of the sugar can also be split, resulting in the accumulation of glyceric, glycolic, and succinic acids.

The complete oxidation of sugars, however, supplies the necessary energy for bacterial growth. The hexose monophosphate pathway is the metabolic pathway for the utilization of sugars. In *Acetobacter*, the tricarboxylic acid cycle (Section 2.2.4) is also used for complete oxidation to $CO_2 + H_2O$, but is absent in *Gluconobacter*. The enzymes for glycolysis either do not exist or only partially exist in acetic acid bacteria.

Oxidation by the hexose monophosphate pathway begins with the phosphorylation of sugar, followed by two successive oxidation reactions. The second is accompanied by a decarboxylation. The xylulose 5-P enters a series of transketolization and transaldolization reactions (Figure 7.2). The overall reaction is the degradation of a glucose molecule into six molecules of CO_2. In parallel, 12 coenzyme molecules are reduced. The transfer of electrons and protons by the cytochromic chains in turn re-oxidizes the coenzymes. The transfer generates three molecules of ATP per pair of H^+ and e^-, i.e. 36 ATPs for the oxidation of a molecule of glucose into CO_2. This metabolic pathway is regulated by the pH of the environment and the glucose concentration. It is significantly inhibited by a low pH (<3.5) and a glucose concentration above 2 g/l. Under these conditions, gluconic acid accumulates in the medium.

From sugars, acetic acid bacteria produce soluble and insoluble exopolysaccharides, variably depending on strains, species, and the nature of sugars in the environment. The synthesis pathway is identical to that of other microorganisms. It starts with the formation of a nucleotide derivative in the sugar (e.g. UDP-glucose), which is assembled with a lipid attached to the membrane. The monomer precursor is repeatedly transported and connected to the polymer. The translocation is done through the periplasmic space and through the external membrane before being released in the environment or organized into capsules. *Komagataeibacter xylinum* (syn. *Gluconacetobacter xylinum*, syn. *Acetobacter xylinum*) are well known above all for producing not only insoluble cellulose from glucose but also, like many others, soluble heteropolysaccharides. Acetan is a soluble exopolysaccharide produced by several species. It is composed of repeated units of cellobiose (its sequence leads to a cellulose skeleton) on which a pentasaccharide side chain, formed of rhamnose, glucose, glucuronic acid, and mannose, is connected (Tayama et al., 1985; Ojinnaka et al., 1996). These polymers could play an important role either in the construction of the microbial biofilm on the surface of the grape berry or later during winemaking where they appear to intervene in viscosity or other physico-chemical actions. Until now, there have been no studies on their effect.

7.4.2 Metabolism of Ethanol

Among the transformations carried out by acetic acid bacteria, enologists are most interested in the transformation of ethanol.

7.4 | Metabolism of Acetic Acid Bacteria

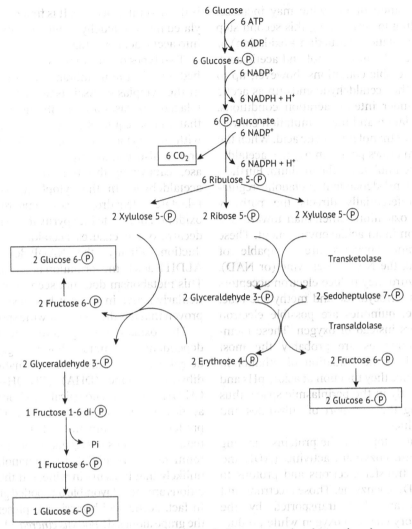

FIGURE 7.2 Degradation of glucose by acetic acid bacteria (hexose monophosphate pathway).

Its oxidation is the source of an increase in volatile acidity in many cases. In fact, the oxidation of ethanol leads to the formation of acetic acid. *Acetobacter* are also capable of oxidizing acetic acid, but this reaction is inhibited by ethanol. It therefore does not occur under winemaking conditions. The formation of acetic acid takes place in two steps; the intermediate is ethanal or acetaldehyde:

$$\underset{\text{ethanol}}{CH_3-CH_2OH} \to \underset{\text{ethanal}}{CH_3CHO} \to \underset{\text{acetic acid}}{CH_3COOH} \quad (7.1)$$

The enzymes involved are, successively, alcohol dehydrogenase (ADH) and acetaldehyde dehydrogenase (ALDH). Their functioning is similar, and they have been identified in the acetic acid bacteria normally found in musts and wines: *Acetobacter*, *Gluconacetobacter*, and *Gluconobacter*.

In the membrane, these two activities are carried out by PQQ prosthetic group proteins, related to the respiratory chain, which transfers electrons and protons through ubiquinone to the final oxygen acceptor. These enzymes operate well at acidic pH, which is not the case for cytosolic enzymes. Acetic acid slows the second step (oxidation of acetaldehyde into acetic acid), when it accumulates in the medium, in which case the acetaldehyde concentration of the wine may increase. According to Asai (1968), this second step is a dismutation (reduction + oxidation) of acetaldehyde into ethanol and acetic acid. Under aerobic conditions, however, up to 75% of the acetaldehyde ends up as acetic acid. Under intense aeration conditions, the oxidation and the dismutation convert all of the ethanol into acetic acid. When the medium grows poorer in oxygen, acetaldehyde accumulates in the medium. Furthermore, a pH-dependent metabolic regulation preferentially directs the pathway toward oxidation rather than toward dismutation in an acidic environment. These membrane enzymes are incapable of reducing the NADP coenzyme (or NAD), but in vitro they reduce electron acceptors such as ferrocyanide and methylene blue. In wine, quinones are possible electron acceptors instead of oxygen. These membrane enzymes are probably the most involved in the oxidation of ethanol in wine, since they function at acidic pHs and act directly in the periplasmic space, thus reducing the transport of substrates and metabolites.

In the cytoplasm, the proteins ensuring these same enzymatic activities (ADH and ALDH) transfer electrons and protons to the NADP coenzyme. Those electrons and protons are then transported by the respiratory chain to oxygen while producing ATP. These proteins function inside the cell and have a high optimum pH of approximately 8.0.

7.4.3 Metabolism of Lactic Acid and Glycerol

In vitro, all of the species of the *Acetobacter* genus oxidize D- and L-lactic acid. Certain strains completely oxidize it into CO_2 and H_2O via the tricarboxylic acid cycle. However, this pathway is repressed in the presence of ethanol, but most species stop at the acetic acid stage. The two isomers are transformed, but the activity is exerted more effectively on the D isomer. Pyruvate is the first intermediate. It is first decarboxylated into acetaldehyde, which is oxidized into acetic acid by ALDH.

Two types of enzymes have been identified, one in the membrane and the other in the cytoplasm (Asai, 1968). The D- and L-lactate oxidases are membrane enzymes that do not require a cofactor but function with the cytochrome chain. The membranes also contain pyruvate decarboxylase, catalyzing the transformation into acetaldehyde. In the cytoplasm, D- and L-lactate dehydrogenases ensure the oxidation of lactate into pyruvate. Pyruvate decarboxylase ensures acetaldehyde production. Finally, the NADP-dependent ALDH leads to the formation of acetic acid. This metabolism does not seem to be particularly active in wine; it has never been proven that it is the source of wine spoilage.

The oxidation of glycerol, by PQQ-dependent membrane-bound glycerol dehydrogenase, leads to the formation of dihydroxyacetone (DHA) ($CH_2OH-CO-CH_2OH$). In an enological environment, acetic acid bacteria, except *A. pasteurianus*, produce this compound. This reaction requires an intense oxygenation of the environment and is inhibited by ethanol. It is unlikely that it occurs in wine, but the conditions are more favorable on spoiled grapes. In fact, acetic acid bacteria are present on the grape alongside *Botrytis cinerea*, glycerol being one of its principal metabolites.

7.4.4 Formation of Acetoin

Acetoin is the compound with the intermediate oxidation level in the group of the three acetoinic molecules present in wine: diacetyl, acetoin, and butanediol. It is formed from pyruvate, itself a metabolic intermediate having different origins in microorganisms.

Acetic acid bacteria produce acetoin from lactic acid, which they first oxidize into pyruvate. For most strains, this pathway is not very significant, but some form up to 74% of the theoretical maximum quantity. Two synthesis pathways are believed to exist (Asai, 1968):

Pyruvate is decarboxylated in the presence of thiamine pyrophosphate (TPP) and leads to the formation of acetaldehyde-TPP. The reaction of acetaldehyde and acetaldehyde-TPP forms acetoin:

$$2CH_3-CO-COOH + 2TPP$$
$$\to 2CH_3CHO-TPP + 2CO_2$$

$$CH_3CHO-TPP \to CH_3-CHO+TPP$$

$$CH_3CHO-TPP+CH_3-CHO$$
$$\to CH_3CHOH-CO-CH_3 + TPP$$
$$\text{acetoin}$$

The other synthesis pathway involves the intermediate step of forming α-acetolactate (identical to the lactic acid bacteria pathway) by the reaction of acetaldehyde-TPP and pyruvate:

$$CH_3CHO-TPP+CH_3-CO-COOH$$
$$\to CH_3-CO-(CH_3)COH-COOH+TPP$$
$$\text{α-acetolactate}$$

$$CH_3-CO-(CH_3)COH-COOH$$
$$\to CO_2 + CH_3CHOH-CO-CH_3$$
$$\text{acetoin}$$

Gluconobacter oxydans can also oxidize butanediol into acetoin (Swings, 1992). The presence of acetoin in wine is less problematic than that of diacetyl, which has a much more pronounced aroma. Moreover, it has never been proven that acetic acid bacteria are capable of oxidizing acetoin into diacetyl.

7.5 Acetic Acid Bacteria Development in Grape Musts

Acetic acid bacteria are present on ripe grapes. The populations vary greatly, in number and species, according to grape soundness. On sound grapes, the population level is low—around 10^2 CFU/ml, and it is almost entirely made up of *G. oxydans*. Moldy grapes, however, are very contaminated: populations can reach upwards of 10^5–10^6 CFU/ml and are mixed, comprising varying proportions of *Gluconobacter* and *Acetobacter* (Lafon-Lafourcade and Joyeux, 1981). Studies conducted with molecular investigation methods confirm these observations and agree on the existence of a greater diversity of species on spoiled grapes (Barata et al., 2012).

This bacterial microflora modifies must composition by metabolizing sugars and sometimes organic acids. *Acetobacter* partially degrades citric and malic acids (Joyeux et al., 1984). However, the most significant activity of these bacteria, especially *G. oxydans*, is producing substances that bind strongly with SO_2 (Sections 8.4.3 and 8.4.6). They transform glucose into gluconic acid and its lactone derivatives, γ- and δ-gluconolactone, which can bind up to 135 mg/l SO_2, in a must containing 24 g/l gluconic acid with a free SO_2 content of 50 mg/l (Barbe et al., 2000). These bacteria also oxidize glycerol to form dihydroxyacetone, which binds with SO_2 in an unstable manner. This compound is metabolized by yeast during fermentation and is no longer a factor in the binding power of wine.

The largest share of SO_2 binding due to *Gluconobacter* involves 5-oxofructose, which is not metabolized by yeast, so it remains unchanged in the wine. It is formed by oxidation of any fructose in the medium, as is the case in grape must. In a botrytized must where the fungus has developed to its most advanced stage, this compound alone accounts for up to 60% of all binding with SO_2. In wines made from this type of must, 5-oxofructose and γ- and δ-gluconolactones are involved in nearly 50% of SO_2 binding, while acetaldehyde and keto acids formed by yeast account for most of the remainder (Barbe et al., 2000).

In addition to acetic acid bacteria, a wide variety of yeasts contaminate grapes. Although alcohol production is limited, these strains do produce small quantities of ethanol directly on extremely rotten grapes or immediately following crushing and pressing. This alcohol is immediately oxidized by acetic acid bacteria, composed in part of *Acetobacter*. Some musts can therefore have a relatively high volatile acidity before fermentation.

Furthermore, acetic acid bacteria produce yeast-inhibiting substances. Lafon-Lafourcade and Joyeux (1981) demonstrated this fact for *Gluconobacter*. Cultured for three, seven, and 14 days in a must that is then inoculated with *S. cerevisiae*, this bacterium stopped alcoholic fermentation. There remained 1.5, 9, and 18 g of non-fermented sugar per liter on average, respectively, as opposed to 0.5 g/l in the control. Gilliland and Lacey (1964) identified the same inhibitory effect of *Acetobacter* toward strains from seven yeast genera possibly present in must, including *Saccharomyces*. As a general rule, however, this effect is very limited. In fact, acetic acid bacteria activity in grape must is normally short-lived: it stops at almost the same time that alcoholic fermentation begins.

In summary, the principal drawback of grape contamination by acetic acid bacteria is the production of volatile acidity and ketone compounds. These molecules are formed from sugars liberated through fissures on the grape caused by fungi during their proliferation on the berry. Some sugars are partially oxidized, mostly by *Gluconobacter*. Others are fermented by yeasts into ethanol, which is used as a substrate for *Acetobacter* (more present on spoiled berries), and lead to production of acetic acid. The conditions required for the biological activity of acetic acid bacteria are ensured in the environment created by the biofilm of mold, yeasts, and bacteria, forming on the surface of berries.

7.6 Evolution of Acetic Acid Bacteria During Winemaking and Wine Aging and the Impact on Wine Quality

The principal physiological characteristic of acetic acid bacteria is their need for oxygen to grow or the availability of suitable electron acceptors. In wine, *A. aceti* or *A. pasteurianus* draw their energy from the oxidation of ethanol, and therefore they produce acetic acid.

Finished wines contain around 0.3–0.5 g of volatile acidity (expressed as H_2SO_4) per liter, resulting from yeast and lactic acid bacteria metabolisms. Above this concentration, acetic acid accumulation most often comes from acetic acid bacteria; this problem is called vinegar taint. This contamination must be avoided not only because of its negative effect on wine quality but also because of the legal limits on the concentration of volatile acidity permitted in wine. The increase in acetic acid is accompanied by an increase in ethyl acetate, which is a real marker for vinegar taint. The perception threshold of this ester is around 160–180 mg/l. Yeasts also form it in concentrations of up to 50 mg/l. An excessive temperature during wine storage accelerates this spoilage (Volume 2, Section 8.3.3).

Acetic acid bacteria multiply easily under aerobic conditions, i.e. in grape must or wine at the surface in contact with air, but this is not the case during fermentation. As soon as alcoholic fermentation begins, the environment grows poor in oxygen and the oxidation–reduction potential falls.

Lafon-Lafourcade and Joyeux (1981) observed the evolution of bacteria during the production of two kinds of wine. In a white grape must parasitized by *B. cinerea*, the initial population of 2×10^6 CFU/ml fell to 8×10^4 CFU/ml, five days after harvest. At this stage, 90 g of sugar per liter was fermented. The population was less than 10^3 CFU/ml on the 12th day, after the fermentation of 170 g of sugar per liter. Similarly in a red grape must, the initial population of 2×10^4 CFU/ml progressively dropped to 20 CFU/ml by the time the wine was drained off the skins. Acetic acid bacteria are therefore not involved during alcoholic fermentation. The same is true during malolactic fermentation. Yet in all cases, they never totally disappear.

During barrel or tank aging, the wine should be protected from air to avoid both chemical and biological oxidation. In addition to oxidative yeasts, acetic acid bacteria still viable after both fermentations are capable of multiplying in the presence of air. To avoid this problem, the containers (tanks or barrels) should be filled as completely as possible. Topping-up should be done with a wine of excellent microbiological quality to avoid contamination. An inert gas, such as carbon dioxide, nitrogen, argon, or a gas mixture, may also be used to replace the atmosphere present at the top of the tanks.

Barrel or tank aging also entails racking for clarifying and aerating the wine—causing limited oxidations that are indispensable to wine evolution. In the absence of air between rackings, acetic acid bacteria remain present in the entire wine mass at concentrations of 10^2–10^4 CFU/ml. During traditional barrel aging, the dissolving of oxygen is more significant than in tanks, due to diffusion across the wood and the bung (when the bung is on top). This slight oxygen dissolution suffices to ensure an oxidation–reduction level compatible with acetic acid bacteria survival.

At the time of racking, the conditions are radically modified. The transfer from one tank or barrel to another is accompanied by the dissolution of 5–6 mg of oxygen per liter in the wine, unless very careful precautions are taken. The gas dissolves more quickly when air contact is favored and the temperature is low. This oxygen is at first rapidly and then more progressively consumed by the oxidizable substances in wine. The oxidation–reduction potential follows the same evolution. Table 7.2 illustrates the evolution of the dissolved oxygen concentration and the acetic acid bacteria population—the growth of which is very active just after racking. Afterward, the bacteria slowly lose their viability until the next racking, several months later, because conditions are limiting. The same phenomenon occurs at each racking during the 18 months of aging.

Between two rackings, bacteria lose their ability to be cultured in culture media. Therefore, they also lose their ability to be counted by classic colony counting methods, thus resulting in the observed population decrease. In reality, viable populations remain high, but they are in a "viable but not culturable" (VBNC) state (Section 6.3.2). They no longer form colonies. This is a sort of dormant state where metabolism is restricted to the maintenance of functions essential to survival. Bacteria return to their culturable state without necessarily increasing their population, when conditions are once again favorable in a context of oxygen dissolution and increase in oxidation–reduction potential (Millet and Lonvaud-Funel, 2000).

TABLE 7.2

Evolution of Dissolved Oxygen and Acetic Acid Bacteria Concentration When Racking Wine from One Barrel to Another

Stage	Dissolved oxygen (mg/l)	Acetic acid bacteria (CFU/ml)
Before racking	0.2	1.0×10^3
During racking	6.0	—
After 3 days	0.8	1.2×10^4
After 20 days	0.6	2.0×10^3
After 60 days	0.3	10

During each population growth period, ethanol is oxidized since it is the preferred source of energy, and as a result acetic acid is produced. Quantities formed depend on the population level. When a red wine is racked, 0.02 g of acetic acid per liter is formed when a population doubles from 3.5×10^4 to 7.0×10^4 CFU/ml. The acetic acid concentration increases by 0.08 g/l when the initial population of 50 CFU/ml grows to 1.5×10^3 CFU/ml. These observations prove that acetic acid bacteria play a key role in the increase in volatile acidity during aging (Millet et al., 1995).

The principal factors affecting acetic acid bacteria development (as with lactic acid bacteria) are alcohol content, pH, SO_2 concentration, temperature, and oxidation–reduction potential. The more the pH and temperature are increased, the more easily the bacteria survive. Their growth is quicker in the case of aeration.

There is no effective method for totally eliminating acetic acid bacteria. Current observations show that even when protected by 25–30 mg of free SO_2 per liter, wines always conserve a viable bacteria population—up to 10^3–10^4 CFU/ml during barrel aging. Only a relatively low temperature of around 15°C can possibly limit this problem.

Furthermore, all other parameters (alcohol, pH, etc.) being equal, the increase of volatile acidity also depends on the wine storage method. In large-capacity tanks, the increase in volatile acidity is lower than in barrels. This is quite logical, since the oxidation–reduction potential is higher in barrels because of the air transfer through the wood. Moreover, the ratio of surface to wine volume exposed to oxygen dissolution from the air is generally greater in a barrel than in a tank.

To avoid spoilage from acetic acid bacteria, the winemaker should first of all concentrate on winery hygiene in order to eliminate potential contamination sources for microorganisms. Similarly, even when the population is around 10^3 CFU/ml at the time of bottling, it decreases slowly but inexorably during bottle aging as the redox potential becomes very restrictive. The total absence of subsequent oxygenation leads to the elimination of these bacteria.

References

Andres-Barrao M.C., Benagli C., Chappuis M., Ortega Perez R., Tonolla M. and Barja F. (2013) Syst. Appl. Microbiol., 36, 75.

Asai T. (1968) *Acetic Acid Bacteria Classification and Biochemical Activities*. University of Tokyo Press, Tokyo.

Barata A., Malfeito-Ferreira M. and Loureiro V. (2012) Int. J. Food Microbiol., 154, 152.

Barbe J.C., de Revel G., Joyeux A., Lonvaud-Funel A. and Bertrand A. (2000) J. Agric. Food Chem., 48, 3413.

De Ley J., Gillis M. and Swings J. (1984) Family VI. Acetobacteraceae. In *Bergey's Manual of Systematic Bacteriology*, pp. 267–268 (Eds N.R. Krieg and J.G. Holt). Williams and Wilkins, Baltimore.

Du Toit W.J., Pretorius I.S. and Lonvaud-Funel A. (2005) J. Appl. Microbiol., 98, 862.

Gilliland R.P. and Lacey J.P. (1964) Nature, 202, 727–728.

Gonzalez A., Hierro N., Poblet M., Mas A. and Guillamon J.M. (2005) Int. J. Food. Microbiol., 102, 295.

Joyeux A., Lafon-Lafourcade S. and Ribéreau-Gayon P. (1984) Appl. Environ. Microbiol., 48, 153.

Komagata K., Iino T. and Yamada Y. (2014) The family *Acetobacteraceae*. In *Prokaryotes*, 4th edition. Alphaproteobacteria and Betaproteobacteria, pp. 3–78 (Eds E. Rosenberg (Editor-in-Chief), E.F. DeLong, S. Lory, E. Stackebrandt and F. Thompson). Springer, Heidelberg.

Lafon-Lafourcade S. and Joyeux A. (1981) Bull. OIV, 608, 803–829.

Mateo E., Torija M.J., Mas A. and Bartowsky E.J. (2014) Int J. Food Microbiol., 178, 98.

Millet V. and Lonvaud-Funel A. (2000) Lett. Appl. Microbiol., 30, 136.

Millet V., Vivas N. and Lonvaud-Funel A. (1995) J. Sci. Tech. Tonnellerie, 1, 123.

Ojinnaka C., Jay A.J., Colquhoun I.J., Brownsey G.J., Morris E.R. and Morris V.J. (1996) Int. J. Biol. Macromol., 19, 149.

Ribéreau-Gayon J., Peynaud E., Ribéreau-Gayon P. and Sudraud P. (1975) *Traité d'Œnologie. Sciences et Techniques du Vin*, Vol. 2. Dunod, Paris.

Saichana N., Matsushita K., Adachi O., Frébort I. and Frébortová J. (2015) Biotechnol. Adv., 33, 1260.

Swings J. (1992) The genera Acetobacter and Gluconobacter. In *The Prokaryotes*, pp. 2268–2286 (Eds A. Balows, H.G. Trüpper, M. Dworkin, W. Harder and K.H. Schleifer. Springer, New York.

Tayama K., Minakami H., Entani E., Fujiyama S. and Masai H. (1985) Agric. Biol. Chem., 49, 959.

Trcek J. and Barja F. (2015) Int. J. Food Microbiol., 196, 137.

Wieme A.D., Spitaels F., Aert M., De Bruyne K., Van Landschoot A. and Vandamme P. (2014), Appl. Environ. Microbiol., 80, 1528.

CHAPTER 8

The Use of Sulfur Dioxide in Must and Wine Treatment

8.1 Introduction
8.2 Physiological Effects
8.3 Chemistry of Sulfur Dioxide
8.4 Molecules Binding Sulfur Dioxide
8.5 Practical Consequences: The State of Sulfur Dioxide in Wines
8.6 Antimicrobial Properties of Sulfur Dioxide
8.7 The Role of Sulfur Dioxide in Winemaking
8.8 The Use of Sulfur Dioxide in the Winery

8.1 Introduction

The widespread use of sulfur dioxide (SO_2) appears to date back to the end of the 18th century. Its many properties make it an indispensable aid in winemaking. Perhaps some wines could be made in the total or near-total absence of SO_2, but it would certainly be presumptuous to claim that all of the wines produced in the various wineries throughout the world could be made in this manner. It must also be taken into account that yeasts produce small quantities of SO_2 during fermentation. In general, the amount formed is rarely more than 10 mg/l, but in certain cases it can exceed 30 mg/l. Consequently, the total absence of sulfur dioxide in wine is rare, even in the absence of sulfiting.

Its principal properties are as follows:

1. Antiseptic: It inhibits the development of microorganisms. It has a greater activity on bacteria than on yeasts. At low concentrations, the inhibition is temporary. High concentrations destroy a percentage of the microbial population. The effectiveness of a given concentration is greater when the initial population has been

---

Handbook of Enology, Volume 1: The Microbiology of Wine and Vinifications, Third Edition.
Pascal Ribéreau-Gayon, Denis Dubourdieu, Bernard B. Donèche and Aline A. Lonvaud.
© 2021 John Wiley & Sons Ltd. Published 2021 by John Wiley & Sons Ltd.

decreased, by filtration, for example. During aging, SO$_2$ hinders the development of all types of microorganisms (yeasts, lactic acid bacteria, and, to a lesser extent, acetic acid bacteria), preventing yeast haze formation, secondary fermentation of sweet white wines (Section 8.6.2), *Brettanomyces* contamination, and the subsequent formation of ethylphenols (Volume 2, Section 8.4.4); the development of mycodermic yeast (flor) (Volume 2, Section 8.3.4); and various types of bacteria spoilage (Volume 2, Sections 8.3.1 and 8.3.3).
Antioxidant: In the presence of catalysts, it binds with dissolved oxygen according to the following reaction:

$$SO_2 + \frac{1}{2}O_2 \rightarrow SO_3$$

2. This reaction is slow. It protects wines from chemical oxidations, but it has no effect on enzymatic oxidations, which are very quick. SO$_2$ protects wine from an excessively intense oxidation of its phenolic compounds and some aroma compounds. It prevents maderization (wine heating and oxidation process). It also contributes to the establishment of a sufficiently low oxidation–reduction potential, favoring wine aroma and taste development during storage and aging.
3. Antioxidase: It instantaneously inhibits the functioning of oxidation enzymes (tyrosinase, laccase) and can ensure their destruction over time. Before fermentation, SO$_2$ protects musts from oxidation by this mechanism. It also helps to avoid enzymatic browning in white and red wines made from moldy grapes.
4. By binding with acetaldehyde and other similar products, it protects wine aromas and makes the oxidized character disappear.

Adding SO$_2$ to wine raises a number of issues. Excessive doses must be avoided, above all for health reasons, but also because of their impact on aroma. High doses neutralize wine aroma, while even larger amounts produce characteristic aroma defects, i.e. a smell of wet wool that rapidly becomes suffocating and irritating, together with a burning sensation on the aftertaste.

However, an insufficient concentration does not ensure the total stability of the wine. Excessive oxidation or microbial development can compromise its presentation and quality.

It is not easy to calculate the precise quantities required because of the complex chemical equilibrium of this molecule in wine. It exists in different forms that possess different properties in media of different composition.

The concentration of sulfur dioxide in wine is habitually expressed in mg SO$_2$ per liter (or ppm), although this substance exists in multiple forms in wine (Section 8.3).

The words sulfur dioxide, sulfurous anhydride, and sulfurous gas can all be used equally, or even sulfurous acid, though the corresponding molecule cannot be isolated. The expression "sulfur," however, is fundamentally incorrect.

Additions made to wine are always expressed in the anhydrous form, in milligram/liter or in gram/hectoliter, regardless of the form effectively employed—sulfur dioxide gas or liquid solution, potassium bisulfite (KHSO$_3$) or potassium metabisulfite (K$_2$S$_2$O$_5$). The effect of the addition to wine is the same, regardless of the form used. The equilibrium established between the various forms is identical. It depends on the pH and the presence of molecules that bind with sulfur dioxide.

Substantial progress in the understanding of the chemistry of sulfur dioxide and its properties has permitted the winemaker to use it rationally in wine. As a result, the concentrations of SO$_2$ employed in wine have considerably decreased. Simultaneously, this technological progress has led

TABLE 8.1

Maximum Sulfur Dioxide Concentrations Depending on Wine Type

EU regulations no: 479/2008 and 606/2009

Types of wine	Sugar content (<5 g/l)	Sugar content (≥5 g/l)
Red wines	150 (+50)[a]	200 (+50)[a]
White and rosé wines	200 (+50)[a]	250 (+50)[a]
Quality sparkling wines	185 (+40)[a]	185 (+40)[a]
Other sparkling wines	235 (+40)[a]	235 (+40)[a]
Fortified wines	150	200
Vins de pays (ABV > 15%; sugar > 45 g/l)		
White AOP wines		
Bordeaux superieur, Graves de Vayres, Côtes de Bordeaux, Saint-Macaire, Premières Côtes de Bordeaux, Sainte-Foy Bordeaux, Côtes de Bergerac (whether followed or not by the denomination Côtes de Saussignac), Haut Montravel, Côtes de Montravel and Rosette, Gaillac		
White DO/DOC wines		
Allela, Navarra, Penedes, Tarragona and Valencia		
Alto Adige, Trentino "passito" "vendemmia tardiva"		
DOC Moscato di Pantelleria naturale and Moscato di Pantelleria		
UK wines (residual sugar > 45 g/l)		
Spätlese		300
Auslese and some Romanian, Czech, Slovakian and Slovenian white wines		350
Beerenauslese, Ausbruch, Ausbruch-wein, Strohwein, Schilfwein, Trockenbeerenauslese, Eiswein		
White AOP wines		
Sauternes, Barsac, Cadillac, Cérons, Louplac, Sainte-Crolx-du-Mont, Graves supérieures, Monbazillac, Jurançon, Pacherenc du Vic Bilh Anjou-Coteaux de la Loire, Bonnezeaux, Quarts de Chaume, Coteaux de l'Aubance, Coteaux du Layon followed by village name, Coteaux du Layon followed by Chaume designation, Coteaux de Saumur		
Alsace and Alsace grand cru followed by "vendanges tardives" or "selection de grains nobles"		
Sweet wines from Greece (sugar ≥ 45 g/l)		
Samos, Rhodes, Patras, Rio Patron, Cephalonia, Limnos, Sitia, Santorini, Nemea, Daphne		
Canadian ice wines		400

EU regulations and OIV recommendations (values expressed in milligram per liter).
Maximum sulfur dioxide concentrations should be adapted in organic and biodynamic wines to comply with specific labeling requirements and certifications.
[a] When required due to weather conditions in certain vineyard areas.

TABLE 8.2

Maximum Acceptable Sulfur Dioxide Limits as a Function of Wine Type

Maximum acceptable OIV limits (*International Code of Winemaking Practices*, 2001)

Types of wine	Sugar content ≤ 4 g/l	Sugar content > 4 g/l
Red wines	150	300
White and rosé wines	200	300
Certain sweet white wines		400

EU regulations and OIV recommendations (values expressed in milligram per liter).

to a decrease in authorized concentrations. In 1907, French legislation set the legal limit in all wines at 350 mg/l, and this was increased, in 1926, to 450 mg/l. Today, French wines are subject to EU legislation (Tables 8.1 and 8.2), which has gradually reduced the permitted level to 160 mg/l for most red wines and 210 mg/l for the majority of white wines. Higher doses may only be used in wines with very high sugar content. They are generally premium wines produced in small volumes and consumed in moderate quantities.

In practice, the concentration used is even lower. For white French wines (excluding special wines), the average concentration is 105 mg/l; for red wines it is 75 mg/l. The International Organisation of Vine and Wine (OIV) recommends slightly higher values than those advocated by the EU in its member countries. In certain countries, the regulations on sulfur dioxide dictate a single limit for all wines. For example, this value is 350 mg/l in the United States, in Canada, in Japan, and in Australia.

Due to the fluctuating equilibrium between free and bound forms of SO_2, in general, the legislation of many countries exclusively refers to the total sulfur dioxide concentration. Certain countries, however, have regulations for the free fraction.

Today, especially for health reasons, there is an effort to reduce the authorized concentrations in different kinds of wines. Such an approach consists in optimizing the conditions and perfecting the methods of using this product. This supposes more in-depth knowledge of the chemical properties of the sulfur dioxide molecule and its enological role. Substitute products can also be considered. Due to the various effects of sulfur dioxide in wine, the existence of another substance performing the same roles without the disadvantages seems very unlikely, but the existence of aids that complement the effect of SO_2 in some of its properties is perfectly conceivable. Enological research has always been preoccupied by the quest for such a substitute product or process (Chapter 9).

In conclusion, sulfur dioxide permits the storage of many types of wine that would not exist without its protection. In particular, it permits extended barrel maturation and bottle aging. In view of its involvement in a wide variety of chemical reactions, it is not easy to determine the optimum dose to obtain all the benefits of SO_2 without any of its unfortunate side effects.

8.2 Physiological Effects

The addition of sulfur dioxide to wine raises health-related objections. These should be taken into account, although this product boasts a long history of use. Its use has always been regulated, and enological techniques have always been aimed at lowering its concentrations. Since the beginning of the 20th century, the possible toxicity of sulfur dioxide has been the subject of much research (Vaquer, 1988).

Acute toxicity has been studied in animals. The absorption of a single dose of sulfites is only slightly toxic. Depending on the animal species, the LD_{50} (lethal dose for 50% of individuals) is between 0.7 and 2.5 g of SO_2 per kilogram of body weight. Sodium sulfite therefore has an acute toxicity similar to inoffensive products such as sodium bicarbonate or potassium chloride.

Chronic toxicity has also been studied in animals. For several generations, a diet containing 1.5 g of SO_2/kg was regularly absorbed. Three kinds of complications resulted: a thiamine deficiency linked to destruction of that compound by sulfur dioxide, a histopathological modification of the stomach, and slowed animal growth. This study permitted the establishment of a maximum nontoxic concentration for rats at 72 mg/kg of body weight. This value led the World Health Organization to set the recommended daily allowance (RDA) at 0.7 mg of SO_2/kg of body weight.

Concerning its toxicity in humans, studies carried out indicate the appearance of intoxication symptoms such as nausea, vomiting, and gastric irritation at high absorbed doses (4 g of sodium sulfite in a single dose). No secondary effects were observed with a dose of 400 mg of sulfur dioxide for 25 days. In humans, its possible toxicity has often been attributed to the well-known destruction of thiamine or vitamin B1 by sulfites, but the corresponding reaction has been observed to be very limited at a pH of around 2, which corresponds to stomach pH.

The main pathway of sulfur dioxide metabolism is oxidation into sulfates by means of sulfite oxidase, which is present mainly in the liver but also in other organs (kidneys, intestine, heart, and lungs). In this form, it is incorporated into the body's reserves of sulfates. The activity of sulfite oxidase varies greatly between species. When taken orally in feed by rats, mice, guinea pigs, and monkeys, bisulfite (sodium or potassium) does not induce toxicity up to a dose of 72 mg/kg/day. Above that dose, stunted growth, weight loss, atrophy of visceral organs, bones and the medulla, stomach inflammation, polyneuritis, and testicular edema can occur. Experiments on humans, including both normal and asthmatic subjects, revealed that short-term inhalation of SO_2 at a concentration of 5–10 ppm can produce bronchial constriction, mostly likely as a reflex, in healthy adults. Subjects suffering from respiratory disorders, in particular asthma, show greater sensitivity to even moderate exposures to SO_2. In asthmatics, the constrictive effect of SO_2 on the bronchial tubes is increased by physical effort at low concentrations of 0.1 ppm. It should therefore be avoided for people suffering from kidney or liver diseases as well as for children. Asthmatics are the most at-risk group. People who are susceptible to conjunctivitis, bronchial infections, emphysema, bronchial asthma, or cardiovascular disease, as well as people with allergies (asthma, achiness, headaches, gastric/digestive or skin irritation, eczema, nausea, diarrhea) or with sulfite intolerance, should pay careful attention to their SO_2 consumption.

In 1973, allergic reactions to sulfites were proven to exist. They occur at very low ingested concentrations (around 1 mg) and primarily concern asthmatics (4–10% of the human population). Asthmatics are therefore often urged to abstain from drinking wine. Although SO_2 sensitivity has not been clearly demonstrated for non-asthmatics, these allergic reactions led the US Food and Drug Administration (FDA) to require the presence of sulfites to be indicated on wine labels in the United States when the concentration exceeds 10 mg/l.

Considering an RDA of 0.7 mg/kg/day, the acceptable dose for an individual is between 42 and 56 mg/day, depending on body weight (60 and 80 kg, respectively). The consumption of half a bottle of wine per day (375 ml) can supply a quantity of SO_2 higher than the RDA. If the total SO_2 concentration is at the maximum limit authorized by the EU (160 mg/l for red wines and 210 mg/l for white wines), the quantity of SO_2 furnished by half a bottle is 60 mg for reds and 79 mg for whites. The average SO_2 concentrations observed in France are much lower: 75 mg/l for red wines and 105 mg/l for white wines. Therefore, the daily consumption of half a bottle furnishes 28 and 39 mg of SO_2, respectively.

In any case, the figures clearly indicate that, with respect to World Health

Organization norms, wines can supply a non-negligible quantity of SO_2. However, the Codex Alimentarius authorizes the addition of SO_2 to various industrial foodstuffs at maximum concentrations of 15 mg/kg (white sugar, powdered sugar, or dextrose, for example), 100 mg/kg (storage of shellfish, such as frozen shrimp) and 1,000 mg/kg (dried fruits and even up to 2,000 mg/kg for dried apricots). It is therefore understandable that national and international health authorities recommend additional decreases in the accepted legal limits.

Experts from the OIV estimate that the concentrations recommended by the EU can be decreased by 10 mg/l, at least for the most conventional wines. Consequently, maximum contents for total SO_2 have been lowered, in EU regulation 606/2009 (for alignment with the OIV limits recognized internationally). The maximum SO_2 content for dry red wines has been dropped from 160 to 150 mg/l and from 210 to 200 mg/l for dry white and rosé wines. For wines with more than 5 g/l of residual sugar, the maximum SO_2 content has gone from 210 to 200 mg/l for red wines and to 250 mg/l for white and rosé wines. The "Contains Sulfites" labeling on bottles and other containers is also mandatory for total SO_2 of 10 mg/l or more. A mandatory logbook must be kept in the winery (including information on handling date, nature, and quantity of products added). EU regulation 606/2009 also stipulates that the mandatory "Contains Sulfites" wording may be accompanied by a pictogram. This pictogram shows the formula for SO_2 surrounded by the words "Allergy Information." This logo is optional. Only the wording is mandatory for applicable wines.

However, in this perfectly justified quest for lower SO_2 concentrations, specialty wines such as botrytized wines must be taken into account. Due to their particular chemical composition, they possess significant binding power with sulfur dioxide. Consequently, their stabilization requires extensive sulfiting. The EU legislation authorizing 400 mg/l is perfectly reasonable, but this concentration is not always sufficient. In particular, it does not guarantee the stability of some batches of botrytized wines and will not prevent them from secondary fermentation. The maximum SO_2 content of sweet wines was not decreased in EU regulation 607/2009. It was thus kept constant for certain specialty sweet wines produced in small batches with exemptions required by their very high sugar content in order to ensure good aging. However, in light of experimental results and scientific studies currently underway on the reduction and substitution of sulfites in wine and on the contribution of sulfites in wine to human nutrition, the maximum values should still be examined again in the future with a view to decreasing them.

8.3 Chemistry of Sulfur Dioxide

8.3.1 Free Sulfur Dioxide

During the solubilization of SO_2, equilibria are established:

TABLE 8.3

Molecular SO_2 and Bisulfite Percentages According to pH (at 20°C) in Aqueous Solution

pH	Molecular SO_2	Bisulfite (HSO_3^-)
3	6.06	94.94
3.10	4.88	95.12
3.20	3.91	96.09
3.30	3.13	96.87
3.40	2.51	97.49
3.50	2.00	98.00
3.60	1.60	98.40
3.70	1.27	98.73
3.80	1.01	98.99
3.90	0.81	99.19
4.00	0.64	99.36

TABLE 8.4

Sulfur Dioxide pK_1 Values According to Alcohol Content and Temperature (Usseglio-Tomasset, 1995)

Alcohol (% vol.)	Temperature (°C)							
	19	22	25	28	31	34	37	40
0	1.78	1.85	2.00	2.14	2.25	2.31	2.37	2.48
5	1.88	1.96	2.11	2.24	2.34	2.40	2.47	2.56
10	1.98	2.06	2.21	2.34	2.44	2.50	2.57	2.66
15	2.08	2.16	2.31	2.45	2.54	2.61	2.67	2.76
20	2.18	2.26	2.41	2.55	2.64	2.72	2.78	2.86

TABLE 8.5

Percentage of Active Molecular SO$_2$ at pH 3.0 According to Alcohol Content and Temperature (Usseglio Tomasset, 1995)

Alcohol (% vol.)	Temperature (°C)		
	19	28	38
0	4.88		
10	7.36	15.40	27.55
20	10.95		

$$SO_2 + H_2O \xrightleftharpoons{K_1} HSO_3^- + H^+ \quad (8.1)$$

$$HSO_3^- \xrightleftharpoons{K_2} SO_3^{2-} + H^+ \quad (8.2)$$

The H$_2$SO$_3$ acid molecule would not exist in a solution. It nevertheless possesses two acid functions whose pK_a's are 1.81 and 6.91, respectively, at 20°C. The neutralization of an acid begins at approximately pH = pK_a − 2. The absence of sulfite ions (SO$_3^{2-}$) at the pH of wine can therefore be deduced. However, the first function is partially neutralized as a function of pH. Knowing the proportion of free acid (active SO$_2$) and bisulfite (HSO$_3^-$) is important, since the essential enological properties are attributed to the former. The calculation is made by applying the mass action law:

$$\frac{[H^+][HSO_3^-]}{[SO_2][H_2O]} = K_1 \quad (8.3)$$

The water concentration can be treated as a constant approaching a value of one:

$$\frac{[H^+][HSO_3^-]}{[SO_2]} = K_1 \quad (8.4)$$

which results in

$$\log \frac{[HSO_3^-]}{[SO_2]} = pH - pK_1 \quad (8.5)$$

Table 8.3 indicates the results for the pH range corresponding to various kinds of wine. The proportion of molecular SO$_2$, approximately corresponding to active SO$_2$, varies by a factor of 1–10. This explains the need for more substantial sulfiting when the must or wine pH is high.

The pK_a value is also influenced by temperature and alcohol content (Table 8.4) and equally by ionic strength, i.e. the concentration of salts. Usseglio-Tomasset (1995) calculated the effect of these factors on the proportion of sulfur dioxide in the form of active SO$_2$ (Table 8.5).

The bisulfite ion (HSO$_3^-$) represents the corresponding acid fraction neutralized by bases, thus almost entirely in the form of ionized salts. With active SO_2 (or sulfurous acid in the free acid state), it represents free sulfur dioxide as defined in enology. The difference between the chemical notion of a free acid and a salified acid should be taken into account.

As a result, the antiseptic properties of a given concentration of free SO_2 toward yeasts or bacteria vary as a function of pH, even though the HSO_3^- form is considered to have a certain level of activity. In the same manner, the more the unpleasant taste and odor of sulfur dioxide, for the same value of free SO_2, increase, the more acidic the wine is. The unpleasant odor of SO_2 is sometimes less the result of excessive SO_2 addition than the nature of the wine—inferior quality, an absence of character and aroma, and very high acidity.

8.3.2 Bound Sulfur Dioxide

Bisulfites possess the property of binding molecules that contain carbonyl groups according to the following reversible reaction:

$$R-CHO + HSO_3^- \rightleftharpoons R-\underset{\underset{H}{|}}{\overset{\overset{OH}{|}}{C}}-SO_3^- \quad (8.6)$$

$$R-\underset{\underset{O}{\|}}{C}-R' + HSO_3^- \rightleftharpoons R-\underset{\underset{OH}{|}}{\overset{\overset{R'}{|}}{C}}-SO_3^- \quad (8.7)$$

These additional forms represent bound sulfur dioxide, or bound SO_2, as it is defined in enology. The sum of free SO_2 plus bound SO_2 is equal to total SO_2. With respect to free SO_2, bound SO_2 has much less significant (even insignificant) antiseptic and antioxidant properties (Section 8.6.1).

In the reactions forming these combinations, the equilibrium point is given by the formula in Equation (8.8) for the reaction in Equation (8.6). This formula presents the molar concentration relationship between the different molecules:

$$\frac{[R-CHO][HSO_3^-]}{[R-CHOH-SO_3^-]} = K \quad (8.8)$$

K is a constant that is characteristic of each substance, with aldehyde or keto functions, able to bind SO_2.

This relation can be written as follows:

$$\frac{[R-CHOH-SO_3^-]}{[R-CHO]} = \frac{[HSO_3^-]}{K} \quad (8.9)$$

For example, a concentration of 20 mg of free SO_2 per liter represents 25 mg of HSO_3^- per liter (molecular weights 64 and 81, respectively). The molar concentration is therefore

$$[HSO_3^-] = \frac{25}{81} * 10^{-3} = \frac{10^{-3}}{3.24} \quad (8.10)$$

The relationship in Equation (8.9) becomes

$$\frac{[B]}{[A]} = \frac{[R-CHO-SO_3^-]}{[R-CHO]} = \frac{10^{-3}}{3.24 \times K} \quad (8.11)$$

It expresses the proportion of carbonyl group molecules bound to SO_2 (B) and those in their free form (A).

First case: K has a low value, less than or equal to 0.003×10^{-3} M, at equilibrium:

$$\frac{[B]}{[A]} = \frac{10^{-3}}{3.24 \times 0.003 \times 10^{-3}} = \frac{1}{0.01} \cong 100 \quad (8.12)$$

In this case, there exists 100 times more of the bound form than the free form. The binding molecule is considered to be almost entirely in the bound form. Free SO_2 can only exist when all of the molecules in question are completely bound. Furthermore, this binding is stable and definitive; the depletion of free SO_2 by

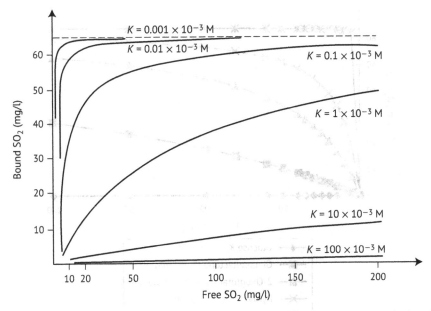

FIGURE 8.1 Sulfur dioxide binding curves as a function of the chemical dissociation constant K (carbonyl concentration = 10^{-3} M (Blouin, 1965).

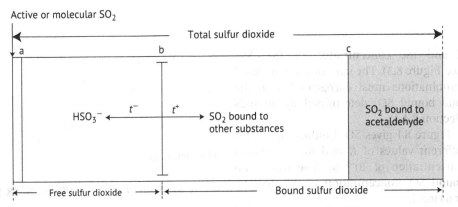

FIGURE 8.2 The different states of sulfur dioxide in wine (Ribéreau-Gayon et al., 1977).

oxidation does not cause an appreciable displacement of the equilibrium.

Second case: K has an elevated value greater than or equal to 30×10^{-3} M:

$$\frac{[B]}{[A]} = \frac{10^{-3}}{3.24 \times 30 \times 10^{-3}} \cong \frac{1}{100} \quad (8.13)$$

In this case, there exists 100 times more of the free form than the bound form. The binding molecule is considered to be slightly bound, and the corresponding binding is not very stable. When free SO_2 is depleted by oxidation, the dissociation of this bound form regenerates free SO_2 in order to reestablish the equilibrium.

Of course, [B] plus [A] represents the total molar concentration of the binding molecule as given by analysis, expressed in millimole per liter. It is therefore possible to establish overall reaction values of bound SO_2 for different free SO_2 values. In fact, by determining the quantity of each carbonyl molecule, the amount of bound SO_2 can be calculated using the value of

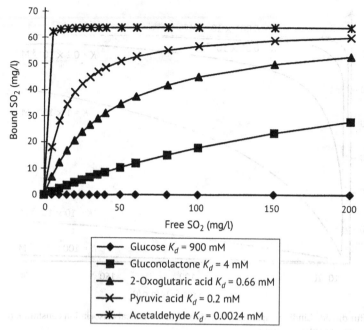

FIGURE 8.3 Sulfur dioxide binding curves for various compounds at a concentration of 1 mM as a function of their K_d (Barbe, 2000).

K and the concentration of free SO_2 (see Figure 8.3). The sum of the individual combinations must correspond with the total bound SO_2 determined by analysis (Section 8.4.3).

Figure 8.1 gives SO_2 binding curves for different values of K and for a carbonyl concentration of 10^{-3} M. The maximum bound SO_2 concentration is also 10^{-3} M, or 64 mg/l.

In conclusion, the different forms of sulfur dioxide existing in wine are summarized in Figure 8.2. Active SO_2 is located on the far left; its separation (a) with HSO_3^- varies according to pH. On the far right, ethanesulfonic acid represents the SO_2 fraction bound with acetaldehyde. Since K is low, this binding is very stable and depends on the acetaldehyde concentration. The (c) separation line is definitive. On the other hand, the (b) separation between sulfur dioxide and sulfur dioxide bound with other substances may shift, moving in one direction or the other according to temperature and the free SO_2 concentration.

8.4 Molecules Binding Sulfur Dioxide

8.4.1 Acetaldehyde

The reaction

$$CH_3-CHO + HSO_3^- \rightleftharpoons CH_3-CHOH-SO_3^- \quad (8.14)$$

generally represents the most significant portion of bound SO_2 in wine. The value of K is extremely low (0.0024×10^{-3}) and corresponds to a bound percentage greater than 99%. Acetaldehyde concentrations between 30 and 130 mg/l correspond to possible bound SO_2 values between 44 and 190 mg/l (Figure 8.3).

In wine no longer containing free SO_2, a weak dissociation of ethanesulfonic acid liberates a trace of acetaldehyde. This acetaldehyde is said to be responsible for the stale character in wine, but the presence of free acetaldehyde is considered to be impossible in wine containing free SO_2.

The binding is rapid. At pH 3.3, 98% of it is bound in 90 minutes and the binding is total in five hours. Within normal limits, binding is independent of temperature. The amount of free SO_2 liberated by raising the temperature is very small. Concentrations in botrytized musts are of the order of 10 mg/l, up to a maximum of 20 mg/l. These concentrations may explain mean binding of under 10 mg/l SO_2.

Alcoholic fermentation is the principal source of acetaldehyde in wine. It is an intermediate product in the formation of ethanol from sugars. Its accumulation is linked to the intensity of the glyceropyruvic fermentation. It principally depends on the level of aeration, but the highest values are obtained when yeast activity occurs in the presence of free SO_2. The formation of ethanesulfonic acid is a means of protection for the yeasts against this antiseptic. Consequently, the level of grape sulfiting controls the concentration of acetaldehyde and of acetaldehyde bound to SO_2.

Considering these phenomena, the addition of SO_2 to a fermenting must should be avoided. It would immediately be bound without being effective. When the grapes are botrytized, the variation in the acetaldehyde content of different wines when 50 mg/l of SO_2 is added to the must accounts for a binding power approximately 40 mg/l higher than that of non-sulfited control wines. When stopping the fermentation of a sweet wine, a sufficient concentration should be added to stop all yeast activity. This concentration can be decreased by initially reducing the yeast population using centrifugation or cold stabilization (−4°C), for example. The highest acetaldehyde concentrations are encountered when successive fermentations occur. The multiple sulfitings required progressively increase bound SO_2 concentrations.

The chemical oxidation of ethanol, by oxidation–reduction in the presence of a catalyst, may also increase the acetaldehyde concentration during storage—for example, during rackings. The binding power of the wine therefore also increases.

8.4.2 Keto Acids

Pyruvic acid and 2-oxoglutaric acid (formerly α-ketoglutaric acid) are generally present in wine (Table 8.6). They are secondary products of alcoholic fermentation. Considering their low K value, they can play an important role in the SO_2 binding rate (Figure 8.3). For example, a wine containing 200 mg of pyruvic acid and 100 mg of 2-oxoglutaric acid per liter has 93 mg of SO_2 per liter bound to these acids for 20 mg of free SO_2.

Those two substances may bind with very different amounts of SO_2. In wines made from botrytized grapes, for a free SO_2 content of 50 mg/l, 2-oxoglutaric acid is likely to bind with an average of 43 mg/l and pyruvic acid with 58 mg/l (Barbe, 2000). The average percentages of pyruvic and 2-oxoglutaric acids in the SO_2 binding equilibrium are 20.7 and 16.7%, respectively.

It is therefore of interest to understand the formation and accumulation condi-

TABLE 8.6

The Keto Acids of Wine (Usseglio-Tomasset, 1995)

Name	Formula	K	Average concentrations in wine
Pyruvic acid	$CH_3-CO-COOH$	0.3×10^{-3} M	10–500 mg/l
2-Oxoglutaric acid	$COOH-CO-CH_2-CH_2-COOH$	0.5×10^{-3} M	2–350 mg/l

TABLE 8.7
Action of Thiamine on Keto Acids and Free Sulfur Dioxide Concentrations (mg/l), Calculated for 250 mg of Total SO$_2$ Per Liter (Ribéreau-Gayon et al., 1977)

Origin of wines	Control			+Thiamine		
	Pyruvic acid	2-Oxoglutaric acid	Free SO$_2$ for total 250 mg/l	Pyruvic acid	2-Oxoglutaric acid	Free SO$_2$ for total 250 mg/l
Monbazillac	10	Traces	136	12	Traces	134
Barsac	Traces	128	104	Traces	107	108
Cerons	Traces	108	113	Traces	82	111
Sauternes	264	121	44	40	73	108
Monbazillac	330	273	20	51	74	109
Sauternes	61	205	52	10	100	88
Cerons	108	72	48	41	70	91

tions of these acids during alcoholic fermentation. They are formed at the beginning of the fermentative process. After initially increasing, their concentration decreases toward the end of fermentation. This explains the higher concentration of these molecules in sweet wines with respect to dry wines. Elevated temperatures and pHs, along with aerations, favor the synthesis and accumulation of keto acids. In numerous fermentations, thiamine (at a concentration of 0.5 mg/l) has been shown to diminish the concentration of these acids and consequently sulfur dioxide binding. The effect of thiamine is not surprising. It is an essential element of carboxylase, which ensures the decarboxylation of pyruvic acid into acetaldehyde. This is an essential step of alcoholic fermentation. The accumulation of keto acids appears to result from a thiamine deficiency.

The figures in Table 8.7 show the effect of thiamine on the accumulation of keto acids and the corresponding binding power. In the first three wines made from slightly botrytized grapes, the sulfur dioxide equilibrium is not modified after the addition of thiamine. In the other cases, the presence of thiamine decreases the keto acid concentration and often improves the sulfur dioxide equilibrium.

To be effective, thiamine needs to be added to clarified and sulfited must sufficiently early. It has no action on the accumulation of acetaldehyde. In certain cases, useful secondary effects are observed: activation of the fermentation and decrease in volatile acidity. On average, in eight cases out of 10, thiamine increases the free SO_2 concentration in sweet wines by 20 mg/l for the same bound SO_2 concentration.

8.4.3 Sugars and Sugar Derivatives

In light of the presence of aldehydic and keto functions in different sugar molecules, they can be expected to have a binding power with sulfur dioxide. Fructose and sucrose, however, practically do not react with SO_2.

In contrast, arabinose binds SO_2 at a rate of approximately 8 mg of SO_2 per gram of arabinose for 50 mg of free SO_2 per liter. Since the concentration of arabinose in wine is low (less than 1 g/l), this binding is not generally taken into account. Glucose has a much lower binding power. One gram binds 0.3 mg of SO_2 for 50 mg of free SO_2 per liter. Due to the high concentration of glucose in musts and sweet wines, this binding should be taken into account, and it is included in the interpretation of the decrease in free SO_2 after sulfiting the grapes or the must.

Burroughs and Sparks (1964, 1973) identified the following substances: 5-ketofructose (5-oxofructose), xylosone, 2-ketogluconic (2-oxogluconic), and 2,5-diketogluconic (2,5-dioxogluconic) acids (Table 8.8). Due to their concentrations in some wines (capable of attaining several dozen milligrams per liter) and their K values, some of them can play a significant role in binding with sulfur dioxide (Figure 8.3). These substances exist naturally in healthy, ripe grapes, and they are also formed in large quantities by *Botrytis cinerea* and acetic acid bacteria (*Gluconobacter*) whose development frequently accompanies various forms of rot.

According to more recent findings (Barbe et al., 2002), among all the previously mentioned compounds, 2- and 5-oxogluconic acids, always present in a ratio of 2.5:1, do not have a significant affinity for SO_2. In contrast, at the pH of botrytized musts and wines, gluconic acid (20 g/l) is in equilibrium with two lactones, γ- and δ-gluconolactone (Figure 8.4), representing about 10% of the concentration of the acid. The affinity corresponds to that of a monocarbonyl compound with a bisulfite binding dissociation constant $K_d = 4.22$ mM. Thus, the lactones of gluconic acid are likely to bind up to 135 mg/l SO_2 for a free SO_2 content of 50 mg/l.

The 5-oxofructose content is also frequently on the order of 100 mg/l in wines

TABLE 8.8
Sulfur Dioxide Binding Sugar Derivatives

	Uronic acids		Sugar oxidation products	
	Galacturonic acid	Glucuronic acid	2-Oxogluconic acid	5-Oxofructose
Formula	CHO H−C−OH HO−C−H HO−C−H H−C−OH COOH	CHO H−C−OH HO−C−H H−C−H H−C−OH COOH	COOH C=O HO−C−H H−C−OH H−C−OH CH$_2$OH	CH$_2$OH C=O H−C−H HO−C−OH C=O CH$_2$OH
K	20×10^{-3} M	20×10^{-3} M	0.4×10^{-3} M	0.3×10^{-3} M
Binding rate[a]	4.4	1.5	66	72

[a] Percentage of bound substances per 50 mg of free SO_2. *Source*: Based on Burroughs and Sparks (1964, 1973).

made from botrytized grapes (Barbe, 2000). Concentrations increase as binding power increases (Figure 8.5). According to Barbe (2000), 5-oxofructose may account for the binding of 4–78% of the sulfur dioxide. Concentrations of this compound are not altered by alcoholic fermentation or by any other aspect of yeast metabolism. Excessive concentrations can, therefore, only be avoided by monitoring grape quality. In the special case of must made from grapes affected by botrytis bunch rot in the advanced stage (*pourri vieux* in French), it contributes, on average, to over 60% of the bound SO_2. This compound is produced from fructose by acetic acid bacteria of the *Gluconobacter* genus (Section 7.5).

8.4.4 Dicarbonyl Group Molecules

In grapes affected by rot, Guillou-Largeteau (1996) identified molecules with two carbonyl groups (Table 8.9). They are probably formed during the development of *B. cinerea* and other microorganisms involved in various types of rot. In view of the fact that concentrations do not exceed 3 mg/l, the contribution of glyoxal to the SO_2 binding balance is practically negligible. Methylglyoxal makes a more significant contribution and may be responsible for binding over 50 mg/l SO_2 for a free SO_2 content of 50 mg/l (Barbe, 2000). Glyoxal and, especially,

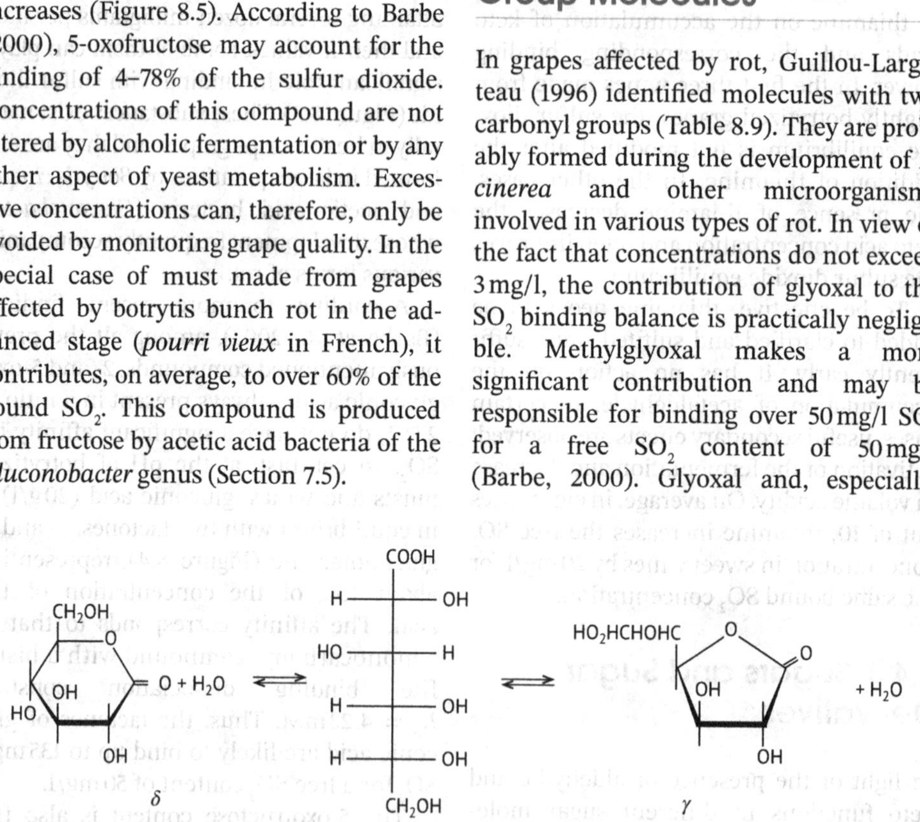

FIGURE 8.4 Formation of γ- and δ-gluconolactone from D-gluconic acid.

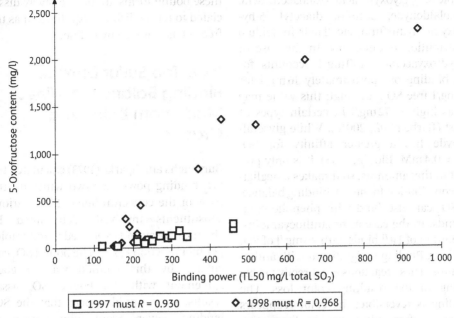

FIGURE 8.5 Changes in 5-oxofructose content as a function of the binding power of the must (Barbe et al., 2000).

methylglyoxal concentrations decrease during alcoholic fermentation, so these two α-dicarbonyl compounds are only responsible for insignificant amounts of bound SO_2 in wine.

8.4.5 Other Bound Forms

Other substances likely to fix small amounts of sulfur dioxide have been identified: glucuronic and galacturonic acids

TABLE 8.9

Some Dicarbonyl Group Molecules Involved in Binding with Sulfur Dioxide

		Contents in musts and wines	
Name	Chemical formula	Healthy grapes	Botrytized grapes
Glyoxal	H-C(=O)-C(=O)-H	Several hundred µg/l	Several mg/l
Methylglyoxal	CH_3-C(=O)-C(=O)-H	Several mg/l	Several dozen mg/l

(Table 8.8), glyoxylic acid, oxaloacetic acid, glycolaldehyde, acetoin, diacetyl, 5-hydroxymethylfurfural, etc. Their individual contribution is quite low. In the case of dihydroxyacetone, 100 mg/l accounts for the binding of approximately 16 mg/l for 50 mg/l free SO_2, although this value may be as high as 72 mg/l in certain types of must (Barbe et al., 2001c). While glyceraldehyde has a greater affinity for SO_2 ($K_d = 0.4$ mM, Blouin, 1995), it is only present in tiny amounts, so it makes a negligible contribution to the SO_2 binding balance.

SO_2 can also bind with phenolic compounds. In the case of proanthocyanidins, a solution of 1 g/l binds with 20 mg/l of SO_2 per liter. Binding is significant with anthocyanins. These reactions are directly visible owing to the resulting color loss. The binding is reversible; the color reappears when the free sulfur dioxide disappears. This reaction is related to temperature (Section 8.5.2) and acidity (Section 8.5.1), which affect the quantity of free SO_2. The SO_2 involved in binding with these compounds is probably titrated by iodine as free SO_2. In fact, due to their low stability, these bound forms are progressively dissociated to reestablish the equilibrium as the free SO_2 is oxidized by iodine.

8.4.6 The Sulfur Dioxide Binding Balance in Wines Made from Botrytized Grapes

Burroughs and Sparks (1973) calculated the SO_2 binding power for two wines on the basis of the concentrations of the various constituents involved, determined by chemical assay and expressed in millimoles per liter (Section 8.3.2). The bound SO_2 calculated by this method was in good agreement with the bound SO_2 assay results, so it would appear that the SO_2 binding powers were fully known in that case.

Blouin (1965) had previously demonstrated the particular importance of keto acids in this type of binding. In spite of all these findings, the sulfur dioxide binding balance cannot be considered complete

TABLE 8.10

Binding Powers of Compounds in a Wine

Wines	CL50 mg/l	Total SO_2 available for binding		SO_2 available for binding with the six compounds accounting for the largest contributions	
		mg/l	% of CL50	mg/l	% of CL50
1	100	92	92	76	76
2	180	175	98	164	92
3	215	217	102	196	92
4	245	228	94	201	83
5	260	261	101	240	93
6	290	258	90	232	81
7	340	306	90	288	85
8	350	339	97	328	94
9	450	449	100	437	98

SO_2 that can be bound by all the compounds assayed or only six of them (acetaldehyde, pyruvic acid, 2-oxoglutaric acid, γ- and δ-gluconolactone, and 5-oxofructose) in nine wines (Barbe, 2000).
CL50; see Section 8.5.1.

TABLE 8.11

Average Quantities (mg/l) of Sulfur Dioxide Bound by the Compounds Under Study in Different Musts (Barbe et al., 2001b)

Compound	Musts (n = 24) with low binding power CL50 = 171 mg/l	Musts (n = 7) with high binding power CL50 = 498 mg/l
5-Oxofructose	24	258
γ- and δ-Gluconolactone	17	55
Dihydroxyacetone	7	60
Glucose	48	45
Methylglyoxal	12	9
Glyoxal	2	2
Acetaldehyde	10	14
2-Oxoglutaric acid	14	15
Pyruvic acid	5	9
Other	32	31

TABLE 8.12

Sulfur Dioxide Binding Balance (in %) in Two Wines with Similar TL50 (Barbe et al., 2001b)

	Wine A[a]	Wine B[b]
5-Oxofructose	2	32
γ- and δ-Gluconolactone	11	16
Trioses (glyceraldehyde + DHA)	1	2
Acetaldehyde	45	16
Pyruvic acid	18	9
2-Oxoglutaric acid	16	12
α-Dicarbonyls (methylglyoxal + glyoxal)	2	1
Glucose	5	6
Other	—	6

[a] Wine A: TL50 = 310 mg/l total SO_2 (i.e. CL50 = 260 mg/l total SO_2).
[b] Wine B: TL50 = 340 mg/l total SO_2 (i.e. CL50 = 290 mg/l total SO_2).

about other compounds, such as dihydroxyacetone, which is in equilibrium with glyceraldehyde (Blouin, 1995; Guillou-Largeteau, 1996), and work on neutral carbonyl compounds in wines (Guillou-Largeteau, 1996). Finally, more recent research by Barbe et al. (2000, 2001a,b,c, 2002) has improved control of sulfur dioxide concentrations by adding to knowledge of the origins of these compounds.

In wines made from botrytized grapes with high or low sulfite binding powers, almost all of this binding is accounted for by the concentrations of 5-oxofructose, dihydroxyacetone, γ- and δ-gluconolactone, acetaldehyde, pyruvic and 2-oxoglutaric acids, glyoxal, methylglyoxal, and glucose (Table 8.10). In contrast, in must made from this same type of grapes, the high binding power is precisely accounted for by the quantities of SO_2 bound by 5-oxofructose, dihydroxyacetone, and gluconic acid lactones (Table 8.11).

Carefully controlled fermentation of botrytized musts minimizes the accumulation of yeast metabolites binding with SO_2, even though stuck fermentations involve much higher concentrations of these compounds than those present in dry

and satisfactory. Progress has been made in establishing the binding balance for wines made from botrytized grapes by finding out

TABLE 8.13

Concentrations Found and K_d Calculated for the Main Molecules Identified in Botrytized Musts and Wines (Burroughs and Sparks, 1964, 1973; Blouin, 1965, 1995; Guillou Largeteau, 1996; Barbe, 2000)

Molecules	Concentrations in wine (min–max) (mg/l)	K_d (mM)	Binding ratio (mg/l)[a]
Acetaldehyde	20–100	0.0024	99.7
Pyruvic acid	20–330	0.3	28
2-Oxoglutaric acid	50–330	0.5	25
Glyoxal	0.2–2.5	—	81
Methylglyoxal	0.7–6	0.017	87
Galacturonic acid	100–700	17	10
Glucuronic acid	Traces to 60	50	1
5-Oxofructose	Traces to 2,500	0.48	22
Dihydroxyacetone	Traces to 20	2.65	16
Glyceraldehyde	Traces to 10	0.4	26
Gluconic acid	1,000–25,000	20	—
2-Oxogluconic acid	Traces to 1,200	1.8	—
5-Oxogluconic acid	Traces to 500	—	—
γ- and δ-Gluconolactone	6 and 4% of the gluconic acid	4.22	5.6
Glucose	±100 g/l	800	0.03

[a] For a 100 mg/l concentration of the compound, the SO_2 binding value for a free SO_2 content of 50 mg/l.

wines. Various technological parameters during fermentation make it possible to reduce the quantities of sulfur dioxide, by affecting only those binding compounds produced by fermentation yeasts. Wines with a lower sulfur dioxide binding power may be obtained by not sulfiting the must, adding 0.5 mg/l of thiamine to must, choosing a yeast strain known to produce little acetaldehyde or 2-oxoacids, and delaying stoppage of fermentation (*mutage*) until the yeast metabolism has been completely shut down (e.g. by filtering or chilling the wine) (Barbe et al., 2001c).

These compounds are produced due to the presence of microorganisms in botrytized grapes. Although yeasts represent a preponderant part of the microorganisms present, acetic acid bacteria, especially those in the *Gluconobacter* genus, are responsible for producing large amounts of these compounds, which act as intermediaries in their metabolism of the two main sugars in botrytized grapes (Barbe et al., 2001a).

The SO_2 binding balance varies considerably between different musts (metabolism of the acetic acid bacteria) and wines (fermentation parameters). Furthermore, the total content of these SO_2 binding compounds in wine may result from both sources, as shown in Table 8.12.

Finally, *B. cinerea* indirectly plays two major roles in the accumulation of substances that bind with SO_2. Firstly, it causes in-depth modifications in the grape skins, which become permeable, thus facilitating access to the various substrates for acetic acid bacteria. Secondly, noble rot causes glycerol to accumulate in the grapes and is thus indirectly responsible for dihydroxyacetone production.

Table 8.13 sums up all the substances that bind with SO_2 identified in musts and wines made from botrytized grapes.

8.5 Practical Consequences: The State of Sulfur Dioxide in Wines

8.5.1 Equilibrium Reactions

In a sulfited wine, an equilibrium exists between the free sulfur dioxide and the bound sulfur dioxide—more precisely the bound sulfur dioxide with a high dissociation constant K. Sulfur dioxide bound to acetaldehyde does not participate in this equilibrium, since its binding has a very low K value and thus is very stable.

Any addition of sulfur dioxide to a wine results in the binding of a part of this sulfur dioxide. Conversely, the depletion of free sulfur dioxide by oxidation results in a decrease of the bound fraction to such a degree that the loss of free sulfur dioxide is less than the amount oxidized. This liberation mechanism is advantageous, since it automatically prolongs the effectiveness of a given concentration of sulfur dioxide.

When the free sulfur dioxide concentration of a wine decreases to a very low level, it rarely falls completely to zero, unless yeasts are involved or other factors modify wine composition. The unbinding of bound SO_2 progressively replaces the missing free sulfur dioxide.

As a result of these equilibria, the total sulfur dioxide concentrations of different wines cannot be compared if they do not have the same free sulfur dioxide concentration. For example, if a sweet wine has a total SO_2 concentration close to the legal limit, the consequence is not at all the same if it only contains 10 mg of free SO_2 per liter (insufficient for ensuring its stability) or 50 mg (largely sufficient).

To remedy this difficulty, Blouin (1965) recommended the use of the expressions "CL20" and "CL50," which represent, respectively, the quantities of bound SO_2 necessary to have 20 or 50 mg of free SO_2 per liter. Known sulfur dioxide additions are used to obtain these numbers experimentally (Kielhöfer and Würdig, 1960). These considerations are most important in the case of sweet wines (Sauternes, Monbazillac, Coteaux de Layon, and Tokay), which require relatively high free SO_2 concentrations to ensure their stability. In practice, the binding power (TL50), or the amount of total SO_2 necessary in a must or wine to obtain 50 mg/l of free SO_2, is calculated by drawing a graph of total SO_2 versus free SO_2 (Barbe, 2000).

8.5.2 Influence of Temperature

The determination of the free sulfur dioxide concentration in samples of a botrytized sweet white wine with a strong binding power varies according to temperature, although the total SO_2 concentration remains constant (Table 8.14). The results for determining free SO_2 concentrations are therefore variable. Depending on the conditions, the results obtained can differ by as much as 20 mg/l.

The storage temperature of the wine must also be taken into account in the

TABLE 8.14

Influence of Temperature on the State of Sulfur Dioxide (mg/l) in a Botrytized Sweet Wine (Sugar 74 g/l; Acetaldehyde 70 mg/l)

Sulfur Dioxide	Temperature		
	0°C	15°C	30°C
Total sulfur dioxide	412	412	412
Free sulfur dioxide	68	85	100
Bound sulfur dioxide (BSO_2)	344	327	312
SO_2 bound with acetaldehyde	104	104	104
SO_2 bound to other substances	240	223	208

evaluation of the effectiveness of sulfiting, at least in the case of sweet wines. Finally, the influence of temperature becomes particularly important when heating wine. The free SO_2 concentration can double, or even more. This release of sulfur dioxide singularly reinforces the effectiveness of heating and helps to explain why wines can be sterilized during bottling at relatively low temperatures (for example, between 45 and 50°C).

8.5.3 Empirical Laws of Binding

For a long time, enology has tried to determine applicable binding rules, both for sulfiting a new wine immediately following fermentation and for adjusting the free SO_2 during storage.

The most satisfactory solution consists in adding increasing concentrations of SO_2 to various samples of the same wine to produce a curve as in Figure 8.3. This operation is long and difficult; consequently, it is not always feasible. Laboratory tests are, however, recommended before the first sulfiting of unknown wines immediately following fermentation. Due to the diversity of harvested grapes, a standard SO_2 concentration can lead to an insufficient free SO_2 concentration for ensuring stability or, on the contrary, an excessively high concentration that would be difficult to lower. The binding curve clearly reveals that bound SO_2 increases with free SO_2. Yet the increase becomes less as the free sulfur dioxide concentration increases.

To increase the free SO_2 concentration of a wine already containing some, the binding of the added amount must be taken into account, especially when free SO_2 is low. As a general rule in standard wines already containing free SO_2, two-thirds of the added amount remains in a free state, and one-third binds with other compounds. As a result, 3 g/hl is necessary to increase the free SO_2 concentration by 20 mg/l. Possible abnormal cases must also be anticipated, corresponding with a much higher binding rate.

In practice, a few days after the addition of SO_2 to wine, the free SO_2 concentration should be verified to ensure that it corresponds with the desired concentration and that the conditions for stabilization are obtained.

8.6 Antimicrobial Properties of Sulfur Dioxide

8.6.1 Properties of the Different Forms

The enological properties of sulfur dioxide were summarized at the beginning of this chapter (Section 8.1). It is essentially a multifaceted antiseptic and a powerful reducing agent that protects against oxidation. Its antifungal and antibacterial activities will be covered here (see also Section 6.2.2); the antioxidant and antioxidase properties will be covered in Section 8.7.2. The various forms of sulfur dioxide do not share these properties to the same extent (Table 8.15).

Its various properties mean that sulfur dioxide is virtually indispensable in winemaking. The goal of enology is not to eliminate this substance completely but rather to establish responsible concentration limits. This supposes a sufficient knowledge of its properties and conditions of use.

8.6.2 Antifungal Activities

The antiseptic action of SO_2 with respect to yeasts can appear in different ways. First, it can be used to stop the fermentation of sweet wines (*mutage*) (Section 14.2.5). It effectively destroys the existing population (fungicidal action). Second, it protects these same sweet wines from possible refermentations—evaluated by the growth of a small residual population. It effectively

TABLE 8.15

Wine Conservation Properties of the Different Forms of Sulfur Dioxide (Ribéreau-Gayon et al., 1977)

Property	SO_2	HSO_3^-	$R-SO_3^-$
Fungicidal	+	Low	0
Bactericidal	+	Low	Low
Antioxidant	+	+	0
Antioxidase	+	+	0
Taste improvement			
Redox potential	+	+	0
Neutralization of acetaldehyde	+	+	+
Sensory role of SO_2	Biting odor, SO_2 taste	Odorless, salty, bitter taste	Odorless, tasteless at normal concentrations

inhibits cell multiplication (fungistatic activity). Moderate sulfiting is also known to inhibit yeast growth temporarily without destroying them totally. The subsequent disappearance of free SO_2 permits the revival of yeast activity. In practice, in the winery, new yeast activity may also come from new contaminations resulting from contact with non-sterile equipment and containers.

For these different reasons, the results concerning the action of SO_2 on wine yeasts cited in various research works and obtained under different conditions are not always easily compared. Moreover, the data on this subject seems incomplete.

Bound sulfur dioxide does not have an antiseptic action on yeasts. Yeasts make use of this binding to inactivate SO_2. HSO_3^- also possesses a low but undetermined antiseptic activity. Table 8.16 indicates the concentrations of free SO_2, titratable by iodine, that must be added to wines (according to their pHs) to have an antiseptic activity equal to 2 mg of active molecular SO_2 per liter. The antiseptic activity of the bisulfite form HSO_3^- is more or less significant, depending on the various hypotheses being considered. According to experience obtained on wine stability, HSO_3^- seems to be 20 times less active than SO_2, notably in wines containing reducing sugars.

Sulfur dioxide is fungistatic at high pHs and at low concentrations, and it is a fungicide at low pHs and high concentrations. The HSO_3^- form is exclusively fungistatic. Each yeast strain probably has a specific sensitivity to the different forms of sulfur dioxide. Romano and Suzzi (1992) considered possible mechanisms that could explain these differences. According to these same authors (Suzzi and Romano, 1982), sulfiting the must before fermentation increases yeast resistance to SO_2. Yeasts from a non-sulfited must, isolated after fermentation, are more sensitive to SO_2 than those coming from the same must that is sulfited before fermentation.

Concerning the *mutage* of sweet wines (fungicidal activity), the fermentation seems to stop abruptly after the addition of 100 mg of SO_2 per liter. The concentration of sugar remains constant, although carbon dioxide continues to be released for about an hour. During this time, the yeasts do not seem to be affected by the sulfiting—they are still capable of multiplying (Table 8.17), whatever the concentration used. It is necessary to wait at least five hours, and more often 24 hours, to observe a decrease in cell viability.

To ensure a complete cessation of fermentation, Sudraud and Chauvet (1985) estimated that 1.50 mg of molecular SO_2 per liter must be added to wine. According

TABLE 8.16

Free Sulfur Dioxide Concentrations Necessary in Wines to Maintain an Antiseptic Activity Equal to 2 mg of Active Molecular SO_2 Per Liter (Ribéreau Gayon et al., 1977)

Wine pH	None	Hypothesis: HSO_3^- activity		
		100 times less than SO_2	20 times less than SO_2	10 times less than SO_2
2.8	22	20	14	11
3.0	34	29	19	14
3.2	54	43	24	16
3.4	87	61	28	18
3.6	134	81	31	19
3.8	200	100	33	20

TABLE 8.17

Sulfiting to Inhibit Yeasts in a Sweet Wine at the End of Fermentation (Values Are Numbers of Viable Cells, Capable of Producing Colonies in Petri Dishes, Per Milliliter; Initial Population 58×10^6/ml) (Ribéreau Gayon et al., 1977)

SO_2 added (mg/l)	Time		
	1h	5h	24h
100	58×10^6	8×10^6	10^5
150	58×10^6	3×10^6	0
300	58×10^6	10^6	0

to the same authors, after the elimination of yeasts by different treatments, 1.20 mg of molecular SO_2 per liter seems sufficient for ensuring the proper storage of wines containing residual sugars (fungistatic activity). Lower concentrations could be recommended for wines stored at low temperatures having a low yeast population.

Romano and Suzzi (1992) reviewed the current understanding of the action of the sulfur dioxide molecule on yeasts. Molecular SO_2 penetrates the cell by either active transport or simple diffusion. Considering the intracellular pH, it must exist in the cell in the form of HSO_3^-. Once inside the cell, it reacts with numerous constituents such as coenzymes (NAD, FAD, FMN), cofactors, and vitamins (thiamine). It would also seem to have an effect on numerous enzymatic systems and on nucleic acids. Finally, a significant decrease in ATP is also attributed to it.

8.6.3 Antibacterial Activities

The activity of free SO_2 on lactic acid bacteria is well known (Section 6.2.2). It is even more influenced by pH than the activity with respect to yeasts. Yet the fraction combined with acetaldehyde or pyruvic acid is also now known to possess an antibacterial activity. The bound SO_2 molecule has a direct effect on bacteria. The mechanism is not explained by the decomposition of the bound form by bacteria, resulting in the release of free SO_2.

Sulfur dioxide combined with acetaldehyde (or pyruvic acid) seems to possess an antibacterial activity 5–10 times weaker than that of free SO_2, even though it can be 5–10 times more abundant.

A large number of bacteria are eliminated by 5 mg of free SO_2 per liter. The same concentration in the bound form lowers the population by 50%. *Oenococcus oeni* is less resistant to sulfur dioxide than *Lactobacillus* and *Pediococcus*.

Significant technical applications for controlling malolactic fermentation and storing wines have resulted from these observations. Sulfiting the grapes does not only act on bacteria in the short pre-fermentation period; it acts by leaving a certain concentration of bound sulfur dioxide that effectively protects the must and retards bacterial growth until completion of alcoholic fermentation. In this manner, the medium that still contains sugar is protected from an untimely bacterial development that could lead to the production of volatile acidity (Section 3.8.1).

When malolactic fermentation is not sought (in dry white wines, for example), it should be noted that wine stability is not due solely to the bactericidal action of free SO_2 but rather to the concentration of bound SO_2 that the wine conserves after fermentation; its action is long lasting during storage. In certain types of wine with too low a pH, bound SO_2 concentrations of 80–120 mg/l can make malolactic fermentation impossible.

Sulfur dioxide is also active on acetic acid bacteria, but additional studies on this subject are needed. These bacteria resist relatively high concentrations. In the winery, acetic acid bacteria are most effectively prevented by avoiding contact with oxygen in the air and controlling temperature in the winery.

8.7 The Role of Sulfur Dioxide in Winemaking

8.7.1 Advantages and Disadvantages

Although the use of sulfur dioxide in the aging and storage of wine seems to be fairly ancient, its use during vinification is more recent. It was recommended at the beginning of the 20th century—essentially for avoiding enzymatic browning. The very appreciable improvement in wine quality by sulfiting rotten grapes was an essential factor in the gain in popularity of this process. Its antiseptic properties and its role in the prevention of bacterial spoilage were discovered later.

Nevertheless, the generalization of sulfiting in winemaking, or at least the establishment of a precise and homogeneous doctrine from one winegrowing region to another, took a long time to come about. Besides its many advantages, sulfiting also presented some disadvantages; therefore, a sufficiently precise understanding of the properties of sulfur dioxide had to be obtained before defining the proper conditions of its use. These conditions permit the winemaker to profit fully from its advantages while avoiding its disadvantages.

When used in excessively high concentrations, this product has an unpleasant odor and a bad taste that it imparts to the wine; the taste of hydrogen sulfide and mercaptans in young wines can also appear when they are stored too long on their lees. The most serious danger of improper sulfiting is the slowing or definitive inhibition of the malolactic fermentation of red wines. Incidentally, for a long time, sulfited grapes were observed to produce red wines with higher acidities. Before the understanding of malolactic fermentation, this observation was attributed to an acidifying effect of sulfur dioxide or to acidity fixation.

8.7.2 Protection Against Oxidation

The chemical consumption of oxygen by SO_2 is slow. It corresponds to the following reaction:

$$SO_2 + \frac{1}{2}O_2 \rightarrow SO_3 \qquad (8.15)$$

In a synthetic medium, SO_2 has been shown to take several days to consume

8.0–8.6 mg of oxygen per liter (this amount corresponds to the saturation of this medium). Such oxidation requires the presence of catalysts, notably iron and copper ions. Yet musts are very oxidizable and should therefore be rapidly and effectively protected against oxidation. Sulfiting accomplishes this. Sulfur dioxide, however, cannot act by its antioxygen effect, that is to say, by combining with oxygen and making it unavailable for the oxidation of other must constituents.

Dubernet and Ribéreau-Gayon (1974) confirmed this hypothesis. The experiment consisted of saturating a white grape must with oxygen and measuring the oxygen depletion rate electrometrically (Figure 8.6). In the absence of sulfiting, the depletion of this oxygen is very rapid and is complete within a few minutes (4–20 on average). This phenomenon demonstrates the extremely high oxidizability of grape must. If at a given moment the must is sulfited, oxygen is no longer consumed, and its concentration remains constant after a given time t, which varies depending on the conditions but is always fairly short. As an initial approximation, the value t varies between one and six minutes when sulfiting varies between 100 and 10 mg/l. The value t is much greater for must obtained from moldy grapes.

In summary, although the antioxygen effect of sulfur dioxide is involved in wine aging and storage, its role is slight during vinification. In this case, SO_2 protects against oxidations by destroying oxidases (laccase) or, at least, blocking their activity, if destruction is not total. Enzymatic oxidation phenomena are inhibited in this manner until the start of fermentation. From that point on, the reductive character of fermentation continues to ensure the protection. However, oxidative phenomena can resume at the end of fermentation insofar as active oxidases remain after the depletion of free SO_2. The enzymatic browning test, or, even better, the determination of laccase activity, permits the evaluation of the risk and the necessary precautions to be taken.

In must, enzymatic oxidations are more significant than chemical oxidations because they are more rapid. In wine, however, chemical oxidations play an unquestionable role, since oxidative enzymes no

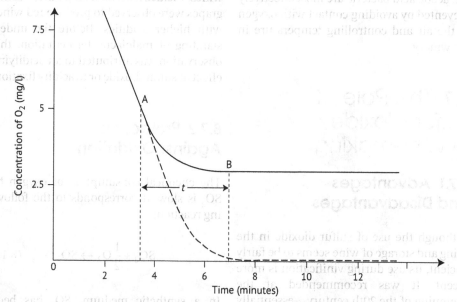

FIGURE 8.6 Oxygen consumption in musts following sulfiting (Dubernet and Ribéreau-Gayon, 1974). (A) Addition of SO_2. (B) Stopping point of oxygen consumption (t = time necessary for oxygen consumption to stop).

TABLE 8.18

Protection of Color of Red Wine Made from Botrytized Grapes by Sulfiting (Sudraud, 1963)

Level of harvest sulfiting	Composition of the wines obtained		
	Total phenolic compounds (index)	Color intensity[a]	Enzymatic browning potential
Without SO_2	32	0.53	++++
+10 g SO_2/hl	41	0.63	++
+20 g SO_2/hl	45	0.3	0

[a] Color intensity = OD 420 + OD 520 (1 mm thickness).

longer exist. In this case, SO_2 reacts with oxygen to protect the wine.

Mold is responsible for the most serious oxidative phenomena. In fact, *B. cinerea* secretes a laccase more active and stable than the tyrosinase from grapes. It is responsible for enzymatic browning in red wines made from moldy grapes. An appropriate sulfiting can protect against this phenomenon to some extent. The figures in Table 8.18 show that intense sulfiting of moldy grapes (as was done in the past) increases total phenols and color intensity while decreasing the risk of enzymatic browning. Progress in phytosanitary vineyard protection has made such situations extremely rare.

From the start of fungal development, oxidase secretion by *B. cinerea* inside the berry can be considerable even when the external signs are barely visible. This situation can be observed in the case of red grapes. The first brown blemishes are more difficult to observe on red grapes than on white grapes. During cold weather, the external vegetation of *B. cinerea* is less developed. These factors must be taken into account when choosing the corresponding sulfiting concentration.

8.7.3 Inhibition, Activation, and Selection of Yeasts

Sulfur dioxide is a general antiseptic with a multifaceted activity on different wine microorganisms. Its mode of action has been described in previous sections.

With respect to yeasts, sulfiting is used first and foremost to ensure a delay in the onset of fermentation, allowing a limited cooling of the grapes. The fermentation is also spread out over a longer period in this manner, avoiding excessive temperatures. More and more often, natural tank cooling is complemented by controlled refrigeration systems.

In the case of white winemaking, the delay in the start of fermentation permits the settling and racking of suspended particles in must.

Sulfiting also makes use of the stimulating effect of sulfur dioxide when used in low concentrations. Consequently, the fermentation speed accelerates, as shown by the curves in Figure 8.7. After an initial slowing of the fermentation, the last grams of sugar are depleted more rapidly. Finally, the fermentation is completed more rapidly in the lightly sulfited must.

When a red wine that still contains sugar is drained off the skins, a light sulfiting (2–3 g/hl) does not block the completion of the fermentation; on the contrary it is known to facilitate it more often than not.

This long-proven effect of sulfiting has been confirmed time and time again. It has been interpreted as the destruction of fungicidal substances by sulfur dioxide. These substances are toxic for the yeast and could come from the grapes, from *B. cinerea*, or even from the fermentation itself. An

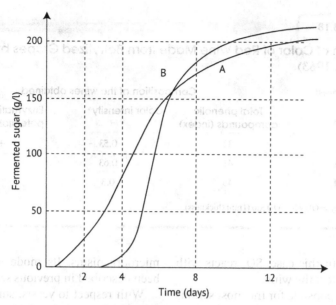

FIGURE 8.7 Effect of moderate sulfiting (5–10 g/hl) on alcoholic fermentation kinetics of grape must. (A) Control must. (B) Sulfited must (5 g/hl).

increase in must protease activity has also been considered. This activity would put assimilable amino acids at the disposal of the yeast (Section 11.6.1). It is more likely that sulfiting acts by maintaining dissolved oxygen in the must (Figure 8.6). Not being tied up in oxidation phenomena, this oxygen is available for yeast growth (Section 8.7.2).

Sulfiting has also been considered to affect yeast selection. Apiculate yeasts (*Kloeckera* and *Hanseniaspora*), which develop before the others, produce lower-quality wines with lower alcohol content. These yeasts are more sensitive to sulfur dioxide. Therefore, moderate sulfiting blocks their development. This has been confirmed by numerous experiments (Romano and Suzzi, 1992), but the research of Heard and Fleet (1988) cast doubts on this generalization. They observed that, in spite of sulfiting, these strains attained an initial population of 10^6–10^7 cells/ml in a few days before disappearing. Moreover, the advisability of eliminating apiculate yeasts and the potential advantages of the successive participation of different yeast species for the production of quality wines are still being considered.

The issue of sterilizing musts, via the total destruction of native yeasts through massive sulfiting, or other processes such as heat treatments, followed by an inoculation using selected yeasts, will be covered elsewhere.

8.7.4 Selection Between Yeasts and Bacteria

Sulfur dioxide acts more on wine bacteria than on yeasts. Lower concentrations are therefore sufficient for hindering bacterial growth or suppressing their activity. No systematic studies have been carried out on this subject, but this fact is well known and is often demonstrated in practice. For example, in the case of a red wine still containing sugar (and thus prone to simultaneous alcoholic and lactic acid fermentation), moderate sulfiting (3–5 g/hl) can initially block both fermentations. Afterward, a pure alcoholic refermentation can take place without the absolute necessity of yeast inoculation.

One of the principal roles of sulfiting in winemaking is to obtain musts that are much less susceptible to bacterial development while undergoing a normal alcoholic fermentation. This protection is most necessary in the case of musts that are rich in sugar, low in acidity, and high in temperature. The risks of stuck fermentations are highest in these cases.

In summary, sulfur dioxide delays, but does not block, yeast multiplication and alcoholic fermentation. The bacteria, supplied by grapes at the same time as the yeasts, are killed or at least sufficiently paralyzed to protect the medium from their development while the yeasts transform the totality of the sugar into alcohol. The serious danger of bacterial spoilage in the presence of sugar is an important factor in wine microbiology (Section 3.8.2).

In white winemaking and for wines in which malolactic fermentation is not sought, sulfiting can be adopted to inhibit bacteria completely. Incidentally, the light sulfiting of white musts undergoing malolactic fermentation can be insufficient to protect effectively against oxidation.

In red winemaking today, malolactic fermentation has become common practice. Excessive sulfiting of the must can cause problems with malolactic fermentation, in some cases stopping it completely.

Generally speaking, the sulfiting of red grape musts enhances wine quality. However, sulfiting must not compromise malolactic fermentation due to its conditions of use and the concentrations employed. To ensure the successful completion of alcoholic fermentation, the amount of sulfite solution added to grape must should be adjusted as a function of pH, temperature, grape soundness, and other factors. Bacterial growth must occur rapidly after the depletion of sugar, because the bacteria must break down malic acid exclusively. The exact SO_2 concentration is difficult to determine, and it varies depending on the region. For red winemaking in the Bordeaux region, 7–10 g/hl seems to be an effective range: below this, the malolactic fermentation is not compromised; above this, it can be considerably delayed (Table 8.19).

8.7.5 Dissolving Power and General Effects on Taste

In red winemaking, sulfiting favors the dissolution of minerals, organic acids, and especially phenolic compounds (anthocyanins and tannins) that constitute the coloring matter of red wines. The solvent activity is due to the destruction of grape skin cells, which yield their soluble constituents more easily in this manner. In fact, the solvent effect of sulfur dioxide seems to have been exaggerated in the case of healthy grapes. The better color of wines derived from sulfited must is probably due to a better protection against enzymatic browning in slightly rotten grapes.

The effectiveness of sulfiting during maceration for extracting grape pigments is indisputable, and this process is used for the industrial preparation of commercial dyes. Yet when rigorous experiments are carried out on healthy red grapes, using classic winemaking techniques, no significant color improvement (anthocyanin and tannin concentration and color intensity value) is observed in the presence of a normal sulfiting level. Since only free SO_2 is active and since this form rapidly disappears in crushed grapes, this effect of

TABLE 8.19

Influence of Must Sulfiting on the Time Necessary (Expressed in Days) for Malolactic Fermentation to Start in Wine After Draining Off the Skins (Ribéreau-Gayon et al., 1977)

Sulfiting	Wine no. 1	Wine no. 2
Control 0	40	30
+2.5 g/hl	45	40
+5 g/hl	70	60
+10 g/l	100	100

sulfiting appears to be exerted for only a brief moment. At the end of fermentation, the effects of maceration time, temperature, and pumping-over are more significant.

Nevertheless, the solvent effect of sulfiting on phenolic compounds is obvious in the case of limited maceration times. This operation is therefore not recommended for crushed white grapes before must extraction by pressing. The sulfiting of grapes also has a clear impact on the color of rosé wines (Section 14.1.4).

Sulfiting also has certain effects on wine quality that still remain poorly defined. The general properties of sulfur dioxide may possibly have indirect consequences (protection against oxidation and the binding of acetaldehyde). In this way, sulfiting often improves the taste of wine—notably in the case of moldy grapes or mediocre varieties. It also protects certain aromas of "new" wines. Moreover, grape sulfiting does not have an obvious impact on the subsequent development of the bouquet of mature wines.

Certain conditions, such as fermentations under strict anaerobic conditions and especially prolonged aging on yeast lees, can lead to the formation of hydrogen sulfide and mercaptans from the added SO_2. The odors of these compounds are unpleasant and can persist in wine.

8.8 The Use of Sulfur Dioxide in the Winery

8.8.1 Winemaking Concentrations

Considering the rapidity of oxidative phenomena, grape and must sulfiting is only effective if the sulfur dioxide is intimately and rapidly incorporated into the total volume before the start of fermentation. If a fraction of the grape must ferments before being sulfited, it is definitively shielded from the action of the SO_2, because the latter immediately binds with the acetaldehyde produced by the fermenting yeasts.

In fact, homogeneous distribution before the start of fermentation is not sufficient. Considering the rapidity of oxygen consumption by grape must, each fraction of the harvested grapes or the must should receive the necessary quantity of sulfur dioxide in the minutes that follow the crushing or pressing of the grapes. This is the only truly effective method of protecting against oxidation. It can be more effective to add 5 g of sulfur dioxide per hectoliter correctly to the harvest than to add 10 g/hl under poor conditions. Poor sulfiting technique is certainly one of the reasons that led to the use of excessive concentrations in the past.

Based on these principles, the only rational sulfiting method for winemaking consists in regularly incorporating a sulfite solution into the white grape must as it is being pressed, or for red grapes as soon as they are crushed. A few successive additions of SO_2 into the tank as it is being filled are not truly effective, even after a homogenization at the end of filling. During homogenization, part of the added sulfur dioxide is already in the bound form and thus inactive.

It is therefore also necessary to use a sufficiently diluted sulfur dioxide solution, capable of being properly incorporated and blended into the must. The direct usage of metabisulfite powder or sulfurous gas in the tank should be avoided. When a tank of red grapes is sulfited by a few additions of a concentrated product during filling, the complete decoloration of certain fractions of the pomace is sometimes observed during draining off the skins. In these cases, the sulfur dioxide was not properly blended, but was instead fixated on certain parts of the grapes, leaving the other parts unprotected.

When choosing the SO_2 concentration to add to the grapes or the must, grape maturity, soundness, acidity (pH), temperature, and any contamination risks must all be taken into account. The choice can

TABLE 8.20
Sulfur Dioxide Doses for Winemaking in Temperate Climate Zones

Status	Sulfur dioxide dose
Red winemaking	
Healthy grapes, average maturity, high acidity	5 g/hl of wine
Healthy grapes, high maturity, low acidity	5–8 g/hl of wine
Moldy grapes	8–10 g/hl of wine
White winemaking	
Healthy grapes, average maturity, high acidity	5 g/hl of must
Healthy grapes, high maturity, low acidity	6–8 g/hl of must
Moldy grapes (gray rot)	8–10 g/hl of must

sometimes be difficult. Table 8.20 gives a few values for vineyards in temperate climates. The generalized use of tank cooling systems and increased hygiene in wineries, combined with a better understanding of the properties of sulfur dioxide, permit the lowering of the concentrations used in winemaking. Today, sanitary practices in the vineyard prevent grape rot, which once justified the intense sulfiting operations indicated in Table 8.20.

During the harvest, progressive increases in the sulfiting concentrations can compensate for the increasingly significant inoculation (notably bacterial) resulting from the development of microorganisms on the equipment—the inner surface of the tanks and the walls of the winery. Fermentation difficulties and microbial spoilage are frequently observed in the last tanks filled. Sufficient must sulfiting should prevent these contaminations.

In white winemaking, excessive concentrations (>5 g/hl) followed by a significant sulfiting at the end of fermentation (>4 g/hl) can be a source of reduction odors and should be avoided. Press wines, however, should be more intensely sulfited—especially in the case of continuous presses that cannot be disinfected regularly.

Concerning the sulfiting technique for red winemaking, the solution should be added after grape crushing to facilitate mixing and to avoid evaporation losses and attacks on metal equipment. If the crushed grapes are transferred with a constant-volume pump, the sulfite solution should be injected into the tube immediately after the pump outlet. The sulfiting is suitably distributed and homogenized in this manner. Of course, the injection pump for the sulfite solution must be properly adjusted and perfectly synchronized with the must pump.

The addition of the sulfite solution after each grape load, by regularly spraying the surface, can only be conducted in small tanks and must also be sufficient in number. Even if it is not completely effective, a pump-over for homogenization purposes is necessary after filling.

In the case of white winemaking, sulfiting must take place after must separation. Sulfiting of the crushed grapes is not recommended since it increases the risk of maceration phenomena and a fraction of the SO_2 is fixated on the solid parts of the grape.

Considering the oxidation speed of white grape must, sulfiting (which ensures the appropriate protection) should be carried out as quickly as possible. Must extraction equipment (the press cage, mechanical drainer, and continuous press) does not supply a constant delivery. Consequently, SO_2 cannot be injected with a pump adapted directly to these outlets. In order to sulfite in this manner, the must has to pass via a small tank through a constant-delivery pump. The corresponding manipulation of the must, in particular the pumping, does not protect the must from a slight oxidation before sulfiting.

The sulfiting of white grape musts can also be calculated from the volume of the juice tray at the outlet of the press. During filling, a homogeneous distribution should be ensured.

The necessary volume of a sulfite solution for sulfiting an entire tank during its filling at the chosen concentration should be prepared in advance. If the system is correctly adjusted, the entire volume of the sulfite solution should have been injected into the tank by the time the tank is full.

8.8.2 Storage and Bottling Concentrations

During storage, sulfiting is, first of all, thought to protect wine from oxidation. As an approximation, oxidative risks are present during prolonged storage below 5–10 mg/l for red wines, 20 mg/l for white wines made from healthy grapes, and 30 mg/l for white wines made from more or less rotten grapes.

At the microbiological level, sulfiting of dry wines must protect against yeast and bacterial development during storage. In dry white wines and red wines having undergone malolactic fermentation, the concentrations used for protection against oxidations are generally sufficient to avoid microbial growth. In red wines that have not undergone malolactic fermentation, the habitual free and total SO_2 concentrations can be insufficient to shield the wine completely from malolactic fermentation—at least a partial one—during storage.

Of course, the sulfiting rules do not apply to certain kinds of wines (red or white, dry or sweet) with qualities due to a certain degree of oxidation or containing acetaldehyde.

Sulfiting also hinders the refermentation of sweet wines, generally provoked by SO_2-resistant yeast strains. The refermentation risks are independent of sugar concentrations but are influenced by alcohol content. Under satisfactory storage conditions, 50 mg of free SO_2 per liter is required to ensure the storage of a sweet wine with a relatively low alcohol content (11%) and 30 mg/l for wines with a high alcohol content (13%).

In practice, carefully adjusted and sufficient concentrations must be used to avoid spoilage risks. The refermentation of a sweet wine can start in the lees of a tank containing a sufficiently high yeast population to ensure the binding of SO_2. Simultaneously, at least for a certain amount of time, all of the liquid remains clear without any refermentation, with 60 mg of free SO_2 per liter. If the fermentation process begins from the lees, refermentation seems possible in spite of the high concentration of free sulfur dioxide.

The size of the yeast population should always be taken into account to evaluate the effectiveness of sulfiting. All operations (fining and filtration) that eliminate a fraction of the yeasts permit the lowering of the free SO_2 concentration necessary for conserving sweet wines.

Free SO_2 concentrations for stabilizing sweet wines can be lowered thanks to steps taken in storing those wines. Clean (if not sterile) conditions have diminished contaminating populations. These criteria for cleanliness should be applied not only to the product but also to the building, the containers, and the material—all contamination sources. Microbiological controls that indicate the number of viable yeast cells are useful tools for adjusting sulfiting.

Table 8.21 indicates free sulfur dioxide concentrations that can be recommended in different situations.[1]

8.8.3 Diminution of Sulfur Dioxide by Oxidation During Storage

The free sulfur dioxide concentration does not remain constant in wines stored in barrels or tanks. There is a continuous loss

---

[1] In all cases, the corresponding active SO_2 content must be taken carefully into account.

TABLE 8.21

Recommended Free Sulfur Dioxide Concentrations (mg/l) in Wines

Dose type	Red wines	Dry white wines	Sweet white wines
Conservation	20–30	30–40	40–80
Bottling	20–30	20–30	30–50
Shipping doses (cask or container)	25–35	35–45	80–100[a]

[a] This type of wine should be bottled at the production site; shipping in bulk should be avoided.

month after month. Over the years, its concentration decreases even in bottled wine.

The decrease in barrels or tanks results from an oxidation catalyzed by iron and copper ions. Although it is very volatile, only a negligible quantity of free sulfur dioxide evaporates during storage in oak barrels. Nor is it bound. A fairly common error is to consider that any decrease in free sulfur dioxide is the result of binding with wine constituents. In reality, after the four or five days following the addition of SO_2, the wine constituents no longer bind with SO_2. An equilibrium is attained, and any decreases occurring afterward are due to oxidation. For more binding to occur, the chemical composition of the wine must be modified. For example, new binding molecules must be formed, such as acetaldehyde, during limited yeast growth or by the oxidation of ethanol when a poorly clarified wine is racked.

The oxidation of sulfurous acid forms sulfuric acid. At the pH of wine, it is almost entirely in the form of sulfate. In botrytized and non-botrytized sweet wines with high free SO_2 concentrations, a considerable amount of sulfate can be formed (0.5 g/l). Less is formed in dry white and red wines, especially those stored in tanks. In the case of barrel-aged wines, the formation of sulfate by the oxidation of free SO_2 is added to the amount resulting from the combustion of sulfur in the empty barrels. This formation lowers the pH and harshens the wine. This phenomenon contributes to the decrease in quality of wines stored in barrels for an excessively long time.

When sulfiting is done without any measurement beforehand, the wine can be excessively sulfited and its taste affected. In general, the characteristic odor appears at or above 2 mg of active molecular SO_2 per liter. Table 8.16 indicates the corresponding free SO_2 concentrations. To lower the concentration of free SO_2 of a wine, the most effective solution, when possible, is to use this wine to increase an insufficient concentration of free SO_2 in a similar wine.

If such an operation is not possible, the most generally recommended method is to aerate the wine. The effectiveness of this method is based on the slow oxidation of sulfur dioxide. During the days that follow, the higher the temperature, the more rapidly the concentration decreases. Aeration has a limited effectiveness, and 16 mg of oxygen per liter is required to oxidize 64 mg of total sulfur dioxide per liter. This approximately corresponds to a decrease of 42 mg of free SO_2 per liter, taking into account the dissociation of bound forms.

The use of hydrogen peroxide is a radical means of eliminating an excess of free SO_2. This method is too severe and is therefore prohibited; it compromises wine quality for a long time.

The impact of aging time and closures on free and total SO_2 concentrations in a Sauvignon Blanc wine was also studied after bottling (Lopes *et al.*, 2009). Forty-eight hours after bottling, free SO_2 concentrations in wines are very similar as a function of cylindrical closure types (natural cork, agglomerated cork, synthetic cork, screw caps, glass stoppers). The total

SO_2 levels are identical but slightly lower for wines closed with synthetic and micro-agglomerated stoppers. Two months after bottling, the free and total SO_2 levels in hermetically sealed bottles drop to 28 and 129 mg/l, respectively. Furthermore, the free and total SO_2 rates drop to 22 and 108 mg/l for wines closed with screw caps, whereas those closed with natural cork and agglomerated and colmated cork have concentrations of 20 and 107–109 mg/l of free and total SO_2. Wines capped with micro-agglomerated and synthetic stoppers have the lowest levels of free and total SO_2. The bottling effect seems to be significant, since free and total SO_2 contents decrease significantly during this step, which is correlated with bottles closed with synthetic and micro-agglomerate stoppers containing the highest levels of dissolved oxygen during bottling. At 12 and 24 months, the hermetically sealed bottle maintains higher levels of free and total SO_2, which are, respectively, 26 and 126 mg/l. Wines closed with screw caps also have high levels of free and total SO_2, which decrease, respectively, from 22 to 19 mg/l and from 108 to 104 and 102 mg/l. Wines capped with synthetic stoppers have lower levels of SO_2, which decrease significantly during aging: free SO_2 concentration after 24 months is less than 10 mg/l, which is considered the lower limit for the protection of white wines. Wines closed with cork stoppers (natural, colmated, agglomerated, and micro-agglomerated) have intermediate free and total SO_2 levels, although absolute concentrations vary. During the first two months of aging, SO_2 levels decrease greatly, in relation with oxygen introduced during bottling, and they continue to drop for the next 22 months. This is particularly important in wines bottled with synthetic closures, which enable continuous entry of oxygen into bottles at high rates (Lopes et al., 2005, 2007). However, the direct reaction of SO_2 with oxygen under wine conditions is very slow and for the most part irrelevant (Waterhouse and Laurie, 2006).

However, SO_2 probably reacts with hydrogen peroxide, aldehydes, and ketones (Danilewicz et al., 2008).

8.8.4 The Forms of Sulfur Dioxide Used

This antiseptic has the advantage of being available in various forms capable of responding to different situations: gaseous state (resulting from the combustion of sulfur), liquefied gas, liquid solution, and crystallized solid.

1. Sulfurous gas (SO_2) liquefies at a temperature of 15°C at normal atmospheric pressure or at normal ambient temperature under a pressure of 3 bar. It is a colorless liquid with a specific gravity of 1.396 at 15°C. Placed in 10–50 kg metal cylinders, this form is used for large-quantity additions that can be measured by weighing the bottle, which is placed directly on a scale. A *sulfidoseur* (or sulfite doser) is used to treat smaller volumes of wine. The graduated container can be precisely filled from the metal cylinder by regulating a pair of needle valves—permitting the addition of precisely measured quantities of the gas. Liquefied sulfur dioxide gas is also delivered in vials containing 25, 50, or 75 g of sulfur dioxide, for example, adapted for sulfiting wine in barrels with capacities of several hundred liters. A special tool perforates these small metal cap-stoppered bottles when they are inside the barrel to be treated.
2. For SO_2 additions to small volumes of wine, or to have a better incorporation, 5–8% solutions prepared in water or must (to avoid dilution) from liquefied sulfurous gas are used. The quantity needed is weighed. The concentration of the

solution is regularly verified by measuring its specific gravity (Table 8.22) or by chemical analysis. It tends to decrease in contact with air. Handling these solutions is unpleasant, since they give off a strong SO_2 odor. Prepared on the premises, they are well adapted to large winemaking facilities such as bulk wineries.

3. Concentrated 10% solutions, or 18–20% potassium bisulfite solutions, are also used. They are more easily handled than the preceding since they are less odorous. Being more concentrated, however, they are less easily incorporated into wine and must. They acidify less than the preceding, since the acidity in these solutions is partially neutralized. Potassium metabisulfite ($K_2S_2O_5$) solutions at 10% diluted in water can also be used. These solutions contain approximately 50 g of sulfur dioxide per liter (5%) and are suitable for limited-volume winemaking. The metasulfite powder should be diluted in water before use. When added directly, it is difficult to blend into the must.

8.8.5 Sulfiting Wines by Sulfuring Barrels

Sulfuring barrels, or small wooden tanks or containers, consists in burning a certain quantity of sulfur in these containers. This is probably the oldest form of sulfur dioxide use in winemaking. It is used for adjusting the free SO_2 concentration of wines at the moment of racking and also for avoiding microbial contamination when storing empty containers. It has a dual sterilizing effect. That effect is exerted at once on the wine and on the internal surface of the container. This practice is part of normal winery operations and could not be replaced by the simple addition of sulfite solution to the wine. Due to the unpleasant odor imparted to the wine cellar by burning sulfur, its usage can be prohibited by the safety and labor legislation of certain countries. Instead of coming from the combustion of sulfur, sulfurous gas can also be delivered from a compressed gas cylinder.

In any case, sulfur combustion is only applicable to wooden containers. In fact, sulfurous gas, coming from the combustion of sulfur, attacks the internal surface of cement tanks and the coating of metal tanks. It also accelerates the deterioration of stainless steel.

TABLE 8.22

Specific Gravity (at 15°C) of Sulfur Dioxide Solutions Prepared by the Dissolution of Sulfur Dioxide Gas in Water

Sulfur dioxide (g/100 ml)	Specific gravity	Sulfur dioxide (g/100 ml)	Specific gravity
2.0	1.0103	6.5	1.0352
2.5	1.0135	7.0	1.0377
3.0	1.0168	7.5	1.0401
3.5	1.0194	8.0	1.0426
4.0	1.0221	8.5	1.0450
4.5	1.0248	9.0	1.0474
5.0	1.0275	9.5	1.0497
5.5	1.0301	10.0	1.0520
6.0	1.0328		

The sulfur is generally supplied in the form of a wick or disc. It may be coated on a cellulose weave or mixed with a mineral base (aluminum or calcium silicate). The units most often used are 2.5, 5, and 10 g of sulfur. Chatonnet et al. (1993) demonstrated a certain heterogeneity in the quantity of SO_2 produced by the combustion of wicks or discs of the same weight as a function of their preparation conditions or storage (moisture absorption).

From the equation

$$S + O_2 \rightarrow SO_2$$
$$32 + 32 = 64 \qquad (8.16)$$

the burning sulfur combines with its weight in oxygen to give double the weight of SO_2. In reality, 10 g of sulfur burned in a 225-l barrel produces only about 13–14 g of SO_2—a 30% loss. One part of the difference is accounted for by the portion of the sulfur that falls to the bottom of the barrel without burning, and the other part by the production of sulfuric acid—a strong acid without antiseptic activity. The sulfiting loss and the acidification of wine (by repeated sulfitings) are explained in this manner.

The combustion of sulfur does not exert its effect by eliminating all of the oxygen from the barrel. The maximum quantity of sulfur that can burn in a 225-l barrel is 20 g, for the maximum production of 30 g of sulfurous gas. At this stage, the combustion stops because the sulfurous gas has the property of hindering its own combustion. It has been determined that approximately 32.5 l of oxygen is present in the barrel at the moment when the combustion stops, compared with 45 l beforehand.

These observations lead to the conclusion that the combustion of sulfur is limited. When a 40 g sulfur wick is burned, not all of the sulfur is consumed, even if the wick is burned to a cinder. About half of it falls to the bottom of the barrel without burning.

All other things being equal, the production of SO_2 is greater in a dry barrel than in a wet one. For instance, the burning of 10 g sulfur in a dry barrel gives 12 g SO_2 and only 5 g in a wet barrel (Ribéreau-Gayon et al., 1977).

The production of SO_2 via the burning of sulfur is thus irregular in a barrel. In addition, the dissolution of the SO_2 formed is generally irregular during the filling of the barrel. Depending on the filling speed and conditions (by the top or bottom, for example), a more or less significant part of the sulfurous gas is driven out of the barrel. Moreover, the distribution of the sulfiting in the wine mass is not homogeneous. The first wine that flows into the barrel receives more SO_2 than the last. In one example, the free SO_2 increased by 45 mg/l at the bottom of the barrel, by 16 mg/l in the middle and not at all in the upper portion. Consequently, the wine should be homogenized after racking—by rolling the barrel, for example. This sulfiting method should only be used for wines stored in small-capacity containers—say, up to 6 hl.

As Chatonnet et al. (1993) stated, the combustion of 5 g of sulfur in a 225-l oak barrel increases the SO_2 in wine by 10 to 20 mg/l. Sulfur wicks are less efficient (10 mg/l) but more consistent than discs (10–20 mg/l). The latter are more sensitive to their external environment, i.e. moisture.

The combustion of sulfur for the storage of empty barrels will be covered in Volume 2, Section 13.6.2.

To decrease the doses of SO_2 in wines during their aging, other key control factors should therefore be carefully followed: strict hygiene and systematic disinfection of winery equipment, control of fast alcohol and malolactic fermentations, working with inert gas and continuous blanketing, limiting the number of transfers, and maintaining suitable aging conditions. As concerns the use of SO_2 for wine aging, research should be encouraged on other aging methods, notably for applications in which the use of SO_2 is responsible for a significant contribution to the RDA, and on testing out alternatives or other solutions (chemical or physical) so that SO_2 use can be decreased.

References

Barbe J.C. (2000) *Les combinaisons du dioxyde de soufre dans les moûts et les vins issus de raisins botrytisés. Rôle des bactéries acétiques*, Thèse de Doctorat, Université Victor Segalen Bordeaux II.

Barbe J.C., de Revel G., Joyeux A., Lonvaud-Funel A. and Bertrand A. (2000) J. Agric. Food Chem., 48, 8, 3413.

Barbe J.C., de Revel G., Joyeux A., Bertrand A. and Lonvaud-Funel A. (2001a) J. Appl. Microbiol., 90, 34.

Barbe J.C., de Revel G., Joyeux A., Lonvaud-Funel A. and Bertrand A. (2001b) 1ère partie, Revue F. d'Œnologie, 189, 26.

Barbe J.C., de Revel G., Perello M.C., Lonvaud-Funel A. and Bertrand A. (2001c) 2ère partie, Revue F. d'Œnologie, 190, 16.

Barbe J.C., de Revel G. and Bertrand A. (2002) J. Agric. Food Chem., 50, 22, 6408.

Blouin J. (1965) *Contribution á l'étude des combinaisons de l'anhydride sulfureux dans les moûts et les vins*, Thèse Docteur-Ingénieur, Université de Bordeaux.

Blouin J. (1995) *CR Journée Technique du CIVB*, Bordeaux, France.

Burroughs L.F. and Sparks A.H. (1964) J. Sci. Food Agri., 3, 176

Burroughs L.F. and Sparks A.H. (1973) J. Sci. Food Agri., 24, 187, 199 and 207.

Chatonnet P., Boidron J.N. and Dubourdieu D. (1993) Œno One, 27, 4, 277.

Danilewicz J.C., Seccombe J.T. and Whelan J.(2008) Am. J. Enol. Vitic., 59, 128.

Dubernet M. and Ribéreau-Gayon P. (1974) Vitis, 13, 233.

Guillou-Largeteau I. (1996) *Etude de substances de faible poids moléculaire combinant le dioxyde de soufre dans les vins blancs issus de vendanges botrytisées. Mise en évidence et importance du rôle de l'hydroxypropanedial*, Thèse Doctorat de l'Université de Bordeaux II.

Heard H. and Fleet G.H. (1988) Aust. N. Z. Wine Ind. J., 3, 57.

International Code of Winemaking Practices (2001) *International Organisation of Vine and Wine, Paris*.

Kielhöfer E. and Würdig G. (1960) Weinberg u. Keller, 7, 313.

Lopes P., Saucier C. and Glories Y. (2005) J. Agric. Food Chem., 53, 6967.

Lopes P., Saucier C., Tesseidre P.L. and Glories Y. (2007) J. Agric. Food Chem., 55, 5167.

Lopes P., Silva M.A., Pons A., Tominaga T., Lavigne V., Saucier C., Darriet P., Tesseidre P.L. and Dubourdieu, D. (2009) J. Agric. Food Chem., 57, 10261.

Ribéreau-Gayon J., Peynaud E., Ribéreau-Gayon P. and Sudraud P. (1977) *Sciences et Techniques du Vin. Vol. 4: Clarification et Stabilisation. Matériels et Installation*. Dunod, Paris.

Romano P. and Suzzi G. (1992) Sulfur dioxide and wine microorganisms. In *Wine Microbiology and Biotechnology*, pp. 373–393 (Ed. G.H. Fleet). Harwood Academic Publishers, Chur.

Sudraud P. (1963) *Etude expérimentale de la vinification en rouge*, Thèse de Docteur-Ingénieur, Faculté des Sciences de Bordeaux.

Sudraud P. and Chauvet S. (1985) Œno One, 19, 1, 31.

Suzzi G. and Romano P. (1982) Vini d'Italia, 24, 138.

Usseglio-Tomasset L. (1995) *Chimie œnologique*, 2nd edition. Tec & Doc, Lavoisier, Paris.

Vaquer J.M. (1988) *L'anhydride sulfureux en œnologie, l'étude de sa toxicité et les méthodes proposées pour en réduire les doses*. Thèse de Doctorat en Pharmacie, Université de Montpellier, France.

Waterhouse A.L. and Laurie F. (2006) Am. J. Enol. Vitic., 57, 306.

CHAPTER 9

Products and Methods Complementing the Effect of Sulfur Dioxide

9.1 Introduction
9.2 Sorbic Acid
9.3 Octanoic and Decanoic Acids (Saturated Short-Chain Fatty Acids)
9.4 Dimethyl Dicarbonate (DMDC)
9.5 Lysozyme
9.6 Destruction of Yeasts by Heat (Pasteurization)
9.7 Ascorbic Acid
9.8 The Use of Inert Gases

9.1 Introduction

Considering the legitimate desire to lower sulfur dioxide concentrations, it is normal to search for aids that complement its action by reinforcing the effectiveness of one of its properties. This chapter covers such chemical products and physical processes that have been or are likely to be authorized by the legislation of different countries. Others are likely to be proposed in coming years.

Sorbic acid, which can be used to increase the antimicrobial properties of sulfur dioxide, is now well known and authorized in many countries. The possibility of using octanoic and decanoic acids will also be covered, though they are not currently authorized. They do not seem to pose any hygiene problems, and they exist naturally in wine; this treatment only reinforces the existing concentrations. Lysozyme from egg white has similar properties. This enzyme is capable of destroying certain bacteria, especially lactic acid bacteria in wine. Its capabilities have been known for a long time, but its practical application in wineries has grown in recent years.

Handbook of Enology, Volume 1: The Microbiology of Wine and Vinifications, Third Edition.
Pascal Ribéreau-Gayon, Denis Dubourdieu, Bernard B. Donèche and Aline A. Lonvaud.
© 2021 John Wiley & Sons Ltd. Published 2021 by John Wiley & Sons Ltd.

Numerous antibiotics and antiseptics known to act on wine yeasts have not yet been authorized, for health reasons. One of them, 5-nitrofurylacrylic acid, is a powerful fungicide but is highly carcinogenic and causes alcohol intoxication. Another, pimaricin, is autodegradable and has a fungistatic effect, without known secondary effects; it is already authorized in the food industry. A proposal requesting the official approval of this product has been filed in France.

In the 1970s, the use of ethyl pyrocarbonate (Baycovin) in wine was proposed and authorized in some countries. This fungicide is very effective for cold-sterilizing wine at the time of bottling. It disappears rapidly, breaking down mainly into ethanol and CO_2, but also releasing tiny quantities of ethyl carbonate that were, nevertheless, significant enough that its use was soon given up altogether. The Bayer pharmaceutical company later marketed a dimethyl dicarbonate (DMDC) product, Velcorin, that is just as effective without any of the health risks.

Among the physical processes capable of complementing the antimicrobial properties of sulfur dioxide, the destruction of germs by heat (pasteurization) can be used. Also the possibility of destroying germs by high pressure has been demonstrated. This method seems to be effective and affects quality less than heat treatment does. The practical conditions of its use in cellars remain to be defined and the appropriate equipment to be designed. Of course, all operations that work toward eliminating microorganisms, even partially (racking, centrifugation, filtration, and pasteurization), facilitate microbial stabilization and permit the winemaker to lower the SO_2 concentration used.

Ascorbic acid is the most used aid, contributing to the antioxidant properties of sulfur dioxide. The storage of wine with an inert gas is another effective means of avoiding oxidation.

9.2 Sorbic Acid

9.2.1 Physical and Chemical Properties

The formula for sorbic acid contains two double bonds:

$$CH_3-CH=CH-CH=CH-COOH$$

Four isomers exist, but only the *trans–trans* isomer is used. Due to its effectiveness and lack of toxicity, its use is authorized in many countries, in particular in the European Union (EU), at a maximum concentration of 200 mg/l. It remains prohibited in a few countries (e.g. Austria and Switzerland).

It exists in the form of a white crystallized powder. In a water-based solution, it can be entrained by steam; for this reason, it is found in wine distillate and increases volatile acidity. It has a slightly acidic taste. Its dissociation constant corresponds to a pK_a of 4.76. In other words, in wine, it is essentially in the form of a free acid.

Sorbic acid is not very soluble in water (1.6 g/l at 20°C; 5.0 g/l at 50°C), but it is soluble in ethanol (112 g/l at 20°C). Its sodium and potassium salts are very water soluble. In this form, concentrated solutions are prepared for treating wine. Potassium sorbate contains 75% sorbic acid. A solution at 200 g of sorbic acid per liter is prepared by dissolving 270 g of potassium sorbate per liter in water. One liter of this solution can treat 10 hl of wine at 20 g of sorbic acid per hectoliter. Sorbic acid can also be dissolved in alkaline solutions, and 200 g will dissolve in a liter of cold water containing 100 g of KOH. These concentrated solutions must be prepared immediately before use. They become yellow with time.

Certain precautions must be taken when incorporating the concentrated solution into the wine being treated. Due to the pH of wine and the pK_a of sorbic acid, the latter will be liberated from its salt as soon as the concentrated solution is introduced in the wine. However, this acid is not very soluble: if its concentration at a given moment is too high, it will precipitate. The concentrated solution must therefore be added slowly while being constantly mixed. The use of a dosing pump is recommended.

9.2.2 Antimicrobial Properties

The fungicidal activity of sorbic acid has been tested on yeasts in different circumstances (Ribéreau-Gayon et al., 1977).

The fungicidal concentration, i.e. to stop a fermentation (*mutage*), is relatively high (5 g/l). From 0.5 g/l on, the fermentation is observed to slow and stop before its successful completion. These elevated concentrations prevent the use of sorbic acid in sweet wine vinification.

The fungistatic concentration, which hinders fermentation in grape must, varies according to must composition (in particular pH), the size of the inoculum, and the nature of the strain. The concentration ranges cited in the literature are between 100 and 1,000 mg/l, with an average value of 300–500 mg/l.

The inhibiting concentration for yeasts in wines containing sugar is lower and depends on the alcohol content and the pH. Concerning the alcohol content, numerous tests are cited in the literature, taking into account the size of the inoculum and the nature of the yeast strains (Table 9.1).

The pH also has a strong influence on the fungicidal activity of sorbic acid, which increases as the pH decreases. In laboratory experiments, a concentration of 150 mg/l of sorbic acid was needed to hinder the refermentation of a sweet wine at pH 3.1. All other factors being equal, 300 mg/l was necessary to obtain the same results at pH 3.5. At a pH > 3.5, the maximum concentration authorized (200 mg/l) may be insufficient to ensure the proper stabilization of a wine.

The effect of pH on the state of sorbic acid has been analyzed. The non-dissociated free acid molecule is known to possess the antiseptic character. Between pH 3.0 and 3.8, the proportion of sorbic acid ($pK_a = 4.76$) in its non-dissociated free acid state passes from 98 to 90%. This difference is too small to explain the much more significant impact of pH. Cell permeability and penetration phenomena of sorbic acid regulated by the pH of the medium may also be involved.

In addition to its action on classic fermentative yeasts, sorbic acid acts against flor yeasts developing on the surface of wine (*Candida*). In capped bottles of red

TABLE 9.1

Sorbic Acid Doses (mg/l) Necessary for Sweet Wine Conservation (Laboratory Tests with *Saccharomyces bayanus*) (Ribéreau-Gayon et al., 1977)

Alcohol content (% vol.)	Yeast population		
	5×10^3 cells/ml	50×10^3 cells/ml	500×10^3 cells/ml
10	150	175	200
11	125	150	200
12	100	150	150
13	75	100	150
14	50	75	125

wine containing 10% vol. alcohol with a relatively significant headspace and stored upright, 150 mg of sorbic acid per liter ensures their storage for three weeks. Higher concentrations may be needed for lower alcohol contents, longer storage periods, higher temperatures, or wines containing residual yeasts.

However, the antibacterial properties of sorbic acid are much less significant. It exerts practically no activity against acetic acid and lactic acid bacteria. Concentrations of 0.5–1 g/l would be necessary to have a significant effect.

Sorbic acid therefore exerts a selective effect on wine microorganisms and opposes yeast development without blocking bacterial growth. It has the opposite effect to that of sulfur dioxide (which favors yeasts at the expense of bacteria). Consequently, sorbic acid must never be used alone but always associated with an antibacterial product (sulfur dioxide). In wines exposed to air, the amount of volatile acidity formed by acetic acid bacteria can be greater in the presence of sorbic acid, due to the absence of an antagonism with yeasts.

In conclusion about these notions concerning its antimicrobial properties, sorbic acid only presents a sufficient effectiveness in practice when associated with a certain concentration of ethanol and free sulfur dioxide. Sorbic acid is an effective complement to sulfur dioxide, since it reinforces its action, but it is not a replacement. Most problems encountered in employing sorbic acid come from its incorrect use. A lack of effectiveness and the appearance of strange tastes are generally reported.

9.2.3 Stability and Sensory Impact

Fresh solutions of sorbic acid have no odor, and this product does not influence wine aromas. Up to a concentration of 200 mg/l, sorbic acid does not modify the taste of correctly stored wines, either immediately after its addition or after several years of bottle aging. For certain wines, its impact on taste is perceptible above this value, but it is distinct only for concentrations above 400–500 mg/l. It does not increase the apparent acidity of wine or thin it out but does accentuate impressions of astringency, bitterness, and harshness, which are perceived in particular in the aftertaste.

In the 1960s, as early as the first treatments, strange odors and tastes were sometimes noted—especially in red wines treated with sorbic acid. For this reason, its authorization was reconsidered. However, when it is correctly used, experience has shown that under normal storage conditions, the evolution and development of bottled wine is not affected. These observations do, however, lead to the problem of the stability of this product in wine.

The sorbic acid molecule, like unsaturated fatty acid molecules, possesses two double bonds. As explained in organic chemistry, these bonds can be oxidized by air to form molecules with aldehydic functions. This reaction explains the unpleasant tastes imparted to fatty substances by oxidation. Concentrated aqueous solutions of sorbate effectively become yellow and take on a pungent odor. This observation is proof of a certain chemical instability of sorbic acid. Yet dilute solutions are noted to be significantly more stable: in wine, in particular, the same quantity initially added is found after three years of bottle aging. Consequently, chemical instability cannot be presented as an explanation for the sensory deviations attributed to sorbic acid.

The appearance of an unpleasant, intense, and persistent odor, similar to the odor of geraniums, in wines treated by sorbic acid, was quickly determined to be related to bacterial development. In fact, this flaw appears at the same time as an increase in volatile acidity or simply during malolactic fermentation. It can also occur in poorly stabilized bottled wines. Lactic acid bacteria are responsible for this spoilage, and numerous isolated wine strains are capable of metabolizing sorbic acid.

The molecule responsible for the geranium-like odor is a derivative of the corresponding alcohol of sorbic acid (Volume 2, Section 8.7.1). This strong-smelling molecule has a perception threshold of less than 1 µg/l.

It is almost impossible to remove the geranium odor from a wine. Since it is still perceptible after significant dilution, blending is not recommended. The most drastic deodorizing treatments fail (fixation on active charcoal, extraction by oil, etc.). The odor passes through during distillation and is concentrated in spirits. Only a severe oxidation with potassium permanganate eliminates it. Fortunately, the necessary conditions to avoid this serious spoilage (the rational use of SO_2) are now well known, and this problem has practically disappeared.

9.2.4 Use of Sorbic Acid

1. Sorbic acid must be used exclusively for the conservation of wines containing reducing sugars. It serves no purpose in dry wines. In the case of red wines in particular, the development of lactic acid bacteria in the presence of sorbic acid can result in the appearance of an extremely serious sensory flaw.
2. Sorbic acid is not a vinification tool. It does not affect the rules of *mutage* for sweet wines. It is incapable of stopping fermentations that are underway. Sorbic acid is exclusively used for the conservation of sweet wines to avoid their refermentation. It can be added to wine after the elimination of yeasts by racking, centrifugation, or filtration.
3. The concentration used is generally 20 g/hl. It can be decreased in wines with little residual yeast, high alcohol content, and/or a low pH. Due to the low solubility of this acid in water, a concentrated solution of the much more soluble potassium sorbate is prepared immediately before treatment. The solution must be introduced slowly into wine and mixed quickly to avoid the insolubilization of the acid.
4. In wines treated with sorbic acid, the free SO_2 concentration must be maintained between 30 and 40 mg/l to protect against oxidation and bacteria and to neutralize the taste of aldehydic substances. This SO_2 concentration alone would be insufficient to protect wine against refermentation.

9.3 Octanoic and Decanoic Acids (Saturated Short-Chain Fatty Acids)

Certain long-chain fatty acids (C16 and C18) activate fermentations. Conversely, other shorter-chain saturated fatty acids, in particular C8 and C10 acids, possess a significant fungicidal action (Geneix et al., 1983). They are formed by yeasts during alcoholic fermentation and can contribute to difficult final fermentation stages. This property, combined with their complete innocuousness, has led to them being proposed as an aid to sulfur dioxide to ensure stability of sweet wines (Larue et al., 1986). Their use as a wine stabilizer could be approved by official regulation.

Different options are possible. Their total concentration added to wine should not exceed 10 mg/l, for example, 3 mg of octanoic acid plus 6 mg of decanoic acid per liter. Octanoic and decanoic acids are prepared for use by being solubilized in ethanol at 60% vol. The concentrations are calculated so that the addition of the solution does not exceed 1% (1 ml/l).

At the indicated concentration, these acids possess a fungicidal effect complementing sulfur dioxide. For the *mutage* (sterilization to stop fermentation) of

TABLE 9.2

Sweet Wine Sterilization at the End of Alcoholic Fermentation by the Use of Octanoic and Decanoic Acids to Complement the Action of Sulfur Dioxide (Larue et al., 1986)

Yeasts conducting fermentation	Sterilization treatment		Viable yeasts 24h after treatment (cells/ml)
	SO_2 (mg/l)	Octanoic acid (3 mg/l) Decanoic acid (6 mg/l)	
Saccharomyces cerevisiae	0	0	10^7
	250	0	4×10^2
	150	9	2×10^2
Saccharomyces bayanus	0	0	7.5×10^6
	250	0	<10
	150	9	<10

sweet wines, 150 mg of SO_2 + 9 mg of fatty acids per liter has the same effectiveness as 250 mg of SO_2 per liter (Table 9.2). The SO_2 saving is significant. The fatty acids should be added 24 hours before sulfiting. Under these conditions, the SO_2 addition acts while yeasts are predominantly (if not completely) inactivated. Also, a fraction of the fatty acids is eliminated by fixation on the yeast cells. After this kind of treatment, sweet wines can be conserved with a free SO_2 concentration of around 40 mg/l.

It should be noted that fatty acids are more effective than sorbic acid, which does not enable a decrease of the SO_2 concentration used for *mutage*.

Due to the aroma intensity of these acids and their esters, the organoleptic effect of such an addition had to be determined. An increase of the four principal wine fatty acids and their ethyl esters has been

TABLE 9.3

Effect of Sterilization Conditions at the End of Alcoholic Fermentation (SO_2 200 mg/l or SO_2 100 mg/l + Fatty Acids 9 mg/l) on Wine Composition (Fatty Acids and Their Ethyl Esters, mg/l) (Larue et al. 1986)

Wine number	Total fatty acids in wine[a]		Total ethyl esters in wine[b]	
	SO_2[c]	SO_2 + fatty acids[d]	SO_2[c]	SO_2 + fatty acids[d]
1	2.6	4.1	0.20	0.56
2	4.7	6.2	0.30	0.90
3	4.6	8.0	0.34	0.95
4	3.3	4.9	0.38	0.58
5	3.0	4.7		
6	12.4	14.3	0.81	1.58
7	4.8	6.2	0.26	0.94
8	2.6	3.2	0.17	0.36
Average	4.95	5.86	0.35	0.76

[a] Total fatty acids = hexanoic acid, octanoic acid, decanoic acid, dodecanoic acid.
[b] Total ethyl esters of fatty acids = ethyl hexanoate + ethyl octanoate + ethyl decanoate.
[c] Sterilization with 200 mg SO_2/l.
[d] Sterilization with 100 mg SO_2/l + 9 mg fatty acids/l.

observed (Table 9.3), but the increase does not represent the totality of the fatty acids that were added. A large portion of them are fixed to the yeast cells during *mutage* and consequently eliminated during clarification. This treatment leads to an increase of a few milligrams (per liter) of the constituents naturally existing in wine. The variations observed are well within the limits caused by the action of certain conventional winemaking methods and operations: aerobic conditions, the addition of ammonium salts, temperature variations, etc.

Numerous sensory analyses have been carried out. This addition is not claimed to be completely without effects, but, provided that the addition does not exceed 9–10 mg/l, the sensory effects are judged to be slight and on the whole positive.

9.4 Dimethyl Dicarbonate (DMDC)

The use of diethyl dicarbonate (DEDC), or ethyl pyrocarbonate (Baycovin), was authorized in the United States and Germany for a few years in the early 1970s. This fungicide is very effective on yeasts; after acting, it is hydrolyzed in a few hours after application, at a rate that varies with temperature, according to the following reaction:

$$C_2H_5-O-\underset{\underset{O}{\|}}{C}-O-\underset{\underset{O}{\|}}{C}-O-C_2H_5 + H_2O$$
$$\longrightarrow 2CH_3-CH_2OH + 2CO_2$$

This product, at doses of a few hundred milligrams per liter, was very effective for cold-sterilizing wine during bottling (Ribéreau-Gayon *et al.*, 1977). It was less useful during wine storage, as it disappeared rapidly and ceased to provide protection from repeat contamination.

It rapidly became apparent, however, that the reaction of ethyl pyrocarbonate in wine was more complex than indicated by the reaction shown above. Ethanol and carbon dioxide were certainly the main degradation products, but small quantities of ethyl carbonate were also formed, and its fruity aroma was perceptible above a certain threshold. Most importantly, ethyl pyrocarbonate is a highly reactive molecule and combines with certain substances in wine (organic acids, polyphenols, and nitrogen-based compounds) to produce urethanes, e.g. ethyl carbamate, which is toxic and carcinogenic. Quantities never exceeded 2–4 µg/l, significantly below the official threshold of 30 µg/l in Canada. However, this risk was sufficient for the product to be completely abandoned.

Bayer, the company that produced Baycovin, replaced it with DMDC (or Velcorin), considered to have the same sterilizing properties in wine without any of its problems (Bertrand, 1999). DMDC breaks down to form methyl carbamate, which is considered to have practically no toxic effects. Ough *et al.* (1988) demonstrated that 100 mg/l DMDC sterilized wine completely at pH below 3.8 in the absence of SO_2, even if the initial yeast population was greater than 10^7 cells/ml.

On the basis of these findings, DMDC was initially authorized in the United States and then in other countries. Like ethyl pyrocarbonate, DMDC is most effective at the time of bottling, although it has also been suggested for use in stopping the fermentation of sweet (botrytized) wines (Bertrand and Guillou, 1999), thus reducing the amount of SO_2 required. In any case, a certain quantity of free SO_2 is always necessary to protect the wine from oxidation.

Adding DMDC to wine has been authorized by the International Organisation of Vine and Wine (OIV) since 2001, with the following objectives: obtaining microbiological stability of bottled wine containing fermentable sugars and preventing the development of undesirable yeasts and lactic acid bacteria. The addition must be conducted very shortly before bottling, and the concentration used cannot exceed 200 mg/l expressed as DMDC. Adding DMDC cannot cause the maximum methanol content in wine to be exceeded, as recommended by OIV. Wine cannot be commercialized as long as DMDC can be detected. Considering these properties, especially the possibility of reducing SO_2 concentrations, this product was registered in the OIV International Code of Winemaking Practices in 2004, after trials. In the EU, the product has been authorized since 2006 at a maximum concentration of 200 mg/l in wine and 250 mg/l for non-fermented drinks (coffee-based beverages, teas, infusions, and flavored beverages), as well as cider. DMDC can also be used to clean and disinfect barrels.

DMDC has proved effective not only on fermentation yeasts but also on those responsible for contamination (*Brettanomyces*; Volume 2, Section 8.4.5), as well as to a lesser extent on bacteria, even though its bactericidal effect is less significant than its anti-yeast effect (Ough *et al.*, 1988).

DMDC decomposes to form mainly methanol and carbon dioxide.

$$CH_3-O-\underset{O}{\overset{\parallel}{C}}-O-\underset{O}{\overset{\parallel}{C}}-O-CH_3 + H_2O$$
$$\longrightarrow 2CH_3-OH + 2CO_2$$

With regard to toxicology, methanol is the most important issue in this case, as it is highly toxic. Theoretically, the breakdown of 200 mg/l of DMDC produces 96 mg/l of methanol. There is no regulated limit for this compound that is present in all wines (Volume 2, Section 2.2.1). In 2004, OIV changed the maximum methanol limits in wines (increased by 100 mg/l), which brings these to 400 mg/l for red wines and to 250 mg/l for white wines. In the United States, the maximum permitted methanol content is 1,000 mg/l. Consequently, the use of DMDC may be considered not to have any serious impact on the methanol content of wine.

The breakdown of DMDC also produces nontoxic methyl carbamate, as well as several methyl, ethyl, and methyl-ethyl carbonates. The latter are aroma molecules but are present in insufficient quantities to modify wine aroma significantly.

Finally, it should be taken into account that handling this undiluted product involves some risk as it is dangerous if inhaled or allowed to come into contact with the skin. It is advisable to use proper equipment to ensure safe handling.

9.5 Lysozyme

9.5.1 Nature and Properties

Ribéreau-Gayon *et al.* (1975, 1977) were apparently the first to describe the winemaking properties of lysozyme. The authors described a tour of the Médoc wineries in the 1950s with Alexander Fleming, winner of the 1945 Nobel Prize for Medicine for his discovery of penicillin. During a fining operation using egg white, Fleming wondered whether lysozyme, which he had discovered in egg white a few years earlier, could play a role in the microbiological stabilization of wine. Lysozyme is a natural polypeptide made up of 129 amino acids, including 21 aspartic acids, five glutamic acids, 12 alanines, 11 arginines, eight cysteines, three phenylalanines, 12 glycines, six isoleucines, one histidine, eight leucines, six lysines, two prolines, two methionines, 10 serines, three tyrosines, seven threonines, six tryptophans, and six valines. It possesses a molecular mass of 14,700 Da.

Lysozyme is an enzyme capable of destroying Gram-positive bacteria (Section 4.3.2), such as lactic acid bacteria in wine. This natural, crystallized substance is capable, at low doses, of causing lysis of the bacteria, i.e. dissolving their cell walls. Egg white contains approximately 9 g/l lysozyme, and this traditional method of fining may introduce 5–8 g/l into the wine. Ribéreau-Gayon *et al.* (1975, 1977) observed that this agent had no effect on acetic acid bacteria in wine, which is not surprising, as they are Gram-negative. However, they confirmed that the use of crystallized lysozyme at doses above 4 mg/l achieved its maximum effect within 24 hours and was capable of destroying almost all the lactic

TABLE 9.4

Effect of Purified Lysozyme on Lactic Acid Bacteria in Wine (Number of Living Bacteria Per Cubic Centimeter) (Ribéreau-Gayon et al., 1975, 1977)

Doses of lysozyme (mg/l)	Red wine 18,000,000		Red wine 520,000		White wine 20,000
	After 4 h	After 24 h	After 4 h	After 24 h	After 24 h
4	2,600,000	2,300	490,000	12,000	30
8	1,430,000	2,250	200,000	8,000	10
12	19,700	1,750	4,200	2,400	40
16	15,800	2,000	4,700	2,700	10

acid bacteria in a wine (Table 9.4). Moreover, the increase of the lysozyme concentration accelerates bacteria lysis, but does not really change the number of resistant cells. Whatever the dose, cells are not all destroyed, and the full stabilization of wine cannot be obtained.

However, it was not possible to detect any effect of lysozyme during fining with egg white, even at very high doses. There was no difference compared to fining with gelatin or bentonite. The lysozyme in egg white is probably not released into the wine during fining and is precipitated, together with the albumin, by contact with tannins.

At that time, the use of lysozyme for the microbiological stabilization of lactic acid bacteria was not envisaged, probably due to the fact that the crystallized product was not widely available and the cost of treatment seemed unacceptable.

The use of lysozyme in the dairy and cheese industries has gradually become widespread, and it has been demonstrated to have no toxic effects on humans. The resulting increase in availability has led to new interest in the use of this product in winemaking since 1990 (Gerbaux et al., 1997; Gerland et al., 1999). Crystallized lysozyme is produced by Fordas (Lugano, Switzerland) and marketed in France, at least for winemaking purposes, by Martin-Vialatte (Epernay, France), under the "Bactolyse" brand. This lysozyme is a protein consisting of 125 amino acids (molecular weight: 14.6 kDa) that acts almost immediately. However, under the conditions prevailing in wine (pH), it is rapidly precipitated out or deactivated (e.g. following bentonite treatment). In contrast to SO_2, the activity of lysozyme increases with pH.

9.5.2 Applications of Lysozyme in Winemaking

Several situations in which lysozyme has useful effects during fermentation and wine storage have been described in the literature (Gerbaux et al., 1999; Gerland et al., 1999), even if stabilization in terms of lactic acid bacteria is not perfect.

Inhibiting malolactic fermentation in white wines

In view of the fact that the dose of SO_2 required in any case to prevent oxidation also has antimicrobial effects, preventing malolactic fermentation in white wines is not usually a problem, except in wines with a high bacteria content and high pH, such as press wines. However, the use of lysozyme makes it possible to achieve the same protection with lower doses of SO_2 (4–5 g/hl). It is necessary to add lysozyme either once (500 mg/l in the must) or twice (250 mg/l in the must and the same in the new wine), as one 250 mg/l dose is not sufficiently effective.

Adding lysozyme does not affect fermentation kinetics. The wines have the same analytical parameters and are unchanged on tasting, provided they are adequately protected from oxidation, but a dose of only 30 mg/l of total SO_2 is required, for example, instead of 50.

As lysozyme is deactivated by bentonite, treatment with the latter should be delayed. For the same reason, wines treated with lysozyme react to the heat test, indicating protein instability. However, protein haze develops at high temperatures (50°C), which do not normally occur during wine shipment and storage.

Delaying the development of lactic acid bacteria and malolactic fermentation in red wines

One case where this applies is the fermentation of whole grape bunches, i.e. carbonic maceration and Beaujolais winemaking methods. It is not unusual for malolactic fermentation to start in this type of medium favorable to the development of lactic acid bacteria before the end of the alcoholic fermentation. Lysozyme (10 g/hl) added to the crushed grapes is at least as effective as sulfiting (5–7 g/hl) and minimizes the risk of an unwanted increase in volatile acidity.

Another situation is that of wines with long maceration times, when bacterial growth increases the risk of malolactic fermentation on the skins. This should always be avoided in view of the risk that there may be trace amounts of residual sugar present, especially if the grapes were not completely crushed prior to fermentation. Lysozyme's effectiveness at this stage is less clear, considering its gradual elimination by precipitation with the phenolic compounds. Adding lysozyme earlier in fermentation is immediately effective, but there is always a risk of recontamination, e.g. during pump-overs, while delaying its addition raises the risk of premature malolactic fermentation.

Use in cases of difficult alcoholic fermentation

If conditions are unfavorable to yeasts, a decrease in their activity causes fermentation to slow down. This may, in turn, promote the proliferation of lactic acid bacteria, which are likely to stop alcoholic fermentation completely (Section 12.7.4), making it extremely difficult to restart (Section 3.8.1). It is well known that bacterial growth in a sugar-containing medium provides the most favorable circumstances for lactic spoilage to occur, together with an increase in volatile acidity.

Sulfiting is the most common technique (Section 8.7.4) for controlling unwanted bacterial growth. A 200–300 mg/l dose of lysozyme provides a useful complementary treatment, especially in white wines, where lysozyme is more stable than in red wines. Lysozyme not only prevents lactic spoilage but is also effective just after spoilage starts: the addition of 25 g/hl reduces the bacterial population very rapidly from several million bacteria to fewer than 100.

Microbiological stabilization after malolactic fermentation

Microbiological stabilization of wines is necessary to avoid the many problems that may be caused by microorganisms. Lysozyme does not protect sweet wines from fermenting again or eliminate contaminant yeasts (*Brettanomyces*), nor is it effective against acetic acid bacteria, which cause volatile acidity in aerated wines.

Lysozyme is, however, effective in eliminating lactic acid bacteria. A dose of 200 mg/l has a similar effect to sulfiting at 50 mg/l. As this level of SO_2 is generally required in any case to benefit from the other properties of this product (preventing oxidation, as well as eliminating yeast and acetic acid bacteria), it seems unnecessary to add lysozyme as well.

It has been observed that delaying sulfiting in red wines by adding lysozyme results in a more intense color, which also remains more stable thereafter (Gerland *et al.*, 1999).

Use and labeling of wines treated with lysozyme

In 1997, OIV authorized lysozyme treatment for must and wine in order to control the growth and activity of bacteria responsible for malolactic fermentation in must or wine. It was also authorized in order to reduce sulfur dioxide additions. The maximum dose of 500 mg/l happens to be sufficient to check the growth and activity of bacteria responsible for malolactic fermentation during alcoholic fermentation. However, lysozyme cannot totally substitute for SO_2, which has antioxidant properties. An SO_2 + lysozyme combination leads to more stable wines. When must and wine are treated with lysozyme, the total concentration cannot exceed 500 mg/l. In 2011, the European Food Safety Authority (EFSA) published its scientific opinion on lysozyme. The panel concluded that wines treated with lysozyme may trigger undesirable allergic reactions in sensitive individuals.

Wines treated with lysozyme were authorized for sale, without specific wording on labels and while stocks lasted, if they were put on the market or labeled before June 30, 2012. Beyond that date, wines treated with lysozyme must be specifically labeled, in application of the provisions of European Commission (EU) regulation No. 1266/2010 of December 22, 2010, which amends directive 2007/68/EC, as regards labeling requirements for wines.

9.6 Destruction of Yeasts by Heat (Pasteurization)

9.6.1 Introduction

The destruction of microbial germs by heat has been known for a long time, but in enology there have not been as many applications of pasteurization as in other food industries. The reasons are easy to understand.

Bottled wine conserves relatively well due to its alcohol content and acidity, provided that it is bottled with a sufficient SO_2 (and possibly sorbic acid) concentration after the satisfactory elimination of microbial populations. "Hot" bottling can contribute to wine stabilization, but, unlike beer, the pasteurization of wine has never been widespread.

"Hot" bottling ensures the stability of bottled red wines with respect to bacterial development and sweet white wines with respect to refermentations. It is generally used with wines of average quality that have microbial stabilization problems. Heating to 45 or 48°C sterilizes the wine and the bottle. The presence of free SO_2 prevents excessive oxidation. Of course, a space appears below the cork after cooling (Volume 2, Section 12.2.4).

Sweet wines are difficult to store during their aging phase between the completion of fermentation and bottling. For this more or less long period, the wine is stored in bulk, in tanks, or in barrels; it normally acquires its stability and limpidity and improves qualitatively, but refermentation risks certainly exist. The destruction of germs by heat should be able to contribute to the required stabilization, while at the same time limiting the sulfur dioxide concentration. The heat conditions necessary for wine stabilization are easy to satisfy without compromising quality. However, the equipment available in a winery makes it difficult to ensure satisfactory sterile conditions for wine handling and storage.

Despite these difficulties, the temperature resistance of wine yeasts is well understood, and the practical conditions for microbial stabilization of bulk sweet wine by heat treatment are also clearly defined (Splittstoesser et al., 1975; Devèze and Ribéreau-Gayon, 1977, 1978). Therefore, the constraints and advantages of these techniques with respect to the desired decrease in sulfur dioxide concentrations can be correctly evaluated.

9.6.2 Theoretical Data on the Heat Resistance of Wine Yeasts

The temperature resistance of microorganisms can be characterized by two criteria: the decimal reduction time (D), which represents the duration of heating, at a constant temperature, required to reduce the population to one-tenth of its initial value, and the temperature variation (z), which permits the multiplication or division of D by 10. The value of D depends on the microorganism, the culture conditions, and the heating environment; the value of z depends almost exclusively on the microorganism.

The following four tested yeast strains are classified according to an increasing resistance to temperature in terms of z and D at 40°C:

Saccharomycodes ludwigii	$z = 3.23°C$	$D_{40} = 10.8 \min$
Saccharomyces bayanus	$z = 3.94°C$	$D_{40} = 8.45 \min$
Zygosaccharomyces bailii	$z = 4.26°C$	$D_{40} = 46.1 \min$
Saccharomyces cerevisiae	$z = 4.34°C$	$D_{40} = 65.7 \min$

Studies of the influence of different factors on the temperature resistance of a *S. bayanus* strain have demonstrated the following:

- Yeasts are half as resistant in a wine at 13% as in a wine at 11% (by volume).
- Yeasts are about three times more resistant in a wine containing 100 g of sugar per liter than in the corresponding dry wine.
- A wine at pH 3.8 must be heated for a period three times longer than the same wine at pH 3.0.
- Yeast resistance is 10 times greater during the final phase of fermentation than during the log growth phase of the cells.

The effectiveness of a heat treatment is expressed in pasteurization units (PU). These units represent the duration of a treatment at 60°C having the same effectiveness on the given microorganism as the treatment being considered. The graph in Figure 9.1 indicates, for wine microorganisms ($z = 4.5°C$), the number of PU corresponding with a given time/temperature combination. Heating operations for three minutes at 55.5°C and for 1.4 minutes at 57°C are equivalent and correspond to 0.3 PU.

The number of PU required to obtain a certain level of destruction can be calculated from yeast temperature resistance criteria in a given medium. Laboratory studies suggest that 0.05 PU would be sufficient for the destruction of yeasts in a dry wine at 12% vol. ethanol. In practice, 0.5 PU are required to sterilize such a wine. The uneven heat supplied by industrial heating equipment and the existence of particularly resistant yeast strains during the final stages of fermentation can explain this difference. Table 9.5 shows the necessary conditions for the destruction of microbes as a function of wine composition.

In addition, moderate sulfiting is required to protect sweet wines from oxidation. This sulfiting increases yeast sensitivity to heat. The effectiveness of a heat treatment can therefore be improved by injecting sulfur dioxide at the inlet of the heat exchanger.

These results show the relative ease of destroying yeasts by heat. They permit the establishment of operating conditions adapted to each particular case. In practice, the principal difficulty arises after the treatment: that is, avoiding subsequent contaminations.

9.6 | Destruction of Yeasts by Heat (Pasteurization)

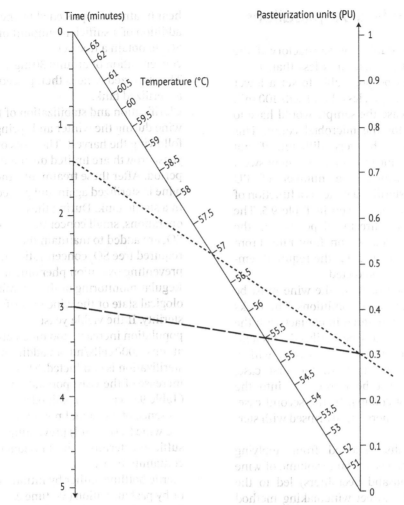

FIGURE 9.1 Pasteurization units supplied by heating at constant temperature according to time and temperature (Devèze and Ribéreau-Gayon, 1977).

TABLE 9.5

Number of Pasteurization Units (PU) Necessary to Sterilize Different Wines (Devèze and Ribéreau-Gayon, 1978)

Alcohol content (% vol.)	Sugar concentration (g/l)		
	0 (PU)	50 (PU)	100 (PU)
10	1.08	1.74	2.83
11	0.73	1.19	1.93
12	0.50	0.81	1.31
13	0.34	0.55	0.90
14	0.23	0.38	0.61

9.6.3 Practical Applications

In practice, stability is satisfactory if the viable yeast population is less than 1 cell/ml. It might be preferable to set a lower limit (for example, less than 1 cell/100 ml), but in this case the sample would have to be filtered for the microbial count. This operation can be very difficult, if not impossible—for example, with new sweet botrytized wines. The number of PU required to sterilize a wine, as a function of its composition, is given in Table 9.5. The heating time directly depends on the industrial pasteurization flow rate. From the graph in Figure 9.1, the required temperature can be predicted.

After pasteurization, the wine must be stored under sterile conditions. The tanks and all of the material in contact with the wine must therefore be sterilized—preferably with steam, and if necessary with a chemical disinfectant. In the first case, sterile air must be introduced into the tanks during cooling. In the second case, they must be energetically rinsed with sterilized water.

The results obtained from applying these techniques to a large volume of wine (several thousand hectoliters) led to the proposal of a sweet winemaking method comprising several steps:

1. A well-adapted heat sterilization at the time of *mutage*, when the sugar/alcohol balance is attained. This heat treatment is preceded by the addition of a sufficient amount of SO_2 to obtain a free SO_2 concentration of around 30 mg/l. The sterilized wine is then placed in a sterilized tank.
2. Clarification and stabilization of the wine during the winter and spring following the harvest. The risks of yeast growth are limited during this period. After these treatments, the wine is sterilized again and placed in a sterile tank. During these operations, small concentrations of SO_2 are added to maintain the required free SO_2 concentration for preventing oxidation phenomena.
3. Regular monitoring of the microbiological state of the wine to verify sterility. If the viable yeast population increases too much and attains 1,000 cells/ml, an additional sterilization is conducted. This increase of the yeast population (Table 9.6) can be explained by the presence of a residual population in the wine (<1 cell/ml) preventing sufficient sterility or by subsequent contaminations.
4. Sterile bottling, either by filtration or by pasteurization (Volume 2, Section 12.2.4), using well-established techniques.

If appropriate equipment (which remains relatively expensive) is used, bottled sweet

TABLE 9.6

Microbiological Control, Over Time, of Yeast Populations in Two Sweet Wine Tanks, Stabilized by Pasteurization (Devèze and Ribéreau-Gayon, 1978)

Tank 1		Tank 2	
Date	Viable yeasts/ml	Date	Viable yeasts/ml
Jan. 5	<1	Mar. 11	<1
Feb. 4	<1	Apr. 18	<1
Apr. 18	100	Jun. 23	70
May 16	880	Jul. 6	100
Jun. 1	1,540	Jul. 20	200

wines can be obtained with the same free SO_2 concentrations as those used for dry white wines. Consequently, the total SO_2 concentration is significantly lowered. Sweet wines stored with only 30 mg of free SO_2 per liter contain on average 60 mg less total SO_2 per liter than those stored with 50 mg of free SO_2 per liter. The latter concentration is indispensable to avoid the refermentation of non-pasteurized wines.

9.7 Ascorbic Acid

9.7.1 Properties and Mode of Action

Ascorbic acid, or vitamin C, exists in fruits and in small quantities in grapes (about 50 mg/l of juice), but it rapidly disappears during fermentation and initial aerations. Wines generally do not contain any.

Ascorbic acid is used in enology (Ribéreau-Gayon et al., 1977) essentially as a reducing agent. Ewart et al. (1987) proposed replacing ascorbic acid with its isomer, erythorbic acid. The latter does not have vitamin properties but possesses the same oxidation–reduction properties. Its industrial production costs are less.

Ascorbic acid was authorized in France as an antioxidant for fruit juices, beers, carbonated beverages, and wines in 1962. Its use does not raise any health-related objections. It is now used in most viticultural countries at a maximum concentration of 150 mg/l, always in association with sulfur dioxide. The recommended concentrations are between 50 and 100 mg/l. As it is completely water soluble (330 g/l), its preparation does not pose a problem. The solution should be prepared at the time of its use. Homogenization should be complete, while avoiding all oxygenation, for example, by mixing it with sulfur dioxide.

The oxidation mechanism of ascorbic acid has incited much research (Makaga and Maujean, 1994). It functions like an oxidation–reduction system. Its oxidized form is dehydroascorbic acid:

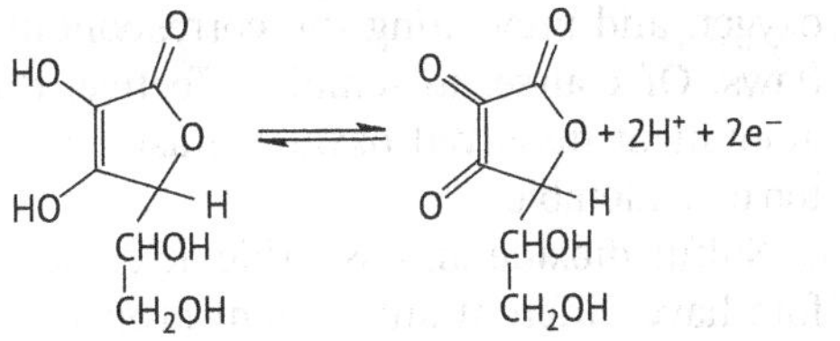

[Chemical structure: Ascorbic acid ⇌ Dehydroascorbic acid + $2H^+ + 2e^-$]

This reaction is theoretically reversible, but, due to its instability, dehydroascorbic acid disappears.

The two electrons that appear in the course of the reaction reduce certain wine constituents, in particular the ferric ion:

$$2Fe^{3+} + 2e^- \leftrightarrow 2Fe^{2+}$$

The effectiveness of ascorbic acid in the prevention of iron haze, which is exclusively caused by Fe^{3+} ions, is explained by the above reactions.

In the presence of oxygen, the oxidation of ascorbic acid leads to the formation of hydrogen peroxide—a powerful oxidant that can profoundly alter wine composition. The presence of a sufficient amount of free sulfur dioxide protects wine from the action of this molecule. It is preferentially oxidized by hydrogen peroxide:

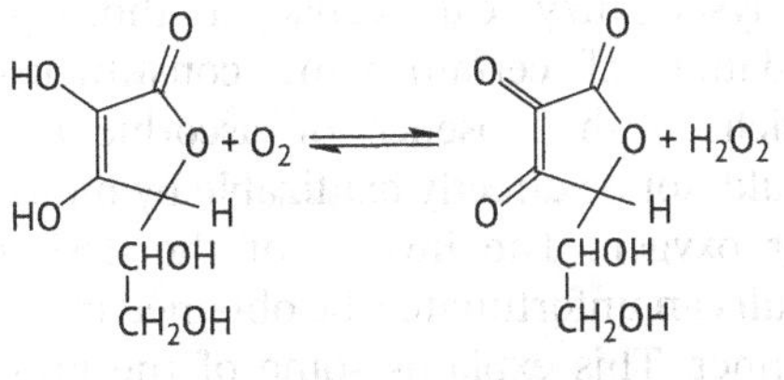

[Chemical structure: Ascorbic acid + O_2 ⇌ Dehydroascorbic acid + H_2O_2]

$$SO_2 + H_2O_2 \rightarrow H_2SO_4$$

The oxidation reaction of ascorbic acid is catalyzed by iron and copper ions. But, contrary to the direct oxidation of SO_2 by molecular oxygen, the reaction is rapid. It constitutes a simple means of almost instantaneously eliminating dissolved oxygen and preventing the corresponding flaws. Of course, to remain effective, the amount of dissolved oxygen must not be too considerable.

Sulfur dioxide and ascorbic acid therefore have different antioxidant properties. The first has a delayed, but stable, effect that continues over time even in the presence of a subsequent oxygenation. It cannot prevent iron casse, which rapidly appears after an aeration. The second has an immediate effect: it can instantaneously offset the damage of an abrupt and intense aeration (iron casse), but it acts only as long as the wine is not in permanent contact with air.

Due to the high oxidation sensitivity of ascorbic acid, its effectiveness is only guaranteed when its contact with air is limited. In other words, it protects well against small, brief aerations but not against intense or continued oxidation. Its role is limited to protecting wine from light aerations, following bottling, for example. It is not effective for prolonged storage in tanks or barrels.

The danger of the oxidation of ascorbic acid, especially in the presence of a large amount of oxygen, should also be considered. Under these conditions, hydrogen peroxide and sometimes other peroxides are formed. Coupled with the presence of catalysts, they can cause a thorough oxidation of certain wine constituents, which in the absence of ascorbic acid would not be directly oxidizable by molecular oxygen. The inverse of the desired result can unfortunately be obtained in this manner. This explains some of the problems that can be encountered when ascorbic acid is used incorrectly. For this reason, ascorbic acid should only be used in wines containing a sufficient concentration of free sulfur dioxide, available for the elimination of the hydrogen peroxide formed in the course of oxidations.

The addition of ascorbic acid to grapes has been authorized by OIV since 2001. It protects the aroma substances of grapes against the influence of oxygen in air, thanks to its antioxidant properties. This is recommended before grape crushing. The concentration used cannot exceed 250 mg/kg. It is preferable to use ascorbic acid with sulfur dioxide. In the case of grape must, the addition of ascorbic acid has been authorized by OIV (2001) for the same reasons as in grapes as well as to limit acetaldehyde formation during alcoholic fermentation by binding with sulfur dioxide. It was also authorized to limit the formation of hydrogen sulfide and volatile thiols during fermentation. In that case, the concentration used, potentially accumulated with the one used on grapes, cannot exceed 250 mg/l. Treating wine with ascorbic acid has also been authorized by OIV since 2001. Adding ascorbic acid is recommended during bottling. Otherwise, it oxidizes on contact with air, and the oxidation product then causes much deeper oxidative spoilage to wine than that resulting from oxygen in the absence of ascorbic acid. The concentration used cannot exceed 250 mg/l. When ascorbic acid has also been used on grapes or in must, the final content, expressed in ascorbic acid plus dehydroascorbic acid, cannot exceed 300 mg/l.

9.7.2 Protection Against Enzymatic Oxidations

The addition of ascorbic acid will limit (if not eliminate) must oxidations catalyzed by tyrosinase or laccase. Moreover, it is a substrate of laccase. It does not act by inhibiting the enzymes, as does sulfur dioxide, but rather by monopolizing oxygen, due to its fast reaction speed. Ascorbic acid is used in this manner in certain countries to complement the protection given by sulfur dioxide against the oxidation of white grape must. Its use is particularly justified for the protection of mechanically harvested grapes, since it does not have an effect on maceration, as does sulfur dioxide (Section 12.4.3). This use is not permitted by EU regulations, which only authorize the addition of L-ascorbic acid in wines up to a maximum concentration of 250 mg/l since 2009.

Effective protection can be obtained in red wines sensitive to enzymatic browning as well as white musts during winemaking. At present, however, the use of ascorbic acid is not widespread in winemaking, probably because the required concentrations are too high and because sulfur dioxide is more effective.

9.7.3 Protection Against Iron Casse

The aeration of wine oxidizes iron. The amount of ferric iron formed (several milligrams per liter) can be sufficient to induce iron casse. Protection against iron casse can be ensured if the wine receives 50–100 mg of ascorbic acid per liter beforehand (Table 9.7).

In parallel, when a wine containing ascorbic acid is aerated, the oxidation–reduction potential slightly increases and then rapidly stabilizes, whereas it continues to increase in the control wine (Figure 9.2). Reciprocally, if ascorbic acid is added to an aerated wine possessing ferric iron, the iron is reduced in the ensuing hours, and the oxidation–reduction potential rapidly decreases. The beginnings of iron casse can be reversed, and the corresponding turbidity eliminated in this manner (Table 9.8).

Ascorbic acid effectively protects against iron casse, which can occur after operations that place wine in contact with air, such as pumping-over, transfers, filtering, and especially bottling. Under the same conditions, sulfur dioxide acts too slowly to block the oxidation of iron. But, if the wine must be aerated again after a treatment following an initial aeration, the ascorbic acid no longer protects the wine. When a wine that has received 100 mg of ascorbic acid per liter a month earlier is aerated, the evolution of iron (III) is identical to that in the control. These results lead to the supposition that the added ascorbic acid has disappeared.

TABLE 9.7

Protection from Iron Casse by the Addition of Ascorbic Acid Before Aeration (Ribéreau-Gayon et al., 1977)

	Fe (III) 48 h after aeration (mg/l)	Limpidity
White wine (total Fe 18 mg/l) control	8	Cloudy
+25 mg ascorbic acid/l	3	Limpid
+50 mg ascorbic acid/l	1	Limpid
+100 mg ascorbic acid/l	0	Limpid
Red wine (total Fe 15 mg/l) control	6	Slightly cloudy
+25 mg ascorbic acid/l	4	Limpid
+50 mg ascorbic acid/l	0	Limpid
+100 mg ascorbic acid/l	0	Limpid

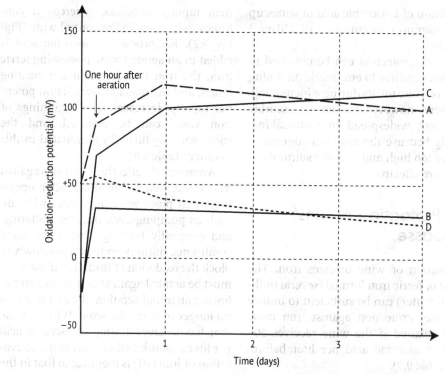

FIGURE 9.2 Evolution of oxidation–reduction potential of reduced wines, aerated after the addition of 100 mg of ascorbic acid per liter, compared with control wine (Ribéreau-Gayon et al., 1977). (A) Red wine control, (B) Red wine + ascorbic acid, (C) White wine control, and (D) White wine + ascorbic acid.

TABLE 9.8
Reduction of Iron Casse by Addition of Ascorbic Acid, 48 Hours After the Aeration Causing the Haze (Ribéreau-Gayon et al., 1977)

	Fe III (mg/l)		Limpidity 48 h after addition
	At time of addition	48 h after addition	
White wine (total Fe 24 mg/l)			
Control	9	7	Very cloudy
+50 mg ascorbic acid/l	9	2	Limpid
Red wine (total Fe 16 mg/l)			
Control	5	4	Cloudy
+50 mg ascorbic acid/l	5	1	Limpid

9.7.4 Organoleptic Protection of Aerated Wines

In certain cases, ascorbic acid improves the taste of bottled wines. Wines generally taste worse when they contain dissolved oxygen and have a high oxidation-reduction potential. Ascorbic acid enables better conservation of wine freshness and fruitiness—especially in certain types of dry or sparkling white wines. It also decreases the length of critical phase that follows bottling, known as "bottle shock." The effect is not as considerable or spectacular for all wines, but wine quality is never lowered by its use.

The taste improvement due to ascorbic acid depends on several factors. The first is the type of wine. Ascorbic acid is of little interest in the case of wines made from certain varieties or very mature wines—for example, barrel-aged wines, oxidized white wines, botrytized sweet wine, and fine red wines. On the contrary, it improves the stability of fresh and fruity wines (generally young wines) that have conserved their varietal aromas.

Another important factor is the concentration of free sulfur dioxide. It should be situated in an intermediate range between 20 and 30 mg/l to ensure a refinement of the wine, which presents a fresher aroma with a floral note. For higher or lower free SO_2 concentrations, the qualitative improvement of the wine is less obvious.

Satisfactory results have been obtained for sparkling wines produced by the traditional method, by the transfer method, or in pressure tanks. The necessary amounts of sulfur dioxide and ascorbic acid are added to the dosage to ensure concentrations of 20–30 and 30–50 mg/l, respectively. The coupled addition of the two substances ensures an optimal aroma and improves the finesse and longevity of the wine.

In sparkling wines, ascorbic acid acts not only by its reducing properties but also by its capacity as an oxidation–reduction buffer. Their potential remains stable at 240 mV for several years. In the absence of ascorbic acid, it varies between 200 and 265 mV, according to the effectiveness of corking. This phenomenon clearly affects the sensory character of wine (Makaga and Maujean, 1994).

The use of ascorbic acid insignificantly modifies the use of sulfur dioxide; it enables a slight lowering of the latter's concentration. However, it possesses other advantages.

9.8 The Use of Inert Gases

9.8.1 Wine Storage Using Inert Gases

Even before the use of antioxidants (sulfur dioxide and ascorbic acid), the first recommendation for protecting wines against the adverse effects of chemical or microbiological oxidations was to limit their contact with air. Wines are thus stored in completely filled containers, sometimes equipped with a system compensating for the effects of expansion. This recommendation cannot always be followed, if the availability of tanks of a satisfactory size is limited or wine is regularly taken from the same tank for several days. Tanks equipped with a floating top, which always remains in contact with the surface of the wine, were introduced, but the seals between the top and the inner surface of the tank are rarely satisfactory, and their effectiveness is questionable.

Satisfactory results are obtained by storing wine in a partially filled tank with an inert gas, in the total absence of oxygen. Wine storage using inert gas also permits the carbon dioxide concentration (lowering or increasing) to be adjusted. Although not directly related to protection from oxidation, this subject will nevertheless be covered in this section.

The following gases are authorized for storage: nitrogen, carbon dioxide, and argon. Argon is rarely used: it is more expensive than the others, and its solubility is limited in wine (4 l/hl). Carbon dioxide is very soluble in wine (107.2 l/hl) and therefore cannot be used alone in partially filled containers. The carbon dioxide concentration of the wine would significantly increase via the dissolution of the gas. It is, however, sometimes used in a mixture with nitrogen (for example, 15% CO_2 + 85% N_2) to avoid the degassing of certain wines that must maintain a moderate CO_2 concentration. Nitrogen is the most commonly used gas. R-grade nitrogen is used; it contains a little oxygen as an impurity but has no impact on wine. It is less soluble in wine than oxygen (1.8 l/hl compared with 3.6 l/hl), but contrary to oxygen, which reacts with wine components by oxidizing them, nitrogen accumulates without reacting. The wine spontaneously becomes saturated in nitrogen during handling in contact with air. Storage in the presence of this gas therefore cannot increase its concentration in wine.

Several principles of inert gas systems for tanks exist, but systems for wineries must be well designed. In particular, they must be perfectly airtight. Maintaining a slight overpressure is recommended in order to monitor for possible leaks. This method is essentially applicable to perfectly sealed tanks.

The gases are stored in compressed gas cylinders. At the cylinder outlet, the gas generally undergoes a double expansion. Initially, it is reduced to a pressure of 2–8 bar and circulates in copper piping up to the storage tank. A second expansion reduces the pressure to 15–20 or 100–200 mbar. This second solution permits easy identification of piping and tank leaks. Each tank has a separate line and a pressure gauge enabling verification of the pressure and thus the airtightness. Finally, a pressure release valve prevents the unfortunate consequences of handling errors.

Facilities have been especially designed for wine storage using inert gas. Metal tanks are connected together by gas lines, but the tanks can be isolated and individually maintained at a slight overpressure (100–200 mbar). This overpressure proves the airtightness of the tanks. It is verified by the pressure gauge reading.

At the beginning of the operation, the tank is completely filled with wine, and the tank top is secured. A hectoliter of wine is then drained from the valve at the bottom of the tank. Simultaneously, nitrogen gas is sparged in the upper portion of the tank—replacing the drained wine and creating a nitrogen atmosphere buffer. Next, the internal pressure is adjusted. The tank is then ready for storage. To remove wine, the nitrogen cylinder should first be opened, then the gas valve for the tank, and finally the wine tank valve. Perfectly clear and stable wines are conserved and protected from oxidation and evaporation for several months in these wine tanks. Nitrogen consumption is extremely limited. The evolution of the taste of these wines is identical to that of wines stored in completely filled tanks.

When there is insufficient wine to fill the tank completely, thus expelling all of the air contained, another solution consists in completely filling the tank with water. It is then emptied under a nitrogen counterpressure before introducing the wine. In many cases, after partially filling the tank with wine, residual air is expelled by sparging nitrogen in the tank headspace. If these operations are carried out in non-airtight tanks, the inert gas must be constantly renewed—resulting in a considerable consumption of gas. These conditions are not recommended and can lead to a false sense of security for wine storage. At the end of a few weeks, the wine is oxidized and has lost its CO_2; its aroma and taste become insipid.

In any case, this storage system does not free the winemaker from using sulfur dioxide or even allow concentrations to be lowered. This antiseptic agent remains indispensable for fighting against yeasts and lactic acid bacteria. It must be used at the same concentrations as in full tanks.

Storage under inert gas can cause either an increase or a decrease in the amount of carbon dioxide naturally existing in wine. The data in Table 9.9 show the impact of the wine volume/gaseous atmosphere ratio. The variations are slight, especially with nitrogen, if the tank is practically full. Storage under a carbon dioxide atmosphere easily leads to an excessive increase of its concentration, with corresponding changes in the organoleptic characters of the wine. Conversely, storage under nitrogen causes a considerable decrease in the CO_2 concentration for half-filled tanks.

The study by Lonvaud-Funel and Ribéreau-Gayon (1977) gives the factors permitting the estimation of the CO_2/N_2 mixture that must be used at a given temperature so that a wine conserves its initial dissolved CO_2 concentration during storage, whatever the tank fill level. Inert gases can also be used for ensuring that wine transfers are protected from oxygen, by injecting nitrogen in the lines while pumping wines, for example.

9.8.2 Adjusting the Carbon Dioxide Concentration

Wine tasters are very sensitive to taste modifications caused by the presence of this gas, even below the sensory perception threshold. For example, in a red Bordeaux wine, more than 50% of the tasters correctly put in order three samples of a wine containing 620, 365, and 20 mg of carbon dioxide per liter (Ribéreau-Gayon and Lonvaud-Funel, 1976). The characteristic pricking sensation of CO_2 was only perceptible in the first sample. The third sample appeared more insipid than the second, which was judged the best. Yet nothing led the tasters to believe that the difference was related to the CO_2 concentration.

Dry white wines tolerate higher carbon dioxide concentrations. Around 90% of the tasters correctly put in order three samples of the same wine containing, respectively, 250, 730, and 1,100 mg of CO_2 per liter. The second sample was preferred overall; the carbon dioxide increased the aroma and the freshness of this wine. Yet the carbon dioxide concentrations should not be exaggerated. Concentrations of 1,000 mg/l are not as appreciated in dry white wines as one might think.

TABLE 9.9

Evolution of the Carbon Dioxide Concentration in Wines Stored in a Carbon Dioxide or Nitrogen Atmosphere, According to Fill Level of Container (Lonvaud-Funel, 1976)

	CO_2 atmosphere		N_2 atmosphere	
Fill level (%)	CO_2 concentration in wine (mg/l)	Increase (%)	CO_2 concentration in wine (mg/l)	Decrease (%)
98	308	7	281	1.5
82	589	106	234	17.8
50	1,132	297	144	49.6
18	1,708	499	51	82.0

Due to its sensory impact, the carbon dioxide concentration should be correctly adjusted. For each type of wine, there is a corresponding optimal concentration. Red wines tolerate less CO_2 (around 200 mg/l) than dry white wines (around 500–700 mg/l). The more tannic and ageworthy the wines are, the less they tolerate CO_2.

It can be useful to eliminate excess carbon dioxide rapidly, by agitation, in young red wines intended for early bottling. Racking in the presence of air can decrease the CO_2 concentration by 10%, but this is not always sufficient.

The injection of fine nitrogen bubbles in wine entrains a certain proportion of dissolved gas (carbon dioxide or oxygen) in a wine. This technique is called degassing by sparging. The wine flow rate, with the device in Figure 9.3, can vary from 30 to 120 hl/h. Temperature plays an important role in the effectiveness of this treatment. Below 15°C, the degassing yield is insufficient. The temperature of the wine should preferably be 18°C. The wine, emulsified with very fine nitrogen bubbles, should then be exposed to air by flowing in a thin film through a shallow tank with a large surface area so that the nitrogen is easily released and entrains the dissolved carbon dioxide.

In tests, simple racking enabled the elimination of 26% of the CO_2. Treating with half a volume of N_2 for a volume of wine eliminated 43% of the CO_2. Four times more nitrogen (two volumes) only eliminated 54%. It is therefore more reasonable to carry out two consecutive treatments with lower volumes of nitrogen.

In certain cases, the carbon dioxide concentration must be decreased; in others, it must be increased. It can be increased by sparging with carbon dioxide gas; the gas can be injected in the winery piping. The same result can be obtained by placing wine in a partially filled tank, its headspace filled with a mixture of N_2 and CO_2. Lonvaud-Funel (1976) has given the mixture required for obtaining a certain CO_2 concentration, according to the respective wine and gas volumes.

These operations are normally carried out at atmospheric pressure. If they were to take place at higher pressures, gasification would occur. The operations would no longer be considered as ordinary wine treatments, since gasified wines are subject to special legislation.

The dissolution of carbon dioxide in wine does not differ much from that in water. It depends on the temperature and ranges between 2.43 g/l at 8°C and 1.73 g/l at 18°C. These values correspond with the maximum amount of CO_2 that can be dissolved in wine.

The sparging of wine by carbon dioxide has been suggested. This method can be useful for avoiding oxygen dissolution during transfers and to ensure protection against oxidation. The wine must be degassed before bottling.

Among the chemical, physical, or biological alternatives to the use of sulfur dioxide, some require more experimentation. For example, high pressure (Lonvaud-

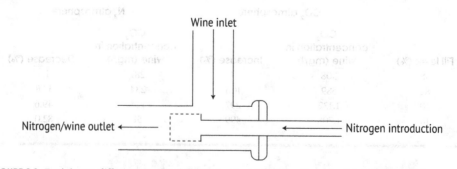

FIGURE 9.3 Gas injector, diffusing very fine bubbles in wine circulating through piping. (Ribéreau-Gayon et al., 1977).

Funel et al., 1994), tangential microfiltration (Poupault et al., 2007), bipolar electrodialysis (Cottereau, 2006), the use of pulsed electric fields, or plasma technology at atmospheric pressure. Glutathione, a sulfur-containing tripeptide, is a compound naturally present in plants and especially in grape berries. Its involvement in enzymatic oxidation mechanisms of white grape must has been clearly described. The effect of adding sulfur-containing peptides or amino acids to preserve the color, taste, or aroma of fruits, vegetables, or fruit juices has been highlighted (Molnar-Perl and Friedman, 1990a,b). Works in the field of wine have shown the beneficial role of glutathione for the prevention of premature aging in dry white wines. More specifically, it preserves their fruity aroma related to the presence of aroma components belonging to the varietal thiol family (Volume 2, Section 8.7). Glutathione content in musts (a few to several dozen mg/l) varies depending on the type of wine and grapevine growing conditions (especially its nitrogen supply). Grape transformation technologies, especially the exposure of must to oxygen during the critical pre-fermentative stages (juice extraction) and aging methods (presence/absence of lees, etc.), are some of the many factors likely to have an impact on residual glutathione content in wine (Lavigne-Cruège et al., 2003; Lavigne and Dubourdieu 2004; Kilmartin et al., 2008). As a result, in addition to the application of techniques likely to preserve glutathione naturally extracted from the grape, its use as a winemaking aid to "dope" original contents and therefore attempt to amplify its antioxidant action has been described (Ugliano et al., 2010).

In 2015, the OIV registered the addition of glutathione in must or in wines (maximum concentration of 20 mg/l) in the International Code of Enological Practices (OIV Resolution 445-2015, 446-2015).

References

Bertrand A. (1999) Vitis, 249, 25.

Bertrand A. and Guillou I. (1999) Bull. OIV, 72, 84.

Cottereau P. (2006) *Winemaking technologies and practices. Code of good organic viticulture and winemaking (Orwine)*, 208–214.

Devèze M. and Ribéreau-Gayon P. (1977) Œno One, 11, 2, 131.

Devèze M. and Ribéreau-Gayon P. (1978) Œno One, 12, 2, 91.

Ewart A.J.W., Sitters J.H. and Brien C.J. (1987) Aust. N. Z. Wine Ind. J. 1, 1, 59.

Geneix C., Lafon-Lafourcade S. and Ribéreau-Gayon P. (1983) C.R. Acad. Sci., 296, *Série III*, 943.

Gerbaux V., Villa A., Monamy C. and Bertrand A. (1997) Am. J. Enol. Vitic., 48, 1, 49.

Gerbaux V., Meistermann E., Cottereau P., Barrière C., Cuinier C., Berger J.L. and Villa A. (1999) Bull. OIV, 72, 819, 349.

Gerland C., Gerbaux V. and Villa A. (1999) Revue des Oenologues, 93S, 44.

Kilmartin P.A., Herbst M. and Nicolau L. (2008) *Austr. NZ Grapegrower Winemaking, 2008*, Annual Tech. Issues.

Larue F., Murakami Y., Boidron J.N. and Fohr L. (1986) Œno One, 20, 2, 87.

Lavigne V. and Dubourdieu D. (2004) *Role du glutathion dans l'élaboration des vins blancs secs*, Actes du Colloque Mondiaviti Bordeaux 2004, 174–180.

Lavigne-Cruège V., Pons A., Chone X and Dubourdieu D. (2003) *Rôle du glutathion sur l'évolution aromatique des vins blancs secs*, Actes du 7ᵉ Symposium d'Oenologie de Bordeaux, 385–388.

Lonvaud-Funel A. (1976) *Recherches sur le gaz carbonique du vin*, Thèse de Doctorat d'Oenologie, Université de Bordeaux II.

Lonvaud-Funel A. and Ribéreau-Gayon P. (1977) Œno One, 11, 2, 165.

Lonvaud-Funel A., Dupont G., Demazeau G. and Bignon J. (1994) J. Int. Sci. Vig. Vin, 28, 57.

Makaga E. and Maujean A. (1994) Bull. OIV, 153, 763.

Molnar-Perl I. and Friedman M. (1990a) J. Agric. Food Chem., 38, 1648.

Molnar-Perl I. and Friedman M. (1990b) J. Agric. Food Chem., 38, 1652–1656.

Ough C.S., Kunkee R.E., Vilas M.R., Bordeu E. and Huang M.C. (1988) Am. J. Enol.Vitic., 39, 279.

Poupault P., Visonneau E. and Benesteau F. (2007) Rev. Fr. Oen., 225, Cahiers techniques, 7–12.

Ribéreau-Gayon P. and Lonvaud-Funel A. (1976) C.R. Acad. Agric., 62, 7, 491.

Ribéreau-Gayon J., Peynaud E., Ribéreau-Gayon P. and Sudraud P. (1975) Sciences et Techniques du Vin. Vol. II: Caractères des vins. Maturation du raisin, levures et bactéries. Dunod, Paris.

Ribéreau-Gayon J., Peynaud E., Ribéreau-Gayon P. and Sudraud P. (1977) Sciences et Techniques du Vin. Vol. IV: Clarification et Stabilisation, Matériels et Installations. Dunod, Paris.

Splittstoesser D.F., Lienk L.L., Wilkinson H. and Stamer J.R. (1975) Appl. Microbiol., 30, 369.

Ugliano M., Kwiatkowski M., Vidal S., Capone D., Solomon M., Dieval J. B., Aagaard O. and Waters E. (2010) *The role of copper and glutathione addition and oxygen exposure in the evaluation of key aroma compounds of Sauvignon blanc, Poster presented at 14th AWITC*, Adelaide.

PART II

Vinification

Reflections on Global Taste and Typicity of Wines

D. Dubourdieu

*Institut des Sciences de la Vigne et du Vin,
University of Bordeaux,
Villenave d'Ornon Cedex, France*

Today's global wine market is characterized by a glut of wine. Global demand has, of course, been steadily increasing for several years, thanks to the growth of several national or regional markets (USA, Asia, Russia, United Kingdom, northern Europe, Brazil, etc.), but this increase in consumption is not sufficient to absorb the glut. There are two essential reasons for this: the growth in surface areas planted to grapevines, particularly in the Southern Hemisphere, and the decline in consumption in traditional wine-drinking countries in Europe (France, Spain, and Italy). This market imbalance further exacerbates competition among wine producers. At the same time, we are seeing more and more uniformity of wine styles, related to worldwide consumption, a small number of "international" varietals (often of French origin), and the industrialization of winemaking processes (sweetening, aroma addition via the use of oak alternatives, total extraction of overripe grapes, etc.). This ampelographic and technological standardization, aimed at pleasing the global palate, is accompanied by the trivialization of distinctive signs of quality: AOC, Château, Grand Cru, 100-point scoring systems, and various awards. The effects of

---

*Lecture given during the Ninth International Enology Symposium, Oeno 2011 (June 16, 2011, Palais de la Bourse, Bordeaux, France).

this competition are nonetheless devastating: lower prices, eroding margins, and, in the longer term, reduction of vineyard areas.

Value creation is therefore the major challenge that world, and particularly European, viticulture must face. Thanks to the diffusion of knowledge, we know how to produce large quantities of wine at lower costs in many regions worldwide. The conditions needed are well known: a warm, relatively dry climate; the possibility to irrigate, if needed; and low-cost labor. Owing to climate change, it is not currently known if all dry-climate wine-growing regions will have sufficient amounts of water in the future. However, for the time being, thanks to irrigation, they can produce large quantities of wine with uniform taste characteristics at low cost. Wine-growing regions that do not have these competitive advantages will always have higher production costs than their competitors in warmer regions. They will be forced to seek maximum value for their products. But producing a wine with high added value is not easy.

Now more than ever, this results from a complex combination of natural and human factors; technical, financial, and commercial know-how; and a great deal of time.

When supply is limited, the value of wine, like that of art, results from the encounter of three communities that are equally demanding in terms of quality: producers, customers, and merchants. Critics, whose role is to guide consumers, also play an essential, but not exclusive, role. Counting too much on good scores can be risky. Wine is, and will remain, "the customer's child." Demanding customers look for demanding producers and demanding merchants. In other words, a demanding retailer needs to find demanding customers. For the others, price will be the only consideration.

To my mind, five, more or less interactive parameters determine a wine's value in the consumer's mind: its image, its price, its taste, its ageworthiness, and its respect for the environment in which it was made. Each of these is necessary, but none is sufficient in and of themselves.

Image is essential: it speaks to the imagination and the emotions. It takes a long time to build, but can be destroyed ever so quickly, particularly when the winemaker does not know how to keep up with the times. A wine's image must absolutely be gratifying for the person who drinks it, buys it, or offers it as a gift. To stay contemporary, that image must evolve. Value is not created in museums.

Price must remain as stable as possible, in line with the quantities on offer and with the target market. Unforeseen variations, whether upward or downward, can destroy a market and ruin a wine's image. A price that drops too low can turn off some consumers, and further decreases can cause demand to drop even more. But to what point? Don't the cheapest wines have a tendency to get cheaper and cheaper?

Typicity is the key factor in wine value. It can be defined as the ability to illustrate a type. It is the quality of a wine that is characteristic of its varietal, its region, or its vineyard. It is not just something that a winery or even a wine region can claim on its own. First and foremost, it is based on how wine lovers think of the wine. Obviously, taste is the linchpin of typicity, provided that the wine has several essential attributes: it is easily recognizable, it is well liked by modern wine drinkers, it is sufficiently complex, and its origin can be located. In other words, it is characteristic of a geographical location and its related winemaking know-how. If it is inimitable or difficult to reproduce elsewhere, it takes on the enviable status of a great wine: an esthetic creation by the winemaker and a delicious experience for the wine lover. The typicity of a wine's taste, just like its image, must be contemporary. Today's Champagnes and top Bordeaux wines are different from those of generations past. That does not impair their typicity in any way. We make the wines for our own times, for the palates of our contemporaries, using current knowledge and resources.

The embodiment of terroir is thus the taste of the wine itself. Such a wine is difficult, if not impossible, to produce anywhere else. As long as this typicity has not been demonstrated, terroir can be no more than virtual, hypothetical, and provisional. At worst, it is just a manner of speaking, an imaginary bulwark against competition both near and far. It is people, of course, who reveal the terroir of a vineyard or region. It takes decades of ongoing interest, effort, and success to establish the existence of a terroir in a given vineyard or appellation. This means the demonstration of a distinctive, characteristic taste that has been associated with this terroir, and with none other, for a long period of time. Sometimes, thanks to the genius of local growers and winemakers, history can move faster. For example, the Marlborough Valley in New Zealand, Priorat and Ribera del Duero in Spain, and Brunello di Montalcino and Maremma in Italy have all been recent revelations. Not long ago, these regions were planted to other crops or produced everyday wines, but in less than 20 years, they have become coveted wine terroirs, thanks to the commercial success and typicity of the wines produced there.

Ageworthiness—the ability of a wine to age while developing its originality—is a decisive factor in a wine's value. Expensive wines are ageworthy wines that are characteristic of their terroir. Selling a wine of a certain value means selling its future, both in terms of its taste and its commercial value. Even the offering of great Bordeaux wines for sale before their release is termed "Futures."

The need to preserve the environment is a requirement for all of humanity, both in terms of biological survival and sustainable economic growth. While they may disagree on the urgency and the nature of solutions to implement, organizations and individuals share the same conviction. Mining operations around the globe cannot be pursued indefinitely in the same way they were done during the Industrial Revolution. Because wine is the most cultural of all beverages, loaded with myths and representations, it must be up to the expectations it elicits. Wine is the symbol of festivity and pleasure, but also conviviality, sharing, communication, and intellectual and mystical communion. In principle, its origin and its conditions of production guarantee its authenticity. However, in the future, it would not be enough to be good or even better than ever. Wine will also need to come from a well-preserved natural environment, from an intact epicurean garden. This will be an integral part of its image and an essential component of its value.

As the sociologist Claude Fischler wrote in 1998, "we are intimately, even magically, convinced that we are what we eat. Our identity is linked to what our bodies absorb. As such, if we no longer know what we eat, there is a great risk of no longer knowing who we are." When we like a well-made wine, whose taste is characteristic of its origin, we soak in the landscapes, the culture, and the history of the place where it was grown. We are suffused by the personality and the philosophy of the people who made it. When we taste a well-aged older vintage, we "absorb" and magically appropriate its wonderful ability to resist time, "to repair the ravages of years irreparable" (Racine, 1873). This is undoubtedly one of the major reasons why high-priced wines have to be ageworthy. In contrast, even the suspicion that the wine may contain environmentally unfriendly exogenous substances, such as pesticides or other chemical contaminants, whether they come from vineyard or winemaking practices, can have dramatic effects. The wine would be irrationally perceived not only as hazardous but also as immoral for our times and would thus be shunned. In other words, we are what we drink, and we want to "drink" a well-preserved natural environment. The environmental ambitions of wine producers must rise to the challenge of consumers' expectations.

Certain viticultural, climatic, and winemaking conditions are required for revelation of the local taste, or terroir. Grape varieties express their characteristics the best when they can attain full ripeness after a long vegetative cycle. Characteristic wines cannot be made from insufficiently ripe grapes or, in contrast, from overripe grapes. While the former always yield poor wines, the latter, even if they are otherwise flawless, are all alike, regardless of their origin, and will have limited ageworthiness. Characteristic, ageworthy wines are, for the most part, obtained from grape varieties grown at their northern limit in the Northern Hemisphere (or southern limit in the Southern Hemisphere). This is the case for Merlot, the Cabernet varietals and Petit Verdot in Bordeaux, Syrah in Tain l'Hermitage, Nebbiolo in the Piedmont, Chardonnay and Pinot Noir in Burgundy, Sangiovese in Tuscany, Tempranillo in Rioja, Sauvignon Blanc in Sancerre or New Zealand, etc. A wine's typicity and value can only be obtained under certain conditions. Terroir is not a privilege, nor is it a gift of nature, as is so often claimed. It is a natural handicap that has been overcome. "There are no pre-ordained great vineyards, there is only the stubbornness of civilizations," the journalist Pierre Veilletet wrote in 1992. In Bordeaux, the handicap for red varietals is the heavy rainfall of the Atlantic climate. Highly colored wines are easier to produce in a drier climate. To satisfy its export markets, Bordeaux has empirically developed high-quality red wine-growing techniques for a mild, humid climate. It has done so by selecting low-fertility soils with low water reserves (gravel, clay, and limestone) and by choosing varietals that are adapted to this water reserve, by promoting grapevine water use with a large leaf surface area, and by restricting yields. Under these conditions, production costs cannot be low. However, when the handicap is overcome and the year's climate conditions are favorable, the inimitable typicity of great Bordeaux wines will be there.

Conclusion

Winemaking, in the noblest sense of the word, consists in guiding, with the least amount of intervention, the natural process of transformation of grapes into wine, in order to reveal the inimitable taste of the place that gave birth to it and which carries its image, in such a way that charms the ever-changing community of wine lovers and wine merchants. To do this, minimalist technology is required, enlightened by precise, knowledgeable enology to support the winemaker's sensibility. It is also absolutely important to develop exemplary viticultural practices that are respectful of the environment. Lastly, we must promote a strong, positive image of wine in the consciousness of consumers, without which there can be no sustainable viticulture.

Bibliography

Fischler C. (1999) *Du vin*. Editions O. Jacob, Paris.

Racine J. (1873) *Athalie [Athaliah], Act II, scene V* (Trans. J. Donkersley). https://www.gutenberg.org/files/21967/21967-h/21967-h.htm.

Veilletet P. (1992) *Le voyage de Bordeaux à Beaune*, p. 34. L'Amateur de Bordeaux, Bordeaux.

… CHAPTER 10

The Grape and Its Maturation

10.1 Introduction
10.2 Description and Composition of the Mature Grape
10.3 Changes in the Grape During Maturation
10.4 Definition of Ripeness—Concept of Vintage
10.5 Impact of Various Other Factors on Maturation and Grape Composition at Ripeness
10.6 *Botrytis cinerea*

10.1 Introduction

The grape constitutes the raw material for producing wines. Its maturity level is the first factor, and certainly one of the most decisive ones, in determining wine quality. It is the result of all of the complex physiological and biochemical phenomena whose proper development and intensity are intricately related to biological and physiological conditions (grape variety, vine age, phenological stage) (Peynaud and Ribéreau-Gayon, 1971; Ribéreau-Gayon et al., 1975; Champagnol, 1984; Huglin, 1986; Kanellis and Roubelakis-Angelakis, 1993; Flanzy, 2000; Roubelakis-Angelakis, 2001; Downey et al., 2006; Dai et al., 2011).

Compared with other fruits, the study of the grape presents many problems. Grapevine fructification is the result of a long and complex reproduction cycle, initiated the year prior to the harvest. The ovary and then the seeds synthesize and attract the growth regulators necessary for their development from the leaves, where they are mainly synthesized. The triggering of the maturation (or ripening) process does not correspond with a true climacteric stage. It is linked to the drop in growth hormone levels because of a strong

Handbook of Enology, Volume 1: The Microbiology of Wine and Vinifications, Third Edition.
Pascal Ribéreau-Gayon, Denis Dubourdieu, Bernard B. Donèche and Aline A. Lonvaud.
© 2021 John Wiley & Sons Ltd. Published 2021 by John Wiley & Sons Ltd.

accumulation of abscisic acid, brassinosteroids, and ethylene in small amounts (Fortes et al., 2015). Consequently, the behavior of the entire plant strongly influences the development of these processes. Certain studies can be carried out on fruiting microcuttings or potted vines under controlled conditions, but the preponderant influence of environmental parameters and growing practices on vine behavior requires that a large number of experiments be carried out in the vineyard. The study of maturation therefore comes up against difficulties due to the extreme variability of berry composition, at any given time and for the same variety.

In spite of these difficulties, the observations made each year by researchers from the *Institut des Sciences de la Vigne et du Vin* (Institute of Science of the Vine and Wine), as well as by other teams in different wine-producing regions, have helped us:

- follow and compare the chemical composition modifications of the grape during maturation;
- compare the maturation kinetics over the years, as a function of weather conditions;
- compare the evolution of different vineyards, in terms of local environmental conditions;
- forecast maturity dates and thus establish the harvest dates.

These preliminary observations directed subsequent research toward a more thorough study of maturation mechanisms and prediction of the maturity date. This chapter will cover the biochemical phenomena characterizing grape maturation and the process of development of mold. It will also focus on the influence of environmental factors on maturation and the impact of the latter in a context of climate change.

10.2 Description and Composition of the Mature Grape

10.2.1 The Berry

The grape is a berry, classified in the group of multi-seeded fleshy fruits. The berries are organized into a cluster. Each berry is attached to the rachis (or stalk or stem) by a small pedicel containing the vessels that supply the berry with water and nutrients (Figure 10.1a). Cluster structure depends on the length of the pedicels: if they are long and thin, the grapes are spread out (Figure 10.1b); if they are short, the bunches are compact, and the grapes are packed together. Varieties used for winemaking often belong to the latter category. Cluster compactness is one of the factors affecting rot sensitivity.

Genetic factors and environmental conditions that characterize berry formation greatly influence its development and its composition at ripeness.

10.2.2 Berry Formation

Obtaining the fruit is a long process, taking place over two consecutive years (Figure 10.2): the first corresponds to the formation of an inflorescence used for flower formation, and the second, the harvest year, corresponds to fruit development. While the first stage takes place in the latent bud the year before the harvest and cannot be seen by the winemaker, the second is visible after bud break. Very early, the potential yield can be estimated, and viticultural practices can be oriented in order to facilitate fruit development.

Fruit development is closely related to the conditions of inflorescence formation and then to the quality of ovule fertilization.

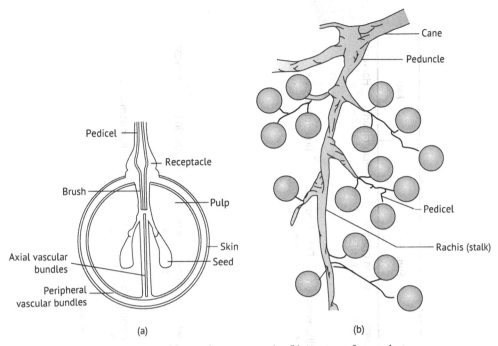

FIGURE 10.1 Fruit of the grape vine: (a) grape berry at maturity; (b) structure of grape cluster

Inflorescence formation processes are complex and correspond to the induction, initiation, and differentiation phases of inflorescence primordia in latent buds. These stages, which occur during the year preceding the harvest, will determine the number of clusters per bud, as well as, in part, the number of flowers per cluster. These early stages are the main yield-building stages and may, depending on weather conditions and grape variety, explain up to 60% in annual yield variation (Guilpart, 2014). These phases are highly dependent on environmental factors (water stress or light conditions) but also on trophic factors such as carbon or nitrogen nutrition, as well as hormonal balances (Li-Mallet et al., 2016) (Figure 10.3).

Just before and just after bud break, the development of inflorescences resumes. The floral induction and initiation stages lead to the formation of all flower parts (Vasconcelos et al., 2009). Starting at the stage with five unfolded leaves, inflorescences are visible at the shoot tip. We have to wait for the stage with 15–20 unfolded leaves to have fully formed flowers. Like for the previous stage, controlling these mechanisms is multifactorial, but the two main factors are apparently temperature (Keller et al., 2010) and carbon availability (Lebon et al., 2008), resulting either from mobilization of reserves or photosynthesis, depending on the stage in question.

Flowering (bloom) corresponds to the opening of the corolla and the ejection of the calyptra (anthesis). The pollen deposited on the stigma will germinate and release a pollen tube reaching the ovary, which is fertilized. Flower opening and the liberation of pollen are facilitated by warm, dry weather. In a cool and humid climate, flowering can be spread out over 10–15 days and sometimes more. This stage is very sensitive to weather conditions. The presence of rain will disturb the ejection of

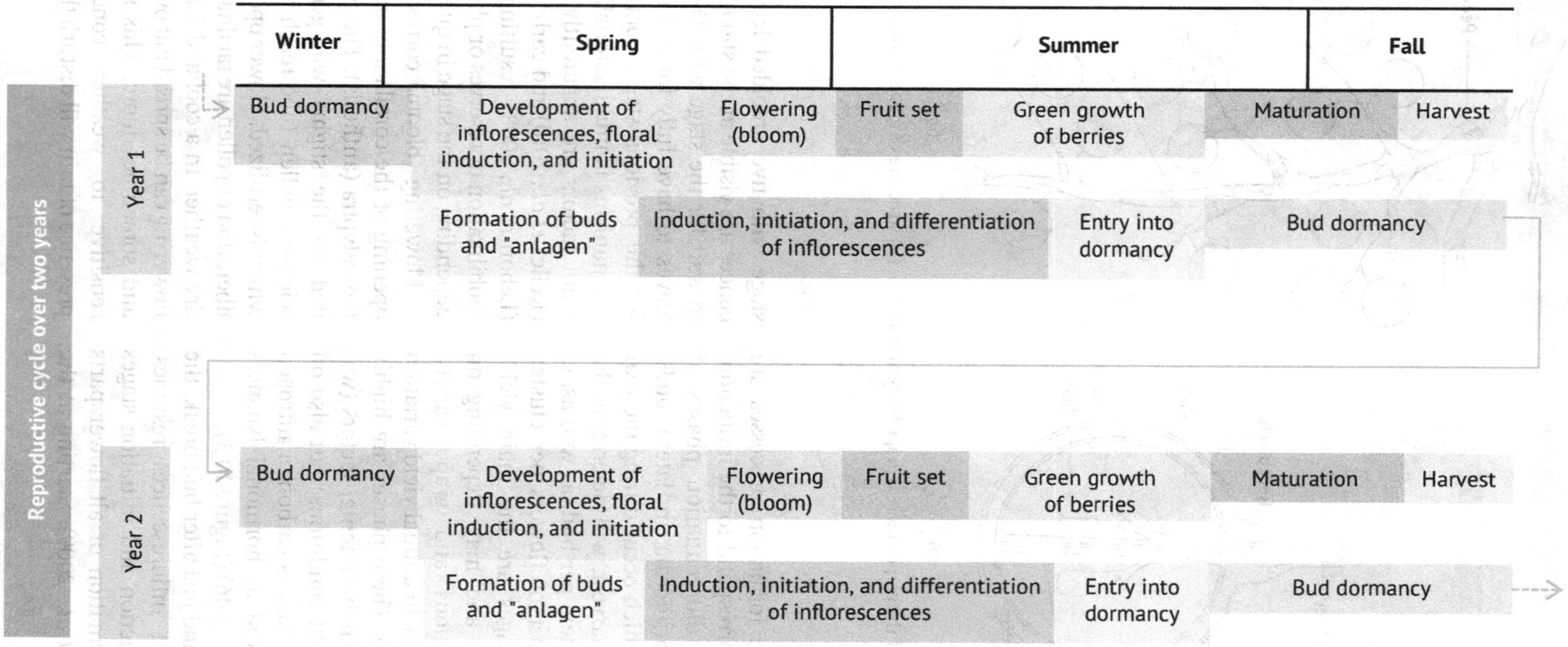

	Winter	Spring	Spring	Summer	Summer	Fall	Fall
Year 1	Bud dormancy	Development of inflorescences, floral induction, and initiation	Flowering (bloom)	Fruit set	Green growth of berries	Maturation	Harvest
		Formation of buds and "anlagen"	Induction, initiation, and differentiation of inflorescences		Entry into dormancy	Bud dormancy	
Year 2	Bud dormancy	Development of inflorescences, floral induction, and initiation	Flowering (bloom)	Fruit set	Green growth of berries	Maturation	Harvest
		Formation of buds and "anlagen"	Induction, initiation, and differentiation of inflorescences		Entry into dormancy	Bud dormancy	

Reproductive cycle over two years

FIGURE 10.2 Grapevine reproductive cycle. *Source*: Adapted from Carmona (2008); Coombe and Iland (2004), and Carbonneau (1992).

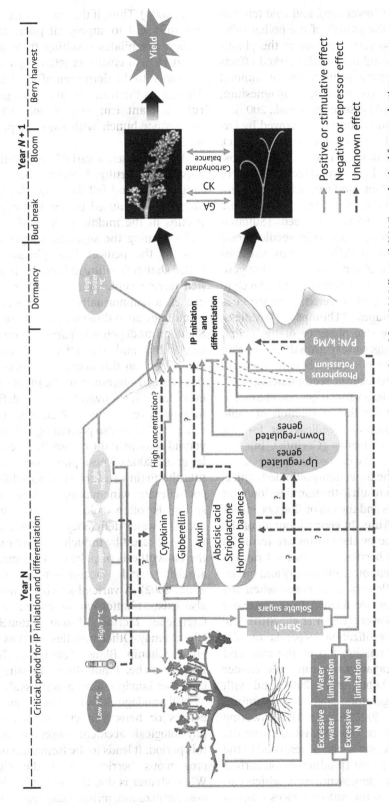

FIGURE 10.3 Factors controlling the induction and initiation phases of inflorescences. *Source:* Adapted from Li-Mallet et al. (2016)—authorized for reproduction.

the calyptra (flower cap), and cold temperatures limit the growth of the pollen tube, which is also very sensitive to the plant's nutritional condition, with marked effects in case of nitrogen limitation or mineral imbalance (potassium, magnesium, molybdenum) (Vasconcelos *et al.*, 2009).

Pollination is normally followed by fertilization of the ovules (one to four), enabling the development of a berry possessing one to four normal seeds. The berry is said to be "set" and can start its growth. However, poor fertilization can lead to the formation of rudimentary seeds (Sultana-type or stenospermocarpic seedlessness) leading to growth deficits in most varieties of grapes. The absence of fertilization generally leads to the berry failing to develop; this is "shatter." However, in some varieties of table grapes (Thompson Seedless, Corinth), the sole presence of the pollen tube may trigger ovary growth and enable the production of seedless berries (Corinth-type or parthenocarpic seedlessness). Seedlessness can be a varietal genetic character, sought after for the production of table grapes (Thompson Seedless) or for the preparation of raisins (Corinth), but not sought after in the case of wine grapes.

Most of the time, non-pollinated, unfertilized ovules result in the non-development of the ovaries and the fall of flowers and/or very young fruits. In general, not all of the flowers borne by the cluster are fertilized and become berries. The fruit set ratio is very dependent on grape variety and on the number of flowers. It is lower when the inflorescences are formed from a greater number of flowers (Keller *et al.*, 2010). The success of fertilization depends on the source–sink relationship in the vine and the plant's capacity to supply the cluster with sugar (Lebon *et al.*, 2008) and with hormone regulators, such as polyamines (Colin *et al.*, 2002; Aziz 2003). Flowering often takes place when switching from the heterotrophic phase (use of reserves) to the autotrophic phase (production of carbon substrates via photosynthesis), which can limit supply to the inflorescences (Zapata *et al.*, 2004). Thus, if the leaf surface area is not sufficient to supply all plant organs, carbon assimilates resulting from photosynthesis will ensure vegetative growth as a priority, to the detriment of fertilization. This limits the fruit set ratio. As a general rule, a plant can only supply 100–200 berries per bunch with sugar, depending on the variety.

After fruit set, a variable proportion of apparently fertilized young berries no longer grow and fall from the plant. This abscission is caused by the hydrolysis of pectins in the middle lamella of the cell walls forming the separation layer at the base of the pedicel. The phenomenon, called shatter (*coulure* in French), is amplified during cold, cloudy weather, which causes an abnormally long flowering—sometimes up to three weeks (Figure 10.4). Shatter thus depends in particular on sugar availability and the effects of weather parameters on this availability (photosynthesis, sugar migration in the plant). However, unfertilized ovaries are also deficient in growth regulators (polyamines) (Colin *et al.*, 2002) and also present significant flavonoid content at the columella level (Cholet *et al.*, 2002). These phenomena constitute the main cause of yield variability in cool-climate winemaking regions where spring is often cold. In warm-climate zones, a water deficiency can bring about the same result by inducing a limitation of the biosynthesis of polyamines because of abscisic acid accumulation (Colin *et al.*, 2002). A varietal-specific sensitivity also exists. Shatter can be complete with Grenache, Merlot, Muscat Ottonel, or Chardonnay. Other varieties, such as Carignan, Chenin Blanc, Sauvignon Blanc, Folle Blanche, Pinot Blanc, Riesling and Cabernet Sauvignon, are much less affected.

Millerandage (also known as shot berries or hens and chicks) is another physiological accident observed during this period. It leads to the formation of heterogeneous berries within the cluster. While shatter is due to non-pollination or non-fertilization, millerandage resembles a

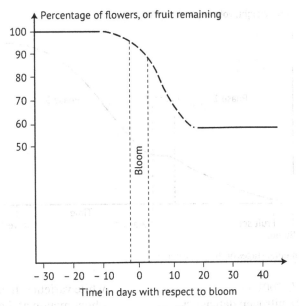

FIGURE 10.4 Flower and fruit evolution during bloom and fruit set (Bessis and Fournioux, 1992).

defective pollination that leads to the formation of an abnormal seed. Therefore, the total growth of the berry cannot be ensured. The limitation of ovarian tissue development could be due to the high activity of an enzyme called polyamine oxidase, which is likely to produce H_2O_2. The latter blocks post-set cell division mechanisms (Colin et al., 2002), thus leading to small berries that remain green.

10.2.3 The Developmental Stages of the Grape

In the course of its development, from ovary to ripe fruit, the grape follows an evolution common to all berries. It is generally divided into three phases (Figure 10.5), taking into consideration parameters such as berry diameter, weight, and volume:

1. An initial rapid growth or green growth phase lasting 45–65 days, depending on grape variety and environmental conditions. The intensity of cell division and then growth depends on the existence of seeds. Growth hormone concentrations (cytokinins and gibberellins) correspond directly with the number of seeds. The application of gibberellic acid on seedless grapes has become a common viticultural practice (Ito et al., 1969). Cell growth begins about two to three weeks after fertilization and continues until the end of the first phase (Ollat et al., 2002). In the course of this first period, chlorophyll is the predominant pigment. The berries have an intense metabolic activity, characterized by an elevated respiratory intensity and a rapid accumulation of acids. The first growth phase is regulated by several factors, including the number of seeds, the source–sink relationship and environmental factors, such as water, temperature, and light. This photosynthetic phase is very sensitive to temperature, which favors cell divisions and growth when it increases. Water and nitrogen availability also affect these two physiological processes

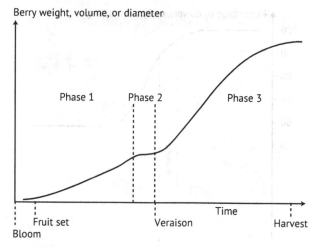

FIGURE 10.5 Developmental stages of the grape berry.

(Ojeda *et al.*, 2001; Ollat *et al.*, 2002): nitrogen deficiency seems to mainly affect cell division, while water stress limits cell expansion.

2. A slower growth phase during which veraison (berry color change) occurs. Veraison is characterized by the appearance of color in colored varieties and a translucent skin in white varieties. It is an abrupt phenomenon at the berry level but takes place over several days when different berries of the same bunch are considered. In a vineyard parcel, this phase lasts 8–15 days or longer if flowering is very slow. It corresponds with the depletion of growth substance synthesis and an increase in the concentration of abscisic

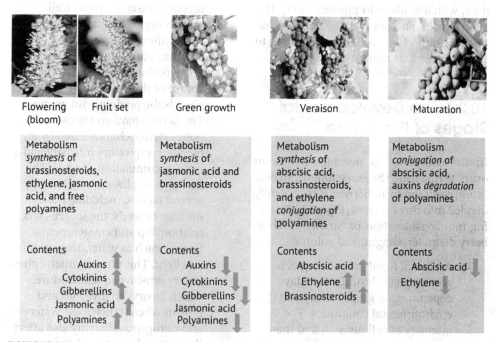

FIGURE 10.6 Hormone balance during berry development. *Source*: Adapted from Fortes *et al.* (2015).

acid, indole-acetic acid, and ethylene (Chervin *et al.*, 2004; Deytieux-Belleau *et al.*, 2007a; Fortes *et al.*, 2015) (Figure 10.6).

3. A second growth phase corresponding to maturation. Cell growth resumes and is accompanied by diverse physiological modifications. The respiratory intensity decreases, whereas certain enzymatic activities sharply increase. This final period lasts 35–55 days, during which the grape accumulates free sugars, cations such as potassium, amino acids, and phenolic compounds, while concentrations of malic acid and ammonium decrease. Grape size at ripeness depends largely on these accumulation processes and is thus very sensitive to carbon availability and source–sink relationships and to all factors affecting photosynthesis (Ollat *et al.*, 2002). The final weight also depends on the number of cells per berry. There is a very close relationship between the dimensions of a ripe grape and the number of seeds it contains (Table 10.1). Finally, the final volume of the berry may vary with the number of berries. Due to significant growth during the first phase, a higher weight can compensate for a low number of berries. A greater assimilation of carbon- and nitrogen-containing assimilates due to low competition between berries enables this phenomenon (Keller *et al.*, 2010).

10.2.4 Grape Morphology

Each grape comprises a group of tissues (the pericarp) surrounding the seeds. The pericarp is divided into the exocarp (the skin), the mesocarp (the pulp), and the endocarp (the tissue that lines the seed receptacles containing the seeds but is not distinguishable from the rest of the pulp) (Figure 10.1a). The fruit is nourished by a branching vascular network of the rachis, which traverses the pedicels. This vascular bundle then branches out in the pulp to form central bundles, which run along the columella; seed bundles, which vascularize the seeds; and peripheral bundles located just under the skin. This network can be observed due to the transparency of certain white varieties at ripeness.

The *skin* of the grape forms a heterogeneous region composed of the cuticle, the epidermis, and the hypodermis (Figures 10.7 and 10.8). The cuticle is a continuous layer whose thickness varies depending on the variety: 1.5–4 µm for certain *Vitis vinifera* varieties and up to 10 µm for certain American vines. It begins to develop three weeks before anthesis. In the course of berry maturation and development, it becomes increasingly disorganized, and its thickness diminishes. The cuticle is generally covered by epicuticular wax (bloom) in the form of stacked platelets, visible by electron microscopy (like leaves piled on top of each other). Wax thickness is relatively constant throughout the course of berry development (about 100 µg of wax/cm² of surface).

TABLE 10.1

Relationship Between Number of Seeds and Berry Size at Ripeness: Merlot Grapes Sampled in 1982 from a Saint-Emilion Vineyard (France)

Number of seeds	Berry weight (g)	Juice volume per berry (ml)	Sugar concentration (g/l)
0–1	1.10	0.75	235
2	1.55	1.01	233
3	1.94	1.12	221

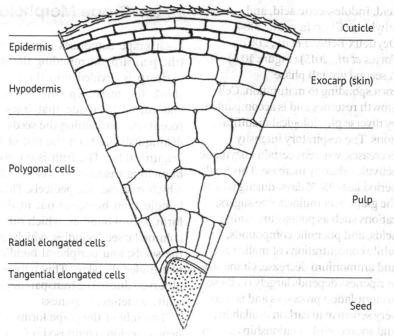

FIGURE 10.7 Different grape berry tissues at maturity.

The epidermis is composed of one or two layers of tangentially elongated cells whose thickness varies depending on the grape variety. The hypodermis comprises several tangentially stretched layers whose cell walls are more or less thick depending on the variety, development stage, and depth. In these cells, anthocyanins and tannins can be found.

The *pulp* is composed of large polygonal cells with very thin, distended cell walls. There are 25–30 cell layers, organized into three distinct regions.

Each normally constituted *seed* has a cuticle, an epidermis, and two coats (external and internal) covering the endosperm and the embryo.

Grape berry consistency depends on the number of skin cell layers and on the cell wall thickness of the pulp and skin. Generally, table grape varieties produce plump, thin-skinned grapes (the pulp having thick cell walls), whereas wine varieties have tough skins and juicy pulp (pulp with thin cell walls).

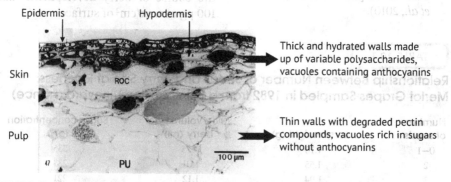

FIGURE 10.8 Structures of the grape berry epidermis, hypodermis, and mesocarp during maturation.
Source: Adapted from Park (1995)—authorized for reproduction.

On the grape surface, there are between 25 and 40 stomata per berry, depending on the variety. After veraison, these stomata no longer function, and they necrotize. Rapid fruit enlargement creates tension, resulting in the development of peristomatic microfissures.

10.2.5 Grape Cluster Composition at Ripeness

Stalks

The stalk, or rachis, represents around 3–7% of the weight of a ripe grape cluster. Its chemical composition is similar to the composition of leaves. It contains little sugar (less than 10 g/kg) and an average acid concentration (180–200 mEq/kg). These acids are in the form of salts, due to the large quantity of cations present.

Stalks are rich in phenolic compounds. They can contain up to 20% of the total phenolic compound concentration of the grape cluster, even though they represent a lower proportion of the total weight. These phenolic compounds are more or less polymerized and have a very astringent taste.

The stalk attains its definitive size around the time of veraison. Although it loses most of its chlorophyll, it remains green during maturation. It often becomes completely lignified well after maturity.

Seeds

The seeds represent 0–6% of berry weight. They contain carbohydrates (35% on average), nitrogen compounds (around 6%), and minerals (4%). An oil can be extracted from the seeds (15–20% of the total weight), which is essentially oleic and linoleic acids. The seeds are an important source of phenolic compounds during red winemaking. Depending on the varieties, they contain between 20 and 55% of the total polyphenols of the berry.

The seeds attain their definitive size before veraison. At this time, they have reached physiological maturity. During maturation, the tannin concentration of the seeds decreases, whereas their degree of polymerization increases. Conversely, the nitrogen compounds are partially hydrolyzed. The seed can yield up to one-fifth of its nitrogen to the pulp while still remaining richer in nitrogen than the other solid parts of the grape cluster.

Skin

Depending on the grape variety, the skin represents from 8 to over 20% of berry weight. Being a heterogeneous tissue, its importance depends greatly on the extraction method used. Separating the skins by pressing the grapes results in the extraction of the pulp and seeds. This method corresponds best to current enological practices. The sugar concentration of skin cells is very low. For the same weight, the skin is as rich in acids as the pulp, but citric acid is predominant. Malic acid, found in significant quantities in the skins of green grapes, is actively metabolized in the course of maturation. The majority of tartaric acid is esterified by phenolic acids (caffeic, coumaric). A significant quantity of cations cause the salification of these acids. The contents of the skin cells always have a higher pH than the pulp.

The skin is especially characterized by significant quantities of secondary products of major enological importance (phenolic compounds and aroma substances). It accumulates these substances during maturation.

The following phenolic compounds are present in the grape skin at ripeness: benzoic and cinnamic acids, flavonols, and tannins. They are distributed in the cells of the epidermis and the first subepidermal layers in both white and red grapes. In addition, the red grape skin contains anthocyanins, essentially located in the hypodermal cell layers. Exceptionally, in certain vintages, the cells adjacent to the pulp can be colored. The pulp itself is colored in the case of Teinturier varieties and some American vines or direct producer hybrids. Anthocyanin composition varies from cultivar to cultivar, depending on the anthocyanidin substitution and their heterosidic nature (see Volume 2, Section 6.2.3).

The ripe grape skin also contains considerable amounts of aroma substances and aroma precursors, corresponding to many chemical families of compounds. Their characterization is constantly getting deeper (monoterpenes, carotene derivatives, cysteine and glutathione S-conjugates, methoxypyrazines, etc.) (Volume 2, Chapter 7).

Finally, the skin is covered by epicuticular wax, essentially composed of oleanolic acid.

All of this information is very important from a winemaking point of view. All methods increasing the solid–liquid exchanges for better color extraction or aroma dissolution should be favored during winemaking.

Pulp

The pulp represents the largest fraction of the berry by weight (from 75 to 85%). The vacuolar contents of its cells contain the grape juice—with the solid parts (cytoplasm, pectin- and cellulose-containing cell walls) constituting less than 1% of this tissue. The juice is a cloudy liquid, generally slightly colored, having a high specific gravity (1.065–1.100 on average), due to the many chemical substances that it contains. Sugars are the primary constituents—essentially glucose and fructose. Fructose is always predominant (the glucose/fructose ratio is around 0.9). Sucrose, which is the migratory form of sugar in the plant, exists in only trace amounts in the grape. Other sugars have been identified in the grape: arabinose, xylose, rhamnose, maltose, raffinose, etc. (Volume 2, Section 3.3.1). The reducing sugar concentration in normal ripe grapes varies from 150 to 240 g/l.

Most of the acids of the metabolism are found in trace amounts in ripe grape pulp (pyruvic, α-oxoglutaric, fumaric, galacturonic, shikimic, etc.). Must acidity, an important element of enological data, is essentially constituted by three acids: tartaric, malic, and citric (Volume 2, Section 1.2.2). It can vary from 3 to 10 g/l expressed as sulfuric acid or from 4.5 to 15 g/l expressed as tartaric acid, depending on the cultivar, the climate, and grape ripeness. Phosphoric acid is the preponderant inorganic anion.

The pulp is particularly rich in cations. Potassium, the principal element, is much more abundant than calcium, magnesium, and sodium. The other cations are present in much lower concentrations, with iron representing 50% of the remaining cations. Concentrations of metal trace elements such as lead are infinitesimal, except in the case of accidental pollution. In spite of this high cation concentration, part of the acids remains unsalified. Must pH commonly ranges between 2.8 and 3.5.

The pulp contains only 20–25% of the total nitrogen content of the berry. The must contains 40–220 mM of nitrogen in its ammoniacal or organic form. The ammonium cation is the most easily assimilable nitrogen source for yeasts and is often present in sufficient quantities (Volume 2, Section 5.2.2). The amino acid fraction varies from 2 to 13 mM in leucine equivalents (2–8 g/l). Most amino acids are found in grape must at variable concentrations, and a few of them (proline, arginine, threonine, and glutamic acid) represent nearly 90% of the total concentration. The relationship between the must amino acid concentration and its organic acid concentration has been known for a long time. The most acidic grapes are always the richest in amino acids. Soluble proteins of the must represent 1.5–100 mg/l.

At ripeness, the grape is characterized by a low concentration of pectic substances with respect to other fruits. Pectins represent from 0.02 to 0.6% of fresh grape weight. Differences between varieties and from year to year can be significant. Only the free pectic fraction, associated with diverse soluble sugars, is likely to be found in must. This fraction also contains small amounts of insoluble proteins.

The skin is considered to be the principal source of aroma substances, but the pulp does contain significant concentrations of these compounds, within a given chemical family (e.g. cysteine *S*-conjugates) (Volume 2, Section 7.5).

There is considerable heterogeneity between different grapes on the same grape cluster. Similarly, the diverse constituents of must are not evenly distributed in the pulp. As a primary technological consequence, the chemical composition of the juice evolves during the course of grape pressing in white winemaking. The peripheral and central zones (near the seeds) are always richer in sugar than the intermediate zone of the pulp (Figure 10.9). In contrast, malic and tartaric acid concentrations increase from the periphery toward the interior of the berry. Potassium is distributed differently within the grape and often causes the salification of the acids, with the precipitation of potassium bitartrate, in the course of pressing. This heterogeneity seems to apply to all must constituents. Finally, the distal half of the grape

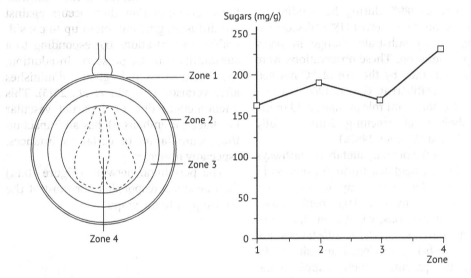

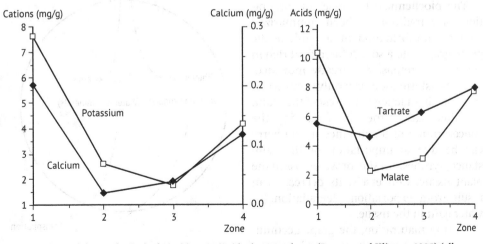

FIGURE 10.9 Breakdown of principal constituents inside the grape berry (Possner and Kliewer, 1985) (all results are expressed in mg per g fresh weight): (a) zones; (b) sugars; (c) cations; (d) acids.

(farthest from the pedicel) is generally richer in sugars and poorer in acids than the proximal half.

10.3 Changes in the Grape During Maturation

10.3.1 General Characteristics of Maturation

As early as 1897, during his studies on grape respiration, Gerber (1898) discovered a respiratory substrate change in berry pulp at veraison. These observations were later confirmed by the use of ^{14}C-marked molecules (Ribéreau-Gayon, 1960). At present, the dominant role of malic acid in the metabolism of ripening fruit is fully established (Ruffner, 1982a).

Most of the primary metabolic pathways have been elucidated through progress in the extraction and study of numerous enzymatic activities. High-performance analytical methods, capable of determining nanograms of volatile substances, are currently being developed and should be able to provide much supplemental information on the secondary metabolism of grapes in coming years.

The biochemical processes of maturation have traditionally been summarized by the transformation of a hard, acidic green grape into a soft, colored fruit rich in sugar and aromas. As already indicated, these transformations can only occur when the grape is attached to the rest of the plant. In this case, the increase in the concentration of a substance in the berry can be due to importation of this substance, synthesis *in situ*, or water loss in the plant tissue. Conversely, its decrease can result from exportation, degradation, or water gain in the tissue.

During maturation, the grape accumulates a significant quantity of solutes, principally sugars. In spite of berry enlargement (cell enlargement), the percentage of solid material increases—indicating that the solutes are imported in greater quantities than water. The amount of water that accumulates each day in the grape is the sum of the phloem (elaborated) and xylem (raw) sap flux minus the water loss due to transpiration (Figure 10.10). At the start of maturation, the berries simultaneously import water with the sugars, but the amount of water transpired rapidly diminishes as the stomata degenerate; then, transpiration occurs exclusively across the cuticular wax. Sugar accumulation then occurs against the diffusion gradient, often up to considerable concentrations corresponding to a substantial osmotic pressure. In addition, the xylem sap supply greatly diminishes after veraison (Knipfer et al., 2015). This phenomenon, due to a partial vascular blockage (or embolism), has an impact on the accumulation of certain substances, especially minerals.

The peripheral network (Figure 10.1a) then becomes responsible for most of the food supply to the grape.

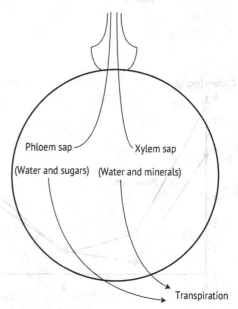

FIGURE 10.10 Grape berry nutrient supply (Coombe, 1989).

The grape is more than an accumulation organ: it maintains an intense activity (respiration and biochemical transformations) during maturation. Veraison also corresponds to the synthesis of new enzyme activities and the release of inhibition on other activities. These variations in gene and protein expression cause profound changes in grape metabolism (Robinson and Davies, 2000; Deytieux-Belleau et al., 2007b).

10.3.2 Sugar Accumulation

The most spectacular biochemical phenomenon of maturation is certainly the rapid accumulation of sugars in the grape from veraison onward. From the start, the inflorescences, due to their growth hormone concentrations, have a strong demand for the products of photosynthesis. However, during the entire green growth phase, the sugar concentration of green grapes does not exceed 10–20 g/kg in fresh weight (around the same as leaves). The sugars imported daily are metabolized at a high intensity for fruit development but in particular for seed growth and maturation. The supply of nutrients toward the grape is even more considerable in the days that precede veraison.

The depletion of growth hormones, notably auxins, and the increase in abscisic acid concentrations correspond with the lifting of the inhibition of the principal enzymatic activities involved in the accumulation of sugars in pulp cell vacuoles. Sucrose phosphate synthase, sucrose synthase, and hexokinase are no longer blocked (Ruffner et al., 1995; Pan et al., 2005). Abscisic acid also stimulates gene expression of sucrose transporters (Cakir et al., 2003). This accumulation occurs against the diffusion gradient. The transport requires energy to counter the growing osmotic pressure as the sugar concentration increases. Up to 30 bar of pressure can be attained toward the end of maturation. An enzymatic complex associated with the tonoplast of the pericarp cells ensures this transport (Figure 10.11). The sugars, synthesized in the leaves, migrate exclusively in the form of sucrose through the phloem to the grapes (Lavee and Nir, 1986). A first invertase, linked to the plasma membrane, hydrolyzes the imported sucrose into glucose and fructose. The free sugars are then phosphorylated by the cytoplasmic hexokinase. After the formation of UDP-glucose, the sugars combine again to form sucrose phosphate with the help of sucrose phosphate synthase. The energy accumulated in this molecule and liberated by a sucrose phosphate phosphatase linked to the tonoplast enables the accumulation of sugars in the vacuole. These enzymes maintain a sufficient activity during the entire maturation process to accumulate a maximum of 2.5 μmol of sugars per hour per berry. However, the direct transfer of cytoplasmic hexoses via tonoplastic transporters cannot be excluded (Robinson and Davies, 2000).

Profound changes in the metabolic pathways also occur at veraison, facilitating the storage of imported sugars. The study of changes in respiration during grape development provides information on these changes. The respiratory intensity increases in proportion to cell multiplication during the first growth phase. It then remains relatively stable until ripeness (Figure 10.12). It does not increase during maturation, as in many other fruits. The most active respiratory sites simply change location. Before veraison, the pulp and in particular the seeds are primarily responsible for respiration, but during maturation the respiratory activity is highest in the skins. The respiratory quotient (ratio between the carbon dioxide released and the oxygen consumed) changes at veraison, indicating a change in the respiratory substrate. During the entire green growth phase, the respiratory quotient remains near one.

In reality, the respiratory quotient of the pericarp of green grapes is slightly higher than one, whereas it is near 0.7 for the

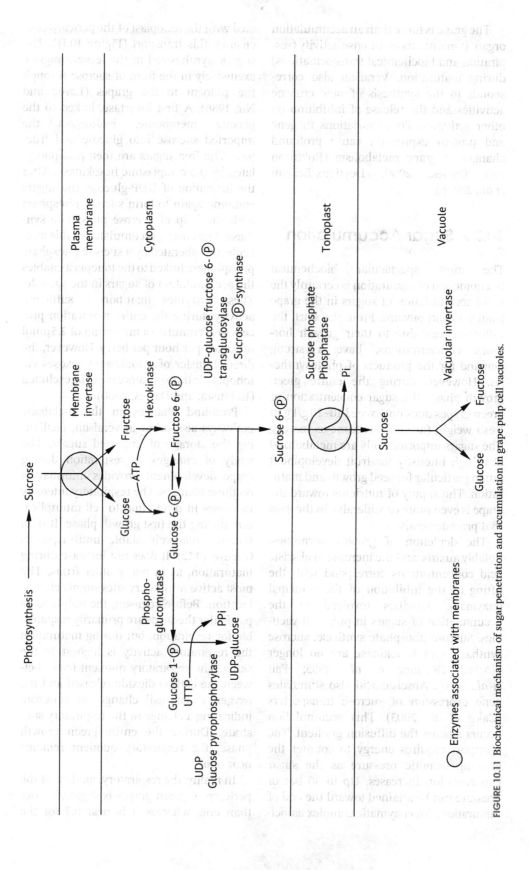

FIGURE 10.11 Biochemical mechanism of sugar penetration and accumulation in grape pulp cell vacuoles.

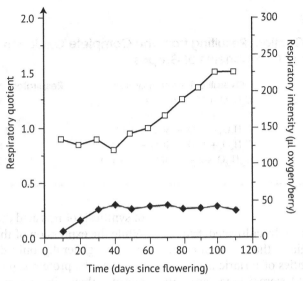

FIGURE 10.12 Evolution of respiration during grape development (Harris et al., 1971). □, respiratory quotient; ♦, respiratory intensity.

seeds. Seeds are rich in fatty acids, which are most likely their respiration substrate. In the pericarp, on the other hand, this quotient results from the combustion of sugars, primarily, but also organic acids (Table 10.2). After veraison, the respiratory quotient increases, reaching 1.5 toward the end of maturation. On the whole, it can logically be considered that the grape essentially uses organic acids as its respiratory substrate during maturation.

Supplementary information on metabolic pathway modifications is provided by observing the evolution of the glucose/fructose ratio during grape development. Sucrose is the principal transport form of photosynthesis products; this ratio should be near one in the grape where these products accumulate. Yet in the green grape at the start of development, glucose predominates and represents up to 85% of grape reducing sugars. This ratio, near five, decreases to two at veraison and then to one at the beginning of maturation. It then remains relatively constant until ripeness (between 1.0 and 0.9). Since glucose is more likely to enter into cell respiration than fructose, the latter preferentially enters into cell synthesis reactions. This phenomenon explains the high glucose/fructose ratio during the green growth phase of the grape and its decrease after veraison, related to a slowing of biosynthetic activity.

10.3.3 Evolution of Organic Acids

From the start to the end of its development, the grape contains most of the acids involved in the glycolytic and shikimic acid pathways as well as in the Krebs and glyoxylic acid cycles. This attests to the functioning of these different pathways. However, their concentrations are generally very low. Tartaric and malic acids represent on average 90% of the sum of the acids. These two acids are synthesized in the leaves and grapes, with a majority produced in the grapes prior to veraison. There is no formal proof of the transport of these diacids from the leaves to the grapes (Ruffner, 1982b).

In spite of their chemical similarity, these two acids have very different metabolic pathways. Their evolution is not identical during grape development and maturation. The malic acid/tartaric acid ratio varies considerably according to the grape variety and the maturation conditions.

TABLE 10.2
Respiratory Quotient Resulting from the Complete Oxidation of the Principal Components of Grapes

Substance	Overall oxidation reaction	Respiratory quotient value
Glucose or fructose	$C_6H_{12}O_6 + 6O_2 \rightarrow 6CO_2 + 6H_2O$	1
Malic acid	$C_4H_6O_5 + 3O_2 \rightarrow 4CO_2 + 3H_2O$	1.33
Citric acid	$C_6H_8O_7 + 4.5O_2 \rightarrow 6CO_2 + 4H_2O$	1.33
Tartaric acid	$C_4H_6O_6 + 2.5O_2 \rightarrow 4CO_2 + 3H_2O$	1.60
Fatty acid	$C_{18}H_{36}O_2 + 26O_2 \rightarrow 18CO_2 + 18H_2O$	0.70

Tartaric acid

The grape is the only cultivated fruit of European origin that accumulates significant quantities of tartaric acid. Specifically, the L-(+) stereoisomer accumulates in the grape, attaining 150 mM in the must at veraison and from 25 to 75 mM in the must at maturity (3.8–11.3 g/l).

This acid is a secondary product of the metabolism of sugars and is synthesized from ascorbic acid. The evolution of tartaric acid is similar to that of ascorbic acid in the berry (Melino et al., 2011; Cholet et al., 2016).

The synthesis of tartaric acid is complex and split into two steps: The first step enables the synthesis of ascorbic acid (Smirnoff–Wheeler pathway), which involves a series of enzymes partially regulated by the cultivar and weather conditions. The second step involves the transformation of ascorbic acid via an enzyme called L-idonate dehydrogenase, regulated by ascorbic acid content and weather conditions (Cholet et al., 2016).

The kinetics of tartaric acid during grape development and maturation are consistent with this dual pathway hypothesis (Cholet et al., 2016). The green growth phase is characterized by a rapid accumulation of tartaric acid, related to intense cell multiplication and equally intense photosynthesis (Melino et al., 2011; Cholet et al., 2016). During maturation, the tartaric acid concentration remains relatively constant in spite of the increase in berry volume. Therefore, we cannot totally refute the hypothesis of a low level of synthesis of this acid during this period. While the expression of the *VvIdnDH* transcript is generally only detectable before veraison, the protein is present in the cytoplasm, then in vacuoles during all development phases of the berry (Wen et al., 2010). Under some weather conditions (rainy late summer leading to growth stoppage), the expression of the *VvIdnDH* transcript may again be detected in relationship with a new accumulation of tartaric acid (Cholet et al., 2016). The small variation in tartaric acid content after veraison seems very dependent on vintage conditions and related to the plant's water supply. Conversely, there is no formal proof of its catabolism during maturation.

Malic acid

Malic acid is a very active intermediate of grape metabolism. The vine contains the L-(−) isomer. The vine assimilates carbon dioxide in the air by a C3 mechanism (Ruffner et al., 1983). In this manner, during the dark phase of photosynthesis, the leaves and young green grapes fix CO_2 on ribulose 1,5-bisphosphate to produce phosphoglyceric acid, which condenses to form hexoses and may also become dehydrated into phosphoenolpyruvic acid. CO_2, catalyzed by PEP carboxylase, is fixed on this acid to form oxaloacetic acid, which is, in turn, reduced into malic acid.

The significant malic acid accumulation during the green growth phase of the grape (up to 15 mg/g fresh weight, or about 95 µmol/berry) is due in part to this mechanism. However, a non-negligible proportion results from its direct synthesis by the carboxylation of pyruvic acid. This reaction is catalyzed by the malic enzyme, whose activity is very high before veraison.

In any case, the imported sugars are the precursors of the malic acid found in grapes. Malic acid is produced by either catabolic pathways (glycolysis, pentose phosphate pathway) or by β-carboxylation.

Grape maturation is marked by a significant change in malic acid metabolism. The malic acid content drops rapidly after veraison to reach levels ranging from 2 to 3 mg/g of fresh weight, i.e. about 1–5 g/l in must at ripeness.

This disappearance corresponds to an increase in the respiratory quotient, which suggests the use of this acid for energy production in the grape (Harris et al., 1971). In fact, during maturation, malic acid takes on the role of an energy vector (Figure 10.13). During the green growth phase, the sugars coming from photosynthesis are

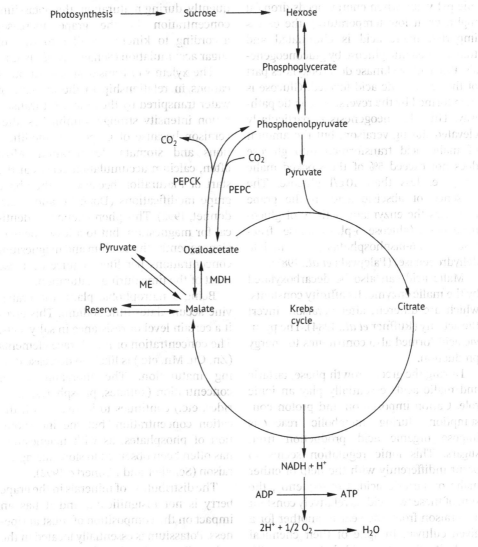

FIGURE 10.13 Role of malic acid in the production of energy (ATP) and the formation of different substrates in the grape (Ruffner, 1982b). MDH, malate dehydrogenase; ME, malic enzyme; PEPC, phosphoenolpyruvate carboxylase; PEPCK, phosphoenolpyruvate carboxykinase.

transformed into malic acid, which accumulates in the pericarp cell vacuoles (the grape being incapable of stocking significant amounts of starch, as many other fruits do). At veraison, due to the severe inhibition of the glycolytic pathway, malic acid importation from the vacuole permits energy production to be maintained. The activation of a specific permease ensures this transport. The *de novo* synthesis of different malate dehydrogenase isoenzymes supports this hypothesis.

In order to maintain a normal cytoplasmic pH value when energy needs drop (at night, or at low temperature), the excess imported malic acid is eliminated and transformed into glucose by gluconeogenesis. PEP carboxykinase decarboxylates part of the oxaloacetic acid formed. Glucose is then formed by the reverse glycolytic pathway. This gluconeogenesis is particularly elevated during veraison, but the amount of malic acid transformed into glucose does not exceed 5% of the stocked malic acid, i.e. less than 10 g/l glucose. The presence of abscisic acid in the grape increases the enzymatic activity of gluconeogenesis (glucose 6-phosphatase, fructose 1,6-bisphosphatase, malate dehydrogenase) (Palejwala *et al.*, 1985).

Malic acid can also be decarboxylated by the malic enzyme. Its affinity constants, which are different after veraison, invert the activity (Ruffner *et al.*, 1984). The pyruvic acid formed also contributes to energy production.

During the green growth phase, tartaric and malic acids essentially play an ionic role. Cation importation and proton consumption during metabolic reactions impose organic acid production from sugars. This ionic regulation seems to occur indifferently with the help of either malic or tartaric acid. Consequently, the sum of these two acids is relatively constant at veraison from one year to another for a given cultivar. In spite of their chemical similarity, these two acids behave very differently during ripening: the tartaric acid content of grapes varies very little, while that of malic acid closely follows the decrease in total acidity. At ripeness, the sum of these two acids is highly variable, depending on vintage conditions.

10.3.4 Accumulation of Minerals

Potassium is one of the rare minerals translocated by the phloem sap. In the phloem, it permits the translocation of sugars derived from photosynthesis. Consequently, during maturation, the potassium concentration in the grape increases according to kinetics similar to that of sugar accumulation (Schaller *et al.*, 1992).

The xylem sap translocates most other cations in relationship to the amount of water transpired by the grape. Yet transpiration intensity strongly diminishes after veraison because of grape skin modifications and stomata degeneration. Most often, calcium accumulation ceases at the start of maturation because of the above grape modifications (Donèche and Chardonnet, 1992). This phenomenon is identical for magnesium, but to a lesser degree. Consequently, the calcium and magnesium concentrations per liter of juice decrease most of the time during maturation.

Being a natrophobic plant, the grapevine accumulates little sodium. This gives it a certain level of resistance in salty soils. The concentration of metal trace elements (Zn, Cu, Mn, etc.) is likely to decrease during maturation. The inorganic anion concentration (sulfates, phosphates, chlorides, etc.) continues to increase with the cation concentration, but the incorporation of phosphates, as with magnesium, has often been observed to slow during veraison (Schaller and Löhnertz, 1992).

The distribution of minerals in the grape berry is not insignificant, and it has an impact on the composition of must at ripeness. Potassium is essentially located in the pulp cell vacuoles, but the skin cells also sometimes contain significant amounts during hot years. In theory, the sum of the acids and cations determines must pH. However, it depends mainly on the tartaric acid and potassium concentrations, according to the following relationship (Champagnol, 1986):

$$pH = f \frac{[\text{tartaric acid}]}{[\text{potassium}]}$$

10.3.5 Evolution of Nitrogen Compounds

The grape nitrogen supply depends on both the phloem and xylem saps. In both cases, nitrates are rarely involved. They are only present in small quantities because of their reduction in the roots and leaves.

Nitrogen transport to the grape essentially occurs in the form of ammonium cations or amino acids. Glutamine represents about 50% of the organic nitrogen imported.

There are two intense nitrogen incorporation phases during grape development: the first following fruit set and the second starting at veraison and finishing at mid-maturation. Toward the end of maturation, the total nitrogen concentration may increase again. As a result, at harvest, half of the nitrogen in the vegetative part of the plant is stored in the grapes (Roubelakis-Angelakis and Kliewer, 1992). In unripe fruit, the ammonium cation represents more than half of the total nitrogen. From veraison onward, the ammonium concentration decreases, whereas the organic fraction increases. The free amino acids increase by a factor of two to five during maturation, attaining 2–8 g/l in leucine equivalents (Volume 2, Section 5.3.2). At ripeness, the amino acid fraction represents 50–90% of the total nitrogen in grape juice.

The incorporation of the ammonium cation on α-oxoglutaric acid appears to be the principal nitrogen assimilation pathway by the grape. It is catalyzed by glutamine synthetase (GS) and glutamate dehydrogenase (GDH) enzymes. Other amino acids are synthesized by the transfer of nitrogen incorporated on glutamic acid.

Research carried out by numerous authors shows that even though the amino acid composition varies greatly depending on conditions, a small number of amino acids predominate: alanine, γ-aminobutyric acid, arginine, glutamic acid, proline, serine, and threonine.

At ripeness, arginine is often the predominant amino acid and can represent from 6 to 44% of the total nitrogen of grape juice. In fact, this amino acid plays a very important role in grape berry nitrogen metabolism (Figure 10.14). A close relationship exists between arginine and diverse amino acids (ornithine, aspartic and glutamic acids, proline). As a result, the proline concentration can increase during maturation by a factor of 25–30 through the transformation of arginine. Moreover,

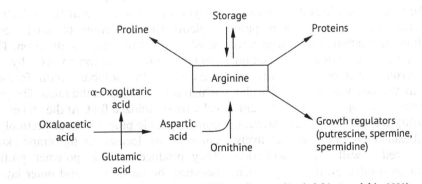

FIGURE 10.14 Role of arginine in the nitrogen metabolism of grapes (Roubelakis-Angelakis, 1991).

aspartic acid constitutes an oxaloacetic acid reserve, which, depending on the demand, can be transformed into malic acid or into sugars during maturation.

Maturation is also accompanied by an active protein synthesis. The soluble protein concentration reaches its maximum before complete ripeness and then diminishes toward the end of maturation. The concentration of grape juice protein can thus vary from 1.5 to 100 mg/l. The concentration of high-molecular-weight insoluble proteins, often attached to the cell wall, is high from the start of development and continues to increase during maturation. During maturation, the protein profile of the skins is highly modified: proteins involved in photosynthesis, primary metabolism, and response to stress are highly overexpressed at the beginning of veraison. However, at the end of veraison, the most common proteins are those involved in anthocyanin synthesis and defense mechanisms (Deytieux-Belleau et al., 2007b).

The juice from ripe grapes contains barely 20% of the total berry nitrogen. The remainder is retained in the skins and seeds, even though the latter are likely to liberate soluble forms of nitrogen (ammonium cations and amino acids) in the pulp toward the end of maturation.

10.3.6 Changes in the Cell Wall

The softening of the grape during maturation is the result of significant changes in cell wall composition—notably in the pulp cells. Cell multiplication and enlargement during grape development and maturation are not accompanied by a proportional increase in the cell wall polysaccharides (Chardonnet et al., 1994; Lacampagne, 2010). Depending on the varieties, either cell wall deterioration or a relatively constant cell wall polysaccharide concentration results, until the approach of ripeness. The pulp texture differences between varieties are explained in this manner.

At the beginning of grape development, the cell walls are primarily composed of cellulose. The veraison period is characterized by considerable pectin synthesis (Volume 2, Section 3.6.2), resulting in the thickening of radial and tangential walls (Lacampagne, 2010). Like cement, pectins ensure cellulose fiber cohesion. They are formed by the polymerization of galacturonic acid and diverse neutral sugars (rhamnose, galactose, and arabinose). A high percentage of the acid functions of galacturonic acid units are methylated.

Maturation is accompanied by a solubilization of these pectins under the influence of several factors. First, pectin methyl esterases (PME) liberate the acid functions of galacturonic acid, resulting in increased grape methanol concentration and acidification of the medium. Cell wall hydration, characterized by swelling, is thus facilitated by increasing the K^+/Ca^{2+} ratio (Possner and Kliewer, 1985). Since the pectins are less chelated by calcium, the free acid functions of the galacturonic residues are the site of attack by other enzymatic activities—polygalacturonases (PG) and pectin lyases. Thus, by demethylating pectins, PME appears to facilitate their digestion by PG, resulting in the degradation of the pectin-based cell wall network and thus the softening of the fruit. These activities are at a maximum between the middle and end of the maturation period (Fasoli et al., 2016). Although PMEs are mostly present in the grape skin, all of these enzymes are also active in the pulp. This explains the reduction of total pectic substances during grape maturation. This phenomenon is accompanied by an increase of the soluble pectin fraction, which is later found in the must. The pulp cells are solubilized first. At the end of ripening, variable proportions of pectinolytic enzymes are located in the grape skins. They produce variable polymer profiles between the most inner and outer layers. The action of these enzymes is strongly regulated by the abscisic acid–auxin hormone equilibrium (Giribaldi et al., 2010).

At ripeness, the grape is characterized by a low pectin concentration with respect to other fruits.

10.3.7 Production of Phenolic Compounds

One of the most remarkable characteristics of maturation is the rapid accumulation of phenolic pigments, which give the red grape its enological importance. These phenolic pigments are secondary products of sugar catabolism. Their biosynthetic pathways are present and partially active right from the start of grape development.

Phenolic compounds derived from a simple unit with a single benzene ring are created from the condensation of erythrose 4-phosphate, an intermediate of the pentose phosphate cycle, with phosphoenolpyruvic acid. This biosynthetic pathway, known as the shikimic acid pathway (Figure 10.15), leads to the production of benzoic and cinnamic acids, as well as aromatic amino acids (PHE, TYR). The condensation of three acetyl coenzyme A molecules, derived from Krebs cycle reactions, also leads to the formation of a benzene ring. The condensation of this second ring with a cinnamic acid molecule produces a molecule group known as the flavonoids. These molecules possess two benzene rings joined by a C3 carbon chain, most often in an oxygenated heterocyclic form. Various transformations (hydroxylation, methoxylation, esterification, and glycosidification) explain the presence of

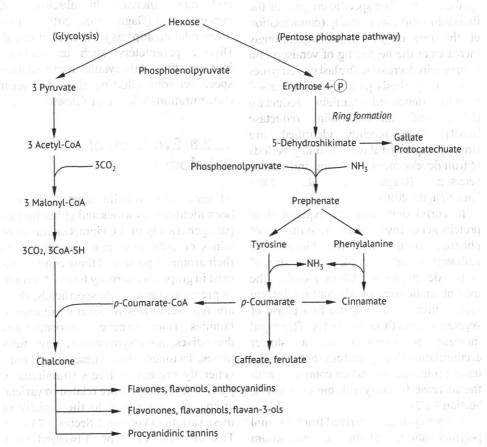

FIGURE 10.15 Biosynthesis pathways of phenolic compounds (Conn, 1986).

many substances from this family in the grape (Volume 2, Section 6.2).

In these metabolic pathways, phenylalanine ammonia-lyase (PAL) is the enzyme that, by eliminating the NH_3 radical, diverts phenylalanine from protein synthesis (primary metabolism) toward the production of *trans*-cinnamic acid and other phenolic compounds. PAL is located in grape epidermal cells as well as in the seeds. Its maximum activity in the seeds occurs during the green growth phase; its activity then decreases after veraison to become very low during maturation. PAL activity contained in the grape skin is very high at the start of development and then decreases up to veraison. In red and black grapes, PAL activity in the skins increases again at the start of veraison. There is a close relationship between its activity and the color intensity of the grape (Hrazdina et al., 1984). Chalcone synthase is the first specific enzyme of the flavonoid synthesis pathway (condensation of the two rings): its activity strongly increases at the beginning of veraison and then rapidly decreases. The last two enzymes of the biosynthesis pathway of condensed tannins (leucoanthocyanidin reductase [LAR] and anthocyanidin reductase [ANR]), both recently identified, are strongly expressed during two key periods of fruit development: flowering and start of veraison (Gagné et al., 2009; Lacampagne, 2010).

In correlation with the expression of protein genes involved in the synthesis of phenolic compounds, the biosynthetic pathways are active as early as the start of grape development. Consequently, the total phenolic compounds continually and rapidly increase during the first phase of vegetative growth of the berry. The rapid increase is followed by a slower accumulation during maturation, owing to lesser protein activity when compared with the increase in berry volume (Volume 2, Section 6.5.2).

The procyanidins, derived from flavanol polymerization, attain a maximum concentration in the seeds before veraison. This then greatly decreases to a lower and relatively stable value when the seeds are ripe. At veraison, the skin tannin concentration is already high—sometimes corresponding to over half of the concentration at ripeness (Figure 10.16).

In white grapes, the concentrations of phenolic acids esterified by tartaric acid, flavan-3-ols, and oligomeric procyanidins are high at the beginning of development. They then diminish to minimal concentrations at ripeness.

In red and black varieties, anthocyanins begin to accumulate in the skins about two weeks before the color is visible. The concentration increases during maturation, but, as with tannins, it attains a maximum and generally diminishes up till ripeness.

This appearance of anthocyanins is linked to sugar accumulation in the grape and rapid increase in abscisic acid concentration (Gagné et al., 2011), but no direct relationship has yet been established. Diverse parameters, such as sunlight, increase the anthocyanin accumulation speed without affecting the skin sugar concentration (Wicks and Kliewer, 1983).

10.3.8 Evolution of Aroma Substances

At least 1,000 volatile compounds have been identified in wines and either participate generically in the vinous character of wines or contribute more specifically to their aroma. A portion of these compounds exist in grapes in a directly fragrant form or in precursor form. More specifically, these are representatives of several compound families (monoterpenes, norisoprenoid derivatives, methoxypyrazines, thiols, furanones, lactones, etc.). These compounds, generally present as traces (hundreds of µg/L to several ng/l), are related to varietal aroma and contribute to the typicity of these varieties (Volume 2, Sections 7.1–7.5). That is the case of 3-isobutyl-2-methoxypyrazine, which was identified for

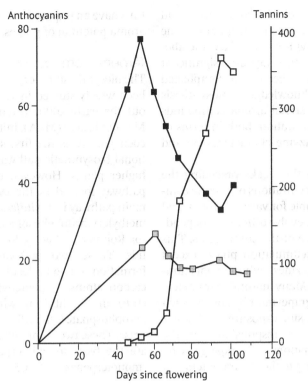

FIGURE 10.16 Evolution of phenolic compounds (Darné, 1991) (results expressed in mg/g dry weight): □, skin anthocyanins; ■, skin tannins; ▫, seed tannins.

the first time in Cabernet Sauvignon grapes (Bayonove et al., 1975) (Volume 2, Section 7.4), the main monoterpenes of Muscat and Muscat-like varieties (linalool, geraniol, hotrienol, rose oxide) (Volume 2, Section 7.2), or volatile thiols of varietal origin, initially found in Sauvignon Blanc and then in many other varieties (Volume 2, Section 7.5). However, all compounds identified up till now can be found in many grape varieties; their concentration level counts for more than specificity. Thus, a trace of terpene alcohols is found in neutral-tasting varieties, whereas their concentration can reach 3 mg/l in certain aromatic varieties (Gewürztraminer, Muscat).

Moreover, aroma potential exists in grapes under various forms, the free odorous form and especially the "bound" form, which is very significant in most grapes of generic varieties. The free form regards odorous volatile substances (monoterpenes, methoxypyrazines, etc.) present in grapes. The "bound" form is represented by aroma precursors. These latter compounds are sometimes volatile, but non-odorous and likely to become odorous through chemical or enzymatic reactions (e.g. terpenes and polyhydroxylated norisoprenoid derivatives (diols, triols, etc.)). Otherwise, these are non-odorous, non-volatile precursors. These compounds are glycosides, more specifically associated with certain varieties, or cysteine S-conjugates and glutathione S-conjugates. Other aroma precursors are present in all grapes, regardless of variety. These include amino acids, which can lead to fermentation aromas (higher alcohols, esters) under the action of yeast metabolism; phenolic acids resulting from volatile phenols; fatty acids; or other lipid compounds.

All of these volatile compounds and precursor forms, present in grapes, come from the grapevine's secondary metabolism. Over the past 15 years, significant progress has been accomplished concerning the knowledge of biosynthesis parameters of varietal aroma compounds during grape maturation, both in terms of biochemical (enzyme characteristics) and genetic aspects.

Therefore, in the weeks preceding the harvest, the concentrations of volatile compounds responsible for varietal aroma, and their precursors, evolve differently depending on the nature of the compounds. This is mainly an accumulation phase, except for methoxypyrazines, whose concentrations decrease. Many natural parameters likely to affect grapevine physiology, such as temperature, sun exposure, and water status, are factors responsible for the observed variations. Grape-growing choices pertaining to these parameters will thus have an impact on the evolution of the aroma potential of grapes.

Terpene compounds

The biosynthesis of terpene compounds has been widely studied in *V. vinifera* throughout the early 2000s (Figures 10.17–10.20). Mevalonic acid (MVA), initiated with acetyl coenzyme A, is the first step of the traditional biosynthetic pathway for terpenes in higher plants. However, this biosynthesis pathway located in the cytosol is not the main pathway in *V. vinifera*. It is, in fact, the methylerythritol phosphate (MEP) pathway (or Rohmer pathway), located in the plastids. These two pathways lead to the formation of two molecules possessing five carbon atoms: isopentenyl pyrophosphate (IPP) in equilibrium with dimethylallyl pyrophosphate (DMAPP) via DMAPP isomerase. These two compounds (IPP, DMAPP) are the basis for all terpene biosynthesis (monoterpenes: 2×C5; sesquiterpenes:

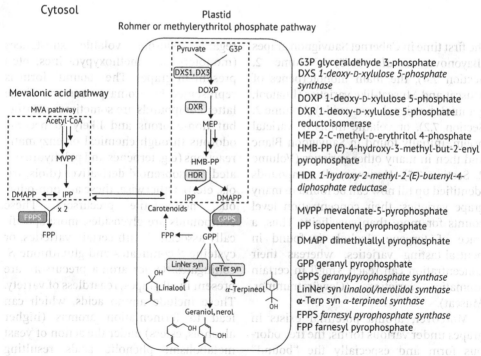

FIGURE 10.17 Biosynthesis pathway of the main grape monoterpenes. *Source*: Adapted from Martin et al. (2012).

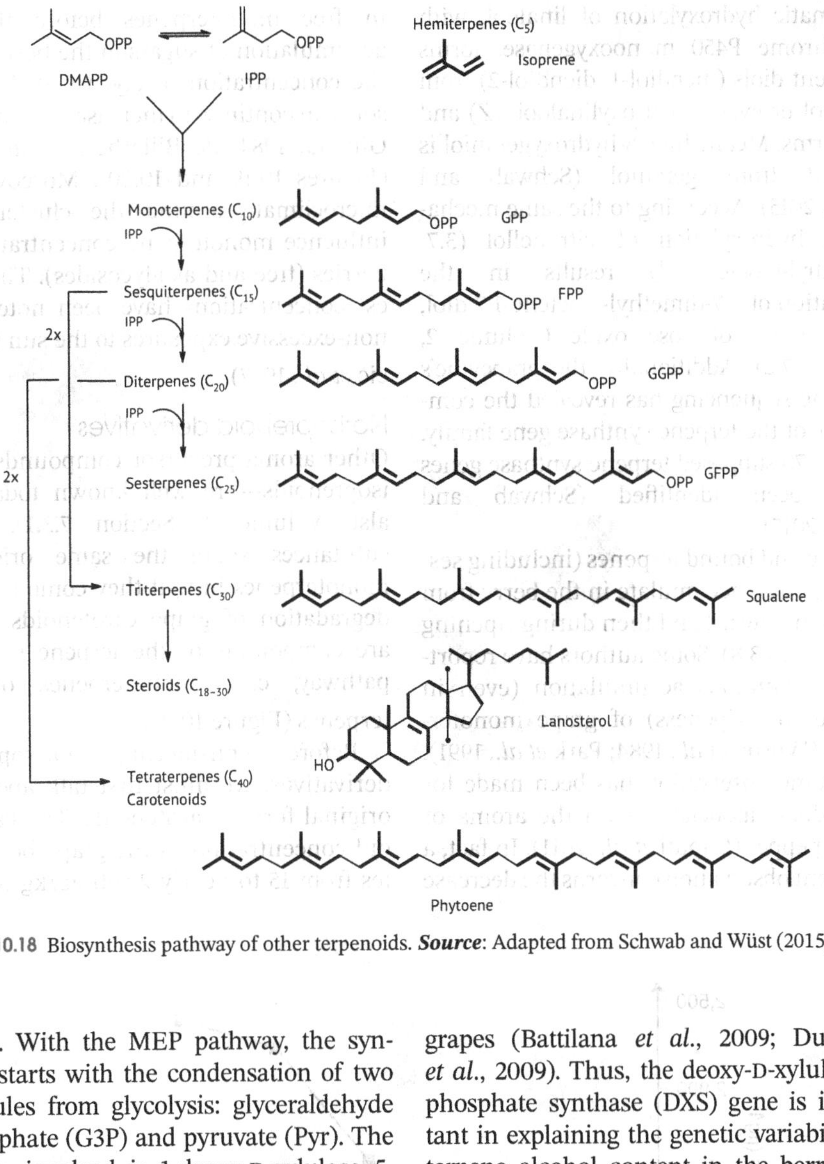

FIGURE 10.18 Biosynthesis pathway of other terpenoids. *Source*: Adapted from Schwab and Wüst (2015).

3×C5). With the MEP pathway, the synthesis starts with the condensation of two molecules from glycolysis: glyceraldehyde 3-phosphate (G3P) and pyruvate (Pyr). The enzyme involved is 1-deoxy-D-xylulose 5-phosphate synthase (DXS), leading to the formation of 1-deoxy-D-xylulose 5-phosphate, which is reduced to 2C-methyl-D-erythritol 4-phosphate by 1-deoxy-D-xylulose 5-phosphate reductase (DXR). Then, after other phosphorylation reactions, it is reduced to IPP, as Rohmer demonstrated (1999). The genetic analysis of monoterpene production in grapes has resulted in the recent identification of two major quantitative loci (QTL) influencing monoterpene alcohol concentration and, more specifically, linalool concentrations, in grapes (Battilana et al., 2009; Duchêne et al., 2009). Thus, the deoxy-D-xylulose 5-phosphate synthase (DXS) gene is important in explaining the genetic variability of terpene alcohol content in the berry. IPP condenses with its isomer, a DMAPP molecule, to produce geranyl pyrophosphate (GPP), a key molecule in terpene synthesis. Luan et al. (2004) proved that the DXP pathway is the dominant one for the biosynthesis of geraniol or linalool in the exocarp and mesocarp of grapes. Once GPP is synthesized, terpene synthases, such as linalool synthase or geraniol synthase, will use this substrate to form linalool or geraniol, respectively. It also produces other polyhydroxylated monoterpenes, such as diendiol, triols, or linalool oxides. For example, enzymatic hydroxylation of linalool, with cytochrome P450 monooxygenase, forms different diols (diendiol-1, diendiol-2) from linalool or even 8-hydroxylinalool, (Z) and (E) forms. Meanwhile, 8-hydroxygeraniol is formed from geraniol (Schwab and Wüst, 2015). According to the same mechanism, hydroxylation of citronellol (3,7-dimethyl-6-octen-1-ol) results in the formation of 3,7-dimethyl-5-octene-1,7-diol, a precursor of rose oxide (Volume 2, Section 7.2). Additionally, the grapevine's genome sequencing has revealed the complexity of the terpene synthase gene family. Nearly 70 supposed terpene synthase genes have been identified (Schwab and Wüst, 2015).

Free and bound terpenes (including sesquiterpenes) accumulate in the berry from veraison onward and then during ripening (Section 10.3.8). Some authors have reported a continuous accumulation (even in case of overripeness) of grape monoterpenes (Wilson et al., 1984; Park et al., 1991). The same observation has been made for rotundone, associated with the aroma of black pepper (Caputi et al., 2011). In fact, a frequent observation concerns the decrease in free monoterpenes before the full accumulation of sugars in the berry, while the concentration of glycosylated precursors can continue to increase (Marais, 1983; Günata, 1984; de Billerbeck et al., 2003) (Figures 10.19 and 10.20). Moreover, the microclimate around the clusters may influence monoterpene concentrations of berries (free and as glycosides). The highest concentrations have been noted with non-excessive exposures to the sun (Belancic et al., 1997).

Norisoprenoid derivatives

Other aroma precursor compounds—norisoprenoids—are well known today (see also Volume 2, Section 7.3.1). These substances share the same origin as monoterpenes in that they come from the degradation of grape carotenoids, which are compounds of the terpene synthesis pathway, called tetraterpenes or C40 terpenes (Figure 10.21).

Before considering norisoprenoid derivatives, we must first talk about the original forms: carotenoids. The carotenoid concentration in the grape berry varies from 15 to nearly 2,500 µg/kg in fresh

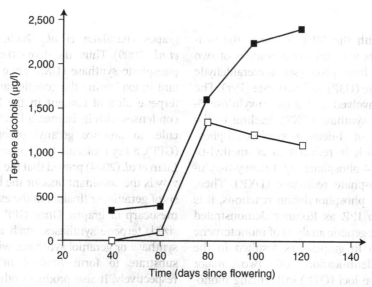

FIGURE 10.19 Average evolution of terpene alcohols during Muscat grape maturation (Bayonove, 1993): □, free terpene alcohols; ■, bound terpene alcohols.

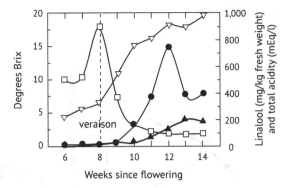

Total acidity (□), degrees Brix (▽), free (●) and bound forms (▲) (glycosylated) of linalool

FIGURE 10.20 Evolution of "free" and "glycosylated" linalool concentrations during maturation of Muscat grapes (de Billerbeck *et al.*, 2003).

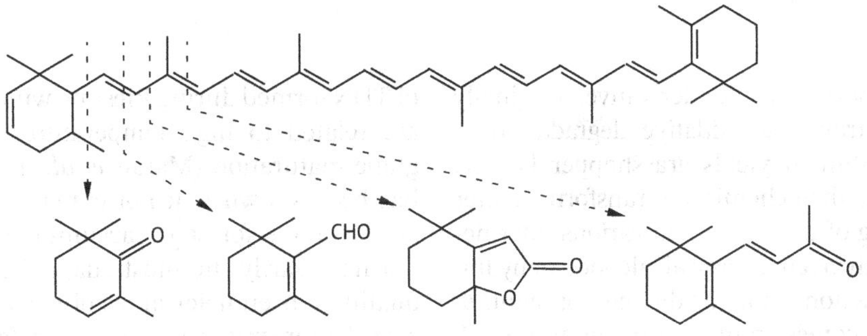

FIGURE 10.21 Degradation of carotenoids, leading to the formation of C9, C10, C11, and C13 norisoprenoids in grapes (Enzel, 1985).

weight (Razungles, 1985). The most important, in decreasing order, are lutein, β-carotene, neoxanthin, and lutein 5,6-epoxide. These molecules, generally enclosed in cell plastids, are essentially located in the solid parts of the berry: the skin is two to three times richer in carotenoids than the pulp. The carotenoids are found in different proportions in the different parts of the berry, depending on their structure. During maturation, a decrease in the carotenoid concentration and an increase in certain carotenoid-derived molecules such as norisoprenoids are observed (Figure 10.22).

The formation of norisoprenoid derivatives in the grape initially results from the oxidative degradation of grape carotenes initiated by regionally selective dioxygenases (Volume 2, Section 7.3). The derivatives formed are then transformed by enzymatic reactions (reductases, glycosylases), resulting in the formation of glycosides. At the end of maturation, and especially during the winemaking process, as well as wine aging, the chemical transformation of these derivatives can sometimes form very strong-smelling compounds (β-ionone, β-damascenone, TDN, TPB, vitispirane, etc.) (Volume 2, Section 7.3). For example, β-damascenone,

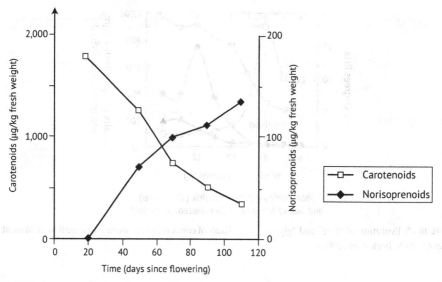

FIGURE 10.22 Average evolution of carotenoids and C13-norisoprenoids during maturation of Muscat grapes (Bayonove, 1993).

a C13-norisoprenoid derivative, originally comes from the oxidative degradation of neoxanthin. It yields grasshopper ketone, which is then chemically transformed after pressing of the grapes, via various intermediates through chemical alcohol dehydration reactions in an acidic environment. β-Ionone comes from β-carotene and TDN, and vitispirane comes from violaxanthin. It is not possible to establish a direct link between the initial concentrations and norisoprenoid derivatives cited, considering the complexity of the intermediate reaction mechanisms.

Grape exposure to sunlight during maturation accelerates carotenoid degradation, especially via the formation of active forms of oxygen, and is accompanied by an increase of C13-norisoprenoid derivative content. This has been observed in Riesling and Syrah grapes (Marais, 1993; Razungles and Bayonove, 1996). Thus, leaf removal from Riesling vines in the cluster zone leads to a greater accumulation of glycosylated C13-norisoprenoid derivatives, except for C13-norisoprenoid glucosides, which lead to β-damascenone (Marais et al., 1992a).

Excessive odors of hydrocarbon in wines, associated with high concentrations of TDN formed during Riesling wine aging, are related to high temperatures during grape maturation (Marais et al., 1992b). It has been shown that hot climates, which are favorable for sugar accumulation, are not necessarily the most adapted to wine quality. Linsenmeier and Löhnertz (2007) also demonstrated that nitrogen fertilization in the vineyard helps limit TDN formation during wine aging. The relationship between the canopy management of Riesling vines and the TDN content of wines is an important subject in the context of climate change.

Methoxypyrazines

The unripe grape contains a high concentration of methoxypyrazines (a few dozen nanograms per liter) in certain varieties, such as Cabernet Sauvignon (Volume 2, Section 7.4). The concentration of these compounds drops significantly in the course of maturation. The highest concentrations are found under the coldest ripening conditions (Lacey et al., 1991) or with very high-vigor grapevines.

In the grape berry, must, and wine, 3-isobutyl-2-methoxypyrazine (IBMP) is the most abundant methoxypyrazine. IBMP is present in the grape at concentrations between a few to several dozen ng/kg. IBMP is characterized by a very low perception threshold, leading to its powerful effect on wine aroma (Volume 2, Section 7.4). Certain varieties such as Cabernet Sauvignon, Carménère, or Cabernet Franc may reach high IBMP concentrations (100 ng/l in extreme cases). Other grape varieties such as Merlot and Sauvignon Blanc present more moderate concentrations between 0.5 and 40 ng/l (Allen et al., 1994; Roujou de Boubée et al., 2000; Poitou et al., 2017).

IBMP is present in free form in grapes. Its origin is not completely known. Nevertheless, several studies have demonstrated that the IBMP precursor is 3-isobutyl-2-hydroxypyrazine (IBHP) (Murray and Whitfield, 1975; Harris et al., 2012). IBHP may be the result of leucine amidation into leucinamide, which is then associated with a glyoxal molecule (ethane-1,2-dione) to form IBHP (Hashizume et al., 2001a) (Figure 10.23). Dunlevy et al. (2013) and Guillaumie et al. (2013) demonstrated the existence of VviOMT3 and VviOMT4 genes in V. vinifera. Both genes code for an O-methyltransferase, responsible for the last stage of the IBMP biosynthesis pathway, i.e. the O-methylation of IBHP into IBMP (Hashizume et al., 2001a,b). IBHP methylation is conducted using S-adenosyl-L-methionine (SAM), which is the methyl donor. A demethylation reaction of IBMP may occur to form IBHP again (Ryona et al., 2010).

Over the last 20 years, many studies concerning IBMP have come out. IBMP appears in clusters several days after anthesis (Ryona et al., 2008). Its concentrations increase at least until the cluster fills out or even until veraison (Hashizume and Samuta, 1999; Roujou de Boubée, 2000; Ryona et al., 2008), reaching sometimes concentrations of 200 ng/kg. This increase is correlated with an increase in the transcripts of VviOMT3 and VviOMT4 genes (Guillaumie et al., 2013). The concentration then decreases during ripening. Ryona et al. (2009) have observed that for Cabernet Franc, the IBMP concentration in ripe grapes is closely correlated with the maximum concentrations determined around the full cluster stage. However, the parameters of plant physiology determining the evolution of IBMP concentrations in fruits (accumulation, then reduction) have not been precisely clarified. IBMP has been detected both in stalks and leaves (Hashizume and Umeda, 1996; Hashizume and Samuta, 1999; Roujou de Boubée et al., 2002). More specifically, authors have found much higher concentration of IBMP in basal leaves, close to the cluster. Experiments using IBMP marked by deuterium have demonstrated that this molecule could migrate from basal leaves to the cluster and then into berries through the phloem (Hashizume and Umeda, 1996; Roujou de Boubée et al., 2000). IBMP marked by deuterium is detected in the stalks but also in seeds and the skin. Additionally, while the

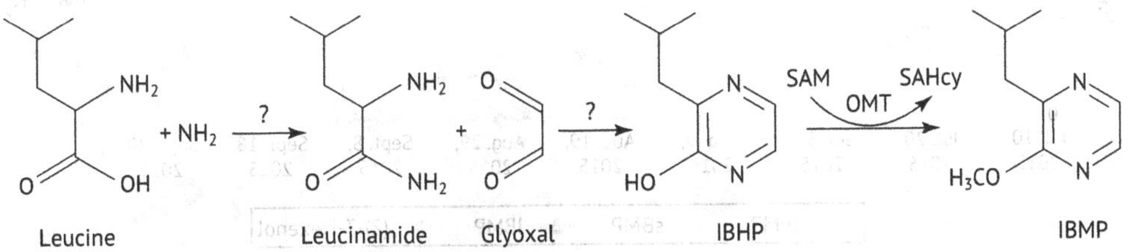

FIGURE 10.23 IBMP biosynthesis pathway in the grape berry. IBHP, 3-isobutyl-2-hydroxypyrazine; IBMP, 3-isobutyl-2-methoxypyrazine; OMT, O-methyltransferase; SAHcy, S-adenosyl-L-homo-Cys; SAM, S-adenosyl-L-methionine. *Source*: Adapted from Hashizume et al. (2001a).

IBMP concentration continually decreases in berries from veraison until ripeness, IBMP biosynthesis continues to take place in basal leaves (Figure 10.24, Roujou de Boubée et al., 2000).

Natural conditions (climate, soil type) and viticultural factors strongly influence the evolution of IBMP concentrations during the vegetative period of the grapevine. Allen and Lacey (1993) noted that in Australia wines made from grapes that ripened at a higher temperature have lower IBMP concentrations. But, on the other hand, Falcao et al. (2007) demonstrated higher concentrations in Brazilian wines resulting from grapes cultivated at higher elevations. In Bordeaux, the analysis of wines from a given vintage, between 1991 and 1995, also showed the impact of climate on methoxypyrazine concentration (Figure 10.25).

Cluster microclimate is thus a major factor in variation of IBMP concentration in grapes (Hashizume and Samuta, 1999; Roujou de Boubée, 2000; Sala et al., 2004; Ryona et al., 2008). Several authors demonstrated that IBMP's sensitivity to UV light could explain the observed variation (Heymann et al., 1986; Maga, 1989; Roujou de Boubée, 2000).

Paradoxically, according to the works of Ryona et al. (2008) and Scheiner et al. (2010), light does not seem to favor the post-veraison degradation kinetics of IBMP. Rather, it limits the accumulation of this compound from fruit set until the full cluster stage or mid-veraison for fruit undergoing veraison. In fact, Roujou de Boubée et al. (2000) indirectly demonstrated the same phenomenon by establishing that the IBMP concentration in grapes is lower when basal leaves are removed early (at fruit set) rather than at veraison. However, because IBMP is abundant in basal leaves, it is possible for the marked decrease in concentration of this compound in grapes to be due also to the elimination of the source (leaves). The work of Gregan et al. (2012) and Gregan and Jordan (2016) supports this hypothesis that leaves can be a source of IBMP in

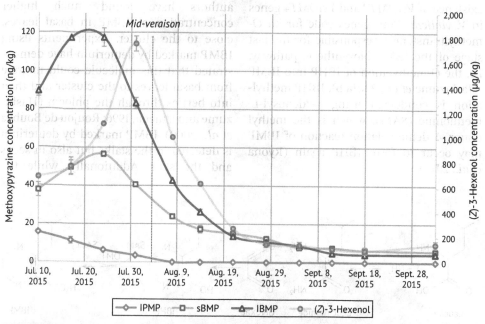

FIGURE 10.24 Kinetics of the evolution of contents of 3-isobuytl-2-methoxypyrazine (IBMP), 3-isopropyl-2-methoxypyrazine (IPMP), 3-sec-butyl-2-methoxypyrazine (sBMP), and Z-3-hexenol in Cabernet Sauvignon grapes (ng/kg and μg/kg) during grape berry development (2015 vintage) (Poitou et al., 2017).

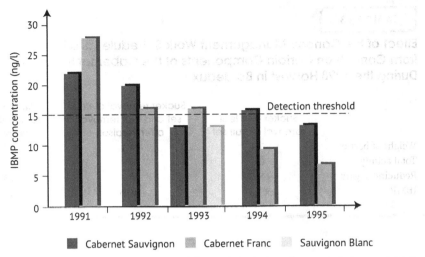

FIGURE 10.25 Comparison of IBMP contents of three different varietal wines, from several vintages, in a given appellation (Roujou de Boubée, 2000).

grapes and that their early removal leads to a lower accumulation of this compound in grapes.

In addition, other physiological parameters of the grapevine, such as yield, water and nitrogen availability (Roujou de Boubée, 2000; Sala et al., 2004), water status, and nitrogen status (Mendez-Costabel et al., 2014a,b, Helwi et al., 2015) may have an impact on IBMP concentrations. Recently, the plant's nitrogen status was shown to have no direct influence on the IBMP concentration of berries or on the expression of the two main genes (*VviOMT3* and *VviOMT4*) involved in its biosynthesis (Helwi et al., 2015). However, if vine vigor is not controlled, a high nitrogen supply may create a cool and shady microclimate in the cluster zone, which limits IBMP degradation, thus favoring its accumulation in grapes. Moreover, nitrogen may increase the plant's leaf area, which is the location of this compound's biosynthesis.

Under the climate conditions of the Bordeaux region, the nature of the soil influences grapevine vigor and plays a major role in methoxypyrazine concentrations of Merlot, Cabernet Franc, and Cabernet Sauvignon wines. In well-drained, gravelly soils, concentrations are lower. In limestone or clay loam soils, Cabernet Sauvignon is richer in methoxypyrazine and is therefore more frequently marked by a vegetal character.

The combined action of leaf removal and elimination of suckers (or suckering) in the cluster zone is also particularly effective in improving the aroma maturation of Cabernet Sauvignon and Sauvignon Blanc (Roujou de Boubée, 2000). As we have seen previously, and in the context of cool years with little sunshine, it is important to conduct this canopy management work between fruit set and the full cluster stage, since late leaf removal (though it can increase sugar accumulation in grapes) is not able to cause a sufficient decrease in IBMP content and thus prevent a marked vegetal aroma in wine due to methoxypyrazines (Table 10.3).

S-Conjugates, precursors of volatile thiols

The main varietal thiols (3SH, 4MSP, 4MSPOH) exist in musts as *S*-conjugates with cysteine (*S*-3-(hexan-1-ol)-L-cysteine (Cys-3SH), *S*-4-(4-methylpentan-2-one)-L-cysteine (Cys-4MSP), *S*-4-(4-methylpentan-2-ol)-L-cysteine (Cys-4MSPOH)) and with glutathione (Gly-Cys-Gly tripeptide,

TABLE 10.3

Effect of the Canopy Management Work Schedule (% Difference from Control) on Certain Components of the Cabernet Sauvignon Berry During the 1998 Harvest in Bordeaux

	Sucker and leaf removal at fruit set	Sucker removal at fruit set and leaf removal after veraison	Sucker and leaf removal after veraison
Weight of berries	−7.4	−4.4	+2
Total acidity	0	0	0
Reducing sugars	+8.5	+6.7	+3
IBMP	−68.4	−10.5	0

Glut-3SH) or as dipeptides (Cys-Gly or Glu-Cys) (Figure 10.26) (Volume 2, Section 7.5).

Biosynthesis pathways of volatile thiol precursors are still far from being fully explained. Recent studies (Kobayashi et al., 2010; Capone et al., 2011; Thibon et al., 2011) have demonstrated the formation of Cys-3SH from Glut-3SH, which is obtained from the conjugation of glutathione (GSH) with *trans*-2-hexenal resulting from the lipoxygenase pathway. The presence of an intermediate aldehyde (Glut-3SHAl) is the result of the conjugation

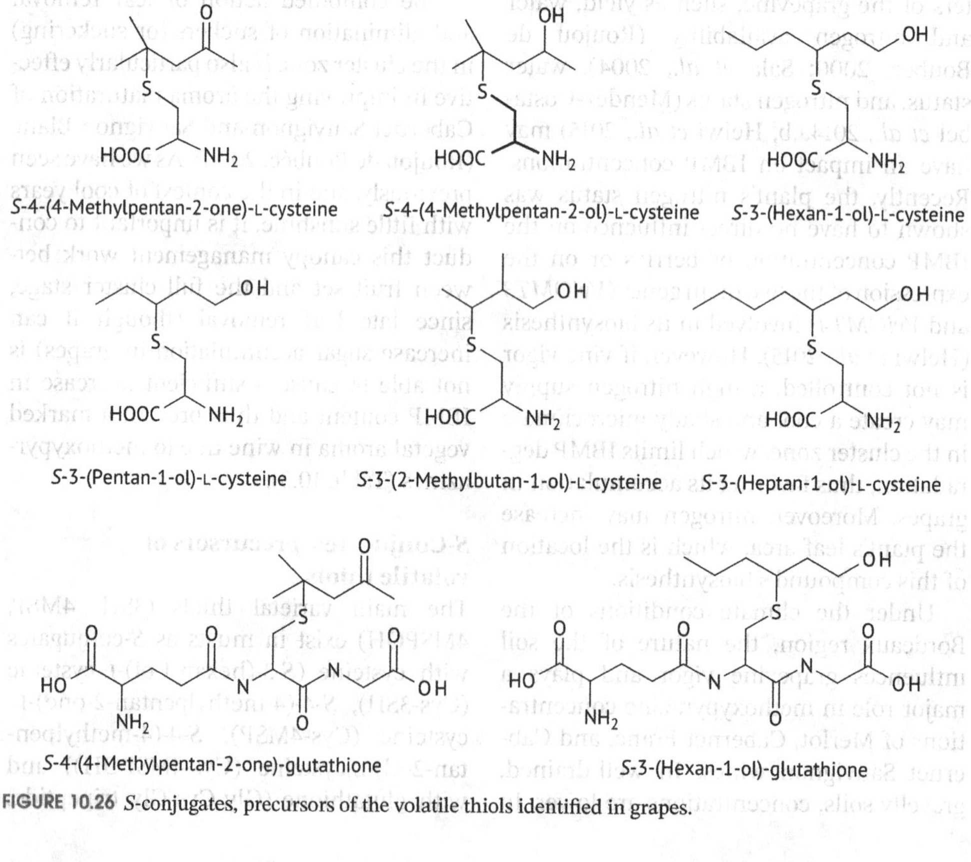

S-4-(4-Methylpentan-2-one)-L-cysteine   S-4-(4-Methylpentan-2-ol)-L-cysteine   S-3-(Hexan-1-ol)-L-cysteine

S-3-(Pentan-1-ol)-L-cysteine   S-3-(2-Methylbutan-1-ol)-L-cysteine   S-3-(Heptan-1-ol)-L-cysteine

S-4-(4-Methylpentan-2-one)-glutathione   S-3-(Hexan-1-ol)-glutathione

FIGURE 10.26 S-conjugates, precursors of the volatile thiols identified in grapes.

of *trans*-2-hexenal and glutathione (Capone *et al.*, 2011; Thibon *et al.*, 2016). Kobayashi *et al.* (2010) reported the existence of the precursor 3SH-cysteine glycine (Cys-Gly-3SH) as an intermediate between the glutathione precursor and the cysteine precursor. Glutathione *S*-transferase (GST) conducts the addition reaction of *trans*-2-hexenal to glutathione in order to generate Glut-3SH. Thus, the synthesis of Cys-3SH from Glut-3SH requires the presence of two other enzymes: γ-glutamyl transpeptidase (GGT), which catalyzes the elimination reaction of glutamic acid, and carboxypeptidase, which is responsible for eliminating glycine to produce Cys-3SH. A study by Kobayashi *et al.* (2011) identified three key genes: *VviGST3*, *VviGST4*, and *VviGGT*. These genes code essential enzymes in the 3SH biosynthesis pathway (Kobayashi *et al.*, 2010). In this study, it was demonstrated that an induction of *VviGST3* and *VviGST4* is followed by an accumulation of Glut-3SH. These same authors also demonstrated *in vitro* that the conjugation of GSH with *trans*-2-hexenal is conducted through *VviGST3* and *VviGST4* activity, thus enabling the formation of Glut-3SH. The latter may be converted into Cys-Gly-3SH by GGT and then into Cys-3SH by a carboxypeptidase (Helwi *et al.*, 2016).

The concentration of the two precursors of 3SH (Glut-3SH and Cys-3SH) increase during maturation in the berry. Their concentrations are considered to reach a maximum 16 weeks after flowering (Peyrot des Gachons *et al.*, 2002; Kobayashi *et al.*, 2010; Cerreti *et al.*, 2015). This increase seems to be dependent on environmental factors, such as climate and soil type (especially water status and nitrogen nutrition), grapevine clone, and rootstock. Moreover, the harvest mode (handpicking versus mechanical), transportation time, and pre-fermentation operations (such as skin maceration, pressing, juice settling, or even stabilization) have been shown to influence precursor concentration in grapes (Peyrot des Gachons *et al.*, 2005; Choné *et al.*, 2006; Thibon *et al.*, 2008; Capone *et al.*, 2010; Roland *et al.*, 2010; Ugliano *et al.*, 2011).

The concentration of the glutathione precursor of 3SH (Glut-3SH) is much higher than that of the cysteine precursor (Cys-3SH) in both the berry and in must (Figure 10.27). It may reach concentrations close to 600 µg/l in Sauvignon Blanc must (Capone *et al.*, 2010; Grant-Preece *et al.*, 2010; Capone *et al.*, 2011; Schüttler *et al.*, 2011; Capone *et al.*, 2012).

The vine's nitrogen status may regulate the precursor concentration of the berry and must independently from the resulting vine vigor. A recent study indicated a significant increase in Glut-3SH concentration in response to an increase in nitrogen supply. However, there was no noticeable effect on the Cys-3SH concentration or on the expression of the *VviGST3*, *VviGST4*, and *VviGGT* genes in its synthesis pathway. RNA sequencing has been used to identify new candidate genes for the 3SH precursor biosynthesis pathway (Helwi *et al.*, 2016).

10.4 Definition of Ripeness–Concept of Vintage

10.4.1 Ripeness

The various biochemical processes just described are not necessarily simultaneous phenomena with identical kinetics. Environmental conditions can modify certain transformation rates, sometimes to the point of upsetting the order of physiological changes in the ripening fruit. In contrast with veraison, which is a fully defined physiological and biochemical incident (Abbal *et al.*, 1992), grape ripeness does not constitute a precise physiological stage. On the contrary, different degrees of ripeness can be distinguished. Biologists consider that the different parts of the berry ripen

344　Chapter 10　The Grape and Its Maturation

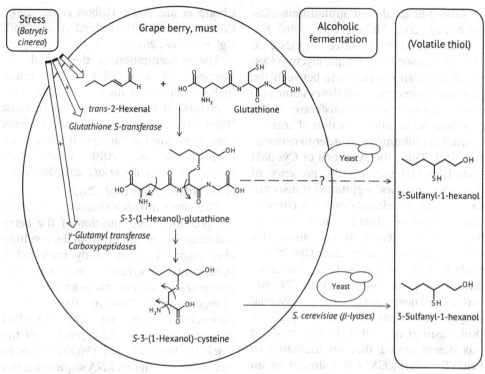

FIGURE 10.27 Biosynthesis of cysteine S-conjugates and then biotransformations during the fermentation process.

successively. The seeds are the first to attain physiological maturity (the ability to germinate) during the period preceding veraison. Over several weeks, the pulp and the skin continue to evolve through a maturation process similar to senescence (alteration of the cell wall, accumulation of secondary metabolites).

In enology, pulp ripeness corresponds to an optimal sugar/acid ratio; skin ripeness is the stage at which the phenolic compounds and aroma substances attain a maximum concentration. These two kinds of ripeness can be distinguished, but the dissociation of the cell wall from the skin must be sufficiently advanced to permit easy extraction of these essential constituents.

Consequently, the definition of ripeness varies, depending on the objective. For example, the production of dry white wines requires grapes whose aroma substances are at a maximum concentration and whose acidity is still sufficient. In certain situations, an early harvest can be desirable. Conversely, when a quality red wine is desired, grape development must be left to continue to obtain the most easily extractable phenolic compounds.

In general, grape maturation results from several biochemical transformations that are not necessarily related to each other. To simplify matters, the increase in sugar concentration and the decrease in acidity are monitored. The accumulation and refinement of white grape aromas and phenolic compounds in red grapes should also be taken into account. The essential property of a quality *terroir* is to permit a favorable maturation. This corresponds to a harmonious evolution of the various transformations in order to reach the optimum point simultaneously at the time of the harvest (Section 13.2.2).

In too cold of a climate, maturation cannot be satisfactory, but in very hot climates, the increase in sugar concentration can impose a premature harvest, even though the other grape constituents are not fully ripe. Of course, environmental conditions (soil, climate) are involved in these phenomena.

10.4.2 Sampling and Study of Maturation

Monitoring maturation poses problems relating to the large variability of berry composition. When precise data are sought in order to compare the diverse constituents of grapes from one vineyard to another, from one week to another or even one year to another, grape sampling methods are of prime importance. Nothing is more heterogeneous than grapes from the same vineyard at a given moment, even if the same variety is considered.

On a grape cluster, the grapes are formed, change color and ripen one after another over a period of up to two weeks, or more under certain difficult conditions. On the same vine, the different grape clusters are never at the same ripeness level. The clusters closest to the trunk contain more sugar than those at the ends of the shoots. The ripest grapes are in general the furthest from the ground, as the sap is preferentially conveyed toward the highest points (apices) and into the longest shoots. These differences are even greater when different vines are considered—some vines always develop more quickly than others. It is therefore risky to determine the harvest date from a single vine sampled at random.

Due to this great heterogeneity, a proper monitoring of the maturation of a given parcel requires regular sampling of a sufficient quantity of grapes: 1.5–2 kg, or about 1,000–2,000 grapes. A large number of samples are required to ensure that the results are representative of the plot (Blouin and Guimberteau, 2000). The most common method consists in gathering, with shears, three or four grape cluster fragments from 100 vines. Grape clusters under the leaves as well as those directly exposed to sunlight should be gathered, taking them alternately from each side of the row at different heights on the vine. When sampling varieties with compact clusters, this method does not generally take into account the berries located at the interior of the cluster. These berries are often less ripe than the others. In this case, whole grape clusters should be sampled to obtain a precise idea of the maturation level of the parcel.

In the laboratory, the berries are separated, counted, and weighed. The juice is extracted with the help of a small manual press or a centrifugal fruit juice separator. The juice volume is measured and the results are expressed per liter of must. The juice sugar and acid concentrations are then determined.

The study of red grape phenolic compounds requires the manual separation of the skins from the seeds of about 500 berries taken at random from the sample. Once separated, the skins and the seeds are dried and lyophilized to facilitate the extraction and the determination of their phenolic content.

Aroma substance monitoring, notably of white grapes, requires the prior maceration of the solid grape parts with the must. After a light crushing of the grapes, this maceration is usually carried out for 16 or 24 hours at a low temperature under a carbon dioxide atmosphere. These techniques require advanced equipment and cannot yet be routinely monitored.

10.4.3 Evaluation of Ripeness: Maturity Index

Grape monitoring during maturation helps vineyard managers and winemakers set the harvest date and maximize the efficiency of their harvest crews according to the ripeness of different cultivars and diverse parcels.

Determining the grape sugar concentration is essential. It is most often done by an indirect physical measure such as hydrometer or refractometer. If the temperature is not at 20°C, a correction is theoretically necessary, but has little effect on the sugar concentration. The results are expressed in various units, depending on the instruments used. This does not facilitate the interpretation of data originating from different wine-producing countries (Blouin, 1992; Boulton et al., 1999).

These assorted measurement scales are compared in Table 10.4. A degree Oechslé corresponds to the third decimal of the apparent specific gravity (SG). The degree Baumé is approximately converted to apparent specific gravity by the following formula:

$$d°Baumé \approx 144.32(1-1/SG).$$

The degree Baumé of a must corresponds fairly well with the percentage alcohol, at least for values between 10 and 12. The degree Brix (or degree Balling) gives the weight of must sugars, in grams, per 100 g of must. In reality, it is a percentage of the

TABLE 10.4
Conversion Table for Various Scales Used to Measure Must Sugar Concentration

Apparent specific gravity (20/20°C)	Degree Oechslé	Degree Baumé	Degree Brix	Refractometric measure (in percentage weight of sucrose)	Sugar concentration (g/l)	Potential alcohol (16.83 g of sugar/l for 1% vol. alcohol)
1.0371	37.1	5.2	9.1	10	82.3	4.9
1.0412	41.2	5.7	10.1	11	92.9	5.5
1.0454	45.4	6.3	11.1	12	103.6	6.2
1.0495	49.5	6.8	12.0	13	114.3	6.8
1.0538	53.8	7.4	13.0	14	125.1	7.4
1.0580	58.0	7.9	14.0	15	136.0	8.1
1.0623	62.3	8.5	15.0	16	147.0	8.7
1.0666	66.6	9.0	16.0	17	158.1	9.4
1.0710	71.0	9.6	17.0	18	169.3	10.1
1.0754	75.4	10.1	18.0	19	180.5	10.7
1.0798	79.8	10.7	19.0	20	191.9	11.4
1.0842	84.2	11.2	20.1	21	203.3	12.1
1.0886	88.6	11.8	21.1	22	214.8	12.8
1.0932	93.2	12.3	22.1	23	226.4	13.5
1.0978	97.8	12.9	23.2	24	238.2	14.2
1.1029	102.9	13.5	24.4	25	249.7	14.8
1.1075	107.5	14	25.5	26	261.1	15.5
1.1124	112.4	14.6	26.6	27	273.2	16.2
1.1170	117.0	15.1	27.7	28	284.6	16.9
1.1219	121.9	15.7	28.8	29	296.7	17.6
1.1268	126.8	16.2	29.9	30	308.8	18.4
1.1316	131.6	16.8	31.1	31	320.8	19.1
1.1365	136.5	17.3	32.2	32	332.9	19.8
1.1416	141.6	17.9	33.4	33	345.7	20.5
1.1465	146.5	18.4	34.5	34	357.7	21.3

dry matter in must, measured by refractometer or hydrometer. This measure is only valid from a certain level of ripeness onward (15° Brix). Before this level, organic acids, amino acids, and certain precursors of cell wall polysaccharides can have similar refraction indexes to sugar and interfere with the measurement.

In the same way, the relationship between must density and alcohol content is always approximate, since sugar is not the only chemical must constituent that affects density. This measurement is more accurate in white winemaking with non-mucilaginous musts having few suspended particles. The values obtained for must from moldy grapes are inaccurately high. Moreover, the estimation of potential alcohol should take into account the sugar/alcohol transformation ratio. The figures in Table 10.4 use the relationship of 16.83 g/l of sugar per liter for 1% alcohol—the official value retained by the EU.

Empirical observation of the inverse variation of sugars and acidity during maturation has led to the development of a sugar/acidity ratio, called the maturity index. This index is very simple, but it should be used with precaution, since there is no direct biochemical relationship between sugar accumulation and acidity loss. More specifically, a given gain in sugar does not always correspond with the same drop in acidity. This ratio is not suitable for comparing different varieties, since varieties exist that are rich both in sugar and in acids. In France, this ratio is calculated from the must sugar concentration (g/l) and the titratable acidity expressed in grams of sulfuric or tartaric acid equivalents per liter. Other modes of expression are used in other countries according to the measurement unit used to express the sugar concentration. In Germany, for example, the ratio obtained by dividing the degree Oechslé of must by the acidity, expressed as tartaric acid, is currently used.

Attempts have been made in the past to describe ripeness, taking into account the respective variations of malic and tartaric acids or the accumulation of cations, but none of the indexes developed have significantly improved the evaluation of the ripeness level. It seems sensible to take into account the individual variations of each berry constituent separately.

In the past 30 years, researchers have focused on the evolution of phenolic compounds during maturation (Volume 2, Section 6.5.4), but the technique of separating the skins from the seeds is awkward and exacting, limiting its practical applicability. There is now a rapid whole-berry grinding technique enabling the evaluation of the color and tannic potential of berries (Glories method, CASV method). The grape grinding is followed by a differential phenolic compound extraction, in either a pH 3.2 buffer (easily extractable compounds) or a pH 1 buffer (total potential of phenolic compounds). The optical density of the solutions obtained is then measured at 280 nm. Information on the total phenolic concentration and their extractability is thus obtained.

Micro-imagery by nuclear magnetic resonance has been shown to give detailed information on the chemical composition and degradation level of grape cell walls (Pope et al., 1993; Lacampagne, 2010), but this technique is still reserved for scientific experimentation. Recently, physical measurements of free water at the surface of the skin enable the evaluation of skin porosity and the estimation of the sensitivity level to *Botrytis cinerea*. It is now possible to assess the skin's texture as well as its hardness using penetrometer measurements (Grotte et al., 2001; Le Moigne et al., 2008). While still limited to the treatment of a small number of samples, these techniques have been developed (Deytieux-Belleau et al., 2009) and make it possible to establish the Skin Permeability Index with respect to skin degradation mechanisms (Lacampagne, 2010).

Fourier transform infrared spectrometry, which has recently been developed, should make it possible to assess grape quality more accurately (Dubernet et al., 2000). This method is easy to implement and only requires prior filtration of the samples. It provides a satisfactory evaluation of the potential alcohol, total acidity, pH, and nitrogen content, as well as the color index for red grapes, in a single operation. In addition to this general analysis of the grapes, it is also possible to detect the presence of mold (gluconic acid, laccase activity, etc.) or fermentation activity (lactic acid, pyruvic acid, etc.). This new technique, however, only gives reliable results after a long, laborious calibration process using samples analyzed by standard methods.

Recently, comprehensive, routine, and nondestructive techniques have been developed to obtain immediate results without sampling berries. These techniques are based on spectroscopy, infrared spectrometry, fluorimetry, or computer vision to evaluate the accumulation of sugars, fruit coloration, or the potential evolution of aroma. However, since these sensors provide direct or indirect measurements that must be stored, transformed, and analyzed, they often require a high level of expertise to understand the results. Getting these innovations to coexist and adapting them to correspond to the real needs of winemakers is still a challenge.

Unfortunately, as of right now, there is no simple method to obtain a maturity index regarding aroma substances (Volume 2, Section 7.7). In fact, even though it is possible to conduct assays in research laboratories much quicker than 20 years ago (in 2000), analyses require at least one hour for sample preparation and for analysis. Grape tasting therefore still remains a pertinent criterion, though it cannot predict the revelation of other aromas at a later date. However, a significant number of berries must be included in this tasting approach at the parcel level to be significant.

10.4.4 Effect of Light on the Biochemical Maturation Process

Three factors have major roles in maturation dynamics: light, heat, and water availability. In general, they affect vine growth and metabolic activity; their action is well known. Yet these also act directly on grapes, and their effects on metabolic pathways translate into changes in grape chemical composition.

In established grape-growing zones, the availability of natural light does not, in general, limit photosynthetic activity and thus the overall functioning of the plant. In fact, photosynthesis is optimum at a sun radiance (expressed in einsteins, E) of about $700 \, E/m^2/s$. Below $30 \, E/m^2/s$, leaf energy consumption is greater than net photosynthetic production (Smart, 1973). In the absence of clouds, sun radiance is greater than $2,500 \, E/m^2/s$. On overcast days, the radiance varies from 300 to $1,000 \, E/m^2/s$. A reduction in photosynthetic activity can thus occur, resulting in a nutrient deficiency in the grape. However, in practice, certain vine trellising methods still cause radiant energy loss. For this reason, wine growers should ensure that the spacing between vine rows is in proportion to the height of the foliage (0.6–0.8) and should avoid leaf crowding by thinning unwanted shoots in the center of the canopy.

Light has a direct effect on floral induction. Grape cultivar fertility depends greatly on bud light exposure during this induction period.

The effects of sunlight on grape composition are even more numerous and complex. In addition to furnishing the energy for photosynthesis and stimulating certain light-dependent metabolic processes, its radiant effect heats not only surfaces but also the air surrounding plant tissue. Grape clusters grown with little light exposure ("shade grapes") always contain less sugar and have a lower pH and higher total acidity and malic acid concentration than grape clusters directly exposed to sunlight. Light is also essential for phenolic compound accumulation, since phenylalanine ammonia lyase (Section 10.3.7) is a photoinductive enzymatic system. Under normal conditions, this photoactivation does not seem to be a factor that limits color formation or phenolic compound concentrations in most varieties. Crippen and Morrison (1986) and Haselgrove *et al.* (2000) showed that the phenolic composition of shaded and light-exposed grape clusters remained the same in Cabernet Sauvignon and Syrah. Only certain sensitive red varieties (Ahmeur Bou Ahmeur, Cardinal, or Emperor) may exhibit color deficiencies when their grape clusters are not exposed to light. Inversely, overexposure may also harm anthocyanin synthesis (Bergqvist *et al.*, 2001). In certain northern vineyards (in the Northern Hemisphere), wines made in climatically unfavorable years are always poorly colored.

The amount of light reaching the grapes also has an impact on the composition and aroma qualities of the grapes. Exposure to the sun accentuates the decrease in methoxypyrazine content during the ripening of Cabernet Sauvignon grapes (Section 10.3.8). Overexposure to light changes the balance of aromas and aroma precursors by increasing volatile compound concentrations related to notes of cooked fruit, particularly some furanones such as furaneol and homofuraneol and lactones such as γ-nonalactone and massoia lactone (Allamy, 2015; Allamy *et al.*, 2018). Conversely, partial shade preserves the floral aromas in Muscat grapes (Belancic *et al.*, 1997).

10.4.5 Influence of Temperature on the Biochemical Processes of Maturation

Temperature is one of the most important parameters of grape maturation and one of the essential factors that triggers it. Temperature affects photosynthetic activity, metabolism, and migration intensity in the vine. Its action is not limited to the period of grape development. Its influence on bud break and flowering dates also has important indirect consequences on grape quality. It is easy to understand that the later the grape develops, the greater the risk that the accompanying maturation conditions will be unfavorable.

Grape growth and development are directly affected by temperature. High temperatures are unfavorable to cell multiplication. During the green growth phase, the optimum temperature is between 20 and 25°C. During maturation, temperature affects migration intensity and thus, indirectly, cell growth. Vine temperature requirements during this period are around 20°C (Calo *et al.*, 1992). Too high of a temperature, even for a short time, can irreversibly alter sugar accumulation. Sepulveda and Kliewer (1986) found that temperatures of 40°C during the day and 20°C at night favored sugar accumulation in other parts of the vine to the detriment of the grapes, which received only a small percentage (about 2.5%), with respect to the control (25°C day/15°C night).

As vines have difficulty growing and producing grapes below 10°C, temperatures above this threshold are known as "active temperatures." A strong correlation exists between the sum of the active temperatures during grape development and the grape sugar concentration in a given location. This measurement permits the evaluation of the climatic potential of a given location to ensure suitable grape maturation. Various bioclimatic indexes have been developed to evaluate this potential:

1. Growing degree-days (Winkler, 1962) are the sum of the average daily temperatures above 10°C from April 1 to October 30, a seven-month period. This sum is often calculated using monthly averages. Initially established for classifying California into different viticultural zones, this index has become widely used in other countries. The climate data for the month of October are not useful: in warm zones, the grape has already been harvested; in cool zones, the average temperature in October is often below 10°C. Furthermore, this index does not take the duration of light exposure into account.

2. The Branas Heliothermic Product (Branas et al., 1946) corresponds to the formula $X \times H \times 10^{-6}$. X is the sum of the average active temperatures above 10°C for the entire year, and H represents the sum of the length of the days for the corresponding period. Vine growing is practically impossible when the product is below 2.6. This index gives the most precise results for vineyards established in cool temperate climates where the end of the period containing active temperatures more or less corresponds with harvest time. In extreme cases of hot climates, this period covers the entire year.

3. In order to obtain a better correlation between bioclimatic data and final grape sugar concentrations, Huglin (1978) proposed a heliothermic index (HI). This index takes into account the maximum daily temperatures over a six-month period from April 1 to September 30. In this relationship

$$HI = \sum_{April\ 1}^{Sept\ 30} \left\{ \left[(ADT - 10) + (MDT - 10) \right] \times K \right\} / 2$$

where ADT represents the average daily temperature, MDT is the maximum daily temperature, and K is the day-length coefficient—varying from 1.02 to 1.06 between latitudes of 40 and 50°. An HI of around 1,400 is the lowest limit for grape growing. This index has made it possible to specify the needs of different varieties in order to attain a given sugar concentration. The comparison of these different indexes (Table 10.5) shows the difficulty of evaluating the viticultural potential of an area based solely on a temperature criterion, even when corrected for light exposure time. These indexes are, however, useful in choosing early- or late-ripening grape varieties to plant in a new vineyard.

In most European viticultural zones, grape varieties are chosen to reach maturity just before the average monthly temperature drops below 10°C. In warmer climates, this drop occurs later. Consequently, maturation takes place during a warmer period. Viticultural zones can thus be classified into two categories, Alpha and Beta, depending on whether the average temperature at grape maturity of a given variety is below or above 15°C (Jackson, 1987). In the Northern Hemisphere, the ideal time for cluster maturation is between September 10 and October 1 (Van Leeuwen and Seguin, 2006).

Temperature also strongly influences many biochemical mechanisms involved in grape maturation. For example, malic acid degradation is considerably accelerated during hot weather: malic enzyme activity (Section 10.3.3) steadily increases between 10 and 46°C. Temperature does not directly influence tartaric acid concentrations. Elevated respiratory quotients, witnessed at temperatures greater than 35°C, were initially interpreted as the respiratory oxidation of tartaric acid, but at

TABLE 10.5
Comparison of Different Methods of Evaluating Climates

Viticultural zones	Sum of degree-days (Winkler, 1962)	Heliothermic product (Branas et al., 1946)	Heliothermic index (Huglin, 1978)
Zone 1 = less than 1,390°C			
Geisenheim	995°C	2.6	—
Geneva	1,030°C	2.5	—
Dijon	1,133°C	—	1,710
Coonawarra	1,205°C	—	—
Bordeaux	1,328°C	4	2,100
Zone 2 = 1,390–1,670°C			
Odessa	1,433°C	—	1,850
Santiago	1,506°C	—	2,290
Napa	1,600°C	—	2,130
Bucharest	1,640°C	—	—
Zone 3 = 1,670–1,950°C			
Montpellier	1,785°C	5.24	2,256
Milan	1,839°C	—	—
Zone 4 = 1,950–2,220°C			
Venice/Verona	1,960°C	—	2,250
Mendoza	2,022°C	2.7–7.8[a]	2,600
Cape Town/ Stellenbosch	2,066°C	—	2,350
Zone 5 = more than 2,220°C			
Split	2,272°C	—	—
Palermo	2,278°C	—	(Bari) 2,410
Fresno	2,600°C	—	3,170
Algiers	2,889°C	—	2,600

[a] Depending on altitude.

such a temperature this activity corresponds more to the initiation of fermentation phenomena in grape pulp—essentially acting on malic acid (Romieu et al., 1989).

Temperature also has an influence on the composition of grape phenolic compounds. Intensely colored wines are known to be difficult to obtain under extreme temperature conditions (too low or high), though the phenomenon involved can at first appear paradoxical. High temperatures stimulate metabolic reactions, whereas low temperatures curb migration. In either case, however, this corresponds with poor grape sugar supply and thus increased competition between primary metabolism (growth) and secondary metabolism (accumulation). Sugar and anthocyanin accumulation is stopped at high temperatures (Sadras and Moran, 2012). The concentration of phenolic compounds is also affected by temperature variations (e.g. day/night) (Kliewer and Torres, 1972; Yamane et al., 2006).

Raising the nighttime temperature from 15 to 30°C while maintaining a daytime temperature of 25°C results in a decrease in grape coloration. The anthocyanins are therefore not blocked metabolites but, on the contrary, are reversible (Mori et al., 2007). Thus, temperature and sun exposure determine phenolic compound accumulation.

Temperature also exerts a considerable effect on aroma substances. It can be difficult however to separate the effects of temperature from those of light. Water stress should also be taken into consideration. The aroma potential of certain white cultivars (Gewürtztraminer, Riesling, Sauvignon Blanc) is known to be fully expressed only in cool climates, where the maturation period is slow and long. By comparing a cool winemaking zone with a warmer zone in South Australia, Ewart (1987) showed that the total volatile terpene quantity increased more slowly in the cool zone but was higher at maturity. In a cool climate, and especially with shaded clusters, methoxypyrazine concentrations can attain more significant levels (Volume 2, Section 7.4). Conversely, warm climates can lead to high concentrations of certain phenolic compounds in white varieties. These compounds confer an excessively astringent character to the wine and lead to the development of a petrol-like odor during aging (Herrick and Nagel, 1985) (Volume 2, Section 7.3). Moreover, empirically, white winemakers know that significant diurnal temperature variation is favorable to the aroma composition of wines.

More so than temperature itself, the hottest period of the growing season changes berry metabolisms in different ways. Thus, a +8°C increase at the full cluster stage will not have the same impact as the same heat shock close to harvest. The work of Lecourieux et al. (2017) demonstrated that heat shock on berries just before veraison delays veraison by shifting the ABA accumulation peak. This shock also limits the appearance of color, slows down fruit texture changes, and reduces the transport of assimilates to the berry (Lecourieux et al., 2017).

Despite the lack of specific experiments, excessive temperatures are known not to be the most favorable conditions for aroma quality. A recent review provided an overview of this subject (Drappier et al., 2019).

10.4.6 Impact of Grapevine Water Supply on Grape Ripening

Effect of water availability on the biochemical processes involved in grape ripening

Unlike most plants, particularly annual crops, vines are generally grown under less than optimum conditions. Various types of environmental constraints are considered to reduce vine vigor and yields while maximizing the winemaking potential of the grapes. Among these constraints, a limited water supply plays a major role in vine behavior and grape composition. A moderately restricted water supply, known as "water deficit," generally has a beneficial effect on wine quality. The expression "water stress" should only be used in situations where an excessive lack of water has a negative impact on grape quality or threatens to kill the vines.

Most high-quality wines are produced in areas where annual precipitation does not exceed 700–800 mm. Evidence indicates that high rainfall and excessive irrigation are detrimental to grape quality.

Before veraison, water is mainly transferred to the grapes via the xylem, and there are close hydraulic relationships between the grapes and the rest of the vine. Any change in the vine's water supply affects raw sap (xylem) circulation and, consequently, grape development. The resulting irreversible reduction in grape size is positive from a qualitative standpoint but also reduces yields. In some countries, the climate may necessitate controlled irrigation of the vines to compensate for water losses via transpiration. After veraison, the deterioration of xylem circulation leads to a concomitant increase in flows via the phloem. At that stage, the phloem provides the main water supply to the grapes. As phloem sap circulation is not directly related to the vine's water supply, grape growth becomes much less dependent on this factor. A minimum water supply is still necessary, however, for the biochemical ripening processes to proceed normally.

Duteau et al. (1981), Matthews and Anderson (1989), and Van Leeuwen and Seguin (1994) showed that water stress caused an increase in the phenolic content of grape juice and skins, with a higher concentration of proline and a lower malic acid content. Inadequate water supply also leads to higher concentrations of terpene compounds (MacCarthy and Coombe, 1984). Conversely, an abundant water supply leads to an increase in grape volume, with a concomitant decrease in phenolic content. Although the acid concentration is often higher, the juice still has a higher pH (Smart and Coombe, 1983). This is due to an increase in imports of tartaric acid and minerals, especially potassium. The aroma compounds are also modified, e.g. excess water gives Semillon grapes a strong herbaceous aroma (Ureta and Yavar, 1982).

While water deficit does not prevent grapes from ripening satisfactorily in terms of their sugar and acid contents, excess water can delay the ripening process and alter the chemical composition of the grapes to a considerable extent. In vineyards where irrigation is used, it should be reduced to a minimum after veraison to maintain a moderate water deficit. Optimal quality is obtained during average to moderate water deficit (Chaves et al., 2010).

Finally, heavy rain when the grapes are close to ripening is likely to cause them to burst due to a sudden absorption of water directly through the skins. This phenomenon is less marked at lower temperatures and depends on respiratory intensity.

Monitoring vine water levels

Studying the vine's response to different levels of water supply requires reliable, easily used indicators of water availability in the soil or the water status of the vine.

The first studies of grapevine reactions to water supply in the late 1960s were based on water balances, carried out using a neutron moisture tester (Seguin, 1970). A probe emitting fast neutrons is inserted in an access tube that stays permanently in the soil. The neutrons are slowed down to a state of thermal agitation when they meet hydrogen atoms. The vast majority of hydrogen atoms in soil are in water molecules; the number of thermal neutrons counted per unit time is thus proportional to the dampness of the soil (humidity by volume). The vines' water consumption between two measurements is calculated by subtracting the second reading from the first and correcting, if necessary, for any precipitation during the interval. Neutron moisture tester studies were used to obtain a detailed view of the water supply in gravel soils in the Haut-Médoc region (Seguin, 1975), clay soils in Pomerol, and asteriated limestone in Saint-Emilion (Duteau et al., 1981). Although this was a highly innovative technique at the time, it had several disadvantages. The water balance calculated using this method does not take into account any horizontal inflows of water through the soil or runoff, which may be significant on sloped vineyards. After a period of time, roots develop around the access tube and distort the results (Van Leeuwen et al., 2001a). Finally, grapevine root systems are often very deep, and vineyard geology (gravel, rocky soil, etc.) may make it particularly difficult to install the access tube. Even though neutron moisture testers are used in some New World countries to control vineyard irrigation, the complexity of this technique prevents it from being used more widely. Using time domain reflectometry (TDR) to establish the vineyard water balance is subject to the same difficulties.

Producing a theoretical water balance by modeling is another approach to determine the grapevine water supply. The aim is to simulate the water reserves remaining in the soil during the summer on the basis of data on the water available at the start of the season, plus any precipitation, minus losses via evapotranspiration. The most advanced model was developed by Riou and Lebon (2000). In this formula, precipitation could be determined accurately and evapotranspiration estimated correctly. The main difficulty with this approach is estimating the water reserves at the beginning of the season, which is particularly complex due to the specific conditions in which vines are grown (deep root systems, rocky soil, etc.).

In view of the difficulty in assessing the vines' water balance on the basis of measurements in the soil or modeling, it seemed more practical to measure water levels in the plants themselves. A water deficit causes several measurable alterations in the vine's physiological functions: variations in xylem sap pressure, closing of the stomata, slowdown in the photosynthesis process, etc. When a plant is used as an indicator of its own water status, we refer to "physiological indicators." Among these indicators, leaf water potential is undoubtedly the most widely used because it is reliable and easy to implement. Water potential is measured by placing a freshly picked vine sample (usually a leaf) in a pressure chamber, connected to a bottle of pressurized nitrogen. Only the leaf stalk remains outside the chamber, via a small hole. Pressure in the chamber is gradually increased, and the pressure required to produce a sap meniscus on the cut end of the stem is noted. This pressure corresponds to the inverse of the water potential: the higher the pressure required to produce the meniscus on the leaf stem, the more negative the water potential and the greater the water deficit to which the vine has been subjected.

There are three applications for water potential measurements using a pressure chamber: leaf potential, basic leaf potential, and stem potential (Choné et al., 2000).

1. Leaf water potential is measured on a leaf that has been left uncovered on a sunny day. This value only represents the water potential of a single leaf. Even if this potential depends on the water supply to the vine, the considerable variability from one leaf to another on the same vine (e.g. due to different sun exposure) leads to a large standard deviation on this measurement, making the value less significant as an indicator.
2. Predawn leaf water potential is measured in the same way as leaf water potential, except it is measured just before sunrise. The stomata close in the dark, and the water potential in the vine comes back into equilibrium with that in the soil matrix. Predawn leaf water potential reflects water availability in the most humid layer of soil in contact with the root system. Therefore, it provides a more stable value that is easier to interpret than leaf water potential measured during the day. It is, however, more difficult to apply, as it requires specific conditions.
3. Stem water potential is measured during the day, on a leaf that has been covered by an opaque, airtight bag for at least one hour before the measurement is made. The leaf stomata close in the dark and the leaf water potential balances with that of the xylem in the stem. This measurement gives a close approximation of the water supply of the whole plant during the day. Provided certain conditions are observed (measuring time and weather conditions), stem water potential is the most accurate of the three pressure chamber applications (Choné et al., 2001a,b).

Carbon 13 isotope discrimination is another physiological indicator of water supply. This isotope represents approximately 1% of the carbon in atmospheric CO_2 and the lighter isotope, ^{12}C, is preferentially involved in photosynthesis. Water deficit causes the stomata to close for part of the day, which slows down CO_2 exchanges between the leaves and the atmosphere and reduces isotopic discrimination. Under these conditions, the $^{13}C/^{12}C$ ratio (known as $\Delta C13$) becomes closer to the ratio in atmospheric CO_2. Measuring $\Delta C13$ in the sugars in must made from ripe grapes (analyzed by a specialized enology laboratory) provides an indicator of the global water deficit to which the vines have been subjected during ripening. $\Delta C13$ is expressed in ‰ in relation to a standard. Values range from −21 to −26‰, where −21‰ indicates a considerable water deficit and −26‰, the absence of water deficit. The advantage of this indicator is that it does not require any field operations other than taking a sample of ripe grapes (Van Leeuwen et al., 2001b; Gaudillère et al., 2002). There is a good correlation between the $\Delta C13$ value measured in must and the stem water potential at maturity.

Impact of water status on grapevine growth and the composition of ripe grapes

A water deficit during the growing season causes profound changes in the physiological functions of the vine. It may progress at varying rates, as shown by the changes in stem water potential measured in the same plot of Saint-Emilion vines in 2000 and 2002 (Figure 10.28). When there is a water deficit, the stomata remain closed for part of the day, increasingly restricting photosynthesis as the deficit becomes more severe. A reduction in water supply tends to stop shoot and grape growth, affecting the grapes especially before veraison (Becker and Zimmermann, 1984). When the soil dries out around the roots, the root tips produce abscisic acid, a hormone that promotes grape ripening. Restricting the water supply to the vine has both negative (restricting photosynthesis) and positive (abscisic acid production, less competition for carbon compounds from the shoot tips, and smaller fruit) effects on grape ripening. If the water deficit is moderate, the positive effects are more marked than the negative factors: the grapes contain higher concentrations of reducing sugars, anthocyanins, and tannins, while the malic acid content is lower (Van Leeuwen and Seguin, 1994). For example, Saint-Emilion wines from the 2000 vintage, when there was an early drop in stem water potential, are better than those from the 2002 vintage (Figure 10.28). In cases of severe water stress, photosynthesis is too severely restricted and ripening may stop completely.

In viticulture, it is essential to know to what extent a water deficit has a positive effect on quality and to locate the threshold of harmful water stress. The answer to this question depends on the type of production, the types of substances considered, and vine yields.

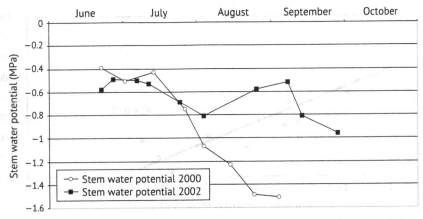

FIGURE 10.28 Comparison of variations in stem water potential in a Saint-Emilion vineyard in 2000 and 2002 (gravelly soil and Merlot grapes). The more negative the values, the more severe the water deficit.

Most studies concerning the link between the grapevine water status and grape composition have dealt with red wine grapes. It is generally accepted that red wine grapes can benefit from more severe water deficits than white grapes. On an estate producing both types of wine, it is therefore logical to plant the red varieties on soils with less plentiful water reserves.

Among the substances that promote red wine quality, sugar accumulation reaches maximum levels when the water supply is moderately restrictive. Grape sugar content is lower both when there is a less restrictive water supply and in cases of severe water stress (Figure 10.29). The anthocyanin content increases in a linear manner over the same range of water deficits, reaching a maximum when the water stress is greatest (Figure 10.30). The quality of a red wine depends more on its phenolic content than on the sugar content of the ripe grapes, so red wine grapes may have the potential to make excellent wine, even if severe water stress has penalized the sugar level of the must.

The issue of the effect of water deficits on quality cannot be settled without discussing yields. The same water deficit may have a positive effect on quality in a vineyard with yields of 30 hl/hectare and lead to blocked ripening with disastrous results at 60 hl/hectare.

Impact of grapevine water status on grape aroma potential

For a long time, the evolution of aroma potential in non-Muscat varieties during maturation could not be studied, due to the

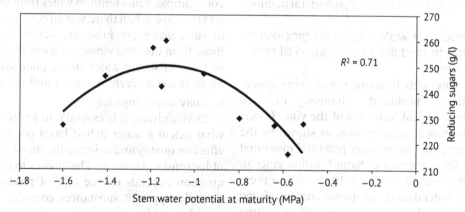

FIGURE 10.29 Correlation between the intensity of water deficit (assessed by the stem water potential at maturity) and the concentration of reducing sugars.

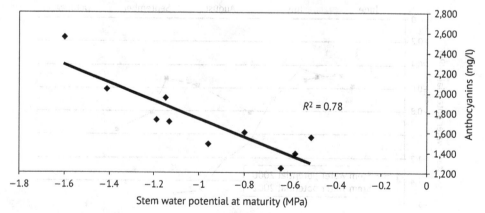

FIGURE 10.30 Correlation between the intensity of water deficit (assessed by stem water potential at maturity) and the anthocyanin concentration.

lack of knowledge about the chemical nature of their aromas and their aroma precursors. Consequently, the study of the influence of soil, climate, and cultural methods on the aroma expression of these varieties could only rely on wine tasting, or even more uncertainly on grape tasting. Nevertheless, winemakers know that, without irrigation, it is difficult to make Sauvignon Blanc wines and more generally aromatic white wines in overly hot or dry climates and/or in water-deficient soils. Knowledge of grape aroma precursors (cysteine and glutathione S-conjugates, glycosides) has helped clarify empirical know-how (Section 10.3.8 and Volume 2, Sections 7.2, 7.5).

The concentrations of cysteine S-conjugate precursors of 4MMPOH in Bordeaux varieties remain stable or increase during grape maturation, while 4MMP and 3MH precursors vary in a more irregular manner (Peyrot des Gachons, 2000). The latter do not follow an accumulation or degradation dynamic. Their evolution depends on the S-conjugate, the type of soil, and climate. In the Bordeaux region, early and intense water stress can manifest itself over a dry summer in certain filtering gravelly soils with a low water reserve. The cysteine precursor concentration of grapes at harvest is lower than in limestone soils. In fact, the latter have a greater water supply because of capillary rise from porous rocks, reducing the water deficit for the vine (Peyrot des Gachons et al., 2005).

Choné et al. (2001b) sought to specify this concept of moderate water deficit of the vine and its favorable effect on the aroma potential of Sauvignon Blanc. Aroma potentials of grapes from Sauvignon Blanc vines exposed to two levels of water deficit were compared at maturity: an unlimited water supply (deep soil) and a moderate water deficit obtained by covering the soil with a waterproof tarp from early June until harvest.

Sauvignon Blanc grapes subjected to a moderate water deficit are richer in cysteine precursors than those for which water supply was not an issue (Table 10.6).

TABLE 10.6

Influence of Grapevine Water Supply on Sauvignon Blanc Composition at Ripeness

Water supply	Unlimited (DS)	Limited (TS)
Ψ_T (MPa) on July 29, 2000 (veraison)	−0.18a	−0.70b
Ψ_T on August 10, 2000	−0.22a	−0.95b
Ψ_T on August 28, 2000 (eight days before harvest)	−0.30a	−1.10b
Harvested weight/vine (kg)	3.6a	2.9b
Primary leaf area (m^2)	3.257a	2.60b
Secondary leaf area (m^2)	3.804a	1.63b
Average berry weight (g)	2.03a	1.81b
Reducing sugars (g/l)	178.4a	201b
Total acidity (g/l)	6.69a	4.21b
Malic acid (g/l)	4.95a	2.44b
Yeast-assimilable nitrogen (mg/l)	172a	225b
P-4MMP (ng of 4MMP Eq/l)	1,263a	2,548b
P-4MMPOH (ng of 4MMPOH Eq/l)	2,226a	2,127a
P-3MH (ng of 3MH Eq/l)	7,254a	24,288b
Phenolic compounds (IPT)	1.6a	2.31b

Ψ_T, stem water potential; DS, deep soil; TS, soil tarped from June 8 till harvest. For the same line, the values followed by a different letter are statistically different.

Moderate water deficits occurring after veraison induce a rise in cysteine precursor concentrations. However, these results do not enable the precise definition of the duration and intensity of the water deficit that would have a positive effect on the aroma potential of Sauvignon Blanc. In all likelihood, the most favorable moderate water deficits are those that appear just after veraison and are accompanied by non-sweltering temperatures. Late ripening is more favorable in that regard than early ripening, which leads to a harvest in hot weather (end of August, beginning of September).

Other studies have confirmed that the impact on the concentration of grape flavors and flavor precursors may vary depending on grapevine water deficit. For example, an Australian study showed that concentrations of (−)-rotundone, a compound associated with the black pepper notes of Shiraz wines, were lower in wines from grapes with the highest water deficit levels, estimated by measuring discrimination by carbon isotopes ($\delta^{13}C$ values) (Scarlett et al., 2014).

Conversely, it has been established that the monoterpene content of Riesling wines from the Rheingau is higher in wines from vines with a greater water deficit (Schüttler et al., 2013).

Impact of water deficit on early ripening
The date when grapes ripen depends both on the phenological cycle, which may be assessed by the date of mid-veraison, and the rate at which they ripen, calculated according to Duteau (1990). The earliness of the phenological cycle depends mainly on the soil temperature, which is related to its moisture content (Morlat, 1989). The ripening rate is largely determined by the vine's water status (Van Leeuwen and Seguin, 1994). A water deficit promotes rapid ripening by keeping the grapes small (thus making them easier to fill with sugar) and reduces the competition between grapes and shoots for carbon-containing substances. Figure 10.31 shows an example of the impact of water availability on the ripening rate and early/late maturity in three plots with very different soils (Van Leeuwen and Rabusseau, unpublished results). To eliminate the impact of temperature on the ripening rate, dates are indicated on the X-axis by the sum of active temperatures starting on August 1—each day is represented by the average

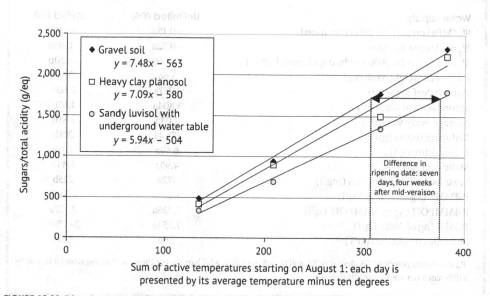

FIGURE 10.31 Ripening rates on three soils in Saint-Emilion (Merlot Noir, 2001).

temperature – 10°C. Grapevines on gravel soils and planosol were subject to water deficit, and the sugar/acid ratio in the pulp evolved rapidly toward ripeness. The water supply on luvisol was not restrictive, and the pulp ripened slowly. Although the mid-veraison dates were very close in all the plots, the difference in ripening date was as much as 70 degree-days, or nearly seven days after four weeks.

The vast majority of damp soils are cool and provide a nonrestrictive water supply. Grapes ripen late on these soils as the phenological cycle is delayed and ripening is slow. By the same reasoning, most dry soils are conducive to early ripening. There are a number of highly reputed estates in Bordeaux, especially in Pomerol, but also a few localized cases in Saint-Emilion and the Haut-Médoc, planted on soils with a high clay content. They are unusual as they have high water contents (and are thus cool) but still cause an early water deficit in the vines. This type of clay (smectite) is unusual in that although it contains large amounts of water, it is unavailable for use by the vines. These soils are conducive to early ripening, and although, for historical reasons, Merlot has been planted on them in Pomerol, Cabernet Sauvignon ripens perfectly on the same type of soil in the Haut-Médoc. This example shows that water deficits play an essential role in the early ripening of grapes and have a greater impact than soil temperature. The choice of a grape variety to suit a particular type of soil should depend mainly on its conduciveness to early ripening and thus on its water status (Van Leeuwen, 2001).

Water status and vintage effect

The water status of a given vintage can be assessed by calculating the water balance. Table 10.7 shows the water status of several vintages in Bordeaux calculated using the method developed by Riou and Lebon (2000). To eliminate the effect of the soil, we introduced a value, 0, for the water reserves at the beginning of the season, which explains the negative values of the water balance. These values indicate a theoretical water deficit, corresponding to the difference between precipitation and real evapotranspiration (in the absence of stomatal regulation). All the lesser quality vintages without exception had only a slightly negative water balance at the end

TABLE 10.7

Correlation Between the Theoretical Water Balance from April 1 to September 30 and the Quality of the Vintage. The More Marked the Negative Water Balance, the Drier the Vintage

Vintage	Theoretical water balance on September 30	Vintage quality (score out of 20)
1990	−306	19
2000	−290	19
1986	−271	18
1998	−256	18
1995	−241	17
1962	−231	17
1964	−220	17
1997	−211	15
1988	−211	17
1970	−210	18
1961	−207	20
1991	−206	13
1989	−204	19
1985	−198	18
1977	−87	11
1993	−80	14
1954	−80	9
1971	−40	17
1956	−39	9
1968	−33	6
1958	−31	12
1969	−14	12
1973	−12	12
1965	−11	3
1963	−8	3
1992	−4	12
1960	−1	12

Source: Based on van Leeuwen and Jaeck (unpublished results).

of September (corresponding approximately to the harvest period). Years in which the vines were subjected to a significant water deficit were all great vintages. Even if ripening may be halted on some plots (especially those with young vines, i.e. shallow root systems) in a very dry summer, which has a detrimental effect on the wine, it is interesting to note that, since 1950, there has been no overall quality loss due to water stress in Bordeaux, at least in red-wine-producing vineyards.

Ways of modifying water supply in a vineyard

The ideal water status for producing grapes to make high-quality wine consists of a moderate water deficit, starting early in the season (before veraison). The grapes will show less winemaking potential if the vines are not subject to water deficit at all, as well as in cases of severe water stress.

Loss of quality is much more commonly due to a plentiful water supply than due to excessive water stress, even if it is generally unnoticed. When summer rains and water reserves in the soil are such that the vines do not regularly suffer a moderate water deficit, leaf surface per hectare must be increased to promote actual evapotranspiration, and vines should be planted on rootstocks that do not take advantage of the plentiful water supply (e.g. Riparia Gloire de Montpellier). Quality can also be maximized by selecting an appropriate grape variety (early-ripening red or white grape varieties; Van Leeuwen, 2001).

In situations where excessive water stress causes a drop in quality in certain vintages (very dry climate and lack of water reserves in the soil), it is possible to minimize the negative impact on the vines by adapting the grapevine training system and vegetative growth (Chone et al., 2001b). The best way to protect vines from the negative effects of water stress is by restricting yields. When yields are low, a relatively small leaf surface area does not negatively affect the leaf/fruit ratio. The most widespread form of adaptation to dry conditions is the use of a drought-resistant rootstock (e.g. 110 Richter). It should also be noted that reducing the vines' nitrogen supply reduces their water requirement by reducing vigor and restricting the leaf area.

Under extreme conditions, grapegrowers may need to irrigate, if permitted by local law. It is considered difficult to grow vines producing viable yields if annual rainfall is under 400 mm. This value may vary, however, depending on the distribution of rainfall throughout the year and the soil's capacity to retain water. In very dry climates, rational irrigation may be a quality factor, while poorly controlled irrigation may also lead to a reduction in winemaking potential. Irrigation should be gradually reduced, so as to produce a moderate water deficit in the vineyard before veraison, while avoiding severe water stress. Monitoring the vines' water status by testing stem water potential is essential to ensure that irrigation is perfectly controlled (Choné et al., 2001b). Other promising monitoring methods are currently in the experimental stage.

For many years, the concept of growing vines under restrictive conditions was purely European, mainly in AOC (controlled appellation of origin) vineyards. It is interesting to observe that this idea is being introduced in some New World vineyards. The Australians have successfully tested two irrigation systems that deliberately restrict the vines' water supply. In regulated deficit irrigation (RDI), a water deficit is deliberately caused after flowering by stopping irrigation for a period of time (Dry et al., 2001). This is particularly aimed at reducing grape size. Partial rootzone drying (PRD) involves irrigating both sides of each row separately, alternating at two-week intervals. Thus, part of the root system is always in soil that is drying out. This has been observed to have very clear impact on the grapes' potential to produce high-quality wine, probably partly due to synthesis of larger amounts of abscisic acid than in vines not subjected to any water deficit (Stoll et al., 2001).

10.4.7 Weather Conditions During the Year: The Idea of Vintage

The three principal climatic parameters (light, heat, and humidity) vary considerably from year to year. Their respective influence on maturation processes is consequently of varying importance and leads to a given grape composition at ripeness. The enological concept of vintage can thus be examined.

Variations in weather conditions do not have the same influence in all climates. The principal European viticultural regions have been classified into different zones (Figure 10.32). Examining only the sugar concentration in the northern continental zone (Alsatian, Champagne, and Burgundian vineyards in France and Swiss and German vineyards for the most part), the length of sun exposure seems to be the principal limiting factor during grape development (Calo *et al.*, 1992).

This factor is also important during maturation in the North Atlantic zone (Loire and southwestern France vineyards), but is less important in the southern zone (Mediterranean vineyards in Spain, France, and Italy). In the latter zone, the water factor interferes with the relative consistency of temperature and sun exposure. High temperatures in this case do not positively affect sugar accumulation, if considerable water stress exists. In the opposite case, they can limit this accumulation by favoring vegetative vine growth when the water supply is not limited.

In the Rioja vineyards of northern Spain, the respective importance (varying from year to year) of the opposing influences of the Atlantic and Mediterranean climate determines wine quality there.

Thus the climate/quality relationship can only be represented approximately. The sum of the temperatures, rainfall, or length of light exposure does not have the

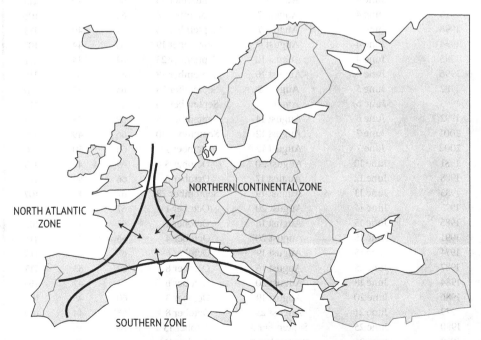

←→ The intermediate zone is more or less subject to the influence of the three neighboring zones depending on weather conditions

FIGURE 10.32 European viticultural climate zones.

most influence on grape quality; rather, it is their distribution in the course of the vine growth cycle.

In northern vineyards, climate conditions favoring an early green growth cycle permit grape maturation during a warmer and sunnier period, thus benefiting grape quality. Recent years have permitted the verification of this simple observation in the Bordeaux region (France).

Among the earliest years for grape development, 1982, 1989, 1990, and 2000 produced wines of outstanding quality (Tables 10.8 and 10.9). The weather conditions of these years are particularly favorable, with warm and sunny days and very little rainfall. At the harvest, Cabernet Sauvignon grapes (main Bordeaux variety) had high sugar and low malic acid concentrations (Table 10.10). A high cation

TABLE 10.8

Recent Phenological Observations on Red Grape Development and Maturation (Merlot and Cabernet Sauvignon) in Bordeaux (France) (Vintages Classified in Order of Early Ripening)

Vintage	Mid-bloom (A)	Mid-veraison (B)	Harvest (C)	Duration in days		
				A–B	B–C	A–C
1997	May 23	July 31	September 15	69	46	115
1990	May 27	August 6	September 24	69	55	124
1989	May 29	August 4	September 16	67	47	114
1999	May 31	August 4	September 20	65	47	112
2000	June 3	August 6	September 24	64	49	113
1976	June 4	August 7	September 18	64	42	106
1998	June 4	August 7	September 25	64	49	113
1994	June 4	August 6	September 19	63	44	107
1995	June 4	August 10	September 23	67	44	111
1996	June 4	August 10	September 28	67	49	116
1982	June 5	August 9	September 23	65	45	110
1993	June 6	August 9	September 26	64	48	112
1992	June 6	August 14	September 28	69	45	114
2001	June 7	August 12	September 30	66	49	115
2002	June 7	August 12	October 1	66	50	116
1981	June 12	August 20	October 5	69	46	115
1988	June 12	August 17	October 6	66	50	116
1983	June 13	August 19	September 28	67	40	107
1975	June 14	August 20	October 1	67	42	109
1985	June 15	August 16	October 1	62	46	108
1991	June 15	August 20	October 4	66	45	111
1974	June 15	August 19	October 6	65	48	113
1987	June 15	August 16	October 8	62	53	115
1984	June 18	August 20	October 6	63	47	110
1986	June 20	August 19	October 3	60	45	105
1979	June 21	August 25	October 8	65	44	109
1980	June 25	September 3	October 13	70	40	110
1978	June 26	September 2	October 12	68	40	108
1977	June 27	September 2	October 12	67	40	107

TABLE 10.9

Comparison Between Recent Vintage Quality and Climate Conditions from April to September in Bordeaux (France) (Vintages Classified in Order of Earliness; See Table 10.8)

Vintage	Sum of average temperatures (°C)	Duration of sun exposure (h)	Number of exceptionally hot days t ≥ 30°C	Rainfall (mm)	Wine quality
1997	3,494	1,216	24	500	—
1990	3,472	1,496	38	319	Exceptional
1989	3,463	1,463	35	364	Exceptional
1999	3,498	1,426	17	523	Very good
2000	3,447	1,454	25	477	Exceptional
1976	3,384	1,430	29	278	Very good
1998	3,373	1,226	25	537	Very good
1994	3,344	1,143	26	620	Very good
1995	3,390	1,149	36	303	Exceptional
1996	3,267	1,207	24	531	Exceptional
1982	3,331	1,262	18	289	Exceptional
1993	3,231	1,086	17	498	Good
1992	3,325	1,219	22	557	Mediocre
2001	3,357	1,505	32	438	Very good
2002	3,309	1,414	18	405	Very good
1981	3,223	1,144	17	289	Very good
1988	3,288	1,249	15	362	Very good
1983	3,354	1,182	24	437	Very good
1975	3,250	1,256	16	362	Very good
1985	3,185	1,326	16	311	Exceptional
1991	3,419	1,370	28	319	Good
1974	3,129	1,279	17	301	Good
1987	3,360	1,200	28	368	Good
1984	3,111	1,308	15	423	Good
1986	3,129	1,300	21	438	Very good
1979	2,938	1,183	3	366	Good
1980	3,057	1,020	9	343	Good
1978	3,029	1,153	12	320	Good
1977	3,044	1,135	2	407	Fairly good

concentration, as shown by high ash content (alkalinity), indicated a suitable circulation of water in the plant and led to relatively high pHs. The long length of maturation in 1990 (55 days on average) resulted in one of the lowest malic acid concentrations in recent years.

Conversely, during late years such as 1980 and in particular 1977, grape development and maturation occurred under unfavorable weather conditions. The grapes sampled from the same parcels were low in sugar and rich in acids—especially malic acid. The importance of an early growing season for grape quality has also been demonstrated by wine-growing regions with similar climate conditions, like the Loire Valley in France and New Zealand.

But in a temperate Atlantic climate, like that of Bordeaux, the moment at which the best or worst climate conditions occur has

TABLE 10.10

Average Composition of Cabernet Sauvignon Grapes (Sampled in Reference Vineyards) at Harvest According to Vintage in Bordeaux (France) (Vintages Classified in Order of Earliness; See Table 10.8)

Vintage	Weight of 100 berries (g)	Sugar concentration (g/l)	pH	Alkalinity of ash (mEq/l)	Total acidity (mEq/l)	Tartaric acid (mEq/l)	Malic acid (mEq/l)
1997	162	196	—	—	82	—	—
1990	113	199	3.38	50	77	92	31
1989	118	208	3.33	48	93	96	45
1999	134	203	3.57	52	80	87	42
2000	148	213	3.63	50	82	83	39
1976	110	196	3.33	48	97	98	44
1998	149	200	3.55	53	79	82	41
1994	140	193	3.31	45	102	84	56
1995	116	194	3.45	47	86	84	42
1996	138	220	3.38	48	102	88	60
1982	116	200	3.41	48	96	94	48
1993	124	181	3.16	43	94	78	51
1992	134	177	3.26	48	103	83	63
2001	146	204	3.59	53	90	86	54
2002	142	205	3.59	—	90	—	—
1981	110	180	3.36	44	103	91	53
1988	120	191	3.31	47	97	83	57
1983	115	195	3.37	48	107	95	59
1975	119	209	3.30	44	96	91	41
1985	117	196	3.48	48	92	99	35
1991	133	185	3.31	46	96	83	58
1974	107	184	3.27	47	98	92	43
1987	143	176	3.35	46	99	88	55
1984	122	185	3.23	44	119	94	66
1986	115	201	3.36	51	86	92	40
1979	116	174	3.18	44	119	98	60
1980	110	181	3.28	49	112	93	71
1978	119	193	3.26	48	120	91	68
1977	118	170	3.29	44	137	90	85

a greater influence on grape quality than the absolute temperature and the total rainfall during the entire vegetative growth cycle (Figure 10.33). Thus, grape quality depends on a period of favorable weather toward the end of maturation. The 1978 vintage in Bordeaux is a paradoxical example of one of the latest years (Table 10.8). Unfavorable weather conditions at the beginning of the growth cycle delayed flowering and grape development. From veraison onward, although the temperature was slightly lower than the seasonal average, a lack of rainfall and considerable sun exposure helped the grapes ripen correctly and attain suitable sugar concentrations (Table 10.10). Among the late vintages of the last 30 years, 1978 is the only year when the grapes reached satisfactory ripeness.

In contrast, the quality acquired at the beginning of development can be

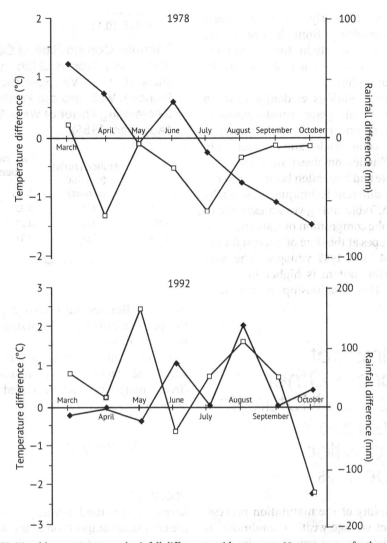

FIGURE 10.33 Monthly temperature and rainfall differences with respect to 30-year average for the period from March to October, 1978 and 1992, at Bordeaux (France): □, temperature difference (°C); ♦ rainfall difference (mm).

compromised by severe bad weather during maturation. In 1992, grape development was initially precocious thanks to high temperatures and moderate rainfall in the months before flowering. After a month of July with normal weather conditions, August was very hot but suffered from extremely high rainfall (about three times the normal rainfall for August in Bordeaux) accompanied by a lack of sunlight. Despite milder weather in September, sugar concentrations remained low (Table 10.10). In fact, sugar accumulation is generally rapid in the weeks following veraison, and migration becomes slower afterward. Sugar concentrations rarely increase rapidly during the days just before harvest.

However, harvest should occur under favorable weather conditions. In 1993, this period was characterized in Bordeaux by heavy rains (more than 180 mm in September). The quality of the vintage dropped substantially during the last days of maturation. Similarly, 1976, an early vintage benefiting from a warm, dry summer, did not attain the exceptional quality hoped for, because of rains at the end of maturation.

Climate conditions evidently have an influence on all grape constituents—in particular secondary metabolites such as phenolic compounds and aroma substances. Studies on these substances are incomplete and have often been carried out with very different techniques—especially extraction. Table 10.11 gives an example of the phenolic composition of Cabernet Sauvignon grapes at the time of harvest for the 1983, 1984, and 1985 vintages. The skin anthocyanin content is higher in quality vintages. This relationship is not valid for tannins.

TABLE 10.11

Phenolic Composition of Cabernet Sauvignon Grapes at Ripeness for Three Different Vintages (Bordeaux France). Vintages are Ranked By Decreasing Order of Wine Quality (Augustin, 1986)

Vintage	Anthocyanins (mg/100 berries)	Tannins (g/100 berries)	
		Skin	Seeds
1985	148	0.35	0.39
1983	132	0.25	0.44
1984	129	0.33	0.52

10.5 Impact of Various Other Factors on Maturation and Grape Composition at Ripeness

The variability of the maturation process, in terms of vintage weather conditions, is also regulated, if not controlled, by other parameters.

Some of these parameters are fixed and exert a constant and permanent action: the nature of the soil, the variety, and possibly the rootstock as well as plant density and trellising methods. All of these factors are established during the creation of the vineyard. Vine age, to a certain extent, can also be placed in this first category.

Other parameters can be continuously changed. Their modification most often corresponds with a desire to adapt plant reactions to vintage conditions: viticultural practices such as pruning, cluster thinning, hedging, leaf thinning, pesticide/herbicide treatments, etc. Fertilizing is also often placed in this second category. However, this practice can change the nature of a soil for a long time.

Finally, vintage conditions can produce accidental factors—both meteorological (frost, hail) and health-related (fungal diseases).

10.5.1 Variety and Rootstock

Rootstock

Rootstocks are used in vine growing when the chemical composition of the soil or the presence of pests (such as phylloxera) prevents the variety from growing on its own roots. The rootstock develops a different root system than the scion would, and this results in changes in the water and mineral salt supply. Grapes of a given variety, grown on the same soil, are known to have a different ionic composition according to their rootstock, but these differences are not sufficiently important to cause a significant variation in must acidity (Carbonneau, 1985).

They are, however, capable of influencing the scion's photosynthetic activity. Vigor may either increase or decrease, depending on the type of soil. In the richest soils, rootstocks such as 110 R, 140 Ru, 1103

P, SO4, and 41 B confer excessive vigor to the scion. This heightened vegetative growth slows and limits the maturation process (Pouget and Delas, 1989). In contrast, the Riparia Gloire and 161-49 C rootstocks create a relatively short vegetative cycle, favoring maturation and respecting the general typicity of the variety.

Grape composition at ripeness differs when a variety has developed on its own roots as opposed to on a rootstock. These differences essentially affect the level of ripeness: they concern the concentrations of sugars, acids, and phenolic compounds. Yet if the rootstock is chosen wisely, the differences that result from diverse rootstocks for a given cultivar are always slight (Guilloux, 1981). Only the nitrogen concentration appears to vary significantly (Roubelakis-Angelakis and Kliewer, 1992).

Variety

Choosing the variety to suit the climate is a decisive factor for obtaining good maturation and high-quality wines. In general, early-ripening varieties are cultivated in cool climates (Chasselas, Gewürztraminer, Pinot) and relatively late-ripening varieties in hot zones (Aramon, Carignan, Grenache). In both cases, ripeness should occur just before the average monthly temperature drops below 10°C. The maturation process should not take place too rapidly or abruptly under excessively favorable conditions.

Quality cultivars such as Cabernet Sauvignon and Pinot Noir lose much of their finesse in terms of aroma substances and phenolic compounds in hot climates. Figure 10.34 indicates the phenological diversity of these two varieties in the different viticultural regions of the world. In a hot climate, characterized by average monthly temperature always above 10°C (for example, Perth, Australia), the development cycle is particularly short—notably the maturation period. Conversely, the cycle grows longer in cool and humid temperate climates (French vineyards and Christchurch, New Zealand).

The choice of variety in a given location must first rely on its ability to produce wines demonstrating originality in taste (aroma, flavor), a richness of aroma composition, harmony, or congruence of flavors. It must also demonstrate aging potential likely to spark the interest of a consumer or a connoisseur.

Choosing a variety for a given area depends greatly on its ability to reach a good acidity balance of 170–200 g/l during maturation. The tartaric/malic acid ratio varies considerably from one variety to another. At maturity, the grapes of most varieties contain more tartaric than malic acid. Some varieties, however, always have a higher concentration of malic acid than of tartaric acid: Chenin Blanc, Pinot Noir, and Carignan. In a hot climate, varieties having a high tartaric/malic acid ratio are preferable.

The most essential element is the aroma and phenolic composition, which is eminently variable from one variety to another (Sections 10.4.1, 10.4.4, 10.4.5, and 10.4.6), all the more so depending on grapevine ecophysiological parameters, as these involve complex combinations of factors, including light exposure level, temperature levels, water deficits, and ripeness (Sections 10.3.6 and 10.4). A great variability of aroma characteristics is also observed within grape varieties from one clone to another.

Vine age

Although many empirical observations note the influence of vine age on wine and grape quality, little scientific work has been devoted to this subject. According to Düring (1994), the young vine develops a root system adapted to its environment during its first years. At the end of its fourth year, a functional equilibrium between the roots and the metabolic activity of the aerial parts of the vine is established. The establishment of an endotrophic mycorrhiza most often facilitates the mineral nutrition of the young vine (Possingham and Grogt Obbink, 1971).

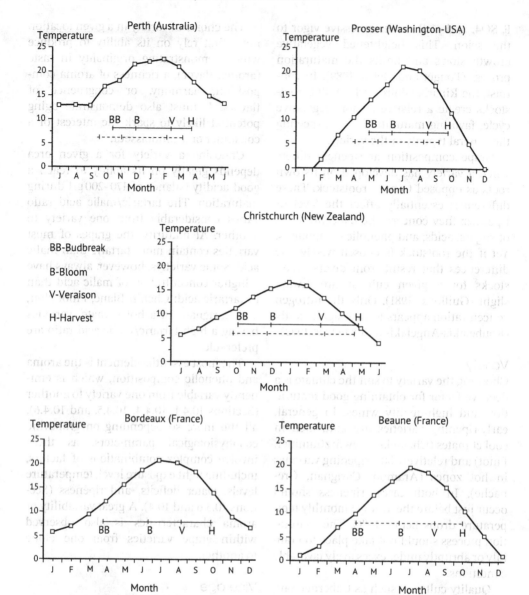

FIGURE 10.34 Phenological behavior of Cabernet Sauvignon and Pinot Noir according to climate. BB, budburst; B, bloom; V, veraison; H, harvest; ———, temperature curve; ———, vegetative cycle of Cabernet Sauvignon; |- - -|, vegetative cycle of Pinot Noir. *Source*: Adapted from Jackson and Lombard (1993).

If the grapegrower then succeeds in respecting the vine–soil–climate equilibrium and in regulating the harvest volume, notably by pruning, the plant can develop sufficient reserves in the old wood to ensure the prerequisites for proper maturation each year. Old vines are thus less sensitive to yearly climate variations and most often produce grapes rich in sugar and secondary products favorable to wine quality.

10.5.2 Soil Composition and Fertilization

The influence of soil on grape composition and wine quality is definitely the most difficult to describe. The soil, by its physical structure and chemical composition, directly affects root system development and consequently the grapevine's water and mineral supplies. It exerts an equally important effect on the microclimate. Soil color and its stone content profoundly modify the minimum and maximum temperatures as well as the light intensity in the lower atmosphere surrounding the grape clusters. Whether with schist plates, limestone pebbles, or siliceous gravel, grapegrowers have long made the most of this "second sun" to improve grape maturation.

Water supply
As mentioned previously (Section 10.4.6), a regular water supply is needed for grape development and maturation, even though a moderate water deficit is favorable for curbing green growth and for the accumulation of secondary metabolites in the grape. This water from the soil transports the minerals that are necessary for growth in the plant. The ion concentration of this solution is related to the nature of the soil and the fertilizers added, but a large amount of the available minerals is the result of biological activity in the soil. A potential imbalance can seriously affect vine growth. The best-known example is the increase in the exchangeable phytotoxic copper concentration in old, traditional vineyards that have received many sulfur- and copper-based treatments to ensure the protection of the vine against fungal diseases. Under the influence of bacteria in the soil, the sulfur is oxidized into sulfates, which accumulate in the soil. The resulting soil acidification causes copper solubilization (Donèche, 1976).

Many synthetic pesticide residues can similarly disrupt certain soil reactions, notably the biological mineralization of nitrogen, but there has been no research on the consequences of these phenomena on grape maturation and composition.

Nitrogen and potassium
Many studies have focused on the influence of different levels of nitrogen and potassium fertilization. The removal of these minerals by the harvest is relatively low, compared with other major crops. Since the roots exploit a large volume of soil, grapevine mineral needs are relatively low.

For example, annual nitrogen fertilization should not exceed 30 kg/ha, which is largely sufficient for meeting the plants' needs. Above this value, nitrogen exerts a considerable effect on vine vigor, and excessive vegetative growth blocks the maturation process. In this case, the grape crop is abundant, but sugar and phenolic compound concentrations are low and the grapes are rich in acids and nitrogen compounds. Excessive addition of nitrogen also increases the concentrations of ethyl carbamate precursors and histamine, which are likely to decrease the healthiness of wine (Ough et al., 1989). The effect of nitrogen on vigor can be limited by water supply deficiencies in warm and dry climates, since all water limitation is also observed through an insufficient assimilation of nitrogen.

Temporary or permanent cover crops between vine rows may lead to a deficit in the vines' nitrogen supply due to competition, but also as a result of mineral nitrogen fixation (or denitrification). Nitrogen deficiency in the grapes may lead to fermentation problems in the must and may also, above all, have a detrimental effect on their synthesis of phenolic compounds and a large number of aroma substances.

The problem of potassium is more complex. This cation predominantly participates in must and wine pH and acidity. Facing the fairly general increase of wine pH in recent years, much research tends to show that the soil is responsible for this high potassium supply, due to excessive soil richness or fertilization. However, a direct relationship between excess potassium fertilization and decreased grape acidity has not been demonstrated definitively in all cases.

Potassium actively participates in grape sugar accumulation. In years with favorable weather conditions, the ripe grape imports large amounts of potassium. Due to the high malic acid degradation characterizing such a maturation, must acidity is principally the result of the high tartaric acid concentration. Insolubilization of tartaric acid salts in the course of winemaking greatly lowers the acidity.

Salty soils
Finally, some vineyards are established on salty soils. High sodium chloride concentrations increase the osmotic potential of the soil solution. As a result, the plant must greatly increase its respiratory intensity to ensure the necessary energy for its mineral nutrition. In hydroponics, this lowers vine vigor and results in earlier ripening. The sugar concentration increases but not the amount of phenolic compounds. In vineyards, these effects are modified by using specific rootstocks (Salt Creek, Dog Ridge and 1613). On salty soils, potassium, magnesium, and organic acid concentrations decrease, whereas calcium and chloride concentrations increase.

Expression of terroir
In conclusion, the primordial influence of soil has been recognized for a long time in the form of viticultural *terroirs*. The soil must create favorable conditions for grape development and maturation (mineral and water supply and microclimate). The temperature above the soil and its water content also have an impact on the earliness or lateness of the growing season (Barbeau *et al.*, 1998; Tesic *et al.*, 2001). These two parameters give an initial indication of the quality of a *terroir*. But a quality *terroir* must also limit the consequences of climate variations from one year to another. Soil study is difficult since all of the factors likely to influence the biochemical processes of maturation should be taken into account.

Our understanding of the role of soil in the intrinsic quality of wine still rests essentially on empirical data. Certain studies, however, shed light on the links between soil composition and the qualitative potential of grapes, especially in terms of aroma and wine composition.

10.5.3 Leaf Treatments, Phytosanitary Protection, and Aroma Composition

Spraying nitrogen and sulfur on leaves during veraison may have a great influence on the aroma potential of volatile thiols and glutathione, as demonstrated by studies on Sauvignon Blanc and other white grape varieties (Peyrot des Gachons *et al.*, 2005; Choné *et al.*, 2006; Charrier *et al.*, 2007; Lacroux *et al.*, 2008).

Nitrogen application has significant consequences on the composition of musts and wines, when applied on the soil only, or sprayed on leaves especially, possibly accompanied by sulfur treatment, on parcels of nitrogen-deficient Sauvignon Blanc vines. An initial study was conducted on low-vigor and highly nitrogen-deficient Sauvignon Blanc parcels (musts containing 30 mg/l of yeast-assimilable nitrogen) (Choné *et al.*, 2006). The fertilization of one section of the parcel, with a mineral nitrogen addition (60 units) in the form of ammonium nitrate, at the beginning of fruit set, leads to a high increase in the vine's nitrogen supply. This translates into a net increase in the yeast-assimilable nitrogen concentration of the must (Table 10.12).

In this experiment, the water supply was not limited, since YT values, as determined from July to the end of August 2000, varied from -0.22 to -0.28 MPa. The increase in the vine's nitrogen supply results in a significantly higher

TABLE 10.12

Effect of the Grapevine Nitrogen Supply on Vigor and Grape Composition

	Deficient	Fertilized with nitrogen (60 units N)
Yeast-assimilable nitrogen (mg/l)	29a	174b
Grape weight/vine (kg)	1.43a	1.58a
Primary leaf surface area (m^2)	2.13a	2.37a
Secondary leaf surface area (m^2)	0.40a	1.44a
Weight of one berry (g)	1.5a	1.9b
Reducing sugars (g/l)	202a	199a
Malic acid (g/l)	2.72a	4.22b
P-4MMP (ng 4MMP Eq/l)	405a	715b
P-4MMPOH (ng 4MMPOH Eq/l)	760a	2,059b
P-3MH (ng 3MH Eq/l)	3,358a	14,812b
Glutathione (mg/l)	17.9a	120b

For a given line, values followed by a different letter are statistically different.

accumulation of cysteinylated precursors and glutathione in berries. However, nitrogen addition to the soil may increase the wine's vigor and its sensitivity to gray rot caused by *B. cinerea*. In a second study, nitrogen was sprayed on the leaves before veraison, as urea in two applications, at a rate of 10 kg/ha, associated or not with micronized sulfur at a dose of 5 kg/ha. The Sauvignon Blanc parcel is still relatively nitrogen deficient, but less so than in the previous experiment and is subject to a slightly limiting water deficiency. In comparison with the control method, yeast-assimilable nitrogen concentrations in musts originating from urea-treated vines are greater (without reaching high concentrations). On the other hand, the number of clusters, the yield and must composition are not significantly different (Table 10.13). Moreover, the increase in nitrogen supply results in an increase of the glutathione concentration in wines. Wines made from vines treated with both nitrogen and sulfur contain higher varietal thiol concentrations.

Consequently, sufficient nitrogen supply for the vine is essential to the aroma expression of the Sauvignon Blanc variety and of many white varieties. First, it favors the synthesis of both the cysteinylated and the glutathione aroma precursors, and second, it apparently limits the synthesis of phenolic compounds (Nikolantonaki and Waterhouse, 2012).

Lastly, certain phytosanitary treatments have unexpected consequences on the varietal aroma of wines. The application of copper-based formulations on Sauvignon Blanc and Cabernet Sauvignon vines leads to decreased aroma in young wines due to the reactivity of thiols with copper (Hatzidimitriou *et al.*, 1996; Darriet *et al.*, 2001). Leaf treatments after veraison have lesser consequences, like those conducted on the entire vegetation up to fruit set. A limitation on copper contamination during the summer season is beneficial to the preservation of this organoleptic component of wines, especially with the current organic farming protocols (Hatzidimitriou *et al.*, 1996; Darriet *et al.*, 2001). Systemic

TABLE 10.13

Effect of Nitrogen Spraying or Nitrogen and Sulfur Leaf Spraying on Sauvignon Blanc Grape and Wine Composition

	Control	N leaf application	N and S leaf application
Number of clusters/vine	9.1 ± 2.1	9.5 ± 2.5	9.2 ± 0.8
Weight of one berry (g)	1.68 ± 0.20	1.72 ± 0.19	1.80 ± 0.18
Must Analysis			
Yeast-assimilable nitrogen (mg/l)	72.7 ± 6.5a	112.3 ± 8.6b	115 ± 1b
Reducing sugars (g/l)	204 ± 5	197 ± 5	199 ± 6
Malic acid (g/l)	4.6 ± 0.44	5 ± 0.38	4.6 ± 0.27
pH	2.97 ± 0.01	2.97 ± 0.03	2.99 ± 0.01
Wine analysis			
Glutathione (mg/l)	13.5 ± 0.2a	20 ± 1b	19 ± 3b
4MMP (ng/l)	53 ± 1a	59 ± 1a	124 ± 1b
3MH	738 ± 10a	690 ± 2a	1,212 ± 90b

For a given line, values followed by a different letter are statistically different.

powdery-mildew treatments (sterol biosynthesis inhibitors) also create an unfavorable effect on the terpene potential of Muscat grapes (Aubert et al., 1997).

10.5.4 Management of Vine Growth

Grape maturation is also influenced by other permanent factors, established during the planting of the vineyard. Planting density, row spacing, and canopy placement (existence and positioning of wire trellising) affect plant physiology through root development with respect to the soil and use of sunlight by the leaves. These factors directly affect vine vigor. Their action on the grape can only be indirect, notably by acting on the microclimate surrounding the grape clusters (temperature and sun exposure).

Rigorous experiments in this domain are difficult to carry out. The existing criteria for establishing a vineyard are primarily empirical, but vine vigor has been shown to increase when plant density decreases, with the risk of delayed maturation.

In northern and temperate regions, tradition (verified by research) recommends relatively high planting densities of around 10,000 vines/ha. In a drier, Mediterranean climate, optimum quality is often obtained with a density of between 3,000 and 5,000 vines/ha. In spite of the water deficit, high density restricts potassium imports and maintains a good acidity level in wines made from these grapes.

Under conditions favoring vegetative growth (irrigation in warm and sunny climates), excessive leaf crowding should be limited by low plant densities (1,000–2,000 vines/ha), and suitable training and pruning methods. Canopy management is of major importance in this case (Carbonneau, 1982).

Alongside climate, soil fertility imposes its own rules. According to Petit-Lafitte (1868), "the poorer the soil, the higher the plant density. . . ." A very compact root system thus enables the maximum exploitation of the soil potential. The same reasoning should be used in dry soils.

10.5.5 Vineyard Practices for Vigor Control

Vine management and growing are characterized by severe measures limiting vegetative growth and the amount of fruit. A certain leaf area is required for nutrient supply to the grapes, and a relationship exists between this area and grape quality at harvest.

The evolution of the leaf area to fruit weight ratio can be used to evaluate grape quality. In an example with Tokay, Kliewer and Weaver (1971) showed that grape sugar concentration diminished sharply when this ratio was lower than $10\,cm^2/g$ (Figure 10.35). Proline and total phenolic compound concentrations were similarly affected. Di Stefano et al. (1983) obtained identical results for terpene compound concentrations in Muscat Blanc. In general, the necessary leaf area for the maturation of 1 g of fruit varies from 7 to $15\,cm^2/g$ for most *V. vinifera* varieties, but increasing this ratio above these values has little effect on grape composition, as shown by the sugar concentration curve in Figure 10.35.

Winter pruning

Winter pruning is the first operation carried out in the vineyard. It consists in controlling vine production by leaving only a certain number of buds capable of producing inflorescences. Grapevine response to pruning is quite variable. Some varieties have such fertile buds at the base of the shoots that pruning is often ineffective for yield control. Bud fertility varies greatly at the same position on the shoot, depending on the variety. Cultivar productivity can also vary with respect to the climate. Many factors must thus be considered. Yet most specialists agree that low yields are needed to obtain proper grape maturation.

Increases in yield have long been known to affect grape sugar concentration negatively (Table 10.14). This problem is especially troubling in vineyards that use varieties close to their limits of cultivation. In very hot regions, the harvest can be delayed to attenuate this phenomenon. In northern vineyards, the high annual variability of climate conditions determines the ratio of grape yield to sugar concentration. Experience shows that great vintages, obtained under particularly favorable climate conditions, most often correspond with abundant grape crops.

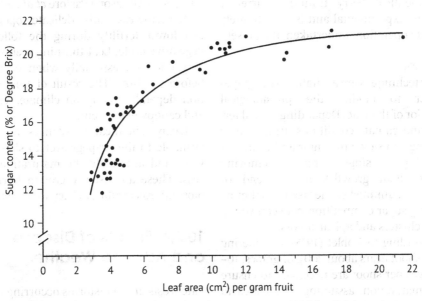

FIGURE 10.35 Influence of leaf area per unit crop weight on sugar content of Tokay berries at ripeness (Kliewer and Weaver, 1971).

TABLE 10.14

Relationship Between Crop Yield and Sugar Concentration

Grape weight per vine (kg)	Sugar concentration (g)	
	Per vine	Per liter of must
0.4	63	235
0.8	126	225
1.2	168	200
1.6	184	165
2.0	182	130

In a warm and sunny climate, increasing the quantity of light energy received by the canopy, combined with controlled irrigation, will often increase yields without lowering grape quality (Bravdo et al., 1985), but several consecutive abundant crops can lead to depletion of vine reserves.

Crop thinning

After fruit set, excess grapes can be removed by thinning. Manual thinning is expensive. Thinning 30% of the crop before veraison results in a 15% increase in the sugar concentration and a 5% drop in acidity. Chemical berry thinning currently remains experimental and is an extremely delicate operation, undertaken at fruit set.

Trimming

Other techniques are available to the grapegrower to modify the physiological behavior of the vine. Depending on soil fertility and climate conditions, trimming or hedging can slow vine vigor and limit leaf crowding. A single topping (removing recent shoot growth), at the end of flowering, diminishes the risk of shatter by limiting sugar competition between young grape clusters and apical shoots.

According to Koblet (1975), considering that a shoot bears about 200 g of grapes, 10–14 leaves per shoot are necessary to ensure their maturation, assuming a leaf area to grape weight ratio of 8–10 cm^2/g. Trimming, which leaves a maximum of 14–16 leaves per shoot, improves the maturation process and increases sugar and secondary metabolite concentrations. Trimming too severely produces the opposite results. Trimming also presents the advantage of lowering the evaporation surface, thus limiting water stress risks under certain environmental conditions (lack of rainfall, soil water deficiencies). However, in case of strong vigor, suckers come out and can alter the microclimate around clusters or compete with young berries for carbon assimilates.

Leaf thinning

Older, less active leaves can also be removed from the base of the shoot. Correct leaf thinning essentially exposes grape clusters to sunlight, improving grape maturation and limiting the risk of mold. Leaf thinning around grape clusters thus reduces malic acid concentrations and the "pyrazine" character while increasing grape anthocyanin concentrations for Cabernet Sauvignon. The vegetal aroma of Sauvignon Blanc can also be lowered in this manner (Arnold and Bledsoe, 1990).

Early leaf thinning before bloom lowers the future crop volume, due to flower-fall (Candolfi-Vasconcelos and Koblet, 1990), and delays veraison (Pastore et al., 2013). Nevertheless, this delicate operation can lower fertility during the following vegetative cycle. Leaf thinning seems to be very effective, especially when used just before veraison. The result of this operation depends greatly on climate, variety, and canopy placement.

Lastly, chemical substances are now available to the grapegrower to slow vine vigor and accelerate the maturation process. These are usually growth hormone biosynthesis inhibitors (Reynolds, 1988).

10.5.6 Effects of Disease and Adverse Weather

Late frosts and hailstorms occurring in the spring often produce the same effects as shoot removal by topping, but latent bud development is disorganized, resulting in bushy vegetation. Flowering is considerably extended, and grape cluster maturation is uneven. The latest grape clusters have difficulty ripening. Damage caused by summertime hail alters nutrient supply to the grape clusters: affected grapes wither or are attacked by parasites, and a rapid harvest may be necessary.

Various causes can result in more or less severe vine defoliation. Maturation is difficult due to insufficient nutrient supply. A late downy mildew attack can cause total leaf loss in certain very sensitive varieties, such as Grenache. Similarly, leaves infected with powdery mildew always lower grape quality. Parasite development leads to significantly reduced crop yields, and very late attacks hinder grape maturation.

A potassium deficiency can cause leaf scorch, leaf yellowing, and premature leaf-drop—and consequently a decrease in grape sugar and phenolic compound concentrations. Black leaf, encouraged by overcropping and soil dryness, often accompanies potassium deficiencies.

Bunch stem necrosis (also known as waterberry in California, shanking in New Zealand, and bunch stem dieback in Australia) also lowers crop quality and can cause crop loss. It is often linked to excessive yields and magnesium deficiencies. Certain varieties are particularly sensitive: notably Gewürztraminer, Sauvignon Blanc, Ugni Blanc, and Cabernet Sauvignon. This necrosis results in a decrease in sugar, anthocyanin, fatty acid, and amino acid concentrations, whereas the grapes remain rich in organic acids (Ureta *et al.*, 1981).

10.6 Botrytis cinerea

10.6.1 Gray Rot and Noble Rot

In addition to the diseases already mentioned (downy and powdery mildew), one of the principal causes of crop quality degradation is grape rot due to the development of various microorganisms (bacteria, yeasts, or other fungi).

The principal microorganism responsible for this is usually *B. cinerea*, which is "a ubiquitous fungus, except perhaps in desert zones" (Galet, 1977). This polyphagous saprophyte can subsist on senescent or dead tissue such as grapevine wood. It is also capable of waiting for favorable conditions in diverse resistant forms (sclerotia or conidia, with a high dissemination capacity). The presence of water on the surface of plant tissue and an optimal temperature of 18°C are ideal conditions for the germination of the resistant forms and mycelial growth. Conidial germination is possible at temperatures between 10 and 25°C. Under these conditions, the contamination area of this fungus is very large and covers a great number of the world's viticultural regions. Gray rot thus remains one of the major concerns of grapegrowers.

In a few areas of the world, particular conditions permit *B. cinerea* to develop on ripe grapes. This process results in an over-ripening that increases the sugar concentration while improving quality (Section 14.2.2). The parasitized grape dehydrates, and the sugars are more concentrated than the acids. Most importantly, the grape acquires the characteristic aromas that permit the production of renowned sweet white wines such as Sauternes-Barsac, Côteaux du Layon (France), Tokaj (Hungary), and Trockenbeerenauslese (Germany and Austria).

Noble rot requires specific environmental conditions. The many studies undertaken have not yet been able to define these conditions precisely, but, in general, *B. cinerea* development in the form of noble rot is thought to be favored by alternating dry and humid periods. Nighttime humidity, dew, and frequent morning fogs in the valleys of certain rivers stimulate fungal development, whereas warm and sunny windy afternoons facilitate water evaporation—limiting fungal growth.

Many factors participate in this phenomenon:

- The soil, by its nature and possibly its drainage, should permit the rapid elimination of rain water.
- The canopy placement and surface should permit a maximum number of grape clusters to be aerated and exposed to sunlight.
- The grape cluster structure should be fairly dispersed.

The nature of the variety also greatly affects grape sensitivity to *B. cinerea*, but no direct relationship seems to exist between variety and noble rot quality. Differences essentially originate from the earliness or lateness of maturation. However, Pucheu-Planté and Leclair (1990) showed the importance of clones on noble rot quality.

10.6.2 Grape Sensitivity to *Botrytis cinerea*

Vine inflorescences can suffer from mold attacks if the climate conditions are favorable to *B. cinerea* development. Peduncular rot causes flowers to fall and consequently a sharp drop in future crop volume. During the entire green growth period, the grape is resistant to this parasite.

Gray rot rarely occurs between fruit set and veraison. In 1983 and 1987, northern European vineyards suffered early attacks, sometimes affecting up to 30% of the berries, but the reasons for the loss of resistance in green grapes are still not known.

In certain cases, i.e. with varieties that have compact grape clusters and high fruit set rates, a few berries can become detached from their pedicel and remain imprisoned inside the cluster. These damaged grapes constitute a direct penetration path for the fungus, circumventing the natural resistance of the grape. More often, the contamination is the consequence of another phenomenon, such as hail or other parasites. After veraison, the grape becomes sensitive to *B. cinerea* more or less rapidly.

These behavioral differences of the grape are due to multiple causes that will now be examined briefly without an in-depth study of the pathology of the grape–*B. cinerea* relationship.

In the first place, the green grape skin, covered by a thick cuticle, constitutes an effective barrier against parasites. Since Bonnet's (1903) initial research, a resistance scale of the principal *Vitis* species has been established based on their respective cuticle thicknesses. American varieties whose cuticle thickness varies from 4 µm (*Vitis rupestris*) to 10 µm (*Vitis coriacea*) have better protected berries than European species (*V. vinifera*), whose cuticle thickness is between 1.5 and 3.8 µm. This observation led to the production of *V. vinifera* and American species hybrids that are effectively more resistant to gray rot, but these hybrids do not usually produce quality grapes on the best *terroirs*.

The same relationship between cuticle thickness and *B. cinerea* resistance has been found in *V. vinifera* varieties but on a smaller scale. The sensitive varieties all have a cuticle thickness of less than 2 µm (Karadimtcheva, 1982).

During maturation, the quantity of cutin and wax per surface unit increases. This accumulation is more intense when the grapes are exposed to sunlight in an environment relatively low in humidity. Contact between berries is always characterized by a lower cuticle thickness.

Although the fungus possesses a cutinolytic activity, the latter is very low. In fact, direct penetration of the grape cuticle by *B. cinerea* via enzymatic digestion has not been proven. Only a developed mycelium produces sufficient amounts of cutinase to attack a neighboring berry cuticle (often less thick if the grape cluster is very compact). The surface of the cuticle has perforations that are a potential point of entry for mycelial filaments (Blaich *et al.*, 1984). Resistant hybrids have fewer perforations than sensitive varieties. The number of these cuticle perforations increases in the course of maturation.

Under the cuticle, the skin also participates in the resistance to *B. cinerea*.

According to Karadimtcheva (1982), the external layer of the hypodermis in certain varieties resistant to gray rot comprises more than seven rows of thin, elongated cells, with a total thickness exceeding 100 μm. In sensitive varieties, it contains only four to six cell rows, with a total thickness of 50–60 μm.

The extent of the thickening of the epidermal cell walls, occurring in the course of maturation, varies depending on the variety. The most sensitive varieties have the thickest cell walls. This phenomenon is caused by the partial hydrolysis of pectic compounds by endogenous grape enzymes (Section 10.3.6). The increase in soluble pectins varies greatly, depending on the variety. Consequently, grape skins exhibit varying degrees of sensitivity to enzymatic digestion by the exocellular enzymes of *B. cinerea* (Chardonnet and Donèche, 1995).

In addition to this mechanical resistance, the grape skin contains preformed fungal development inhibitors. All epidermal cells possess tannin vacuoles. These phenolic compounds exert a weak fungistatic effect on the pathogen.

As in many fruits, an inhibitor of the endo-polygalacturonase of *B. cinerea* is also contained in the grape skin cell walls. This glycoprotein is released during cell wall degradation by the pathogen. The concentration and persistence of this inhibitor in the skin vary according to the stage of development and the variety considered.

The green grape skin is also capable of synthesizing phytoalexins in response to an infection (Langcake and Pryce, 1977). These stilbene derivatives (*trans*-resveratrol and its glucoside, *trans*-piceid, ε-viniferin dimer, α-viniferin trimer, β-viniferin tetramer) have fungicidal properties (Table 10.15).

Resveratrol is obtained by the condensation of *p*-coumaroyl-CoA with three malonyl-CoA units in the presence of stilbene synthase (Figure 10.36). Viniferin formation is then ensured by grape peroxidases.

At the time of a parasite infection, the normal flavonoid metabolism (Section 10.3.6) is diverted toward stilbene derivative production by the action of stilbene synthase. Remarkably, after veraison, this capacity is very rapidly lost, even in pathogen-tolerant American vines (Figure 10.37).

TABLE 10.15

Principal Stilbene Derivatives Identified in the Grapevine (Langcake and Pryce, 1977)

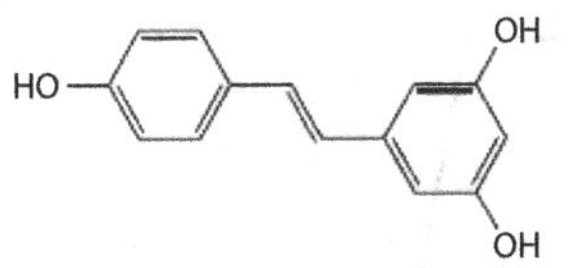

trans-Resveratrol

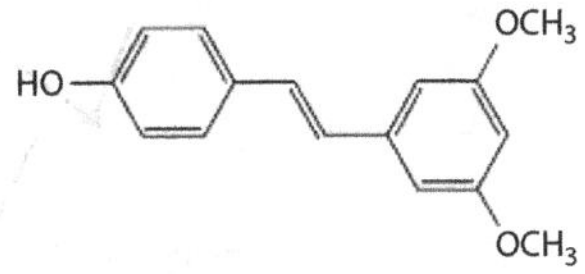

trans-Pterostilbene

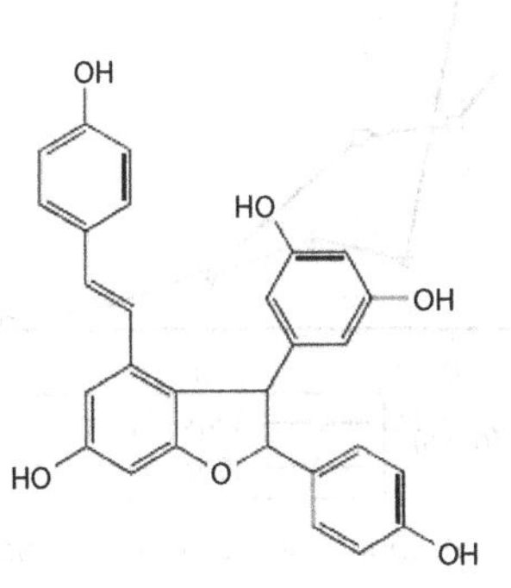

ε Viniferin

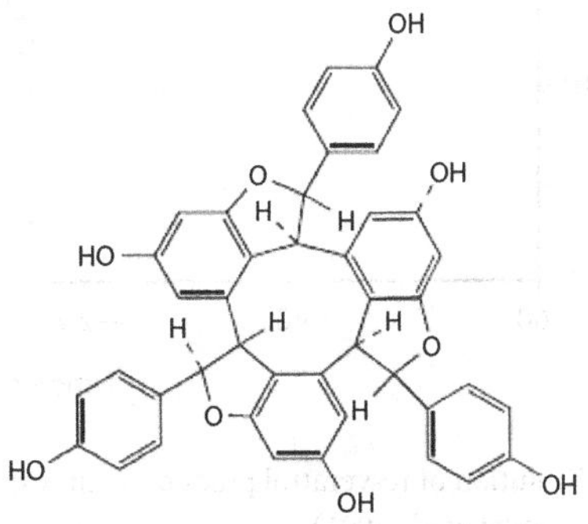

α Viniferin

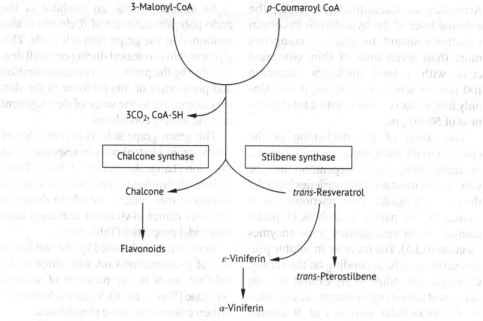

FIGURE 10.36 Biosynthesis pathways of stilbene phytoalexins in the grapevine (Jeandet and Bessis, 1989).

Thus during maturation, the grape loses most of its physical and chemical defenses. The differences in parasite sensitivity observed for diverse varieties and clones essentially result from differences in their grape development time. Microperforations of the cuticle and stomatic fissures (Section 10.2.3 and 10.3.7) also constitute a passageway for the efflux of grape exudate, which is indispensable for conidial germination

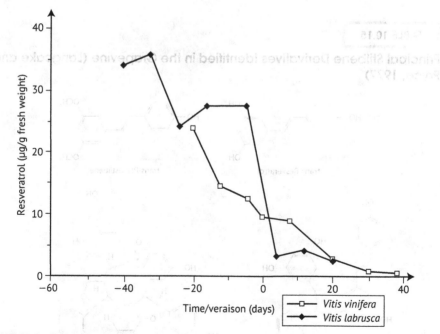

FIGURE 10.37 Evolution of resveratrol production in response to infection by a parasite during berry development (Jeandet et al., 1991).

and the proliferation of *B. cinerea* (Donèche, 1986). Changes in the chemical composition of the grape during maturation—notably an increase in sugar, amino acid, and soluble pectin concentrations—furnish the fungus with essential nutrients for mycelial growth.

10.6.3 Noble Rot Infection Process

Some observations have led to the conclusion that *B. cinerea* is sometimes present inside grapes as early as fruit set (Pezet and Pont, 1986). When it comes out of the latency phase, it develops preferentially toward the skin, due to the presence of antagonistic enzymes (chitinase and β-glucanase) in the pulp cells. However, the infection is most often caused by the germination of a conidium on the grape's surface. An exogenous nutrient supply facilitates the elongation of the germ tube. When the mycelial hypha reaches a microfissure, it penetrates the grape. Thus *B. cinerea* development occurs mainly in the grape's superficial cell walls. More precisely, the mycelial filaments are located in the middle lamella of the pectocellulosic cell walls. The latter are degraded by the enzymes of the fungus (pectinolytic, cellulase complex, protease, and phospholipase enzymes).

When the mycelium has totally overrun the pectocellulosic cell walls of the epidermal and immediate subepidermal cells, white grapes take on a chocolate-brown color that is characteristic of the *pourri plein* stage (fully developed rot). The mycelium then produces filaments that emerge from the skin surface by piercing the cuticle or by taking advantage of the diverse fissures used for the initial penetration. The filament extremities differentiate by producing conidiophores, whose conidia later detach and contaminate nearby berries.

The cell walls of the plant tissue are so greatly modified that they can no longer ensure their functions. In particular, berry cell hydration is no longer regulated. It can vary as weather conditions change and, under ideal conditions, should lead to a characteristic desiccation accompanied by the cytoplasmic death of the epidermal cells. The sugar concentration of these cells is considerable. Due to the high osmotic pressure, the fungus can no longer subsist and stops developing. This shriveled *confit* stage, known as *pourri rôti*, is used for making sweet white wines (Table 10.16).

The infection process, from healthy grape to *pourri rôti* grape, lasts from 5 to 15 days, depending on environmental conditions. This overripening period must be characterized by an alternation between short humid periods (three to four days, with 10 to 30 mm water rainfall), favoring conidial germination, and longer dry periods (about 10 days), enabling grape concentration and chemical transformations to occur.

For high-quality noble rot development, the phenomenon must be rapid and occur near maturity; otherwise the grape titratable acidity would decrease, but also the berries must reach that stage intact.

Many years of observation in Sauternes vineyards (France) have shown that the first symptoms of attack appear 15–20 days before maturity. Regardless of weather conditions, *B. cinerea* development is slow until maturity. At this stage, the parasite spreads rapidly, and its growth is explosive. At a certain moment, a high percentage of grapes simultaneously reach the *pourri plein* stage (Figure 10.38).

Grape maturation in a vineyard, in a parcel, or even on a given cluster is never absolutely synchronous (Section 10.4.1). As soon as a berry approaches maturity, it is contaminated by conidia from a nearby moldy berry. This asynchronism makes successive sorting necessary during the harvest ("successive sorting" is a local term that means that successive handpicking is used to ensure that only noble rotted grapes are harvested). The use of different varieties with varying degrees of earliness is beneficial in practice: for example, Muscadelle, Semillon, and Sauvignon Blanc in Sauternes (France) or Fürmint, Hárslevelü, and Muscat Ottonel in Tokaj (Hungary).

TABLE 10.16

Principal Modifications of Chemical Composition of Grapes Undergoing a Noble Rot Attack

Constituent	Per liter of must		Per 1,000 berries	
	Healthy berry	Noble rotted berry (*pourri rôti*)	Healthy berry	Noble rotted berry (*pourri rôti*)
Weight (g)	—	—	2,020	980
Volume of must (ml)	—	—	1,190	450
pH	3.33	3.62	3.33	3.62
Sugar concentration (g)	247	317	294	143
Glycerol (g)	0	7.4	0	3.4
Alkalinity of ash (mEq)	33	81	39	36
Total acidity (mEq)	123	112	146	50
Tartaric acid (mEq)	71	33	85	15
Malic acid (mEq)	81	117	96	54
Citric acid (mEq)	2.7	3.5	3.2	1.6
Acetic acid (mEq)	5.4	6.9	6.4	2.4
Gluconic acid (mEq)[a]	0	10.6	0	4.8
Ammonium (mg)	85	56	101	25
Amino acids (mg)	1,282	1,417	1,526	638
Proteins (mg)	2,815	3,795	3,350	1,708

[a] Musts from infected grapes contain from 1 to 2.5 g gluconic acid/l.

10.6.4 Changes in the Chemical Composition of Noble Rot Grapes

Physical data (juice volume, berry weight) effectively express the grape desiccation phenomenon. Grapes can be concentrated by a factor of two to five, depending on weather conditions.

The enological profile of must obtained from botrytized grapes is specific (Table 10.17). This juice is very rich in sugars, but its acidity is similar to that of juice obtained from healthy grapes. The tartaric acid concentration is often even lower and the pH higher (from 3.5 to 4.0), attesting to the concentration of other substances such as potassium ions. Compounds not present or in negligible concentrations in healthy grapes are encountered in considerable quantities in botrytized grapes (*pourri rôti* stage). For example, glycerol and gluconic acid can reach concentrations of 15 g/l and over 3 g/l in botrytized must, respectively.

But this concentration phenomenon masks profound chemical composition changes, resulting from the biological activity of *B. cinerea*. As Müller–Thurgau demonstrated (1888), these changes affect sugars and organic acids.

Glucose and fructose

Botrytis cinerea does not use much of the pectocellulosic cell wall residues for its development. For example, the contaminated grape becomes rich in galacturonic acid derived from the degradation of pectic compounds. It prefers to assimilate glucose and fructose accumulated in the pulp cells, and up to 50% of the sugars are lost in the production of these noble rot wines.

Metabolic studies *in vitro* have shown that the young mycelium of *B. cinerea* possesses the enzymes of the Embden–Meyerhof pathway, the hexose monophosphate

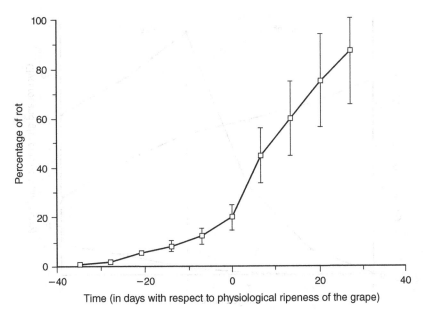

FIGURE 10.38 Evolution of a noble rot attack in a Sauternes vineyard (France) (average of 10 years of experimentation on different parcels).

shunt, and the tricarboxylic acid cycle (Donèche, 1989). It directly oxidizes glucose into gluconic acid. The latter, according to a process identical to the Entner–Doudoroff pathway, helps the young mycelium synthesize substantial quantities of cellular material. However, when the fungus is partially deprived of oxygen, mycelial growth is low, and the complete oxidation of glucose is accompanied by the liberation of glycerol in the medium.

Thus, the initial fungal development under the grape skin is marked by considerable glycerol accumulation (Figure 10.39). When *B. cinerea* emerges on the outside of the grape and reaches its stationary growth phase, changes occur in the sugar assimilation pathways, and remarkably, it can no longer assimilate gluconic acid. This acid, which accumulates in the grape, is a characteristic secondary product of significant sugar degradation. The gluconic acid concentration depends on the duration of the external development of the fungus, varying from 5 to 10 mEq/berry.

The glycerol concentration of contaminated grapes also varies according to the duration of the respective internal and external fungal development phases (Section 14.2.3). The glycerol concentration is between 50 and 60 µmol per berry at the *pourri plein* stage. Despite the concentration phenomenon, only 10–40 µmol exist per berry at the *pourri rôti* stage. Part of the

TABLE 10.17

Relationship Between the Level of Rot Determined By Visual Inspection and Laccase Activity Measured By the Oxidation of Syringaldazine (Redl and Kobler, 1992)

Level of rot (%)	Laccase activity (units/ml)
<1	0.39
1–5	0.78
6–10	2.25
11–25	6.56
26–50	8.12
51–100	15.86

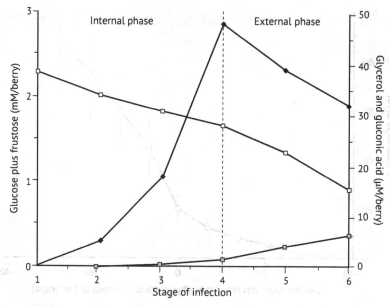

FIGURE 10.39 Sugar assimilation and secondary product formation during a noble rot attack (Donèche, 1987): □, glucose + fructose; ♦, glycerol; ▨, gluconic acid. Stages of infection: (1) healthy berry; (2) spotted berry (spot diameter less than 5 mm^2); (3) spotted berry (spot diameter greater than 5 mm^2); (4) fully rotted berry (completely spotted or *pourri plein*); (5) appearance of mycelial hyphae on the berry surface; (6) *pourri rôti*.

accumulated glycerol is oxidized by a glycerol dehydrogenase in the course of the external growth phase of *B. cinerea*. Musts obtained from botrytized grapes normally contain 5–7 g/l of glycerol.

The ratio of glycerol to gluconic acid represents the length of the internal and external growth phases of the parasite. It constitutes a noble rot quality index (Figure 10.40) (Section 14.2.3). In vintages with favorable weather conditions (for example, 1981, 1982, and 1985), rapid grape desiccation from the *pourri plein* stage onward leads to a high glycerol to gluconic acid ratio.

During a botrytis attack, other polyols (mannitol, erythritol, and meso-inositol) are formed, and their concentrations increase in the grape (Bertrand *et al.*, 1976). *Botrytis cinerea* also produces a glucose polymer, which accumulates in contaminated grapes. Its concentration can attain 200 mg/l in must. This glucan is often at the root of subsequent wine clarification difficulties (Dubourdieu, 1978).

Acids

Botrytis cinerea growth is always accompanied by the degradation of the principal acids of the grape. This biological deacidification lowers initial grape acidity by 70% on average. Tartaric acid degradation is progressive in the course of the infection process, and it stimulates sugar assimilation. Malic acid is generally less degraded: it occurs especially at the end of a botrytis attack and corresponds to a strong energy demand in order to accumulate reserve substances in the developing conidia.

The other acids are less degraded. This sometimes leads to their increased concentrations in *pourri rôti* grapes. Citric acid is an example and can be synthesized by certain *B. cinerea* strains, but, in spite of the concentration phenomenon, its concentration rarely exceeds 8 mEq/l. Acetic acid behaves similarly.

Mucic acid has also been observed to accumulate. It is a product of the oxidation of galacturonic acid (Wurdig, 1976). This acid is capable of precipitating in wine in

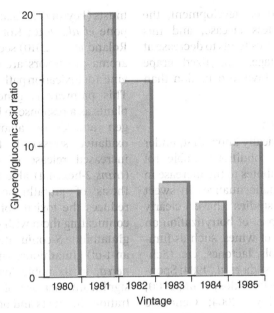

FIGURE 10.40 Influence of vintage on noble rot quality, expressed by the glycerol/gluconic acid ratio (Donèche, 1987).

the form of calcium salts, but this phenomenon is rare and seems limited to northern winemaking regions. More generally, it affects wines made from grapes that were insufficiently ripe when Botrytis set in.

Nitrogen substances

Botrytis cinerea growth is also to the detriment of grape nitrogen compounds. The fungus degrades grape proteins and liberates the nitrogen in amino acids with the help of proteases and amino oxidases diffused in the grape. *Botrytis cinerea* then assimilates this nitrogen and synthesizes the metabolic proteins necessary for its growth. The grape thus becomes rich in exocellular fungal proteins. Musts obtained from *pourri rôti* grapes contain less ammonium and more complex forms of nitrogen than musts from healthy grapes.

Phenolic compounds

Like many other fungi, *B. cinerea* produces an exocellular laccase: *p*-diphenol oxygen oxidoreductase (Dubernet *et al.*, 1977). This enzyme oxidizes numerous phenolic compounds. It is involved in the pathogenetic process, and its synthesis is induced by two groups of substances. The first group comprises phenolic compounds (gallic and hydroxycinnamic acids) that are most likely toxic to the fungus. The second group consists of pectic cell wall substance degradation products (Marbach *et al.*, 1985). The fungus adapts the molecular structure of this exocellular laccase to the pH of the host tissue and the nature of the phenolic compounds present. The quantity of the enzyme produced is also regulated.

Laccase and other oxidases produced by *B. cinerea* transform the principal white grape phenolic compounds (caffeic and *p*-coumaric acids—both free forms and forms esterified by tartaric acid) as well as flavanols (catechin, epicatechin) into quinones (Salgues *et al.*, 1986; Carbajal-Ida *et al.*, 2015; Stamatopoulos *et al.*, 2015). These quinones tend to polymerize, forming brown compounds. These compounds are most likely responsible for the characteristic chocolate color of *pourri plein* grapes.

Toward the end of development, the fungus produces less laccase, and this enzymatic activity thus tends to decrease at the *pourri rôti* stage. Botrytized grape musts are less sensitive to oxidation than one might expect.

Aroma substances

Winemakers empirically know that, under favorable weather conditions, noble rot development contributes to the increase in characteristic aromatic nuances in sweet wines. Various studies have clearly established the impact of botrytization on aroma compounds of wines, such as furanones, volatile thiols, lactones, etc. (Schreier et al., 1976; Masuda et al., 1984; Sponholz and Hühn, 1994; Miklósy et al., 2000; Miklósy and Kerényi, 2004; Genovese et al., 2007; Sarrazin et al., 2007, 2010; Stamatopoulos et al., 2014, 2015, 2016; Wang et al., 2017).

These studies show that the *B. cinerea* fungus acts on several levels to increase the concentration of key aroma compounds or precursors. Of course, *B. cinerea* accelerates the *passerillage* phenomenon (drying of grapes on the vine) and favors the concentration of volatile compounds or their precursors, especially furanones with cooked sugar notes. It also modifies the composition of berries by producing aroma compounds, phenylacetaldehyde (Sarrazin et al., 2008) and lactones, some of which are specifically related to noble rot like 2-nonen-4-olide (Volume 2, Section 7.7). Moreover, it creates an increase in thiol precursor concentrations related to zesty notes. These include cysteinylated and glutathionylated precursors (Sarrazin et al., 2008; Thibon et al., 2009, 2011). For all of these phenomena, a phase of increasing concentrations in the berry is noted, reaching a maximum at the *pourri plein* stage or the *pourri rôti* stage.

Volatile thiol precursor concentrations (*S*-conjugates of cysteine and glutathione) increase with botrytization (Thibon et al., 2009). The simultaneous presence of cysteine and glutathione *S*-conjugates in musts (Peyrot des Gachons et al., 2002; Capone et al., 2010; Kobayashi et al., 2010; Roland et al., 2010) seems to indicate that aroma precursors are synthesized by the vine's detoxication pathway (Section 10.3.8). This pathway is generally activated by plants as a response to biotic stress (pathogen attack) or abiotic stress (injury, oxidative stress). It translates into an increased release of reactive aldehydes (*trans*-2-hexenal) able to induce the synthesis of glutathione transferase. This reduces the toxicity of these reagents by conjugating them with glutathione to form glutathione *S*-conjugates, such as *S*-3-(hexan-1-ol)-glutathione. The presence of *B. cinerea*, in its noble form, enables a very significant increase of precursor concentrations in musts and of the corresponding aromas in wines (up to 100 times more with respect to the same grape before botrytization). Recently, a simplified *in vitro* model was developed, and its use has clarified the role of *B. cinerea* and helped determine the origin of cysteinylated and glutathionylated precursors in the grape cell (Thibon et al., 2011). This model has confirmed that *S*-3-(hexan-1-ol)-cysteine is produced from *S*-3-(hexan-1-ol)-glutathione, itself formed by the conjugation of glutathione with *trans*-2-hexenal (Figure 10.27). This biogenesis pathway of aroma potential in the grape is greatly amplified by the development of *B. cinerea* as noble rot on the berry.

Lactone concentrations also evolve significantly with botrytization (Schreier et al., 1976; Miklósy et al., 2000; Miklósy and Kerényi, 2004; Genovese et al., 2007; Sarrazin et al., 2008; Stamatopoulos et al., 2014). The evolution of these concentrations in Semillon grapes was also measured during their botrytization over several vintages. Monitoring demonstrates that their evolution varies depending on the molecular structure of the lactones. The concentration of certain "unsaturated lactone" forms was lower after the *pourri rôti* stage. Meanwhile, "saturated lactone" concentrations (less associated with noble rot quality, such as γ-octanolactone, γ-nonadecalactone, etc.) continue to increase during the last stage (*pourri rôti* + 15 days). Therefore, "unsaturated lactone" concentrations, i.e. 2-nonen-4-olide and massoia lactone, reach their maximum during the *pourri roti* stage and then decrease (Figure 10.41). In the same grapes, the concentration of volatile thiol precursors (cysteine and glutathione *S*-conjugates) presents similar evolution kinetics (Figure 10.42). Differences in the evolution of 2-nonen-4-olide, massoia lactone, and γ-nonalactone concentrations could be due to a process that limits unsaturated lactone synthesis as noble rot development progresses. Considering the sensory impact of 2-nonen-4-olide and of massoia lactone, these may be considered as molecular markers for the quality of noble rot. Moreover, this study also demonstrated differences in lactone concentration between various vintages (2012 and 2014) and between parcels as a function of terroir (Stamatopoulos et al., 2015).

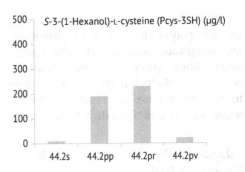

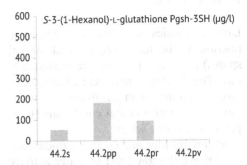

FIGURE 10.42 Evolution of *S*-conjugates of cysteine and glutathione in Semillon grapes during botrytization (2014 vintage) (Sauternes, parcel 44.2). H: Healthy; pp: *pourri plein*; pr: *pourri rôti*; pv: *pourri vieux* (old rotted).

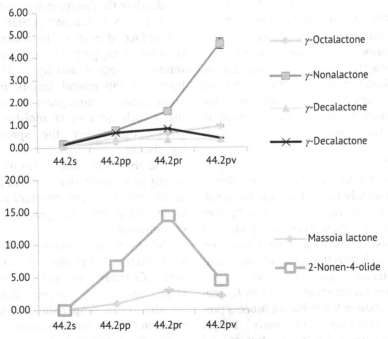

FIGURE 10.41 Evolution of saturated lactones (γ-octa-, nona-, deca-, dodeca-) and unsaturated lactones (massoia lactone, 2-nonen-4-olide) in Semillon grapes during botrytization (2014 vintage) (Sauternes, parcel 44.2). S: Sound grapes; pp: *pourri plein*; pr: *pourri rôti*; pv: *pourri vieux* (old rotted).

After a botrytis attack, grapes and must contain polysaccharides with phytotoxic and fungistatic activities. *Botrytis cinerea* also produces diverse antibiotic substances: botrydial, norbotryal acetate, and botrylactone. Some of these substances can be the source of fermentation difficulties.

10.6.5 Gray Rot and Other Kinds of Rot

Large quantities of a variety of epiphytic microflora (bacteria, yeasts, and fungal spores) are present on the grape skin surface. The development or reproduction of a given microorganism is, above all, determined by environmental conditions (temperature and free water).

When healthy grapes ripen under dry conditions, the low level of water activity on the skin surface promotes the proliferation of osmophilic microorganisms, especially yeasts (Rousseau and Donèche, 2001). Average water activity (from condensation or fog) is required for fungi to develop, while bacteria need large quantities of free water, generally from heavy rainfall, before they can multiply (Figure 10.43).

Furthermore, the various microorganisms interact (antagonism and competition for nutrients). Indeed, a number of yeast strains capable of restricting the development of *B. cinerea* are used in organic disease control.

Gray rot

Noble rot is a regular phenomenon developing uniformly throughout the vineyard. Gray rot attacks, however, are usually very heterogeneous. Partially or totally infected grape clusters are often encountered on one plant, whereas the grapes on the neighboring vine are totally untouched.

The gray rot infection process by *B. cinerea* is identical to the noble rot process previously described, but early fungal development is difficult to detect on red grapes. The external development of the fungus is certainly the most characteristic

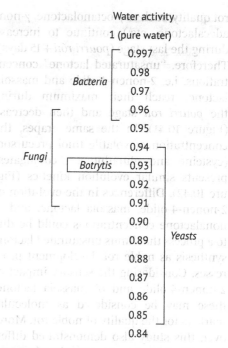

FIGURE 10.43 Minimum values of water activity for the growth of epiphytic microorganisms in the grape.

trait of gray rot. The conditions leading to the death of the fungus in the case of noble rot do not occur. A mycelial "felt" forms on the surface of grapes. The contamination of neighboring grapes is facilitated by the intense biological activity of this mycelium. All viticultural factors increasing grape cluster compactness and maintaining a high amount of moisture on the grapes thus favor the spreading of the disease.

The chemical composition of grapes is greatly modified in the course of a gray rot attack. All of the intermediary forms between noble rot and gray rot can be encountered.

Botrytis cinerea consumes grape sugars while accumulating glycerol and gluconic acid. Contrary to noble rot, sugar concentration by grape dehydration remains low in comparison with sugar degradation (Figure 10.44). Consequently, the sugar concentrations of musts obtained from grapes infected by gray rot rarely exceed 230 g/l. The fungus also accumulates

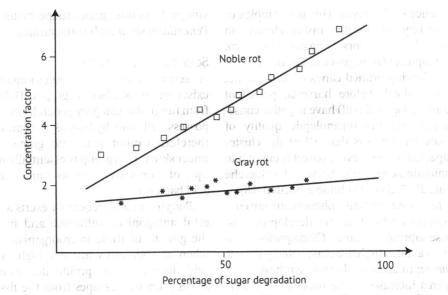

FIGURE 10.44 Relationship between sugar degradation and berry dehydration according to the type of rot (Donèche, 1992).

large quantities of gluconic acid during its external development phase (more than 10 μEq/berry or more than 3 g/l of must and up to 10 g/l).

Malic and tartaric acid degradation is more significant than in the case of noble rot. Up to 90% of the initial concentrations present in healthy grapes can be degraded. The fungus also accumulates higher amounts of citric and acetic acids in the contaminated grape, but the acetic acid concentration rarely exceeds 5 μEq per berry.

The differences between the two kinds of rot are even more pronounced when considering phenolic compounds. These are much more oxidized by laccase, especially in red grapes whose skin is rich in phenolic substrates. Laccase activity increases as mycelium grows and indicates age of gray rot. The risk of color alteration, known as enzymatic browning, is considerable when the must is exposed to air after crushing (Section 11.6.2).

In contrast to noble rot, which gives sweet white wines their specific qualitative aromas, gray rot often causes aroma flaws (Volume 2, Section 8.8). The grapes and wines obtained are often marked by characteristic "moldy" or "undergrowth" odors. The responsible compounds are derivatives of cuticular fatty acids (1-octen-3-one, 1-octen-3-ol) or terpene compounds (2-methylisoborneol) formed during skin maceration by the mycelial biomass (Bock *et al.*, 1988; La Guerche *et al.*, 2006).

Other fungi are often simultaneously present with *B. cinerea*. As a result, the rot takes on varied colors: black (*Aspergillus niger*), white, blue, or green (*Penicillium* sp., *Cladosporium* sp.). These fungi develop less mycelial biomass than *B. cinerea*; glycerol and gluconic acid accumulation is less substantial. The contaminated grapes are often extremely bitter and possess aroma flaws. In wines, these compounds give iodine (Ribéreau-Gayon, 1982), damp earth, and fungal odors (La Guerche *et al.*, 2006).

Some grapes have a strong "damp earth" smell due to the accumulation of geosmin (Volume 2, Section 8.8). This compound, derived from the biosynthesis of terpenoids, is formed by several strains of *Penicillium* (mainly *Penicillium expansum*), in the presence of *B. cinerea*. This is a complex or secondary rot, located inside clusters in particular. Factors aggravating rot development (resurgence of humidity in the vine, leaving pruned canes on the ground, heavy rainfall before harvests, persistent morning fog, and hail) have negative consequences on the organoleptic quality of wines. In fact, less than 1% of the cluster impacted in a grapevine parcel is enough to contaminate an entire harvest (La Guerche et al., 2005, 2007) (Volume 2, Section 8.8).

Moreover, oxidative phenomena are consequences related to the development of these saprophytic fungi. *Cladosporium* possesses a much higher laccase activity than *B. cinerea*; in addition, laccase synthesis by *B. cinerea* increases in the presence of *Aspergillus* or *Penicillium* (Kovac, 1983).

All fungi have similar water requirements, and the most decisive parameter in their selection is certainly temperature. For this reason, the toxicogenic strains of *Aspergillus* are most widespread in hot-climate vineyards, as this species adapts better than *Penicillium* sp. at high temperatures.

Sour rot and acid rot

A second category of microorganisms exists on the surface of grapes. Differing from fungi, this category generally does not possess cell wall hydrolysis enzymes and therefore cannot penetrate grapes with intact skins. This group is essentially made up of oxidative yeasts and acetic acid bacteria.

Botrytis cinerea frequently exerts a powerful antagonistic influence and hinders the growth of these microorganisms, but when temperatures are too high, acetic acid bacteria can proliferate, utilizing sweet juice that escapes from the fissures created by the emergence of *B. cinerea* on the exterior of the grape. The evolution of the grape from the *pourri plein* stage onwards is thus different and leads to sour rot (*pourriture aigre*) (Figure 10.45) (Section 13.2.1).

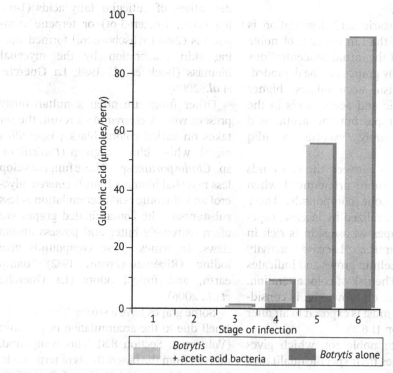

FIGURE 10.45 Evolution of gluconic acid concentrations in grapes during development of *B. cinerea*, or of *B. cinerea* followed by development of acetic acid bacteria.

These bacteria transform the glycerol, formed beforehand by *B. cinerea*, into dihydroxyacetone. Among these acetic acid bacteria, *Gluconobacter* species oxidize glucose with the help of membrane dehydrogenases. Sugar degradation is thus substantial and is accompanied by the accumulation of gluconic, 2-oxogluconic, 5-oxogluconic, and 2,5-dioxogluconic acids in the grape. The production of these ketone compounds substantially increases the binding potential of the must with sulfur dioxide.

The development of these acetic acid bacteria is also characterized by acetic acid production. The musts obtained from grapes infected with sour rot can contain more than 40 g of acetic acid and up to 25 g of gluconic acid per liter. Since these bacteria only slightly degrade grape acids, the musts obtained have extremely low pHs. This is the worst form of rot.

Yeasts may also be involved in grape contamination, either alone or associated with acetic acid bacteria. In fact, yeasts have been identified as responsible for certain acid rot attacks in Mediterranean vineyards. This disease is caused by oxidative yeast development (*Candida*, *Kloeckera*, *Hanseniaspora*). It is known that certain phytopathogenic strains are likely to cause lesions in plant tissue. Tartaric acid in the grape is not attacked. The formation of gluconic, acetic, and galacturonic acids greatly increases acidity. These yeasts produce a small amount of ethanol, and the must possesses high concentrations of ethyl acetate and acetaldehyde.

Damaged grapes follow the same spoilage pathway. Grape skin lesions can occur at any stage of maturation. The causes are diverse: bursting due to a rapid water flux, insect bites, or skin degeneration due to overripeness.

Exceptionally, *B. cinerea* can be observed to develop exclusively, forming a crescent-shaped mycelial mass to obstruct the fissure (Donèche, 1992). Changes in the chemical composition of the grape are thus also characteristic of a gray rot attack.

Usually, however, the sweet juice seeping out of the damaged grape favors the multiplication of oxidative yeast and acetic acid bacteria. These microorganisms are generally transported by the insects responsible for the lesions. The grape thus inevitably evolves toward vulgar rot.

Grapes infected with vulgar rot (sour or acid) cannot be used to make wine. Their presence in the vineyard must be detected as soon as possible, and the grape clusters should be eliminated to limit the spreading of the disease.

10.6.6 Evaluating the Soundness of the Harvest

Botrytis cinerea development, alone or associated with other microorganisms, lowers potential grape quality. The enological consequences are serious in wines made from spoiled grapes: oxidation, degradation of color and aromas, and fermentation and clarification difficulties. The objective measurement of the soundness (absence of rot) of the harvest is therefore of obvious interest.

For a long time, only visual evaluation methods were available to the grapegrower for judging the extent of contamination. This technique takes only external fungal development on the berries into account. However, *B. cinerea* has already partially altered the grape before emerging on the surface of the grape. To complicate the matter further, the infection spots are difficult to see on the surface of a red grape.

At present, a grape selection criterion is based on monitoring the laccase activity within the grape. This enzyme is secreted early on by *B. cinerea*. Due to the natural presence of another oxidoreductase (tyrosinase) in healthy grapes, a specific substrate must be used to measure the laccase activity in must. Two measurement methods exist.

The first method is based on a polarographic measurement of the must oxygen consumption in the presence of a laccase-specific substrate. This method is not very sensitive, and the elimination of must phenolic compounds is not indispensable. It has the advantage of taking into account all of the oxidase activities likely to exist in must obtained from contaminated grapes. However, this method may not be able to differentiate between slightly contaminated and perfectly sound grapes. A machine based on this method has been developed, which automates this analysis (Salgues et al., 1984).

The other process includes a colorimetric measurement that makes use of syringaldazine, a laccase-specific substrate (Harkin and Obst, 1973). This colorless ortho-diphenol is stable with respect to chemical oxidation as well as the presence of tyrosinase and polyphenol oxidase, in sound grapes. The quinone formed in the presence of laccase has an intense rose-mauve color (Figure 10.46). The speed at which it appears is measured by spectrophotometry (Dubourdieu et al., 1984). The reaction must be carried out on a non-sulfited must. The phenolic compounds must also be eliminated by percolation on a polyvinylpolypyrrolidone (PVPP) column to avoid their interference in the analysis. The results are expressed in laccase activity units per milliliter of must. A laccase activity unit is defined as the quantity of the enzyme capable of oxidizing a nanomole of syringaldazine per minute under the conditions of analysis. Manual analysis by this method is relatively quick (5–10 minutes, depending on the percolation time for the sample). It is simple to conduct, since ready-to-use kits exist, including PVPP cartridges and ready-to-use reagents as well as a color chart indicating the corresponding laccase units. A semi-quantitative determination is thus possible in the winery without using colorimetry. An automated analyzer also exists based on the same principle and adapted to large-volume wineries. The results, obtained in two minutes, are given in laccase units, as with the manual colorimetric method.

A strong correlation has been demonstrated between visually determined grape contamination levels, and laccase activity (Table 10.17). However, Redl and Kobler (1992) emphasized that the laccase concentration does not permit the estimation of the total phenolic decomposition level of moldy grapes. Notably, there is no correlation between laccase activity and the total phenolic index, determined by spectrophotometry at 280 nm.

Cagnieul and Majarian (1991) developed an immunological method for detecting B. cinerea, but no diagnostic kits are currently being developed. In this method, polyclonal antibodies are used, which recognize the presence of specific polysaccharides secreted by the fungus.

Other microorganisms present on the grape, such as *Aspergillus*, *Penicillium*, *Cladosporium*, acetic acid bacteria, and

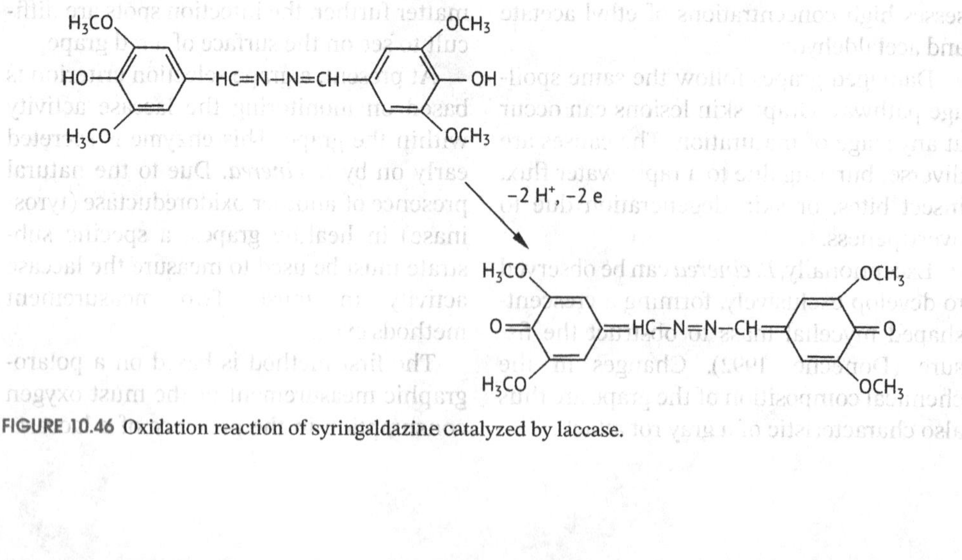

FIGURE 10.46 Oxidation reaction of syringaldazine catalyzed by laccase.

oxidative yeasts, do not interfere with this immunological test. Thanks to its sensitivity, the fungus can be detected 20–30 days before the harvest, providing adequate time for applying fungicides. This method is also capable of differentiating between wines made from botrytized grapes and healthy grapes at any given stage of the winemaking process (Fregoni *et al.*, 1993).

Fourier transform infrared spectrophotometry can be an invaluable tool for assessing the condition of harvested grapes, if calibration is done regularly and with precision. This method is already capable of detecting the presence of rotten grapes, but there is no close correlation with the measurement of laccase activity.

In conclusion, progress in grape-growing and disease prevention has greatly improved wine quality, not only by diminishing wine flaws but also by letting the harvest be delayed until optimal ripeness.

New risks have arisen from this progress. Improved vineyard practices can result in excessive plant vigor. Or on the contrary, too much control over yields and over the leaf area may result in wilting. Above a certain level, increased or decreased vegetative growth is always detrimental to the biochemical maturation processes. The start of these processes is at least delayed, and thus maturation may occur under unfavorable weather conditions. Excess production is another risk: diluted grapes are obtained, producing wines with little structure, color, and aroma.

Climate change will have to be taken into even greater account in the years to come. The best production conditions must be conserved to maintain quality and typicity.

The grape is more than a reserve storage organ. At harvest time, it still possesses an intense metabolic activity. Particular attention should therefore be given to grape handling, especially when a percentage of the grapes are infected by bunch rot. In this case, the grapes contain many additional enzymes of fungal origin.

References

Abbal P., Boulet J.C. and Moutounet M. (1992) Œno One, 26, 231.

Allamy L. (2015) *Recherches sur les marqueurs moléculaires de l'arôme de "fruits cuits" des raisins et des vins rouges issus des cépages Merlot et Cabernet-Sauvignon. Approche sensorielle, analytique et agronomique*, Thèse de doctorat, Université de Bordeaux.

Allamy L., Darriet P. and Pons A. (2018) Food Chem., 266, 245.

Allen M.S. and Lacey M.J. (1993) Vitic. Enol. Sci., 48, 211.

Allen M.S., Lacey M.J. and Boyd S. (1994) J. Agric. Food Chem., 42, 1734–1738.

Arnold R.A. and Bledsoe A.M. (1990) Am. J. Enol. Vitic., 41, 74.

Aubert C., Baumes R., Günata Z., Lepoutre J.P., Cooper J.F. and Bayonove C. (1997) Œno One, 31, 57.

Augustin M.M. (1986) *Etude de l'influence de certains facteurs sur les composés phénoliques du raisin et du vin*, Doctorat d'Université, Bordeaux.

Aziz A. (2003) J. Exp. Bot., 54, 355.

Barbeau G., Asselin C. and Morlat R. (1998) Bull. OIV, 805–806, 247.

Battilana J., Costantini L., Emanuelli F., Sevini F., Segala C., Moser S., Velasco R., Versini G. and Grando M.S. (2009) Theor. Appl. Genet., 118, 653–669.

Bayonove C. (1993) Les composes terpéniques. In *Les acquisitions récentes en chromatographie du vin* (Ed B. Donèche). Tec & Doc, Lavoisier, Paris.

Bayonove C., Cordonnier R. and Dubois P. (1975) C. R. Acad. Sci., 281, D, 75.

Becker N. and Zimmermann A. (1984) Bull. OIV, 641–642, 584.

Belancic A., Agosin E., Ibacache A., Bordeu E., Baumes R., Razungles A. and Bayanove C. (1997) Am. J. Enol. Vitic., 48, 181–185.

Bergqvist J., Dokoozlian N. and Ebisuda N. (2001) Am. J. Enol. Vitic., 52, 1.

Bertrand A., Pissard R., Sarre C. and Sapis J.C. (1976) Œno One, 10, 427.

Bessis R. and Fournioux J.C. (1992) Vitis, 31, 9.

Blaich R., Stein U. and Wind R. (1984) Vitis, 23, 242.

Blouin, J. (1992) *Techniques d'analyses des moûts et des vins*. Dunod, Paris.

Blouin J. and Guimberteau G. (2000) *Maturation et maturité du raisin*. Féret Ed, Bordeaux.

Bock G., Benda I. and Schreier P. (1988) Z. Lebensm., 186, 33.

Bonnet A. (1903) Ann. Ecol. Nat. Agric. Montpellier, 3, 58.

Boulton R.B., Singleton V.L., Bisson L.F. and Kunkee R.E. (1999) *Principles and Practices of Winemaking*. Food Science and Nutrition. Springer.

Branas J., Bernon G. and Levadoux L. (1946) *Eléments de viticulture générale*. Published by the authors, printed by Delmas, Bordeaux.

Bravdo B., Hepner Y., Loinger C., Cohen S. and Tabacaman H. (1985) Am. J. Enol. Vitic., 36, 125.

Cagnieul P. and Majarian W. (1991) *Diagnostic kit for grape grey mold. Comptes Rendus de la 3ème Conférence Internationale sur les maladies des plantes (Bordeaux, 1991)*. ANPP, Paris.

Cakir B., Agasse A., Gaillard C., Saumonneau A., Deirot S. and Atanassova R. (2003) Plant Cell, 15, 2165.

Calo A., Costacurta A., Tomasi D., Becker N., Bourquin H.D., de Villiers F.S., Garcia de Lujan A., Huglin P., Jaquinet L. and Lemaitre C. (1992) Riv. Vitic. Enol., 3, 3.

Candolfi-Vasconcelos M.C. and Koblet W. (1990) Vitis, 29, 199.

Capone D.L., Sefton M.A., Hayasaka Y. and Jeffery D.W. (2010) J. Agric. Food Chem., 58, 1390–1395.

Capone D.L., Pardon K.H., Cordente A.G. and Jeffery D.W. (2011) J. Agric. Food Chem., 59, 11204–11210.

Capone D.L., Black C.A. and Jeffrey D.W. (2012) J. Agric. Food Chem., 60, 3515–3523.

Caputi L., Carlin S., Ghiglieno I., Stefanini M., Valenti L., Vrhovsek U. and Mattivi F. (2011) J. Agric. Food Chem., 59, 10, 5565–5571.

Carbajal-Ida D., Maury C., Salas E. and Siret R. (2015) Eur. Food Res. Technol., 242, 117.

Carbonneau A. (1982) Prog. Agric. Vitic., 99, 290.

Carbonneau A. (1985) Trellising and canopy management for cool climate viticulture. In *Proceedings of the International Symposium on Cool-Climate Viticulture and Enology*, pp. 158–183 (Ed D.A. Heatherbell, P.B. Lombard, F.W. Bodyfelt and S.F. Pryce). OSU Agricultural Experimental Station Technical Publication No. 7628. Oregon State University, Corvallis.

Cerreti M., Esti M., Benucci I., Liburdi K., de Simone C. and Ferranti P. (2015) Aust. J. Grape Wine Res., 21, 411–416.

Champagnol F. (1984) *Eléments de physiologie de la vigne et de viticulture générale*. Montpellier. Published by the author.

Champagnol F. (1986) Prog. Agric. Vitic., 103, 361.

Chardonnet C. and Donèche B. (1995) Vitis, 34, 95.

Chardonnet C., Gomez H. and Donèche B. (1994) Vitis, 33, 69.

Charrier F., Dufourcq T. and Guérin-Schneider R. (2007) *Comptes Rendus du 8e Symposium International d'œnologie de Bordeaux*, Vol. 1, p. 50. Vigne et Vin Publications Internationales, Bordeaux.

Chaves M.M., Zarrouk O., Francisco R., Costa J.M., Santos T., Regalado A.P., Rodrigues M.L. and Lopes C.M. (2010) Ann. Bot., 105, 661.

Chervin C., El-Kereamy A., Roustan J.P., Latché A., Lamon J. and Bouzayen M. 2004 Plant Sci., 167, 6, 1301–1305.

Cholet C., Mondolot L. and Andary C. (2002) Aust. J. Grape Wine Res., 8, 126.

Cholet C., Claverol S., Claisse O., Rabot A., Osowsky A., Dumot V., Ferrari G. and Gény L. (2016) BMC Plant Biol., 16, 144.

Choné X., Tregoat O., Van Leeuwen C. and Dubourdieu D. (2000) Œno One, 34, 4, 169.

Choné X., Tregoat O. and Van Leeuwen C. (2001a) Œno One, 35, Hors Série, 47.

Choné X., Van Leeuwen C., Dubourdieu D. and Gaudillère J.P. (2001b) Ann. Bot., 87, 4, 47.

Choné X., Lavigne-Cruège V., Tominaga T., Van Leewen C., Castagnède C., Saucier C. and Dubourdieu D. (2006) Œno One, 40, 1–6.

Colin L., Cholet C. and Geny L. (2002) Aust. J. Grape Wine Res., 8, 101.

Conn E.E. (1986) *Recent Advances in Phytochemistry, Vol. 20: The Shikimic Acid Pathway*. Plenum Press, New York.

Coombe B.G. (1989) Acta Hortic., 239, 149.

Crippen D.D. and Morrison J.C. (1986) Am. J. Enol. Vitic., 37, 1986.

Dai Z.W., Ollat N., Gomès E., Decroocq S., Tandonnet J.P., Bordenave L., Pieri P., Hilbert G., Kappel C., Van Leeuwen C., Vivin P. and Delrot S. (2011) Am. J. Enol. Vitic., 62, 4.

Darné G. (1991) *Recherches sur la composition en anthocyanes des grappes et des feuilles de vignes*, Thèse Doctorat ès Sciences, Bordeaux.

Darriet P., Bouchilloux P., Poupot C., Bugaret Y., Clerjean M., Sauris P., Medina B. and Dubourdieu D. (2001) Vitis, 40, 93.

De Billerbeck G.M., Ebang-Oke J.P., Ambid C., Lonvaud A., de Revel G. and Darriet P. (2003) *Oenologie 2003*, pp. 89–93. Tec & Doc, Lavoisier, Paris.

Deytieux-Belleau C., Gagné S. L'Hyvernay A., Donèche B. and Geny L. (2007a) Œno One, 41, 3, 141.

Deytieux-Belleau C., Geny L., Lapaillerie D., Claverol S., Bonneur M. and Donèche B. (2007b) J. Exp. Bot., 58, 7, 1851.

Deytieux-Belleau C., Geny L., Roudet J., Veyssière V., Donèche B. and Fermaud M. (2009) Eur. J. Plant Pathol., 125, 4, 551.

Di Stefano R., Corina L. and Bosia P.D. (1983) Riv. Vitic. Enol., 36, 263.

Donèche B. (1987) *Etude biochimique de la relation hôte-parasite dans le cas du raisin et de Botrytis cinerea*, Thèse Doctorat ès Sciences, Université de Bordeaux II.

Donèche B. (1976) *Effets du mancozèbe sur la microflore des sols de vignobles et participation des microorganismes à sa dégradation*, Thèse Doctorat de 3ème cycle, Université de Bordeaux II.

Donèche B. (1986) Agronomie, 6, 67.

Donèche B. (1989) Can. J. Bot., 67, 2888.

Donèche B. (1992) Botrytized wines. In *Wine Microbiology and Biotechnology*, pp. 327–351 (Ed G.H. Fleet). Harwood Academic Publishers, Chur.

Donèche B. and Chardonnet C. (1992) Vitis, 31, 175.

Downey M.O., Dokoozlian N.K. and Krstic M.P. (2006) Am. J. Enol. Vitic., 57, 3.

Drappier J., Thibon C., Rabot A. and Geny-Denis L. (2019) Crit. Rev. Food Sci. Nutr., 59, 14.

Dry P., Loveys B., McCarthy M. and Stoll M. (2001) Œno One, 35, 3, 129.

Dubernet M., Ribéreau-Gayon P., Lerner H.R., Harel E. and Mayer A.M. (1977) Phytochemistry, 16, 191.

Dubernet M., Coulomb S., Lerch M. and Traineau I. (2000) Rev. Fr. Oenol., 185, 18.

Dubourdieu D. (1978) *Etude des polysaccharides sécrétés par Botrytis cinerea dans la baie de raisin. Incidence sur les difficultés de clarification des vins de vendanges pourries*, These Doctorat de 3ème cycle, Université de Bordeaux II.

Dubourdieu D., Grassin C., Deruche C. and Ribéreau-Gayon P. (1984) Œno One, 18, 4.

Duchêne E., Butterlin G., Claudel P., Dumas V., Jaegli N. and Medinoglu D. (2009) Theor. Appl. Genet., 118, 541–552.

Dunlevy J.D., Dennis E.G., Soole K.L., Perkins M.V., Davies C. and Boss P.K. (2013) Plant J., 75, 4, 606–617.

Düring H. (1994) Am. J. Enol. Vitic., 45, 297.

Duteau J. (1990) *Actualités oenologiques 89*. Dunod, Paris.

Duteau J., Guilloux M. and Seguin G. (1981) Œno One, 15, 3, 1.

Enzel C.R. (1985) Pure Appl. Chem., 57, 5, 693.

Ewart A.J.W. (1987) Influence of vineyard site and grape maturity on juice and wine quality of *Vitis vinifera*, cv. Riesling. In *Proceedings of Sixth Australian Wine Industry Conference*, pp. 71–74 (Ed T.H. Lee), Adelaide.

Falcao L.D., de Revel G., Perello M.C., Moutsiou A., Zanus M.C. and Bordignon-Luiz M.T. (2007) J. Agric. Food Chem., 55, 3605.

Fasoli M., Dell'Anna R., Dal Santo S., Balestrini R., Sanson A., Pezzotti M., Monti F. and Zenoni S. (2016) Plant Cell Physiol., 57, 6, 1332.

Flanzy C. (2000) *Œnologie: fondements scientifiques et technologiques*. Tec & Doc, Lavoisier, Paris.

Fortes A.M., Teixeira R.T. and Agudelo-Romero R. (2015) Molecules, 20, 9326.

Fregoni M., Perino A. and Vercesi A. (1993) Bull. OIV, 745–746, 169.

Gagné S., Lacampagne S., Claisse O. and Geny L. (2009) Plant Physiol. Biochem., 47, 282.

Gagné S., Cluzet S., Merillon J.M. and Geny L. (2011) J. Plant Growth Regul., 30, 1.

Galet P. (1977) *Les maladies et les parasites de la vigne*. Paysan du Midi, Montpellier.

Gaudillère J.P., Van Leeuwen C. and Ollat N. (2002) J. Exp. Bot., 53, 369, 757.

Genovese A., Gambuti A., Piombino P. and Moio L. (2007) Food Chem., 103, 1228.

Gerber C. (1898) Ann. Sci. Nat. Bot., 4, 1.

Giribaldi M., Geny L., Delrot S. and Schubert A. (2010) J. Exp. Bot., 61, 9, 2447.

Grant-Preece P.A., Pardon K.H., Capone D.L., Cordente A.G., Sefton M.A., Jeffery D.W. and Elsey G.M. (2010) J. Agric. Food Chem., 58, 1383–1389.

Gregan S.M. and Jordan B. (2016) J. Agric. Food Chem., 64, 2200–2208.

Gregan S.M., Wargent J.J., Liu L., Shinkle J., Hofmann R., Winefield C., Trought M. and Jordan M. (2012) Aust. J. Grape Wine Res., 18, 227.

Grotte M., Cadot Y., Poussier A., Loonis D., Pietri E., Duprat F. and Barbeau G. (2001) Œno One, 35, 2, 87.

Guillaumie S., Ilg A., Brette M., Decroocq S., Claudel P., Jaegli N., Trossat-Magnin C., Pons M., Butterlin G., Léon C., Baltenweck R., Fischer M., Delrot S., Duchêne E., Meyer S., Gomès E., Darriet Ph. and Hugueney P. (2013) Plant Physiol. 162, 604–615.

Guilloux M. (1981) *Evolution des composés phénoliques de la grappe pendant la maturation du raisin. Influence des facteurs naturels*, Thèse Doctorat Université, Université de Bordeaux II.

Guilpart N. (2014) *Relations entre services écosystémiques dans un agroécosystème à base de plantes pérennes: compromis entre rendement de la vigne et régulation de l'oïdium*, Thèse Doctorat, SupAgro, Montpellier.

Günata Z. (1984) *Recherches sur la fraction liée de nature glycosidique de l'arôme du raisin: Importance des terpénylglycosides, action des glycosidases*, Thèse Docteur-Ingénieur, Université Sciences et Techniques du Languedoc, Montpellier.

Harkin J.K. and Obst J.R. (1973) Experientia, 29, 381.

Harris J.M., Kriedemann P.E. and Possingham J.V. (1971) Vitis, 9, 291.

Harris S.A., Ryona I. and Sacks G.L. (2012) J. Agric. Food Chem., 60, 11901–11908.

Haselgrove L., Botting D., van Heeswijck R., Høj P.B., Dry P.R., Ford C. and Iland P.G. (2000) Aust. J. Grape Wine Res., 6, 141.

Hashizume K. and Samuta T. (1999) Am. J. Enol. Vitic., 50, 194.

Hashizume K. and Umeda N. (1996) Biosci. Biotechnol. Biochem., 60, 802.

Hashizume K., Tozawa K., Endo M. and Aramaki I. (2001a) Biosci. Biotechnol. Biochem., 65, 795–801.

Hashizume K., Tozawa K., Hiraga Y. and Aramaki I. (2001b) Biosci. Biotechnol. Biochem., 65, 2213–2219.

Hatzidimitriou E., Bouchilloux P., Darriet P., Clerjeau M., Bugaret Y., Poupot C. and Dubourdieu D. (1996) Œno One, 30, 3, 133.

Helwi P., Habran A., Guillaumie S., Thibon C., Hilbert G., Gomes E., Delrot S., Darriet P. and Van Leeuwen C. (2015) J. Agric. Food Chem., 63, 9789–9802.

Helwi P., Guillaumie S., Thibon C., Keime C., Habran A., Hilbert G., Gomes E., Darriet P., Delrot S. and Van Leeuwen C. (2016) BMC Plant Biol., 16, 173.

Herrick I.W. and Nagel C.W. (1985) Am. J. Enol. Vitic., 36, 95.

Heymann H., Noble A.C. and Boulton R.B. (1986) J. Agric. Food Chem., 34, 268.

Hrazdina G., Parsons G.F. and Mattick L.R. (1984) Am. J. Enol. Vitic., 35, 220.

Huglin P. (1978) C. R. Acad. Agric. Fr., 64, 1117.

Huglin P. (1986) *Biologie et écologie de la vigne*. Payot, Lausanne.

Ito H., Motomura Y., Konno Y. and Hatayama T. (1969) Tohoku J. Agric. Res., 20, 1.

Jackson D.I. (1987) Vinifera Winegrowers J., 14, 144.

Jackson D.I. and Lombard P.B. (1993) Am. J. Enol. Vitic., 44, 4, 409.

Jeandet P. and Bessis R. (1989) Bull. OIV, 703–704, 637.

Jeandet P., Bessis R. and Gautheron B. (1991) Am. J. Enol. Vitic., 42, 41.

Kanellis A.K. and Roubelakis-Angelakis K.A. (1993) Grape. In *Biochemistry of Fruit Ripening*, pp. 189–234 (Eds G.B. Seymour, J.E. Taylor and G.A. Tucker). Chapman & Hall, London.

Karadimtcheva B. (1982) Bull. OIV, 613, 246.

Keller M., Tarara J.M. and Mills L.J. (2010) Aust. J. Grape Wine Res., 16, 3, 445.

Kliewer W.M. and Torres R.E. (1972) Am. J. Enol. Vitic., 23, 71.

Kliewer W.M. and Weaver R.J. (1971) Am. J. Enol. Vitic., 22, 172.

Knipfer T., Fei J., Gambetta G., McElrone A.J., Shackel K.A. and Matthews M.A. (2015) Plant Physiol., 168, 1590.

Kobayashi H., Takase H., Kaneko K., Tanzawa F., Takata R., Suzuki S. and Konno K. (2010) Am. J. Enol. Vitic., 61, 176–185.

Kobayashi H., Takase H., Suzuki Y., Tanzawa F., Takata R., Fujita K., Kohno M., Mokizuchi M., Suzuki S. and Konno, T. (2011) J. Exp. Bot., 62, 1325.

Koblet W. (1975) Wein Wiss., 30, 241.

Kovac V. (1983) Bull. OIV, 628, 420.

La Guerche S., Chamont S., Blanchard D., Dubourdieu D. and Darriet P. (2005) Antonie Van Leeuwenhoek, 88, 139.

La Guerche S., Dauphin B., Pons M., Blanchard D. and Darriet P. (2006) J. Agric. Food Chem. 54, 9193.

La Guerche S., Dauphin B., Pons M., Blancard D. and Darriet P. (2007) Antonie van Leeuwenhoeck, 92, 331.

Lacampagne S. (2010) *Localisation et caractérisation des tannins dans la pellicule du raisin: étude de l'impact de l'organisation physico-chimique des parois cellulaires sur la composante tannique, la qualité du fruit et la typicité des raisins de Bordeaux*, Thèse de Doctorat, Université de Bordeaux.

Lacey M.J., Allen M.S., Harris R.L.N. and Brown W.V. (1991) Am. J. Enol. Vitic., 42, 103.

Lacroux F., Treogat O., Van Leeuwen C., Pons A., Tominaga T., Lavigne V. and Dubourdieu D. (2008) Œno One, 42, 125.

Langcake P. and Pryce R.J. (1977) Phytochemistry, 16, 1193.

Lavee S. and Nir G. (1986) Grape. In *CRC Handbook of Fruit Set and Development*, pp. 167–191 (Ed S.P. Monselix). CRC Press, Boca Raton, FL.

Le Moigne M., Maury C., Bertrand D. and Jourjon F. (2008) Food Qual. Prefer., 19, 2, 220.

Lebon G., Wojnarowiez G., Holzapfel B., Fontaine F., Vaillant-Gaveau N. and Clément C. (2008) Botany, 59, 10, 2565.

Lecourieux F., Kappel C., Pieri P., Charon J., Pillet J., Hillbert G., Renaud C., Gomès E., Delrot S. and Lecourieux D. (2017) Front. Plant Sci., 8, 53, 1.

Li-Mallet A., Rabot A. and Geny L. (2016) Botany, 94, 3, 147.

Linsenmeier A.W. and Löhnertz O. (2007) S. Afr. J. Enol. Vitic., 28, 17.

Luan F., Hampel D., Mosandl A. and Wust M. (2004) J. Agric. Food Chem., 56, 7, 1371–1375.

MacCarthy M.G. and Coombe B.G. (1984) Acta Hortic., 171, 447.

Maga J.A. (1989) Sensory and stability properties of added methoxypyrazines to model and authentic wines. In *Flavors and Off-Flavors*, p. 61–70 (Ed G. Charalambous). Elsevier, Amsterdam.

Marais J.S. (1983) S. Afr. J. Enol. Vitic., 4, 2, 49.

Marais J.S. (1993) C.R. du Symposium international "Connaissance aromatique des cépages et qualité des vins", Montpellier, February 9–10, 1993, Rev. Fr. Œnologie Ed.

Marais J.S., Van Wyck C.J. and Rapp A. (1992a) S. Afr. J. Enol. Vitic., 13, 33.

Marais J.S., Versini G., Van Wyck C.J. and Rapp A. (1992b) S. Afr. J. Enol. Vitic., 13, 71.

Marbach I., Harel E. and Mayer A.M. (1985) Phytochemistry, 24, 2559.

Martin D.M., Chiang A., Lund S.T. and Bohlmann J. (2012) Planta, 236, 3, 919–929.

Masuda M., Okawa E., Nishimura K. and Yunome H. (1984) Biol. Chem., 48, 2707.

Matthews M.A. and Anderson M.M. (1989) Am. J. Enol. Vitic., 40, 52.

Melino V.J., Hayes M.A., Soole K.L. and Ford C.M. (2011) J. Sci. Food Agric., 9, 1712.

Mendez-Costabel M.P., Wilkinson K.L. Bastian S.E.P., Jordans C., McCarthy M., Ford C.M. and Dokoozlian N.K. (2014a) Aust. J. Grape Wine Res., 20, 100–110.

Mendez-Costabel M.P., Wilkinson K.L., Bastian S.E.P., Jordans C., McCarthy M., Ford C.M. and Dokoozlian N.K. (2014b) Aust. J. Grape Wine Res., 20, 80–90.

Miklósy E. and Kerényi Z. (2004) Anal. Chim. Acta, 513, 177–181.

Miklósy E., Kalmar Z., Pölös V. and Kerényi Z. (2000) Chromatographia 51, 305.

Mori K., Goto-Yamamoto N., Kitayama M. and Hashizume K. (2007) J. Exp. Bot., 58, 8, 1935.

Morlat R. (1989) *Le terroir viticole: contribution à l'étude de sa caractérisation et de son influence sur les vins. Application aux vignobles rouges de la moyenne vallée de la Loire*, Thèse Doctorat ès Sciences, Université de Bordeaux II.

Müller-Thurgau H. (1888) Landwirt. Jarbücher, 17, 83.

Murray K.E. and Whitfield F.B. (1975) J. Sci. Food Agric., 26, 973–986.

Nikolantonaki M. and Waterhouse A.L. (2012) J. Agric. Food Chem., 60, 34, 8484–8491.

Ojeda H., Deloire A. and Carbonneau A. (2001) Vitis, 40, 3, 141.

Ollat N., Diakou-Verdin P., Carde J., Barrieu F., Gaudillere J. and Moing A. (2002) Œno One, 36, 3, 109.

Ough C.S., Stevens S.D. and Almy J. (1989) Am. J. Enol. Vitic., 40, 219.

Palejwala V.A., Parikh H.R. and Modi V.V. (1985) Physiol. Plant., 65, 498.

Pan Q.H., Li M.J., Peng C.C., Zhang N., Zou X., Zou K.Q., Wang X.L., Yu X.C., Wang X.F. and Zhang D.P. (2005) Physiol. Plant., 125, 157.

Park H.S. (1995) *Le péricarpe des baies de raisin normales and millerandées: ontogenèse de la structure et évolution de quelques constituants biochimiques, notamment des tannins*, Thèse de Doctorat, Université Bordeaux I.

Park S.K., Morrisson J.C., Adams D.O. and Noble A.C. (1991) J. Agric. Food Chem., 39, 514.

Pastore C., Zenoni S., Fasoli M., Pezzotti M., Tornielli G.B. and Filippetti I. (2013) BMC Plant Biol., 13, 30.

Petit-Lafitte A. (1868) *La vigne dans le Bordelais*. J. Rothschild, Paris.

Peynaud E. and Ribéreau-Gayon P. (1971) The Grape. In *The Biochemistry of Fruits and Their Products*, Vol. 2, pp. 171–205 (Ed A.C. Hulme). Academic Press, London.

Peyrot des Gachons C. (2000) *Recherches sur le potentiel aromatique des raisins de Vitis vinifera L. Cv. Sauvignon blanc*. Thèse de doctorat. Université de Bordeaux 2.

Peyrot des Gachons C., Tominga T. and Dubourdieu D. (2002) J. Agric. Food Chem., 50, 4076–4079.

Peyrot des Gachons C., Van Leeuwen Tominaga T., Soyer J.P., Gaudillère J.P. and Dubourdieu D. (2005) J. Sci. Food Agric., 85, 73–85.

Pezet R. and Pont V. (1986) Rev. Suisse Vitic. Arboric. Hortic., 18, 317.

Poitou X., Thibon C. and Darriet P. (2017) J. Agric. Food Chem., 65, 383–393.

Pope J.M., Jonas D. and Walker R.R. (1993) Protoplasma, 173, 177.

Possingham J.V. and Grogt Obbink J. (1971) Vitis, 10, 120.

Possner D.R.E. and Kliewer W.M. (1985) Vitis, 24, 229.

Pouget R. and Delas J. (1989) Œno One, Hors Série, 27.

Pucheu-Planté B. and Leclair P. (1990) *Comptes Rendus du 4ème Symposium International d'Œnologie (Bordeaux, 1989)*, Dunod, Bordas, Paris.

Razungles A. (1985) *Contribution à l'étude des caroténoïdes du raisin: teneur et localisation dans la baie, évolution au cours de la maturation*, Thèse Ecole Nationale Supérieure Agronomique, Montpellier, France.

Razungles A. and Bayonove C. (1996) *La Viticulture à l'aube du IIIe millénaire, Œno One*, Hors série, p. 85 (Ed J. Bouard and G. Guimberteau).

Redl H. and Kobler A. (1992) Mitteilungen Klosterneuburg, 42, 25.

Reynolds A.G. (1988) Hortic. Sci., 23, 728.

Ribéreau-Gayon G. (1960) Vitis, 2, 113.

Ribéreau-Gayon P. (1982) Bull. OEPP, 12, 201.

Ribéreau-Gayon J., Peynaud E., Ribéreau-Gayon P. and Sudraud P. (1975) *Traité d'Œnlogie: Sciences et Techniques du vin. Vol. 2: Caractères des vins, maturation du raisin, levures et bactéries*. Dunod, Paris.

Riou C. and Lebon E. (2000) Bull. OIV, 73, 837–838, 755.

Robinson S.P. and Davies C. (2000) Aust. J. Grape Wine Res., 6, 175.

Rohmer M. (1999) Nat. Prod. Rep., 16, 565–574.

Roland A., Vialaret J., Razungles A., Rigou P. and Schneider R. (2010) J. Agric. Food Chem., 58, 4406–4413.

Romieu C., Robin J.P., Nicol M.Z. and Flanzy C. (1989) Œno One, 23, 165.

Roubelakis-Angelakis K.A. (1991) *Proceedings of the International Symposium on Nitrogen in Grape and Wine*, Seattle, WA.

Roubelakis-Angelakis K.A. (2001) *Molecular Biology and Biotechnology of the Grapevine*. Kluwer Academic Publishers, Dordrecht.

Roubelakis-Angelakis K.A. and Kliewer W.M. (1992) Hortic. Rev., 14, 407.

Roujou de Boubée D. (2000) *Recherches sur la 2-méthoxy-3-isobutylpyrazine dans les raisins et les vins. Approches analytique, biologique et agronomique*, Thèse de Doctorat, Université de Bordeaux 2.

Roujou de Boubée D., Van Leeuwen C. and Dubourdieu D. (2000) J. Agric. Food Chem., 48, 4830–4834.

Roujou de Boubée D., Cumsille A.M., Pons M. and Dubourdieu D. (2002) Am. J. Enol. Vitic., 53, 1–5.

Rousseau S. and Donèche B. (2001) Vitis, 40, 2, 75.

Ruffner H.P. (1982a) Vitis, 21, 247.

Ruffner H.P. (1982b) Vitis, 21, 346.

Ruffner H.P., Brem S. and Rast D.M. (1983) Plant Physiol., 73, 582.

Ruffner H.P., Possner D., Brem S. and Rast D.M. (1984) Planta, 160, 444.

Ruffner H.P., Hürlimann M. and Skrivan R. (1995) Plant Physiol. Biochem., 33, 25.

Ryona I., Pan B.S., Intrigliolo D.S., Lakso A.N. and Sacks G.L. (2008) J. Agric. Food Chem., 56, 10838–10846.

Ryona I., Pan B.S. and Sacks G.L. (2009) J. Agric. Food Chem., 57, 8250.

Ryona I., Leclerc D. and Sacks G.L. (2010) J. Agric. Food Chem., 58, 9723–9730.

Sadras V. and Moran M. (2012) Aust. J. Grape Wine Res., 18, 115.

Sala C., Busto O., Guasch J. and Zamora F. (2004) J. Agric. Food Chem., 52, 3492–3497.

Salgues M., Olivieri C., Chabas M. and Pineau J. (1984) Bull. OIV, 638, 308.

Salgues M., Cheynier V., Gunata Z. and Wylde R. (1986) J. Food Sci., 55, 1191.

Sarrazin E., Dubourdieu D. and Darriet P. (2007) Food Chem., 103, 536–545.

Sarrazin E., Tominaga T. and Darriet P. (2008) *Proceedings of the 12th Weurmann Flavour Research Symposium in Expression of Multidisciplinary Flavour Science. Blank I, Wüst M and Yeretzian C (Eds.). Zürcher Hochschule für Angewandte Wissenschaften, Winterthur, Zürich*, 72.

Sarrazin E., Shinkaruk S., Pons M., Thibon C., Bennetau B. and Darriet P. (2010) J. Agric. Food Chem., 58, 10606–10613.

Scarlett N., Bramley R. and Siebert T. (2014) Aust. J. Grape Wine Res., 20, 214.

Schaller K.O. and Löhnertz O. (1992) Vitic. Enol. Sci., 47, 202.

Schaller K.O., Löhnertz O. and Chikkasubbanna V. (1992) Vitic. Enol. Sci., 47, 36.

Scheiner J.J., Sacks G.L., Pan B., Ennahli S., Tarlton L., Wise A., Lerch S.D. and Vanden Heuvel J.E. (2010) Am. J. Enol. Vitic., 61, 358–364.

Schreier P., Drawert F. and Junker A. (1976) J. Agric. Food Chem., 24, 331.

Schüttler A., Gruber B.R., Thibon C., Lafontaine M., Stoll M., Schultz H.R., Rauhut D. and Darriet P. (2011) *Oeno2011-Actes de colloques du 9e symposium international d'oenologie de Bordeaux*, p. 65. Dunod, Paris.

Schüttler A., Fritsch S., Hoppe J.E., Schüssler C., Jung R., Thibon C., Gruber B.R., Lafontaine M., Stoll M., De Revel G., Schultz H.R., Rauhut D. and Darriet P. (2013) Rev. Œnol., 149S, 36.

Schwab W. and Wüst M. (2015) J. Agric. Food Chem., 63, 10591–10603.

Seguin G. (1970) *Les sols de vignobles du Haut-Médoc. Influence sur l'alimentation en eau de la vigne et sur la maturation du raisin*, Thèse Doctorat ès Sciences, Université Bordeaux 2.

Seguin G. (1975) Œno One, 9, 1, 23.

Sepulveda G. and Kliewer W.M. (1986) Am. J. Enol. Vitic., 37, 20.

Smart R.E. (1973) Am. J. Enol. Vitic., 24, 141.

Smart R.E. and Coombe B.G. (1983) Water relations of grapevines. In *Water Deficits and Plant Growth, Vol. VII: Additional Woody Crop Plants*, pp. 137–196 (Ed T.T. Kozlowski). Academic Press, New York.

Sponholz W.R. and Hühn T. (1994) Wein Wiss., 49, 37.

Stamatopoulos P., Frérot E., Tempère S., Pons A. and Darriet P. (2014) J. Agric. Food Chem., 62, 2469–2478.

Stamatopoulos P., Thibon C., Redon P. and Darriet P. (2015) *CR Actes de conferences Oeno2015 10ième Symposium Oenologie Bordeaux*, p. 75. Vigne et Vin Publications Internationales, Bordeaux.

Stamatopoulos P., Brohan E., Prevost C. Siebert T.E., Herderich M. and Darriet P. (2016) J. Agric. Food Chem., 64, 8160–8167.

Stoll M., Loveys B. and Dry P. (2001) J. Exp. Bot., 51, 1627.

Tesic D., Wolley D.J., Hewett E.W. and Martin D.J. (2001) Aus. J. Grape Wine Res., 8, 15.

Thibon C., Shinkaruk S., Tominaga T., Bennetau B. and Dubourdieu D. (2008) J. Chromatogr. A, 1183, 150–157.

Thibon C., Dubourdieu D., Darriet P. and Tominaga T. (2009) Food Chem., 114, 4, 1359–1364.

Thibon C., Cluzet S., Mérillon J.M., Darriet P. and Dubourdieu D. (2011) J. Agric. Food Chem., 59, 1344–1351.

Thibon C., Böcker C., Shinkaruk S., Moine V., Darriet P. and Dubourdieu D. (2016) Food Chem., 199, 711–719.

Ugliano M., Kwiatkowski M., Vidal S., Capone D., Siebert T., Dieval J.B., Aagaard O. and Waters E.J. (2011) J. Agric. Food Chem., 59, 2564–2572.

Ureta C.F. and Yavar O.L. (1982) Œno One, 16, 187.

Ureta C.F., Boidron J.N. and Bouard J. (1981) Am. J. Enol. Vitic., 32, 90.

Van Leeuwen C. (2001) Œno One, Hors Série, 97.

Van Leeuwen C. and Seguin G. (1994) Œno One, 28, 2, 81.

Van Leeuwen C. and Seguin G. (2006) J. Wine Res., 17, 1.

Van Leeuwen C., Choné X., Tregoat O. and Gaudillere J.P. (2001a) Aust. Grapegrow. Winemak., 449, 18.

Van Leeuwen C., Gaudillère J.P. and Treogat O. (2001b) Œno One, 35, 4, 195.

Vasconcelos M.C., Greven M., Winefield C.S., Trought M.C.T., Raw V. (2009) Am. J. Enol. Vitic., 60, 4, 411.

Wang X.J., Tao Y.S., Wu Y., An R.Y. and Yue Z.Y. (2017) Food Chem., 226, 41–50.

Wen Y.Q., Li J.M., Zhang Z.Z., Zhang Y.F. and Pan Q.H. (2010) Biosci. Biotechnol. Biochem., 74, 12, 2413.

Wicks A.S. and Kliewer W.M. (1983) Am. J. Enol. Vitic., 34, 114.

Wilson B., Strauss C.R. and Williams P.J. (1984) J. Agric. Food Chem., 32, 919.

Winkler A.J. (1962) *General Viticulture*. University of California Press, Berkeley, CA.

Wurdig G. (1976) Wein wiss., 112, 16.

Yamane T., Jeong S.T., Goto-Yamamoto N., Koshita Y. and Kobayashi S. (2006) Am. J. Enol. Vitic., 57, 1, 54.

Zapata C., Deleens E., Chaillou S. and Magne C. (2004) J. Plant Physiol., 161, 9, 1031.

CHAPTER 11

Harvest and Pre-fermentation Treatments

11.1 Introduction
11.2 Improving Grape Quality by Overripening
11.3 Harvest Date and Operations
11.4 Acidity Adjustments of the Harvested Grapes
11.5 Increasing Sugar Concentrations
11.6 Enzymatic Transformations of the Grape After Harvest
11.7 Use of Commercial Enzymes in Winemaking

11.1 Introduction

The definition of maturity, the biochemical transformations of grapes during ripening, and related subjects have been described in Sections 10.3 and 10.4. Grape maturity varies as a result of many parameters and is not a precise physiological state.

Under certain conditions (for example, dry white winemaking in warm climates), grapes are sometimes harvested before complete ripening (Section 13.2.2). In other conditions (in temperate climates, for example), the natural biochemical phenomena may be prolonged when unfavorable weather conditions have disrupted normal ripening kinetics, and this has become a tradition in certain northern vineyards ("late harvest," *beerenauslese*, etc.). In other regions, *Botrytis cinerea* in the noble rot form causes overripening (Sections 10.6 and 14.2.2).

All on-vine overripening methods increase the ratio of sugar to acid. Grapes accumulate sugar while breaking down malic and/or tartaric acid. In all cases, this natural drying process lowers crop volume due to water loss.

Similar results are sought by exposing picked grapes to sunlight or storing them

---

Handbook of Enology, Volume 1: The Microbiology of Wine and Vinifications, Third Edition.
Pascal Ribéreau-Gayon, Denis Dubourdieu, Bernard B. Donèche and Aline A. Lonvaud.
© 2021 John Wiley & Sons Ltd. Published 2021 by John Wiley & Sons Ltd.

in ventilated buildings. These techniques are used in grape-growing regions as varied as Jerez (Spain) and Jura (France).

Must quality can also be improved after the harvest, but these wine adjustments, whether physical or chemical, should not be made simply to compensate for basic viticultural inadequacies. These processes are strictly regulated to avoid potential abuse and to prevent them from becoming standard practices with the objective of replacing the work of nature.

The grape remains a living organism after it is picked. Many enzymes maintain sufficient activity to ensure various biochemical processes. Enzymatic activity is regulated by cellular compartmentation, which continuously limits available substrates. Enzymatic activity is higher in moldy grapes, in which case they should be maintained intact for as long as possible. Great care should be given to their harvest and transport.

Pre-fermentation practices at the winery destroy the structure of grape cells. Enzymes are placed in direct contact with abundant substrates, resulting in explosive enzymatic reactions. Laccase activity is the best-known reaction: secreted by *B. cinerea*, it alters phenolic compounds in the presence of a sufficient amount of oxygen. Not all of the enzymes present are harmful to quality—pectinolytic enzymes, for example, favor must clarification by hydrolysis of cell wall components. For several years, enologists have sought to amplify these favorable reactions by using more active, industrially produced enzymes.

Thanks to an increased fundamental understanding of grape constituents and their enzymatic transformation mechanisms, manufacturers have greatly improved equipment design, treatments, and technological processes. Many methods for maintaining, increasing, and, if necessary, correcting grape quality are currently available to the enologist. Enology will likely bring about other improvements in future years, notably with respect to phenolic compounds and aroma substances.

Nevertheless, two significant constraints persist. International trade imposes increasingly strict regulations, encouraged by the consumer's desire for natural products. Also, certain technological methods require costly equipment, and this level of investment is restricted to large wineries.

As a bridge between viticultural practices and winemaking methods, pre-fermentation treatments demand great care from the enologist.

11.2 Improving Grape Quality by Overripening

Overripening is a natural extension of the ripening process but differs from it on a physiological level. The aging of vascular tissue in the grape stems progressively isolates the grapes from the rest of the plant. As a result, crop volume generally diminishes since evaporative water loss is no longer compensated for by an influx from the roots. Overripening is also characterized by an increase in fermentative metabolism and alcohol dehydrogenase activity (Terrier et al., 1996). Noble rot is also a process that improves grape quality by overripeness; it is described in Sections 10.6 and 14.2.

11.2.1 On-vine Grape Drying

The grapes are left on the vine for as long as possible with this natural drying method (*passerillage* in French)—sometimes after twisting of the grape cluster peduncle. The berries progressively shrivel, losing their water composition. They produce a naturally concentrated must, richer in sugar and aroma substances. Acidity does not increase in the same proportions and can even decrease by malic acid oxidation. Other biochemical ripening phenomena also occur; notably, the skin cell walls deteriorate. This method should therefore be used only with relatively thick-skinned varieties to limit the risk of *B. cinerea* development.

11.2.2 Off-vine Grape Drying

In certain regions, this method can be limited to simply exposing grapes to sunlight for a variable length of time. In the Jerez region, grapes of the Pedro Ximenez variety are exposed to the sun on straw mats for 10–20 days before pressing (Reader and Dominguez, 1995). The grapes are turned over regularly and covered at night to protect them from moisture. During this process, locally called *soleo*, must density regularly attains 1.190–1.210 but sometimes exceeds 1.235. The juice yield is very low (250–300 l/metric ton), and only vertical hydraulic presses are capable of extraction. The resulting juice is extremely viscous and dark, with pronounced grape aromas. The Pedro Ximenez variety is particularly rich in organic acids. The heat of the Andalusian sun provokes the formation of a significant quantity (50–75 mg/l) of hydroxymethylfurfural from fructose.

In this same region of Spain, as well as in many other Mediterranean winegrowing regions (Greece, Cyprus, Italy, Turkey, etc.), this sun-drying method is applied to Muscat grape varieties (Muscat of Alexandria, for example). More than simply concentrating grape sugar, sun drying in particular increases the aroma typicity of the must. These musts attain high free terpene alcohol concentrations, thus contributing to wine aroma.

In the Jura region of France, healthy grapes are sorted on the vine, and the different varieties (in particular, Savagnin) are carefully picked to make a blend. The selected grape clusters are hung to dry, spread out on wooden grids covered with straw, or suspended on wires in well-ventilated storage rooms. This drying method results in extensive losses, not only by desiccation but also by rot. Moldy grapes are removed regularly. This operation lasts two to four months. The grapes are generally pressed after Christmas, producing musts containing 310–350 g of sugar per liter (legal minimum = 306 g/l) with a higher-than-normal volatile acidity. The yield is approximately 250 l/metric ton.

11.2.3 Artificial Drying

Natural grape drying (*passerillage*) is a difficult operation to master, due especially to the risks of rot-induced grape spoilage. Since the beginning of the 20th century, enologists have been trying to replace this natural process with a suitable technology. The principles of an industrial dryer are simple. The equipment circulates hot and dry air over the grapes, which are placed in small boxes inside the heated compartment. The ventilation system circulates 2,500–5,000 m^3 of dry air (below 15% relative humidity) per hour, at a temperature varying from 25 to 35°C.

According to recent experiments that confirm earlier tests (Ribéreau-Gayon et al., 1976), the apparatus reduces the grape crop mass by 10–15% in eight to 15 hours and increases potential alcohol by 1.5% vol. The decrease in acidity, via the oxidative degradation of malic acid, varies according to air temperature.

Other biochemical phenomena probably accompany this artificial overripening. Wines obtained from these treated grapes are richer in color and tannins and are always preferred in tastings. This treatment is exclusively for red winemaking, since the resulting increased phenolic compound concentrations are detrimental to quality white winemaking.

Equipment costs and utilization constraints limit the use of this technique, but its effectiveness is proven.

11.3 Harvest Date and Operations

First and foremost, the grapes should be protected from attacks and contaminations such as European grapevine moth and mold, right up to the harvest.

Optimal enological maturity depends on grape variety, environmental conditions, and wine type (Section 10.4.1). Thus, a perfect knowledge of veraison conditions and veraison midpoint dates will enable the grower to organize the harvest according to the various maturity periods. Maturity analysis monitoring complements this information (Sections 10.4.2 and 10.4.3).

The grape crop should be harvested under favorable weather conditions. After rainfall, the grape clusters retain water that is likely to dilute the must. The grapes should therefore be allowed to dry out, at least partially. The most recent research measuring water activity on the surface of grapes has shown that at least two hours of drying are necessary. Morning fog can also cause must dilution. Harvesting should begin after the sun has dried the vines. In contrast, if harvest is done during the warmest hours of the day, the prolonged maceration of harvested grapes in the juice of the inevitably burst grapes should be avoided.

Mechanical harvesting facilitates compliance with the above recommendations. Progress in harvest machine technology has helped to avoid berry damage and excessive leaf debris. Thanks to its speed and ease of use, the harvester enables a rapid harvest of grapes at their optimal quality level and at the most favorable moment. Manual grape-picking can be even more selective and qualitative, but its cost is not justifiable in all wineries.

Whatever the harvest method, the grower's principal concern should be the maintenance of grape quality.

11.3.1 Grape Harvest

From ancient times to recent years, harvest methods have barely evolved—other than slight improvements in tools for grape cutting and gathering.

In certain appellations (Champagne, for example) and vineyards, quality concerns prohibit mechanized harvesting. In noble rot (Section 10.6.3) vineyards, it cannot be implemented because the harvester is not capable of selecting grapes that have reached the proper stage of noble rot. Everywhere else, since the beginning of the 1970s, mechanical harvesting has undergone spectacular growth as a result of increased production costs and the greater scarcity of manual labor (Vromandt, 1989).

Lateral or horizontal strike harvesting techniques are easily adapted to traditional vine training methods. The grapes are shaken loose by two banks of flexible rods that straddle the vine row. The banks of rods transmit an alternating transversal oscillation to the vines (Figure 11.1). This

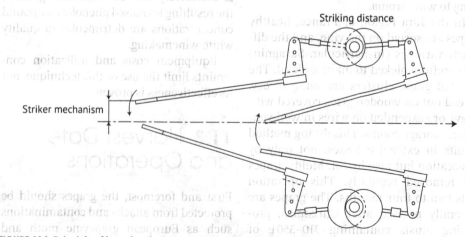

FIGURE 11.1 Principle of lateral strike harvesters (Jacquet, 1983).

movement transmits a succession of accelerations and decelerations to the grape clusters that results in individual grapes, partial grape clusters, or entire grape clusters falling. The adjustment of these machines is complicated and requires a complete mastery of the technique. The number, position, and angle of the rods or rails in the banks must be chosen in accordance with the training and pruning system used. Finally, the striking frequency must be adapted to the forward speed of the harvester. The frequency is adjustable from 0 to 600 strikes/min on conventional harvesters; some more recent models attain up to 1,400 strikes/min.

Harvesters should be adjusted for not only the variety (the ease of grape dislodgement depends on the variety) but also the pruning method employed and the canopy density at the time of the harvest. For example, a harvester with an insufficient striking frequency in thick foliage does not harvest the entire grape crop. In contrast, with a sparser canopy, an excessive oscillation amplitude and striking speed will transmit a lot of kinetic energy to the rod banks, so the rod strikes burst the poorly protected berries.

New striking methods are responsible for most of the recent improvements in mechanical harvesting. On conventional harvesters, the rod ends are not attached; thus, their inertia is not controlled. In more recent models, manufacturers have eliminated this free rod end. Their solutions vary depending on the machines (Figure 11.2). In some cases, two rod ends are connected by an articulated link, and these flexible rails strike by bending. In other

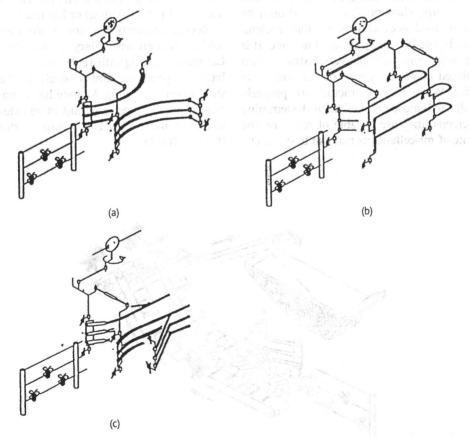

FIGURE 11.2 New lateral strike harvester models (Vromandt, 1989): (a) flexible striker, (b) dual-drive rigid striker, and (c) differential rod bank striker.

cases, striker shafts at the front and back of the machine oscillate semirigid rods at different amplitudes.

The dislodged grapes are gathered on an impermeable mobile surface made up of a series of overlapping plastic elements: in general, there are two rows of plastic elements shaped like fish scales. These elements yield to vine trunks and trellis posts by moving on their rotation axis. Lateral grill or perforated belt transporters then drain the grapes as they are conveyed. Upon reaching the extremity of the machine, the grapes are transferred to a shelf or bucket elevator (Figure 11.3). This type of machine permits high harvest speeds to the detriment of crop quality. Juice losses may represent up to 10% of total weight.

Striker mechanisms inevitably entrain leaves, leaf fragments, and other vine parts (material other than grapes [MOG]) with the grape clusters. The MOG should be eliminated as quickly as possible, preferably before becoming covered in juice. It is usually removed by blowers. Extractor fans placed above the conveyor belt are effective. When these extractors are properly adjusted and combined with destemming screens, they are capable of reducing the rate of miscellaneous rubbish to 0.5%. The crop is generally stored temporarily in one or two hoppers with a capacity of 8–20 hl. These hoppers are capable of dumping the crop directly into the transport containers.

Ribéreau-Gayon *et al.* (1976) had already indicated that rigorous comparative studies between manual and mechanical harvest quality were practically impossible. The two methods would have to be examined on a sufficiently large and homogeneous parcel, and the harvest reception and winemaking equipment would have to be capable of identically handling a large immediate grape supply from mechanical harvesting or a progressive grape supply from manual harvesting.

An experienced enologist is capable of examining the grape crop visually to compare mechanical harvest and manual harvest quality, including the proportion of burst or rotten grapes and the presence of leaves and petioles (intact or lacerated).

Recent research has confirmed those initial observations (Clary *et al.*, 1990). Careful harvesting with a correctly adjusted harvester produces similar results to classic manual harvesting. Manual harvesting, however, continues to permit more extensive, but more expensive, grape sorting (Section 11.3.3).

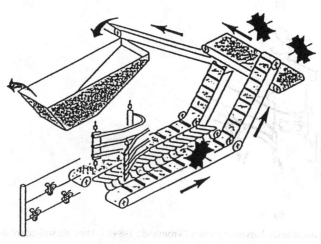

FIGURE 11.3 Lateral strike harvester (Vromandt, 1989).

11.3.2 Harvest Transport

Choosing and implementing harvest transport equipment is a complex issue. It is linked to the organizing of harvest work and the winery reception installation and is subject to certain enological and economic constraints.

From an enological viewpoint, grapes should arrive at the winery intact. More precisely, the container should transport the grapes in the physical or biochemical state obtained after picking and transfer them to the reception bin. Any excessive bruising and crushing of grapes can be avoided by:

- using shallow transport containers (not exceeding a depth of 0.8 m);
- using easily cleaned material to ensure proper hygiene;
- limiting the number of grape transfers and the load and dumping height.

Growers in certain viticultural regions (Champagne, for example) are required to follow these strict rules. Small perforated containers, preferably plastic, are used for grape-picking to ensure grape quality from the first step of the harvest. These containers are stacked on an open trailer and gently emptied at the winery.

Mechanical harvesting produces a different rate of grape supply compared with manual picking:

- Hourly crop volume is considerable (from 4 to 10 metric tons/h) and the daily harvesting time is often 12 hours.
- The harvest is partially destemmed and crushed—sometimes with a lot of juice (10–30% of total juice).
- The harvest is full of MOG (leaves, petioles, shoot fragments) and sometimes small animals.

Mechanical harvest transport does not follow the same rules. The harvested grapes should be brought rapidly to the winery after the juice has been separated from the solid parts to the extent possible. Sulfur dioxide should not be added to the unseparated harvest; it favors the maceration of the solid material during transport. Similarly, carbon dioxide protects must from oxygen only when separated from the grapes.

Transport equipment can be grouped into two categories (Figure 11.4):

1. Removable containers are placed on a transport chassis—sometimes several at once, depending on their size (0.2–10 hl). In some cases, the containers have large capacities (10–150 hl) and are used for transporting the grape crop over long distances. This operation is not recommended from an enological viewpoint. Medium-sized grape containers should be used to avoid crushing the grapes and to reduce the number of times they need to be transferred.
2. Fixed containers may be used, corresponding to one transport unit. Within this category, we can distinguish between dumping containers, which instantaneously empty their entire crop into the reception bin, and containers equipped with screw pumps, which progressively empty the crop. The latter do not require any reception facility at the winery.

Harvest trailers are transport containers attached to a tractor. The vineyard therefore requires at least one additional trailer for other harvest operations. These trailers vary in terms of their capacity and dumping method. The smallest (from 20 to 30 hl) can pass between vine rows, thus eliminating intermediate crop transfers. They are gravity-tilt trailers and often require a costly recessed reception area. High-capacity bin trailers are shallower, which limits grape bruising and crushing, but they are too large to pass between vine rows. These containers empty the grapes into the reception bin using a hydraulic lift system. Elevator bin trailers also exist and are capable of lifting their containers up to 1.5–2 m

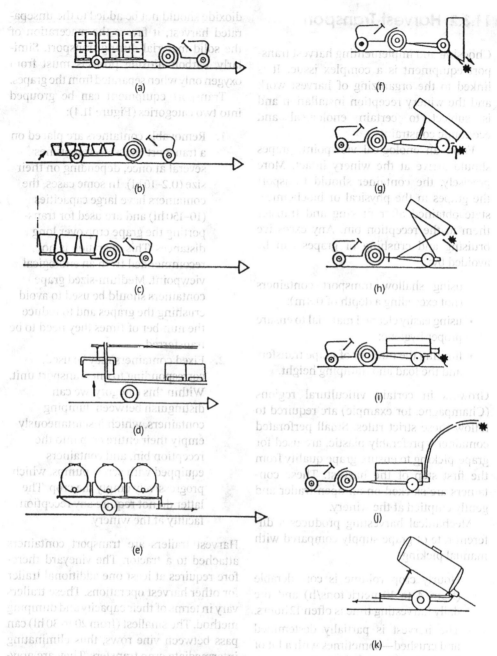

FIGURE 11.4 Different harvest transport containers (Ribéreau-Gayon et al., 1976). On the left, mobile containers: (a) 20–90-l stackable boxes; (b) 60–100-l plastic or wooden containers; (c) 600–800-l harvesting bins; (d) 1,000–2,000-l high-capacity containers, transported by truck; and (e) 15–20-hl mobile tanks. On the right, containers attached to trailer: (f) 15–25-hl gravity dumping bin; (g) 20–30-hl mechanical dumping bin; (h) elevator dumping bin; (i) screw-driven bin; (j) screw- and pump-driven bin; and (k) pumper tank.

before emptying, depending on the model. This system does not require a recessed reception area. The grapes can be directly fed to the first step of the winemaking process (destemmer-crusher or press). Similarly, screw-bin trailers equipped with pumps (Figure 11.4) eliminate the need for a reception facility. Although these systems are very practical, the mechanism of some screw pumps is too brutal and can decrease crop quality.

Mechanically harvested grape crops, especially white, require rapid draining. This is most often done with grills forming a double floor in the bottom of bin trailers.

11.3.3 Cleaning and Sorting the Grape Crop

These operations include eliminating MOG (leaves and stems) and damaged, unripe, or rotten grapes.

During manual harvesting, MOG is eliminated (at least for red grapes) at the same time as the crushing and stemming process. The same equipment is used for cleaning mechanically harvested grapes, but the different appearance of these grapes could justify adapted machinery, which will certainly be developed in the future.

Sorting the harvest to eliminate bad grapes is only possible with intact grapes. This operation is difficult with machine-harvested grapes. Cutters, however, may precede the harvester to eliminate most of the damaged, spoiled, or unripe grapes. However, better results have been obtained using automatic sorting systems with vibrating screens.

Manually harvested grapes have the undeniable advantage of being able to undergo an effective sorting. This can be carried out as the transport bin is being filled. At present, sorting tables are used, and workers, placed around these tables, remove bad grapes.

Among the various sorting tables proposed by manufacturers, there are models adapted to the back of the tractor that directly feed the bin trailer. Usually, however, the sorting table is an independent, detachable unit installed between the reception area and the first piece of winemaking equipment. A sorting table essentially consists of a conveyor belt on a metal frame. The belt is driven by rollers powered by an electric motor. The belt speed should be slow (less than 5 m/min) to limit workers' eye fatigue. The belt is often made of food-quality rubber and is sometimes perforated. An articulated plastic belt is also used. The sorting tables are often slightly inclined (5–10%) to facilitate draining. Only intact grapes should be supplied to the sorting table. Screw pumps often damage grapes and therefore compromise sorting effectiveness. Ideally, small-capacity containers should be emptied directly on the table, in which case the grapes are spread out as the containers are emptied. Vibrating tables are now used to spread the grapes out, although they are rather noisy.

11.3.4 Grape Selection and Selective Must Extraction by Low-Temperature Pressing

Generally, in a homogeneous vineyard parcel containing only one grape variety, ripening intensity can vary from one grape cluster to another and even from one grape to another on the same cluster (Section 10.4.2). During manual grape-picking, and even more so during machine harvesting, differences in ripeness are difficult to distinguish. At best, the incoming grape crop can be sorted according to its sugar concentration and to eliminate any moldy grapes.

In noble rot regions, grapes and cluster fragments are selected by successive harvesting (sorting). Only the grapes having reached the *rôti* (roasted berry) stage are picked (Sections 10.6.3 and 14.2.4). Even in this case, the grape-picker cannot always precisely evaluate the degree of grape concentration. Furthermore, the condition of the grapes does not necessarily make the appropriate choice possible.

In white winemaking, a method called controlled-temperature pressing currently permits selective must extraction from grapes richest in sugar. These grapes freeze at a lower temperature than those that are less rich in sugar. This modern technique, which is authorized as a harvest selection method by European legislation, originates from a traditional grape-picking method in certain viticultural regions. In the northern vineyards of Germany, Austria, and Canada, the winemaker benefits from the severe climate by harvesting and pressing white grapes while they are partially frozen. A must particularly rich in sugar is thus obtained and is used to make highly prized ice wines (*eiswein*). This pressing method greatly accentuates the concentrations of sugar and aroma substances of these overripe grapes.

The method of "selective cryoextraction" consists in cooling grapes until only those richest in sugar remain normal. The others are frozen solid and are not compressible. The grapes are then immediately pressed. Only the must from the grapes richest in sugar and therefore highest in quality is extracted (Chauvet *et al.*, 1986).

The method is carried out differently depending on the harvest method. With manual harvesting, grapes in small containers are placed in a freezing chamber. After cooling, the boxes are manually emptied into the press. When grapes are harvested mechanically, this method is less effective. The grapes must be drained before cooling and are then frozen in bulk by a freezing device. Liquid nitrogen is often the cooling source.

Noble rot grape winemaking has found particularly interesting applications for this technique (Section 14.2.4), but it can also be applied to healthy grapes for making dry white wines. In addition to sugars, other elements of grape chemical composition are modified. While the total acidity and malic acid concentrations are higher in the selected must, the tartaric acid concentration generally decreases. The increased potassium concentration resulting from cryoextraction makes a portion of this acid insoluble. Total phenol concentrations remain stable, indicating that cryoextraction acts essentially on the pulp without altering the skin. The wines produced are richer and more complex.

Supraextraction (Section 13.3.6) is derived directly from cryoextraction. It consists in subjecting the grapes to freeze–defrost cycles and then pressing them at room temperature. The ice crystals tear the cell walls, so pressing extracts even more compounds from the grapes.

11.4 Acidity Adjustments of the Harvested Grapes

Generally, in temperate climates and in traditional winemaking regions, regular grape sampling, maturity assessment, and the correct choice of grape varieties ensure a proper level of harvest acidity. In the Bordeaux region (France), the low acidity of Merlot is often compensated by the higher acidity of the Cabernet varieties and possibly Malbec, planted together in the same vineyard. But acidity may need to be corrected when making single-variety wines or in case of extreme weather conditions.

Acidity adjustments consist in either increasing (acid additions) or decreasing (acidity reduction) total must acidity. Various products are used for this purpose. In contrast with overripening techniques, acidity adjustments are strictly regulated in the European Union and in other countries that comply with International Organisation of Vine and Wine (OIV) recommendations.

11.4.1 Acidification

In hot winegrowing regions and during exceptional ripening years in temperate zones, a considerable amount of malic acid is degraded during maturation. To maintain the freshness and firmness of wines desired by consumers, especially in whites, the acidity should be increased by adding an acid. High acidity also enhances the protective effects of sulfur dioxide.

Strong inorganic acids, such as hydrochloric, phosphoric, and sulfuric acid, may not be added to wine and are prohibited in all countries. European legislation and the OIV only recommend tartaric acid, but this acid tends to harden wines and should be added with caution.

For a total acidity between 3.0 and 3.5 g/l, expressed as H_2SO_4 (4.6–5.4 g/l as tartaric acid), 50 g of tartaric acid should be added per hectoliter. If the total acidity of the grapes is below 3 g/l as H_2SO_4, then 100 g of tartaric acid should be added per hectoliter. In any case, legislation limits this acidification to 1.5 g of tartaric acid per liter (Table 11.1). The need for acid additions can also be determined with respect

TABLE 11.1

Grape Acidity Correction Regulations[a] in the European Union (Except Portugal)

Schematic delimitation of viticultural zones	Acidification[b,c] Normal years	Acidification[b,c] Exceptional years[d]	Deacidification[e]
(a) Vineyards of Belgium, Luxembourg, the Netherlands, Great Britain, Austria, and Germany (except the Baden region)	—	—	Authorized
(b) Vineyards of Baden (Germany) and the northern half of France (Loire valley included)	—	—	Authorized
(C Ia) Vineyards of the southern half of France, except the southernmost, and northwest of Spain	—	Authorized: max. 1.5 g/l	Authorized
(C Ib) Alpine vineyards of northern Italy	—	Authorized: max. 1.5 g/l	Authorized
(C II) Southernmost vineyards of France except zone (C III); vineyards of the northern half of Spain; Italian vineyards except zones (C I) and (C III)	Authorized: max. 1.5 g/l	—	Authorized
(C III) Coastal vineyards of the eastern Pyrénées, Var, and Corsica in France; vineyards of the southern half of Spain; Italian vineyards of Basilicata, Puglia, Calabria, Sicily, and Sardinia; the vineyards of Greece	Authorized: max. 1.5 g/l	—	Authorized

[a] Grape musts, partially fermented grape must and "nouveau" wines that are still fermenting.
[b] By addition of tartaric acid exclusively.
[c] In finished wines, acidification permitted up to 2.5 g/l.
[d] The EU declares exceptional years.
[e] No limit.

to must pH. For example, they are necessary for pH above or equal to 3.6. In low-acidity musts, the production of succinic and lactic acid during fermentation tends to increase acidity in greater proportions, and this should be considered when determining the need for an acid addition.

In both red and white winemaking, tartaric acid should be added before or, preferably, toward the end of fermentation. Allowing for precipitation, the addition of 200 g of tartaric acid per hectoliter increases the acidity by 1 g/l expressed as H_2SO_4 (1.5 g/l as tartaric acid). Acid additions, however, should never be calculated to bring acidity up to normal levels. As adding tartaric acid insolubilizes potassium, it tends to have more effect on pH and flavor than on total acidity levels. Harvesting a portion of the crop before full maturity or using grapes from the second crop can provide natural acidification, but these methods are not recommended as they are detrimental to quality.

The addition of citric acid is not a useful solution, since the total citric acid concentration limit is set at 1 g/l as citric acid. This limit, imposed by European legislation, does not noticeably increase the total acidity. Furthermore, lactic acid bacteria can break down this acid during winemaking, increasing volatile acidity.

Calcium sulfate (plaster of Paris or gypsum) was traditionally used in certain viticultural regions (Jerez in Spain). Its addition at doses of 1.25–2.25 g/l lowered must pH by precipitating calcium tartrate. This method has now been practically abandoned. Although effective, it is not advisable; moreover, its use must be mentioned on the wine label (Benitez et al., 1993) (Volume 2, Section 12.4).

Cation exchange treatments in must and wine are authorized in some countries. When used for stabilizing tartrate precipitation, these treatments acidify the resulting product; but under normal treatment conditions, the pH must not be lowered by more than 0.2. Eliminating potassium by electrodialysis (Volume 2, Section 12.5) also causes a slight decrease in pH.

11.4.2 Deacidification

When grapes do not reach complete maturity in northern vineyards, grape acidity can be high. Under these conditions, malic acid concentrations are almost always greater than those of tartaric acid. When the biological degradation of malic acid is not desired due to the organoleptic changes that it causes, the juice must be chemically deacidified. This practice is authorized in many countries, and various products are available.

The two main compounds, calcium carbonate and potassium bicarbonate, react according to the same mechanism (Blouin and Peynaud, 2001). When they combine with tartaric acid, H_2T, the carbonate is broken down into carbonic acid, released as CO_2, and the calcium or potassium forms an insoluble salt with the tartrate, which then precipitates out.

• With calcium carbonate

$$H_2T + CaCO_3 \rightarrow CaT\downarrow + CO_2\uparrow + H_2O$$
$$\text{150 g} \quad \text{100 g}$$

• With potassium bicarbonate

$$H_2T + KHCO_3 \rightarrow KHT\downarrow + CO_2\uparrow + H_2O$$
$$\text{150 g} \quad \text{100 g}$$

Theoretically, in both cases, 1 g/l of deacidifying agent neutralizes 1.5 g/l of H_2T, giving a decrease in acidity of 20 mEq/l or 1 g/l expressed in H_2SO_4 (or, of course, 1.5 g/l expressed in tartaric acid). In practice, the reaction is less efficient, especially in the case of $KHCO_3$, where higher doses are recommended to achieve the same level of deacidification (Table 11.2). As a result, this product should be reserved for minor acidity corrections. In any case, it should always be borne in mind that these deacidifying agents act exclusively on tartaric acid, so they should not be used to try to adjust acidity to normal levels, which would necessitate the elimination of excessive

TABLE 11.2
Product Doses for the Deacidification of Musts and Wines

Initial total acidity (g H_2SO_4/l)	$CaCO_3$ (g/hl)	$KHCO_3$ (g/hl)
Less than 7.0	50	75
From 7.0 to 7.5	75	110
From 7.5 to 8.0	100	—
Greater than 8.0	125	—

amounts of tartaric acid and cause an unacceptable increase in pH.

In the case of calcium carbonate, deacidification results directly from the salt's chemical reaction, so maximum deacidification is rapidly achieved and is relatively predictable. However, if, for any reason, the wine has a high calcium content, there is a risk of further precipitation at a later date, even in bottle.

In contrast, the deacidification mechanism involving potassium bicarbonate is more complex. Following an initial decrease in acidity due to this salt's reaction with tartaric acid, the formation of potassium bitartrate upsets the ion equilibrium and precipitates, producing a secondary deacidification that is not, theoretically, predictable. It is advisable to conduct laboratory trials to determine the appropriate dosage (Flanzy, 1998).

For all these reasons, it is advisable to use calcium carbonate to deacidify musts with excessive acidity levels before or during fermentation. Potassium bicarbonate should be reserved for slight corrections (after laboratory trials), during the final preparation phase. It is, in many cases, preferable to implement deacidification in two stages.

Another authorized product, dipotassium tartrate, is rarely used, due to its cost and its low deacidifying power. To lower the total acidity by 1g/l as sulfuric acid requires 2.5–3 g of dipotassium tartrate per liter. This product deacidifies by precipitating potassium bitartrate, which possesses an acid function.

All of these products act solely by precipitating tartaric acid, since the potassium and calcium salts of malic acid are soluble. However, insufficiently ripe grapes contain excess malic acid.

Wucherpfennig (1967) recommended a deacidification technique in Germany based on the precipitation of a double calcium malate and tartrate salt, insoluble above pH 4.5. With this method, a fraction of the must to be treated is completely neutralized by calcium carbonate, containing a small amount of calcium malate and tartrate seed crystals. After precipitation of the double salt, this strongly deacidified volume is filtered before being blended back into the untreated fraction. This procedure was developed and perfected by Haushofer (1972). Where M (in hl) is the total volume of must to deacidify, the fraction to be treated (M_d) by calcium carbonate is calculated by using the following formula:

$$M_d = \frac{(T_1 - T_2)}{T_1} \times M \quad (11.1)$$

T_1 is the titratable must acidity in grams of tartaric acid per liter; T_2 is the desired acidity after the treatment. At a must volume M_d, a quantity of calcium carbonate is added, given by the formula Q (g) = $0.67(T_1 - T_2)M$.

When there is too much surplus malic acid in the initial must, part of the calcium carbonate added to the treated volume produces soluble calcium malate. At the time of final blending, the surplus calcium reacts with the tartaric acid of the untreated must. The deacidification thus occurs in two steps and acts essentially on the tartaric acid. The tartaric acid concentration therefore has to be adjusted in order to decrease malic acid concentrations substantially by double salt formation (Usseglio-Tomasset and Bosia, 1992). This result can be obtained by using a mixture of calcium carbonate and calcium tartrate, also containing a small amount of the double calcium salts of malic and tartaric acid to favor calcium tartromalate crystallization. The deacidification power of this mixture depends on the proportion of calcium tartrate used. In this case, the use of this deacidification process becomes complex. It only reaches final equilibrium after a long time.

In white winemaking, this deacidification should be done after must clarification but before fermentation. Aromatic ester production by yeasts is facilitated by a moderate pH. Conversely, this treatment permits a more precise deacidification of red wines when performed at the end of alcoholic fermentation, when the wine is drained off the skins. It can also help to trigger malolactic fermentation.

European legislation does not impose must deacidification limits, but there is a limit of 1 g/l of the total acidity, expressed as tartaric acid, for wine. Table wines must have a minimum total acidity of 4.5 g/l as tartaric acid. In any case, this treatment must be declared and cannot be combined with an acidification.

11.5 Increasing Sugar Concentrations

Certain regions have difficulty in producing quality wines. Chapter 10 discussed the climatic and soil conditions that excessively favor vine growth and grape development. These conditions lead to musts with high sugar concentrations, but the resulting wines lack finesse and aroma complexity.

In the best *terroirs* of northern vineyards, unfavorable weather conditions during difficult vintages often hinder ripening. Many parameters (reduced photosynthesis, continued vegetative growth, excessive crop yields, etc.) limit grape sugar accumulation; thus, adjusting the natural sugar concentration can be useful. Of course, these adjustments must be limited and cannot replace complete ripening. In particular, their use should not incite premature harvesting or exaggerated crop yields.

Various subtractive techniques increase sugar concentrations by eliminating part of the water found in grapes—similarly to natural overripening. The crop yield is consequently lowered. Although some of these techniques are still in the experimental stage, they are largely preferred by international authorities over additive techniques (Tinlot, 1990). Additive techniques, despite their long length of use, always have the drawback of increasing crop volumes by the addition of an exogenous product—sugar or concentrated must (Dupuy and de Hoogh, 1991).

11.5.1 Subtractive Techniques

These techniques, similar to drying and cryoextraction, consist in eliminating part of the water contained in grapes or in must. Two physical processes can be used: water evaporation or selective separation across a semipermeable membrane (reverse osmosis). European legislation has set the limits for these treatments: a 20% vol. maximum decrease and a 2% vol. maximum potential alcohol increase.

Quite apart from equipment costs, the considerable crop yield loss resulting from the use of these methods has hindered their proliferation, and chaptalization was preferred for a long time. In recent years, a focus on increasing wine quality has renewed interest in these methods. In red wines in particular, tannin concentrations are simultaneously increased. Of course, these methods should never be used with the intent of correcting excessive crop yields.

Vacuum concentration

Heat concentration is a long-standing method often used in the food industry. For more than 40 years, it has been used to make concentrated must, but potential wine quality must not be compromised during the heating process. The denaturing of heat-sensitive must constituents and the appearance of organoleptic flaws (hydroxymethylfurfural, for example) must therefore be avoided. Earlier equipment, operating at atmospheric pressure, required relatively high temperatures, which produced off-aromas. For this reason, certain French appellations prohibited their use. Today, vacuum evaporators have lowered the evaporation temperature to 25–30°C, and interesting qualitative results have been obtained. Even lower evaporation temperatures are possible, but the must delivery rate becomes too low.

In addition, horizontal shell-and-tube heat exchangers permit continuous high-speed treatment of must. This system limits the risk of heat damage from the prolonged contact between a fraction of the must and the hot exchanger surface. A thermocompressor, acting as a heat pump, extracts part of the water vapor coming from the must and mixes it with the steam produced by the steam generator. This system has the two-fold advantage of lowering the treatment temperature and saving energy (by favoring must evaporation). Current equipment can treat from 10 to 80 hl of must per hour, with an evaporation capacity of 150–1,200 l/h. In controlled appellation (e.g. AOC) areas, this treatment should be done in a closed circuit directly linked to the fermentor. The evaporation occurs at a low temperature (25–30°C), and under these conditions the concentration factor is always less than two. The sugar concentration of must coming out of the concentrator remains practically constant during its operating cycle. Iron and malic acid are concentrated to the same degree as the sugars, but potassium and tartaric acid concentrations are lower, due to their partial precipitation during the treatment (Table 11.3). This technique, however, is not recommended for concentrating musts made from grape varieties with marked varietal aromas.

Reverse osmosis

Peynaud and Allard (1970) used reverse osmosis to eliminate water from grape must at ambient temperatures. The results obtained were satisfactory from an enological viewpoint, but problems in regenerating the cellulose acetate membranes stopped this technique from being developed further. The usefulness of reverse

TABLE 11.3

Must Concentration with a Vacuum Evaporator (Cabernet Sauvignon, Bordeaux, 1992)

Constituent	Initial must	Concentrated must	Condensed vapors
Sugar concentration (g/l)	179	204	0
pH	3.26	3.39	3.80
Total acidity (g/l H_2SO_4)	5.1	5.6	0.05
Malic acid (g/l)	4.2	4.8	0
Tartaric acid (g/l)	6.2	6.8	0
Potassium (g/l)	1.7	1.9	0
Iron (mg/l)	1.8	2.1	0

osmosis for obtaining wines of superior quality was confirmed by Wucherpfennig (1980), but it was the development of composite membranes that gave rise to new experiments with reverse osmosis (Guimberteau et al., 1989).

Figure 11.5 illustrates the principle of reverse osmosis. An appropriate membrane is used to separate a concentrated saline solution (A) from a more diluted solution (B). The difference in chemical potential tends to make the water pass from the low potential compartment to one with the higher potential (direct osmosis); the latter is diluted. The diffusion of water stops when the internal pressure of compartment A (osmotic pressure) counterbalances the pressure that diffuses the water across the membrane. If a pressure greater than this osmotic pressure is exerted on compartment A, the direction of the diffusion of the water is reversed (reverse osmosis) and solution A is concentrated.

A pressure of at least twice the osmotic pressure must be exerted on the must to force the water in the must to cross the membrane. This concentrates the various must constituents. During water transfer, molecules and ions retained by the membrane are apt to accumulate on its surface, increasing the real concentration of the treated must and thus the required pressure. To limit this concentration polarization phenomenon, the filtering side of the membrane must be cleaned to minimize the thickness of this accumulating boundary layer and to facilitate the retrodiffusion of the retained solutes. Reverse osmosis can be placed in the same category as tangential hyperfiltration. The equipment is comparable and only differs by the nature of the membrane.

Several models exist, depending on membrane layout. Plate modules, derived from plate filters, were the first to be used (Figure 11.6). The fluid to be treated circulates between the membranes of two adjacent plates, thus ensuring the mechanical support of the membrane and the draining of the permeate. The systems currently in

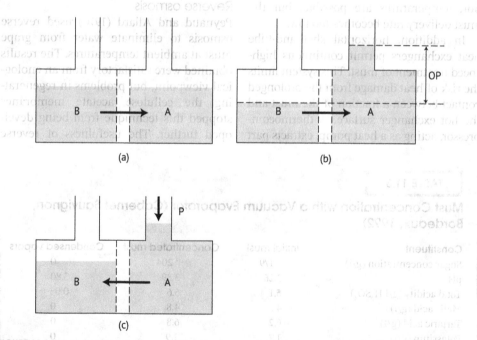

FIGURE 11.5 Principle of reverse osmosis (Guimberteau et al., 1989). A, concentrated saline solution. B, diluted saline solution. (a) Direct osmosis, (b) osmotic equilibrium (OP = osmotic pressure), and (c) reverse osmosis (P = pressure greater than osmotic pressure). Arrow, migration of water.

11.5 | Increasing Sugar Concentrations

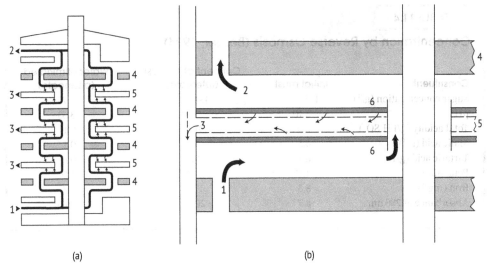

FIGURE 11.6 Plate membrane module for reverse osmosis: (a) schematic diagram of entire module and (b) detail of cell. (1) Must to be treated, (2) concentrated must, (3) permeate, (4) intermediate plate, (5) membrane support, and (6) membrane.

use are equipped with spiral or tubular modules. Membrane surface is often maximized to compensate for the low delivery rate of these systems. Several modules must be installed in parallel to have a satisfactory delivery rate.

Due to the extremely thin flow stream, the must undergoes an intensive clarification beforehand. Depending on the equipment, the must should be settled, partially clarified, or perfectly limpid before this process. In practice, for red wine grapes, the must is taken from a tank, cooled, clarified by settling or filtration to a turbidity of around 400 NTU, concentrated by reverse osmosis, and returned to its original tank. In certain cases, to avoid the formation of potassium bitartrate crystals, the use of metatartaric acid has been suggested. Depending on the modules, the working pressure varies from 60 to 120 bar, the temperature is regulated between 15 and 25°C, and the permeate delivery rate (the quantity of water removed over a period of time) is from 0.5 to 5 l/h/m² of filter surface.

Several findings have been confirmed in experiments over the last 10 years (Berger, 1994). The membranes have a solute retention rate of over 99.5%, but losses increase when the number of modules in parallel is increased in order to attain high continuous treatment delivery rates. In this case, traces of sugars and minerals are found in the permeate (Table 11.4). Tastings reveal that the permeate sometimes gives off sulfurous odors reminiscent of low tide. The retentate or enriched must is of good organoleptic quality. Generally, the concentrations of malic acid, metals, phenolic compounds, and macromolecules such as proteins and polysaccharides increase proportionally with the sugar concentration. Salification phenomena cause a limited change in the concentrations of tartaric acid and potassium and the pH.

Subtractive techniques increase tannin concentrations by decreasing juice volume with respect to pomace volume and are therefore more often applied to red winemaking than white. In this manner, they act like *saignée*, or tank bleeding (Section 12.5.8), but without the loss of sugar. These enrichment methods have the distinct advantage of being self-limiting for technical (concentration of bad tastes and flaws) and especially economic (operating costs and volume losses) reasons. They

TABLE 11.4
Concentration by Reverse Osmosis (Berger, 1994)

Constituent	Initial must	Concentrated must (retentate)	Water eliminated (permeate)
Sugar concentration (g/l)	175	434	1.75
pH	3.10	2.93	3.53
Total acidity (g/l H_2SO_4)	5.9	11.5	0.1
Malic acid (g/l)	4.5	9.7	0.1
Tartaric acid (g/l)	6.8	9.2	0.6
Potassium	1.17	1.43	0.03
Iron (mg/l)	8.3	17	1.3
Absorbance at 280 nm	8.7	20	0.05

represent a considerable cost: in equipment investment, operating costs, and crop yield loss. The last drawback is not applicable if the crop yield is above authorized limits. The main interest of these techniques is to treat must made from grapes soaked by heavy rain during the harvest.

11.5.2 Additive Techniques

Contrary to subtractive techniques, increasing the sugar concentration by the addition of an exogenous product is not directly limited by technical constraints. These techniques are therefore very strictly regulated. Such strict regulations have led the technical branches of regulatory organizations to develop sophisticated analytical methods to verify that the regulations are indeed being followed: magnetic resonance of the deuterium in ethanol, the isotopic ratio of ^{13}C by mass spectrometry, etc. (Martin et al., 1986). Must sugar concentration can be increased by directly adding pure sugar, concentrated must, or rectified concentrated must (RCM). It is generally legally limited to a 2% increase in alcohol content by volume—more under certain conditions. In practice, however, this increase should be limited to 1–1.5% vol. to avoid throwing the wine out of balance by excessive "heat," which would mask wine fruitiness. Within responsible limits, sugar addition is an effective means of improving the sensory quality of wine by increasing body and harmony.

Chaptalization

Sugar addition, better known as chaptalization since the end of the 18th century, consists in adding refined white sucrose to must. The sucrose must be at least 99% pure but can be derived from any plant (sugar cane, sugar beet, etc.). The quantity of sucrose required to increase a wine by 1% alcohol by volume varies from 16 to 19 g/l, depending on the yeast strain, must oxygenation, and the initial sugar concentration. European legislation has established precise doses: 18 g/l in red winemaking and 17 g/l in white winemaking. The sucrose is dissolved in a fraction of must. This operation should be done during the first one-third of alcoholic fermentation and during a pump-over. The sucrose dissolves more quickly in warm must, and simultaneous aeration stimulates the fermentative activity of the yeasts. The increase of alcohol content by chaptalization modifies some of the constituents of the corresponding wine. Total acidity decreases by 0.1–0.2 g/l (as H_2SO_4) for 1% vol. of added alcohol. This drop is caused by increased potassium bitartrate precipitation. In red winemaking, phenolic compound extraction increases by approximately 5% for each additional 1% vol. of ethanol. The increase in volatile acidity is negligible as long as chaptalization is not excessive. For alcohol contents greater than 13% vol., it increases by approximately 0.05 g/l (as H_2SO_4). Glycerol and dry extract increase in lesser proportions than ethanol. For a long time, the differences in concentrations of the various wine constituents, expressed in ratios, were the only methods available for enforcing chaptalization regulations. Today, isotopic methods can distinguish the sugar and thus the ethanol formed in terms of its botanical origin (grape, sugar beet, or sugar cane).

Various atoms have isotopes: deuterium ^2H for hydrogen ^1H and ^{13}C for carbon ^{12}C, for example. In a multiple atom molecule of natural origin, a very small but variable quantity of these atoms can be replaced by the corresponding isotopes. The ethanol molecule possesses several hydrogens. An isotopomer is formed if one or more of its hydrogens, for example, are replaced by deuterium. The proportion of these different isotopomers depends on the origin of the fermented sugar. Indeed, the grapevine's photosynthesis mechanism is different from that of other crops (beets, sugar cane), and the deuterium content varies according to latitude. Isotopomers are determined by nuclear magnetic resonance (NMR). This method is capable of detecting the addition of beet sugar (chaptalization) by comparing a sample wine with control wines made in the same geographical region during the same vintage. To evaluate a mixed addition of beet sugar and cane sugar during fermentation, this first analysis must be complemented by a carbon isotope determination using mass spectrometry.

Rectified concentrated must

The addition of RCM is similar to chaptalization. This colorless liquid contains an equimolar mixture of glucose and fructose. The RCM is obtained by dehydration. All compounds other than sugar are eliminated by ion exchange resin treatment (Table 11.5). European legislation has precisely defined the characteristics of this product. RCM must have a refractometric index greater than or equal to 61.7%. RCM, diluted to 25° Brix, must not have a pH greater than 5, an optical density greater than 0.1 at 425 nm, or a conductivity greater than 120 μS/cm. The legal limits are set at less than or equal to the following concentrations: titratable acidity, 15 mEq/kg (of sugar); sulfur dioxide, 25 mg/kg; total cation concentration, 8 mEq/kg; and hydroxymethylfurfural, 25 mg/kg.

TABLE 11.5

Analytical Comparison Between Concentrated White Must and the Same Concentrated Must Rectified (Brugirard, 1987)

Constituent	Concentrated must	Rectified concentrated must
Specific gravity (20°C/20°C)	1.3620	1.3535
Sugar concentration (g/l)	871	852
Potential alcohol (% vol.)	51.23	50.11
Total acidity (g/l H_2SO_4)	12.5	0.25
Iron (mg/l)	20.3	0
Copper (mg/l)	1.1	0
Ashes (mg/kg of sugars)	177	<1.2
Total phenols (mg/kg of sugars)	478	152

RCM is used in the same manner as pure sucrose and leads to the same chemical modifications in the corresponding wines. It has several drawbacks with respect to sucrose. First of all, RCM is more expensive. Equally important, it is not in crystallized form; therefore, its purity and stability over time depend on preparation and storage conditions. Attempts have been made to separate the glucose and fructose in order to crystallize them separately, but production costs are too high. In addition, RCM increases the dilution of the product. Finally, a similar product prepared by the hydrolysis of sucrose can be used to adulterate and sometimes counterfeit RCM. The only advantage of RCM is that it is derived from grapes. This reference to its origin is secondary, since the purest possible product is desired and obtained through appropriate treatments.

Concentrated must

Must constituent modifications are even more pronounced when non-purified concentrated must is used. These concentrated musts are usually obtained by indirect heating. All organic and mineral elements in the musts are concentrated. They have very low pHs (often lower than 3) despite the insolubilization of a portion of the tartaric acid in the form of potassium and calcium salts. Sulfur dioxide is used in high concentrations to conserve the must before concentration. The SO_2 is oxidized into sulfates during the treatment. Concentrated musts are also very rich in iron and are apt to cause iron casse (iron-induced haze). They are always very dark in color, due to the decomposition of sugars and nitrogen compounds in this hot and acidic environment. The resulting products intensify the color produced by phenolic compounds. The minimum sugar concentration in concentrated must is 582 g/l (51° Brix). Concentrated must greatly modifies the composition of the wine obtained—especially white wines: a slight increase in total and volatile acidity and a more significant increase in glycerol, dry extract, and phenolic compounds in red wines. Concentrated must often comes from the same grape varieties as the must to be enriched and sometimes from the same geographical zone.

Due to the diversity of its viticultural zones, the European Union has developed very elaborate legislation on this subject (Table 11.6). In many appellations, notably in France, the legislation is even more restrictive.

11.6 Enzymatic Transformations of the Grape After Harvest

When conserved intact after its harvest, the grape still maintains an intense physiological respiratory activity—which is utilized during drying. The lack of oxygen and the depletion of available respiratory substrates rapidly provoke the onset of fermentative processes in the berry.

The destruction of grape cell structures during pre-fermentation treatments results in oxygen dissolution, despite the precautions taken. Two enzyme categories, oxidoreductases and oxygenases, are responsible for many grape constituent transformations. They often harm grape quality.

Depending on winemaking techniques, the duration of grape solid maceration in must varies. During this period, hydrolase-type enzymes act on grapes and must. These enzymes are responsible for the hydrolysis of diverse macromolecules such as proteins, polysaccharides, heterosidic derivatives, and various esters. Their action often improves the grape/must mixture. This maceration phase should therefore sometimes be extended.

TABLE 11.6

Sugar Additions and Limits for Quality Wines Produced in Certain Regions According to European Legislation (Values Expressed in % vol.)

	European viticultural zones[a]					
	A	B	C I		C II	C III
			(a)	(b)		
Potential alcohol before addition	6.5	7.5	8.5	9	9.5	10
Addition limits in potential alcohol (% vol.)						
Normal year	3.5	2.5	2	2	2	2
Exceptional year	4.5	3.5	2	2	2	2
Authorized volume increase (%)[b]						
Normal year	11	8	6.5	6.5	6.5	6.5
Exceptional year	15	11	6.5	6.5	6.5	6.5
Total alcohol content (% vol.)						
Minimum commercial	9	9	9	9	9.5	10
Maximum commercial	Variable in France according to *appellation controleé*. Not limited in most other countries					

[a] See Table 11.1 for the definition of zones.
[b] By addition of concentrated or rectified concentrated must.

11.6.1 Hydrolytic Enzymes

Proteases

The active protein synthesis that characterizes ripening is responsible for the high protein concentration in mature grapes. In must, proteins often represent 50% of the total nitrogen. In white winemaking, part of these proteins are insoluble and are eliminated during clarification. Endogenous grape proteases are apt to hydrolyze these proteins into soluble forms. These forms are more easily assimilated by yeasts during fermentation. The enzymes catalyze the hydrolysis of the peptide bond between two amino acids (Figure 11.7).

The grape possesses a low, constant protease activity during its green growth phase. From veraison onward, this activity greatly increases. In a ripe grape, the protease activity is essentially located in the pulp (Table 11.7). But the proteases are generally bound to cell structures. Healthy grape juice thus has relatively few proteases (30% of total protease activity).

All physical treatments of grapes (mechanical harvesting, stemming, crushing) increase the proportion of soluble proteases. The higher free amino acid concentration of these musts attests to this (Cantagrel *et al.*, 1982).

Grape proteases are acidic, with an optimum pH near 2.0. In the pH range of must, 40–60% of the potential protease activity exists. Protein hydrolysis activity during the pre-fermentation phase varies greatly,

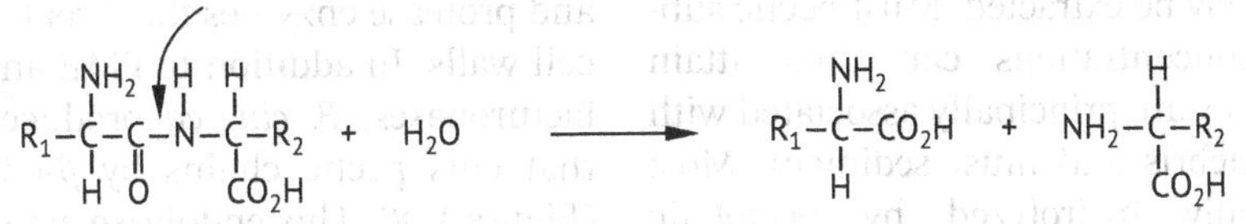

$$R_1-\underset{\underset{H}{|}}{\overset{\overset{NH_2}{|}}{C}}-\underset{\underset{O}{\|}}{C}-\underset{\underset{CO_2H}{|}}{\overset{\overset{H}{|}}{N}}-\underset{\underset{H}{|}}{C}-R_2 + H_2O \longrightarrow R_1-\underset{\underset{H}{|}}{\overset{\overset{NH_2}{|}}{C}}-CO_2H + NH_2-\underset{\underset{CO_2H}{|}}{\overset{\overset{H}{|}}{C}}-R_2$$

FIGURE 11.7 Protease mode of action by hydrolysis of peptide bonds.

TABLE 11.7

Distribution of Protease Activity in the Different Areas of a Healthy Grape (Expressed as % of Total Activity per Berry) (Cordonnier and Dugal, 1968)

Grape area	Protease activity
Pulp + juice	81.6
Skin	17
Seeds	1.4

depending on grape maturity and harvest treatments. This certainly affects fermentation kinetics, but the relationship has never been established. A slight sulfur dioxide addition (around 25 mg/l), however, has been confirmed to stimulate protease activity. This explains, at least partially, its activation effect on fermentation (Section 8.7.3).

Finally, botrytized grapes also contain fungal proteases. In contrast with grape proteases, these are soluble and pass entirely into the must. *Botrytis cinerea* aspartate proteinase has an optimum pH in the vicinity of 3.5. Whatever their origin, proteases are thermostable; they increase soluble nitrogen, even during thermovinification.

Pectolytic enzymes

Among fruits, the grape is one of the least rich in pectic substances (Volume 2, Section 3.6.2). These substances are predominately located in skin cell walls (Section 10.2.5). Must consequently contains a small amount of these compounds in the soluble form (0.5–1 g/l expressed as galacturonic acid). Depending on harvest treatments, some insoluble skin compounds may be extracted. Must pectic substance concentrations can thus attain 2.5 g/l. They are principally associated with cellular debris and must sediment. Most are rapidly hydrolyzed by pectolytic enzymes of the grape. Ethanol subsequently precipitates the rest of them at the end of fermentation. Wine is therefore practically devoid of pectic substances from the must (Usseglio-Tomasset, 1978).

The ripe grape is rich in pectin methyl esterase (PME) and polygalacturonases. Ripe grapes have high PME and polygalacturonase contents but contain no pectin lyase. PME is not a hydrolytic enzyme but rather a saponification enzyme. It liberates the acid functions of galacturonic units, resulting in the accumulation of methanol in the must (Figure 11.8). Grape PME is heat stable and has an optimum pH of 7–8. Its activity is reduced at the pH of must, but its action beforehand in the grape results in a significant decrease in the degree of esterification of pectic compounds released in the must. This action is essential, because polygalacturonases can only act on the free carboxylic functions of the galacturonic units. Two types of polygalacturonase exist in the grape: exo-polygalacturonases exert their hydrolytic action sequentially, beginning at one end of the polygalacturonic chain, and endo-polygalacturonases act at random on the interior of the chains. In the latter case, although pectin chain hydrolysis is very limited, the endo-polygalacturonic activity leads to a rapid and significant decrease in must viscosity. The viscosity is reduced by one-half when fewer than 5% of the glycolytic bonds are broken. Grape polygalacturonases retain a significant activity at the pH of must (optimum activity pH is between 4 and 5). The homogalacturonan zones are susceptible to rapid hydrolysis. The rhamnogalacturonan zones are more resistant, due to the presence of side chains of arabinose and galactose.

When the *B. cinerea* fungus infects grapes, it synthesizes cellulase, pectinase, and protease enzymes that break down the cell walls. In addition to PME and polygalacturonases, *B. cinerea* produces a lyase that cuts pectic chains by β-elimination (Figure 11.8). This endolyase activity is not

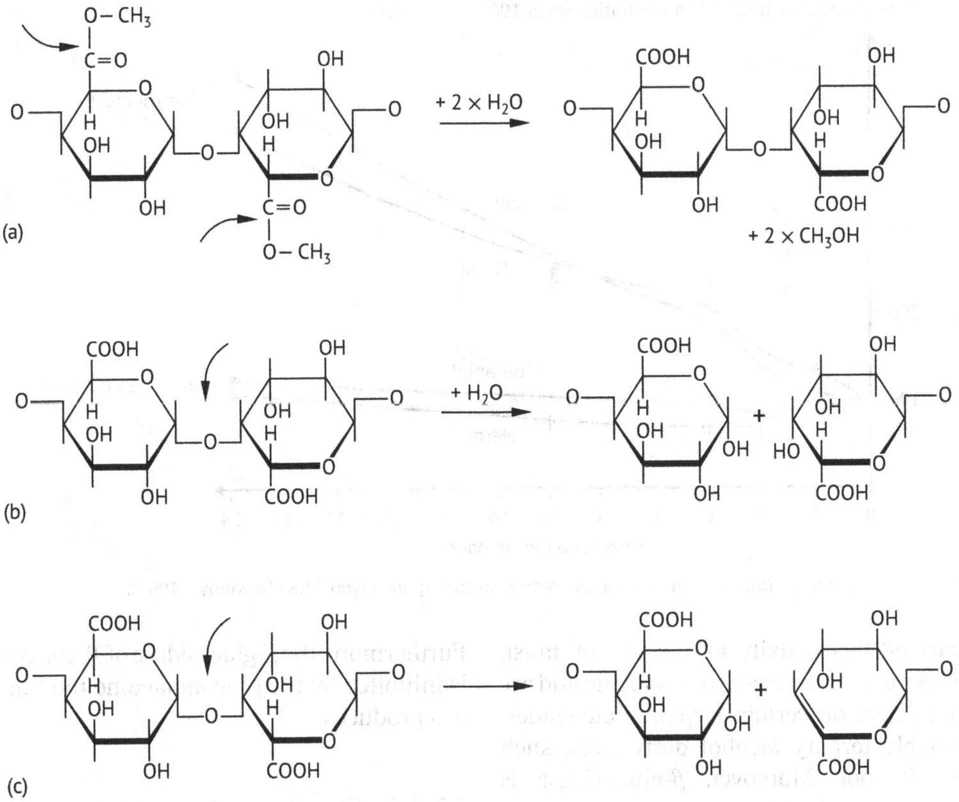

FIGURE 11.8 Mode of action of different pectolytic enzymes: (a) pectinesterases, (b) polygalacturonases, and (c) pectin lyases.

influenced by the esterification level of the carboxylic functions of the galacturonic units. All pectolytic enzymes of this fungus have an optimum pH near 5 and are therefore very effective during a botrytis attack. Musts made from contaminated grapes consequently contain very few pectic substances. Glucan, a polysaccharide secreted by *B. cinerea*, is mostly responsible for their viscosity (Volume 2, Section 3.7.2).

The pectolytic enzymes produced by the grape or the fungus are fairly resistant to sulfur dioxide. Their activity is, however, reduced at temperatures below 15°C and above 60°C.

Glycosidases

In Muscat-type aromatic varieties, a considerable proportion of their aromatic potential is in the form of terpene heterosides—non-odorous in ripe grapes (Section 10.3.8). During pre-fermentation treatments, enzymatic hydrolysis of these compounds increases must aroma intensity (Figure 11.9). This phenomenon is enhanced by maceration of grape solids because of the high concentration of bound terpene compounds in skins. These compounds, called diglucosides, are composed of a glucose associated with another sugar, i.e. rhamnose, arabinose, or apiose. The hydrolysis of these heterosides requires two sequential enzymatic activities. An α-L-rhamnosidase, an α-L-arabinosidase, or a β-D-apiosidase must act on the molecule before the β-D-glucosidase is able to exert its action (Figure 11.10). In practice, this hydrolysis is relatively limited. Grape glycosidases have an optimal activity at a pH between 5 and 6, and they only retain

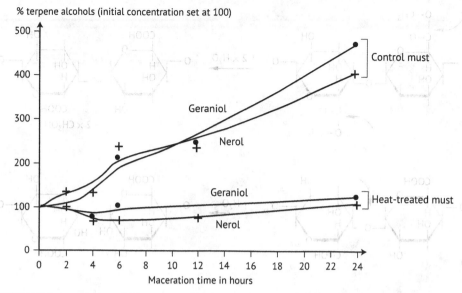

FIGURE 11.9 Demonstration of the enzymatic hydrolysis of terpene glycosides (Bayonove, 1993).

part of this activity at the pH of must. These glycosidases are very specific and are not active on certain terpene heterosides, notably tertiary alcohol derivatives, such as linalool. Moreover, β-glucosidase is strongly inhibited by free glucose (Bayonove, 1993). In moldy grapes, the glycosidases secreted by *B. cinerea* are more active. But the fungus totally degrades the aroma potential of the grape (Section 10.6.3). Furthermore, the β-glucosidase of *B. cinerea* is inhibited by the gluconolactone the fungus produces.

11.6.2 Oxidation Enzymes

Oxygenases

A green leaf-type odor or herbaceous note is produced when plant tissue (especially

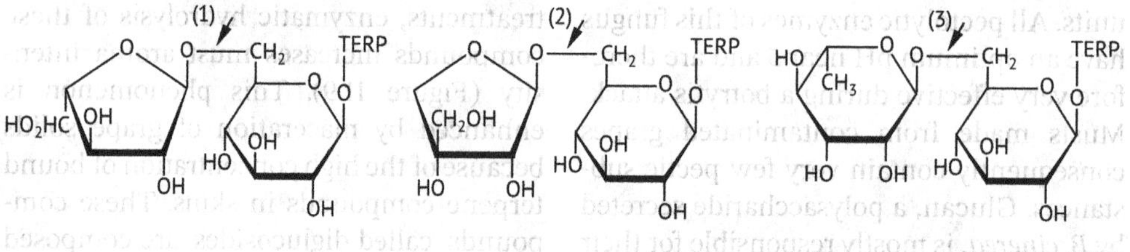

FIGURE 11.10 Enzymatic hydrolysis mechanism of terpene glycosides (Bayonove, 1993). (1) α-L-Arabinofuranosidase, (2) β-D-apiosidase, (3) α-L-rhamnosidase, and (4) β-D-glucosidase. TERP-OH: terpene alcohol.

leaves, but also fruit) is crushed. This phenomenon was shown to exist in grapes by Rapp et al. (1976). Four enzymatic activities are sequentially involved (Figure 11.11). First, an acylhydrolase frees the fatty acids from membrane lipids. Next, lipoxygenase catalyzes the fixation of oxygen on these C18 unsaturated fatty acids. This enzyme preferentially forms hydroperoxides at C13 from linoleic and linolenic acids. The peroxides obtained are then cleaved into C6 aldehydes. Some of the latter are reduced to their corresponding alcohols by the alcohol dehydrogenase of the grape (Crouzet, 1986). These alcohols are responsible for their corresponding odors. Since the cleavage enzymes are linked to membrane fractions, the aldehyde concentrations are proportional to the intensity of solids maceration. To limit

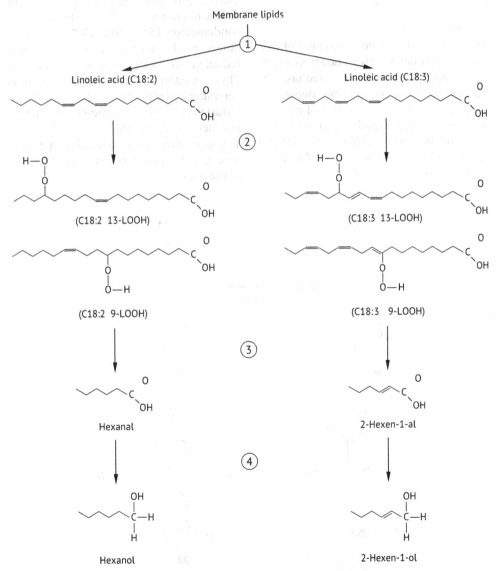

FIGURE 11.11 Enzymatic formation mechanism of aldehydes and C6 alcohols, responsible for grassy flavors (Crouzet, 1986). (1) Acylhydrolase, (2) lipoxygenase in the presence of oxygen, (3) peroxide cleavage enzyme, and (4) alcohol dehydrogenase.

their concentration during white winemaking, a sufficiently clear must should be obtained as quickly as possible (less than 200 NTU) (Dubourdieu et al., 1986).

The breakdown of the grape cellular structure during pre-fermentation treatments is also accompanied by other enzymatic oxidations. The oxygen consumption rate thus varies from 0.5 to 5 mg/l/min, depending on must origin. This variation is caused for the most part by the oxidation of phenolic compounds.

Oxidoreductases

Ripe grapes contain an ortho-phenol oxygen oxidoreductase, also known as cresolase, catechol oxidase, or tyrosinase. Its activity is extremely variable, depending on the grape variety and degree of ripeness (Dubernet, 1974). Tyrosinase consists of a group of isoenzymes differing in inductor nature and catalyzed activities (Mayer and Harel, 1979).

In white grape must (Section 13.4.2), this enzymatic activity preferentially oxidizes tartaric derivatives of hydroxycinnamic acids (1 and 2), which are the most prevalent phenolic compounds in grape pulp (Figure 11.12). The quinones produced (2) are unstable and likely to enter into two different reactions (Figure 11.13) (Rigaud et al., 1990). First, these very reactive quinones can condense with other phenolic compounds (flavonoids), forming polymerized products. Their color evolves from yellow to brown according to the degree of condensation (Singleton, 1987). The quinones are also apt to react with a strongly reductive molecule such as glutathione. This reaction produces a colorless derivative, S-glutathionyl-2-$trans$-caffeoyl tartaric acid, known as the grape reaction product (GRP) (3) (Salgues et al., 1986). This derivative is not oxidizable by tyrosinase and thus does not modify the color of the must.

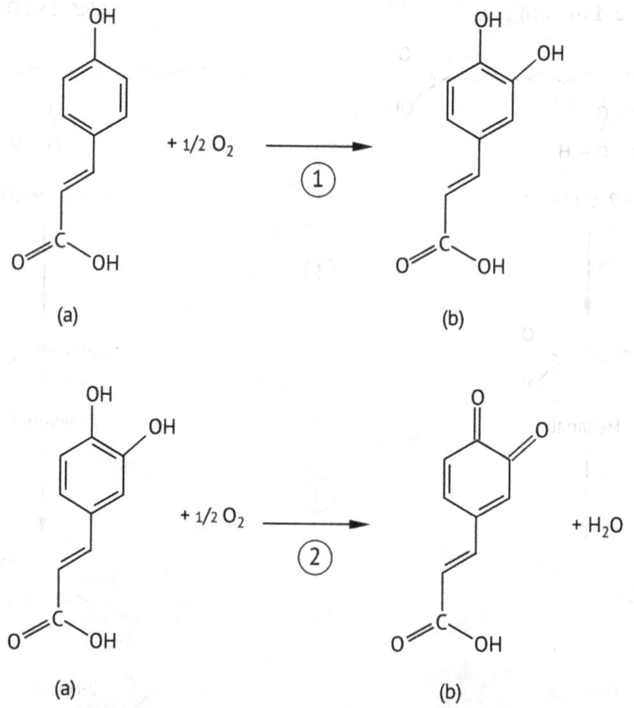

FIGURE 11.12 Mode of action of grape tyrosinase on hydroxycinnamic acids (Mayer and Harel, 1979). (1) Cresolase activity: (a) coumaric acid and (b) caffeic acid. (2) Catecholase activity: (a) caffeic acid and (b) quinone.

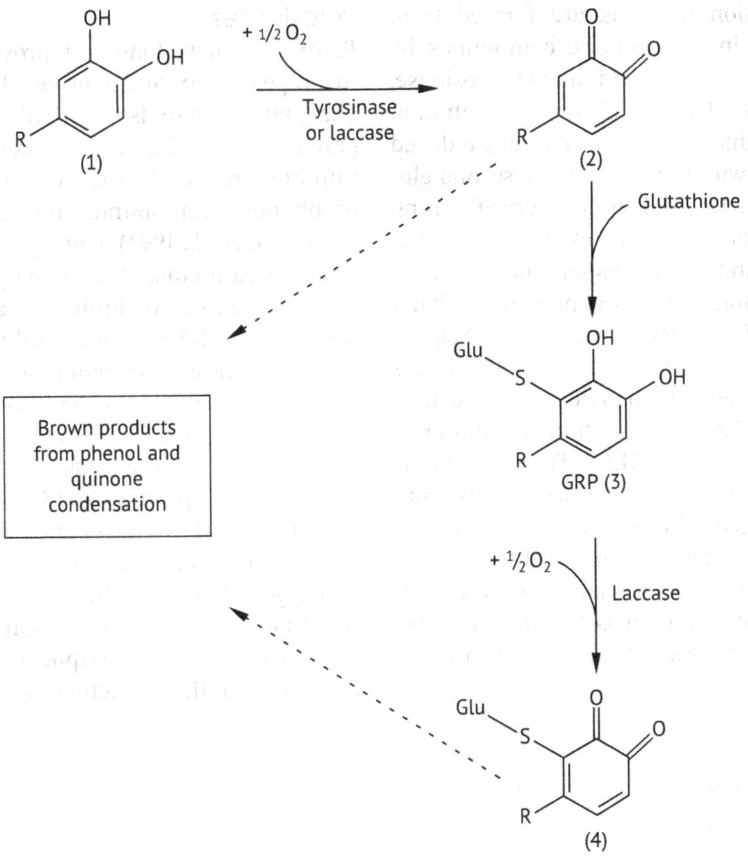

FIGURE 11.13 Oxidation mechanisms of healthy grape must by tyrosinase and botrytized grape must by laccase (Salgues *et al.*, 1986).

Must browning depends, of course, on the flavonoid concentration and consequently on mechanical treatments that favor maceration of the solid parts of the cluster. These operations are also involved in the solubilization of the tyrosinase bound to the chloroplast membranes.

Yet the trapping of quinones by glutathione limits oxidation phenomena. Must browning is therefore also dependent on the glutathione concentration.

The tyrosinase from grapes is active but unstable at the pH of must (optimum activity at pH 4.75). Temperatures above 55°C or the addition of more than 50 mg of sulfur dioxide per liter is necessary to denature this enzymatic activity. Lower sulfur dioxide concentrations only delay the oxidation phenomena. In fact, the bisulfite ions regenerate the potential enzyme substrates by reducing the quinones formed. Finally, treating must with bentonite reduces the soluble fraction of tyrosinase.

Phenolic compound oxidation is much more dangerous when the grapes have been attacked by *Botrytis*. Botrytized grapes contain a para-phenol oxygen oxidoreductase known as laccase (Dubernet, 1974). Contrary to tyrosinase, this fungal enzyme is stable at the pH of must and is more resistant to sulfur dioxide. It is also able to oxidize a greater number of phenolic substrates and molecules belonging to other chemical families. Laccase is thus capable of oxidizing the phenol-glutathione (3) complex to quinone (4) (Figure 11.13). Glutathione, therefore, can no longer trap quinone (Salgues *et al.*, 1986). More brown condensation products are formed from the same initial phenolic compounds by laccase than during oxidation by tyrosinase.

However, if the glutathione concentration is high, quinone (4) can be partially reduced to phenol with the fixation of a second glutathione molecule. This new derivative is no longer oxidizable by laccase. Oxidation phenomena and the corresponding browning are thus limited. This second reaction is not likely in botrytized grape musts (Salgues et al., 1986). The oxygen consumption rate is not much higher than in healthy grape must (Section 8.7.2), but the action of sulfur dioxide is slower (Figure 11.14). The moldy grape contains many other oxidases that also consume oxygen (glucose oxidase, amine oxidase, etc.). A temperature of 50°C destroys laccase more quickly than tyrosinase. This thermal denaturation is the only possible treatment, since adding bentonite only very slightly decreases laccase activity.

Peroxidases

Peroxidases have long been proven to exist in grapes (Poux and Ournac, 1972). This enzymatic activity is essentially located in grape cell vacuoles. It most likely plays an important role in the oxidative metabolism of phenolic compounds during ripening (Calderon et al., 1992). During pre-fermentation treatments, the activity of this enzyme seems to be limited by a peroxide deficiency in the presence of sulfur dioxide.

An increased understanding of these oxidation phenomena has spurred the development of a pre-fermentation technology called white must hyperoxygenation (Müller-Späth, 1990) (Section 13.4.1). As soon as cellular structures are broken down, a sufficient and controlled addition of oxygen provokes the denaturation of tyrosinase during the oxidation reactions that it catalyzes. The disappearance of the enzyme and the depletion of oxidizable

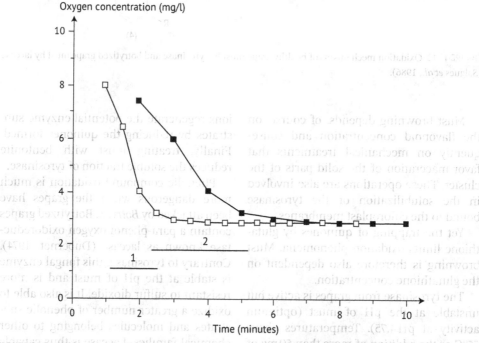

FIGURE 11.14 Effect of sulfur dioxide on oxygen consumption in musts made from healthy and moldy grapes. (1) Time necessary to stop oxygen consumption after the addition of sulfur dioxide in must made from healthy grapes. (2) Time necessary to stop oxygen consumption after the addition of sulfur dioxide in must made from moldy grapes. ■, oxygen consumption in must made from healthy grapes; □, oxygen consumption in must made from moldy grapes.

phenolic substrates thus make the must stable with respect to oxidation. The condensation products responsible for browning should be eliminated before fermentation. Due to its possible impact on aroma compounds, this technique seems better adapted to certain cultivars. It is not applicable to botrytized grapes due to the resistance of laccase.

11.7 Use of Commercial Enzymes in Winemaking

The beneficial action of diverse hydrolytic enzymes from grapes is often limited by must pH or an insufficient activity due to the limited duration of pre-fermentation treatments. Manufacturers have developed better adapted enzymatic preparations, essentially from diverse species of fungi (*Aspergillus*, *Rhizopus*, and *Trichoderma*). Research in this field is very active (van Rensburg and Pretorius, 2000). The enzymatic profile of currently available commercial preparations is still unclear, and users must develop their own experimentation. Many countries permit the use of these preparations. They are added as early as crushing to increase juice extraction or to finished wine to improve filterability. These methods can also be used to improve color extraction and must quality (settling, fermentability, and aroma intensity) in red and white winemaking, respectively.

11.7.1 Juice Extraction

The addition of pectolytic enzymes to crushed grapes can improve juice extraction for certain varieties that are very rich in pectic substances (Muscat, Sylvaner, etc.). Commercial preparations contain diverse enzymatic activities that are active at a low pH: PME, polygalacturonases, pectin lyases, and hemicellulases. At a concentration of 2–4 g/hl, 15% more juice can be obtained during a settling period of 4–10 hours; even a shorter settling period (one to two hours) increases the proportion of free-run juice (Table 11.8). Effectiveness varies according to the nature of the grapes.

These pectolytic preparations can also contain diverse glycosidases (Cordonnier *et al.*, 1989) and proteases (Schmitt *et al.*, 1989), which are responsible for secondary transformations. Their degree of purity must therefore be ensured.

11.7.2 Must Clarification

Pectolytic preparations lower white must viscosity and thus accelerate sedimentation (Figure 11.15). In less than an hour, the colloidal equilibrium is destabilized, resulting in rapid sedimentation and increased must limpidity. A more compact must deposit, facilitating static settling, is rare. This treatment can lead to excessive juice clarification. Its use should be determined according to must composition. The enzymatic degradation of pectic compounds is subsequently demonstrated by a distinct improvement in the filterability of the musts and wines obtained. These wines are often better prepared for tangential filtration (Volume 2, Section 11.9).

In red winemaking, these preparations are used in particular for press wines and heat-treated grapes and must (Section 12.8.3). In the latter case, the must is very rich in pectic compounds and

TABLE 11.8

Enzymatic Treatment of Crushed Grapes (Kadarka Variety, Hungary; Enzymatic Preparation: Vinozyme 2 g/hl) (Canal-Llaubéres, 1989)

Juice	Control (%)	With enzyme (%)
Free run	66	93
Press	34	7

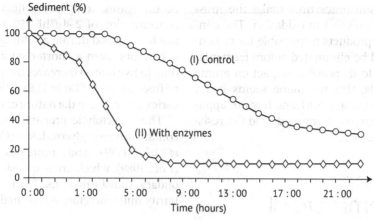

FIGURE 11.15 Effect of pectolytic enzymes on the sedimentation speed of white must lees (Canal-Llaubères, 1989).

devoid of endogenous grape enzymes. These are destroyed by heat (Martinière and Ribéreau-Gayon, 1973). Pectolytic enzymes can also be used when the wine is drained off the skins after a traditional maceration.

In musts from botrytized grapes, the pectic compounds are degraded for the most part and replaced by a fungal polymer, glucan (Section 10.6.3). A glucanase (Volume 2, Section 11.5.2) is industrially prepared from *Trichoderma* sp. fungus cultures (Dubourdieu *et al.*, 1981). The enzyme is preferably added (1–3 g/hl) after fermentation. Its action takes 7–10 days and must occur at a temperature equal to or greater than 10°C. Higher doses are required in red winemaking since phenolic compounds partially inhibit glucanase. Industrial glucanase also affects the yeast cell walls and improves the wine's colloidal stability.

11.7.3 Color Extraction and Stabilization

Red wine color results from maceration of grape solids (skins, seeds, and sometimes stems) during alcoholic fermentation. Phenolic compound extraction thus depends on many factors: grape variety, grape maturity, length of maceration, number of pumpovers, temperature, etc. (Section 12.5).

Adding pectolytic enzymes at the start of maceration can facilitate this extraction (Table 11.9). The resulting wine is richer in tannins and anthocyanins with a higher color intensity and redder hue.

This treatment also improves the organoleptic characters (notably structure) of the wine (Canal-Llaubères, 1992). It apparently favors color stabilization by forming

TABLE 11.9

Influence of Pectolytic Enzymes on Color Extraction in Red Winemaking (Merlot in Bordeaux, France, 1988; Vinozyme 3 g/hl at Filling) (Canal-Llaubères, 1990)

Wine after draining off the skins (20 days of maceration)	Control tank	Enzyme-treated tank
Absorbance at 280 nm	64	66
Tannins (g/l)	3.5	3.8
Anthocyanins (mg/l)	768	895
Color intensity	1.58	1.68
Hue	0.44	0.40
Absorbance at 420 nm (%)	27.8	26.2
Absorbance at 520 nm (%)	63.0	65.3
Absorbance at 620 nm (%)	9.2	8.5

polymerized pigments (Parley *et al.*, 2001). Further research is needed to evaluate the stability of these changes during aging, especially to define the types of wine that may benefit from such treatment. These preparations also contain β-D-glucosidase, likely to hydrolyze anthocyanin glucosides (van Rensburg and Pretorius, 2000).

11.7.4 Freeing of Aromas

The glycosidases contained in commercial pectolytic enzymes are capable of partially hydrolyzing terpene glycosides (Table 11.10). The first tests of these enzymes were conducted on dry wines because of the inhibiting effect exerted by glucose on β-glucosidase. These enzymes may also act on other aromatic compounds present in the form of odorless precursors in certain grapes.

This treatment is likely to complement the terpene compound transformations effected by yeasts during fermentation (Volume 2, Section 7.2). However, it releases all the terpene alcohols too rapidly. The pleasant-smelling monoterpenes, such as linalool, nerol, and geraniol, may be converted into more stable forms during aging, including terpineol, which has a less attractive aroma (Park, 1996).

TABLE 11.10

Liberation of Terpene Alcohols by Enzymatic Hydrolysis (Gewürztraminer 1985; Novoferm 12 = 15 ml/hl, one month of incubation at 18°C) (Canal-Llaubéres, 1990)

Terpene alcohols (µg/l)	Control wine	Enzyme wine
Linalool	141	151
Terpineol	74	75
Citronellol	45	52
Nerol	53	104
Geraniol	216	358
Total	529	740

In any case, care should be taken to avoid enzyme preparations containing cinnamate decarboxylase, as it may lead to the development of ethylphenols with a highly unpleasant musky odor (Volume 2, Section 8.4.3).

References

Bayonove C. (1993) Les composes terpéniques. In *Les Acquisitions récentes en Chromatographie du Vin*, pp. 99–120 (Ed B. Donèche). Tec & Doc, Lavoisier, Paris.

Benitez J.G., Grandal-Delgado M.M. and Martin J.D. (1993) Am. J. Enol. Vitic., 44, 400–404.

Berger J.L. (1994) In *Les Acquisitions récentes dans les traitements physiques du vin*, pp. 65–80 (Ed B. Donèche). Tec & Doc, Lavoisier, Paris.

Blouin J. and Peynaud E. (2001) *Connaissance et travail du vin*, 3rd edition. Dunod, Paris.

Brugirard A. (1987) Rev. Œnol., 44, 9–12.

Calderon A.A., Garcia-Florneciano E., Munoz R. and Ros Barcelo A. (1992) Vitis, 31, 139–147.

Canal-Llaubères R.M. (1989) Rev. Œnol., 53, 17–22.

Canal-Llaubères R.M. (1990) Rev. Fr. Œnol., 122, 28.

Canal-Llaubères R.M. (1992) Enzymes in winemaking. In *Wine Microbiology and Biotechnology*, pp. 477–506 (Ed G.H. Fleet). Harwood Academic Publishers, Chur.

Cantagrel R., Symonos P. and Carles J. (1982) Sci. Aliments, 2, HS 1, 109–142.

Chauvet S., Sudraud P. and Jouan T. (1986) Vitis, 101, 11–38.

Clary C.D., Steinhauer R.E., Frisinger J.E. and Peffer T.E. (1990) Am. J. Enol. Vitic., 41, 176–181.

Cordonnier R. and Dugal A. (1968) Ann. Technol. Agric., 17, 3, 189–206.

Cordonnier R.E., Gunata Y.Z., Baumes R.L. and Bayonove C.L. (1989) Œno One, 23, 7–23.

Crouzet J. (1986) Rev. Fr Œnol., 102, 42–49.

Dubernet M. (1974) *Recherches sur la tyrosinase de Vitis vinifera et la laccase de Botrytis cinerea. Applications technologiques*, Thèse de doctorat, Université de Bordeaux II, France.

Dubourdieu D. Ollivier C. and Boidron J.N. (1986) Œno One, 20, 53–76.

Dubourdieu D., Villetaz J.C., Desplanques C. and Ribéreau-Gayon P. (1981) Œno One, 15, 161–177.

Dupuy P. and De Hoogh J. (1991) *The Enrichment of Wine in the European Community—Wageningen Agricultural University*. Report EUR 13239 EN, 1-151, Commission of the European Communities.

Flanzy C. (1998) *Oenologie. Fondements scientifiques et technologiques*. Tec & Doc, Lavoisier, Paris.

Guimberteau G., Gaillard M. and Wajsfelner R. (1989) Œno One, 23, 2, 95-118.

Haushofer H. (1972) Œno One, 6, 373.

Jacquet P. (1983) Vititechnique, 71, 27.

Martin G.J., Guillou C., Naulet N., Brun S., Tep Y., Cabanis J.C., Cabanis M.T. and Sudraud, P. (1986) Sci. Aliments, 6, 385-405.

Martinière P. and Ribéreau-Gayon J. (1973) Ann. Technol. Agric., 22, 1-20.

Mayer A.M. and Harel E. (1979) Phytochemistry, 18, 193-211.

Müller-Späth H. (1990) Rev. Fr. Œnol., 124, 5-12.

Park S.K. (1996) Food Biotechnol., 5, 280-286.

Parley A., Vanhanen L. and Heatherbell D. (2001) J. Grape Wine Res., 7, 146-152.

Peynaud E. and Allard J.J. (1970) CR Acad. Agric., 56, 18, 1476-1478.

Poux C. and Ournac A. (1972) Ann. Technol. Agric., 21, 1, 47-67.

Rapp A., Hastrich H. and Engel L. (1976) Vitis, 15, 29-36.

Reader H.P. and Dominguez M. (1995) Fortified wines: Sherry, Port and Madeira. In *Fermented Beverage Production*, pp. 159-207 (Eds A.G.H. Lea and J.R. Piggott). Blackie Academic and Professional, Glasgow.

Ribéreau-Gayon J., Peynaud E., Ribéreau-Gayon P. and Sudraud P. (1976) *Sciences et Techniques du vin Vol. 3: Vinifications, transformations du vin*, pp. 3-28. Dunod, Bordas, Paris.

Rigaud J., Cheynier V., Souquet J.M. and Moutonet M. (1990) Rev. Fr. Œnol., 124, 27-31.

Salgues M., Cheynier V., Gunata Z. and Wylde R. (1986) J. Food Sci., 51, 1191-1194.

Schmitt A., Köhler H., Miltenberger A. and Curschmann K. (1989) Dtsch. Weinbau, 11, 408-414.

Singleton V.L. (1987) Am. J. Enol. Vitic., 38, 69-77.

Terrier N., Sauvage F.X. and Romieu C. (1996) *Œnologie 95 (Compte Rendu du 5ème Symposium International d'Œnologie)*, pp. 24-28. Tec & Doc, Lavoisier, Paris.

Tinlot R. (1990) Vignes Vin, 5, 51-54.

Usseglio-Tomasset L. (1978) Ann. Technol. Agric., 27, 261-274.

Usseglio-Tomasset L. and Bosia P.D. (1992) Bull. OIV, 731-732, 5-13.

Van Rensburg P. and Pretorius J.S. (2000) S. Afr. J. Enol. Vitic., 21, 52-73.

Vromandt G. (1989) Vitis 137, 83-103.

Wucherpfennig K. (1967) *Compte Rendu du 2ème Symposium International d'Œnologie*, pp. 411-433. INRA.

Wucherpfennig K. (1980) Bull. OIV, 583, 186-205.

CHAPTER 12

Red Winemaking

12.1 Generalities
12.2 Mechanical Processing of the Harvested Grapes
12.3 Tank Filling
12.4 Controlling Alcoholic Fermentation
12.5 Maceration
12.6 Draining Off the Skins and Pressing
12.7 Malolactic Fermentation
12.8 Automated Red Winemaking Methods
12.9 Carbonic Maceration

12.1 Generalities

Red wine is made by means of maceration, or contact with the grape skins. The extraction of solids from grape clusters (specifically from skins, seeds, and possibly stems) accompanies the alcoholic fermentation of the juice. In conventional red winemaking, extraction of grape solids involves maceration, which occurs during must fermentation. Other methods exist that dissociate fermentation and maceration, such as thermovinification.

The localization of red pigment exclusively in skins, at least in the main red varieties, allows for a slightly tinted or white wine to be made from the colorless juice obtained from a delicate pressing of red grapes. Base wines for the production of sparkling wines are a good example. The designation *blanc de blancs* was created to distinguish white wines derived from white varieties and those from red (*blanc de noirs*). Finally, varietal nature is not sufficient for characterizing the origin of a red wine. Maceration intensity is of prime importance.

The length and intensity of maceration are adjusted according to grape variety and the type of wine desired. In fact, maceration is a means by which the winemaker can personalize the wine.

Handbook of Enology, Volume 1: The Microbiology of Wine and Vinifications, Third Edition.
Pascal Ribéreau-Gayon, Denis Dubourdieu, Bernard B. Donèche and Aline A. Lonvaud.
© 2021 John Wiley & Sons Ltd. Published 2021 by John Wiley & Sons Ltd.

Nouveau-type wines are made to be drunk young: their aromas and fruitiness are inversely related to their phenolic content, but premium wines require a sufficient tannin concentration to develop properly during aging.

Grape quality directly influences grape skin contact quality in red winemaking and is thus of the greatest importance. In fact, the grape skin is more affected than the juice by grape variety, maturation conditions, and the condition of the grapes (absence of mold). Vintage and vineyard (growth) rankings are therefore much more clearly defined with red wines than whites. In the Bordeaux region, anthocyanin and tannin concentrations in the same parcel can so much as double, from one year to another, depending on maturation conditions. Must acidity and sugar concentrations can fluctuate by 50 and 15%, respectively. These numbers are not surprising, since the plant requires a lot of energy to synthesize anthocyanins. For this reason, the northernmost vineyards produce only white wines. In any case, when phenolic contents are examined in relation to environmental conditions, their nature, properties, and location in the tissues must also be considered. Enologists readily define "good" tannins as those that give wines a dense structure without aggressiveness and "bad," or harsh, tannins as those characterized by vegetal flavors and herbaceous astringency. The nature and chemical properties of these various phenolic compounds are covered elsewhere in this series (Volume 2, Sections 6.2 and 6.3). They highlight the need to wait until the grapes reach full phenolic maturity, which may occur later than general ripeness. Similarly, high levels of methoxypyrazines in insufficiently ripe grapes of certain varieties (especially Cabernet Sauvignon) are responsible for a herbaceous, green bell pepper aroma in must and wine that is considered a defect above certain levels (Volume 2, Section 7.4).

Grape composition and quality variability result in heterogeneous grape crops. Grape selection can compensate for this heterogeneity, and tanks should be filled with a homogeneous single-variety grape crop that has the same degree of soundness (absence of rot) and level of maturity. *Terroir*, quality, vine age, rootstock, fruit loads, and a number of other factors should also be taken into consideration. Appropriate vineyard management methods are increasingly being applied to achieve the low yields essential to ensure perfect grape ripeness and high quality. This batch selection, conducted at tank filling time, must be maintained during the entire winemaking process until the definitive stabilization after malolactic fermentation (MLF). The best batches are then blended together to make a wine of superior quality. The complementary characteristics of the various batches often produce a blended wine that is superior in quality to each of the batches before blending.

The grape crop should also be carefully sorted to eliminate damaged, moldy, or unripe grapes. This operation can be conducted in the vineyard during picking or in the winery at harvest reception. At the winery, the grapes are spread out on sorting tables. A conveyor belt advances the crop, while workers eliminate bad grapes. A concern for perfection in modern winemaking has led to the spread of such practices. Their effectiveness is even more pronounced when they are applied to grape crops of superior quality.

Red grape crop heterogeneity requires specific winemaking techniques to be adapted according to the crop. Much remains to be learned in optimizing the various grape specifications.

The generalized use of MLF is another characteristic of red winemaking. This phenomenon has been recognized since the end of the last century, but, until the last few decades, it was not a consistent component of red winemaking. For a long time, a slightly elevated acidity was considered to be an essential factor in microbial stability and thus in wine quality. Moreover, red wine must acidification was a widespread practice. It has currently disappeared for the most part, since it is only justified in particular situations. Today, on the contrary, MLF is known to produce a more stable wine by eliminating malic acid, a molecule easily biodegraded.

It was in temperate regions that MLF first became widespread. These wines, which are rich in malic acid, are distinctly improved, becoming more round and supple. MLF was then progressively applied to all red wines, even those produced in warm regions already having a low acidity. This type of fermentation may not be advisable in all regions, and another method of stabilizing red wines containing malic acid should be sought.

The classic steps in red winemaking are:

- mechanical processing of the harvested grapes (crushing, destemming, and tank filling);
- *cuvaison* (primary alcoholic fermentation and maceration)—draining off the skins;
- separation of wine and pomace by dejuicing and pressing—completion of fermentation, exhaustion of the last grams of sugar by alcoholic fermentation, and MLF.

There are currently many variations on each stage in traditional winemaking, but the operations described in this chapter constitute the basic method for producing high-quality red wines. It does, however, require considerable tank volume capacity and many demanding handling operations. Consequently, other techniques have been developed. The standard order of certain operations has been changed to make a certain level of automation possible—for example, in continuous winemaking and heat extraction (Section 12.8).

Lastly, fermentation with carbonic maceration takes advantage of the special aroma qualities produced by fermenting whole grapes under anaerobic conditions (Section 12.9). This special fermentation gives these wines specific organoleptic characters.

12.2 Mechanical Processing of the Harvested Grapes

12.2.1 Harvest Reception

Diverse methods, adapted to each winery, are used to transport the harvest from grapevine to winery. In world-renowned vineyards, small-capacity containers are carefully manipulated by hand. In most vineyards, the harvest is transported in shallow bed trailers or trucks. Whatever the container capacity, the grapes should be transported intact without being crushed. Transport containers should also be kept clean. If the transport time is even somewhat long, the grapes should be transported during the cooler hours of the night.

Red grapes are certainly less sensitive to maceration and oxidation phenomena than white grapes, but microbial contamination is likely to occur in a partially crushed crop, left in the vineyard, especially in the presence of sunlight. These risks must be avoided.

During mechanical harvesting, the grapes are transported in high-capacity containers. Speed and hygiene are even more important in this case, since the grapes are inevitably partially crushed with this method.

Small-volume containers are emptied manually. More generally, a dumping trailer is used, which empties its load into the receiving hopper (Figure 12.1). In high-capacity facilities, the bins are placed on a platform that dumps the grapes sideways—thus avoiding excessive truck or tractor maneuvering. If the winery is equipped, the grape crop may pass over a sorting table (Section 11.3.3) before reaching the receiving hopper. Manual sorting is only effective if the grapes are whole. It is almost impossible to combine it with mechanical harvesting: at best, obviously damaged grapes can be removed from the vines before the harvester arrives. Otherwise, sorting may

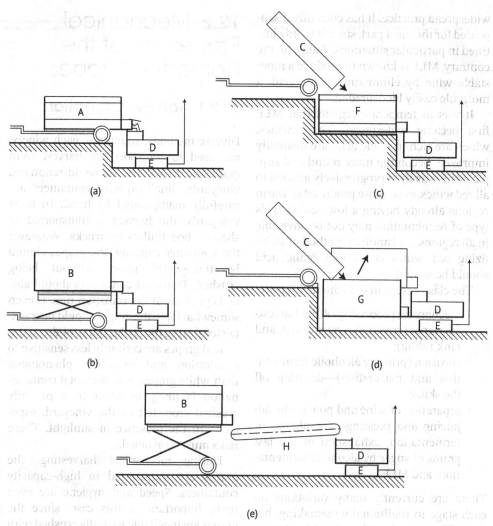

FIGURE 12.1 Examples of harvest receiving equipment for red winemaking. (a) Gondola with screw conveyor, gravity dumping. (b) Elevator gondola with screw conveyor, adjustable dump height. (c) Gravity-dump gondola. (d) Gravity-dump gondola, elevator hopper designed to feed destemmer-crusher by gravity. (e) Sorting table between trailer and destemmer-crusher. Key: (A) trailer with a screw conveyor; (B) same trailer with an elevator system; (C) dumping trailer; (D) destemmer-crusher; (E) grape pump; (F) stationary hopper feeding the destemmer-crusher by gravity; (G) elevator hopper with adjustable height to feed the crusher by gravity; (H) sorting table. (*Source*: P. Jacquet, Bordeaux, personal communication.)

take place in the vineyard, immediately after the grapes have been picked, or when the grapes arrive at the winery. In the latter case, the grapes should be transferred to the sorting table manually to spread them evenly. Transfer screws should not be used, as they crush the grapes and make sorting impossible.

Two sorting tables may be necessary at wineries producing high-quality red wine.

The grapes are initially sorted when they arrive from the vineyard. The second sorting operation, after destemming, removes any small fragments of stems and leaves that were missed during the first sorting operation and is followed by crushing. There are now increasing numbers of machines, based on various techniques, available to do this operation automatically. Vibrating tables are quite successful.

Receiving hoppers are available in various designs. In small wineries, they may be installed directly above the crusher-stemmer and filled directly from the transfer vehicle. In general, a worm screw in the bottom of the hopper regulates throughput, and it should turn slowly to avoid excessive crushing of the grapes. Throughput may be increased by using a larger-diameter hopper.

When buying grapes according to weight and sugar concentration, these values must be determined at the time of reception. Grape crop heterogeneity complicates the determination of the sugar concentration. The sample should therefore be taken after crushing and homogenization. The soundness of the grapes may also be assessed at this stage by analyzing their laccase activity (Section 10.6.6).

At the outlet of the destemmer-crusher, a pump distributes the grapes to a given tank. Sulfur dioxide is added at this time (Sections 8.7 and 8.8.1), as can any other necessary additions (Section 12.3.1).

Grape handling should be minimized, limiting transfer distances and maximizing the use of gravity. Rough handling is likely to shred or lacerate stem tissues, so sap is liberated from green plant tissue and later found in wine. The suspended solid concentration simultaneously increases; in fact, this measurement may be used to evaluate equipment quality. The most quality-oriented solution consists in sorting and destemming the grapes by hand, then crushing them, if necessary, through a wooden screen, thus eliminating the need to crush them mechanically. Finally, the must is transferred without pumping. Of course, only the most prestigious estates can afford the high cost of these techniques.

In partially botrytized grapes, the brutal mechanical action on grapes disperses a colloidal carbohydrate (glucan) in the must. Glucan is produced by *Botrytis cinerea* (Volume 2, Section 11.5.2) and is located between the pulp and the skin inside the berry. The wine obtained is difficult to clarify. When the same grapes are carefully handled, the wine can be clarified much more easily. Wine clarification difficulties with botrytized grapes are always observed in the same wineries. The type of equipment used is often responsible.

12.2.2 Crushing

Grapes are traditionally crushed to break the skin in order to release the pulp and the juice. This operation is probably one of the most ancient harvest treatments. Partial crushing can be obtained by the traditional technique of stomping the grapes. High-speed centrifugal crusher-destemmers ensure an energetic crushing. There are also many other systems between these two extremes.

The consequences of crushing are as follows:

1. The juice is aerated and is simultaneously inoculated by yeasts. The fermentation is quicker, and the temperature higher. In certain circumstances, a slower fermentation speed and lower temperatures can be obtained by not crushing the grapes (carbonic maceration; Section 12.9.4).
2. Aeration can be harmful. In partially rotted grapes, it can provoke enzymatic browning.
3. Crushed grapes can be pumped, and sulfiting is more homogeneous.
4. All of the juice is fermented; at the time of draining off the skins, the press wine does not contain sugar.
5. Crushing has a significant effect in facilitating maceration and accentuating anthocyanin and tannin dissolution. An energetic crushing intensifies this effect. Tannin concentrations proportionally increase more rapidly than the color. This increased maceration can be an advantage in certain cases, but it tends to increase the harsh astringency and unpleasant tastes of average varieties.

Premium wine grapes are traditionally lightly crushed to burst the berries without lacerating the solid parts. Crushing is used to facilitate fermentation and avoid residual sugar in press wines. Methods other than crushing should be used to increase maceration (maceration time, pump-over operations, punchdowns, temperature). They better respect wine quality. Even when carbonic maceration is not strictly used (Section 12.9), winemakers may wish to avoid crushing the grapes for great wines with long maceration times to avoid brutal damage to the plant tissues.

Two kinds of crushers exist. Roller crushers (Figure 12.2) are coated with plastic; the opposing rollers turn in opposite directions, and their spacing is easily adjustable. This system works well, but delivery rates are limited. High-speed rotary crushers can be either horizontal or vertical. Beaters project the grape clusters against a perforated cage, and the burst grapes pass through the perforations. These machines simultaneously destem the grapes (Figure 12.3). As they are rough on the grapes, they are not recommended, especially in making high-quality wines.

12.2.3 Destemming

This operation, also known as destalking, is now considered indispensable (after much discussion of its advantages and disadvantages for a long time).

Destemming has a number of consequences:

1. A primary and financially important advantage of this operation is the reduction of the required tank capacity by 30%. In addition, the pomace volume to be pressed is greater with a grape crop containing the stems. Although the stem facilitates juice extraction during pressing, a higher-capacity press is required.
2. Fermentations in the presence of stems are always quicker and more complete (Figure 12.4). The stems facilitate fermentation not only by ensuring the presence of air but also by absorbing heat, thus limiting temperature increases. Fermentation difficulties are rarely encountered with non-destemmed grapes.
3. The stems modify wine composition. They contain water and very little sugar, thus lowering alcohol content. Moreover, stem sap is rich in potassium and not very acidic. Destemming therefore increases must acidity and alcohol content.
4. With moldy grapes, stems protect wine color from enzymatic browning. The laccase activity of *B. cinerea* is most likely fixated or inhibited.
5. Destemming most significantly affects tannin concentrations. Table 12.1 indicates the approximate proportion of phenolic compounds

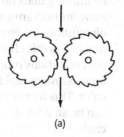
(a)

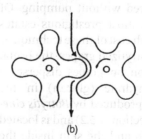
(b)

FIGURE 12.2 Crusher roller design: (a) spiral ribbed rollers and (b) grooved rollers with interconnecting profiles.

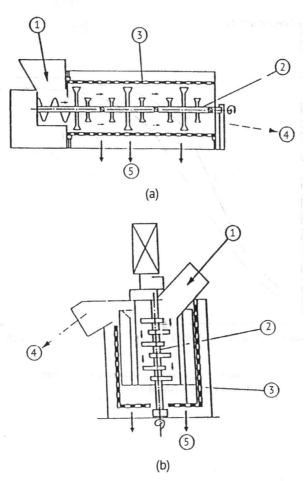

FIGURE 12.3 Operating principle of (a) a horizontal destemmer and (b) a vertical centrifugal destemmer. Key: (1) hopper; (2) shaft with arm and paddles; (3) perforated cylinder; (4) stem outlet; (5) destemmed grape outlet.

supplied by the various parts of the grape cluster. In this experiment, 54% of the total tannins come from grape skins, 25% from seeds and 21% from stems. Results may vary according to grape quality and variety. More precise details on the nature and concentrations of phenolic compounds from the various parts of the grape cluster are discussed elsewhere in this series (Volume 2, Section 6.5). Table 12.2 summarizes the principal changes in wine composition caused by destemming. Despite the increase in total phenolic compounds in the presence of stems, color intensity (CI) diminishes. This long-observed fact is interpreted as the adsorption of grape skin anthocyanins on the ligneous surface of the stems. This interpretation has been confirmed in a model solution containing anthocyanins and tannins; either a stem extract or the stems themselves are added. In the first case, the tannin concentration increases considerably, while the CI slightly increases. In the second case, the tannins increase, but the CI decreases. Tannins play an

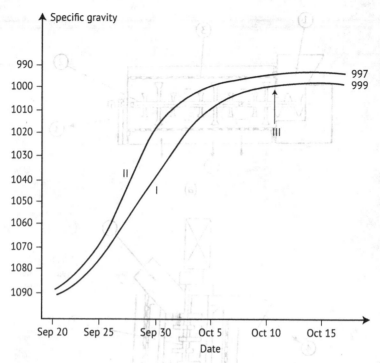

FIGURE 12.4 Influence of stems on alcoholic fermentation (Ribéreau-Gayon et al., 1976). I: destemmed grapes. II: non-destemmed grapes. III: date of draining off the skins.

TABLE 12.1

Influence of Different Parts of the Grape Cluster on Phenolic Compounds and Wine Color (Malbec, Fermentation at 25°C, Maceration Time 10 days) (Ribéreau Gayon et al., 1976)

Component Color[a]:	Juice	Juice + skins	Juice + skins + seeds	Juice + skins + seeds + stems
Intensity	—	1.81	1.40	1.17
Hue	—	0.39	0.43	0.48
Anthocyanins (g/l)	—	0.98	0.94	0.85
Tannins (g/l)	—	1.75	2.55	3.25
Total phenolic compounds (permanganate index)	5	32	47	56

[a] Intensity = OD 420 + OD 520.
Hue = OD 420/OD 520.
(OD 420 and OD 520 = optical density, 1 mm thickness, at 420 and 520 nm).

TABLE 12.2
Influence of Stems on Wine Composition (Ribéreau-Gayon et al., 1976)

Component	Destemmed	Not destemmed
Alcohol content (% vol.)	13.2	12.7
Total acidity (mEq/l)	86	78
Volatile acidity (mEq/l)	11.4	11.4
Total phenolic compounds (permanganate index)	38	58
Color[a]:		
Intensity	1.28	1.18
Hue	0.51	0.57

[a] See Table 12.1.

important role in the color of mature wines. Although wines made from non-destemmed grapes have less color when young, they become more colored than their destemmed counterparts in the course of aging.

6. The increased tannin and phenolic compound concentration of wines made from non-destemmed grapes can increase wine quality in certain cases, e.g. for young vines and wines with insufficient structure without the stems. However, grape stems are likely to give unpleasant vegetal and herbaceous flavors to wines. In general, when finesse is desired, destemming is indispensable. In any case, the decision on total or partial destemming must take into account stem quality, which is related to grape variety and ripeness.

In the past, the grape crop was destemmed by hand directly in the vineyard or, more generally, in the winery, by rubbing the grape clusters with rakes against a wooden screen. Today, this operation is carried out mechanically (Figure 12.3). Destemmers comprise a perforated cylinder, with a shaft equipped with paddle-like arms running through its center. When the shaft turns, it draws in the grape clusters and expels the stems out the other end. The juice, pulp, and grape skins pass through the perforations. The continual quest for higher output has led to increased rotation speeds and replacement of the paddle-shaped rods with beaters. The beaters apply sufficient force to burst the grapes without the need of a crusher. Vertical shaft destemmers can process 20–45 metric tons/h. They operate at 500 rpm, and the centrifugal force evacuates the stems by the top of the machine. These machines have a brutal action on the grapes and produce fine suspended solids, imparting vegetal and herbaceous flavors to wines. Their use should be avoided, at least for the production of premium wines (Section 12.2.2).

Crushing and destemming are generally done by the same piece of equipment, but in certain cases it is desirable to have the option of not destemming. For a long time, with conventional crusher-destemmers, crushing preceded destemming. Today, there are an increasing number of machines that eliminate the stems before crushing the grapes. The stems do not pass between the crusher rollers. In this manner, the risk of shredding the stems is lowered. This order of operation increases must quality, since stem shredding releases vegetal vacuolar sap, which is bitter and astringent.

A good destemmer should not leave any berries attached to the stems. Reciprocally, the stems should not be impregnated with juice. The stems should also be entirely eliminated, with no broken fragments remaining. The laceration of ligneous stem tissue by the machine can seriously affect quality, and in these instances, destemming should be avoided. Attempts have been made to eliminate residual stem waste after the destemmer and before the crusher. Quite large quantities are removed in this way.

12.3 Tank Filling

12.3.1 Tank Filling and Related Operations

In the case of fermentation with carbonic maceration (Section 12.9), tanks must be filled directly from the top with uncrushed grapes, which obviously requires a very complex system. Otherwise, grapes are usually received at a single winery location and transferred to the fermentation tanks after destemming and crushing. Transfer pumps must do as little damage as possible to grape tissue, and distances should be kept to a minimum, with as few bends as possible in the hoses. This operation can be carried out manually, without pumping. As the must increases in volume during fermentation, about 20% empty space should be left in each tank. If anti-foaming agents (Section 3.2.5) are used, less headspace is required.

A considerable volume of gas is released during fermentation, approximately 50 l of carbon dioxide per liter of must fermented. Before one enters a tank, it is essential to check for oxygen, in view of the danger of asphyxiation and death from carbon dioxide. The traditional method is to check that a flame stays alight inside the fermentation vessel. Today, handheld O_2, CO_2, and multiple-gas meters can be used to determine if the tank is safe to enter.

The grapes must be sulfited adequately and homogeneously during transfer to the tank (Section 8.8.1).

Several operations may be carried out during transfer of the grapes/must or in the following few hours. Firstly, they may be inoculated (Section 3.5.4) with a fermenting must (a few percent corresponding to 10^6 cells/m) or active dry yeast (ADY), *Saccharomyces cerevisiae*, chosen from among the various commercial strains (over 100 on the market). The qualities required are the aptitude to complete fermentation successfully and heat resistance. The impact of yeast strains on the character of red wines is less marked than in the case of white wines (Section 13.7.2). Winemakers must still ensure, however, that the selected strain is suitable for the type of wine being made. Recommended doses of 10–25 g/hl correspond to inoculation with $2 \times 10^6 - 10 \times 10^6$ cells/m. Native yeasts must be inhibited by appropriate doses of SO_2 to ensure effective inoculation. Dry yeasts must be reactivated for 20 minutes by mixing them into a mixture of must and water (1:1) at 40°C. The reconstituted yeast must be distributed evenly through each tank.

Acidity can be corrected (Section 11.4) during the initial transfer into tanks or at a later time. If sugar levels need to be increased (chaptalization; Section 11.5.2), this is best done when the must is warm at the beginning of fermentation. The sugar dissolves more easily, and the subsequent aeration stimulates fermentation, though this does generate heat, which must be taken into account. In addition, relatively low sugar levels during the growth phase promote reproduction of the yeast cells.

If an assay (Section 3.4.2) indicates a nitrogen deficiency, ammonium sulfate (10–30 g/hl) may be added as soon as the tank has been filled, or, preferably, once fermentation has started.

Adding tannin during red winemaking was abandoned for a long time, but the quality of products now available, particularly those made from white grape skins or fresh grape seeds, has revived interest in this procedure. These products are not only considered capable of improving body and tannin structure but also of stabilizing color by promoting condensation of anthocyanins and tannins (Volume 2, Section 6.3.10).

It is thus useful to add tannin early in the fermentation process, when the tannins have not yet been extracted from the grape seeds, so that they can react with the anthocyanins released early in maceration. According to some authors, the results are uneven due to the low solubility of the tannins and the difficulty of mixing them into the must (Blouin and Peynaud, 2001), so it is preferable to add the product after draining off the skins. High doses (20–50 g/hl) are required to raise the initial tannin levels by approximately 10%.

Another operation currently attracting some interest is the addition of pectolytic enzymes to promote extraction of phenolic compounds (Sections 11.7.3 and 12.5.1), for the purpose of obtaining wines with a higher tannin content, but less astringency and bitterness (Blouin and Peynaud, 2001).

Glycosidases may also be used to promote extraction of terpene aromas—particularly useful in making Muscat wines. Care must be taken in traditional red winemaking to avoid producing off-aromas. The use of enzymes in winemaking requires further scientific research.

Some juice can also be bled off at this stage (Section 12.5.8), mainly to eliminate rainwater and juice that has not yet absorbed compounds from the skins. Decreasing the quantity of must facilitates concentration of the phenolic compounds during maceration. This operation is generally carried out after the tank has been filled and the juice has been separated from the pomace. Water can be eliminated (e.g. by reverse osmosis or vacuum evaporation) (Section 11.5.1) under the same conditions. The results are very similar, but these methods maintain the natural grape sugars. These techniques are capable of concentrating the must by 5–10% or even as much as 20%. Excessive concentration of the must changes the flavor balance of the wine completely, and it is preferable to adapt vineyard management methods and reduce yields on the vine to achieve similar results.

12.3.2 Principal Tank Systems

Various types of fermentation tanks exist. They differ in the aeration level supplied to yeasts, which helps to ensure a complete fermentation, and in the specific behavior of pomace as regards maceration and phenolic compound extraction.

Fermentation releases gas within the must. The bubbles rising toward the surface of the fermentor entrain solid particles, which unite and agglomerate, forming the pomace or skin cap. The skin cap is maintained at the top of the fermentor by the pressure of the released gas.

Pomace plays an important role. First and foremost, during maceration, it yields its constituents (anthocyanins and tannins). These compounds are indispensable components of the character of red wine. Yeast reproduction is also particularly intense within the pomace: $10-50 \times 10^6$ cells/ml have been observed in the juice at the bottom of the tank and $150-200 \times 10^6$ cells/ml in the juice impregnating the pomace.

Although no longer recommended, open floating-cap fermentors are still used in small-scale wineries. They were used in the past because the extended contact with air enabled successful fermentations, even in musts containing high concentrations of sugar. Moreover, temperature increases are less significant. Yet, the drawbacks are undeniable. Alcohol losses can attain and sometimes exceed 0.5% by volume (Section 12.6.1; see Table 12.10). The risk of enzymatic browning with moldy grapes is also significant. Additionally, as soon as the active fermentation stops, the pomace cap surface is no longer protected from aerobic germ development. Bacterial growth is facilitated, and contamination risks are high due to the large surface area of this spongy matter. As soon as the fermentation slows, the pomace cap should be regularly submerged to drown the aerobic germs. This operation, known as cap punching (*pigeage*), can only be carried out manually in small-capacity fermentors. If necessary, it can be done mechanically with a jack or other equipment. Submerging the pomace cap also contributes to the extraction of its constituents. It also aerates the must and homogenizes the temperature. But this type of fermentor does not permit an extended maceration. The tanks must be run off before the carbon dioxide stops being released. Afterward, spoilage risks in the pomace cap are certain, and the resulting press wine would have an elevated volatile acidity. Manually removing the upper layer of the most contaminated part of the pomace cap is not sufficient, nor is covering the tank with a tarpaulin after fermentation.

To avoid cap spoilage and to eliminate the laborious work of regularly punching down the cap, systems have been developed that maintain the cap submerged in the must, for example, under a wooden screen fitted to the tank after filling. The must in contact with air is permanently renewed by the released gas. Acetic acid bacteria have more difficulty developing in this environment. The compacting of the pomace against the wooden screen does not facilitate the diffusion of its constituents, and several pump-overs are therefore recommended to improve maceration.

Today, most red wines are fermented in tanks that can be closed when the carbon dioxide release rate falls below a certain level. The complete protection from air enables maceration times to be extended, almost as long as desired. The tank can be hermetically sealed by a water-filled tank vent (Figure 12.5) or simply closed by placing a cover on the tank hatch. In the latter case, the CO_2 that covers the upper part of the tank disappears over time, and the protection is not permanent. The tank should therefore be completely filled with wine, or a slight pumping-over operation should be carried out twice a day to immerse the aerobic microbes.

For a long time, the major drawbacks of the closed fermentor were a considerable temperature increase and the absence of oxygen. As a result, fermentations were often long and difficult, and stuck fermentation occurred frequently. Today, these two drawbacks are mitigated by temperature control systems and pump-overs with aeration, permitting the dissolution of the necessary oxygen for a successful fermentation.

In addition, this fermentor design avoids alcohol loss by evaporation. Press wine

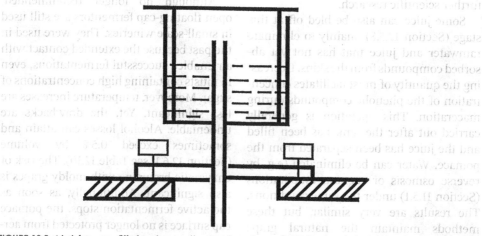

FIGURE 12.5 Airtight water-filled tank vent allowing the release of CO_2 from tank during fermentation without air entry.

quality is greatly increased, while the laborious work of cap punching is eliminated. This kind of tank has also been empirically observed to facilitate MLF.

12.3.3 Fermentor Construction

Red wine fermentors have been successively made of wood, concrete, and steel, and on occasion plastic.

Wood is a noble material, and wooden tanks have long been part of the tradition in great winemaking regions. New wood releases aroma compounds into the wine during fermentation, but after a few years of use, this phenomenon no longer occurs during fermentation. Disadvantages are that wooden tank maintenance is difficult, old wooden tanks are a source of contamination and off-aromas, and wooden tanks are not completely airtight. They must sometimes be expanded with water before use, with all of the corresponding risks of microbial contamination. In addition, the flat ceiling of a truncated cone-shaped tank is rarely airtight—this kind of tank is not suitable for prolonged wine conservation. Wood is also a poor heat conductor. Wooden tanks are subject to considerable temperature increases that must be compensated by appropriate refrigeration systems; yet when the fermentation is completed, they retain the heat generated for a long time, thus favoring post-fermentation maceration.

Concrete enables good use of available space, since the tanks are manufactured on site, but the acids in wine attack concrete. The inner tank walls must therefore be protected. The tanks can be coated with a 10% solution of tartaric acid applied three times at intervals of several days. Under these conditions, the inner tank walls are coated with calcium tartrate. The wine in the tank contributes to maintaining the coating. It is, however, preferable to coat the inner tank walls with an innocuous and chemically inert lining such as epoxy resins or Araldite®. Whatever lining is used, the coating of these tanks requires continuous maintenance. Concrete is a better heat conductor than wood, but refrigeration systems are still indispensable. These tanks are completely airtight and can be used for wine storage.

Steel, particularly stainless steel, is the material most often used today for manufacturing fermentors. Two categories of stainless steel exist: one contains molybdenum; the other does not. Chrome–nickel–molybdenum steel is more resistant to corrosion and is necessary for the long-term conservation of sulfited white wines, especially in partially filled tanks. In the humid atmosphere above the wine, sulfur dioxide gas is concentrated, and the condensation formed on the tank walls is corrosive. For red winemaking and storage in completely filled tanks, the less expensive chrome–nickel steel is sufficient.

Stainless steel tanks have the significant advantage of being airtight and easily fitted with various types of equipment. Their internal and external maintenance is also facilitated; their inner walls are impregnable. Stainless steel also has good heat exchange properties, avoiding excessive temperature increases. In certain cases, red wine fermentations can occur at 30°C without cooling. In any case, cooling is simplified: a cool liquid is circulated within the double jacket of the tank or in an integrated heat exchanger. When a sufficient amount of cool water is available, running water over the exterior of the tank can be sufficient. In the 1960s and 1970s, stainless steel tanks represented a considerable advance in temperature control compared with wooden and concrete tanks, and this superior control was much appreciated by winemakers. Today, however, it has been observed that these tanks insufficiently warm the fermenting must when the ambient temperature is too low. This phenomenon is accentuated in cases where the tanks have been placed outside to lower the cost of investment. As soon as fermentation stops, the tank temperature rapidly decreases to the ambient temperature; as a result, maceration phenomena, which are influenced by temperature, are slowed.

To master red winemaking, temperature-controlled (heating and cooling) stainless steel tanks are necessary. However, in any case, as heat inertia is limited, it is difficult to obtain homogeneous temperatures in the post-fermentation phase. Recent developments in cooling equipment have led to renewed interest in wooden or concrete fermentation tanks, where homogeneous temperatures are easier to maintain.

Fermentor capacity must also be considered when designing a winery. High-capacity tanks are of course economical, but tank size should not be too large and should be adapted to the winery (50–350 hl). It is difficult to control the various winemaking steps in tanks containing over 350 hl. Tanks of limited capacity permit superior batch selection and skin extraction due to increased skin contact. The tank should be filled before the start of fermentation, and for this reason the filling time should not exceed 12 hours.

Tank shape is important for red winemaking. The exchange surface between the pomace and the juice should be sufficient. The dimensions of sheet steel used during manufacturing sometimes result in tanks that are too high with respect to their diameter. Reciprocally, tanks should not be too wide. In this case, pump-overs lose their effectiveness in covering the entire skin cap, and air contact can also be excessive. Tank height should slightly exceed tank diameter. High-performance pump-over systems can compensate for disproportionately high tanks to a certain extent.

12.3.4 Fermentor Equipment

Basic red winemaking tanks should be equipped with the following:

- Two juice evacuation valves on the lower part of the tank, placed at different heights to facilitate racking (elimination of sediment). There should also be an opening at the lowest point of the tank for emptying and cleaning.
- One or two doors: one, a bit higher is used to empty the skins from the tank after draining; the second, lower door is less essential but can be useful for cleaning the tank.
- A gauge to indicate the filling height.
- A tasting spigot to take samples.
- A thermometer.
- An airtight lid at the top of the tank. A water-filled tank vent (Figure 12.5) makes the tank completely airtight, while allowing the liquid to expand.

This basic configuration has often been complemented with additional more specific add-ons. Steel tanks lend themselves particularly well to additional equipment. Nowadays, several manufacturers offer tanks specifically designed for fermenting red wines. They are equipped with complex attachments that turn them into complete systems for monitoring and controlling fermentation. For example, Selector System (Gimar Tecno 15040, Occimiano, A.I., Italy) has an automated tank cleaning system, programmable pump-overs with or without aeration and/or spraying of the pomace cap, temperature control, and management of fermentation kinetics according to changes in specific gravity so that the entire fermentation cycle can be programmed and controlled directly by the system.

Temperature control systems are generally the first add-on. Automatic temperature monitoring, sometimes continuous, is standard. Temperature probes, which must be properly placed to ensure correct measurements, are often part of a more elaborate temperature control system. Initially, these systems simply consisted in running cool water over the exterior of the tank. Today, cooling fluids (water or a dilute glycol solution) are often circulated through a double jacket in the tank or an internal heat exchanger. The latter is more efficient but makes tank cleaning more difficult.

Tanks not only need to be cooled but also heated on occasion. An identical system is therefore used to circulate a heated liquid (hot water). The liquid can be sent through the same heat exchanger or preferably another pipeline. Since the liquid (must) and the solid (pomace) are separated in the tank, a pump-over is required, at the same time as cooling or heating, to homogenize the temperature.

In high-performance facilities, when a tank reaches a maximum preset temperature, a pump-over is automatically performed to homogenize the temperature. If the temperature remains too high after this first operation, the tank is cooled. Automated temperature control systems have been designed that regulate the temperature throughout the entire fermentation process.

Automatic pump-over systems have also been developed to facilitate skin extraction (Section 12.5.5) and permit the aeration of fermenting must (Section 12.4.2). In the past, the carbon dioxide released during fermentation and the resulting pressure were used to pump the fermenting must to an upper tank for aeration, after cooling if necessary. Opening a valve releases the pressure, causing the must to cascade on the pomace cap. Due to the complexity of this system, the pump should be specifically adapted to the tank. This pump-over operation can be done with or without aeration in a tank located below the fermentor. Aeration could be better controlled by injecting a predetermined quantity of oxygen in the lines. Pump-over frequency and duration should of course be adapted to the must. Pumping-over too often may make the wine excessively hard and astringent. Various systems are available to improve cap submersion and intensify maceration.

Injections of pressurized gas at 3 bar (nitrogen, CO_2, or even air) can replace conventional pump-over operations, with installation of a probe on the internal piping of the tank. Results that would normally take over an hour with a traditional pump-over (rotating irrigator, stream breaker, etc.) are obtained in a few minutes. This system has been combined with standard pump-overs at the beginning and end of maceration using food-grade gas.

Other methods complementing or replacing pump-overs can be employed to improve pomace extraction. Hydraulically controlled pistons have been developed, which immerse the pomace cap, replacing the traditional punchdown method (*pigeage*). Various systems break up the pomace cap inside the tank, particularly rotating cylindrical tanks. Due to the continuing evolution of these systems, a more detailed description is difficult (Blouin and Peynaud, 2001), but their use is covered in Section 12.5.8.

Another piece of equipment in high demand is an automatic pomace removal system, which evacuates the skins from the tank toward the press, replacing the arduous task of manually emptying the tank.

A number of self-emptying tanks have been proposed. Models with inclined floors (20° slope) improve pomace removal. The worker removes the pomace with an adapted rake without having to enter the tank. Automated self-emptying tanks are also available, and several models have been proposed (Figure 12.6). The simplest are cylindrical with an extremely inclined floor (45° slope) ending in a large door. The slope of the tank floor and the door dimensions should be appropriate to the nature and viscosity of the grape crop; for example, extended maceration times dry out the pomace, making removal more difficult. With this type of evacuation system, the entire tank contents should be emptied in one go; a sufficiently large receiving system linked to the press must therefore be placed below the tank. Progressively emptying the tank may cause the pomace to get stuck: it forms an increasingly compacted arch, which is very difficult to break.

Hydraulic systems evacuate pomace by inclining the tank. Screw conveyor systems inside tanks enable a controlled and regular pomace evacuation; they are used in rotating tanks.

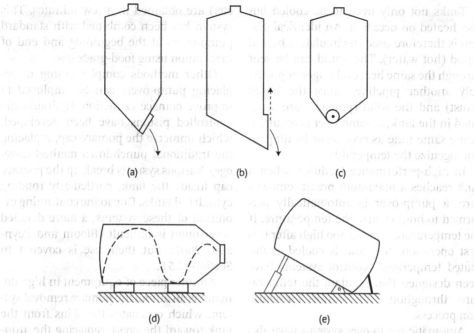

FIGURE 12.6 Self-emptying tanks: (a, b) self-emptying by gravity, (c) screw conveyor self-emptying tank (screw conveyor is incorporated into the chute at the bottom of the tank), (d) rotating tank for cake breakup and pomace evacuation, and (e) hydraulic dump tank.

12.4 Controlling Alcoholic Fermentation

12.4.1 Effect of Ambient Conditions

In the past, a distinction was made between red winemaking methods in warm and cool climates. In certain variable-climate vineyards, warm and cool vintage winemaking techniques were also distinguished. Problems linked to fermentation temperature control and grape composition were responsible for these distinctions. These differences are less important today. The necessary conditions for successful winemaking are known: they are adapted to the nature of the grape crop and are not difficult to carry out, as long as the appropriate equipment is available.

A cool year or cool climate is characterized by late and often insufficient ripeness. Grape acidity is high, and the musts are thus relatively protected against bacterial attack. However, there is a risk of *Botrytis* attacks and the formation of enzymatic browning, since cool climates often correspond to rainy climates. In addition, grape crops arriving at the winery are often characterized by relatively low temperatures in cool years. As a result, the initiation of fermentation can be difficult, even more so when the grapes have been washed off by rain; the native yeast population can be insufficient.

Ferré (1958) observed in Burgundy vineyards that fermentation was activated in 12 hours at 25°C, in 24 hours at 17–18°C, and in five to six days at 15°C; it was nearly impossible at 10°C. These numbers are of course approximate and depend on many other factors, in particular the yeast inoculation concentration. Tanks should not be left at insufficient temperatures. The resulting fermentations are often slow and incomplete, with a risk of mold development beforehand. Tanks are also

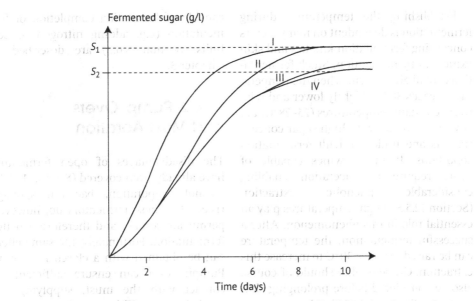

FIGURE 12.7 Effect of momentary aeration (by pump-over operations at different times) on fermentation kinetics. I: open tank, permanent aerobic conditions—all sugar in must (S_1) is fermented. II: closed tank, aeration on second day by pump-over—the fermentation is accelerated, with respect to total anaerobic conditions, and goes to completion. III: closed tank; aeration on sixth day—the acceleration of fermentation is insignificant. IV: closed tank under total anaerobic conditions—the fermentation stops when fermented sugar (S_2) is lower than S_1.

immobilized for extended periods, which can create problems in wineries that use each tank several times during the harvest.

The must should therefore be warmed as quickly as possible to 20°C. If the fermentation does not begin shortly after warming, the temperature rapidly drops down to its initial value. A simultaneous yeast inoculation is required to avoid this problem; it also accelerates the fermentation and thus provokes a more considerable temperature increase. If the temperature becomes too high, cooling may be required after these operations, which accelerate the fermentation. Aeration also remains useful, as long as the must is not susceptible to enzymatic browning.

In contrast with a cool year, a warm year or warm region produces an early harvest. The resulting must is rich in sugar, and so a complete fermentation can be difficult to obtain. The low acidity also increases bacterial risks and requires appropriate sulfiting (Section 8.8.1). All of the harvest conditions combine to produce a high fermentation temperature, and temperature control systems are therefore indispensable. In such a situation, the risks of a stuck fermentation and consequently bacterial spoilage are at a maximum. Paradoxically, the highest-quality wines can be made under these winemaking conditions. In temperate climates, the greatest vintages have long been known as the most difficult ones to produce, but winemaking methods adapted to these conditions are relatively recent. Temperature control and refrigeration have been essential. Although a moderate temperature (20°) is necessary to initiate fermentation correctly, the temperature should not be excessive. Yeasts in their growth phase are particularly heat sensitive: when the initial temperature is between 26 and 28°C, the increase in temperature during the yeast growth phase makes stuck fermentation more difficult and increases the risk of the yeast producing excessive volatile acidity. Initial cooling of the grape crop is therefore recommended.

Establishing the temperature during fermentation is dependent on many factors concerning fermentation kinetics and skin extraction by maceration. Stuck fermentations are likely to occur when the temperature exceeds 30°C. Slightly lower and relatively constant temperatures (25–28°C) are advised for musts with high sugar concentrations and under difficult fermentation conditions. Premium wines capable of aging require a maceration enabling considerable phenolic extraction (Section 12.5.5). High temperatures play an essential role in this phenomenon. After a successful fermentation, the temperature can be raised to above 30°C to increase this extraction. Grape quality should of course also be considered before prolonging skin contact (Section 12.5.8). *Nouveau-type* wines, however, are made to be drunk young and retain the fruitiness of the grape; lower fermentation temperatures are recommended for these wines (25°C).

Under difficult fermentation conditions linked to excessive temperatures, several palliatives were formerly recommended. Limited crushing slowed the fermentation process and thus produced less heat. Another method consisted of simultaneously filling several fermentors over several days in the hope that the regular addition of fresh grapes would moderate the fermentation process. In the latter case, sulfiting is not sufficient to avoid the increased risks and resulting consequences of stuck fermentation and hydrogen sulfide production. Lastly, high temperatures justify early draining off the skins. This process, which separates the juice from the pomace, lowering bacterial risks, is employed in hot regions. Long maceration times, however, are conducted in more temperate zones.

Today, must aeration or, more specifically, aeration of yeasts during their growth phase (Section 3.7.2), along with yeast inoculation (Section 3.5.4) and temperature control, is the most effective way of helping difficult fermentations. It is carried out during pump-overs, or, possibly, by means of micro-oxygenation. Other processes capable of facilitating completion of fermentation (e.g. adding nitrogen or cell hulls; Section 3.6.2) are described in Chapter 3.

12.4.2 Pump-Overs and Must Aeration

The disadvantages of open fermentors have already been covered (Section 12.3.1): alcohol evaporation, bacterial spoilage risks, etc. These fermentors do, however, permit air contact and therefore a better fermentation. Fortunately, the same effect can be obtained with a closed fermentor. Pumping-over can ensure sufficient air contact with the must, supplying the needed oxygen. This operation consists in letting fermenting must flow in contact with air and then pumping it back into the upper part of the tank (Figure 12.8). The effectiveness of this method has been known in the Bordeaux region since the end of the 19th century. In-depth research was carried out in the 1950s. Due to its simplicity, its use has been widespread.

The numbers in Table 12.3 (Ribéreau-Gayon et al., 1951) demonstrate the effectiveness of pump-overs for improving the fermentation process. They also show that yeasts better resist high temperatures when aerated, but there is a certain amount of confusion as to the most opportune time to aerate. The same authors demonstrated that correct timing of oxygenation is essential. Early aerations at the beginning of fermentation help to prevent stuck fermentations: the yeasts are in their growth phase, and oxygen is utilized to improve their growth and produce survival factors (Section 3.7.2). Early pump-overs have the additional advantage of avoiding alcohol loss by evaporation.

An aeration carried out on the second day of fermentation is the most effective. The effectiveness of later aerations, in the presence of fermentation difficulties, is greatly diminished (Figure 12.7), sometimes to the point of being nonexistent. In

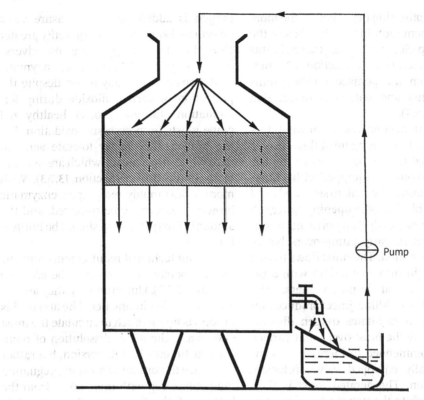

FIGURE 12.8 Pump-over operation, showing must aeration below and skin extraction above (P = pump).

TABLE 12.3

Effect of Aeration by Pump-overs on Fermentation Kinetics (Ribéreau-Gayon et al., 1951)

	Tank aerated by pump-over		Non-aerated tank (without pump-over)	
Time	Temperature (°C)	Specific gravity	Temperature (°C)	Specific gravity
Day 1	22	1.088	23	1.088
Day 2	26	1.084	26	1.084
Day 4	32	1.047	29	1.073
Day 6	20	0.996	27	1.045
Day 10	—	—	27	1.020
Day 20	—	—	20	1.002

the final stages of fermentation, the yeast does not make use of oxygen, since ethanol and other toxic metabolites hinder its nitrogen assimilation. A nitrogen addition in the final stages of fermentation, therefore, does not help to reestablish fermentation activity, even after aeration.

Pumping-over with must aeration is only beneficial at certain moments, but the pump-over in general has other effects. It increases the tank's maximum temperature and thus requires refrigeration. It also homogenizes the temperature, sugar concentrations, and yeast population of the fermentor, offsetting the effects of the more active fermentation in and just below the pomace cap (Section 12.3.1). Above all, this operation facilitates extraction of compounds from the pomace (anthocyanins and tannins) and enhances maceration (Section 12.5.4).

The pump-over process is shown in Figure 12.8. The fermenting must flows from a valve located in the lower part of the fermentor. It should be equipped with a filtering system inside the tank to stop seeds and skins from blocking the opening, since the obstruction of this opening presents a serious problem in early pump-overs before skin cap formation. The must flows from a certain height into a container with a capacity of several hundred liters. The pressure of the falling juice produces an emulsion that facilitates oxygen dissolution. Running the must over a flat surface is also recommended to increase air contact. Specially equipped valves intensify the emulsion. The aerated must is then pumped back to the upper part of the fermentor, soaking the pomace cap. Aeration may be eliminated by pumping must in a closed system, directly from valve to pump to the upper portion of the fermentor. Cap submersion may also be avoided by placing the pipe, in the upper part of the fermentor, through the pomace cap into the liquid.

Admittedly, the system in Figure 12.8 and its use are based on empirical data. The quantity of oxygen dissolved in must with this system cannot even be approximated. However, the quantity of oxygen dissolved in must exposed to air is on the order of 6–8 mg/l and varies according to temperature. The quantities necessary to avoid stuck fermentation are approximately 10–20 mg/l, which can be obtained by pump-overs with aeration two or three times, 24 hours apart. Nevertheless, a system permitting controlled oxygen addition (from a compressed gas cylinder, for example) would be preferable. Indeed, this is the aim of a process known as micro-oxygenation. The precise amount of oxygen necessary must also be determined. Of course, oxygen is added simply to ensure yeast growth and survival, and a quantity greater than the optimum dose has no adverse effects on yeasts. Nevertheless, enzymatic oxidations in must may occur, despite the protection of carbon dioxide during fermentation. Tannins protect healthy red grape musts from excessive oxidation. For this reason, they better tolerate aeration than white grape musts, which are not generally pumped over (Section 13.7.3). With more or less moldy red grapes, enzymatic browning can easily be triggered, and the amount of oxygen added should be limited, if not nil.

Certain technical requirements must be met for pomace extraction to be effective (Section 12.5.5). Unfortunately, they are not always satisfied in practice. The aim of this process is not so much to circulate the must and ensure the direct dissolution of pomace constituents by submersion, but rather to replace the saturated must impregnating the pomace cap with must taken from the bottom of the fermentor. Approximately two-thirds of the pomace is immersed in the fermenting must, and one-third floats above the liquid as a cap. All of the must should be pumped over, and the entire pomace cap should be soaked to obtain satisfactory results. These conditions can be difficult to achieve in narrow rectangular fermentors, especially if the lid is not located in the center of the tank. The same limited must fraction participates in this pump-over operation. Draining (possibly with aeration) a third to a half of the tank volume and then brutally releasing the must from the top of the tank enables the submersion of the entire pomace cap (*délestage* or rack-and-return). During this process, the cap descends in the tank. Various systems (cables) permit the cap to be broken up and reformed.

Various types of rotating irrigators exist. They must be placed at the center of the fermentor and ensure the thorough soaking of the entire surface. To be fully effective, their pump delivery rate must be sufficient. Increasing the flow rate during a pump-over can suffice for modifying tannin concentrations and consequently wine style. Even under ideal conditions, the liquid may pass through preferred channels in the pomace cap. Depending on operating conditions, pump-over effectiveness with respect to pomace extraction is extremely variable. Close monitoring is indispensable.

A volume of juice corresponding to one-third to one-half of the tank volume should generally be pumped over. The number of pump-overs should be increased but not their duration. In any case, the frequency of pump-over should be modulated. This operation contributes to the tannic structure of wine and favors the extraction of the highest-quality tannins, making wine rich and supple, but an excessive tannin concentration can lead to hard, aggressive, unpleasant wines. Other techniques (e.g. punching down the cap) also give similar results (Section 12.5.5).

Due to its simplicity and its favorable effects, pumping-over is an essential operation in red winemaking. Inspired by the recommendations of Blouin and Peynaud (2001), the following steps are applicable to Bordeaux-style winemaking:

1. As soon as the tank is filled, a homogenization pump-over blends the grapes and evenly distributes the sulfur dioxide. Yeast may be inoculated at this time, together with any other additives, but aeration is unnecessary.
2. Pumping-over with aeration is essential as soon as fermentation starts, as well as the following day and, possibly, the day after that.
3. During active fermentation, pump-overs are conducted for extraction of phenolic compounds. The number of pump-overs should be adapted to wine type and grape quality. The fermenting must should be pumped over every one to two days. "Free-run" wine and "press" wine are also homogenized during this operation.
4. After fermentation, pump-over operations should be discontinued in airtight, closed tanks to avoid oxygen exposure. In open tanks, pump-overs can be continued. In case of extended maceration in partially filled tanks, it is recommended to saturate the top of the tank with carbon dioxide to prevent the development of aerobic microorganisms. Even light pump-overs during extended maceration increase the astringency and bitterness of wines.
5. Pump-over operations rarely suffice to restart slow or stuck fermentations.

12.4.3 Monitoring the Fermentation Process: Determining Its Completion

Monitoring temperature and fermentation kinetics via specific gravity measurements has already been described (Sections 3.2.2 and 3.2.3). It is indispensable in winemaking. Other controls are also recommended to complement this data.

Fermentation rates have been observed to vary under apparently identical conditions (temperature, sugar content, amount of yeast inoculated, etc.). Sluggish fermentations may be completed successfully, but they are always a cause for concern. Besides specific factors in the must, one explanation is that several yeast strains are involved, and fermentation kinetics may be affected by antagonism between them (killer effect; Sections 1.7 and 3.8.1).

At the end of alcoholic fermentation, malic acid concentrations should be determined and monitored if necessary. MLF normally occurs after the complete depletion of sugars. An early onset of MLF is generally linked to alcoholic fermentation difficulties and insufficient sulfiting. In certain cases, the two fermentations take place simultaneously, even though the antagonistic phenomena between yeasts and bacteria tend to inhibit alcoholic fermentation.

Volatile acidity concentrations can be monitored to identify bacterial contamination. This is indispensable when MLF is initiated during slow alcoholic fermentations, before the complete depletion of sugar (Section 3.8). On rare occasions, apparently normal alcoholic fermentations produce excessive volatile acid concentrations. In this case, the fermenting must has most likely been contaminated by the fermentor or poorly maintained equipment. Spoilage bacteria can also produce volatile acidity before the start of alcoholic fermentation. After fermentation, its origin is more difficult to identify. Renovating the winery and replacing old equipment generally eliminates this contamination risk.

Yet the production of volatile acidity does not always indicate the presence of bacteria. Yeasts may also produce volatile acidity. In certain, as yet poorly defined, cases, excessive amounts of volatile acidity are produced (0.4–0.6 g/l expressed as H_2SO_4 or 0.5–0.7 g/l as acetic acid). By determining the concentrations of the two lactic acid isomers (L(+)-lactic acid and D(−)-lactic acid), the origin of the acetic acid can be identified. During fermentation, yeasts produce no more than a few dozen mg/l of the former and not more than 200 mg/l of the latter (Lafon-Lafourcade and Ribéreau-Gayon, 1977). Higher values indicate the involvement of lactic acid bacteria (LAB) in the production of volatile acidity, but the possibility of relatively high volatile acidity production levels by yeasts should always be considered. Moreover, the standard methods used to protect against bacterial spoilage have no effect on yeasts. Additionally, wine acidification tends to increase yeast-based volatile acidity production. Certain yeasts have an increased capability for production of volatile acidity, which attains a maximum in the course of fermentation, tending to decrease toward the end. In red winemaking, an excessive temperature (28°C) at the onset of fermentation contributes to elevated volatile acidity levels.

The final stages of fermentation should be closely monitored. When the specific gravity drops below 1.000, this measurement is no longer sufficient to measure the evolution of the fermentation precisely. Moreover, the relationship between any residual sugar and specific gravity is complex. When fermentation is complete, wine specific gravity can vary between 0.991 and 0.996, according to alcohol content. In addition, free-run wines always have a lower specific gravity than press wines, which are rich in extracted constituents.

The completion of fermentation is verified by chemically measuring the sugar concentration. For a long time, the reducing property of sugars was exclusively used to determine their concentration, but methods based on this characteristic also measured other substances in addition to fermentable sugars (glucose and fructose). Due to this "interference," fermentations were considered complete when these methods indicated less than 2 g of sugar per liter. This value signified the presence of less than 2 g of reducing agents per liter, including, among other substances, glucose and fructose. Due to a higher quantity of reducing agents in press wines and wines made from moldy grapes, fermentations are considered complete at approximately 3 g/l in these cases. Today, glucose- and fructose-specific analysis methods are available. Their overall concentration should not exceed a few hundred milligrams per liter when the fermentation is complete. This is necessary to avoid spoilage due to the development of contaminant yeasts (*Brettanomyces*) during barrel aging (Volume 2, Sections 8.4.5 and 8.4.6).

12.5 Maceration

12.5.1 The Role of Maceration

Red wines are macerated wines. Maceration is responsible for all of the specific characteristics of sight, smell, and taste that differentiate red wines from white wines. Phenolic compounds (anthocyanins and tannins) are primarily extracted, participating in the color and overall structure of wine. Aromas and aroma precursors, nitrogen compounds, polysaccharides (in particular, pectins), and minerals are also released in the must or wine during maceration.

The corresponding chemical elements come from the skins, seeds, and sometimes the stems. Flavor-wise and chemically speaking, each of these organs supplies different phenolic compounds. The aroma and taste differences are confirmed by tasting wines made in the presence of one or more of these grapevine organs. Stems give wine herbaceous flavors. Skin contact alone produces a supple but incomplete wine with light body. Skin and seed contact makes a more balanced wine that is capable of longer maceration times and is softer in the mouth. The phenolic compounds of each organ also vary according to variety, maturation conditions, and other factors. Furthermore, in the same organ (for example, in the grape skin), herbaceous, vegetal, and bitter substances along with leafy and grassy substances are located alongside phenolic compounds favorable to wine quality. Fortunately, the latter substances are extracted before the others.

Consequently, the maceration should be modulated and fractionated. Only useful grape constituents should be dissolved—those positively contributing to wine flavor and aroma. The extraction of these desirable substances should be total.

The concentration of substances in grape tissues detrimental to wine quality increases as grape quality diminishes. This phenomenon can be verified by chewing a grape skin after the pulp and seeds have been removed by pressing the berry between the thumb and index finger. Initially mild flavors evolve toward mellow tannins. Afterward, vegetal sensations become increasingly bitter and aggressive. The transition rate from pleasant to unpleasant sensations varies according to grape quality. The evolution of tannin quality can be evaluated during maturation in this manner. The same experiment conducted on seeds leads to similar results. Harshness and astringency diminish during maturation, while sensations of body and harmony increase.

An abundance of pleasant-tasting substances useful for winemaking and a lack of unpleasant ones characterize the grapes from top-ranked vineyards. These characteristics are most pronounced in years with good ripening conditions, i.e. great vintages. Such wines are capable of undergoing the most intense extractions and extended maceration times. The resulting high tannin concentrations are necessary to ensure their long-term aging. Lesser quality red wines, made for immediate consumption, have relatively short macerations—more flaws than qualities would result from longer macerations.

The extraction of pomace constituents during maceration should therefore be modulated according to grape variety and quality and also the style of wine desired (Volume 2, Section 6.6). Yet each grape crop is capable of producing a given type of wine, depending on natural factors (i.e. *terroir*).

For example, rotundone (pepper aroma) has been shown to accumulate exclusively in the exocarp of the grape berry skin (Vespolina variety). Skin contact could modulate the peppery character of red wines produced from the grapes containing this compound. Rotundone extraction after vinification is relatively low. Only 10% of rotundone present in grape clusters is extracted during a nine-day fermentation/maceration at 20°C, and only 6% will then be found in the bottled wine (Caputi et al., 2011).

Premium wines require a tannin structure that should not compromise finesse and elegance. These wines are difficult to produce and require grapes of superior quality benefiting from great *terroirs* and great vineyards within those *terroirs*.

Light, fruity red wines are relatively easy to obtain—grape quality is not essential, but if grape quality (variety, ripeness, soundness, etc.) is insufficient, tannic red wines rapidly become heavy, coarse, and without charm. A limited maceration lessens the occurrence of unpleasant characteristics.

A number of methods are available to the winemaker to adjust extraction levels during maceration. They essentially influence tissue destruction and favor the dissolution of phenolic compounds (Volume 2, Section 6.6.2). Techniques are continually evolving, and engineers regularly propose new solutions. Current methods will be described later in this chapter, each one probably having preferential effects on one or more groups of extracted substances. For example, brutal crushing promotes the extraction of bitter and herbaceous substances. Percolation of must during pumpovers, on the contrary, favors soft, supple tannins. Extractability of the constituents of the various organs varies with several factors (variety, ripeness, etc.).

Enzymatic reactions, activated by grape enzymes, are involved in cell wall degradation. They favor the dissolution of their vacuolar contents (Section 11.7.3). Commercial enzymatic preparations have been developed to activate these phenomena; they have pectinase, cellulase, hemicellulase, and protease activities of diverse origins (Amrani-Joutei, 1993). These enzymes seem to favor the extraction of skin tannins over skin anthocyanins. They act on the tannins bound to the polysaccharides of the cell wall, giving the enzyme-treated wine a more full-bodied character than the control wine.

Preparations containing pectolytic enzymes with polygalacturonase, pectinesterase, pectin lyase, and rhamnogalacturonase activities may change the polysaccharide and oligosaccharide composition of wines. In Merlot wines from the Bordeaux region, oligosaccharides in control wines are partially methylated homogalacturonans, corresponding to the smooth regions of pectins. However, the enzyme-treated wines have rhamnogalacturonan-like structures linked to neutral side chains. Enzymatic preparations may cleave the rhamnogalacturonan backbone of hairy regions (rich in side chains) and demethylate and hydrolyze the smooth regions (Ducasse et al., 2011).

Touzani et al. (1994) obtained encouraging results from an enzymatic pool produced from laccase-free *B. cinerea* cultures. This preparation attacks cell walls and favors anthocyanin extraction over tannin extraction. It could therefore be interesting for making *nouveau-type* wines to be drunk young, fruity, and rich in color but not very tannic.

Future research will be required to determine the selective effect of this enzymatic extraction on the various phenolic compounds, according to conditions. Its effect, with respect to standard practices for regulating maceration, also needs to be explored. Regardless of the mechanisms involved, tasting has confirmed the interest of using enzymatic preparations for maceration during red winemaking.

These various results demonstrate the utility of a better chemical and organoleptic understanding of the molecules involved in maceration phenomena. The extraction of a specific combination of these molecules could thus be obtained depending on the maceration technique used. Certain practices may be beneficial to wine quality in some situations but not in others. Only fundamental research on the chemistry of phenolic compounds will be capable of giving definitive answers. These studies are complicated by the extreme complexity and the high reactivity of the molecules involved (Volume 2, Section 6.3).

In traditional winemaking, maceration occurs in the fermentor (*cuvaison*), while the pomace soaks in the juice. Alcoholic fermentation occurs in the juice, producing ethanol and raising the temperature. Both ethanol and temperature participate in the dissolution of pomace constituents.

12.5.2 Different Types of Maceration

There is a current trend distinguishing between the various types of maceration, in addition to standard extraction during fermentation:

1. High-temperature extraction prior to fermentation, as used in thermovinification (Section 12.8.3), either followed by normal fermentation or separate fermentation of the juice.
2. Low-temperature extraction (cold soaking) prior to fermentation, aimed at enhancing aroma complexity. The start of fermentation is postponed by maintaining low temperatures and an appropriate level of SO_2, as well as by delaying inoculation with active yeasts.
A more elaborate form of this technique consists in cooling the grapes to around 5°C, by injecting liquid CO_2 or dry ice and maintaining this temperature for 5–15 days. The temperature shock bursts the grape skin cells and releases intensely colored juice (Blouin and Peynaud, 2001). Once the must has been heated to normal temperature, fermentation proceeds as usual. The purpose of this technique is to obtain wines with high concentrations of phenolic and aroma compounds. The results of this rather laborious method are not universally appreciated. Further research is required to identify the conditions required to produce deep-colored, aromatic wines without any rustic or herbaceous character. Satisfactory results have been achieved with Pinot Noir (Flanzy, 1998), producing finer, fruitier wines. In any case, the results are better than those obtained with cold maceration, following microbial stabilization with considerable doses of SO_2, which gave deep-colored wines that were lacking in varietal character and tended to dry out on the end of the palate (Feuillat, 1997).
3. Post-fermentation maceration is required for the best premium quality red wines to extend skin contact after the end of fermentation, sometimes combined with an increase in temperature (final, high-temperature maceration; Section 12.5.5).

Different types of maceration have been compared: traditional, TM; cold soaking prior to fermentation CS, *délestage* (rack-and-return), DEL; Ganymede systems, GAN (use of fermentative carbon dioxide to dislodge the pomace cap); addition of enzymes and tannins, E+T; or addition of oak chips during alcoholic fermentation, OC. They were compared with regard to the phenolic composition and color characteristics of young red wines made from Mencia grapes in Galicia, Spain (Soto Vasquez et al., 2010). In this case, adding pectolytic enzymes (2.4 g/hl) and tannins (ellagitannins and grape procyanidins at 35 g/hl) results in red wines forming many pigmented polymers that are stable over time. Cold soaking prior to fermentation (10°C for four days), *délestage*, the Ganymede system and the addition of fresh oak chips (300 g/hl) may also be used to make early-drinking red wines. After alcoholic and malolactic fermentations, anthocyanin concentrations from the highest to the lowest in mg/l are 643 (TM), 628 (E+T), 612 (GAN), 578 (CS), 578 (OC), and 565 (DEL). Moreover, proanthocyanidin concentrations in g/l are 2.07 (E+T), 1.90 (OC), 1.70 (DEL), 1.59 (CS), 1.54 (TM), and 1.49 (GAN). In these tests, the variations explain the sensory differences observed in favor of enzyme and tannin use or the addition of oak chips with respect to the other methods.

12.5.3 Principles of Maceration

The passage of pomace constituents, particularly phenolic compounds (anthocyanins and tannins), into fermenting juice depends on various basic factors. The results constitute overall maceration kinetics. The phenomena involved are complex and do not cause a regular increase in extracted substances. In fact, among these various factors, some tend to increase phenolic compounds, while others lower concentrations. Moreover, they do not necessarily always act in the same manner on the various constituents of this family.

Maceration is controlled by several mechanisms (see also Volume 2, Section 6.6.1):

1. The extraction and dissolution of different substances. Dissolution is the passage of cell vacuole contents from the solid parts of the pomace into the liquid phase. This dissolution depends first of all on grape variety and ripeness levels. This is especially important for anthocyanins. In certain cases, strongly colored musts are obtained immediately after crushing. In other cases, a period of 24–48 hours is required. Tissue destruction through enzymatic pathways or crushing facilitates dissolution. The more intense the crushing, the more dissolution is favored. Finally, dissolution depends on the various operations that participate in tissue destruction: sulfiting, anaerobic condition, ethanol, elevated temperatures, and contact time.
2. Diffusion of extracted substances. Dissolution occurs in the pomace, and the impregnating liquid rapidly becomes saturated with extracted substances; exchanges therefore stop. Further dissolution is dependent on the diffusion of the extracted substances throughout the mass. Pumping-over or punching down the pomace cap renews the juice impregnating the pomace cap. This diffusion is necessary for suitable pomace extraction. It homogenizes the fermentor and reduces the difference between the phenolic concentrations of free-run wine and press wine.
3. Refixation of extracted substances on certain substances in the medium: stems, pomace, and yeasts. This phenomenon has been known since Ferré's (1958) observations (Section 12.2.3).
4. Modification of extracted substances. This hypothesis still requires further theoretical interpretations. Anthocyanins may temporarily be reduced to colorless derivatives (Ribéreau-Gayon, 1973). The reaction appears to be reversible, since the color of new wines exposed to air for 24 hours increases, with the exception of those made from moldy grapes. Anthocyanin–Fe^{3+} ion complex formation may be involved in this color increase in the presence of oxygen. Ethanol may destroy tannin–anthocyanin associations extracted from the grape (Somers, 1979). Under the same environmental conditions, free anthocyanins are less colored than tannin–anthocyanin combinations, which are formed again during aging and ensure color stability.

The quantity of anthocyanins and tannins found in wine depends first of all on their concentration in the grape crop. Ripe grapes are the first condition for obtaining rich and highly colored wines. However, only a fraction of the phenolic potential of the grape is found in wine. Their concentration depends not only on the ease of phenol extraction but also on the extraction methods used. The phenol concentrations of various components of grape clusters and wine have been compared. Approximately 20–30% of the phenolic potential of grapes is transferred to wine. The loss is significant, and efforts have been made to improve this yield, but, due to the complexity of this phenomenon and the molecules involved, a simple solution is difficult to find.

Finally, "bleeding" a tank (*saignée*) (Section 12.5.9) is a way of raising tannin levels by reducing volume. Eliminating water by other techniques (Section 11.5.1) achieves similar results but keeps all the sugar in the must.

In red winemaking, maceration must be adapted to suit the nature and composition of the grapes (Volume 2, Section 6.6.2).

12.5.4 Influence of Maceration Time

The dissolution of phenolic compounds from solids into fermenting must varies according to maceration time, but no proportional relationship between maceration time and phenol concentration exists. CI has even been observed to diminish after an initial increase during the first 8–10 days (Ferré, 1958; Sudraud, 1963). The graphs in Figure 12.9 depict the evolution of a maceration extended well beyond normal conventions. In this laboratory experiment, the CI passes through a maximum on the eighth day and then diminishes. The evolution of total phenolic compounds is different: during an initial phase lasting a few days, their increase in concentration is rapid and then slows afterward. This behavioral difference is due to tannin concentrations (skins and seeds) in the grape crop being 10 times greater than

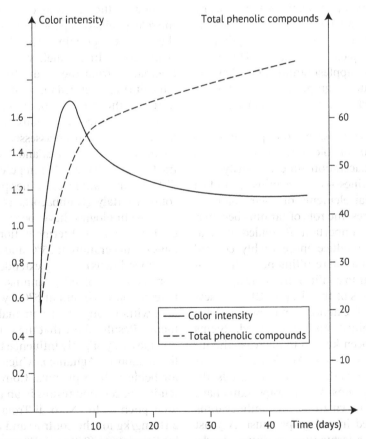

FIGURE 12.9 Evolution of color intensity and phenolic content of red wines as a function of maceration time (Ribéreau-Gayon *et al.*, 1970). Color intensity is defined as the sum of the optical densities at 420 and 520 nm at 1 mm thickness, (CI = OD 420 + OD 520). Total phenolic compounds are determined by the permanganate index.

anthocyanin concentrations. In both cases, the concentrations increase during the first few days. Afterward, tannin losses are proportionally less significant, and an overall increase is always observed. Certain varieties have very low tannin concentrations. In this case, tannins evolve similarly to color (Figure 12.9). Other experiments have shown that the nature and properties of tannins vary according to maceration time.

Similarly, the various grape organs (skins, seeds, and stems) contain specific phenolic compounds. Their extraction varies according to diverse conditions. Skin anthocyanins are extracted first; ethanol is not required for their dissolution. Skin tannin extraction begins soon after, facilitated by the increasing presence of ethanol during fermentation. A relatively long maceration is necessary for seed tannin extraction. The presence of ethanol is required to eliminate lipids beforehand. The skins contain the supplest tannins, but they can become bitter if grape ripeness is insufficient. Seed tannins are harsher but less bitter.

Of course, these notions pertaining to the evolution of color and anthocyanins in terms of maceration time primarily concern new wines—anthocyanins are in fact the essential elements of their color. As wine matures, the role of tannins becomes increasingly important. Extended maceration times produce more highly colored wines, even if the resulting new wines initially appear to confirm the contrary.

The causes of this drop in CI, after several days of extended maceration, have been examined and interpreted. Stems have long been known to decrease the intensity of wine color (Section 12.2.3); this phenomenon is the result of their adsorbing anthocyanins. White grape skins have also been shown to adsorb anthocyanins when placed in red grape must. A yeast biomass in a fermenting medium adsorbs both anthocyanins and tannins.

Chemical reactions also diminish CI. Grape tannin–anthocyanin combinations are destroyed, and anthocyanins are reduced to colorless forms during these reactions (Section 12.5.1).

These facts lead to an important conclusion. On approximately the eighth day of maceration, wine CI is at its maximum and tannin concentrations are limited, permitting fruity sensations to be conserved. This maceration time (*cuvaison*) is best adapted to wines for early drinking. In contrast, extended maceration times produce rich, tannic wines capable of extended aging, but the resulting high tannin concentrations require grapes of high quality.

Recent works (Cretin, 2016) have shown that post-fermentation maceration at 30°C leads to increased smoothness (perception of sweetness) and decreased astringency in red wines. This phenomenon could be explained by the autolysis of yeast, causing the release of sweet peptides resulting from the yeast protein Hsp12 (Marchal *et al.*, 2011). In parallel, sweet molecules originating from the seed diffuse into the wine during extended maceration and contribute to the sweet flavor of dry wines.

The influence of various winemaking technologies has been assessed in terms of chemical characteristics and the phenolic fraction of Aglianico, Montepulciano, Nero di Troia, and Sangiovese wines produced in southern Italy (Gambacorta *et al.*, 2011). Four technologies have been compared: control (traditional fermentation with five days of maceration at 25°C and three days of punchdowns), 10-day extended maceration, addition of ellagitannins, and cryomaceration (24 hours at 5°C by using dry ice), without any other winemaking treatments. Results show that the various technologies only slightly influence the phenolic fraction of Aglianico, which is known for being rich in phenols. Conversely, extended maceration results in an increase in total phenols for Nero di Troia (2,592 vs. 2,115 mg/kg for the control) and a decrease for Sangiovese (869 vs. 1,013 mg/kg for the control). Adding tannins leads to an increase in total phenols for Montepulciano (1,358 vs. 1,216 mg/kg for control) and to a decrease for Sangiovese (916 vs. 1,013 mg/kg). Finally, cryo-maceration results in decreased anthocyanins for all varieties (about 15% less).

12.5.5 Influence of Pump-overs (*Remontage*) and Punchdowns (*Pigeage*)

Pumping-over is important in red winemaking, at least for certain varieties such as Merlot and Cabernet Sauvignon. Section 12.4.2 describes the objectives and steps of this operation. In addition to introducing oxygen, it plays a major role in extracting compounds from the pomace and homogenizing the contents of the tank. When the tanks are emptied at the end of maceration, the free-run wine and press wine have more similar tannin concentrations when pump-over operations were carried out. The numbers in Table 12.4 show the dual effect of pump-overs. Increasing the number of such operations accentuates these effects. An effective pump-over should thoroughly wet the pomace cap. Precise experiments have shown that phenolic contents and CI vary significantly according to pump-over conditions. Rack-and-return (*délestage*) can be recommended and consists in draining off part of the fermenting must, then reintroducing it all at once over the broken-up pomace.

Pumping-over has an additional advantage that is more difficult to evaluate. This operation does not affect tissue integrity and promotes the extraction of higher-quality tannins, at least in some grape varieties. It imparts a rich and full-bodied structure to wines, without bitterness and vegetal characteristics.

The timing of pump-overs influences the selective extraction of skin and seed tannins. Skin tannins are more easily released and are sufficient for making *nouveau-type* wines, but seed tannins are necessary to obtain a premium wine. Pump-overs in the final stage of maceration are used to extract these compounds. The advantages of this technique have led to its general use. However, in certain cases, it is abused, producing aggressive and unpleasantly tannic wines; the resulting press wines are thin and unusable.

Gas injection systems have been proposed to replace traditional pumping (Section 12.4.2, Figure 12.8) for the extraction of pomace constituents. These systems

TABLE 12.4

Influence of Pumping-over on Color and Tannin Dissolution in Open Fermentors (Sudraud, 1963)

Time	No pumping-over		Two pump-overs, on days 1 and 3	
	Tannin (permanganate index)	Color intensity[a]	Tannin (permanganate index)	Color intensity[a]
Maceration for 3 days	39	0.83	46	0.93
Maceration for 6 days	43	0.87	48	0.98
Maceration for 10 days	45	0.89	52	1.04
Free-run wine at end of maceration	48	0.93	56	1.16
Press wine at end of maceration	102	1.35	95	1.30

[a] See Table 12.1.

consist in introducing a specially adapted pipe into the lower part of the tank. The pipe injects carbon dioxide, nitrogen, or filtered air into the fermentor. The tank contents are churned up briskly. This operation is much more rapid than traditional methods, and a period of 1 minute/100 hl of grape crop at a pressure of 3 bar is recommended. The technique should not be used during the final stage of fermentation (when the specific gravity falls below 1.010). Continuing injections beyond this value would lead to the definitive disintegration of the pomace cap.

In certain cases, punchdowns can replace pump-overs. As its name suggests, a punchdown consists in plunging the skin cap completely into the liquid. This immersion results in the disintegration of the cap and increases maceration phenomena. The process promotes seed tannin extraction and thus increases the tannic structure of wine. For certain varieties (e.g. Pinot Noir), punching down the cap produces better results than pumping-over, but there is some concern that it may give a slightly "rustic" character to Cabernet Sauvignon and Merlot wines, especially if used to excess. Various types of tanks can be equipped with automatic punchdown systems (Blouin and Peynaud, 2001). We can also use various mechanical stirring processes (Section 12.5.8). Finally, nitrogen injection simultaneously conducts pumping-over and punching.

12.5.6 Influence of Temperature

Heat is a means of degrading tissues. It increases the dissolution of pomace constituents and accelerates maceration. The technique of heating crushed red grapes has been used for a long time and is particularly important in thermovinification (Section 12.8.3).

In addition to this extreme case, temperature is an essential factor in standard maceration. It should be sufficiently high to ensure satisfactory extraction of phenolic compounds. The experiment (Sudraud, 1963) in Table 12.5 clearly illustrates its impact.

Both the average and maximum temperatures affect extraction. The results of a laboratory experiment in Table 12.6 indicate the simultaneous influence of maceration time and temperature. When maceration is extended, an elevated temperature can exacerbate the drop in anthocyanin concentrations and CI.

Moderate maceration temperatures (25°C) are preferred for the production of *nouveau-type* wines. These are made to be drunk young, and this approach gives them a good color while conserving their fruity aroma. Moderate temperatures are also recommended when there is a risk of fermentation difficulties (high sugar concentration). High temperature (30°C) extracts the tannins required to produce a premium

TABLE 12.5

Influence of Fermentation Temperature on Dissolution of Phenolic Compounds (Sudraud, 1963)

Temperature (°C)	Total phenols (permanganate index)	Color intensity[a]
20	44	0.71
25	48	0.87
30	52	0.96
20–37 (average 29.5)	52	1.21
25–37 (average 32.6)	60	1.43

[a] See Table 12.1.

TABLE 12.6

Influence of Maceration Temperature on Dissolution of Phenolic Compounds (Ribéreau-Gayon et al., 1970)

Time and temperature	Hue[a]	Color intensity[a]	Anthocyanins (g/l)	Tannins (g/l)	Total phenol (permanganate index)
Maceration for 4 days					
Temperature: 20°C	0.54	1.04	0.54	2.2	39
25°C	0.52	1.52	0.63	2.4	45
30°C	0.58	1.46	0.64	3.3	55
Maceration for 8 days					
Temperature: 20°C	0.45	1.14	0.59	3.0	43
25°C	0.56	1.62	0.61	3.2	48
30°C	0.56	1.54	0.62	3.6	55
Maceration for 14 days					
Temperature: 20°C	0.53	1.16	0.49	2.5	48
25°C	0.51	1.36	0.59	3.5	58
30°C	0.56	1.44	0.58	3.8	59
Maceration for 30 days					
Temperature: 20°C	0.56	1.45	0.38	3.5	63
25°C	0.67	1.20	0.39	3.7	67
30°C	0.80	1.47	0.21	4.3	72

[a] See Table 12.1.

wine capable of long aging; higher temperatures would promote further extraction but would also compromise yeast activity—they should thus be used with caution.

In stainless steel tanks, especially if they are installed in the open air, excessively low fermentation temperatures may cause problems in certain years in some climates. These temperatures do not permit sufficient maceration. Moreover, as soon as the fermentation stops, such tanks can rapidly drop in temperature and no longer produce heat. The temperature in wooden tanks changes differently.

Tank temperature control systems have permitted an almost perfect regulation of the maceration temperature during fermentation. Cool grape crops and excessively cool fermenting juice can be warmed to a suitable temperature, but since an integrated heat exchanger warms only a portion of the tank contents, a pump-over is required to homogenize the tank. Temperature control must of course be set according to fermentation kinetics. For example, the initiation of fermentation requires a moderate temperature (20°C), since yeast is relatively heat sensitive during its growth phase (Sections 3.7.1 and 3.8.1), and there is a greater risk of stuck fermentation at that stage. A moderate temperature must be maintained (about 25°C in the liquid phase) during alcoholic fermentation. After fermentation, maceration may be extended at a higher temperature (30°C) to accelerate enzymatic mechanisms of yeast autolysis. This causes a better extraction of mannoproteins, which participate in the tartaric stabilization of wines, as well as sweet peptides resulting from the Hsp12 protein.

Glories *et al.* (1981) developed a technique called hot post-fermentation maceration. The objective of this method is to separate the fermentation phase at a moderate temperature from the maceration phase. After fermentation, the tank contents are warmed to between 35 and 40°C for several days. In this process, only the wine is directly heated, with the pomace cap being heated indirectly. The wine is heated to between 50 and 60°C during the entire maceration period. This process has not been observed to alter the taste of wine. In addition, MLF takes place normally. As long as this method is conducted in the absence of residual sugar, there is no risk of bacterial spoilage resulting in increased volatile acidity. The method most significantly affects phenol concentrations and CI, which are greatly increased by hot maceration after fermentation.

In the experiment in Table 12.7, tasting reveals wine A to be diluted because of high crop yield and insufficient ripeness. Wine B is more full bodied and has more taste but no vegetal character. Wine B is much superior when compared with wine A, which is clearly lower in quality. Applying hot post-fermentation maceration to wines that are already naturally rich greatly intensifies their concentration, which could be interpreted as an improvement, except that this concentration is often at the expense of finesse in the mouth: the tannins rapidly become hard and rustic, and this tannic astringency tends to increase during aging. The repeated pump-overs required to maintain the temperature during this process can magnify these flaws. Suspended solids are often produced, and press wines often become unusable. The above observations emphasize the need to carefully consider use of hot post-fermentation maceration.

Temperature drops in steel tanks should be avoided after the completion of fermentation. A post-fermentation maceration at 30°C for several days often favors wine quality. This approach emulates the temperature conditions in wooden tanks following fermentation. Hot post-fermentation maceration must not be confused with thermovinification (Section 12.8.3).

Temperature control, coupled with pump-overs and extended maceration time (another very sensitive tool at the winemaker's disposal), is another means of modulating extraction during maceration. Temperature regulation can profoundly modify the tannic structure of wine. But there is a risk of also increasing the rustic character of the tannins. The nature of the grape crop and the type of wine desired are criteria that should be considered when determining optimum temperature regulation.

12.5.7 Effect of Grape Sulfiting and Alcohol Produced by Fermentation

The impact of must sulfiting on pigment extraction is covered in Section 8.7.5. Sulfur dioxide destroys cell tissue and promotes the dissolving of pomace constituents, but in traditional winemaking with

TABLE 12.7

Effect of "Hot Post-fermentation Maceration" on Wine Composition

Component	Traditional winemaking (wine A)	Hot post-fermentation maceration (4 days at 45°C) (wine B)
Color intensity[a]	0.45	0.67
Hue[a]	0.82	0.75
Total phenols (Folin–Ciocalteu index)	29	38
Anthocyanins (mg/l)	273	329

[a] See Table 12.1.

ripe, healthy grapes, pumping-over, temperature, and maceration time have a greater impact. All things considered, SO_2 has little influence on the CI and phenol content of wines.

However, the solvent effect of SO_2 is evident in rosé winemaking, since the phenol content is low in this case. In fact, SO_2 can be detrimental to white winemaking. This solvent effect may also affect red grapes if they are insufficiently ripe, and pigment extractability is poor. In that case, sulfiting facilitates anthocyanin extraction in the early stages, especially during cold soaking (Section 12.5.2).

With moldy grapes, sulfiting does not improve the extraction of pigments; instead, it prevents the laccase activity of *B. cinerea* from destroying them. The numbers in Table 12.8 show that high sulfur dioxide concentrations increase the total phenol content and CI in the case of highly contaminated grapes. The elevated hue value in the control sample is due to a yellow component, characteristic of enzymatic browning.

The impact of the ethanol produced by fermentation seems to be complex. According to Somers (1979), it is involved in the decrease in CI observed during maceration. The mechanism (Section 12.5.2) seems to correspond to the destruction of tannin–anthocyanin combinations, resulting in the liberation of free anthocyanins, which are less colored. At the same time, alcohol is considered to participate in tissue destruction and, as a result, in the dissolution of pomace constituents (Table 12.9). In large wineries containing many fermentors, with relatively homogeneous grape crops, the wines with the highest tannin concentrations and color intensities are often observed to have the highest alcohol content.

TABLE 12.8

Influence of Sulfiting Moldy Grapes on Phenolic Compounds of the Resulting Wines (Sudraud, 1963)

	Total phenols (permanganate index)	Color intensity[a]	Hue[a]
Control	32	0.53	0.76
Sulfited at 10 g/hl	41	0.63	0.42
Sulfited at 20 g/hl	55	0.83	0.43

[a] See Table 12.1.

TABLE 12.9

Influence of Alcohol on Extraction of Pomace Phenolic Compounds in a Model Solution (10 days of Maceration at 20°C, pH 3.2) (Canbas, 1971)

Alcohol (% vol.)	Tannins (g/l)	Total phenols (permanganate index)	Anthocyanins (mg/l)	Color intensity[a]
0	0.66	12	169	1.95
4	0.96	16	214	3.60
10	1.32	20	227	6.35

[a] See Table 12.1.

12.5.8 Impact of Various Mechanical and Physical Processes Acting Directly on the Pomace (Flash Détente) and Pulsed Electric Fields

The impact of crushing grapes has already been covered in Section 12.2.2. Energetic crushing increases the diffusion of solid tissue components, but, as a general rule, the corresponding tissue destruction promotes the extraction of inferior-quality tannins. These tannins impart vegetal and herbaceous tastes to the must and resulting wine. Unfortunately, no measures are currently available to confirm this fact. Yet when compared with wines made from energetically crushed grapes, pump-over operations have been observed to favor the extraction of soft and more pleasant tannins. The extent of the negative effect caused by excessive crushing depends on the grape variety and its degree of vegetal character. However, all things being equal, quality grape crops are much more sensitive to crushing intensity than ordinary grapes.

Similarly, the breaking up and punching down of the pomace cap has long been used to increase maceration while protecting the wine from bacterial growth. In the past, these operations were common and carried out manually in small-capacity open fermentors. They are no longer possible in today's large fermentors, but tanks can now be equipped with mechanical devices (screw, helix, jack, and piston based), which ensure the breaking up, reshaping, and punching down of the pomace cap (Blouin and Peynaud, 2001). The pomace cap can also be broken up by a mechanical claw, which is also used for emptying pomace from the tank for pressing. When the claw is not functioning, the tank is closed by a removable cover. The operation of these various systems is fairly complex, but they are effective in terms of extracted substances. They can be applied to certain varieties (Pinot Noir; Section 12.5.4), but with other varieties, these methods can rapidly lead to excessively hard and unpleasant tannins.

Rotating cylindrical fermentors have similar constraints. A fixed device inside the tank (for example, a helical fin) breaks up the cap and is also used for discharge the skins from the fermentor. This kind of fermentor is relatively expensive. It has the same advantages and drawbacks as the equipment described in the above paragraph. Grape crops with low concentrations of phenolic compounds are extracted rapidly. In other situations, a satisfactory extraction is also obtained rapidly, permitting the fermentor to be used several times during the same harvest. This system does not necessarily give satisfactory results for producing premium wines.

Another developed process (IMECA DIF, 34800 Clermont-L'Hérault, France), aimed at intensifying maceration of pomace in grape must, is known as *Flash Détente* (Boulet and Escudier, 1998). It consists in bringing the crushed grapes rapidly to a high temperature and then chilling them almost instantaneously in a strong vacuum. It may be considered a variation on thermovinification (Section 12.8.3). Under these conditions, the skin cell tissue structure is completely degraded, and components essential to wine quality (phenolic compounds, polysaccharides, aromas, etc.) are rapidly released during later maceration.

The grapes are destemmed and part of the juice separated out. As soon as the crushed grapes come into the heating chamber, it is brought up to a high temperature (70–95°C) very rapidly by direct injection of saturating steam at 100°C, produced using the reserved juice. The must is then transferred by means of a positive displacement pump into a high vacuum chamber connected to a vacuum pump coupled with a condenser. When the hot must enters the vacuum chamber, the water it contains is vaporized, cooling it rapidly to the boiling point of water under the vacuum conditions used, i.e. 30–35°C.

According to Boulet and Escudier (1998), this sequence has three consequences: the crushed grapes are cooled in less than one second, the grape cell walls are broken, and there is only a very low concentration of residual oxygen in contact with the must.

The water recovered from the vacuum chamber is condensed and represents 7–12% of the total volume of the must. Part of this is used to produce the hot steam used in the system, in addition to that made from the separated juice. The rest may be added back into the must, making it more concentrated. This is only possible if permitted by regulation. This technique does not dilute the must, as the steam used to heat the must is made from grape juice; however, it may concentrate the must, as part of the water is eliminated.

If the treated grapes are pressed immediately, the resulting wine is similar to that produced by fermentation combined with standard techniques for heating the must (Section 12.8.3). If, however, the must is left on the skins at a suitable temperature after *Flash Détente* treatment, the pomace extraction kinetics are much faster than in a normal winemaking situation, reaching maximum anthocyanin and polyphenols levels after three to four days.

This high-temperature treatment destroys laccase, so it is extremely suitable for grapes affected by *B. cinerea*. Pectolytic enzymes are also partially destroyed, so the must becomes viscous after *Flash Détente* and is difficult to drain. Pressing is facilitated by mixing treated must with a sufficient quantity of untreated juice.

Generally speaking, according to Boulet and Escudier (1998), wines treated by *Flash Détente* contain 30–60% more polyphenols than controls, even after the traditional three-week skin contact time.

The resulting wines have more intense color and a better tannic structure. They have different aromas, but do not lose their varietal character. This winemaking technique is obviously better suited to some grape varieties than others. A better understanding of the nature and properties of the substances extracted under different maceration conditions will lead to the development and use of mechanical techniques enhancing high-quality maceration phenomena. In this sector, as in many others, empiricism has preceded research. There is currently no theoretical knowledge of these phenomena that permits the explanation and prediction of the results observed.

The effects of applying a continuous treatment using pulsed electric fields (PEF treatment) to Cabernet Sauvignon pomace on the color evolution and phenol content of red wines have been studied (Puértolas et al., 2010). After crushing and destemming, grapes were treated by PEF, which consisted in 50 pulses from a 5 kV/cm electric field (total specific energy: 3.67 kJ/kg) at a frequency of 122 Hz. The temperature is measured at both the inlet and the outlet of the treatment chamber. In all experiments, the temperature increase after treatment never exceeds 2°C. At the end of alcoholic fermentation, wines produced from PEF-treated grapes present higher CI, total polyphenols (TPI), and total anthocyanin content (TAC) than the control wine. This effect has been observed even if the maceration time for wines from PEF-treated grapes is 48 hours less than the control wine. Differences remain during MLF and aging. After four months of bottle aging, the CI, TPI, and TAC of wine produced from PEF grapes were, respectively, 38, 22, and 11% more than the control. Phenolic profiles of wine were qualitatively similar, without detection of any selective effect on different phenols (with increases for some markers: malvidin 3-glucoside, malvidin 3-acetylglucoside and delphinidin 3-O-(6-p-coumaroyl)glucoside, catechin and epicatechin monomers, and gallic acid). Besides astringency that appears to be intensified, no other significant differences have been detected on sensory characteristics between the wines. Results indicate that PEF represents a future technology of interest for red winemaking. It helps reduce maceration time and increase CI and phenolic extraction.

12.5.9 The Maceration Process: Grape Quality and Tannin Concentrations in Wines

Grape quality

In Section 12.1, the importance of grape quality was emphasized. It directly influences the consequences of maceration. The quantity and quality of phenolic compounds are directly related to grape variety, *terroir*, ripeness, soundness, etc.

Proper maturation conditions are essential to the accumulation of phenolic compounds. The climate plays a major role in phenolic compound production, since this requires a considerable amount of energy. Viticultural techniques also affect maturation. Moreover, phenolic compound accumulation is limited in young vines; therefore, relatively old vines are necessary for premium wine production.

Crop yields also greatly affect the accumulation of tannins and anthocyanins, but this factor must be interpreted carefully. In some cases, the same climate criteria that favor quality can also favor quantity. In vintage-dependent, temperate vineyards, the best quality years are sometimes also the most abundant. Reciprocally, low-yield vintages do not necessarily produce the best quality grapes. When considering crop yields, plant density should also be taken into account: must sugar concentrations have long been known to diminish when per vine production increases. Vineyards with a long tradition of quality choose to maintain their plant density at 10,000 vines/ha in poor soils. In this manner, satisfactory production is ensured while respecting grape quality. In richer soils, lower plant densities decrease cultivation costs (as low as 2,000 vines/ha, and sometimes even lower). As a result, to have the same production per hectare, higher yields per vine are required. Under satisfactory climate conditions, the grapes on these vines ripen normally, producing a relatively large harvest, but under less satisfactory maturation conditions, crop volume is more apt to delay ripening with low plant densities than with high plant densities.

The relationship between vine production and maturation conditions (in particular, phenol content) is complex and difficult to interpret. Practices that increase vine vigor (fertilizing, rootstock, pruning, etc.) are known to delay maturation. Phenolic compounds are the first substances affected. When production is excessive, wines rapidly become diluted and lack color. Some grape varieties (e.g. Cabernet Franc) are more susceptible to flavor dilution due to excessively high yields than others (e.g. Cabernet Sauvignon). A carefully measured equilibrium between an acceptable selling price and optimal wine quality will determine crop yields and consequently the future of premium red wines. The future of great red wines no doubt depends on maintaining a balance between producing enough wine to achieve reasonable profitability and keeping yields low to optimize quality. A few Bordeaux winegrowers have started keeping their yields extraordinarily low—about half the normal level (20–30 hl/hectare)—to produce extremely concentrated wines that find a market at unusually high prices.

Excessively vigorous vines and excess rain, leading to berry swelling, also cause abundant harvests. Various techniques are available to mitigate the resulting defects. The first is cluster thinning, which consists in eliminating a portion of the grape clusters between fruit set and veraison. Cluster thinning should preferably be carried out near veraison. At this time, grapes manifesting physiological retardation can be eliminated; the vegetation is also less affected. This green harvest, however, is difficult work, and its effectiveness is limited. The retained grapes swell to compensate for the thinning. When 30% of the grapes are removed, crop yields generally decrease by only 15%. Vine vigor and pruning should preferably be regulated to ensure a crop yield corresponding to quality grapes.

Eliminating a fraction of the must can also increase tannin concentration of dilute red grape must. This method increases the ratio of skins and seeds (pomace) to juice. A few hours after filling the fermentor, as soon as the juice and solids can be separated, some of the juice is drawn off (approximately 10–20% of the total juice volume). This operation significantly affects tannin concentrations and CI. The method should be used with caution, as excessive concentration of the must can lead to exaggeratedly aggressive tannins. The volume of must to be drawn off depends on skin quality, ripeness, the absence of vegetal character, and the soundness (lack of mold) of the grapes. The drawn-off juice can be used to make rosé wine. It is not advisable to throw away the excess must or rosé wine in order to avoid pollution. In any case, it constitutes a loss of production, which must be compensated by producing a better-quality wine capable of fetching a higher price.

Siegrist and Léglise (1981) obtained data illustrating the importance of the solid parts of the grape. In their study, a Pinot Noir must containing 60% juice and 40% pomace is higher in quality than a similar must containing 80% juice and 20% pomace.

Instead of drawing off juice, it is now possible to eliminate water directly from the grape must (Section 11.5.1). Two methods currently exist: the first circulates the must across membranes that retain water by reverse osmosis (Degremont); the second evaporates water in a low-temperature (20–24°C) forced vacuum (Entropie). These techniques have the additional advantage of increasing the sugar concentration, thus eliminating the need for chaptalization in some cases.

In addition to the purely quantitative aspect of phenol content in grapes, the nature and properties of these substances also play an essential role in the maceration process and its consequences.

The potential dissolution of skin pigment varies, especially according to ripeness. Phenolic maturity corresponds to a maximum accumulation of phenolic compounds in the berry. Cellular maturity is defined with respect to the level of cell wall degradation (Volume 2, Sections 6.5.3 and 6.5.4). Extraction of phenolic compounds increases with this degradation level (Amrani-Joutei, 1993). Augustin (1986) defined the anthocyanin extraction coefficient (AE) as follows:

$$AE = (\text{Wine anthocyanins} / \text{Grape anthocyanins at ripeness}) \times 100$$

This coefficient varies from one year to another. For example, the following values were obtained for Merlot and Cabernet Sauvignon: 46.5 in 1983, 26.5 in 1984, and 39.0 in 1985. Moreover, these values correspond fairly well with ripeness, expressed by the ratio (sugar concentration)/(total acidity). The same coefficient varies less for tannins (26.8 in 1983, 23.2 in 1984, and 30.7 in 1985). It is not correlated with juice maturity.

The organoleptic quality of tannins is directly related to maturation conditions. Enologists have defined this quality in terms of "good" tannins and "bad," or harsh, tannins (Section 12.5.1). The chemical understanding of these phenolic compounds has made it possible to make a better choice of winemaking techniques that optimize the quality of various kinds of grapes. A perfect state of phenolic maturity not only supposes a maximum tannin concentration but also corresponds to soft, nonaggressive, nonbitter tannins.

Environmental conditions (*terroir* and climate) and grape variety determine this phenolic maturity, which can be illustrated by Cabernet Sauvignon. In cool climates, its insufficiently ripe tannins take on a characteristic vegetal note. The same flaw can occur in excessively hot climates: the rapid sugar accumulation forces harvesting before the tannins reach their optimum maturity. A harmonious maturation of the various constituents of the grape characterizes great *terroirs* and great vintages. When conditions permit, grapes should never be harvested before complete phenolic maturity. Harvest dates based on sugar/acid ratios should be delayed, when necessary, so that tannins may soften. To ensure this maturation, several more days are sometimes needed before harvesting. During this period, grapes should be protected against *Botrytis* attacks in certain situations; in other situations, excessively high sugar concentrations should be avoided by close monitoring.

Wine tannin concentration

By taking into account the previously discussed concepts, general red winemaking principles can be improved for the better control of maceration time and intensity.

If grapes have low anthocyanin and tannin concentrations, only light red wines should be made. These wines, however, should be fresh and fruity. A limited concentration of grape phenolic compounds nevertheless deserves an explanation. It can be a varietal characteristic, which must be taken into account. Vine cultivation conditions, favoring crop yields over quality, can also be responsible. Suitable winemaking techniques are necessary in these cases. Techniques for compensating a phenolic deficiency are palliative and are not a substitute for perfect grape ripeness.

Grapes rich in phenolic compounds are capable of making premium wines enabling many years of aging. Tannins play at least as important a role in wine aging potential as alcohol or acidity. However, tannin quality also contributes to aging potential. For example, common varieties, incomplete ripeness, and unsound grapes contribute aggressive phenolic elements that should be limited, if not totally avoided. Viticultural traditions have led to the establishment of the longest maceration times in the best *terroirs*. Reciprocally, rosé wines should be made from grape crops whose quality does not improve with maceration. Intermediate techniques can also be used.

In the 1970s, the great Bordeaux wines were considered to have insufficient tannic structure. The young wines did not taste well, and there was concern that they would not age as gracefully as older vintages. There were certainly significant changes in vineyard management practices during this period, leading to higher yields and less concentrated must. However, at the same time, progress in winemaking improved management of the fermentation process. The resulting clean, fruity wines no longer needed many years' aging for certain defects to be attenuated. Nowadays, thanks to recent developments in vineyard management and winemaking techniques, Bordeaux wines have good structure and are already enjoyable immediately after vinification. It is possible to judge these wines and evaluate their quality when they are still young. The commercial value of these wines is often established within a few years of their production, when offered to the market. An unpleasant-tasting young wine would be difficult to sell in today's market by simply arguing that it should improve with considerable aging.

Despite the pleasant taste of present-day premium wines immediately following fermentation, they are still capable of long-term aging. Additionally, the number of well-made wines is much higher than in the past. Yet not all wines lend themselves to long-term aging. *Terroir* and vintage also participate in a wine's aging potential. Vineyard management conditions leading to high crop yields also limit the proper development of grape constituents. Truly great vintage wines, however, are fruity and enjoyable when young, although they have sufficiently high levels of good-quality tannins to age well for a remarkably long time. Although as pleasant as lighter new wines, these wines are capable of long-term aging. They are made from the grapes that best support extended maceration, resulting in a harmonious tannic structure.

Thus, in Bordeaux in the 1990s, winegrowers reverted to more quality-oriented vineyard management practices. In particular, yields have been reduced to produce wines that are both more complex and more intense, as well as fruity and well balanced. In certain cases, winemakers aim for extreme concentration by keeping vine yields very low (20–30 hl/hectare) and emphasizing extraction (*saignées*, pumpovers, and long maceration times). The resulting rich flavors are reinforced by marked oakiness. Of course, these wines must be sold for sufficiently high prices to justify these expensive techniques. A number of these wines have been commercially successful, indicating that their quality has been recognized.

These wines are appreciated for their deep color, their rich aromas, featuring oak as an essential element, and their powerful structure and complex flavors. They stand out from other wines in blind tastings and are real "competition wines." As accompaniments to a meal, however, they are less enjoyable due to their aggressiveness, which may dominate to the point of being barely acceptable. It is easy to understand the variable appreciation of these wines.

Another consequence of this type of production is a standardization of quality that is more due to winemaking techniques than natural factors. In general, these wines are made with noble grape varieties from well-known winegrowing areas. However, successful wines have been produced by these methods from *terroirs* that had never been recognized as top quality, as well as from others that certainly had been recognized. Finally, there is no information available as yet on their aging potential. It is understandable that there should be some doubt concerning the long-term future of this type of production and its attendant prestige.

In conclusion, only the best grape varieties grown on the best *terroirs* produce wines that combine the high tannin content indicative of aging potential with aromatic finesse and complexity. On tasting, these wines are not only superbly concentrated but also well balanced and elegant. Winemakers today are aware that excessive tannin extraction tends to mask a wine's fruit and that perfect balance is the sign of a well-made wine.

12.6 Draining Off the Skins and Pressing

12.6.1 Choosing the Moment for Draining Off the Skins

Choosing the optimal maceration time after fermentation is a complicated decision with many possible solutions. It depends on the type of wine desired, the preferred characteristics (tannin intensity and harmonious structure are not always compatible), and the nature of the grape. This decision also depends on winemaking conditions. For example, only closed fermentors permit extended maceration times. In open fermentors, the must, in full contact with air, ferments easily, but the risks of bacterial spoilage and alcohol loss make short maceration times necessary (Section 12.3.1).

In the 1950s in France, maceration times after fermentation tended to be shortened from the traditional three or even four weeks. The goal of this approach was to produce more supple and less tannic wines, but the major reason was the preoccupation with avoiding bacterial spoilage. Ferré (1958) was a principal advocate of short maceration times for quality wines: "Postfermentation maceration times can be reduced to five or six days without affecting wine quality; maceration times longer than eight days should be avoided, if only to reduce the alcohol loss occurring in open fermentors." The data in Table 12.10 are important; they indicate the amount of alcohol loss that can occur.

Other techniques (aeration through pumping-over, temperature control, etc.)

TABLE 12.10

Alcohol Loss as a Function of Maceration Time in an Open Tank (Ferré, 1958)

Maceration time (days)	Density at tank draining	Temperature at tank draining (°C)	Alcohol content of wine (% vol.)	Alcohol loss (% vol.)
0	1.082	15	11.1	—
1	1.074	17	11.1	0
2	1.071	19	11.0	0.1
3	1.030	23	10.9	0.2
4	1.017	28	10.6	0.5
5	0.999	24	10.5	0.6
6	0.997	23	10.4	0.7
7	0.997	21	10.3	0.8

have made it possible to extend maceration in closed tanks without risking spoilage. Winemakers also aim to achieve greater concentration in many types of wine. Today, premium wines often have extended maceration times of two to three weeks. Extended maceration times are chosen to increase tannin concentrations, but, according to analysis, the third week does not significantly increase this concentration. The extended maceration time nevertheless has a "maturing effect" on the tannins. This maturation softens the tannins and improves the organoleptic quality of wines. The chemical transformations during this phenomenon are not known precisely, but they can be appreciated by tasting macerating wines between their 8th and 20th day of extended maceration. The oxidation of tannins is a possible explanation of these transformations. The oxygen introduced during pump-overs could be responsible for this oxidation. Controlling this phenomenon would represent a considerable advance in winemaking.

Wineries producing ordinary wines often extend maceration of their wines for only two to four days. This technique is often used in hot climates, because short post-fermentation maceration times eliminate the risk of significant bacterial spoilage. Additionally, longer times (and thus greater extraction) risk increasing organoleptic flaws to the detriment of finesse. In fact, maceration intensity should be established in accordance with grape quality. Maceration is shortened for ordinary varieties and in poor-quality *terroirs*; noble grape varieties in quality viticultural regions are capable of extended maceration.

Adjusting the skin contact time is a simple method for modifying the maceration, and it is therefore one of the most variable aspects of red winemaking from one region to another. Its duration should be chosen by the winemaker according to grape quality and cannot be generalized: it varies from one vineyard to another, one year to another, and even one fermentor to another, since grape quality is never homogeneous. This quality depends on the ripeness of the grapes (resulting from vine exposure and age) and their soundness. Winemaking equipment should never be the determining factor for deciding maceration times, but unfortunately too many wineries do not have sufficient tank capacity. Winemakers are therefore sometimes forced to stop maceration prematurely in order to free up tank space. In such cases, maceration times can be too short.

Three types of maceration techniques are summarized below:

1. Draining off the skins before the end of fermentation—the wine still contains sugar, and the must specific gravity is between 1.020 and 1.010. This short maceration time of three to four days is generally recommended for average-quality wines coming from hot climates. This method is adopted for producing supple, light, fruity wines for early drinking, but it can also be used to attenuate excessive tannin aggressiveness due to variety or *terroir*.
2. Draining off the skins immediately after fermentation, as soon as the wine no longer contains sugar—approximately the eighth day of maceration. Under these conditions, a maximum CI with a moderate tannin concentration is expected (Section 12.5.3, Figure 12.9). The organoleptic equilibrium of new wines is optimized. Their aromas and fruitiness are not masked by an excessive polyphenol concentration. This maceration method is recommended for premium wines that are to be rapidly released on the market. The resulting wines are not harsh or astringent and can be drunk relatively young. When the grape crop is exceptionally ripe and thus very concentrated, premium wines may also be made in this manner. Finally, open fermentors must be drained off the skins immediately following the end of fermentation.
3. Draining off the skins several days after alcoholic fermentation. Maceration times may exceed two to three weeks. This method is often used to produce premium wines. The tannins assuring the proper aging of the wine are supplied during this extended maceration (Figure 12.9). After several years, free anthocyanins have all but disappeared. Wine color is essentially due to bound forms of anthocyanins and tannins. When making premium wines, successful winemaking requires a compromise. On the one hand, the tannin concentration must be sufficient to ensure long-term aging. On the other, the wine should remain fairly soft and fruity. These criteria are important, since wines are often judged young.

In fact, maceration times do not follow precise rules. They depend on the kind of wine desired and on grape quality.

12.6.2 Premature Fermentor Draining Due to External Factors

Sometimes fermentors must be drawn off before the ideal tannin concentration has been attained. This operation is recommended for stuck fermentations (Section 3.8.1). For reasons already mentioned (Section 3.8.3), there is the risk of development of LAB in sugar-containing musts with inactive yeasts. The volatile acidity would consequently increase dramatically. Drawing off the juice is a means of eliminating the majority of the bacterial population located in the pomace. Light sulfiting can be conducted at the same time (3 g/hl). This operation may, of course, delay MLF, but the sulfur dioxide concentration should be calculated to allow the alcoholic fermentation to restart while blocking bacterial activity.

Various grapevine diseases alter grape crops. As a result, unpleasant tastes often appear in wine. Early draining may help to lessen the severity of these alterations. Bunch or gray rot (*B. cinerea*) is a typical example. Certain vineyards are susceptible to gray rot, since the maturation period coincides with the rainy season.

Fortunately, the vineyard treatments currently available have greatly reduced the frequency of this disease. *Botrytis* has multiple effects on grape constitution and wine character (Section 10.6.5). Their impact influences maceration decisions.

First of all, the various forms of rot impart mushroom-like, iodine-like, and moldy odors to wine (Volume 2, Section 8.8). A short maceration time is recommended to avoid accentuating them.

Moreover, *Botrytis* secretes laccase. This enzyme has a very high oxidative activity (Section 11.6.2), and it can rapidly alter a red wine exposed even briefly to air.

In this case, laccase analysis or, more simply, an enzymatic browning test is advised before draining off the skins. This test consists of filling a wine glass halfway and leaving it in contact with air for 12 hours. The wine is considered to have a browning risk if (1) it changes color, (2) it is turbid, (3) there is sediment in the glass, (4) its hue is less brilliant, (5) there is an iridescent film on the surface, or (6) the color becomes a brownish yellow.

If the results of these tests are positive, an extended maceration is not necessary. A longer maceration time would in fact intensify the flaws due to grape rot. In this case, the wine should also be sulfited when the fermentor is drained. MLF will of course be more difficult, but all oxidative risks are avoided during draining.

Certain measures should be taken when wines made from moldy grapes are drained. First of all, sulfur dioxide has a high binding rate in these wines. The sulfur dioxide concentrations must therefore be relatively high (5 g/hl, or more). In the presence of SO_2, enzymatic activity is instantly inhibited, but the complete destruction of laccase activity is slow. At concentrations of 20–30 mg of free SO_2 per liter, several days are required to destroy this enzyme completely. Fortunately, during this time, the free sulfur dioxide protects the wine against enzymatic browning. After the complete destruction of laccase, the protection is definitive and independent of the presence of free SO_2.

The data in Table 12.11 demonstrate the effectiveness of SO_2 in destroying laccase. Three different wines prone to enzymatic browning receive 0, 50, and 100 mg of sulfur

TABLE 12.11

Laccase Destruction (Expressed in Arbitrary Units) and Enzymatic Browning Protection as a Function of Sulfiting (Free-Run Wine) (Dubernet, 1974)

SO_2 added (mg/l)	Corresponding free SO_2 (mg/l)	Initial laccase activity	Laccase activity after 6 days	Laccase activity after 15 days
Wine no. 1				
0	0	0.16	0.16	0.16
50	16	0.16	0.02	0.00
100	28	0.16	0.00	0.00
Wine no. 2				
0	0	0.13	0.13	0.12
50	18	0.13	0.10	0.07
100	34	0.13	0.03	0.01
Wine no. 3				
0	0	0.16	0.16	0.16
50	28	0.16	0.11	0.09
100	56	0.16	0.08	0.03

dioxide per liter. The second column indicates the binding rate, which is particularly high in wine 1. This wine also has the lowest residual free SO_2 concentrations. The following columns indicate the decrease in laccase activity (expressed in arbitrary units) after sulfiting. In wine 3, the enzyme has not totally disappeared within 15 days of sulfiting at a high concentration (10/g/hl).

Figure 12.10 indicates the role of sulfiting in protecting against enzymatic browning in wine. The wine contains laccase and is exposed to air. Figure 12.10a corresponds to the aeration of a non-sulfited sample. The laccase activity diminishes with time but does not disappear. In the first phase, the red color component (OD 520) and the yellow color component (OD 420) increase. In the second phase, enzymatic browning appears with an increase in the yellow color component and a decrease in the red color component. In phase 3, enzymatic browning causes a precipitation of colored matter.

Figure 12.10b corresponds to the evolution of a sample of the same wine, exposed to air, after being sulfited at 36 mg/l. The free SO_2 disappears after 24 hours, but the laccase activity is not entirely destroyed. As long as the wine contains free sulfur dioxide, it is protected against enzymatic browning (the red color component and the yellow color component increase). The enzymatic browning occurs afterward, lowering the red color component.

Lastly, Figure 12.10c indicates the evolution of the same wine exposed to air, after sulfiting at 55 mg/l. When the free SO_2 concentration falls to zero after 48 hours, the laccase activity has been completely destroyed. The wine is thus definitively protected from enzymatic browning. The yellow color component and especially the red color component increase with exposure to air.

12.6.3 Draining the Free-Run Wine into Tanks or Barrels

The draining operation consists in recovering the wine that spontaneously flows out of the fermentor by gravity. The wine is then placed in a container where alcoholic and malolactic fermentations are completed.

In the traditional, quality-orientated European vineyards, the free-run wine was collected in small wooden barrels. The wooden fermentors were not hermetic enough to protect wine from contact with air. Concrete and stainless steel tanks have been recommended since their development for wine storage during the final phase of fermentation. This final phase precedes barrel aging. The tanks must of course be completely full and perfectly airtight.

When wines are barreled down directly, without blending beforehand, the wine batches may be heterogeneous. Yeasts and bacteria participate in these differences, and they govern the completion of the fermentations. As a result, wine composition (residual sugar, alcohol, and tannin concentrations) may be affected. The less the grapes are crushed and the fewer the pump-over operations, the greater the difference between barrels of wine.

Temporarily putting the wine in tanks, a technique that came into general use in Bordeaux in the 1960s, offers four advantages. First, it presents an opportunity to blend the wines. Second, yeast and bacteria cells may be evenly distributed; fermentations are thus more easily completed. Third, abrupt temperature drops occurring in small containers are avoided (they can hinder the completion of fermentation). Fourth, the daily analysis of fermentation kinetics, including the completion of alcoholic and malolactic fermentation, is easier and more rigorous with a limited number of large tanks than with a great number of small barrels.

New wines nevertheless evolve differently according to the storage method used. Slow final fermentation stages (up to several months) accentuate these differences. Wine clarification in tanks occurs more slowly and is more difficult to obtain than in barrels. Carbon dioxide

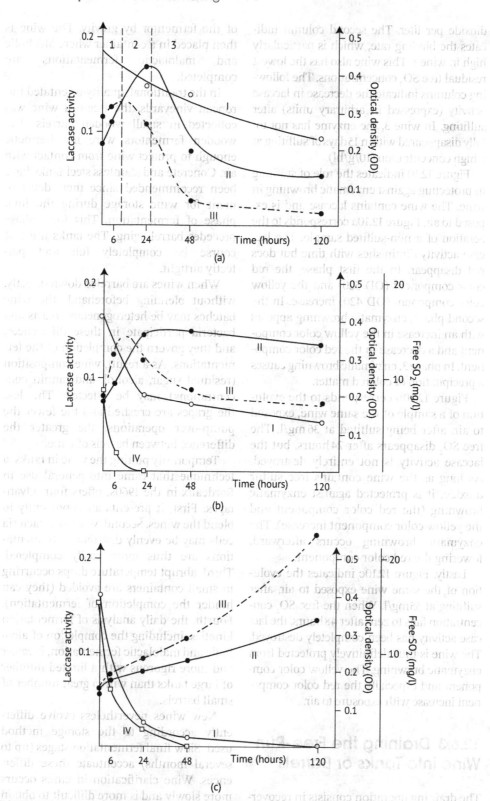

FIGURE 12.10 Evolution of laccase activity, red color (OD 520), yellow color (OD 420), and free SO_2 concentration upon air contact (Dubernet, 1974): (a) non-sulfited sample, (b) sample sulfited at 36 mg/l, (c) sample sulfited at 55 mg/l. I: laccase activity (arbitrary units). II: OD 420 (optical density at 420 nm, 1 mm thickness), yellow. III: OD 520 (optical density at 520 nm, 1 mm thickness), red. IV: free SO_2.

concentrations are also maintained over a longer time, negatively impacting wine taste. Tanks are also known to generate reduction odors from lees, such as hydrogen sulfide or mercaptans.

In the late 1990s, it became increasingly popular to run the wine off into barrel immediately (Section 12.7.2), as MLF in oak barrels has been shown to enhance aroma complexity as well as the finesse of oak character. In fact, it is not known whether this undisputed improvement is due to the effect of bacteria on molecules released by the oak or the fact that the new wine is still warm when it is put into barrel. The fact remains that, if red wines are to be barrel-aged, they should be run off into barrel as soon as possible. We now have all the necessary techniques to avoid the problems that led to the abandonment of barrel aging in the past: blending, temperature control, analytical monitoring of fermentation in individual barrels, etc.

12.6.4 Pressing the Pomace

After the wine is run off from the fermentor, the drained pomace is emptied from the tank and pressed. Self-emptying or automatic-discharge fermentors are capable of executing this operation automatically (Section 12.3.3). Pomace unloading can also be carried out manually, but this is laborious.

These automated alternatives do not always respect quality criteria. In fact, fermented skins are more sensitive to the shredding and sometimes grinding effect of mechanical solutions than fresh grape crops. As a result, suspended solids are formed, and press wines are turbid, bitter, and sometimes colorless. Furthermore, pressing must be rapidly conducted, due to pomace sensitivity to oxidation phenomena. Finally, in some cases, the addition of a fraction of press wine to free-run wine can improve overall wine quality. The goal of obtaining quality press wines is therefore completely justifiable.

To simplify pomace unloading, a method was developed consisting of energetically mixing fermentor contents to disperse the pomace and homogenize the tank. A pump transfers the mixture to the press; the juice and skins are then separated in the press cage. This method, however, is detrimental to wine quality: the brutal mechanical action on the pomace induces vegetal and herbaceous tastes; furthermore, these flaws are not limited to the press wine—they are distributed to all of the wine.

To ensure wine quality, pomace unloading should be carried out manually, and the pomace extracted carefully. A screw or, even better, a conveyor belt system is used to transport the pomace out of the tank. Clearly, a worker must be inside the tank to feed the pomace transport system (the absence of carbon dioxide must be verified before a worker enters the tank). Ideally, the extracted pomace should fall directly from the transport system into the press. The press should therefore be mobile and capable of being placed in front of each tank. This, however, is not always possible. Moreover, this kind of pressing operation affects winery cleanliness. For this reason, pomace pumps are used to transfer drained grape skins to the press through a pipeline. The press is immobile and generally located outside of the tank room, favoring winery cleanliness, but this setup jeopardizes wine quality, making it unsuitable for premium wines. Since the appearance of these pumps on the market 20 years ago, their operation has been much improved. Current models have less of an impact on tissue integrity, especially with short pipelines, but the high pressure required to displace the pomace through the pipeline, especially through its bends, is detrimental to wine quality. This system therefore always affects the quality of press wine. The addition of wine to the pomace to facilitate its transport further diminishes overall wine quality.

When the press cage cannot be placed in front of the tank door, a conveyor belt system can be used as long as the tank and the press are not too far from each other. Another possibility is to fill containers (several hundred liters each) directly in front of the tank. These containers can then be transported to the stationary press and emptied into it.

Oxidation should be avoided during all pomace handling (unloading, transport, and pressing). All equipment and receiving tanks should also be perfectly clean. Good hygiene prevents the possible development of acetic acid bacteria. In fact, these bacteria may already be present in the pomace, if the maceration time was long and the fermentor not completely airtight.

Presses currently used are illustrated in Figure 12.11. They are also used in white winemaking. But fermented skins are pressed more easily than fresh skins. In fact, a smaller press capacity is needed for red winemaking. When the pomace is

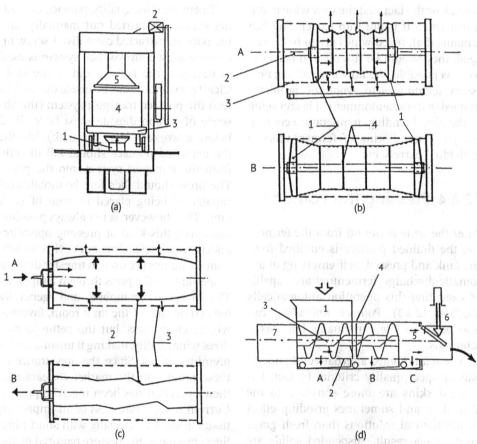

FIGURE 12.11 Different types of press. (a) Vertical hydraulic press. The shaft (1), driven by an electric motor (2), raises the mobile press bottom (3) and presses the pomace in the basket (4) against the fixed head (5). (b) Moving-head press (two heads) (A). (1) In the pressing phase, the heads advancing toward each other with the movement of the screw (2), inside the cage (3), press the pomace. (B) During head retraction, the chains and the hoops (4) break up the cake, while the cage is being rotated. (c) Bladder press. (A) The injection of compressed air (1) is responsible for pressing by inflating the bladder (2) against the press cage (3). (B) After decompression, the rotation of the cage (crumbling) breaks up the cake. (d) Continuous press: (1) hopper; (2) perforated cylinder, filtering the juice; (3) screw; (4) anti-rotation system preventing the pomace from rotating with the screw; (5) compression chamber; (6) restriction door; (7) motor. Two or three levels of juice selection (A, B, C) are possible along the length of the screw. (*Source:* P. Jacquet, Bordeaux, personal communication.)

pressed, the solids must be broken up between each pressure increase–decrease cycle so that more juice can be obtained. Each of these operations has an impact on the quality of the press wine, so it should be kept in separate batches.

Vertical hydraulic presses are the oldest design. They produced good-quality press wine, but loading, unloading, and breaking up the pomace between pressure cycles were awkward, labor-intensive operations. There has been renewed interest in this type of press, as it can easily be moved in front of the tank door for filling. Breaking up the pomace has been simplified, or even eliminated altogether, by inserting efficient drains through the pomace, which makes it possible to extract a large volume of good-quality press wine in a single operation without applying excessive pressure.

Moving-head presses have the significant advantage of being automatic and sometimes programmable. Chains and press rotation are used to break up the cake after decompression. This operation produces suspended solids and can lead to off-aromas.

Pneumatic presses comprise a horizontal press cage and an inflatable membrane. Air forced into the membrane crushes the pomace against the cage. After decompression, cage rotation breaks up the cake. The lack of a central shaft, as opposed to moving-head presses, increases the press capacity of a cage of the same size. This type of press produces the highest-quality results. Another option is a pneumatic press continually fed by an axial pomace pump, but this practice decreases wine quality and is not recommended. Many of the advantages of a pneumatic press are lost with this technique. In any case, the pomace should be transported as short a distance as possible with a minimum number of bends in the pipeline.

Quite a few years ago, continuous screw presses were fairly popular, due to their ease of use and high pressing speed. Yet, even with a large-diameter screw turning slowly, these presses have a brutal action on pomace; press wine quality is affected. Screw presses always produce lower quality wines than other presses. Due to pressing variations along the length of the screw press, the tank receiving press wine should be divided for separate collection of the batches corresponding with the first pressing, second pressing, etc. of discontinuous presses. The small volume of the last batch is generally very low in quality: it should be eliminated and distilled.

12.6.5 Composition and Use of Press Wines

The wine impregnating the pomace constitutes the press wine. Its volume during winemaking depends on the level of pulpiness of the grapes. In the Bordeaux region, it represents approximately 15% of the finished wine on average. Press wine is composed of interstitial wine. This wine is easy to separate from the skins and relatively similar to free-run wine, when the fermentor has been well homogenized by pump-over operations and the grape correctly crushed. Press wine is, however, also made up of a wine that saturates the pomace tissues. This wine is very different from free-run wine and much more difficult to extract. Following this principle, the two kinds of press wine are generally separated. The first press wine (approximately 10% of the finished wine or two-thirds of the press wine) is obtained through a direct pressing. When pomace handling and pressing are done correctly, the first press is of a good quality. The second press wine (approximately 5% of the total quantity of wine and one-third of the press wine) is not as good in quality, as it is obtained at high pressure after the press cake has been broken up. This damages grape tissues that have become more fragile during fermentation, releasing substances with bitter, herbaceous overtones and accentuating the characteristic astringency of press wines, due to their high tannin content.

Grape quality primarily affects press wine quality. Varieties of ordinary quality and grapes ripened in hot climates can produce press wines containing a high concentration of aggressive and vegetal tannins. Press filling conditions and pressing methods also affect press wine quality, i.e. both the number of pressings with cake breakup and the maximum pressure. Finally, a slow and regular pressure increase, even between two pressings, is beneficial to press wine quality. A single pressing, without pomace cake breakup, is recommended for premium wines, of course, while assuring a slow and regular pressure increase up to the maximum. This method produces less press wine but of a superior quality.

All elements, except for alcohol, are concentrated in press wine. Table 12.12 gives an example of this phenomenon. The alcohol content decreases by 40% and in some cases by even more. The presence of reducing agents is most likely responsible for the higher sugar concentration. Uncrushed grapes may also liberate unfermented sugar during pressing. The volatile acidity of press wine is always higher than in free-run wine—indicating an increased bacterial risk in the pomace. Total wine acidity is generally also a little higher, but the higher mineral concentration also increases the pH of press wine. More phenolic compounds (anthocyanins and tannins) are present, reflected in the extract values. Press wines also contain more nitrogen compounds. In certain hot climates, high maturity levels lead to extremely concentrated grapes: the resulting press wines are so rich in tannins that it can make them too bitter and astringent when tasted. Although there is little analytical data on this, press wines also contain polysaccharides and other colloids that add body to blends.

Until recently, very few studies had examined the aroma characteristics of red press wines. By focusing first on sensory aspects, studies have shown the existence of a specific sensory space for these wines (Poitou, 2016; Poitou et al., 2017). From an olfactory standpoint, red press wines could be very significantly differentiated from free-run wines and commercial wines. The panel of experts and professionals selected "bell pepper," "straw," and "herbaceous" as the main descriptors related to the vegetal aroma family. These aromas are not correlated with increased contents of 2-methoxy-3-alkylpyrazine nor with C6 alcohols in press wines.

Characterization of related compounds has enabled the identification of (Z)-4-heptenol. This compound presents green notes, and its contents in press fractions are well correlated with pressing intensity (Figure 12.12). In several series of press wines, the (Z)-4-heptenol content exceeded the sensory detection threshold of this

TABLE 12.12

Composition of Free-Run and Press Wine (Ribéreau-Gayon et al., 1976)

Component	Free-run wine	Press wine
Alcohol content (% vol.)	12.0	11.6
Reducing sugars (g/l)	1.9	2.6
Extract (g/l)	21.2	24.3
Total acidity (g/l H_2SO_4)	3.23	3.57
Volatile acidity (g/l H_2SO_4)	0.35	0.45
Total nitrogen (g/l)	0.28	0.37
Total phenols (permanganate index)	35	68
Anthocyanins (g/l)	0.33	0.40
Tannins (g/l)	1.75	3.20

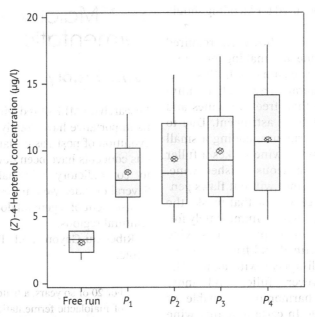

FIGURE 12.12 Distribution of (Z)-4-heptenol concentrations (μg/l) in samples of red wines ($n = 63$ wines). Increasing pressure levels: P_1–P_4.

compound, estimated at 12.6 μg/l in red wine. Moreover, a quantification method for carbonyl compounds (alkanals, alkenals, alkadienals, and alkadienones) presenting green odors has been developed to show trends in terms of concentrations between these compounds and pressing intensity. When added to wines at the levels found in press wines, some of these compounds in the mixture ((E,E)-2,6-nonadienal, (E,Z)-2,6-nonadienal, (E,E)-2,4-nonadienal, decanal, and (Z)-1,5-octadien-3-one) modify the aroma expression of wines. Furthermore, MLF, addition of sulfur dioxide, and blanketing of the press with inert gas all limit the concentrations of these compounds in press wines, since the sensory impact of pressing under a blanket of neutral gas is generally significant when wines have not undergone MLF (Poitou, 2016; Poitou et al., 2017).

Press wine quality also depends on winemaking conditions. Repeated pump-overs, high maceration temperatures, and other techniques that increase maceration will deplete the pomace of desirable phenolic compounds. The resulting press wines lack body and color and are dominated by astringent and vegetal flavors. These inferior-quality wines cannot be blended with free-run wines to improve overall wine structure and quality, and must sometimes be distilled or eliminated, representing a considerable loss in wine volume. Press wine quality must therefore be ensured by avoiding excessive maceration and extraction techniques.

The decision to blend press wines with free-run wines is complicated. It not only depends on both free-run and press wine quality but also on the type of wine desired. In general, press wines are not added when making nouveau-style wines for early drinking, except when the free-run wine is excessively light in body. Moreover, press wines should not be incorporated into wines made from ordinary and rustic varieties. Premium wines made from very concentrated grapes are often very tannic, even after short maceration times. In all these cases, there is the question of what to do with the press wines. When possible, press wines that are not used for blending should be distilled.

Press wines are, however, required most of the time for making premium wines in temperate climates. In this case, press wines generally have higher tannin concentrations than free-run wines and are often excessively astringent. Due to their colloidal structure, adding a small percentage of press wine makes a fuller and more homogeneous finished wine. But even press wines without flaws generally have a heavy odor that masks the fruitiness of new wines. Immediately following the addition of press wines, wine aroma is less refined and fruity, but this flaw tends to disappear with aging. The wine is, however, fuller and more balanced and harmonious—capable of long-term aging. In certain quality wine regions, press wines are generally considered indispensable to wine quality.

Delaying the addition of press wines so that the clarification process may take place can be beneficial to wine quality. The press wines may undergo fining or pectolytic enzyme treatments (0.5 g/hl) at draining before MLF. Too long of a delay (for example, until the spring following the harvest) causes the free-run wine to evolve. As a result, it may not blend well with the typical flavors of press wine. The best solution is to decide whether to add a certain proportion of press wine soon after the completion of MLF. The press wine percentage (5–10%) must be determined according to the anticipated aging potential of the wine, based on precise tasting tests on samples. At this stage, a certain level of tannic aggressiveness should be sought. These tannins improve the barrel and bottle aging potential of the wine. The press wine may also be progressively blended in during the months following fermentation to compensate for thinning (which always accompanies the first stage of wine maturation). With this process, the wine is at its optimum quality during the period when it is judged and sometimes sold.

12.7 Malolactic Fermentation

12.7.1 History

Research on MLF of red wines, its role, and its importance have greatly influenced the evolution of post-Pasteurian enology. Various concepts have been developed leading to contradictory winemaking methods. Several decades were necessary for the establishment of a general doctrine in all viticultural regions.

Ribéreau-Gayon and Peynaud (1961) wrote:

> For 20 or so years, a better understanding of malolactic fermentation, its agents, its mechanism and its factors has permitted considerable progress in research in this area. An ever-increasing number of observations and studies in relevant winemaking regions have also participated in a better understanding of this process. Yet viticultural regions are slow to apply this information. Its practice seems to spread slowly from one region to another and is difficult to establish. The fact that these concepts have made winemaking methods more complex when compared with certain simplistic theories has created a certain amount of resistance. The need to modify outdated, but generally accepted, doctrines has also slowed progress. It is surprising that solidly established and widely confirmed concepts have encountered so many obstacles.

MLF is both relatively simple and extremely important in practice, and all sensible winemaking and aging techniques take its existence and laws into account. It is an important element in premium red wines, even in fully ripe years. In addition, it regulates wine quality from year to year. The less ripe the grapes and therefore the higher the malic acid concentration, the more MLF lowers wine acidity. The differences in acidity of wines from the same region are much smaller than those of the corresponding musts.

Another less readily accepted consequence of MLF is an improvement in biological stability caused by bacteria; though it raises pH, it also eliminates highly unstable malic acid.

The existence and importance of MLF were not easily recognized. It occurs under variable conditions that make proving its existence difficult. If it takes place during or immediately following alcoholic fermentation, it can be completed without being noticed, but it can also occur several weeks or months after alcoholic fermentation. Since little carbon dioxide is released, the phenomenon is sometimes almost imperceptible. The decrease in total acidity observed can also be interpreted as a potassium bitartrate precipitation. Additionally, the chemical analysis of malic acid, especially in the presence of tartaric acid, was difficult in the past. The determination of malic acid concentrations by paper chromatography was the first simple and significant method (Ribéreau-Gayon, 1953); it could be used in the winery and permitted the disappearance of malic acid to be monitored. This greatly contributed to establishing the concept of MLF.

MLF is nevertheless part of winemaking tradition. It occurred irregularly but did exist in past red wines. The data in Table 12.13 are significant in this respect. It was not until the 10-year period from 1963 to 1972 that MLF became systematic. Better control of microbial spoilage simultaneously permitted the lowering of volatile acidity concentrations, essentially affecting maximum values. Bordeaux was the forerunner with this systematic control of MLF, which occurred much later in many other viticultural regions throughout the world. Although not pertaining directly to this chapter, the figures in Table 12.13 concerning alcohol content are interesting: they show that chaptalization has led to

TABLE 12.13

Analyses of Different Vintages of Red Wines from Medoc and Graves Vineyards (Analyses Performed in 1976; (Ribéreau-Gayon, 1977))

Period	Number of samples	Levels	Alcohol content (% vol.)	Total acidity (g/l H_2SO_4)	Volatile acidity (g/l H_2SO_4)	Total SO_2 (mg/l)	Malic acid (g/l)	Sugar (g/l)
1906–1931	12	Min	10.0	3.72	0.53	27	0	2.0
		Max	12.3	5.59	0.90	83	3.5	4.2
		Av.	*11.1*	*4.40*	*0.71*	*41*	*0.8*	*2.5*
1934–1942	11	Min	9.9	3.92	0.46	27	0	2.0
		Max	13.1	5.39	1.10	105	3.5	3.8
		Av.	*11.2*	*4.32*	*0.75*	*53*	*0.6*	*2.4*
1943–1952	17	Min	10.3	3.43	0.45	28	0	2.0
		Max	12.7	5.29	1.05	70	2.0	3.8
		Av.	*11.3*	*4.03*	*0.70*	*45*	*0.2*	*2.3*
1953–1962	18	Min	11.1	3.19	0.45	29	0	2.0
		Max	13.3	5.29	0.72	80	1.5	2.5
		Av.	*11.9*	*3.55*	*0.57*	*52*	*0.1*	*2.1*
1963–1972	18	Min	11.8	3.09	0.37	51	0	2.0
		Max	12.7	3.53	0.50	89	0	2.5
		Av.	*12.0*	*3.38*	*0.42*	*67*	*0*	*2.0*

For each period, we provide minimum, maximum, and average values. Italics denote the average values vs. min/max.

greater regularity in the alcohol content of recent Bordeaux wines in comparison with past vintages, but maximum values have remained similar over the years.

The first observations of MLF date back to the end of the 19th century in Switzerland and Germany and to the beginning of the 20th century in France. The data in Table 12.14, pertaining to winemaking in 1896, give characteristic examples of MLF. Researchers at that time were not capable of correctly interpreting the information: they focused in particular on the volatile acidity increase, following the bacteria population increase observed under the microscope. They attributed the lowering of total acidity to potassium bitartrate precipitation—the disappearance of malic acid was not even considered. This situation was thought to be the beginning of a serious microbial contamination that should absolutely be avoided.

Past researchers also noticed that sulfiting of must resulted in wines with higher acidity. This phenomenon was interpreted as a greater dissolution of the acids in the pomace in the presence of sulfur dioxide. The idea that bacteria were inhibited and that malic acid was not degraded was not even considered.

The general existence of this phenomenon was established from 1922 to 1928 in Burgundy by the research of L. Ferré (1922) and from 1936 to 1938 by the studies of J. Ribéreau-Gayon in Bordeaux (both authors cited in Ribéreau-Gayon et al. (1976)). The importance of this second fermentation was demonstrated to be an essential step in making premium red wines. Yet the apparently simple corresponding concepts were difficult to accept. For a long time in enology textbooks, MLF was described in the chapters covering diseases and spoilage. Certain enology schools contested both the existence and especially the value of this second fermentation.

It took time to establish the importance and utility of MLF because of the involvement of LAB, which were considered to be contaminating agents (Table 12.14). Their frequent presence in red winemaking was thought to correspond with the beginning of spoilage that should be avoided at all costs. Pasteur once said: "Yeasts make wine, bacteria destroy it." It seemed pretentious at the time to go against the beliefs

TABLE 12.14

Alcoholic and Malolactic Fermentations in Tank; Results Obtained in 1896 (Gayon, 1905)

Tank no.	Duration of fermentation (days)	Remaining sugar (g/l)	Total acidity (g/l H_2SO_4)	Volatile acidity (g/l H_4SO_4)	Number of bacteria under microscope
1	2	116	4.35	0.12	0
	5	45	4.25	0.12	0
	6	23	4.25	0.13	1–2
	7	6	3.19	0.21	25–30
	10	3	3.15	0.35	30–35
	13	1	3.15	0.36	35–40
2	2	104	4.25	0.10	0
	5	29	4.25	0.10	0
	6	4	4.47	0.12	0
	7	1	4.20	0.17	4–5
	8	1	3.35	0.23	20–25
	13	1	3.40	0.45	50–60

of a great scientist. Finally, the idea that the same bacteria could be beneficial when they degrade malic acid and detrimental when they attack other constituents was difficult to accept.

Furthermore, a slightly elevated acidity was considered in the past to be a sign of quality. A low pH effectively opposes bacterial development and thus can limit the production of volatile acidity. However, malic acid is a highly biodegradable molecule, and its disappearance results in a biological stabilization of the wine, even though the pH increases. When a red wine containing malic acid is bottled, there is always a risk that MLF will start in the bottle after a few months, resulting in spoilage due to gassiness and an increase in volatile acidity.

The diagram in Figure 3.8 (Section 3.8.2) summarizes the principles of correct present-day red winemaking. LAB should only be active when all of the sugar has been fermented. Interference between the two fermentations should be avoided (Section 3.8.1). It compromises the completion of alcoholic fermentation and can result in a considerable increase in volatile acidity, if the bacteria decompose the remaining sugar. When there is no longer any sugar, the bacteria develop mainly by degrading malic acid, the most easily biodegradable molecule. The high biodegradability of malic acid is one of the reasons why it must be eliminated. The bacteria are thus beneficial during this process.

As soon as the MLF is completed, the same bacteria can rapidly become detrimental, and certain precautions are necessary to avoid this unwanted evolution. The bacteria are apt to decompose pentoses, glycerol, tartaric acid, etc. These transformations cause common wine diseases (lactic spoilage, *amertume* or acrolein spoilage, tartaric acid degradation, etc.), which increase volatile acidity and lactic acid concentrations to a variable degree. Total acidity is thus also increased. The second fermentor in Table 12.14 gives an example of this phenomenon. Between the eighth and 13th day, the bacteria population, total acidity, and volatile acidity increase considerably. These increases indicate that the transformation is no longer a pure MLF.

Fortunately, LAB have a preference for malic acid—otherwise, present-day premium red wines would not exist (at least as we know them)—but care must be taken to ensure useful microbial transformations while avoiding harmful ones. Despite its acidity and alcohol content, wine is susceptible to spoilage, but luckily not too susceptible.

The errors committed in certain French wineries during the 1950s due to winemaking principles at the time are understandable. Wine must was massively sulfited to be absolutely sure of avoiding bacterial contamination. On the one hand, the wine did not benefit from the advantages of MLF. On the other hand, since the wine was not stored under sterile conditions, it remained susceptible to subsequent contamination. Untimely and uncontrolled MLF could therefore occur at any moment.

In view of the gradual decrease in total acidity observed in many vineyards today, there may be some doubt as to the absolute need for MLF. In the future, steps may be taken to prevent it in certain specific cases. Of course, for the moment, that is only a hypothesis as, according to our present understanding of these phenomena, MLF is still an indispensable stage in red winemaking.

Current technology should lead to the development of stabilization methods preventing uncontrolled MLF. The first step is to avoid excessive contamination, even though absolute sterility is difficult (if not impossible) to obtain. Physical methods such as heat treatments are the most effective methods for eliminating LAB. Various sterile bottling techniques exist that make use of either filtration or heat treatments. Among chemical methods, sulfiting is effective due to the antibacterial effect of bound sulfur dioxide. LAB inhibitors also exist: egg white lysozyme (Section 9.5.2), fumaric acid, and nisin. These substances are not always completely effective, nor are they perfectly stable. In any case, the high resistance of certain strains in wine should be taken into account, especially when wine pH is low.

12.7.2 Wine Transformations by Malolactic Fermentation

This section provides further details on the chemical and flavor changes that occur in wine during MLF (Section 6.3.3). The reaction mechanisms involved are described in Section 5.3.1.

The overall reaction of this phenomenon is shown below.

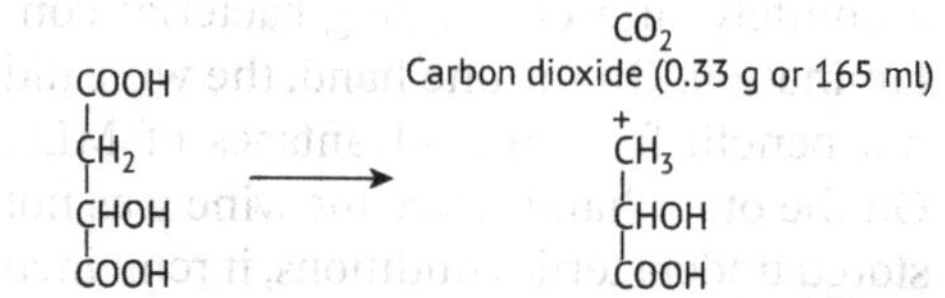

$$\begin{array}{c} \text{COOH} \\ | \\ \text{CH}_2 \\ | \\ \text{CHOH} \\ | \\ \text{COOH} \end{array} \longrightarrow \begin{array}{c} \text{CO}_2 \text{ Carbon dioxide (0.33 g or 165 ml)} \\ + \\ \text{CH}_3 \\ | \\ \text{CHOH} \\ | \\ \text{COOH} \end{array}$$

Malic acid (1 g)   Lactic acid (0.67 g)

This reaction is a simple decarboxylation, explaining the loss of an acid function. In practice, at the pH of wine, malic acid is partially neutralized in the form of dissociated salts, thus in an ionic form, but the overall phenomenon described remains the same. Each time that a molecule of malic acid, in the free acid or ionic form, is degraded, a free acid function is eliminated. Only a limited amount of carbon dioxide is released, but it is perceptible if the cellar is quiet. It can, in fact, be the first sign of the onset of MLF.

The decrease in acidity following MLF varies according to the malic acid concentration and thus grape ripeness. This decrease in acidity can range from 2g/l (expressed as H_2SO_4) to sometimes 3g/l (3–4.5g/l as tartaric acid). Total acidity decreases from 4.5–6.5g/l as H_2SO_4 (6.75–9.75g/l as tartaric acid) to 3–4g/l H_2SO_4 (4.5–6g/l as tartaric acid). The fermentation of 1g of malic acid per liter lowers the total acidity by approximately 0.4g/l as H_2SO_4 (0.6g/l as tartaric acid).

The preceding reaction does not explain acetic acid production, but volatile acidity always increases during MLF. This production is due, at least in part, to citric acid degradation (Section 5.3.2). Although one molecule of citric acid produces two acetic acid molecules, this degradation is always limited because grapes do not contain large quantities of citric acid.

Bacteria also produce volatile acidity from the degradation of pentoses. In fact, these sugars might be used as energy sources. Malic acid degradation alone does not seem sufficient to ensure cell energy needs (Henick-Kling, 1992).

Observations show that volatile acidity increases at the end of this phenomenon, when malic acid is almost entirely depleted. Moreover, this increase is even greater when MLF is facilitated (low-acidity musts, for example).

Table 12.15 shows the main chemical transformations in wine during MLF. In this case, it is incomplete, as the wine still contains 0.5g/l malic acid that has not been degraded.

The results (mEq/l) reflect the consequences of MLF. The lactic acid formed corresponds to half of the malic acid transformed. The drop in fixed acidity corresponds approximately to the difference between the loss in malic acid and the gain in lactic acid.

The chemical transformations of wine by MLF are much more complex in reality. MLF also produces ethyl lactate, which contributes to the sensation of body in wine (Henick-Kling, 1992). Additionally, other secondary products have been identified, the most important being diacetyl, produced by bacteria (a few milligrams per liter). The amount of diacetyl formed enters a complex pool of production and degradation mechanisms. At moderate concentrations, acetyl contributes to aroma complexity, but above 4mg/l its characteristic buttery aroma dominates.

Another transformation attributed to LAB is the decarboxylation of histidine into histamine, a toxic substance (Section 5.4.2).

TABLE 12.15

Analysis of a Wine Before and After Malolactic Fermentation (Ribéreau-Gayon et al., 1976)

Acids	Concentrations (g/l)[a]		Concentrations (mEq/l)		
	Before	After	Before	After	Difference
Total acidity	4.9	3.8	100	78	−22
Volatile acidity	0.21	0.28	4.3	5.6	+1.3
Fixed acidity	4.7	3.6	96	73	−23
Malic acid	3.2	0.5	48	8	−40
Lactic acid	0.12	1.8	1.4	20	+19

[a] Total acidity, volatile acidity, and fixed acidity are expressed as H_2SO_4.

Wine color change always accompanies MLF. CI decreases, and the brilliant red hue diminishes. This modification is due to the decolorization of anthocyanins when the pH increases, but condensation reactions between anthocyanins and tannins are probably also involved. These reactions modify and stabilize wine color.

The organoleptic character of the wine is also greatly improved. First, wine aromas become more complex. Wine bouquet is intensified, and the character and firmness of the wine are improved, as long as the lactic notes are not excessive. MLF conditions (bacterial strains and environmental and physical factors) certainly influence results, and this fact is illustrated by conducting MLF on white wines, which are, of course, simpler and thus more sensitive to changes brought about by MLF. These transformations merit further study. Harmful off-odors may occur, especially with difficult MLF.

The taste of the wine is also considerably improved, especially when the initial malic acid concentration of wine is high and thus deacidification is significant. The softening of wine is due first of all to a decrease in acidity. The substitution of the malic anion by the lactic anion also contributes. In fact, malic acid corresponds to the aggressive green acid of unripe apples. Lactic acid is the acid found in milk; it has a much less aggressive effect on the taste buds. Additionally, the association of the flavor of malic acid with the astringency of tannins is not harmonious. MLF causes red wines to lose their acidic and hard character. They become softer, fuller, and fatter—essential elements for a quality wine.

Attempts have been made to determine the influence of bacterial strains and operating conditions on organoleptic changes in red wines brought about by MLF. At present, no definitive results have been found.

According to certain recent theories, red wine quality is even more greatly improved when MLF takes place in barrels (Section 12.6.3). Oakiness is better integrated; tannins are fuller and more velvety. These differences are already present at the end of MLF, but they are often less flagrant at the time of bottling. When this technique is applied correctly, it has no detrimental effects, but requires a lot of extra work in the cellar, especially in handling and inspecting large numbers of barrels.

Lastly, concerning the transformations in wine, the degradation of the malic acid molecule leads to a biological stabilization, even though the pH increases (Section 12.7.1).

12.7.3 Monitoring Malolactic Fermentation

It is essential to determine the onset of MLF and to monitor the disappearance and complete depletion of malic acid in each tank. For a long time, malic acid concentration was difficult to determine chemically—it could only be extrapolated through comparing total acidity before and after MLF. In this case, simultaneous potassium bitartrate precipitation could also lower total acidity, falsifying the estimate of malic acid concentrations. In due course, paper chromatography appeared (Ribéreau-Gayon, 1953). This analytical tool, which permitted a simple visual method for monitoring the disappearance of malic acid (Figure 12.13), represented a considerable advance and greatly contributed to the general use of MLF in wineries. Although not a very precise method for analysis of malic acid, it also permits the taste characteristics of a wine to be compared according to the stage of MLF.

Paper chromatography is easy to use and is still widely employed in wineries to monitor this second fermentation, though the method is sometimes slightly varied.

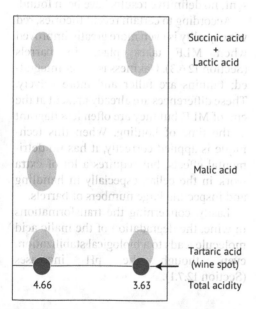

FIGURE 12.13 Separation of organic acids found in wine by paper chromatography (Ribéreau-Gayon, 1953). Left: wine that has not undergone malolactic fermentation. Right: the same wine after malolactic fermentation. Total acidity is expressed in g/l H_2SO_4.

The Bordeaux region, for example, produces 4–6 million hl red wine per year, representing some 40,000–60,000 tanks that have to be tested at least twice, requiring a considerable number of analyses. Today, there is also a more accurate enzymatic method for assaying malic acid. The reagents are expensive, but the analysis may be automated and is especially suitable for checking the completion of MLF in a large number of tanks.

MLF is normally monitored after the wine has been drained and pressed off the skins—in other words, after the completion of alcoholic fermentation. Yet MLF may begin prematurely, when musts are insufficiently sulfited or are inoculated with bacteria before alcoholic fermentation. In this case, the two fermentations may be slow to complete, with no true end point. The overlapping of the two fermentations can lead to a stuck alcoholic fermentation, and in this situation the bacteria are also apt to produce volatile acidity from sugars (Sections 3.8.1 and 3.8.2). This must be closely controlled to avoid serious spoilage, and the monitoring of both the malic acid concentration and volatile acidity is recommended. When the alcoholic fermentation is not complete, the sugar concentration should also be closely followed. If bacterial spoilage begins to occur, the wine should be immediately sulfited (3g/hl).

Experience has shown that, even in sugar-containing media, LAB during their growth phase do not produce acetic acid and decompose only malic acid (Section 3.8.3). The complete depletion of malic acid, however, greatly increases the risk of serious spoilage when the wine still contains sugar. For example, MLF sometimes occurs on pomace before the wine has been run off during the post-fermentation phase. In this case, the wine may still contain some residual sugars, especially in the case of slightly crushed grapes.

After skin and juice separation and pressing, the tanks or barrels are completely filled with wine. The MLF process is then monitored routinely using paper chromatography. This second fermentation generally takes a few days to a few weeks (maximum). An excessive delay of its onset is generally due to winemaking errors: too much sulfiting or too low a pH or wine temperature. In the past, certain vineyards and sometimes entire winemaking regions claimed that MLF was impossible. A winemaking error, most often excessive sulfiting, was generally responsible.

If MLF does not commence, the fear of an increase in volatile acidity does not justify prematurely sulfiting the wine. In fact, in the absence of sugar, LAB begin to develop by degrading malic acid into lactic acid. However, an excessive delay in the start of the fermentation does require other precautions to be taken. To avoid excessive oxidation, the wine should be sulfited. Although the lees contain bacteria, it may be necessary to remove them to prevent reduction odors. MLF has been observed (especially in the past) to occur normally in the spring or summer following the harvest.

The disappearance of malic acid is also monitored in order to choose the ideal moment for definitively stabilizing the wine. Sulfiting at 3–6 g/hl permits this stabilization. Before sulfiting, the wine is racked to eliminate a fraction of the bacteria with the coarse lees. This operation should follow the complete depletion of malic acid. After its onset, malic acid degradation is normally completed in several days. In rare cases, bacteriophages destroy bacteria (Section 6.5): they stop MLF, and several hundred milligrams of malic acid per liter may remain undegraded. In this situation, the wine must be reinoculated to restart the fermentation.

The residual malic acid content should be under 100 mg/l when the wine is stabilized by sulfiting, and it would be dangerous to bottle a wine containing over 200 mg/l. This stabilization should not be excessively delayed: bacterial spoilage, such as increased volatile acidity, is most likely to occur in the final stages of MLF. The possibility of strains highly resistant to sulfur dioxide, especially at high pHs, should also be considered, but they are not a significant problem if the wine no longer contains sugar, malic acid, and citric acid and if its conservation temperature is relatively low. The evolution of wines should nevertheless be closely monitored.

12.7.4 Conditions Required for Malolactic Fermentation

Bacterial growth conditions are described in Chapter 6. During the first hours after the fermentor is filled, the bacteria originating on the grape develop rapidly. As soon as alcoholic fermentation is initiated and ethanol is formed, the bacteria population greatly decreases, and environmental conditions become increasingly hostile—only the most resistant strains are capable of surviving. During draining and pressing, the contact of the wine with contaminated equipment may increase the population of these resistant strains. They remain latent for a variable period and then start MLF when the population reaches levels on the order of 10^6 cells/ml.

In good winemaking practice, the latent phase should be sufficiently long to avoid the undesirable overlapping of the two fermentations (Section 12.7.3). This phase should also be short enough so that the malic acid may be degraded within a reasonable amount of time.

Techniques may be employed to influence MLF, but the ideal conditions for this phenomenon remain ill defined. Malolactic fermentability of wine varies according to the region, vineyard, and year. Wine also tends to ferment better in large containers than in small ones. These facts are difficult to interpret, but environmental and nutritive conditions of bacteria appear to play an important role. Bacterial growth is limited by alcohol and acidity in wine; in addition, bacteria are incapable of synthesizing certain essential substances (nitrogen compounds, amino acids, and growth factors). Specific deficiencies may therefore make certain MLF difficult. Yet, in practice, modifying wine composition does not improve malolactic fermentability to any significant extent, except by increasing its pH. Each bacterial strain probably has an optimum nutrient medium.

Optimum growth conditions for bacteria of grape origin are described in this section. Inoculation will be covered in Section 12.7.5.

Alcohol is the first limiting factor on MLF. Malic acid concentrations often decrease fastest in tanks containing the lowest alcohol concentrations. *Leuconostoc oenos* (now known as *Oenococcus oeni*) is predominantly responsible for MLF in red wines, and it cannot grow in alcohol concentrations exceeding 14% volume. Some lactobacilli can resist 18–20% (vol.) alcohol and are apt to cause spoilage in fortified wines. Besides alcohol production, the wine yeast strain responsible for alcoholic fermentation affects bacterial growth and MLF in other ways. It yields macromolecules (polysaccharides and proteins) to the medium. The enzymatic systems of the bacterial cell wall hydrolyze these substances.

The following factors participate in the control of MLF: acidity, temperature, aeration, maceration time, and sulfiting.

Influence of acidity (Section 6.2.1)

As acidity increases, a growing number of bacterial species are inhibited. MLF becomes increasingly difficult, but, simultaneously, it is increasingly pure. In other words, malic acid is predominantly degraded. The degradation of other wine components is slight, thus limiting the increase in volatile acidity.

LAB growth is optimum at a pH between 4.2 and 4.5. In the pH range of wine (3.0–4.0), MLF speed increases with the pH. The pH limit for growth is 2.9, but even at 3.2 bacterial growth is very limited. MLF becomes possible at a pH of 3.3 or higher.

MLF is necessary with insufficiently ripe grapes, but the high malic acid concentrations in these grapes hinder this fermentation. With very ripe grapes having a low acidity, the impact on wine taste is less significant, but MLF occurs easily and the risk of bacterial spoilage is much higher.

When the pH is excessively low, wine can be deacidified to facilitate the initiation of MLF: for example, 50 g of $CaCO_3$ per hectoliter can be added to the wine (Section 11.4.2). The role of this deacidification is to rectify the pH without removing an excessive amount of tartaric acid. This operation must take into account the decrease in total acidity brought about by MLF, depending on the malic acid concentration in the wine. This operation should be conducted on a fraction of the total volume (20–30%, for example). This deacidified fraction is used to initiate natural deacidification reactions (MLF followed by potassium bitartrate precipitation).

Influence of temperature (Section 6.2.4)

The effect of temperature is twofold. First, an elevated temperature (above 30°C) during maceration can affect bacteria. Second, bacterial growth and the onset of MLF require a certain temperature range. The impact of temperature on bacterial growth depends on the alcohol content of the wine. For 0–4% (vol.) of ethanol, the optimal growth temperature is 30°C, as opposed to 18–25°C for an alcohol content of 10–14% by volume (Henick-Kling, 1992). In practice, the optimal temperature for MLF is between 20 and 25°C. The fermentation is slowed outside these ranges.

New wines should therefore be maintained at a temperature of at least 18°C. Temperature is the simplest means of influencing MLF. In the past, wine cellars were not temperature controlled, and the cold autumn air was the principal factor blocking MLF.

The fermentation of malic acid is slow at 15°C, whereas it is complete in a few days at 20°C. When initiated at a suitable temperature, MLF generally goes to completion, even when the temperature drops to 10°C. Its initiation in winter is unlikely if the temperatures are unfavorable: fermentation will most likely occur the following spring, when the temperature rises naturally. Sulfiting should be carefully timed to avoid blocking this phenomenon.

The fermentation should be conducted at as low of a temperature as possible (18–20°C, for example). The low temperature makes MLF slower but limits the risk of bacterial spoilage—in particular, excessive volatile acidity production due to the transformation of substances other than malic acid.

Influence of aeration (Section 6.2.5)

Each bacterial species has its specific needs. In practice, these different needs are not known. MLF is possible in a large range of aeration levels. This factor does not seem to be decisive. Wine oxygenation by contact with air generally accelerates the onset of MLF, but saturating wine with pure oxygen delays but does not completely block it. In practice, moderate aeration is often beneficial to MLF (Blouin and Peynaud, 2001).

Influence of sulfiting (Section 6.2.2)

Bacteria are known to be highly sensitive to sulfur dioxide (Section 8.6.3). They are much more sensitive to it than yeasts (Section 8.7.4). Moderate concentrations of sulfur dioxide ensure a pure alcoholic fermentation without bacterial contamination—always dangerous in the presence of sugar. Both free and bound sulfur dioxides have an effect on bacteria. Over several months of storage, wine is regularly sulfited; this operation increases the total SO_2 concentration, and MLF becomes difficult, if not impossible, even with a low concentration of free SO_2.

Sulfiting particularly affects MLF in two circumstances: sulfiting the crushed grapes during tank filling and sulfiting wine at tank emptying.

In normal winemaking, the wine should not be sulfited at draining/pressing to avoid compromising MLF. There are two exceptions corresponding to accidental factors: moldy grapes and stuck fermentations (Section 12.6.2). In the first case, a light sulfiting (2–5 g/hl) protects against enzymatic browning; in the second case, it prevents lactic spoilage. In both cases, the situation is serious enough to justify making MLF more difficult. After sulfiting at 2.5 g/hl at draining/pressing, MLF has been reported to be delayed until the following summer. Sulfiting at 5 g/hl can definitively block fermentation. In these exceptional cases, the wine should be massively inoculated with a wine that was not sulfited at draining/pressing and that is undergoing a normal MLF.

New red wines should not be sulfited immediately. This can pose a problem if the onset of MLF is slow. Bacterial spoilage is, of course, unlikely because LAB initially degrade malic acid, but oxidation can be detrimental to wine quality. In general, premature sulfiting can make MLF impossible.

Adding sulfur dioxide when the must is put into tank also has an impact on MLF. The fermentation may be delayed to a variable extent, depending on the concentration of sulfur dioxide used and the way it is mixed into the must (Section 8.8.1) and may, in extreme cases, even be permanently inhibited. The concentration chosen must be sufficient to retard MLF to avoid its interference with alcoholic fermentation and the associated risks, but not so excessive that the MLF cannot be completed within a reasonable time period.

The action of sulfur dioxide depends not only on the concentration chosen but also on grape composition. The pH and soundness of the grapes, in particular, influence the SO_2 binding rate. The ambient temperature is also a factor. Under Bordeaux conditions, 5 g/hl has little effect on delaying MLF, 10 g/hl clearly slows it, and at 15 g/hl or higher, it becomes impossible. In northern/cool-climate vineyards, 5 g/hl can be sufficient to stop it, but in hot regions, MLF may still occur at 20 g/hl.

Determining sulfiting levels to add at harvest is difficult. Sulfiting, however, remains a particularly sensitive method for controlling the MLF process. Proper sulfiting permits a complete alcoholic fermentation, while avoiding spoilage phenomena, without compromising or excessively delaying the MLF. The use of lysozyme (Sections 9.5.1 and 9.5.2) has been recommended to supplement the effect of SO_2 in delaying the development of indigenous bacteria and, thus, the start of MLF.

12.7.5 Malolactic Fermentation Inoculation

The optimum conditions for obtaining MLF in new wines were recommended in the last section. Most often, this fermentation begins within a reasonable amount of time, but it is not always initiated spontaneously. At optimum conditions in a winery where alcoholic fermentation has already occurred, MLF often occurs in at least a few of the fermentors. It can therefore be propagated throughout the winery by massively inoculating the other fermentors. For example, a third of the volume of a fermentor with a completed MLF can be mixed with two-thirds of the volume of a fermentor with a difficult MLF. Fermentors are sometimes inoculated with the lees from a nearly completed MLF tank. The fermentation is thus (hopefully) completed within a few weeks instead of a few months. The elimination of malic acid no longer poses significant practical constraints. Better control of winemaking conditions, especially temperature control and sulfiting, has led to the progressive resolution of past difficulties.

Mixing wines to inoculate other fermentors, however, may run contrary to the legitimate desire of selecting batches according to grape quality. Definitive blending is generally carried out several weeks after MLF, when tasting permits a more exact judgment of wine quality.

The temperature must be maintained while waiting for spontaneous MLF, and this can become costly. In addition, a significant delay at the end of alcoholic fermentation may enable the development of *Brettanomyces* sp. before MLF starts (Volume 2, Section 8.4). Finally, indigenous bacteria are not necessarily higher in quality than commercial strains: an inoculation with selected strains could be preferable.

As a result, research has been focused for a long time on developing commercial, selected bacterial strains that can be inoculated into wine to ferment malic acid. The possibility of implanting genuine MLF starters in wine is definitely interesting, but it has also posed many difficulties, which have been progressively resolved, though these solutions are not definitive. Such inoculation is general practice in many wineries around the world but is not used systematically in all of them. In general, *O. oeni* strains are used; this bacterial species is best adapted to MLF.

For a long time, direct inoculation of new wines after alcoholic fermentation consistently failed. The bacteria introduced could not develop, due to the environmental conditions (pH and alcohol content) encountered in wine. Bacterial populations were observed to regress rapidly, resulting from cell death. The interpretation of this situation can be summarized simply. To have a sufficient biomass, commercial bacteria are initially cultivated in an environment promoting their growth and are therefore adapted to these specific environmental conditions. When placed in wine, a much less favorable environment, they must adapt in order to multiply and initiate MLF. This adaptation becomes increasingly difficult the more the composition of the two media differs. Indigenous bacteria from the grape, however, undergo a progressive selection according to their ability to adapt to changing environmental conditions. They more easily ensure MLF than commercial strains. The difficulty in using commercial strains led to the experimental development of the techniques described below, even if some are no longer in use:

a. Inoculating must before alcoholic fermentation, when the alcohol-free

environment is most favorable to bacterial growth;
b. Using a sufficiently large non-proliferating bacterial biomass to degrade malic acid without cell growth;
c. Inoculating wine after alcoholic fermentation with a commercial biomass that has undergone a reactivation phase just before use;
d. Inoculating wine with a commercially prepared biomass that is already adapted to wine.

Inoculating must before alcoholic fermentation

In traditional winemaking, bacteria from the harvested grapes grow in the sugar-containing must before the onset of alcoholic fermentation (Section 6.3.1). From this initial population, a progressive selection, during alcoholic fermentation, results in a reduced population, which is, however, relatively well adapted to environmental conditions. This reduced population capable of carrying out MLF can be increased by inoculating the must with *O. oeni* at the start of alcoholic fermentation.

To avoid the risk of inhibiting yeasts by a bacterial inoculation, it is advised to inoculate simultaneously with yeasts and bacteria. Current commercial preparations, whether freeze-dried or frozen, contain 10^{11}–10^{12} viable cells per gram. An inoculation of 1 g/hl corresponds to 10^6–10^7 cells/ml. Bacteria can be directly added to the must without preparation beforehand.

First proposed in the 1960s, this method seems to be a satisfactory solution to the problem of inoculation for MLF. This is especially true of the risk of development of spoilage yeast (e.g. *Brettanomyces* sp.). Inoculating this starter either 24 hours or 48 hours after yeasting is called early co-inoculation.

The results, however, are not always as satisfactory as supposed. In practice, three situations can occur:

1. After a significant population decline, bacterial growth occurs toward the end of alcoholic fermentation; MLF starts simultaneously and completes rapidly. This is the ideal situation.
2. The population decline leads to complete disappearance of the inoculated bacteria. MLF kinetics are not improved. The inoculation has no effect. Nevertheless, this situation is less and less frequent with the starters currently available.
3. A difficult alcoholic fermentation, accentuated by antagonistic phenomena between yeasts and the high bacterial population, leads to a stuck alcoholic fermentation and premature growth of LAB in a sugar-containing medium. Volatile acidity is produced. *O. oeni* is the best adapted bacterium for MLF. It is, however, a heterofermentative coccus that forms acetic acid from sugars. This risk is very rare, however, with the inoculation of malolactic starters. The production of volatile acidity remains limited when LAB starters are used.

As a result, the early inoculation of LAB starters (common in some US and Australian wineries) has grown a lot over the past decade. It helps avoid the "ecological vacuum" between the end of alcoholic fermentation and the start of MLF, which favors the growth of *Brettanomyces* sp. yeasts.

More recently, another attempt to inoculate must before alcoholic fermentation was made using a *Lactobacillus plantarum* starter (Prahl et al., 1988). This process uses non-proliferating cells. It is described below.

Inoculating with non-proliferating bacteria

Having witnessed the difficulty of obtaining LAB growth in wine, Lafon-Lafourcade (1970) studied the possibility of obtaining malic acid degradation by using a biomass sufficiently abundant and rich in malolactic enzyme so that the reaction can occur without cell reproduction.

When the evolution of an inoculated bacteria population in wine is studied, an abrupt drop in the number of viable cells is observed in the first few hours. Afterward, the decline is slower. After several days, bacterial growth may occur, but this growth is too uncertain to be used as a technique for initiating MLF. Yet, during the decline of the population, the malolactic enzyme supplied by the bacteria induces the partial degradation of malic acid. In this case, the bacteria do not act as a fermentation starter but rather as an enzymatic support.

Despite efforts to establish the necessary conditions, an *O. oeni* biomass inoculated in wine is not capable of degrading all of the malic acid present. The complete reaction can only be obtained by massive inoculations (1–5 g/l), which are not feasible in practice. In general, when the population has completely disappeared, the reaction stops, leaving malic acid. In addition, the malolactic activity of commercial preparations rapidly diminishes during conservation, even at low temperatures.

The kinetics of the reaction could possibly be improved by fixing cells or even enzymatic preparations on solid supports. The resulting protection with respect to the medium could increase the average duration of the enzymatic activity. Wine would circulate in these reactors to have malic acid converted into lactic acid. At present, this research has not led to practical applications.

The reaction would of course be easier in the must before alcoholic fermentation, but this technique is not feasible with heterofermentative *Oenococcus* strains. The risk of these bacteria developing in a sugar-containing medium cannot be taken, since volatile acidity production would be significant (see above).

However, a non-proliferating *L. plantarum* biomass could be introduced in the must. This homofermentative strain exclusively produces lactate from sugars. Prahl et al. (1988) demonstrated that a *L. plantarum* preparation could be inoculated into the must when filling the fermentor in order to degrade malic acid. The preparation contains 5×10^{11} viable cells/g. A concentration of 10 g/hl is used, corresponding to 5×10^7 cells/ml. Malic acid degradation is initiated rapidly; it then continues slowly and is completed during alcoholic fermentation. These bacteria are not resistant to ethanol. As a result, their activity progressively diminishes; sugar assimilation is negligible, and no volatile acidity production is observed. This method is simple and has no adverse organoleptic effects, but its use is limited, due to the risk of the bacteria population completely disappearing before the end of the reaction. Furthermore, the bacteria are sensitive to free sulfur dioxide. For these various reasons, the general application of this technique is not possible for the moment.

Inoculating with commercial *Oenococcus oeni* preparations after reactivation

Oenococcus oeni is the best adapted strain for MLF in wine. It is involved in practically all spontaneous fermentations. Due to the presence of ethanol, adding this strain in wine after alcoholic fermentation results in a significant decline in its population. Part of the malic acid may be degraded, but the cell growth necessary for ensuring a complete MLF does not consistently occur.

Lafon-Lafourcade et al. (1983) were the first to show that bacteria survival could be improved during their transfer to wine, as long as the population is brought to a suitable physiological state beforehand. These authors proposed using the expression "reactivation" to designate this operation. In fact, this is not a simple pre-cultivation. The population increase that accompanies this operation is a beneficial side effect but is not the primary objective sought.

Many authors have since used this idea of reactivation. Although many different procedures have been proposed, that of Lafon-Lafourcade et al. (1983) is the most commonly used. Non-sulfited grape juice is diluted to half its original concentration (80 g/l of sugar per liter); a commercial yeast autolysate is added (5 g/l); and the pH is adjusted to 4.5 with $CaCO_3$. After several hours, commercial biomasses inoculated at 10^6 cells/ml at 25°C produce fermentation starters rich in malolactic enzymes. These starters are also more resistant in wine than non-reactivated starters. Populations increase to 10^6, 10^7, and 10^9 cells/ml after two hours, 24 hours, and six days of reactivation, respectively.

The starters prepared in this manner are inoculated into wine after alcoholic fermentation. Table 12.16 attests to the effectiveness of this operation. In all cases, wine is inoculated at 10^6 cells/ml. By the end of the 12 days, cell growth has occurred, and MLF is nearly complete, if the starter has undergone a reactivation of 24 hours or six days. A two-hour reactivation is insufficient. Without reactivation, the population declines, and MLF is still not initiated after 12 days.

In practice, the reactivated starter preparation added to wine should not exceed a concentration of 1/1,000, since the yeast autolysate is highly odorous. To obtain a cell concentration of 10^6–10^7 in wine, its concentration must be between 10^9 and 10^{10} in the reactivated medium. This result is obtained by inoculating the reactivation medium at 10^8 and 10^9 cells/ml with a reactivation time of 48–72 hours. Commercial starter preparations contain 10^{10}–10^{11} viable cells per gram. The reactivation medium must therefore be inoculated at 10 g/l.

This method is effective, but it does require a certain level of knowledge of microbiological methods—not always possible in wineries. This constraint limits its more widespread use. Many wineries prefer spontaneous MLF, even though that requires more time.

Inoculating with commercial *Oenococcus oeni* preparations not requiring a reactivation phase: Direct inoculation of starters in wine
For a long time, attempts were made to inoculate commercial biomasses directly into wine after alcoholic fermentation, but these failed. Bacteria populations had difficulty adapting to the physicochemical conditions of wine.

The reactivation procedure previously described could be assumed to confer an indispensable characteristic to bacteria. It would therefore be very difficult (if not impossible) to obtain commercial preparations ready for use in wine. However, in the last two decades, starters that can be inoculated directly into wine immediately after alcoholic fermentation, or even before the end of MLF, have been developed. Success rates are high. Experimental results obtained with this preparation in the laboratory and in the winery have shown that

TABLE 12.16

Effect of Bacteria Reactivation Conditions on Malolactic Fermentation (Lafon-Lafourcade *et al.*, 1983)

Measurement on 12th day	Inoculation with non-reactivated biomass	Inoculation with reactivated biomass: duration of reactivation		
		2 hours	24 hours	6 days
Population (cell/ml)	10^5	3×10^7	4.4×10^7	9.4×10^7
Malic acid degraded (g/l)	0	2.3	3.7	3.6

Initial malic acid concentration: 4.5 g/l.
Bacteria inoculation: 10^6 cells/ml.
Temperature: 19°C.

bacterial growth and MLF can be obtained 15 days early, with respect to a control (Figure 12.14). No organoleptic flaws are observed.

The effectiveness of this preparation is based on selecting a suitable strain, in terms of its resistance to alcohol, pH, SO_2, and various other limiting factors in wine. It also depends on the particular preparation conditions of the commercial biomasses. This preparation includes a progressive adaptation to the limiting environmental conditions of wine.

12.8 Automated Red Winemaking Methods

12.8.1 Introduction

In red winemaking, the complexity of the operations linked to controlling pomace extraction lends itself to the development of manufacturing processes permitting the automation of winemaking steps. Equipping fermentors to provide a certain level of automation has already been discussed (Section 12.3.3). In the 1960s, the development of two particular winemaking techniques focused on automation: continuous winemaking and thermovinification (heating the harvested grapes). The monograph *Sciences et Techniques du Vin* (Ribéreau-Gayon et al., 1976) was published while these techniques were being developed. It gave a detailed description of these methods. They were subject to the same popularity that all innovations were causing at that time. Some techniques were being generalized that, in reality, were best suited to highly specific applications. Today, the use of these techniques is on the decline; they are still worth mentioning but no longer justify a detailed description.

12.8.2 Continuous Winemaking

Initially, the development of continuous winemaking was based on the advantages of continuous fermentation. This method was adopted in certain industries using

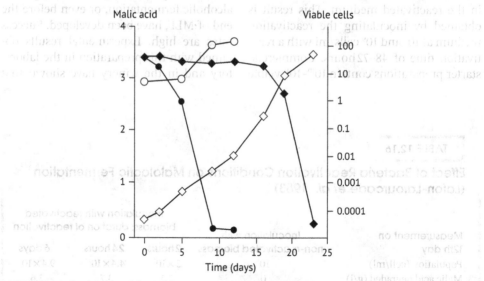

FIGURE 12.14 Induction of malolactic fermentation in red wine (Cabernet Sauvignon, Graves, 1992) by direct inoculation with a freeze-dried preparation (*Viniflora oenos*, Chr. Hansen's Laboratory, Denmark H/S). Temperature = 20°C; pH = 3.5; alcohol = 11.8% vol.; total SO_2 = 5 mg/l; glucose = 0.30 g/l; fructose = 0.45 g/l. Malic acid (g/l): — ● — (inoculated medium); — ♦ — (control). Viable cells (10^6/ml): — ○ — (inoculated medium); (10^6/ml) — ◊ — (control).

fermentation because of its rapidity and regularity. Continuous fermentation is generally conducted in a battery of communicating fermentors. Fresh must enters at one end, and the fermented product flows out from the other. Under these conditions, yeast growth is well controlled, and their population and fermentation activity are at their maximum. The same conditions may be reproduced by regularly supplying a single fermentor with must at the bottom and extracting the fermented product from the top at the same rate as the supply.

In actual red winemaking, fermentation and maceration are sought simultaneously. For this reason, continuous fermentation cannot provide the full benefits of traditional winemaking techniques.

Continuous winemaking uses rational equipment and permits rigorous operating control and good work organization. It is best applied to high-volume winemaking to produce wines of the same quality and style.

Continuous fermentors comprise a 400- to 4,000-hl stainless steel tower. A 4,000-hl system can handle 130 metric tons of harvest per day, and it can produce approximately 23,000 hl of wine in three weeks. An annual wine production of 40,000 hl is necessary to justify the costs of such a system. These fermentors (Figure 12.15) enable the daily reception of fresh grapes and the evacuation of an equivalent amount of partially fermented wine and pomace. In the upper part, a rotating rake removes the skins toward a continuous press. A quarter of the total volume of the tower is renewed each day. This corresponds with a four-day average maceration time. The seeds that accumulate at the bottom of the tank are regularly eliminated;

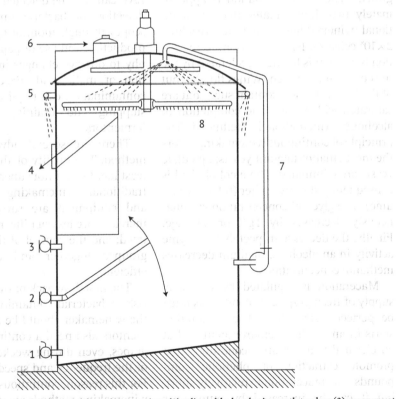

FIGURE 12.15 Continuous fermentor: (1) seed evacuation; (2) adjustable-level wine outlet valve; (3) grape must inlet valve; (4) pomace evacuation; (5) cooling nozzles; (6) expansion dome for wine storage; (7) pumping-over conduit; (8) pomace extraction rake (Ribéreau-Gayon et al., 1976).

seed maceration for long periods in the presence of alcohol can confer herbaceous flavors and excessive astringency to wine. The weight of the seeds thus eliminated depends on tank dimensions and sometimes attains one metric ton per day.

The winemaker has all of the equipment required for controlling operations at his or her disposal. Before being transferred to the fermentor, the harvest is automatically sulfited by a dosing pump. Pre-fermentation adjustments such as modifying acidity and chaptalization can be performed. Temperature control and pump-over operations are automated. The daily supply of fresh grapes minimizes temperature increases: under similar conditions, the temperatures in continuous fermentors are 5–7°C lower than in traditional batch fermentors.

In continuous winemaking, environmental conditions are favorable to yeast growth. The yeast population is approximately two times greater than in traditional winemaking, sometimes reaching 2×10^8 cells/ml. For this reason, fermentation is rapid. It is further accelerated by the introduction of oxygen. Wines flowing out of the fermentor still contain sugar but are saturated with yeasts. The completion of alcoholic fermentation is facilitated. The principle of continuous winemaking favors the most ethanol-tolerant yeasts; apiculate yeasts are eliminated. The alcohol yield is consequently slightly higher (0.1–0.2% volume). The glycerol concentration simultaneously decreases by 1g/l on average. Finally, the decrease in pectolytic enzyme activity in an alcoholic medium decreases methanol concentrations.

Maceration is regulated by the daily supply of fresh grapes. Its conditions must be perfectly controlled. The maceration starts in an alcoholic environment and at an elevated temperature—conditions that promote extraction of phenolic compounds. The maceration is relatively short, but it can be increased by pump-over operations. The concentration of phenolic compounds in the wine is related to the frequency of draining and addition of fresh grapes.

When this method is correctly applied, the resulting wines have no significant organoleptic differences with respect to traditionally made wines.

Continuous fermentors present a particularly high risk of bacterial contamination. Their operating conditions lend themselves to LAB development, and MLF can be initiated since the fermentor is continuously supplied with fresh grapes. In this sugar-containing environment, lactic spoilage may occur inside the fermentor. To avoid this dangerous contamination, a homogeneous sulfiting is recommended. The SO_2 concentrations should be slightly higher than in traditional winemaking. Lactic acid isomer analysis is particularly effective for detecting bacterial contamination in continuous fermentors (Section 12.4.3) (Peynaud et al., 1966). Contamination by LAB can thus be detected (well before the detection of bacteria under the microscope) through monitoring of acetic acid production and use of paper chromatography to observe changes in the concentration of malic acid. Excessive microbial contamination can require the immediate stopping and draining of continuous fermentors.

There are several advantages to this method. The quality of the products is at least identical, if not superior, to that from traditional winemaking; space, labor, and equipment are saved; temperature increases are less significant; MLF is facilitated; and the control of the operations is grouped together and therefore more efficient.

The first drawback of this method is the risk of bacterial contamination, to which the winemaker should be alert. These fermentors also need a continuous supply of grapes, even during weekends, regardless of the frequency and speed of the harvest. For this reason, continuous and traditional winemaking methods should be employed simultaneously to adapt to varying conditions.

The primary disadvantage is the need to mix grapes of different origins and quality. Grapes cannot be selected, nor can their diversity be expressed in the wine: a single type is produced. This approach is contrary to current winemaking concepts—the diversity of grape origins is now emphasized. For this reason (at least in France), after a period of growth, this technique has lost popularity.

In the first half of the 20th century, various winemaking methods using continuous fermentation were studied, in particular in the Soviet Union. The first industrial continuous fermentors appeared in Argentina in 1948 and were later developed in Algeria and in the south of France. The largest expansion of this method was in the 1960s and 1970s, when about 100 of these facilities were built. Today, their use is on the decline.

12.8.3 Thermovinification: Heating the Harvest

Heating whole or crushed grapes promotes the diffusion of phenolic compounds from the skins. More highly colored musts are thus obtained. This phenomenon has been known for a long time; it was referred to even in the 18th century. Attempts have long been made to increase red wine color by heating.

Until fairly recently, heating methods remained very empirical. Only part of the harvest was heated; it was then blended with the rest of the tank and underwent traditional winemaking methods.

The idea is not new but, during the last 30 years, industrial heating processes have been developed. They permit large volumes of grapes to be heated rapidly to high temperatures (65–75°C). Various techniques are used, although heating the grapes directly with steam has been almost entirely abandoned (Blouin and Peynaud, 2001). Destemmed, crushed grapes may be heated directly in a tubular heat exchanger, heated by steam or, preferably, hot water, or plunged into juice that has been separated from the solids and heated.

The pressed juice may be cooled before fermentation, but if the must is to be fermented on the skins, the solid and liquid components must be cooled together, which is a much more complex operation, requiring special equipment.

Products based on this method were developed with two distinct objectives. In one application, the method was integrated into traditional winemaking to increase concentrations of phenolic compounds, especially anthocyanins (color). In the other, it was used to automate red winemaking.

Heating the grapes to extract more color is not currently in favor, at least in controlled appellation (AOC, DOC, DO, etc.) vineyards. First of all, fermentors are now preferably equipped with temperature control systems, which permit a more flexible use of heat to promote the extraction of phenolic compounds. Excessive heating of the entire crushed grape crop, combined with a traditional maceration, might cause excessive tannic bitterness—the wine is consequently without finesse. The increased must color obtained through heating the crushed grapes has also been shown to be unstable, disappearing during fermentation (Table 12.17).

In addition, even if young thermovinification wines are more highly colored than traditionally made wines, they progressively lose this advantage during maturation.

Thermovinification units were developed above all with the goal of automating winemaking. The destemmed, crushed grapes are heated to between 65 and 75°C and then transferred to a stabilization tank for several minutes. The results depend on the temperature used and on the length of time the heat is applied. The grapes are then cooled and pressed. The highly colored juice is then fermented. During this time, it loses part of its color. All of the operations can be automated, which results in substantial savings in labor costs. Moreover, this system significantly decreases the total fermentor volume needed. This

TABLE 12.17

Evolution of Anthocyanin Content and Color Intensity in a Heated and Pressed Must, Compared with Traditional Winemaking, During Alcoholic Fermentation (Ribéreau-Gayon et al., 1976)

Fermentation time (hours)	Traditional winemaking		Thermovinification of red grape must	
	Anthocyanins (mg/l)	Color intensity	Anthocyanins (mg/l)	Color intensity
0	252	0.37	816	3.08
3	248	0.45	810	2.98
6	244	0.47	824	3.35
10	200	0.48	936	3.36
24	260	0.51	596	1.23
72	302	0.81	508	1.00
96	400	1.16	540	1.20
End of fermentation	468	0.75	476	0.92

alone can justify the installation of a thermovinification unit.

Whatever the heating method used, it is recommended that the must or crushed grapes should be cooled before the onset of fermentation, which must take place at approximately 20°C. Excessive production of volatile acidity by yeasts can hopefully be avoided in this way.

Tasting results are not always homogeneous and depend on grape composition and on heating and hot maceration conditions. The participation of these factors is poorly understood. In certain cases, the wines obtained have more color and are better than the traditionally made control wines. They can be rounder and fuller bodied, while still having a fruitiness that gives them personality. In other cases, they have abnormal tastes, a vegetal aroma dominated by amyl notes, a loss of freshness, and a bitter aftertaste.

Figure 12.16 shows that the temperature should be higher than 40°C for 15 minutes to obtain significant color extraction, but the extraction is not increased for temperatures above 80°C. Identical results are observed for the tannins. For this reason, a temperature of 70°C for 10 minutes corresponds to a standard thermovinification treatment.

Heating grapes destroys the natural pectolytic enzymes of the grape, and so spontaneous clarification of new wines is difficult. This circumstance intensifies potential sensory flaws. Adding commercial pectolytic enzymes can resolve this problem, but their effectiveness varies.

Destruction of oxidases and protection against oxidation are favorable consequences of thermovinification. Rotten grapes benefit the most from this treatment as they contain laccase, which has a significant oxidizing activity. However, enzymes are only destroyed at temperatures over 60°C, while their activity increases with temperature up to that point, so the must has to be heated very rapidly.

Also, it has been noticed that heating Cabernet Sauvignon must attenuates the green bell pepper character due to 3-alkyl-2-methoxypyrazines in insufficiently ripe grapes.

Heating also affects fermentation kinetics. At temperatures well above those that kill yeasts, heated musts ferment easily. Yeast activity continues at temperatures that yeasts generally do not support. However, this heating destroys nearly all of the yeasts originating from the grapes. A second natural inoculation occurs during the subsequent handling of juice and skins, and this new

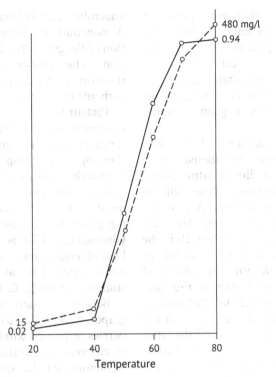

FIGURE 12.16 Anthocyanin extraction and evolution of color intensity according to temperature (Ribéreau-Gayon et al., 1976). ——— Color intensity (OD 520 + OD 420) - - - - Anthocyanins (mg/l). T = heating temperature (°C). Heating time = 15 minutes.

population rapidly becomes significant. Thus, manual inoculation is unnecessary. Heating is therefore not a viable method for killing the indigenous yeast population, which should be eliminated when using selected yeast strains. This activation of fermentation is not due to a natural selection of thermoresistant yeasts; it is most likely caused by the dissolution, or at least dispersal, of activators in the grape must belonging to the steroid family. These activators come from the grape skins. Flash pasteurization, i.e. rapid heating to a high temperature, has also been suggested as a means of restarting a stuck fermentation (Section 3.8.3).

Nitrogen compounds may also be involved in the improvement of fermentation kinetics. Heating the crushed grapes increases not only the total nitrogen and amino compound concentrations but also the consumption of nitrogen during fermentation.

Heating grapes puts many complex chemical and microbiological phenomena into play. Yet until these phenomena are better understood, separating the maceration and fermentation phases has no distinct advantage. In addition, the performance of temperature control systems used in traditional batch fermentors is continually improving. These systems often produce higher-quality wines. For this reason, thermovinification techniques no longer present the same interest as they did not so long ago.

12.9 Carbonic Maceration

12.9.1 Principles

Like all plant organs, the grape berry has an aerobic metabolism. Respiration produces the necessary energy to ensure its vital functions. In this complex chain of reactions, the grape makes use of oxygen from air to decompose sugar into water and carbon dioxide. Yet when many plants are deprived of air, they adopt an anaerobic metabolism and produce ethanol from sugars. *Saccharomyces cerevisiae* is the classic example of this phenomenon. The anaerobic metabolism is significant because this yeast has a good tolerance to ethanol.

The whole, uncrushed berry also develops an anaerobic metabolism when placed in a carbon dioxide atmosphere. During this phenomenon, various chemical and physicochemical processes occur, especially ethanol production. They are linked to the functioning of the cells in the whole berry, but, in contrast to yeasts, grape berry cells are not very tolerant of ethanol. Ethanol production is therefore limited: it varies from 1.21 to 1.89% volume for the Carignan variety, regardless of the must sugar concentration, when the latter is between 184 and 212 g/l (Flanzy et al., 1987). The intensity of anaerobic metabolism depends on variety, vintage, and maceration temperature and duration.

Enzyme systems in the grape cells, particularly alcohol dehydrogenase, cause the phenomena that give carbonic maceration wines their specific character.

Anaerobic metabolism occurs whenever the oxygen concentration is low, in either a gaseous or liquid environment, but in a liquid environment the intensity of the phenomenon diminishes (Flanzy et al., 1987). Whole grapes immersed in must undergo a less intense anaerobic metabolism than the same grapes placed in a carbon dioxide atmosphere. This diminution is due to exchanges between the berry and the ambient environment, which are greater in the liquid phase than in a gaseous atmosphere. The diffusion of sugars, phenolic compounds, and malic acid across the grape skin toward the solution lowers the concentration of anaerobic metabolism substrates in the berry. In addition, the diffusion occurs in both directions. When intact berries are placed in a medium containing alcohol, their ethanol concentration is increased, thus inhibiting anaerobic metabolism. This observation demonstrates the importance of the condition of the grape crop for carbonic maceration. The higher the proportion of uncrushed grapes, the more effective is carbonic maceration.

Pasteur is credited with first noting the taste modification of whole berries during fermentation. He confirmed his observations by placing grapes in a bell jar filled with carbon dioxide. These grapes took on a vinous odor and taste reminiscent of fermented grapes. He concluded that crushing grapes has an essential impact on red winemaking. This is the basic principle behind fermentation with carbonic maceration, initiated by M. Flanzy (1935) and studied in detail by C. Flanzy (1998).

Before mechanized crushing, when grapes were still crushed by foot, many berries remained whole. A certain degree of carbonic maceration occurred in the fermentor. At the same time, the juice of uncrushed grapes was progressively released by the weight of the harvest, thus fermenting more slowly. Consequently, the fermentor temperature was moderate. In warm climates, winemakers directly benefited from this phenomenon in the past.

Carbonic maceration comprises two steps:

- The fermentor is filled with whole grapes under a blanket of carbon dioxide and kept at a moderate temperature (20–30°C) for one to two weeks. This is the pure carbonic maceration phase. Anaerobic metabolism reactions modify grape composition. Substances from the solid tissue disintegrated by anaerobic conditions are also diffused in the juice and the pulp.
- The fermentor is emptied and the pomace is pressed. The free-run and press wines are usually blended together prior to normal alcoholic and malolactic fermentation.

In practice, it is impossible to fill a fermentor with only whole berries. Some are crushed and their juice undergoes a normal alcoholic fermentation. During maceration, additional grapes continue to be crushed as the processes occurring under anaerobic conditions weaken cell tissue. The fermentation of completely crushed grapes and pure carbonic maceration occur simultaneously to varying degrees. The condition of the grapes influences the amount and intensity of carbonic maceration. In practice, during the first step of winemaking, yeast-based fermentation always accompanies the anaerobic metabolism of the berry. The winemaker should take steps to minimize this interference.

12.9.2 Gaseous Exchanges

During the first hours of anaerobic conditions, the berry tissues absorb carbon dioxide. Metabolic pathways make use of this dissolved CO_2. Using CO_2 marked with ^{14}C, it has been demonstrated that the gas is integrated not only into various substrates, such as malic acid and amino acids, but also into sugar and alcohol. The volume of carbon dioxide dissolved in the berry in this manner is temperature dependent. It represents 10% of berry volume at 35°C, 30% at 20°C and 50% at 15°C (Flanzy et al., 1987).

The berry metabolism simultaneously releases CO_2, which eventually attains an equilibrium with the amount absorbed. In a closed system, the equilibrium is established in six hours at 35°C, 24 hours at 25°C, and three days at 15°C (Flanzy et al., 1987).

The initial CO_2 concentration controls the intensity of the anaerobic maceration phenomena, reflected by ethanol production. Under certain experimental conditions, for a given time and temperature, this production can vary by a factor of two, depending on the CO_2 concentration in the atmosphere (20–100%).

12.9.3 Anaerobic Metabolism

It has long been known that the grape berry is capable of producing ethanol. This production is always low and depends on the variety. According to different authors, it varies from 1.2 to 2.5% vol. or from 0.44 to 2.20% vol. The speed and limits of ethanol production are governed by temperature (Figure 12.17). Maximum production is obtained earlier at higher temperatures than at lower temperatures, but a higher maximum is obtained at lower temperatures.

Temperature is consequently a major factor in the intensity of anaerobic metabolism. Raising the temperature of excessively cool grapes is therefore recommended. Heating conditions also have an impact.

The yield of the transformation of sugar to alcohol is difficult to determine. It seems to be similar to the alcoholic yield of yeasts—18.5 g of sugar per 1% vol. of ethanol. Various secondary products are simultaneously formed: 1.45–2.42 g of glycerol, 21–46 mg of acetaldehyde, approximately 300 mg of succinic acid, and 40–60 mg of acetic acid per liter. The presence of all of these compounds indicates the existence of a mechanism similar to yeast-based alcoholic fermentation. Yet in this case, the share of glyceropyruvic fermentation would seem to be greater, since the average glycerol/ethanol ratio×100 is 18–20% instead of 8%.

During anaerobic metabolism, total berry acidity diminishes. Peynaud and Guimberteau (1962) demonstrated in rigorous laboratory experiments that tartaric and citric acid concentrations remained constant, while malic acid concentrations dropped sharply. The degree of this decrease depends on the variety: 32% for Petit Verdot, 42% for Cabernet Franc, 15% for Grenache Gris, and 57% for Grenache Noir. As with ethanol production, temperature affects malic acid degradation (Figure 12.18). It regulates the speed and limit of the phenomenon.

Decrease in malic acid is a major effect of carbonic maceration. Ethanol is produced after a double decarboxylation. Yeasts use an identical mechanism (Figure 12.19). Two enzymes have been confirmed as being involved in these

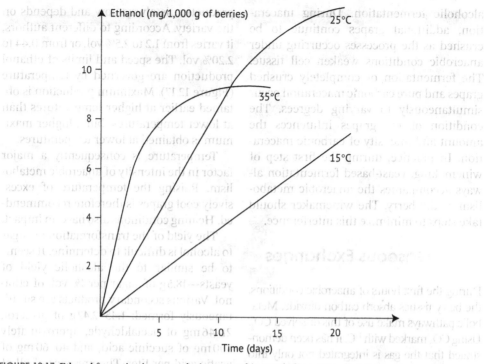

FIGURE 12.17 Ethanol formation in grapes during anaerobic metabolism, according to temperature (Flanzy et al. 1987).

reactions—they are even considered as markers of an anaerobic metabolism. The specific activity of the malic enzyme reaches its maximum between the third and fourth days of anaerobic conditions at 35°C. During the same period, alcohol dehydrogenase is progressively inactivated. This is probably linked to the accumulation of ethanol (Flanzy et al., 1987).

Traces of succinic, fumaric, and shikimic acids, but not lactic acid, are generally considered to be formed by this anaerobic metabolism. Ascorbic acid concentrations decrease. There are also significant changes in the nitrogen compounds, including an increase in the amino acid content, probably dissolved from the solids in the must, as well as a decrease in protein nitrogen.

The anaerobic metabolism also causes a breakdown of the cell walls, with hydrolysis of the pectins leading to an increase in the methanol content, which may reach levels up to 80 mg/l, corresponding to the hydrolysis of approximately 500 mg of pectin.

Finally, within 30 minutes, the development of an anaerobic metabolism in the grapes leads to a significant decrease in the ATP and ADP molecules responsible for energy transfers in biological systems. After an initial decrease of approximately 20% at the time of passing into anaerobic conditions, the energy charge (EC = $(ATP + \frac{1}{2}ADP)/(ATP + ADP + AMP)$) stabilizes for about one week before decreasing again. Under anaerobic conditions, the capacity to regenerate energy-rich bonds (ATP, ADP) is limited (Flanzy et al., 1987).

An important result of carbonic maceration during red winemaking is the characteristic aroma produced. The nature and origin of the molecules involved in this aroma remain rather obscure. According to Flanzy et al. (1987), the formation of aspartic acid from malic acid, along with the formation of succinic and shikimic acids, may be the source of aroma precursors. These researchers also noticed differences in

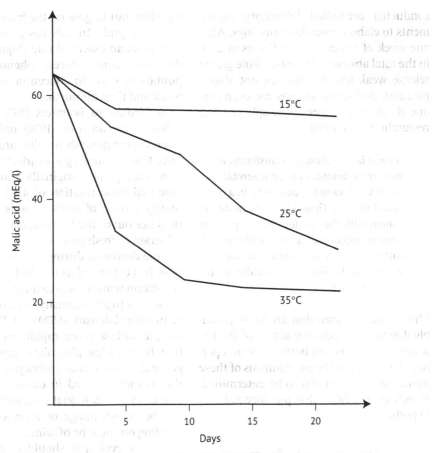

FIGURE 12.18 Malic acid degradation during anaerobic metabolism, according to temperature (Flanzy et al., 1987).

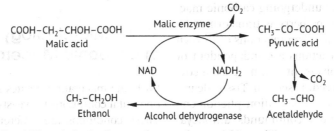

FIGURE 12.19 Malic acid degradation by the grape berry under anaerobic conditions.

higher alcohol and fermentation ester concentrations with respect to wines that did not undergo carbonic maceration. However, the principal difference is the increase in various aromatic derivatives: vinylbenzene, 2-ethylphenyl acetate, benzaldehyde, 4-vinylguaiacol, 4-vinylphenol, 4-ethylguaiacol, 4-ethylphenol, eugenol, and methyl and ethyl vanillate. Ethyl cinnamate, in particular, was proposed as an indicator of carbonic maceration wines.

Peynaud and Guimberteau (1962) pointed out that the simultaneous action of the intracellular berry and yeast cell metabolisms were responsible for the pleasant aroma produced in fermentors. They conducted controlled laboratory experiments to elaborate on these findings. After one week of anaerobic conditions at 25°C in the total absence of yeasts, whole grapes release weak aromas that are not always pleasant. Reduction aromas are even produced in a nitrogen atmosphere. The researchers concluded:

> If there is a biochemical transformation of essential substances under anaerobic conditions, it does not appear to be in a favorable direction. These observations do not concur with the development of pleasant aromas noted during winemaking with carbonic maceration. The aroma improvement may be due in particular to the action of yeasts.

The carbonic maceration aroma is probably due to the successive action of the anaerobic metabolism of berry, yeast and perhaps bacteria, but the mechanisms of these transformations remain to be determined. Flanzy et al. (1987) also put forward this hypothesis.

12.9.4 Grape Transformations by Carbonic Maceration

In a fermentor undergoing carbonic maceration, the grape berry is transformed by anaerobic metabolism reactions of its own cells. These reactions are independent of any yeast involvement and have been covered in the preceding section. Tissue degradation favors the maceration phenomena involved. Phenolic compounds, anthocyanins, nitrogen compounds, and other components of the solid parts of the berry are diffused in the juice of the pulp.

The data in Table 12.18 express the consequences of these phenomena. A slight increase in nitrogen and possibly mineral concentrations is observed. There is also a systematic increase in total polyphenol concentrations. The dissolution of anthocyanins also results in an increase in CI. This increase is considerable for certain varieties, but in general the juice obtained is simply pink. In this case, temperature also plays an essential role (Figure 12.20). At high temperatures, phenolic compounds increase in concentration for one week and then diminish.

According to Bourzeix (1971, cited in Ribéreau-Gayon et al., 1976), only 0.7 g of phenolic compounds per liter are found in juice that has undergone carbonic maceration from grapes originally containing a potential concentration of 4 g/l. Approximately 150 mg of anthocyanins pass into the juice out of the 1,650 mg contained in a kilogram of fresh grapes.

Hue decreases during carbonic maceration. It is expressed as the degree of yellow coloration with respect to the degree of red coloration (optical density at 420 nm divided by optical density at 520 nm). The anthocyanins diffuse more rapidly in the juice than the colorless phenolic compounds. In general, tannin and anthocyanin extraction is more limited in carbonic maceration than in traditional winemaking. This can be an advantage or a drawback, depending on the type of wine.

Aroma evolution should of course be taken into account when considering grape transformations by carbonic maceration, but no experimental results currently exist.

12.9.5 Microbiology of Carbonic Maceration

Carbonic maceration creates specific developmental conditions for yeasts and bacteria. These conditions are different in the two stages of the process. They are also different with respect to traditional winemaking.

In the first stage, the yeasts originate from the grape or are added as a yeast starter. They develop in juice produced by the progressive crushing of a portion of the grapes in a carbon dioxide atmosphere (containing no oxygen) and a non-sulfited or lightly sulfited environment. At the time of draining and pressing, yeast populations attain 10^8 and sometimes 2×10^8 cells/ml.

TABLE 12.18
Changes in Composition of Cabernet Franc and Petit Verdot Grape Juice Under Anaerobic Conditions (Eight Days at 25°C in a CO_2 or Nitrogen Atmosphere) (Peynaud and Guimberteau, 1962)

Component	Cabernet Franc			Petit Verdot		
	Control	CO_2	Nitrogen	Control	CO_2	Nitrogen
Reducing sugars (g/l)	200	162	162	145	109	104
Ethanol (g/l)	0	15.9	15.1	0	17.5	18.3
Glycerol (g/l)	0.23	2.65		0.60	2.05	
Acetaldehyde (mg/l)	12	47	54	16	58	37
Methanol (mg/l)	0	50	50	0	70	50
Total nitrogen (mg/l)	532	588	574	490	588	588
Permanganate index[a]	9	12	10	9	19	18
Color intensity	0.20	0.30	0.21	0.09	0.72	0.65
Hue	0.90	0.71	0.79	1.00	0.56	0.60
pH	3.25	3.40	3.40	3.00	3.35	3.30
Titratable acidity (mEq/l)	96	80	84	134	98	102
Ash alkalinity (mEq/l)	52	52	50	49	55	52
NH_4^+	8.4	5.2	7.2	6.4	5.2	5.6
Sum of cations (mEq/l)	**156**	**137**	**141**	**189**	**158**	**160**
Tartaric acid (mEq/l)	92	92	94	110	96	98
Malic acid (mEq/l)	50	29	34	65	44	42
Citric acid (mEq/l)	2.5	2.3	2.1	3.0	2.5	2.9
Phosphoric acid (mEq/l)	2.1	2.1	2.1	3.1	3.1	3.1
Acetic acid (mEq/l)	0.6	1.8	1.8	0.6	2.4	1.8
Succinic acid (mEq/l)	0	5.0		0	5.0	
Sum of anions (mEq/l)	**147**	**132**	**134**	**182**	**153**	**148**

Bold denotes the sum of cations/anions values.
[a] Total phenol index.

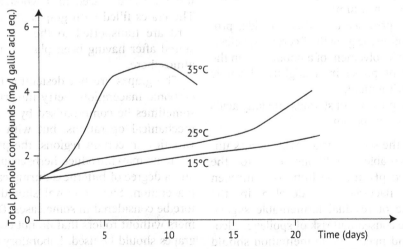

FIGURE 12.20 Influence of temperature on phenolic compound diffusion in juice pulp during anaerobic metabolism (Flanzy et al., 1987).

This significant yeast growth is explained by the low ethanol concentration and the presence of compounds from grape bloom (oleic and oleanolic acids). These unsaturated fatty acids, like sterols, activate the fermentation and compensate for the absence of oxygen to a certain degree. This large and active population ensures a rapid fermentation of sugar during the second stage of this process. The increase in yeast-assimilable nitrogen during anaerobic metabolism certainly favors fermentation. However, if high temperatures (35°C) are attained during the carbonic maceration phase, all or part of the yeast population may be destroyed and alcoholic fermentation may become stuck, creating an opportunity for bacterial growth. In this case, both volatile acidity and fermentation inhibitors such as ornithine are produced. These inhibitors reinforce fermentation difficulties and increase the risks of stuck fermentations (Flanzy et al., 1987).

Carbonic maceration consequently facilitates the development of LAB and the MLF process. The risk of bacterial spoilage is also greater, especially during difficult alcoholic fermentations.

Many factors promote bacterial development:

- the absence of sulfiting or at least the irregular sulfiting of a heterogeneous medium;
- the presence of carbon dioxide, promoting the growth of certain species;
- the involvement of a population in the latent phase in a slightly alcoholic environment;
- the presence of steroids and fatty acids from the bloom.

During the second stage, MLF occurs under favorable conditions due to the increased pH and the improved nitrogen supply. Bacteria may develop in the presence of residual fermentable sugars, with the consequent risk of spoilage. Alcoholic and malolactic fermentation should never overlap (Section 3.8.2). Freshly picked grapes must be sulfited prior to carbonic maceration, even if it is difficult to distribute the sulfur dioxide evenly among the berries. The various phases of the microbiological processes should also be rigorously monitored.

12.9.6 Using Carbonic Maceration

Different systems can be developed to make use of carbonic maceration. Its success depends on anaerobic metabolism intensity, which itself depends on fruit integrity, degree of anaerobic conditions, possible traces of oxygen, and time/temperature.

Harvesting, transport and maceration methods must ensure the integrity of the grapes and grape clusters. Flanzy et al. (1987) have described various methods, including placing picked grapes in small containers to avoid crushing and using a system to fill the fermentors gently, after possibly weighing the grapes, to limit the bursting. Pumps should never be used to transfer grapes from the receiving area to a fermentor. Conveyor systems are preferred since they maintain tissue integrity better than worm screws.

Anaerobic conditions are generally obtained in an airtight fermentor, but grapes can also be wrapped in an airtight plastic tarpaulin and placed in a wooden crate. The crates filled with grapes at the vineyard are transported to the winery and stored after having been placed in a CO_2 atmosphere.

The grapes are not destemmed before carbonic maceration. Berry integrity may sometimes be compromised by necessary mechanical operations, but with certain varieties, in certain regions, the presence of stems may introduce herbaceous notes and a degree of bitterness during carbonic maceration. Stem removal should therefore be considered in some cases. Destemmers without rollers that do not crush the grapes should be used. Laboratory experiments have shown that the metabolism is less intense for berries that have been detached from their peduncle than for whole grape clusters. Similarly, current mechanical harvesting methods do not permit carbonic maceration to be conducted under satisfactory conditions. In the future, new equipment combined with suitable viticultural methods will perhaps lead to a mechanical harvest better suited to the needs of carbonic maceration.

Whatever the precautions taken while filling the fermentor, some grapes are inevitably crushed and release juice. During the anaerobic phase, other grapes are progressively crushed, increasing juice volume. In an experiment at 25°C with the Carignan variety, 15% of the total free-run juice is released in 24 hours, 60% by the fifth day, and 80% by the seventh day of maceration. Variety, maturity and fermentor height are factors influencing the formation of this free-run juice. A homogenizing pump-over, when conducted, is also a factor. It is sometimes used for even distribution of the sulfur dioxide (3–8 g/hl), which is necessary to avoid microbial spoilage. The use of lysozyme has been considered (Section 9.5.2) to prevent the premature development of LAB.

A fermentor undergoing carbonic maceration therefore contains the following:

1. Whole grapes immersed in a carbon dioxide atmosphere, poor in oxygen. They are the grapes most affected by anaerobic metabolism. In addition, they are located in an environment with an increasing ethanol concentration. At a certain partial pressure, ethanol diffuses into the berries under anaerobic conditions. At pressing, the press juice has a higher alcohol content than the juice from a solely anaerobic metabolism.
2. Whole grapes immersed in must from crushed grapes. They undergo a less intense anaerobic metabolism than (1).
3. Must from certain crushed grapes undergoing yeast-based alcoholic fermentation. Crushed grapes macerate in this juice, whose fermentation occurs at the base of the fermentor. It must be carefully monitored to avoid bacterial spoilage. Acetic acid bacteria may develop when the fermentation starts slowly. The addition of a fully active yeast starter helps to avoid this problem. It also gives protection from the untimely development of LAB in the case of slowed yeast activity. When the pH is excessively high (pH 3.8), tartaric acid may be added to the juice at the bottom of the fermentor (up to 150 g/hl, taking into account the total must volume anticipated at the end of the anaerobic maceration phase). Sulfiting is also indispensable for inhibiting LAB (3–8 g/hl). Homogenizing pump-overs must be minimized when making these additions; otherwise, free-run juice volume is increased. During this fermentation phase, microbial activity must be monitored via the disappearance of sugar and malic acid, the increase in volatile acidity, and possibly the analysis of L(+)-lactic acid, whose presence indicates bacterial activity.

When the addition of sugar (chaptalization) is judged necessary, it is conducted after tank emptying, at the start of the second fermentation stage.

Anaerobic conditions are obtained by filling an empty fermentor with carbon dioxide from an industrial gas cylinder or a fermenting tank. After filling, the carbon dioxide supply must be continued for 24–48 hours to compensate for possible losses and dissolution in the grape. After this period, fermentation emissions compensate for losses. The extinguishing of a candle flame when placed in the tank verifies anaerobic conditions.

The temperature and duration of the anaerobic phase are essential parameters of carbonic maceration. The increase in temperature is less significant with carbonic maceration than with crushed grapes, which have more active fermentations. In hot climates, this fact was used to the winemaker's advantage, when controlling temperatures was more difficult than today. The anaerobic metabolism, however, must take place at a relatively high temperature (30–35°C) for this method to be fully effective. Yet the temperature must not exceed 35°C, above which this metabolism is affected. Maceration for six to eight days at 30–32°C is recommended. An insufficient temperature can be compensated for by prolonging maceration time—for example, 10 days at 25°C or 15 days at 15°C—but the result is not necessarily identical. In some regions, excessively low temperatures (15–20°C) restrict the use of carbonic maceration, as the reactions are slowed down. Systems have been devised to warm the grape crop; this operation is always complex. Immersing the grape crop in warm must or wine is laborious and several days are needed to obtain a perfect homogeneity of heat. Attempts at heating the grapes directly on the conveyor belt, using microwave technology, have met with limited success.

The moment of tank emptying should be chosen according to the style of wine desired. This difficult decision is based on experience but takes into account the evolution of specific gravity and temperature, color and tannic structure, aroma, and juice flavor along with the degree of grape degradation and pulp color. Pumping the must over once or twice before tank emptying enhances the must's aroma intensity and tannic structure.

After tank emptying, the grapes should be carefully pressed using a horizontal moving-head press or a pneumatic press. These presses do not affect tissue structure. Due to the presence of whole grapes, pressing capacity must be considerable (one-third to one-half higher than in traditional winemaking). Pressing is also slower. Grapes may be crushed just before pressing to simplify this operation. Since the press wine is potentially organoleptically richer than the free-run juice, the press/free-run ratio should be as high as possible.

At the time of tank emptying, the free-run juice has a specific gravity between 1.000 and 1.010 and the press juice between 1.020 and 1.050.

Table 12.19 compares free-run and press juice composition for traditional winemaking (crushed grapes) and carbonic

TABLE 12.19

Free-Run and Press Wine Analysis Comparing Traditional Winemaking with Carbonic Maceration (Flanzy et al., 1987)

Component	Winemaking using crushed grapes		Carbonic maceration	
	Free-run	Press	Free-run	Press
Alcohol content (% vol.)	12.5	10.96	11.15	13.00
Specific gravity at 15°C	0.9949	0.9991	0.9966	0.9920
Glycerol (g/l)	9.29	9.75	9.10	7.91
Dry extract (g/l)	23.8	32.0	25.5	19.2
Total acidity (g/l H_2SO_4)	3.30	3.50	3.50	2.80
pH	3.90	4.05	3.93	3.90
Total nitrogen (mg/l)	154	425	144	123
Color intensity[a]	388	912	510	487
Tannic matter[b]	1,342	2,550	1,582	1,440

[a] Sum of optical densities at 420 and 520 nm.
[b] Sum of optical densities at 260 and 280 nm.

maceration. In carbonic maceration press wines, the alcohol content is higher (caused by ethanol fixation) and the acidity lower (due to malic acid degradation). These wines also have lower concentrations of phenolic compounds and other extracted components; their dissolution is lower.

Due to their complementary composition, free-run wines and press wines should be blended immediately after pressing before the completion of alcoholic and malolactic fermentation. Bacterial contamination in the free-run wine, leading to premature MLF and the risk of an increase in volatile acidity, is the only reason for fermenting the free-run and press wine separately. Microbiological analysis should be systematic at this stage, followed by sulfiting and reseeding the must with fresh yeast to complete the alcoholic fermentation, if necessary.

During the second fermentation phase, the complete transformation of sugar into alcohol is generally very quick. It is carried out at 18–20°C to preserve aroma components. Afterward, the favorable conditions permit the easy initiation of MLF. Despite the existence of two distinct phases, carbonic maceration requires less time than traditional winemaking when done properly. This method is therefore well adapted for wines that are quickly put on the market.

12.9.7 Characteristics of Wines Made by Carbonic Maceration

Table 12.20 (Flanzy *et al.*, 1987) compares the composition of traditionally made wines (crushed grapes) and wines having undergone carbonic maceration (conducted at 25 and 35°C). The importance of temperature in anaerobic metabolism is shown. At 35°C, this technique permits the same tannic structure as traditional winemaking. In general, density and dry extract, fixed acidity, and phenolics are lower with carbonic maceration than with traditional winemaking. This winemaking technique produces a lighter wine, containing fewer substances from the solid parts of the grape. The method has advantages when used with rustic grape varieties—it avoids the excessive extraction of aggressive aroma and taste components lacking finesse. In other cases, carbonic maceration may result in insufficient structure and an impression of thinness or results anywhere in between.

In Table 12.20, volatile acidities were observed to be relatively high—often greater than 0.5 g/l (expressed as H_2SO_4) and sometimes attaining 0.69 g/l (0.65 and 0.84 g/l when expressed as acetic acid). These numbers indicate the inherent bacterial risk associated with this winemaking method.

The structural difference of a wine having undergone carbonic maceration, with respect to a traditionally made wine, as shown by laboratory analysis, is reflected in its organoleptic characters. Carbonic maceration produces supple, round, smooth, and mellow wines. For this reason, they are often used in blends to improve wine quality. However, this positive characteristic in certain situations can be negative in others: wines can become thinner and more fluid, and, depending on the variety and ripeness, the less abundant tannins can also be more bitter, probably due to the presence of the stems.

Carbonic maceration is certainly most interesting from an aroma viewpoint. It produces wines with a unique aroma. Some have accused this technique of producing uniform wines and of masking the aromas of premium varieties (Ribéreau-Gayon *et al.*, 1976). Other authors (Flanzy *et al.*, 1987) find that the aromas of certain varieties (Muscat and Syrah) are intensified including the observation of increased concentrations of ethyl cinnamate in wines (Versini and Tomasi, 1983). This technique has also been observed to increase the aroma intensity of relatively neutral varietal wines (Aramon, Carignan).

Changes in the concentrations of secondary products of alcoholic fermentation

TABLE 12.20

Analytical Comparison of the Composition of Wines Having Undergone Carbonic Maceration at 25°C (CM25) and 35°C (CM35) and of the Same Wines Made from Crushed Grapes (CG) (Analysis Carried Out Four Months After the End of Malolactic Fermentation, 1,983 Vintage) (Flanzy et al., 1987)

Component	Carignan			Mourvèdre		
	CG	CM25	CM35	CG	CM25	CM35
Alcohol content (% vol.)	11.6	11.4	11.4	12.25	12.25	12.35
Ash (g/l)	3.0	2.7	3.0	3.4	2.6	2.8
Ash alkalinity (mEq/l)	34.5	32.5	34.5	41.0	33.0	33.3
Glycerol (g/l)	8.0	7.0	7.3	9.0	7.7	8.5
Total nitrogen (mg/l)	196	146	238	179	120	129
Total acidity (g/l H_2SO_4)	3.10	3.10	3.00	3.00	3.30	3.30
Volatile acidity (g/l H_2SO_4)	0.41	0.34	0.51	0.43	0.54	0.69
Tartaric acid (mEq/l)	24.9	24.0	25.7	22.2	19.3	19.9
Malic acid (mEq/l)	0	0	0	0	0	0
pH	3.71	3.61	3.74	3.85	3.71	3.80
Potassium (mEq/l)	35.3	32.8	34.8	42.5	34.8	38.4
Total SO_2 (mg/l)	38	57	27	44	45	57
Optical density 520×10^3 (red)	394	393	384	570	410	505
Optical density 420×10^3 (yellow)	101	94	108	146	109	142
Optical density 280×10^3 (tannins)	810	770	903	1,250	980	1,240
Total polyphenols (g/l gallic acid)	1.573	1.436	1.755	2.690	2.120	2.736
Anthocyanins (g/l)	0.509	0.518	0.474	0.738	0.527	0.624

have been reported. In particular, aroma substances specific to this winemaking method seem to be produced. Yet the nature and origin of the corresponding molecules are not always clear, in spite of the considerable research that this technique has incited. The typical aroma seems to be acquired during the anaerobic metabolism phase, but yeasts seem to be involved in its expression.

The description of the specific aromas of carbonic maceration wines is confronted by the well-known difficulties of tasting vocabulary. According to experts, carbonic maceration wines have a dominant fruitiness with notes of kirsch, cherry, plum, and fruit brandy, whereas traditionally made wines have a dominant vinous character with notes of wood, resin, and licorice. In addition, the various aroma components may be more harmoniously integrated in carbonic maceration wines.

Carbonic maceration is best applied in making nouveau-type wines for early drinking. Experts, however, do not agree on the aging potential of these wines. For some, carbonic maceration wines lose their specificity after two years of aging but are always better than the equivalent traditionally made wines. For others, these wines evolve poorly after a year of aging: they lose their characteristic aromas and do not undergo the harmonious sensory development of traditional wines. When evaluating the differences of opinion regarding this technique, the variety should be considered, and carbonic maceration conditions should also be taken into account. Harvest temperature is particularly important, since it determines the intensity of the anaerobic metabolism.

Carbonic maceration is essentially used for red winemaking. It is best adapted to certain varieties, such as Gamay. Reservations have always been expressed about using this technique in regions known for their fine wines with aging potential, owing to concerns that varietal character may be lost. The technique is also used for rosé wines (Section 14.1.1) and *vins doux naturels* (sweet fortified wines). It has been used experimentally to produce white wines and base wines for sparkling wines and spirits but has not been further developed on a production scale. Contaminated harvests (more than 15% rotten grapes) and mechanically harvested grapes should not undergo carbonic maceration.

References

Amrani-Joutei K. (1993) *Localisation des anthocyanes et des tanins dans le raisin. Etude de leur extractibilité*, Thèse Doctorat de l'Université de Bordeaux II. Mention (Oenologie-Ampélologie).

Augustin M. (1986) *Etude de l'influence de certains facteurs sur les composés phénoliques du raisin et du vin*, Thèse Doctorat de l'Université de Bordeaux II.

Blouin J. and Peynaud E. (2001) *Connaissance et travail du Vin*, 3rd edition. Dunod, Paris.

Boulet J.C. and Escudier J.L. (1998) Flash-détente. In *OEnologie. Fondements scientifiques et technologiques* pp. 797–805 (Ed C. Flanzy. Tec & Doc, Lavoisier, Paris.

Canbas A. (1971) *Les facteurs de dissolution des composés phénoliques au cours de la vinification*, Thèse Doctorat de 3ème cycle, Universitéde Bordeaux II.

Caputi L., Carlin S., Ghiglieno I., Stefanini M., Valenti M., Vrhovsek U. and Mattivi F. (2011) J. Agric. Food Chem., 59, 5565.

Cretin L. (2016) *Recherches sur les déterminants moléculaires contribuant à l'équilibre gustatif des vins secs*, Thèse de Doctorat, Université de Bordeaux.

Dubernet T.M. (1974) *Recherches sur la tyrosinase de Vitis vinifera et la laccase de Botrytis cinerea*, Thèse Doctorat de 3ème cyle, Universitéde Bordeaux II.

Ducasse M.A., Williams P., Canal-Llauberes R.M., Mazerolles G., Cheynier V. and Doco T. (2011) J. Agric. Food Chem., 59, 6558.

Ferré L. (1958) *Traité d'Œnologie Bourguignon*. INAO, Paris.

Feuillat M. (1997) Rev. Œnol., 82, 29.

Flanzy M. (1935) C.R. Acad. Agric., 21, 935.

Flanzy C. (1998) *Œnologie. Fondements scientifiques et technologiques*. Tec & Doc, Lavoisier, Paris.

Flanzy C., Flanzy M. and Benard P. (1987) *La vinification par macération carbonique*. INRA, Paris.

Gambacorta C., Antonacci D., Pati S., la Gatta M., Faccia M., Coletta M. and La Notte E. (2011) Eur. Food Res. Technol., 233, 1057.

Gayon U. (1905) *Préparation et conservation des vins*. Pech éditeur, Bordeaux.

Glories Y., Ribéreau-Gayon P. and Ribéreau-Gayon J. (1981) C.R. Acad. Agric., 67, 623–627.

Henick-Kling T. (1992) Malolactic fermentation. In *Wine Microbiology and Biotechnology*, pp. 289–326 (Ed G.H. Fleet). Hartwood Academic Publishers, Chur.

Lafon-Lafourcade S. (1970) Ann. Technol. Agric., 19, 2, 141–154.

Lafon-Lafourcade S. and Ribéreau-Gayon P. (1977) C.R. Acad. Agric., 551.

Lafon-Lafourcade S., Carre E., Lonvaud-Funel A. and Ribéreau-Gayon P. (1983) Œno One, 17, 55–71.

Marchal A., Marullo P., Moine V. and Dubourdieu D. (2011) J. Agric. Food Chem., 59, 2004.

Peynaud E. and Guimberteau G. (1962) Ann. Physiol. Vég., 4, 2, 161–167.

Peynaud E., Lafon-Lafourcade S. and Guimberteau G. (1966) Am. J. Enol. Vitic., 4, 302.

Poitou X. (2016) *Contribution à la connaissance aromatique des vins rouges: approche sensorielle et moléculaire des nuances « végétales, vertes » en lien avec leur origine*. Thèse de Doctorat, Université de Bordeaux.

Poitou X., Achaintre V.-C., Redon P., Noilet P. and Darriet P. (2017) Rev. Œnol., 165, 42–44.

Prahl C., Lonvaud-Funel A., Korsgaard S., Morrison E. and Joyeux A. (1988) Œno One, 22, 197–207.

Puértolas E., Hernández-Orte P., Sladaña G., Álvarez I. and Raso J. (2010) Food Res. Int., 43, 761.

Ribéreau-Gayon P. (1953) C.R. Acad. Agric., 39, 807.

Ribéreau-Gayon P. (1973) Vitis, 12, 144.

Ribéreau-Gayon P. (1977) C.R. Acad. Agric., 63, 120.

Ribéreau-Gayon J. and Peynaud E. (1961) *Traité d'Œnologie*, Vol. II. Béranger, Paris.

Ribéreau-Gayon J., Peynaud E. and Lafourcade S. (1951) Ind. Agric. Alim., 68, 141.

Ribéreau-Gayon P., Sudraud P., Milhe J.C. and Canbas A. (1970) Œno One, 2, 133.

Ribéreau-Gayon J., Peynaud E., Ribéreau-Gayon P. and Sudraud P. (1976) *Sciences et Techniques du Vin, Vol 2: Vinification. Transformations du Vin*. Dunod, Paris.

Siégrist J. and Léglise M. (1981) C.R. Acad. Agric., 67, 300.

Somers T.C. (1979) J. Sci. Food Agric., 30, 623–633.

Soto Vazquez E., Rio Segade S. and Orriols Fernandez I. (2010) Eur. Food Res. Technol., 231, 789.

Sudraud P. (1963) *Etude expérimentale de la vinification en rouge*, Thèse Docteur-Ingénieur, Faculté des Sciences de Bordeaux.

Touzani A., Muna J.P. and Donèche B. (1994) J. Int. Sci. Vigne et Vin, 28, 1, 19.

Versini G. and Tomasi T. (1983) L'Enotecnico, 19, 595.

CHAPTER 13

White Winemaking

13.1 Distinctive Characteristics of White Winemaking
13.2 White Grape Quality and Picking Criteria
13.3 Juice Extraction
13.4 Protecting Juice from Oxidation
13.5 Clarification
13.6 Juice Treatments and the Advisability of Bentonite Treatments
13.7 Fermentation Operations
13.8 Making Dry White Wines in Barrels
13.9 Controlling Reduction Off-Aromas During White Wine Aging

13.1 Distinctive Characteristics of White Winemaking

13.1.1 The Essential Role of Pre-fermentation Operations in Dry White Winemaking

While red wines are obtained by the alcoholic fermentation of musts in the presence of the solid parts of the berry (skins and seeds), white wines are produced by the fermentation of grape juice alone. Thus, in the production of white wines, must extraction and varying degrees of clarification always precede alcoholic fermentation. It is the absence of skin contact in the alcoholic phase, and

Handbook of Enology, Volume 1: The Microbiology of Wine and Vinifications, Third Edition.
Pascal Ribéreau-Gayon, Denis Dubourdieu, Bernard B. Donèche and Aline A. Lonvaud.
© 2021 John Wiley & Sons Ltd. Published 2021 by John Wiley & Sons Ltd.

not the color of the grape, that distinguishes white winemaking from red winemaking. White wines can be made from red grapes having white juice, if the grapes are pressed under conditions that prevent grape skin anthocyanins from coloring the must. This is the case of *blancs de noirs* from Champagne, made from Pinot Noir grapes.

That is not to say that white winemaking does not include any maceration. If this term designates solubilization of solid components in juice, a certain degree of maceration is inevitably associated with white winemaking. It occurs in the absence of alcohol during the pre-fermentation phase at the time of juice extraction and clarification.

Varietal aromas and aroma precursors are located in the grape skin or in the underlying cell layers in most quality cultivars (Volume 2, Chapter 7). Yet these zones are also the richest in grassy-smelling and bitter-tasting substances, especially when the grapes are not completely ripe, are infected with rot, or are from a *terroir* less favorable for producing quality wines. The taste of a dry white wine, made from a given grape, therefore depends greatly on the conditions of various pre-fermentation operations: harvest, crushing, pressing, and clarification.

All winemaking includes a selective extraction of grape components; white winemaking is not an exception to this general principle. Winemaking not only consists of carrying out the alcoholic fermentation of must or grapes but also, and especially, extracting the best part of the grape berry while limiting the diffusion of substances in the liquid phase capable of generating aroma and taste flaws.

In red winemaking, fractional extraction of grapes occurs primarily during alcoholic fermentation and maceration. The winemaker influences the future taste of a red wine by adjusting maceration, or skin contact, times (Section 12.5). By adjusting various operations during maceration, the winemaker approaches day by day, over a period of two to three weeks, the desired tannin, color, and aroma concentrations for the wine. During maceration, time is the winemaker's ally.

In white winemaking, conditions for the extraction of berry components are radically different, since maceration phenomena occur before alcoholic fermentation. In this case, pre-fermentation treatment conditions control the passage of compounds responsible for the qualities and flaws of grapes into must. The quality of a dry white wine, made from given grapes, depends above all on grape and must handling during production. In other words, the art of making dry white wines lies in knowing how to press the grapes and clarify the musts in a manner that simultaneously extracts and preserves potential grape quality. For certain varieties (Sauvignon Blanc, Muscat, etc.), limited skin contact (pre-fermentation skin maceration) before pressing can be useful in facilitating the diffusion of varietal aromas and their precursors in the juice. The winemaker has only a limited amount of time to extract components from the grape skins before the juice begins to ferment—generally a few hours to a few days maximum. In addition, the choices that are made during the pre-fermentation phase are definitive: pressing time and program, juice selection, possible skin contact, blending free-run juice and press juice, and degree of must clarification.

Therefore, in dry white winemaking, the important choices are made before alcoholic fermentation. Afterward, corrections and adjustments are practically impossible. When alcoholic fermentation is initiated, the taste of the dry white wine is already largely determined. Even the decision to ferment in tank or barrel is made fairly early. In fact, barrel-fermented wines should be barreled at the onset of fermentation to avoid the wood dominating the wine later (Section 13.8). This decision must be made as early as possible so that barrel purchases can be planned properly. It is too late to barrel down a wine after tank fermentation, even if wine quality would have justified barrel aging. Juice clarification is another example of definitive pre-fermentation decisions: it is impossible to mitigate its consequences subsequently during fermentation. In fact, no satisfactory methods exist to stimulate sluggish fermentations of overclarified juice or to eliminate vegetal and reduced odors that appear during the fermentation of poorly clarified juice.

Botrytized sweet winemaking constitutes an extreme case of the fundamental decisions involved concerning fractional berry component extraction. The most important decisions made by the winemaker essentially concern picking grapes at the ideal noble rot stage (Section 10.5.2). Noble rot is not only an overripening by water loss, like raisining (*passerillage* or *passito*), but also and especially an intense enzymatic skin maceration paired with a *Botrytis*-specific metabolism. The decisive part of botrytized sweet winemaking occurs in the grape on the vine (Sections 14.2.2 and 14.2.3).

In conclusion, each type of winemaking contains a key phase during which the decisions of the winemaker have a decisive and almost irreparable effect on wine taste: maceration time for red wines, pre-fermentation operations for dry white wines, and noble rot development and picking conditions for botrytized sweet wines.

13.1.2 White Wine Diversity and Current Styles

White wines are generally thought to present a greater diversity of styles than red wines (Ribéreau-Gayon et al., 1976). In fact, apart from still and sparkling wines, white wines can be dry or contain a varying amount of residual sugars (from several grams to several dozen or, sometimes, even more than 100 g/l). These differences can occur in wines made from the same varieties and parcels—the grapes being picked at different maturity levels. This is the case of great German, Austrian, and Alsatian Rieslings and Gewürztraminers, Hungarian Furmints, Semillons and Sauvignon Blancs from Bordeaux, Loire Valley Chenin Blancs, etc. The fixed acidity of dry and sweet white wines can also vary greatly (from 3 to 6 g/l expressed as H_2SO_4). Moreover, dry wines may or may not undergo malolactic fermentation, which, in contrast, is universally used for red wines. Intense or discreet, predominantly marked by the grape variety or only by secondary products of alcoholic fermentation, white wines also seem to have more diverse aromas than red wines. This relatively rich typology characterizing white wines may be divided into two general categories, which also concern red wines:

1. Premium wines improve during bottle aging by developing a bouquet.
2. *Nouveau*-type wines, incapable of aging, are to be drunk young.

In addition, oak-aged white wines are partially or totally made in new barrels. In other wines, the organoleptic character supplied by oak is not sought—these wines are made in neutral vessels (tanks or used barrels). Finally, like certain red wines, some white wines are distinguishable by their oxidative character (sherry and *vin jaune*). Yet most are made in the virtual absence of oxygen and under the protection of antioxidants such as sulfur dioxide and ascorbic acid to preserve their fruitiness.

The diversity of white wine types and winemaking methods has strongly diminished during the last 20 years, due to the trend toward a world market, a standardization of consumer tastes, and a general trend of producers to imitate a few universally appreciated models. The growing influence of the wine critic on the market has certainly amplified and accelerated this convergence of white wines toward a few widely recognized types. Four categories currently distinguish international dry white wines: neutral, Chardonnay, Sauvignon Blanc, and aromatic white wines.

Neutral white wines

Neutral white wines do not possess a particular varietal aroma. They only contain young wine fermentation aromas—essentially due to ethyl esters of fatty acids and acetates of higher alcohols produced by yeasts (Section 2.3.5), when the fermentation of clarified juice occurs at relatively low temperatures (16–18°C). These wines are appreciated especially for their thirst-quenching character, due to their refreshing acidity possibly reinforced by the presence of carbon dioxide (0.6–1 g/l). They should be low in alcohol and without bitterness. Their fleeting aroma rarely lasts for more than a year of storage; these white wines are generally bottled a few months after the completion of fermentation and should be drunk within the year that follows the harvest. A particular varietal aroma is not sought nor is an expression of *terroir* expected in these "quaffable" white wines. They are generally made from high-yielding, slightly, or nonaromatic varieties such as Ugni Blanc (Trebbiano in Italy), Macabeu (Viura in Spain), Airen (also of Spanish origin and the world's mostly widely planted varietal in terms of surface area), Grenache Blanc, Clairette, etc. Unfortunately, neutral white wines are sometimes produced with noble varieties, due to excessively high crop yields and unfavorable soil and climate conditions. This is often the case of Semillon at yields greater than 60 hl/ha or Sauvignon Blanc grown in hot climates regardless of yields. In the 1970s, these simple and inexpensive white wines sold well, especially when promoted by a strong brand name. Today, the demand for them has dropped, as the market orientates itself toward more expressive white wines, particularly in English-speaking countries. Oxidized wines possibly containing several grams of sugar and nonoxidized wines containing several grams of sugar have all but disappeared.

Chardonnays

Chardonnay is the principal current international white wine standard. White Burgundy wines supplied the original model (Meursault, Chassagne-Montrachet, Chablis, etc.). The top estates from this region are among the best dry white wines in the world. Their wines are powerful, firm, aromatically intense, and "sweet," although they do not contain residual sugar. The great white Burgundies are distinguished by their aging potential. During the aging process, they develop a remarkable bottle bouquet. In its area of origin, the Chardonnay variety produces grapes rich in both sugar and acid, often reaching 13% potential alcohol for an acidity between 6 and 7 g/l (expressed as sulfuric acid) and remarkably low pH (3.1–3.3). The traditional Burgundy winemaking method, with barrel fermentation and on-lees aging, has profoundly influenced current white winemaking methods. Today, enological research has shed light on and justified these empirical Burgundian practices, now put to use worldwide and not just for Chardonnay.

During the last 30 years, Chardonnay has been widely planted in Mediterranean-type climates in Europe and the New World. Along with Cabernet Sauvignon for red wines, it is certainly the varietal best adapted to climatic conditions that are warmer than its original *terroir*. All Chardonnay producers try to attain the Burgundy archetype, just as makers of Cabernet Sauvignon wines strive to attain the classified growth model of the Medoc. Excellent Chardonnays are found in many viticultural regions throughout the world, but the diversity of expression of this variety in different Burgundy terroirs, or *climats* (Burgundian term for precisely delineated vineyard parcels), still remains more fascinating for the wine buff.

Sauvignon Blancs

Inspired by the wines of Central France (Sancerre and Pouilly-Fumé), Sauvignon Blancs constitute another important world standard for dry white wines. Their often intense and complex varietal aroma is easily recognized. Certain volatile substances responsible for this aroma as well as their precursors in the grape have recently been identified (Volume 2, Chapter 7). The Sauvignon Blanc aroma is more sensitive to climatic conditions during maturation than the Chardonnay aroma. It is therefore less constant and stable and more difficult to reproduce. The aroma expression of Sauvignon Blanc is often disappointing in Mediterranean climates. It has consequently been less universally successful than Chardonnay. Due to its cool climate, New Zealand without a doubt produces one of the most aromatic Sauvignon Blancs in the New World. On average, Sauvignon Blanc wines have a lesser aging potential than Chardonnay wines, except in very particular situations. Sauvignon Blanc also originated in the Bordeaux region and is nearly always blended with Semillon in this area. Sauvignon Blanc contributes fruitiness, firmness, and acidity, while Semillon gives the wine body, richness, and bouquet during aging. These two varieties are particularly complementary. During recent years, Sauvignon Blanc winemaking methods have undergone many changes—including a return to barrel fermentation of musts originating in the best *terroirs* as well as on-lees aging of new wines, whatever the fermentation method (barrel or tank). Current Chardonnay and Sauvignon Blanc winemaking methods are very similar, but malolactic fermentation is rarely used on Sauvignon Blanc wines (Section 13.7.6).

Aromatic white wines

Various aromatic white wines compose the fourth group. Sometimes, these wines are famous and made from premium varieties. Their geographical territory has remained limited to their original regions. An exhaustive list of these wines is not included in this text, but a few examples will be given.

Within this group, the dry white, premium wines of Germany and Alsace are worth mentioning. These wine styles are also made in Austria and continental Europe. The most notable varieties are Riesling, Pinot Gris, and Gewürztraminer. Late harvesting of these varieties produces premium sweet wines capable of considerable aging. They have a characteristic aroma reminiscent (at least in part) of their grape or juice aroma. These floral or Muscat varieties are distinguishable from nonaromatic varieties such as Sauvignon Blanc, Chardonnay, Chenin Blanc, etc. The juice of nonaromatic varieties is not very fragrant, but their wines have a characteristic varietal aroma essentially derived from odorless precursors located in the grape. The role of volatile terpene alcohols and certain norisoprenoids in the aroma of Muscat varieties has been widely studied and proven (Volume 2, Chapter 7). The specific aroma of these different varieties, however, is far from being totally elucidated.

Several regional varieties also exist that, for diverse reasons, have until now only produced characteristic wines in relatively limited zones. Some of them have never been planted outside their region of origin, while others lose their character in warmer climates. French varieties of this type include Chenin Blanc (Savennières, Loire Valley), Viognier (Condrieu), and Petit and Gros Manseng (Jurançon). The same is true of Albariño in the north of Spain and the remarkable and rare Petite Arvine in the Swiss Valais.

13.2 White Grape Quality and Picking Criteria

Varietal aroma finesse, complexity, and intensity are the primary qualities sought after in a dry white wine. Its personality is due to varietal expression or, more precisely, its particular aroma profile on a given *terroir*. Fermentation aroma components are present in all wines and are not very stable over time. These esters and higher alcohols produced by yeasts are not sufficient to give a white wine an aroma specificity, but they were the first to be measured by gas chromatography because of their relatively high concentrations in wines. Consequently, in the past, the importance of their contribution to the aroma quality of dry white wine was exaggerated. It has been widely accepted that aroma quality is mainly due to the primary aroma—the aroma originating in the grape—even though the volatile compounds responsible are far from being identified and the production mechanisms from grape to wine remain unknown. The handicap of neutral varieties is thus explained: no winemaking method can compensate for this neutrality. For all that, the varietal aroma is not the only characteristic of a dry white wine. The balance of acidity and softness, mouthfeel, body, structure, and persistence and the impression of density and concentration also play an important role in quality appreciation. Healthy ripe grapes must be used to obtain a wine with all of these characteristics. Grape soundness and maturity level, in particular in terms of aromas, are the essential harvest selection criteria for making quality dry white wines. Harvest time and methods (mechanical or manual) influence these two essential parameters and are therefore very important.

13.2.1 Grape Soundness

White grape varieties are susceptible to gray rot due to *Botrytis cinerea* development on the grape (Section 10.6). In a given region, the earlier white varieties are more subject to this disease than red varieties. Obtaining healthy grapes with Sauvignon Blanc, Semillon, and Muscadelle grapes is much more difficult than with Merlot and Cabernets. Muscats in Mediterranean climates, Chardonnay in Champagne, and Chenin Blanc in the Loire Valley are also affected.

From early contamination of the grape cluster (latent since bloom), *Botrytis* can develop explosively near harvest time. Feared by winegrowers, this pathogen is triggered by severe rains near *veraison* and during maturation. The fungus contaminates both green and burst berries, degrading the skin—the site of aromas and aroma precursors.

Even a relatively small percentage of botrytized grapes in the crop always seriously compromises the aroma quality of dry white wines. Gray rot on white grapes results in a decrease in varietal aroma, a greater instability of fermentation aromas, and the appearance of off-aromas. These consequences of gray rot on the aroma of white wines are much more serious than enzymatic browning itself. The latter is a direct manifestation of the laccase activity on wine color, especially with red and rosé wines (Section 12.6.2), but it can be observed in certain white wines—in particular bottled sparkling wines—even several years after bottle fermentation.

Although empirically witnessed with all aromatic varieties, the harmful effects of gray rot on the primary grape aroma have only been quantified with Muscat varieties by measuring monoterpene alcohol concentrations in musts (Boidron, 1978). When *Botrytis* contaminates 20% or more of a grape crop, the total terpene alcohol concentration of Muscat Blanc or Muscat of Alexandria drops by nearly 50% with respect to concentrations in healthy grape must (approximately 1.5–3 mg/l). The most fragrant terpene alcohols (linalool, geraniol, and nerol) are the most affected. These alcohols are partially transformed into less fragrant compounds, such as linalool oxides, α-terpene alcohol, and other compounds (Rapp *et al.*, 1986), themselves original components of healthy juice. This rapid degradation of terpenes by *B. cinerea* can be observed in the laboratory in a fungus culture on a medium supplemented with monoterpene alcohols.

Gray rot is also observed to affect the specific aromas of other varieties. A relatively small percentage of gray rot (less than 10%) diminishes the Sauvignon Blanc varietal aroma in wine. This aroma is due at least in part to very fragrant volatile thiols, existing in trace amounts (a few nanograms or a few dozen nanograms per liter) in wines (Volume 2, Chapter 7). The reaction of fragrant thiols with quinones formed by the oxidation of grape phenolic compounds can help to trap them. *Botrytis* laccase activity in a must containing phenolic compounds inevitably leads to the formation of quinones. The quinones trap Sauvignon Blanc varietal aromas as they are formed during alcoholic fermentation. When Sauvignon Blanc must is insufficiently sulfited during the pre-fermentation phase, it is oxidized. The resulting combination of thiols and quinones produces wine with a slight or nonexistent varietal aroma (Section 13.4.1).

Paradoxically, noble rot does not destroy the specific aroma of white varieties used to make great botrytized sweet wines (Section 14.2). In the Sauternes region, the lemon and orange fragrances of Semillon and Sauvignon Blanc are even enhanced, as is the mineral character of Riesling or the lychee aroma of Gewürztraminer in Alsatian or German late-harvest wines. The bouquet of dry wines made from healthy grapes and sweet wines made from botrytized grapes of the same variety and from the same *terroir* has even been observed to converge during bottle aging. In the ideal noble rot case, the intense skin maceration of the ripe grape under the action of fungal enzymes promotes the diffusion of free and bound varietal aromas in the must. These aromas are concentrated without being degraded. These complex phenomena have recently received an interpretation in which *B. cinerea* acts as a "stimulator" of the biosynthesis of the fruit's aroma potential (Thibon *et al.*, 2009, 2011) (Volume 2, Sections 7.2, 7.5, and 7.7). This process is different from *passerillage*, in which the grapes are concentrated by the sun that burns the grape skin. Most of the varietal-specific aromas are lost, and a character peculiar to raisins is acquired, varying little from one white variety to another. Theoretically, a small proportion of noble-rotted grapes could be added to a grape crop intended for dry white winemaking, but in practice this is difficult. In effect, at the time of the healthy white grape harvest, most of the rot-infected grapes on the vine correspond to early *Botrytis* sites developed on the unripe grape and thus gray rot.

Gray rot also greatly diminishes the intensity of fermentation aromas of dry white wines. Among the exocellular enzymes liberated by *Botrytis* in the infected grape, esterases exist whose activity persists in juice (Dubourdieu *et al.*, 1983). They are capable of catalyzing the rapid hydrolysis of esters produced by yeasts during alcoholic fermentation. Figure 13.1 shows the hydrolysis kinetics of these different aroma compounds in a dilute alcohol medium in the presence of a *Botrytis* enzymatic extract. The occurrence of gray rot is more detrimental to neutral varietal wines, which are characterized essentially by fermentation aromas.

Finally, gray rot seriously affects the aroma sharpness of dry white wines. Varietal aromas are masked, while dusty, dirty, and moldy aromas appear. It also promotes the development of rancid, camphor, and waxy odors, appearing later during aging, especially in the bottle. This type of off-aroma is comparable with premature oxidative aging of white wines (Volume 2, Section 8.2.3). Gray rot is not solely responsible, and white wines made from healthy grapes can also contain the flaw. The responsible compounds and their formation mechanisms have not yet been described.

The level of *B. cinerea* contamination of a grape, in the form of gray rot, therefore constitutes a decisive criterion for evaluating grape quality, whether red or white. But white grape crops, and consequently dry white wine quality, are more affected at lower levels of gray rot contamination than red grapes. The visual examination of the percentage of botrytized berries, despite its insufficiencies, was previously the only method available to winegrowers. A dozen or so years after the pioneering research of Dubernet (1974) on *B. cinerea* laccase, new methods of analyzing this enzymatic

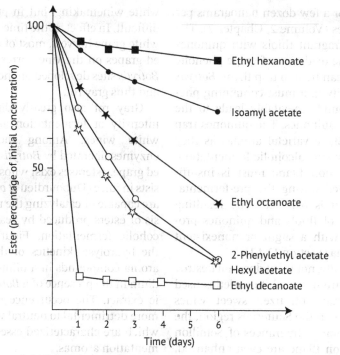

FIGURE 13.1 Enzymatic hydrolysis (pH 3.4 at 18°C 11% ethanol) of different esters by an exocellular protein extract from *B. cinerea* (Dubourdieu et al., 1983).

activity in juice (Section 10.6.6) were developed. This new way of quantifying *Botrytis* development on the grape appears promising. Two methods have been proposed: the first, a polarographic method, measures the oxygen consumption in a must sample with a Clark electrode (Salgues et al., 1984), and the second, a more sensitive colorimetric method, uses syringaldazine as a specific reagent that produces a pink quinone in the presence of laccase (Dubourdieu et al., 1984; Grassin and Dubourdieu, 1986). Results are expressed in laccase units (Section 10.6.6).

Juice from healthy grapes is evidently devoid of laccase. Juice from infected grapes can contain from one to several dozen units per milliliter (U/ml), depending on the fungal development stage. Variety and weather conditions also influence the concentration contained in the grapes. Before the appearance of *Botrytis* conidiophores, the infected berries contain little laccase activity (1–2 units/ml). Activity considerably increases with sporulation (15–20 units/ml) and continues to grow, due to a concentration effect, during the shriveling of the grape (Table 13.1). Universal enological tolerance thresholds are

TABLE 13.1

Development of *B. cinerea* on the Grape Berry and Laccase Activity of Juice (Grassin and Dubourdieu, 1986)

Development of *Botrytis*	Laccase activity (units/ml)[a]
Healthy grape	0
Full rot without conidiophores	1–2
Appearance of conidiophores	15–20
Shriveled rotted grapes	20–70

[a] Laccase activity unit corresponds to the quantity of enzymes capable of oxidizing 1 nmol of syringaldazine per minute under laboratory conditions.

always difficult to establish for a grape defect: they depend on the level of quality or perfection desired for a wine. Ideally, a white grape crop should not contain any botrytized berries; the laccase activity should be zero or at least less than 1 unit/ml. The enzymatic browning threshold for red wines is greater than 3 units/ml, and the sensitivity limit of the laccase measurement by the colorimetric method is 0.5 units/ml. In the event of a gray rot attack, the grapes must be manually sorted in the vineyard; this is the sole means of maintaining the quality of the healthy portion of the harvest (Section 13.2.3).

Although less widespread than gray rot, sour rot (Section 10.6.5) can seriously affect the soundness of grape crops, particularly when ripening occurs in a warm and humid climate. In the Bordeaux region, white grapes, in particular Sauvignon Blanc, are more sensitive to this disease than red grapes. It has not been extensively studied and is poorly known, despite its seriousness. The grapes take on a brick-red color within a few days, while letting some of the juice flow out, and they simultaneously give off a strong acetic acid odor. The microbial agents responsible for this acetic acid fermentation are a combination of aerobic yeasts (*Hanseniaspora uvarum*) and acetic acid bacteria. Fruit flies are known to be the contamination vector (Bisiach *et al.*, 1982; Guerzoni and Marchetti, 1987), but the exact causes of the berry contamination by the microorganisms that habitually make up the microfauna of the grape surface have not been elucidated. The development of sour rot is encouraged (like gray rot) by excessive swelling of the berries following heavy precipitation during maturation. The pressure of surrounding grapes can often detach certain grapes from the pedicel. Contamination can occur, beginning at this rupture zone. Microscopic epidermal fissures permitting juice flow, invisible to the naked eye, may also be a cause. The evolution of the grape crop toward gray rot or sour rot from these situations depends on environmental conditions. When excessive temperatures (higher than 30°C) block the development of *B. cinerea*, sour rot quickly appears and is capable of destroying the entire harvest in a few days. While the development of gray rot in humid climates ceases with the return of hot and dry conditions, sour rot continues its growth inexorably—whatever the weather conditions. Young, vigorous vines with a superficial root structure planted in well-drained soils are the most sensitive to sour rot. This phenomenon is aggravated by bird and insect damage in vines located in urban areas and zones that are well lit at night.

Sour rot obviously damages dry white wine quality more than gray rot. Musts made from partially sour grapes can contain more than 1 g of acetic acid and several grams of gluconic acid per liter (Section 10.6.5). They have very high sulfur dioxide binding rates caused by ketone substances formed by the acetic acid bacteria metabolism. Their propensity for premature fermentations makes natural settling particularly difficult. Finally, to combat the spreading of sour rot, the harvest must often be started before complete maturity, and the grapes must be rigorously sorted in the vineyard.

The presence of even a small proportion of grapes infected by powdery or downy mildew in the harvest leads to the appearance of characteristic off-aromas, having an earthy and moldy smell. These odors can adversely affect the aroma of dry white wine. Fortunately, these types of grape spoilage have become rare.

13.2.2 Maturity and Setting the Harvest Date

The need for picking ripe grapes to make good wine is well understood. Yet optimum grape crop maturity (whether red or white) is difficult to define, and there is no universally agreed concept of grape maturity. It depends on the latitude of the growing region, climate, vintage, variety, vineyard parcel, and the type of wine desired.

Must sugar concentration and acidity do not solely define the maturity of grapes destined to produce dry aromatic white wines. Aroma and aroma precursor concentrations are also decisive factors. No systematic relationship, however, exists between optimum grape aroma composition and maximum grape sugar concentrations—no more than between the latter and optimum grape acidity levels for a given type of wine. The characteristics of Chardonnay maturity are not the same in Meursault and Champagne or from a *grand cru* terroir and a generic appellation. It is therefore impossible to establish a general rule for this subject. While the concept of aroma maturity is often used by certain enologists and winemakers, this language can be misleading. Optimum maturity can only correspond to a level of grape maturity that produces the best wine from a grape crop of a given parcel. Furthermore, the optimal aroma composition of a grape is not easy to define. In fact, the grape, like all fruits, progressively loses its vegetal and grassy aromas during maturation to acquire fruity aromas that are more or less stable toward the end of maturation (Section 10.3.6).

The formation of these different aromas in wine is also relatively complex. Some exist in a free state in the grape; others are formed from precursors located in the must, during the pre-fermentation phase under the action of grape enzymes, or during alcoholic fermentation through yeast metabolism. The grape has potential for both undesirable grassy flavors and sought-after fruity aromas. These two potentials evolve in the opposite direction during maturation. Theoretically, such changes should be measurable. According to the theoretical representation in Figure 13.2, the grape has an optimum composition

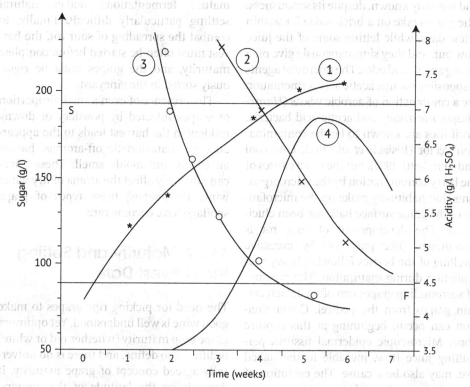

FIGURE 13.2 Theoretical diagram of the evolution of Sauvignon Blanc grapes during the seven weeks following veraison. 1: evolution of sugar concentration (S = minimum concentration in France). 2: evolution of acidity (optimum: between 5 and 6 g/l H_2SO_4). 3: evolution of herbaceous aroma (arbitrary units) (F = minimum perception threshold). 4: evolution of potential fruitiness (arbitrary units).

during the fifth week following veraison. It is now becoming possible to monitor the aroma evolution of the principal varieties analytically. However, these analyses are complex, and the analysis times are long. As a result, winemakers must often set the harvest date based on conventional ripeness tests, mid-bloom and mid-veraison dates, and grape tasting, which helps them to evaluate the aroma maturity of the grapes.

Standard maturity assessment, from *veraison* until harvest, follows the evolution of three principal parameters: berry weight, sugar concentration, and total acidity. It is also useful to measure malic acid concentration and pH, but these analyses are not often carried out by winemakers. A maximum sugar concentration without loss of berry weight indicates the completion of maturation. Overripening, which begins when berry weight diminishes, is generally shown by an additional increase in sugar and possibly acid concentrations. Overripening is rarely sought for white grapes used for the production of dry white wines, due to the accompanying aroma losses.

A minimum concentration of must sugar has been empirically determined for each variety, within a particular region, for producing dry white wines of satisfactory quality. For example, in the Bordeaux region, Sauvignon Blanc and Semillon must have at least 190 and 175 g of sugar per liter, respectively. Below these thresholds, regardless of winemaking methods, the wines obtained have a vegetal aroma. They lack finesse and rarely express the personality of the *terroir*. Similarly, the optimum acidity of ripe white grape musts is specific to both the location of the vineyard and the variety used. In Bordeaux, the optimum acidity at the time of the harvest is between 5 and 6 g/l (expressed as H_2SO_4) for Sauvignon Blanc and 4–5 g/l for Semillon. These threshold values correspond to average acidities and sugar concentrations of samples taken over several years at the time of ideal maturity (Table 13.2).

When grapes have reached their minimum sugar concentration, harvest is possible, but several other conditions must also be satisfied: grape tasting indicates the disappearance of herbaceous aromas, whereas fruity aromas, characteristic of

TABLE 13.2

Average Sauvignon and Semillon Must Composition at Harvest in Representative Parcels in Classified Growths of Graves (Bordeaux)

Year	Sauvignon blanc		Sémillon	
	Sugar (g/l)	Acidity (g/l H_2SO_4)	Sugar (g/l)	Acidity (g/l H_2SO_4)
1987	212	5.0	193	4.3
1988	217	5.2	198	4.0
1989	213	5.9	195	4.7
1990	210	4.3	193	3.2
1991	188	6.8	160	6.3
1992	161	6.1	155	4.7
1993	200	5.2	168	4.2
1994	201	5.8	183	4.5
1995	192	5.2	186	3.6
1996	193	7.2	180	5.2
Average	198	5.7	180	4.5
Potential alcohol (%)	12.0		10.9	

the variety, are present; approximately 40 days have passed since mid-veraison (±10%); and the acidity is within in the optimum range for the given variety and vineyard. In general, the slower the rate of decrease in acidity during maturation, the longer the harvest can be delayed without fear of losing fruity varietal aromas. In the best terroirs for making aromatic dry white wines capable of considerable aging, the grape remains fruity and sufficiently acidic in the final stages of maturation. These terroirs have slow and complete maturation conditions. Conversely, excessively hot climates, early harvests, and excessive water stress in the summer are unfavorable to the aroma evolution of white grapes.

Table 13.3 gives an example of Sauvignon Blanc maturation on two different Bordeaux soils: a sandy-gravely soil (G) and a sandy-clay soil on compact limestone (C). Soil G filters well and has a low water reserve. The water supply is limited, making the vine ripen earlier than in soil C, which has less water stress. The grapes on soil C may be picked about 10 days later than those on soil G and are slightly overripe. Soil C grapes undergo a slower maturation. At the same maturity level in the two soils, soil C grapes are slightly more acidic with less malic acid and a lower pH. Over the years, soil C grapes are observed to remain fruitier during the maturation process than soil G grapes. The harvest date can consequently be set to correspond with practically the maximum sugar concentration desired. On the gravely-sandy soil G, the characteristic Sauvignon Blanc aroma can be almost totally lost in the course of a week. Early harvesting on this type of soil is necessary, not only due to the earliness of the terroir but also due especially to the instability of the varietal aroma. In a given vineyard, an understanding of the earliness and behavior of parcels greatly influences the reasoning used in setting harvest dates.

TABLE 13.3

Maturation Characteristics of 1993 Bordeaux Sauvignon Blanc on Sandy-Gravel Soil (G) and Sandy-Clay Soil on Limestone (C) (Sergent, 1996)

Characteristic	Sample dates						
	Aug. 11	Aug. 19	Aug. 25	Sep. 1	Sep. 9	Sep. 16	Sep. 20
Soil (G)							
Berry weight (g)	0.98	1.17	1.34	1.44	1.47		
Veraison percentage	81	97	100				
Sugar concentration (g/l)	112	139	149	174	197		
Total acidity (meq/l)	290	200	150	128	106		
Malic acid (meq/l)	212	156	102	73	55		
pH	2.57	2.84	3.06	3.15	3.16		
Soil (C)							
Berry weight (g)	1.03	1.25	1.33	1.62	1.54	1.67	1.55
Veraison percentage	66	97	100				
Sugar concentration (g/l)	81	121	138	160	184	193	193
Total acidity (meq/l)	424	266	202	172	142	124	110
Malic acid (meq/l)	301	173	120	85	62	54	32
pH	2.43	2.64	2.82	2.86	2.89	3.01	3.05

13.2.3 Harvest

White grape harvesting for quality wine production has long been known to be more difficult and require more precautions than red grape harvesting. More sensitive to oxidation, easily masked by off-aromas, white wines have a more fragile aroma than red wines. The aroma can be partially lost or altered as early as the harvest, if certain rules are not followed. Harvest conditions must be such that the grapes picked are healthy and their enological maturity (sugar, acidity, and aroma concentrations) is as uniform as possible. Leaves, petioles, dirt, and assorted debris should be avoided in the harvest. From harvest to their arrival at the winery, the grapes must be as intact as possible to limit must oxidation and stem maceration.

The grapes should be harvested at a temperature below 20°C. In warm climates, harvesting may have to occur at night or in the early morning; but moisture on the grape clusters should be avoided, as it can be a significant cause of dilution.

The choice of harvesting method depends on grape maturity and soundness on the one hand and on economic constraints on the other. White grapes can be harvested manually or mechanically, all at once or in several stages, with or without sorting in the vineyard or on sorting tables at the winery.

In cool-climate vineyards sensitive to gray rot, selective manual harvesting in multiple passes optimizes dry white wine quality. Only healthy grapes reaching the desired maturity level are picked. Infected grapes are eliminated in the vineyard. This is the most effective sorting method. Spoiled grapes and grape clusters are left at the foot of the vine, and unripe grapes are picked at a later date. Leaf removal and cluster thinning, carried out during the year, combine to avoid cluster crowding and promote sun exposure. In this manner, ripe grape clusters are more easily identified by the grape-picker. Well-planned pruning, the early elimination of suckers and laterals, leaf removal near veraison, and grape cluster thinning should all be carried out with the objectives of promoting a healthy state and homogeneous grape maturity. These efforts not only improve wine quality but also facilitate harvesting.

Selective harvesting in multiple passes has long been recognized to increase grape quality. Chaptal (1801), restating established principles, wrote:

> Only healthy and ripe grapes should be picked; all infected grapes should be discarded with care and unripe grapes should be left on the vine. The harvest is carried out two to three times in places where wine quality is a great concern. In general, the first cuvée is the best. Some regions nevertheless harvest all grapes at the same time. The characteristics of the good and bad are expressed at the same time. A much lower-quality wine is thus produced compared with the potential of the grapes, if more precautions were taken during the harvest.

In certain years, grapes may have a homogeneous maturity and perfect health. In these situations, all of the grapes can be harvested at the same time—multiple passages through the vineyard not necessary.

Due to its lower cost, its speed, and its simplicity, mechanical harvesting has been increasingly adopted over the last 20 years. Its effect on dry white wine quality can be negligible under optimum health and maturity conditions, but mechanical harvesting of a heterogeneous crop always sacrifices wine quality. In this case, it is of economic and enological interest to have the infected grapes removed by a picking team before harvesting the healthy portion of the crop.

Mechanically harvested white grapes must be protected against oxidation. Sulfiting, however, must be avoided since it promotes the extraction of phenolic compounds. The addition of dry ice to the crop is a preferable alternative. Some countries use ascorbic acid, but this antioxidant is authorized for treating wines only in the European Union (EU).

Whether manually or mechanically harvested, the grapes should be transported rapidly to the winery in containers that minimize berry crushing.

13.3 Juice Extraction

13.3.1 General Principles

In dry white winemaking, pre-fermentation operations (grape and juice handling and treatments) are decisive factors in final product quality (Section 13.1.1). Their role is multiple. They have to extract and clarify juice in a relatively limited amount of time while minimizing juice loss. In addition, the diffusion of certain grape skin substances in the juice, in particular fruity aromas and aroma precursors, must be promoted during these operations. The dissolution of herbaceous and bitter-tasting compounds, associated with the solid parts of the berry, must simultaneously be limited. The formation of substances capable of decreasing the stability of extracted fruit aromas must also be avoided. Oxidized or oxidizable phenolic compounds in particular are able to trap certain aromas (Section 13.2.1).

Before describing the different techniques used and their consequences for juice and wine composition, the principles of proper juice extraction should be discussed.

The fermentation of juice containing too many suspended solids (resulting from juice extraction) does not produce quality dry white wine. In fact, high concentrations of suspended solids in juice are known to have detrimental effects on wine quality (Section 13.5). The first criterion of a juice extraction method, therefore, is its ability to produce clear juice with a turbidity as near as possible to desired levels (200 NTU). All winemaking is a series of basic operations, and each one must be conceived with the idea of facilitating the others that follow. The lower the concentration of suspended solids in the juice after draining or pressing, the easier it is to accomplish juice clarification. Conversely, clarification becomes impossible after a poorly adapted pressing that produces excessive suspended solids. When designing winery equipment, this criterion is not always sufficiently taken into account. The production of suspended solids during juice extraction has other disadvantages: it indicates that the grape has undergone severe mechanical treatment and consequently a greater amount of grassy aroma substances are diffused in the juice.

Proper juice extraction should also minimize oxidation phenomena, the dissolution of phenolic compounds from the skins, seeds, and stalks, and pH increases, linked to potassium extraction from the solid parts of the grape. The resulting oxidation phenomena and juice browning can be evaluated by measuring the absorbance of filtered juice at 420 nm. To evaluate the extraction of phenolic compounds from the skins, particularly flavan-3-ols, represented mainly by catechin and epicatechin in white musts, analysis by HPLC coupled with fluorescence detection is required (Nikolantonaki, 2010). It is not sufficient to estimate the phenolic content of white grape juice and white wines by direct measurement of the total phenols index (TPI) (absorbance at 280 nm), even when subtracting the absorbance of the must at 280 nm after percolation on polyvinylpolypyrrolidone (PVPP). The increase in pH during pressing is an indirect way of monitoring this operation more systematically in practice. The evolution of juice electrical conductivity during pressing can also provide interesting information on pressing kinetics.

The objectives described above are more easily attained when the following conditions are met:

1. Low pressing pressure;
2. Limited mechanical action capable of grinding the grape skins;

3. Slow and progressive pressure increases;
4. High volume of juice extracted at low pressure;
5. Juice extraction at a temperature lower than 20°C;
6. Limited press-cake breakup during pressing;
7. Minimum air contact—rapidly protected from air exposure and sulfited.

The transformation of grapes into juice can be obtained by various methods. Juice extraction can be immediate or preceded by a skin maceration phase. It can be continuous or in batches, with or without crushing and destemming. Continuous and immediate juice extraction processes (very widespread until recently in high-volume wineries, despite the disastrous consequences on juice quality) are fortunately being abandoned; they will therefore be covered only briefly. Immediate batch pressing (whole clusters or after crushing) and skin contact will be described in more detail.

13.3.2 Immediate Continuous Extraction

In this process (Figure 13.3), the grapes, crushed by rollers beforehand, fall by gravity into a continuous inclined drainer containing a helical screw and are transferred into the continuous press placed below. Continuous drainers are capable of processing large amounts of grape crops (several hundred kilograms per minute), releasing a high proportion of free-run juice (70%). They have the disadvantage of producing juice with a high concentration of suspended solids and elevated turbidity (1,000–10,000 NTU). The percentage of lees after natural settling is between 30 and 50%. Clarifying this volume of suspended solids is problematic: it requires costly large-scale filtration or centrifugation

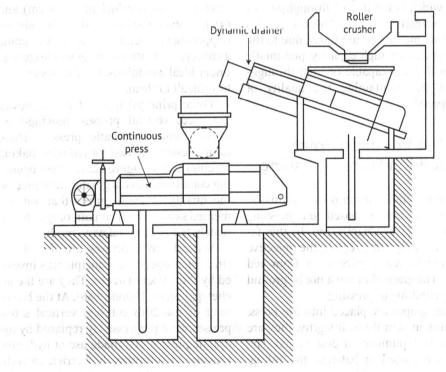

FIGURE 13.3 Continuous juice extraction process.

equipment with poor yields. Moreover, the high speed of continuous drainers limits the diffusion of aroma components from the skins into the juice. Sometimes even at the same turbidity after clarification, juice having undergone continuous juice draining has more difficult fermentations than juice from batch pressing, which is more apt to extract compounds indispensable to yeasts.

Continuous presses extract the remaining 30% of juice contained in the skins after continuous draining. The skins are pushed into a cylinder by a large helical screw against a stop to compact the skins and form a plug. Due to the tearing and grinding of the grapes caused by the screw, the juice obtained with this equipment, regardless of any improvements made by manufacturers, is bitter, vegetal, highly colored, and high in tannins and has an elevated pH (Peynaud, 1971; Maurer and Meidinger, 1976). The wines obtained can never be used in a premium blend (e.g. *Appellation d'origine contrôlée* wine in France). Speed is the only advantage of this press, which is capable of throughputs of up to 100 metric tons/h. The use of this extraction method is rare today, due to the development of high-capacity pneumatic presses that are capable of high throughputs while maintaining the quality of batch pressing.

13.3.3 Immediate Batch Extraction Without Crushing

Also called whole-cluster pressing, this extraction process is based on pressing methods used in Champagne. In this famous sparkling wine region, the objective is to obtain white juice even from red grapes. The grape skin must not be ground during handling or pressing.

Intact grapes are placed into the press. To maintain skin tissue integrity, they are not crushed, pumped, or destemmed. The juice is extracted in batches: the filling, pressing, and emptying operations are carried out successively and make up a cycle. In small facilities that process premium grapes, the small containers (wooden tubs or crates) used to transport the whole grapes are also used to fill the presses. They are easily maneuvered on pallets by means of a forklift.

For large-scale production, if trailers are used to transport grapes from the vineyard to the winery, they must be capable of emptying their contents into the press without the need for a must pump, which inevitably crushes the grapes. Various techniques are effective. Trailers can dump by gravity into a screw-driven hopper feeding a conveyor belt leading to the press. Hydraulic lift trailers capable of elevating themselves to the level of the press may also be used. The grapes are then transferred into the press with a helical screw. A helical screw-based system does not result in a significant amount of burst berries and does not adversely affect wine quality as long as (1) the transfer distance does not exceed 4–5 m, (2) the height of grapes above the screw is kept to a minimum (a few dozen centimeters), (3) the screw diameter is sufficient (30–40 cm), and (4) its rotation speed is slow. Belt-driven hoppers have recently appeared, containing a conveyor belt: they transfer whole clusters under ideal conditions, but this system can be difficult to clean.

Three principal types of batch presses are used: vertical presses, moving-head presses, and pneumatic presses. These same presses are used for red winemaking (Section 12.6.4, Figure 12.11). Press operating conditions have a greater influence on the quality of fresh grapes than on fermented skins, which contain only approximately 15% of the total wine produced.

Vertical screw presses are the oldest, since their operating principle was invented by the ancient Greeks. They are the archetypal press (Hiéret, 1986). At the beginning of the 20th century, vertical screw presses were progressively replaced by hydraulic presses that made use of hydraulic pressure to compress the berries. In vertical presses with a mobile basket, the hydraulic jack lifts the basket and compresses the berries from top to bottom in the direction of the fixed pressure plate. In fixed basket presses (traditional Champagne presses), the top-to-bottom compression is produced by a mobile pressure plate equipped with a hydraulic jack that lowers itself.

The quality of juice extracted by vertical hydraulic presses is indisputable, since the pressure is exerted without grinding the grapes. The juice has a low concentration of suspended solids due to the filtration resulting from the cake thickness. This type of press requires elevated pressures, from 4 to 5 bar during the first pressing to 14 bar for the last. The extended pressing time and the percolation of juice through the skins increases the concentration of aroma compounds from the skins in the must. The primary disadvantages of these vertical hydraulic presses are their slow throughput and the labor-intensive operation of breaking the press cake. In most facilities, they have been replaced by rotating horizontal presses, either moving head or pneumatic, permitting the cake to be broken up mechanically.

In threaded-axle moving-head presses, depending on the direction of basket rotation, the plates (heads) approach from each end (compression phase) or separate from each other (decompression phase). The separation of the heads provokes the breakup of the cake. The press is filled and emptied through central openings in the basket. These presses generally contain internal hoops connected by stainless steel chains fixed to the heads. This setup effectively breaks up the cake but also sharply increases the formation of suspended solids. For this reason, models designed for the production of Champagne and other sparkling wines have no hoops and chains. In the most popular horizontal head press (Vaslin), two basket rotation speeds make rapid pressure increases and cake breakup possible. Rotating-axle horizontal presses are preferable to fixed-axle presses. Rotating the axle in the opposite direction of the basket moves the heads more rapidly and limits the number of basket rotations necessary between pressing cycles. Moving-head presses generally have five to six pressure levels (up to 9 bar). Pressing can be controlled manually or automatically—the program is adjusted according to the nature of the grape. The pressing quality obtained with this type of press depends a lot on the choice of pressing cycle. Increasing the pressure too quickly and excessive, too rapid, and ill-timed cake breakup lead to vegetal and oxidized juice with suspended solids. Slow manual pressing, while monitoring throughput and juice turbidity, yields the best results.

Figure 13.4 gives an example of a pressing cycle with a horizontal moving-head press (Vaslin, 22 VT) in manual mode using whole, healthy, and ripe Sauvignon Blanc. All pressings, as well as the first three cake breakups, are executed at a slow basket rotation speed. Each time after retracting the heads, must extraction is normally resumed at a pressure lower than before breaking up the cake. Table 13.4 states the various pressing times and juice volume and turbidity for this diagram. The drained-off juice and the juice from the first pressing are the most turbid. As soon as the pomace cake is formed, it acts as a filter, and 50% of the juice is extracted without breaking up the cake. The juice from the first two pressings (Π_1 and Π_2) constitutes the free-run juice. The press juice is composed of the last three pressings and represents approximately 15% of the total extracted volume. Total pressing time exceeds three hours, but the overall turbidity of the pressings (about 500 NTU) is satisfactory with respect to the 200 NTU preferred for a juice before fermentation. Consequently, the percentage of solids obtained through natural settling is generally less than 10%.

When correctly carried out to obtain quality juice, pressing with a horizontal moving-head press is necessarily slow. Additionally, due to mechanical constraints imposed by the basket, presses larger than 60 hl cannot be manufactured; thus their

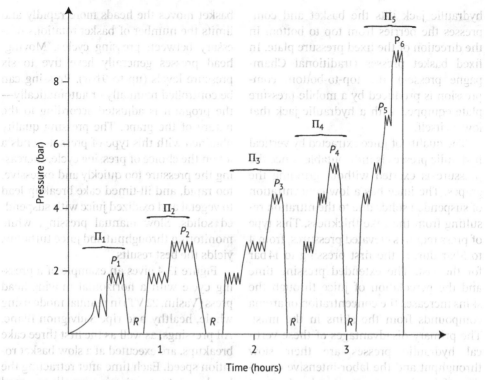

FIGURE 13.4 Whole-cluster pressing cycle in manual mode with a moving-head press (Vaslin 22 VT). P_1 to P_6 = pressure level. Π_1 to Π_5 = pressure step. C = crumbling, drying, or cake breakup.

TABLE 13.4

Evolution of Must Turbidity During Whole-Cluster Pressing of Sauvignon Blanc Grapes in a Moving-Head Press (Vaslin 22 VT)

Pressing	Time (min)	Volume (hl)	Juice (%)	Turbidity (NTU)
Draining	15	0.6	4	630
P_1	60	6.9	46	690
$P_1 + P_2$	50	5.0	34	290
$P_1 + P_2 + P_3$	40	1.3	8	580
$P_2 + P_3 + P_4$	30	0.9	2	350
$P_3 + P_4 + P_5$	15	0.3	2	310
Total	210	15.0	100	550

use is limited to quality white wines and small wineries.

In pneumatic presses, the pressure exerted to extract the juice is applied to the grape clusters by an internal membrane that is inflated with compressed air. The maximum pressure attained by a pneumatic press is 2 bar. Different models exist: perforated basket or closed tank, equipped with drains, with an axial bladder or side-mounted membrane, and filled axially or through doors. They can function manually or automatically with more or less sophisticated programs.

Closed tank presses are preferable to perforated basket presses, since the grapes can be more easily protected with a blanket of carbon dioxide during filling. These presses can also be used for skin maceration under proper conditions (Section 13.3.5). Additionally, the tank press has a greater mechanical resistance than the perforated basket press for the same metal thickness. Juice collection by the drains in tank presses also limits juice oxidation. In basket presses, the juice flows out in a thin layer, increasing oxidation risks. Membrane tank presses are currently the most popular, especially for high-capacity presses. The membrane is located on the half of the tank opposite the drains. The largest membrane presses currently have a 350 hl capacity. Filling, pressing, crumbling, and emptying is depicted in Figure 13.5. During the pressing phase, the tank is immobile with the drains facing the bottom. During crumbling, the membrane is deflated, and the basket rotates.

Axial filling is only an option when the press is filled with crushed grapes (Section 13.3.4). This method leads to an increase in suspended solid concentrations after pressing.

As with horizontal moving-head presses, the pressing quality of pneumatic presses depends on the chosen pressing method and cycle. The general rules are the same. The maximum volume of juice must be extracted at the lowest possible pressure, and crumbling must be limited. Crumbling generates less suspended solids than in moving-head presses, but the oxidation promoted must be taken into account. Figure 13.6 is a representation of the standard pressing program used by Bucher presses. Time and pressure parameters are adjustable. Table 13.5 gives an example of a Sauvignon Blanc pressing with a Bucher 22 hl press, which is interesting to compare with Table 13.4 (moving-head press).

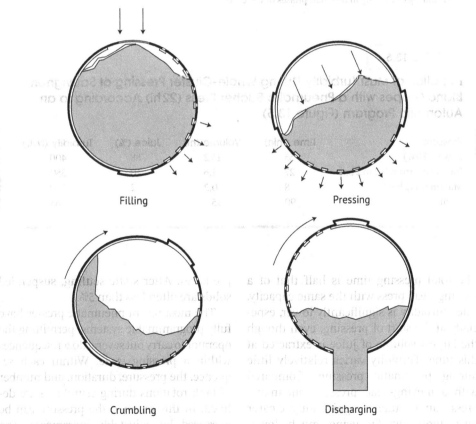

FIGURE 13.5 Operation of a closed tank membrane press.

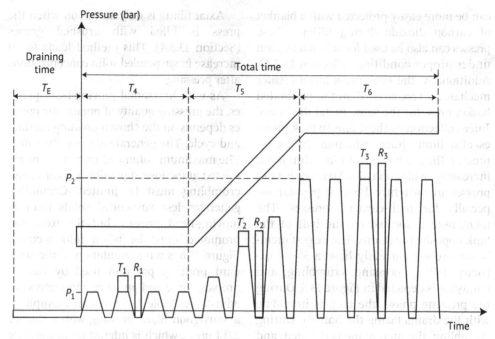

FIGURE 13.6 Standard pressing program of a Bucher pneumatic press. T_E, T_4, T_5, T_6: times at low pressure, increasing pressure and maximum pressure. T_1, T_2, T_3: times at constant pressure. R_1, R_2, R_3: number of rotations during crumbling in different phases of the cycle.

TABLE 13.5

Evolution of Must Turbidity During Whole-Cluster Pressing of Sauvignon Blanc Grapes with a Pneumatic Bucher Press (22 hl) According to an Automatic Program (Figure 13.5)

Pressure	Time (min)	Volume (hl)	Juice (%)	Turbidity (NTU)
Low (0.2 bar)	55	13.2	88	400
Increasing from 0.2 to 2 bar	27	1.6	10	350
Maximum (2 bar)	8	0.2	2	300
Total	90	15	100	463

The total pressing time is half that of a moving-head press with the same capacity. Juice turbidity is significantly lower, especially at the start of pressing, even though the largest volume of juice is extracted at this time. Turbidity varies relatively little during pneumatic pressing. Compared with a moving-head press, a pneumatic press can liberate a significantly clearer juice more rapidly using much lower pressures. After static settling, suspended solids are often less than 5%.

The most recent pneumatic presses have fully programmable systems, permitting the operator to carry out several dozen sequences within a pressing cycle. Within each sequence, the pressure, duration, and number of tank rotations during crumbling are defined. In this manner, the pressure can be increased by adjustable increments. The juice produced has a low turbidity and sometimes does not require clarification. The pressing programs that limit the number of crumblings and thus juice oxidation are by far the best. By measuring juice flow rate, the system determines the pressure level at which the press must operate and the amount of time that this pressure should be maintained. The press incrementally increases the pressure in steps and determines the time and intensity of crumbling.

The most important characteristic of pneumatic pressing is the low increase in the concentration of phenolic compounds in juice during the pressing cycle (Maurer and Meidinger, 1976). The juices of the final pressings, or at least a greater percentage of final press juices, may be blended into the finished product.

In conclusion, different batch pressing systems have been successively used in the making of quality white wines. Clear juice extraction and slow total pressing times were two inherent characteristics of vertical presses. With this kind of press, even if the winemaker had wanted to press more quickly and less carefully, this was impossible. With a moving-head press, clear juice extraction is dependent on pressing slowly and correctly. Yet these presses have often been incorrectly used to work more quickly, producing juices of inferior quality with suspended solids. The creation of pneumatic presses radically changed the constraints on white grape pressing. Clear juices are obtained with much reduced pressing times and a lower proportion of inferior quality press juice. The total pressing time, however, must be sufficiently long to permit the diffusion of characteristic aromas of the grape into the juice.

13.3.4 Advisability of Crushing and Destemming with Immediate Extraction

Crushing consists of breaking the grape skin for immediate liberation of pulp and part of the juice. Grooved rollers with adjustable spacing have long been used for crushing. Turning in opposite directions, they seize the grape clusters and crush the berries. The machinery should be adjusted so that the stems and seeds are maintained intact during this operation. The crusher should be placed above the press to permit gravity filling. This setup eliminates the need for a must pump, which always generates suspended solids. This principle is not always respected. In high-volume wineries, pneumatic presses are frequently filled axially—using a must pump. Draining is facilitated by the periodic rotation of the basket. In this case, the presses are used as dynamic drainers before pressing. These operations always increase the formation of suspended solids.

Crushing before pressing has the advantage of permitting a more significant draining of the grape crop during the filling of the press. Press capacity can be increased by 30–50%. Additionally, crushed grapes are pressed slightly more quickly than whole clusters. On the negative side, the juice obtained is at least two times more turbid than after whole-cluster pressing. Lees volume after natural settling is approximately 20% if a moving-head press is used and slightly lower with a pneumatic press. The suspended solids are liberated in the drained juice before the press cake is capable of playing its role as a filter. Consequently, when a closed tank press is filled with crushed grapes, instead of draining immediately, the drains should be closed during the first half of filling to limit the formation of suspended solids.

The mechanical action exerted on the skins by crushing seems to promote the diffusion of aroma components into the juice. Nevertheless, by rapidly liberating the juice, crushing curtails skin contact. The two phenomena work against each other. The role of crushing on primary aromas located in the skins is uncertain.

Crushing is generally thought to increase grassy and herbaceous aromas (hexanol, *cis*-3-hexenol, and *trans*-2-hexenol) in juice and wine, especially in the

TABLE 13.6

Influence of Semillon Grape Ripening and Crushing on Concentrations (mg/l) of C6 Compounds (Hexanol + Hexenols) in Semillon Wines

State of grapes	Harvest A[a]	Harvest B[a]
Crushed	2.1	1.5
Not crushed	1.4	1.0

[a] Harvests A and B are separated by 10 days.

case of insufficient grape maturity (Table 13.6). Even after adjusting juice to suitable turbidity levels before fermentation, the wines obtained have considerably higher C6 alcohol concentrations when early harvested grapes are crushed.

Destemming white grapes intended for immediate pressing also presents certain disadvantages. The stems act as a draining aid during pressing. Removing them increases draining time and the number of crumblings required. To facilitate the pressing of mechanically harvested grapes, certain manufacturers (Bucher) have equipped their pneumatic presses with complementary drainage systems. The presence of stems during pressing also limits the concentration of thermolabile proteins in the juice, causing protein haze in white wines. Wines are therefore stabilized with lower bentonite concentrations when produced from juice extracted from non-destemmed grapes (Volume 2, Section 6.6.2).

If press capacity permits, crushing and destemming should be avoided, and the grapes should be hand-harvested and pressed immediately. These operations are only necessary when the grape undergoes skin maceration before pressing.

13.3.5 Maceration or Skin Contact

Experience and cautiousness have led to the creation of general white winemaking principles recommending as little maceration as possible with the solid parts of the grape cluster. The diffusion of substances from these solids into the juice leads to various flaws in the wine: vegetal aromas from unripe grapes; astringency and bitterness due to phenolic compounds from seeds, skins, and stems; and moldy, earthy, and fungal odors from spoiled grapes. It should therefore be avoided. With grapes of heterogeneous maturity levels and health status, immediate and rapid juice extraction followed by rigorous selection of press juice is indispensable. Draining chambers, used in the 1950s to store crushed grapes before pressing, were abandoned for this reason (Ribéreau-Gayon et al., 1976). They provoke oxidation and an uncontrolled maceration in the presence of stems.

However, with certain varieties, when soil and climatic conditions combine to produce perfectly ripe and healthy grapes, skin maceration can be sought for the better extraction of grape skin components that participate in the aroma, body, and aging potential of dry white wines. In this case, the positive elements largely outweigh the negative elements linked to insufficient maturity or poor grape soundness.

Slow pressing contributes to the extraction of aroma compounds from the grape skin. In fact, press juice is the result of a certain degree of maceration. In certain cases, adding it to wine is desirable; in other cases, it should be avoided. The higher the sugar concentration and aroma intensity and the lower the pH, adding more press juice improves wine quality. In vineyards in the Bordeaux region, incorporation of Sauvignon Blanc or Semillon press juice before fermentation is systematic when the grapes come from old vines and the best parcels. It is avoided with juice from young vines, insufficiently ripe grapes, and excessively high-yielding vines (Section 13.2.2).

Skin maceration consists of voluntarily permitting a contact phase between the skins and the juice under controlled conditions. A suitable tank is filled with moderately crushed, destemmed grapes. Several hours later, the drained juice is collected and the drained pomace is pressed.

Results reported in the literature, as well as winemakers' opinions on skin contact and the quality of wine obtained by this method, are sometimes contradictory. This is not surprising, since the nature of the grape (variety, soundness, and maturity) and maceration conditions (temperature, tank, and grape handling) greatly influence its effect. According to certain authors (Ough, 1969; Ough and Berg, 1971; Singleton et al., 1975), skin contact lasting for more than 12 hours results in coarse, phenolic wines of inferior quality. Others (Arnold and Noble, 1979) find that the skin maceration of Chardonnay significantly improves aroma quality and wine structure without increasing bitterness and astringency. The best results, in this particular case, are obtained with relatively long maceration (16 hours). Shorter maceration has been recommended for Austrian varieties (Haushoffer, 1978).

Skin maceration grew increasingly popular in France in the mid-1980s (Dubourdieu et al., 1986; Ollivier, 1987). This operation produces satisfactory results with white Bordeaux varieties (Sauvignon Blanc, Semillon, and Muscadelle) as well as with Muscats, Chardonnay, and Gros Manseng—as long as it is carried out with the proper equipment on healthy grapes (negative response to laccase activity test) at the same level of ripeness.

The fully destemmed grapes are transferred to the maceration tank with a must pump, and the tank is then blanketed with a layer of carbon dioxide to avoid oxidation. Sulfiting is avoided to minimize the extraction of phenolic compounds. Various facilities can be used.

The first solution consists in carrying out the skin maceration in a pneumatic press, if it is airtight. When maceration is complete, the juice is drained and the skins are pressed. This system has enological advantages and is simple. The grapes are only transferred once, thus eliminating oxidation. The primary disadvantage is that it takes the press out of service.

Skin maceration is generally carried out in a tank equipped with a system permitting the drained juice (70%) to be removed and the drained skins to be transferred to the press by gravity. The volume of this tank must be triple that of the press.

Skin maceration in a membrane tank (Elite-Pera) is a process situated between maceration in a pneumatic press and tank maceration. At the end of maceration, the juice is collected first by natural draining and then by inflating the tank membrane, incrementally increasing pressure (0.1–0.25–0.4 bar). Ninety percent of the juice can be collected by this method. The drained skins are transferred by gravity into the press with the help of a screw conveyor. The juices obtained by this method do not contain many suspended solids (200–300 NTU) and are particularly well protected from oxidation.

Grape temperature must be maintained below 15°C during maceration. Direct refrigeration of the tank by circulating cold fluid through the jacket or cooling coil is not possible without agitating the grapes, but this operation is not recommended because it promotes the formation of suspended solids and the extraction of phenolic compounds. The crushed grapes may also be cooled with a tube heat exchanger. This process requires considerable cooling capacity and draws the grapes through small-diameter piping with many bends. Increased production of suspended solids may result. Another method consists in incorporating liquid carbon dioxide into the grape crop during filling at the outlet of the must pump. The grapes are cooled without any supplemental mechanical treatment. In addition, the oxygen dissolved in the must during crushing is eliminated by the flow of CO_2. The grapes are also transferred to the tank under an inert atmosphere. It requires 0.8 kg of CO_2 to cool 120 kg of destemmed grapes by 1°C. This technique is costly.

Maceration times vary from 12 to 20 hours, depending on the winery. At controlled temperatures (10–15°C) and in the absence of oxygen, this time period seems to permit a suitable extraction of aroma compounds from the skins without the risk of significant dissolution of phenolic compounds.

Pressing macerated grapes does not pose any particular problems. Due to the destruction of the pectic structure by grape enzymes, the grapes can be pressed at low pressures with only one to two crumblings being necessary. The first pressing is immediately reincorporated with the free-run juice. The final pressings are left separate. The decision to incorporate them with the free-run juice and other pressings is made after clarification.

Skin maceration results in a decrease in must acidity and an increase in pH (Table 13.7). These changes are linked to the liberation of potassium from the skins and the resulting partial salt formation with tartaric acid. The acidity can decrease by as much as 1–1.5 g/l (expressed as H_2SO_4), but the degree of these changes depends on the variety and the terroir. Acidity and pH often vary less in Chardonnay than in other varieties such as Grenache Blanc (Cheynier et al., 1989).

With respect to whole-cluster pressing, skin maceration also provokes an increase in phenolic compounds in the juice. Table 13.8 presents recently obtained values for flavan-3-ols. Maceration increases the amino acid concentration in juice, resulting in an improved fermentation speed, which is often observed in practice. Macerated grapes also produce juice and wine that is richer in neutral polysaccharides (Table 13.9) and proteins than in the case of whole-cluster pressing. Wines made from macerated grapes require higher bentonite concentrations to be stabilized (Volume 2, Section 5.6.2).

Skin maceration makes the most of the aromatic potential of the grapes, and in general it significantly enhances varietal aroma without increasing grassy flavors. In Muscat wines, these sensory differences can be analytically interpreted by measuring free and bound terpene alcohols. Baumes et al. (1989) observed increases in free terpenes from 576 to 742 μg/l and in bound terpenes from 689 to 1,010 μg/l. Measuring S-cysteine and S-glutathione conjugates indicated a significant increase of these aroma precursors in Sauvignon Blanc in relation with skin contact (Peyrot des Gaschons, 2000).

13.3.6 Cryoselection and Supraextraction

Chauvet et al. (1986) initially developed these techniques to improve the quality of juice intended for sweet winemaking (Section 14.2.4), but they are also of inter-

TABLE 13.7

Influence of Pre-fermentation Maceration (18 hours at 20°C) on Total Must Acidity Before Clarification (1985 Harvest) (Dubourdieu et al., 1986)

Variety	Control[a] Total acidity (g/l H_2SO_4)	pH	Pre-fermentation maceration Total acidity (g/l H_2SO_4)	pH	Duration (in hours)
Sauvignon Blanc 1	5.6	3.05	4.05	3.35	8
Sauvignon Blanc 2	5.3	3.15	4.00	3.35	12
Sauvignon Blanc 3	6.05	3.43	4.75	3.53	12
Sauvignon Blanc 4	6.9	2.98	5.50	3.30	18

[a] Must obtained by whole-cluster pressing.

TABLE 13.8

Influence of Pre-fermentation Maceration (10 hours at 20°C) Under Practical Conditions on Must Phenolic Compounds (2010 Vintage) (Nikolantonaki, 2010)

	Unmacerated grapes	Macerated grapes
Variety	Sum of flavan-3-ol contents (catechin and epicatechin) (mg/l)	Sum of flavan-3-ol contents (catechin and epicatechin) (mg/l)
Sauvignon Blanc	12.5	17.9
Sauvignon Blanc	9.6	15.4

TABLE 13.9

Influence of Pre-fermentation Maceration on Total Polysaccharide Concentrations (mg/l) in Wine (Dubourdieu et al., 1986)

Variety	Whole-cluster pressing	Pre-fermentation maceration[a]
Sauvignon Blanc 1	389	469
Sauvignon Blanc 3	356	547
Sauvignon Blanc 4	385	520
Sauvignon Blanc 6	266	359
Semillon 1	362	435
Semillon 2	228	442
Muscadelle 1	290	373

[a] Maceration: 12 hours for Sauvignon Blanc and 18 hours for other varieties.

est for dry white winemaking. The process consists in chilling whole grape clusters in small crates for 20 hours or so in a walk-in freezer at a temperature of −2 to −3°C. Two phenomena—cryoselection and supraextraction—are at the origin of these changes in juice composition observed with respect to traditional pressing.

Cryoselection involves pressing grapes at low temperature. Only the sweetest grapes remain unfrozen and release their juice. High-quality juice is obtained, with the volume increasingly limited as the temperature is lowered. After thawing, a second pressing releases lower-quality juice, coming from grapes with a lower sugar concentration.

Supraextraction corresponds to pressing whole grapes after they have been thawed. The freezing and thawing of the skins and lower epidermal layers results in modifications in tissue ultrastructure. In certain aspects, it produces an effect comparable with skin maceration. Notably, aromas and aroma precursors are released more easily from the grape. Extraction of skin phenolic compounds is, however, lower than with skin maceration and even immediate whole-cluster pressing (Ollivier, 1987). Sugar extraction from the skins is also increased during supraextraction, leading to a 0.3–0.6% bump in potential alcohol. Despite its slowness and high cost, supraextraction promotes the aroma expression of certain premium white varieties.

13.4 Protecting Juice from Oxidation

13.4.1 Traditional and Current Techniques

Oxygen is often said to be the enemy of white wines. In fact, except for rancio (oxidized) wines, whose flavor results from intense oxidation during production, white wines are protected from oxygen (or at least from oxidative phenomena) during winemaking and aging. These precautions are taken to protect the fruity aromas of young wine and to avoid browning. They also promote the later development of bottle bouquet in premium wines during bottle aging.

The oxidation of substances in white wine can occur at any time during winemaking and aging. While the need to protect white wine from oxidation after fermentation is generally accepted, protecting must from oxidation is not unanimously considered necessary.

Most winemakers prefer limiting air contact with crushed grapes and white juice as much as possible. A suitable sulfur dioxide addition to juice blocks the enzymatic oxidation of phenolic compounds. This philosophy is based on empirical observation: juices of many grape varieties must conserve a green color during the pre-fermentation phase to be transformed into fruity white wines. Oxidation phenomena must consequently be avoided as much as possible.

Other enologists believe, on the contrary, that musts too well protected from oxygen give rise to wines that are much more sensitive to oxidation. Furthermore, pressing experiments carried out in an air-free environment have shown that these wines brown quicker in contact with air than wines made from traditionally pressed grapes (Martinière and Sapis, 1967). In addition, these wines are more difficult to stabilize with sulfur dioxide.

Müller-Späth (1977) was the first to contest the need to sulfite white juice before alcoholic fermentation. His research clearly showed that adding pure oxygen to non-sulfited juice before clarification improves the stability of white wine color without producing oxidation-type flaws. This process, called hyperoxidation or hyperoxygenation, consists in oxidizing juice polyphenols to precipitate them during clarification and eliminate them during alcoholic fermentation.

Must oxidation results in a varying degree of color stabilization of white wines, depending on variety (Cheynier et al., 1989, 1990; Schneider, 1989; Moutounet et al., 1990). Hyperoxygenation has also been used successfully on an experimental basis to remove color and improve the quality of second-pressing Pinot Noir and Pinot Meunier juice in Champagne (Blank and Valade, 1989). The impact of this technique on the aroma quality of the wine varies according to the variety and the tasting judges. The effect on aroma is sometimes judged favorable or neutral for Alsatian and German varieties, Chardonnay and Chasselas (Fabre, 1988; Müller-Späth, 1988; Cheynier et al., 1989), but hyperoxygenation, or simply not protecting musts from oxidation, considerably affects the aroma of Sauvignon Blanc (Dubourdieu and Lavigne, 1990). The 4MMP concentration decreases when the must is less well protected from oxidation (Figure 13.7). The mechanisms of this phenomenon will be discussed in Section 13.4.2. Juice oxidation also decreases the aroma intensity of other varieties, such as Semillon and Petit and Gros Manseng, whose aroma similarity to Sauvignon Blanc is due to the participation of sulfur-containing compounds (Volume 2, Chapter 7). The best Chardonnay wines also seem to be made by limiting juice oxidation. Boulton et al. (1995) shared this opinion, believing that juice hyperoxidation harms the varietal aroma of wines.

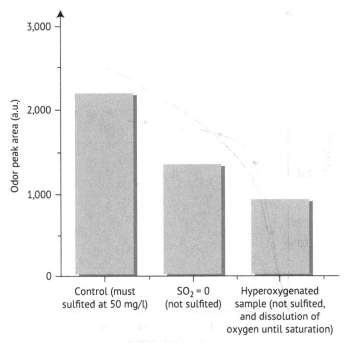

FIGURE 13.7 Influence of must oxidation on 4MMP concentration in Sauvignon Blanc wines (Dubourdieu and Lavigne, 1990).

13.4.2 Mechanisms of Juice Oxidation

Oxygen consumption in juice is essentially due to the enzymatic oxidation of phenolic compounds. Two oxidases (Section 11.6.2) are involved (Dubernet and Ribéreau-Gayon, 1973, 1974): tyrosinase in healthy grapes and laccase from *B. cinerea*, which only exists in juice from botrytized grapes. Laccase activity can be specifically measured in order to evaluate grape soundness analytically (Sections 10.6.6 and 13.2.1).

The substrates of tyrosinase are almost exclusively cinnamic acids and their esters with tartaric acid (caftaric and coutaric acids). It transforms caftaric acid into quinones (Section 11.6.2, Figure 11.12). Owing to cresolase activity of the enzyme, coutaric acid yields the same quinone as caftaric acid. These oxidation reactions are extremely quick. The oxygen consumption rate in juice, when first put into contact with air, can exceed 2 mg/l/min, whereas it is around 1–2 mg/l/day in wine. A certain degree of oxidation in juice inevitably results during white winemaking before protection by sulfur dioxide. The decrease in the speed of oxygen consumption during successive oxygen saturations is caused much more by the depletion of the substrate, caftaric acid, than by the inhibitive effect of the oxidation products formed. Adding caftaric acid reestablishes the initial consumption rate (Moutounet *et al.*, 1990), but laccase is capable of catalyzing the oxidation of a large variety of substances. It not only acts rapidly but also continues over a much longer time period (Figure 13.8).

The quinone formed from caftaric acid has several possible destinations:

1. It can combine with "quinone traps" in juice—in particular glutathione (GSH) (Section 11.6.2, Figure 11.13), a highly reductive tripeptide with a free sulfhydryl (–SH) group, found in concentrations up to 100 mg/kg in certain varieties. The product formed is

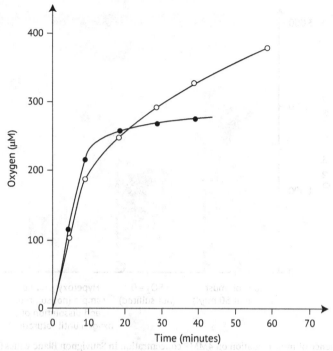

FIGURE 13.8 Oxygen consumption kinetics of a healthy Colombard must (— ● —) and of one with 10% contamination by *B. cinerea* (— ○ —) (Moutounet et al., 1990).

S-glutathionyl-2-caftaric acid, initially called the grape reaction product (GRP) (Cheynier et al., 1986). This oxidation–reduction reaction regenerates the o-diphenol function. Tyrosinase has no action on this GSH derivative, but it can be oxidized by laccase. The caftaric acid quinone can also combine with other juice-reducing agents, such as ascorbic acid. The coupled oxidation regenerates caftaric acid. As long as GSH and ascorbic acid concentrations are high, consumption of must oxygen does not result in quinone accumulation or juice browning.

2. The caftaric acid quinone can also enter into coupled oxidations with flavonoids as well as with GRP. Flavonoid and GRP quinones are formed. These then combine with GSH to form di-S-glutathionyl caftaric acid or GRP2.

3. The caftaric acid quinone is capable of condensing with o-diphenols, first with caftaric acid itself. The color and insolubility of the products formed increases with their degree of condensation. Flavonol quinones also enter into condensation reactions, resulting in strongly colored and later insoluble products.

The oxygen consumption rate of white juice and the nature of the products formed therefore depend on initial concentrations of caftaric acid, GSH, ascorbic acid, and flavonoids in juice. Variety and, most likely, grape maturation conditions influence the proportion of caftaric acid and GSH found in juice. The differences in how two juices react to the presence of oxygen are illustrated in Figure 13.9 (Rigaud et al., 1990). Colombard juice is rich in GSH and ascorbic acid, and its oxidation frees few caftaric acid quinones. In an initial phase, as caf-

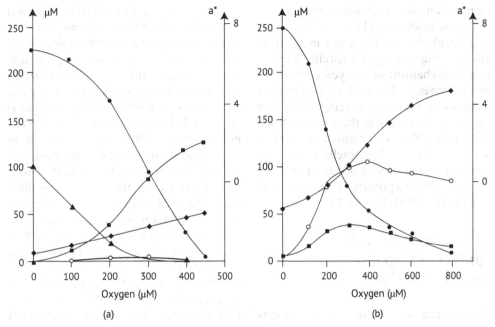

FIGURE 13.9 Oxidation of (a) Colombard and (b) Ugni Blanc must (Rigaud *et al.*, 1990). ○ caftaric acid quinone; ● caftaric acid; ■ GRP; ▲ ascorbic acid; ♦ a* (tristimulus coordinates corresponding to the green-red color range).

taric acid is formed, it is reduced by ascorbic acid. When ascorbic acid is depleted, caftaric acid combines with GSH to form GRP, which accumulates. Juice color changes from green to beige. There is no browning and few reactions coupled with the flavonoids, when quinones are not available. Ugni Blanc juice contains relatively little GSH and no ascorbic acid: as a result, it behaves differently. Oxygen consumption is more rapid, and a large quantity of quinones are formed, resulting in a pronounced browning of the juice with a continued increase in orange color. The caftaric acid quinone enters into reactions coupled with flavonoids and GRP, whose concentration decreases.

The oxidation phenomena linked to the properties of tyrosinase and laccase are rapid and are present as early as crushing and pressing. Unsulfited juices exposed to air consume a variable quantity of oxygen depending on their caftaric acid and flavonoid concentrations.

The effects of juice hyperoxygenation on the stability of white wine color are variable, due to the existence of several reaction mechanisms.

If color stabilization by oxidation is sought, other juice components must not be adversely affected—especially aroma. The difficulty of obtaining fruity Sauvignon Blanc wines from oxidized or brown-orange tinted juices remained unexplained for a long time, but it is now known that various thiols, playing a role in Sauvignon Blanc aroma (Volume 2, Chapter 7), combine very rapidly with quinones in accordance with Michael's addition mechanism. Furthermore, the forms of addition of a very aromatic thiol, 3-sulfanylhexanol (3SH), with quinones from the main phenolic compounds of grapes have recently been characterized, including flavanols [(−)-catechin, (+)-epicatechin] (Nikolantonaki *et al.*, 2012). The presence of GSH and sulfur dioxide helps to protect these aromas by preventing quinone formation, by reacting with or reducing them into phenols in order to minimize addition mechanisms with thiols. However, we can note that there is a different aroma composition of wines made from musts that were oxidized and then sulfited. This is particularly true for wines made from late pressings, owing to a specific fermentation mechanism of the yeast *Saccharomyces cerevisiae*. The fatty acid ethyl ester contents, particularly acetates of higher alcohols, are doubled in these wines (Nikolantonaki, 2010). The content of ethyl 2-sulfanylacetate, a bad-smelling thiol, is greater in white wines made from oxidized juice from the last pressings (Nikolantonaki and Darriet, 2011) (Section 7.5).

13.4.3 Techniques for Protecting Juice from Oxidation

The winemaker can implement various complementary techniques to limit juice oxidation:

1. Sulfiting—antioxidant and antioxidase activity;
2. Protection from oxygen by using carbon dioxide or nitrogen gas;
3. Adding ascorbic acid for its antioxidant effect;
4. Cooling grapes and musts to slow oxidation reactions;
5. Heating musts to 60°C for several minutes to destroy oxidases;
6. Handling grapes in the absence of air to limit the dissolution of oxygen;
7. Clarification to eliminate a portion of the tyrosinase activity associated with solids and to limit the oxidase activity of juice.

Sulfiting is the first, most simple and effective method of protecting juice from oxidation. To destroy tyrosinase, 50 mg/l of sulfur dioxide must be added to the juice. If the grapes are healthy, this addition definitively blocks enzymatic oxidation mechanisms. Sulfiting must be carried out with higher concentrations (70 mg/l) to inactivate tyrosinase in highly colored press juices, containing quinones.

The entire sulfur dioxide addition should be made at the same time. It should be fully homogenized into the juice. Sulfiting at concentrations below 50 mg/l should be avoided, since this only delays oxidation phenomena and juice browning; in time, all of the oxygen contained in the juice is consumed. The worst method consists of progressively adding small quantities of SO_2. The total amount of oxygen consumed under these conditions by a juice in contact with air is greater than in an unsulfited juice. The final color of the two juices, in terms of oxidation, is practically equal.

Sulfiting grapes promotes the extraction of skin phenolic compounds. Protecting with ascorbic acid (10 g/hl) does not have this disadvantage, but this powerful reducing agent does not act on oxidase enzymes. Like low sulfur dioxide concentrations, it only limits browning by reducing quinones but does not limit oxygen consumption. The effectiveness of sulfur dioxide is enhanced when musts are kept in a neutral gas atmosphere (nitrogen, carbon dioxide) during pre-fermentation skin contact and pressing.

Grapes must be in limited contact with oxygen when using ascorbic acid. For example, they should be handled in the presence of dry ice when filling maceration tanks or pneumatic presses. Juices not protected by sulfur dioxide should not be allowed to stagnate in contact with air in press pans during pressing. Their large surface area promotes rapid oxidation.

Cooling grapes and juices is extremely effective in slowing juice oxidation, and it should be used systematically. Oxygen consumption is three times faster at 30°C than at 12°C (Dubernet and Ribéreau-Gayon, 1974). Cooling with liquid CO_2 while filling tanks and presses combines the handling of crushed grapes in an inert atmosphere with the effects of cooling.

Cryoextraction or supraextraction (pressing whole grapes at temperatures below 0°C) considerably limits oxidation. This technique enhances the fruity character of dry white wines, compared with pressing at ambient temperatures. Not only are aromas and aroma precursors freed by the freezing and thawing of the skins, but oxidative phenomena are also limited during pressing at a temperature close to 0°C.

Clarification (Section 13.5) limits oxidase activity but does not prevent juice from browning. A sufficient soluble tyrosinase activity remains in fresh juice unprotected from oxygen, allowing rapid browning. Clarification is a means of eliminating oxidation products—in particular condensed flavonoids formed during coupled oxidations.

Heating juice theoretically destroys oxidases, but it must occur quickly after extraction. The heating process must also be rapid. It is rarely used in practice.

13.5 Clarification

13.5.1 Formation and Composition of Suspended Solids and Lees

Freshly extracted grape juice is more or less turbid. It contains suspended solids of diverse origin: soil, skin and stem fragments, cellular debris from grape pulp, insoluble residues from vineyard treatment products, etc. Macromolecules in colloidal solution or in the course of precipitating are also involved in juice turbidity. Among them, grape pectic substances play an essential role (Volume 2, Section 3.6). With rotted grapes, juice turbidity is also caused by the presence of polysaccharides, especially (1,3)-(1,6)-β-D-glucan, produced by *B. cinerea* in the berry. A few milligrams of these substances are enough to provoke serious clarification difficulties (Volume 2, Sections 3.7 and 11.5.2). These carbohydrate macromolecules influence juice turbidity through the Tyndall effect (Volume 2, Section 9.1.2). Acting as protective colloids (Volume 2, Section 9.4), they also hinder clarification by limiting or blocking particle flocculation and sedimentation phenomena as well as clogging filter surfaces. By acting on the colloidal structure of the juice, natural grape pectinases (or those added by the winemaker) facilitate natural settling. After several hours, the juice separates into two phases: a more or less opalescent clear juice and a deposit varying in thickness. The latter contains different-colored successive strata: greenish brown in the lower portion of the deposit and green to light beige in the upper portion. Some winemakers distinguish between the heavy deposit that forms first during natural settling and the light deposit that accumulates more slowly. Clarification consists in separating (by racking, for example) the clear juice from the lees before alcoholic fermentation.

The quantity of suspended solids formed during juice extraction and the speed of sedimentation depend on variety, grape disease status, maturity, and especially winemaking methods (crushing, draining, pressing, etc.) (Section 13.3).

Under normal conditions, juice turbidity generally decreases during grape maturation (Hadjinicolaou, 1981). This evolution results from the hydrolysis of pectic substances in the berry by pectic enzymes of the grape (endo-polygalacturonase and pectin esterase). Under dry weather conditions, the grape remains pulpy, and the juice is more difficult to extract and clarify, due to a lack of pectinase activity. Toward the end of maturity, the soluble acidic polysaccharide (pectin) concentration in juice generally evolves in parallel with juice turbidity. This is a good indicator of clarification potential (Robertson, 1979; Dubourdieu et al., 1981; Ollivier, 1987). When the pectin concentration in juice steadily drops during maturation, the juice is generally easy to clarify. In the opposite case, clarification is more difficult, and exogenous pectic enzymes must be used.

High levels of rot in the harvested grapes increase juice turbidity and make clarification difficult, due to the protective colloidal effect of glucan produced by *Botrytis*. A low concentration of rot (less than 5%) tends to facilitate juice clarification, due to the pectinase activity in moldy grapes, which is nearly 100 times higher than in healthy grapes.

Juice extraction methods have a prime influence on the formation of suspended solids. Slow batch pressing, while minimizing cap breakup, leads to the clearest juices (Section 13.3.3).

The exact physical structure and chemical composition of the suspended solids remain unknown. They are made up of varying-sized particles of less than 2mm. They are generally observed (Alexandre et al., 1994) to contain essentially insoluble polysaccharides (cellulose, hemicellulose, pectic matter) and relatively few nitrogen compounds, essentially insoluble proteins not utilizable by yeasts (Table 13.10). They also contain mineral salts and a significant amount of lipids—most likely from cell membranes. This lipid fraction contains a slightly higher proportion of unsaturated than saturated fatty acids. The principal fatty acids are linoleic acid (C18:2), palmitic acid (C16:0), and oleic acid (C18:1) (Table 13.11). They enter into the composition of membrane phospholipids of grape cells, but a small proportion also exists in a free state, adsorbed on solid particles (Lavigne, 1996). These particles are definitely utilizable by yeasts. Phytosterols, mainly represented as β-sitosterol, also are associated with suspended solids and contribute to *S. cerevisiae* yeast metabolism during alcoholic fermentation (Casalta et al., 2016).

TABLE 13.11

Total Fatty Acid Composition of Suspended Solids (%) (Alexandre et al., 1994)

Component	%
Lauric acid C12:0	8.3
Palmitic acid C16:0	25.0
Palmitoleic acid C16:1	5.5
Stearic acid C18:0	22.2
Oleic acid C18:1	22.2
Linoleic acid C18:2	25.0

13.5.2 Influence of Clarification on Dry White Wine Composition

Winemakers have long observed an improvement in dry white wine quality resulting from proper juice clarification. Clarifying juice improves wine quality more dramatically, when compared with the unclarified juice, when there is a high concentration of suspended solids in the initial juice. Wines made from juices containing too many suspended solids have heavy, green aromas and bitter tastes. They are also higher in color and richer in phenolic compounds, and their color is less stable to oxidation. At the end of fermentation, they often contain reduction odors, which are more or less difficult to eliminate by aeration and racking. Inversely, the fruity character of the variety is more distinct and stable in wines made from clear juice. Some of these empirical observations based on tasting have been confirmed by analysis of wines.

Since the 1960s, clarification has been known to improve the fermentation aromas of dry white wines (Crowell and Guymon, 1963; Bertrand, 1968; Ribéreau-Gayon et al., 1975). Wines made from clarified juices have lower concentrations of

TABLE 13.10

Suspended Solids Composition (%) (Alexandre et al., 1994)

Component	%
Total neutral polysaccharides	71.9
Total nitrogen	2.6
Ashes	5.5
Acidic polysaccharides	5.2
Lipids	7.8
Total	93

pungent higher alcohols and higher concentrations of ethyl esters of fatty acids and higher alcohol acetates, which have more pleasant aromas.

Clarification also limits the concentration of C6 alcohols in wines (Table 13.12) (Dubourdieu et al., 1980). Before fermentation, juices essentially contain C6 aldehydes (hexanal, *cis*-3-hexenal, and *trans*-2-hexenal) formed by enzymatic oxidation of linolenic and linoleic acids during pressing. The detailed mechanisms of these reactions are described in Section 11.6.2. These compounds are not very soluble in juice and most likely remain partially associated with the grape solids. During alcoholic fermentation, they are systematically reduced into the corresponding alcohols by the yeast and pass into solution in the wine. The elimination of suspended solids from the must therefore helps to lower vegetal aromas in dry white wines. The influence of clarification increases when pressing and handling of grapes become more brutal and the harvested grapes are less ripe.

Initial research demonstrating the enological value of clarification generally reported the percentage of clear juice obtained by different clarification methods but rarely specified the turbidity of the clarified juice. Yet relatively small turbidity variations have been shown to have a decisive influence on alcoholic fermentation kinetics and wine composition.

More recent research has focused on the influence of the degree of clarification on the production of unpleasant sulfur-containing compounds during alcoholic fermentation and the more or less stable reduction off-odors that result (Lavigne-Cruège, 1996; Lavigne and Dubourdieu, 1997) (Volume 2, Section 8.6.2). Some heavy sulfur-containing compounds produced by yeasts increase juice turbidity (Table 13.13). However, in light of the perception threshold and sensory descriptors of these various compounds (Volume 2, Section 8.6.2), only methionol (3-(methylthio)-1-propanol), with an unpleasant odor of cooked cabbage, is significantly involved in the off-odor observed when juice turbidity exceeds 250 NTU. Methionol is stable in wine, and it cannot be eliminated by racking and aeration. This troublesome consequence of insufficient clarification is definitive. The degree of juice turbidity must therefore be adjusted with precision to maintain the aroma finesse of dry white wine.

Increased methionol formation in insufficiently clarified juices cannot be interpreted as a methionine enrichment of juice by grape solids. In fact, grape solids do not contain soluble amino acids, and the acid protease they contain is incapable of freeing them by hydrolyzing proteins in the juice. Furthermore, even prolonged contact between the juice and lees does not result in an increased concentration of amino acids.

Some experiments demonstrate the role of the lipid fraction of the grape solids in

TABLE 13.12

Influence of Must Clarification on C6 Alcohol Concentrations (Hexanol + Hexenols) in Wine (Dubourdieu et al., 1980)

Must treatment	Must turbidity (NTU)	C6 alcohols in wine (mg/l)
Non-clarified must	400	2.0
Clarified must	260	1.0
Deposited solids	ND[a]	2.1
Deposited solids after filtration	8	0.9

[a] ND = not determined.

TABLE 13.13

Influence of Must Turbidity on Concentration of Heavy Sulfur-Containing Compounds in Wines (µg/l) (Lavigne-Cruège, 1996)

Substances	Must turbidity			Perception threshold of substance (model solution)
	120 NTU	250 NTU	500 NTU	
2-Mercaptoethanol	113	140	179	130
2-Methyltetrahydrothiophenone	102	131	191	70
2-(Methylthio)ethanol	61	61	66	250
Ethyl 3-(methylthio)propanoate	1	2	2	300
3-(Methylthio)-1-propanol acetate	5	6	6	50
3-(Methylthio)-1-propanol (methionol)	1,097	1,958	3,752	1,200
4-(Methylthio)-2-butanol	35	66	60	80
Dimethyl sulfoxide	363	728	1,448	Odorless
Benzothiazole	28	26	29	50
3-(Methylthio)propionic acid	85	178	310	50

the excessive production of methionol during alcoholic fermentation. This fraction most likely promotes the incorporation of methionine in yeasts that is transformed into methionol via the Ehrlich reaction (Section 2.4.3). The lipid fraction of the grape solids has also been shown to be involved in limiting acetic acid production by yeasts. The practical consequences of these phenomena are obvious. If juice turbidity is too low, an insufficient concentration of long-chain unsaturated fatty acids could induce excessive production of acetic acid by yeasts. If the turbidity is too high (greater than 250 NTU), an excess of these same fatty acids promotes excessive methionol formation.

Juice turbidity also influences the production of volatile sulfur-containing compounds by yeasts (Volume 2, Section 8.6.2): H_2S, methanethiol, dimethyl disulfide, and carbon disulfide. Wine made from juice fermented at 510 NTU has a pronounced reduction off-odor (Figure 13.10). In the same wine fermented at 270 NTU, dimethyl disulfide is present in concentrations above the sensory threshold but does not produce a noticeable reduction flaw. However, when the methanethiol concentration exceeds its sensory threshold (0.3 µg/l), which is the case with a less clarified must, the quality of wine aroma is immediately lowered. Methanethiol thus plays a major role in reduction flaws in dry white wines following insufficient clarification. Elemental sulfur and certain pesticide residues in the grape solids explain the effect of juice turbidity on the formation of light sulfur-containing compounds.

At equal turbidity levels, sulfiting also influences the production of heavy and light volatile sulfur-containing compounds by yeasts. Methionol and hydrogen sulfide, for example, increase greatly with the sulfur dioxide concentration used. This concentration must not exceed 5 g/hl, added in a single dose as soon as the juice is received. The free sulfur dioxide concentration should not be adjusted before alcoholic fermentation, or during or after clarification, since this practice does not provide greater oxidation protection for the juice and systematically promotes the production of sulfur-containing compounds by yeasts.

The role of clarification in fruity varietal aromas is not well known, and winemakers' observations in this respect are sometimes contradictory. Insufficiently clarified juices,

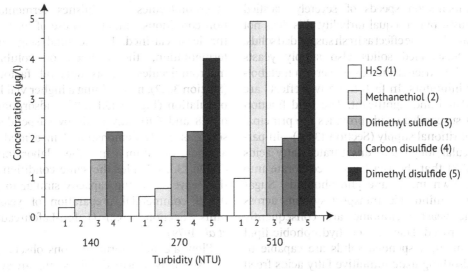

FIGURE 13.10 Influence of must turbidity on the formation of light volatile sulfur-containing compounds by yeasts (Lavigne, 1996).

especially those containing insoluble products of phenolic compound oxidation, can produce wines with decreased varietal aromas. In contrast, overclarification (less than 50 NTU) also decreases the fruitiness of dry white wines. This phenomenon has been observed with Muscat, Chardonnay, Sauvignon Blanc, Semillon, Gros and Petit Manseng, etc. It is exacerbated by difficult fermentation conditions and by excessively slow fermentations with increased volatile acidity production. The varietal aroma of wines made from excessively clarified juices is sometimes masked by an artificial, banana, amyl, or soapy aroma, linked to the presence of a greater quantity of esters.

It is difficult to recommend an optimum turbidity that is valid for all varieties. A range between 100 and 250 NTU is generally used, since it is a suitable compromise between a good alcoholic fermentation and aroma finesse of dry white wines.

13.5.3 Effect of Clarification on Fermentation Kinetics

Slow and stuck fermentations of dry white wines are well-known consequences of clarification. This phenomenon, varying in intensity according to juice composition and clarification methods, has incited much research and been interpreted in different manners in the past.

Clarification depletes must microflora. Inoculating juice with yeasts after clarification has long been known to hasten the onset of fermentation but does not noticeably change its duration or the quantity of residual sugar present upon its completion (Ribéreau-Gayon and Ribéreau-Gayon, 1954). Clarification is thus not simply an "unyeasting" of the juice.

A variety of physical actions also contribute to the stimulating effect of suspended solids on alcoholic fermentation. By providing nucleation sites for gas bubbles, it has been suggested that suspended solid particles promote the elimination of CO_2 from the fermentation medium—thus limiting its inhibitive effect on yeasts. This effect is very limited at tank pressures found in dry white winemaking. Suspended solids are also thought to promote yeast growth by serving as a support. As a matter of fact, the addition of various supports such as diatomaceous earth (Schanderl, 1959), bentonite (Groat and Ough, 1978), and cellulose (Larue et al., 1985) improves the fermentation speeds of severely clarified musts, but, at equal turbidity, this does not have the same effect as fresh suspended solids.

Suspended solids also supply yeasts with nutrients and adsorb certain metabolic inhibitors. In fact, these two effects are related and significant. The lipid fraction of suspended solids provides the principal nutritional supply (Section 13.5.1)—in particular long-chain unsaturated fatty acids (C18) that the yeast can incorporate into its own membrane phospholipids. Sugar and amino acid transport systems across the yeast membrane are consequently improved. Due to their hydrophobic lipid content, suspended solids are capable of adsorbing toxic inhibitive fatty acids freed in the juice during alcoholic fermentation (C8, C10, C12). The combination of these two effects (lipid nutrition and toxic fatty acid adsorption) produces a survival factor effect for yeasts (Section 3.5.2) (Ollivier et al., 1987; Alexandre et al., 1994).

In Table 13.14 and Figure 13.11, six lots corresponding to different clarification levels were constituted from fresh Muscadelle juice. Lowering juice turbidity prolongs alcoholic fermentation. In the case of the most clarified juice (lot CN), this can lead to a stuck fermentation. Supplementing lot CN with either colloids or soluble macromolecules reestablishes fermentation conditions similar to those of lot CS, the least clarified. In the final stage of fermentation, the colloids or soluble macromolecules act as survival factors (Section 3.5.2), maintaining a higher viable population (Figure 13.11). The adsorption of C8 and C10 fatty acids by suspended solids is easily demonstrated in a model alcohol medium in the laboratory (Table 13.15). Under the same conditions, they have a fixating capacity similar to a 0.5 g/l commercial preparation of yeast hulls (Section 3.6.2) (Lafon-Lafourcade et al., 1979).

Slow alcoholic fermentations observed in extremely clarified juices are always accompanied by increased acetic acid concentrations in wines (Section 2.3.4). The degree of this effect varies depending on the yeast strain used (Table 13.16). It is intensified by high sugar concentrations.

13.5.4 Clarification Methods

The most simple and effective juice clarification method is natural settling—the natural settling of suspended solids followed by a careful racking. Free-run juice and the first pressing on the one hand and subsequent pressings

TABLE 13.14

Influence of the Colloidal Composition of Muscadelle Must Clarified Under Different Conditions on the Length and Completion of Alcoholic Fermentation (Ollivier et al., 1987)

	Treatments[a]					
Component	CC	CS	CE	CN	CN+HF	CN+SM
Turbidity (NTU)	280	62	1.5	2.6	120	6.8
Proteins (mg/l)	—	506	412	356	382	548
Total polysaccharides (mg/l)	344	323	218	318	323	540
Length of alcoholic fermentation (days)	18	25	33	54	25	31
Residual sugar at the end of fermentation (g/l)	1.2	2.0	2.0	2.7	2.0	1.9

[a] CC, coarse clarification; CS, cold settled must; CE, cold settling + enzymatic clarification; CN, centrifuged must; HF, colloidal haze; SM, soluble macromolecules.

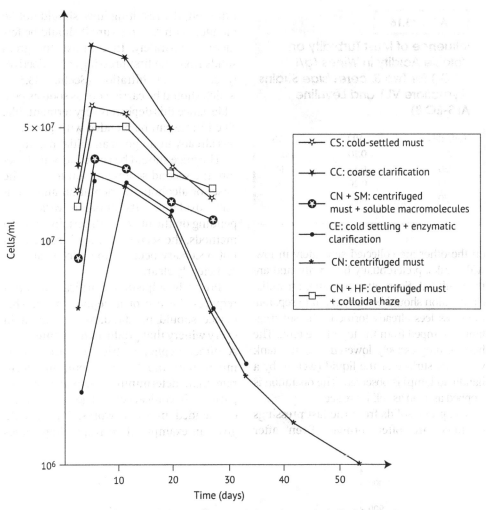

FIGURE 13.11 Influence of must turbidity on yeast populations during alcoholic fermentation.

TABLE 13.15

Influence of the Amount of Colloidal Haze on Fatty Acid Adsorption (mg/l) After 24 Hours of Contact (Ollivier et al., 1987)

Turbidity (NTU)	Hexanoic acid	Octanoic acid	Decanoic acid
1	5.6	10.8	4.1
15	5.6	11.0	4.3
31	5.0	10.2	3.3
62	4.4	7.5	1.5
124	5.3	8.7	2.0

TABLE 13.16

Influence of Must Turbidity on Volatile Acidity in Wines (g/l H_2SO_4) for Two *S. cerevisiae* Strains (Zymaflore VL1 and Levuline ALS-EC 8)

Turbidity	VL1	EG8
500	0.10	0.14
250	0.12	0.14
100	0.20	0.40
50	0.30	0.55

on the other are collected separately in low, wide tanks, preferentially by gravity, and are then sulfited. Soon after pressing, an initial clarification should be carried out to separate the gross lees already formed. The supernatant is pumped from the top of the tank. The hose is progressively lowered into the tank while the surface of the liquid (well lit by a handheld lamp) is observed. The operation is stopped as soon as solids are seen.

The grape solids from the last pressings of juice are often brown. Even after filtration, the resulting juice should not be blended with the free run: it should be fermented separately. In contrast, the gross solids from the first pressings are effectively clarified by filtration (Section 13.5.5), which should be carried out as soon as possible, since this deposit is very fermentable. The filtrate can be blended with juice that has already undergone an initial racking.

The juice should be cooled to 5–10°C before the second sedimentation to slow the onset of alcoholic fermentation and minimize oxidation. Its duration varies, depending on the juice. With certain pressing methods, the second racking is sometimes not necessary because the juice is already sufficiently clear.

Precisely adjusting clarification levels requires the use of a nephelometer. This device should be standard equipment in every winery that produces dry white wine. A direct nephelometric measurement is much more rapid, convenient, and accurate than determining the percentage of particles in conical centrifuge tubes, as recommended in some works. Figure 13.12 gives an example of corresponding values

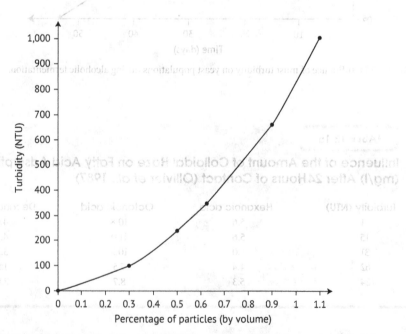

FIGURE 13.12 Example of the correlation between turbidity expressed in NTU and percentage of solid particles in must.

between turbidity and percentages of solids. The optimum turbidity range of 100–250 NTU corresponds to 0.3–0.5% of particles. The nephelometric measurement is much more precise than the solids percentage. In addition, the nephelometric reading is direct, whereas the solids percentage measurement requires at least five minutes of centrifugation.

Samples are taken from the middle of the decanting tank to monitor evolution of juice turbidity during clarification. When optimum juice turbidity is attained, a second racking is carried out as indicated above. Once this operation is accomplished, the turbidity of the clear racked juice must be verified. If it is too high due to error, additional settling is necessary. If, on the contrary, it is too low, fine solids must be added to the clear juice.

To obtain the highest-quality juice from final pressings, suitable for blending with juice from first pressings, commercial pectinases should be used to maximize clarification. Very low turbidity, from 10 to 15 NTU, should be obtained. The light solids of these press juices are highly colored by oxidized phenolic compounds and should be eliminated. Like the heavy solids from press juice, they are not worth filtering. Optimum press juice turbidity (100–200 NTU) is obtained by adding the appropriate quantity of fine solids from the natural settling of the corresponding free run.

When juice clarification is too slow, due to insufficient activity of natural grape pectinase, settling can be accelerated by using commercial pectinases from *Aspergillus niger*. These preparations should be pure and not contain cinnamate esterase activity (Volume 2, Section 8.4.3) to minimize vinylphenol production by yeasts. Like grape pectinases, commercial pectinases have several activities—notably a pectin esterase activity that demethylizes pectic chains and an endo-polygalacturonase activity that hydrolyzes glycoside bonds between galacturonic residues. The use of exogenous pectinases can result in excessive juice clarification. Juice turbidity must often be adjusted after settling by adding fine solids. Juice clarity must be fastidiously adjusted to assure a complete fermentation and allow the expression of aroma quality of dry white wines.

After settling, the juices from the last pressings must be evaluated to determine if they can be blended with first pressing juice. There are no general rules to guide this decision. Tasting, evaluation of visual color (the least oxidized possible), phenolic index, sugar concentration, and especially pH are taken into account. For example, for a Sauvignon Blanc from Bordeaux, press juices with a pH greater than or equal to 3.5 are not blended, but the addition of quality press juice to free run intensifies the varietal aroma of certain Sauvignon Blanc wines. Press juices should be blended before alcoholic fermentation.

Other juice clarification methods exist using more or less expensive equipment: centrifugation, filtration, tangential microfiltration, and carbon dioxide or nitrogen flotation. For various reasons, these techniques do not produce as high-quality wine as natural settling. Centrifugation always causes a certain amount of oxidation, and filtration systems generally produce wines lacking aroma intensity. These techniques are even less justified today, since pneumatic presses extract relatively clear juice requiring minimal settling. Replacing continuous extraction systems with pneumatic presses in large winemaking facilities has made all of the mechanical clarification systems practically obsolete.

13.5.5 Clarification Methods for Grape Solids Deposits

The volume of grape solids deposits obtained after natural settling represents a sizable proportion of the harvest.

Two types of filters are used for solids filtration: diatomaceous earth rotary vacuum filters and plate-and-frame filters (using 1–1.5 kg of perlite/hl of solids filtered).

These two methods extract clear juice (less than 20 NTU), without clogging, at a rate of 1–2 hl/h/m². Their recovery rate is near 90% when the deposit to be filtered contains 10% solids. There is practically no juice loss (Dubourdieu et al., 1980; Serrano et al., 1989). The juice obtained can be blended with clear juice from natural settling without a quality difference detectable by tasting or analysis.

Plate-and-frame filters are easier to use than rotary vacuum filters—especially for small wineries. They also have the advantage of not exposing the juice to air.

13.6 Juice Treatments and the Advisability of Bentonite Treatments

Sugar and acidity adjustments are described in Sections 11.4 and 11.5. They should be carried out after clarification.

In the past, bentonite treatments were also recommended to eliminate proteins in the juice—responsible for instability of juice clarity (Milisavljevic, 1963; Ribéreau-Gayon et al., 1976) (Volume 2, Sections 5.5 and 5.6). The use of bentonite in juice before fermentation offers long-known advantages but also has some more recently discovered drawbacks.

Treating juice with bentonite is recommended for wines that are to be clarified shortly after the completion of alcoholic fermentation. Additional handling of the wine is avoided in this manner at a time when the wine is thought to be more fragile. However, protein stability tests carried out on juice are imprecise and therefore not practical. Within a given winery, the same bentonite concentration is generally used on all musts. In spite of this treatment, white wines are sometimes unstable at bottling and require an additional bentonite treatment.

If white wines are to undergo barrel or tank aging on the lees (Sections 13.8 and 13.9), bentonite treatments are not recommended for two reasons:

1. Maintaining wine for several months on lees containing bentonite, with weekly stirring, has been observed to damage sensory quality.
2. Lees aging naturally stabilizes white wines with respect to protein precipitation (Volume 2, Section 5.6.4). The mechanisms of this phenomenon have long occurred unnoticed. Yeast autolysis progressively releases different mannoproteins in wine, including one with a strong stabilizing effect on the proteins responsible for protein haze. Lees aging of white wine decreases the bentonite concentration necessary for stabilization by a factor of two to four.

After barrel or tank aging, lees-aged white wines are treated with bentonite at relatively low concentrations, determined by precise and reliable protein stability tests.

13.7 Fermentation Operations

13.7.1 Tank Filling

Fermentors are filled with clarified juice. Approximately 10% of the tank volume is left empty to avoid the overflowing of foam (Section 3.2.5) produced during the tumultuous phase of alcoholic fermentation.

Different clarified juices must often be blended together when filling a high-capacity fermenting tank. This operation requires several basic precautions. Before blending the clarified juices from different tanks, fine solids that have settled after racking must be reincorporated into the juice. In addition, juice that has not started fermentation should not be blended with fermenting juice, since the yeasts fermenting one juice produce H_2S in the presence of free SO_2 from the other juice. The onset of fermentation must occur after the blend is made.

13.7.2 Yeast Inoculation

Within the last 40 years, the use of active dry yeast (ADY) in winemaking has increased considerably. It has replaced the traditional practice of "pied de cuve" (native yeast starter) in many wineries (except for those choosing to use spontaneous fermentation, for example, to be certified as biodynamic). In this method, juice is obtained from not fully ripened grapes and then sulfited at 3 g/hl to eliminate spoilage yeasts and promote the growth of wine yeasts. It is then inoculated into newly filled fermentors at a concentration of 2–5% after several days of spontaneous fermentation.

The kinetics of spontaneous dry white wine fermentation are sometimes difficult. The speed and degree of completion of fermentation vary, depending on the native strains present. In addition, clarification "deyeasts" juice to a certain degree. A slow fermentation due to a low yeast population can result. Sometimes quality wild microflora can produce spontaneous fermentations with excellent results. Selection of wine yeast strains began by isolating strains with successful spontaneous fermentations.

Today, hundreds of ADY strains belonging to *S. cerevisiae* are used in white winemaking. They have been selected based on more or less empirical criteria of their enological aptitudes in different winemaking regions of the world.

The ADY strain chosen for white winemaking has significant consequences on fermentation kinetics and the development of varietal aromas. A difficult fermentation always gives rise to dull wines lacking aroma definition and intensity. The most important quality of a yeast strain intended for dry white winemaking is the ability to ferment completely a juice with a turbidity of between 100 and 200 NTU containing up to 220 g of sugar per liter, without excessive production of volatile acidity. This ability to ferment clarified juice is not widespread among wild yeasts, and a wild yeast inoculum from spontaneous fermentations sometimes does not contain such strains.

Some strains produce high ester concentrations—in particular higher alcohol acetates contributing to the fermentation aromas of dry white wines. Their use is only recommended for neutral grape varieties—they mask the varietal aroma of noble varieties.

Other strains were selected for their low vinylphenol production. These compounds possess rather unpleasant pharmaceutical aromas. Above a certain concentration, they dull the aroma of dry white wines (Volume 2, Sections 8.4.2 and 8.4.3). These strains have low cinnamate decarboxylase activity. During alcoholic fermentation, this enzyme catalyzes the partial transformation of p-coumaric and ferulic acids found in juice into 4-vinylphenol and 4-vinyl guaiacol. Since this enzyme is inhibited by phenolic compounds, only white wines can contain quantities of vinylphenols likely to affect their aroma. The use of strains with low cinnamate decarboxylase activity is recommended—particularly for white juices containing high concentrations of hydroxycinnamic acid.

The role of yeasts in the varietal aroma of wines is poorly understood. With the exception of the terpene aromas of Muscat varieties and the characteristic Sauvignon Blanc aroma, little research has been dedicated to the aroma characteristics of other varieties. Yeasts have been shown to release only small amounts of free terpene alcohols from terpene glycosides present in juice. The yeast strain used for fermenting Muscat juice, therefore, does not greatly influence the terpene alcohol composition in wine. However, many winemakers have observed the particular aptitude of certain *S. cerevisiae* strains (EG8, 2056, VL3, X5, R2, TXL, Delta, etc.) to intensify the varietal aroma of certain grape varieties (Sauvignon Blanc, Gewürztraminer, Mansengs) (Volume 2, Chapter 7).

As a result, the winemaker must understand not only the fermentative behavior of the yeast strain used but also its effect on the specificity of the wine made. The composition of aroma precursors in the grapes is responsible for the aroma intensity of the wine. The role of the yeast is to transform this grape aroma potential into free aromas. Ipso facto, the yeast reveals the grapes' polymorphism depending on the vintage and terroir. A good yeast strain enables the expression of the finesse and complexity of the aromas in the grape, but it does not detract from this character by revealing flaws, masking it with excessive fermentation aromas or caricaturing it by revealing only some of the particular nuances. If yeast strains are chosen wisely, the use of selected yeasts will not lead to a standardization of dry white wine aromas.

White juice should be inoculated with ADY at a concentration of 10–15 g/hl or 10^6 cells/ml of juice, immediately after clarification. The cells are reactivated beforehand for 20 minutes in a water and must mixture (1:1) at 40°C. If the must was clarified at low temperatures (10–12°C), it is not necessary to wait for the must temperature to rise before inoculating, since early inoculation guarantees the implantation of the starter.

It is essential to homogenize the tank or barrel contents at the time of inoculation. In this manner, the suspended solids are properly distributed in the juice during yeast growth. In high-capacity tanks without an agitator, the blending operation is difficult, and the starter should be pumped into the fine solids at the bottom of the tank, as opposed to over the possibly overclarified must at the top of the tank.

13.7.3 Addition of Nitrogen Sources and Juice Aeration

The general mechanisms that link the nitrogen and oxygen needs of yeasts during the alcoholic fermentation process (Sections 2.4.2, 3.4.2, and 3.5.2) explain the advisability of the addition of ammonium salts and the necessity of aeration during dry white winemaking.

Yeast-assimilable nitrogen concentrations (ammonium cation and amino acids except for proline) in white juice from cool-climate vineyards (northern or Atlantic type) are generally sufficient to ensure normal yeast multiplication, but, even in these climates favorable to white varieties, an insufficient nitrogen supply to the grapevine or excessive summer dryness can sometimes result in juices deficient in yeast-assimilable nitrogen. Viticultural conditions favoring this situation are varied but always foreseeable: superficial root systems of young vines, winter root asphyxiation in poorly drained soils, light soils with an insufficient water reserve, and ground cover strongly limiting water and nitrogen supplies. The winemaker should pay close attention to these potential problems, since nitrogen-deficient white juices almost always produce heavy white wines with little fruit and aging potential.

White juices with concentrations of less than 140–160 mg of yeast-assimilable nitrogen per liter should be supplemented with ammonium sulfate. Yeast-assimilable nitrogen concentrations in suspected nitrogen-deficient white juices should be systematically analyzed using various analytical methods (Casalta et al., 2016). The most efficient is likely the sum of alpha-amino nitrogen measured by the N-o-phthaldialdehyde (NOPA) method and ammonium using an enzymatic or chemical protocol or FTIR. Some white musts can contain less than 40 mg of yeast-assimilable nitrogen per liter. Nitrogen can be added to the must in various forms. It can be an inorganic form of ammonia nitrogen (ammonium sulfate or diammonium phosphate), since organic winemaking tolerates diammonium phosphate exclusively. Supplementation may also be carried out with forms of organic nitrogen, derived from yeast breakdown products, depending on the specific authorizations for each certification type (organic, biodynamic, etc.). The maximum dose of nitrogen (in all its forms) that can be added to the must is 100 g/hl in the EU. Chronic nitrogen deficiencies can be corrected in the vineyard by appropriate viticultural practices.

Nitrogen is added either all at once at the time of inoculation or in two additions, the second occurring at the same time as the aeration on the second or third day of alcoholic fermentation. The second method sometimes permits a more rapid fermentation.

In the past, due to fears of aroma loss through oxidation, white juices were not aerated during alcoholic fermentation in high-capacity tanks. Opinions today are less categorical. The varietal aroma of white wines is affected by oxidation during pressing and draining (Section 13.4.2), but an aeration in the first half of fermentation has no effect on the fruity aroma of aromatic varieties. At this stage, the considerable reducing power of yeasts very effectively protects the aromas from oxidation. If the addition of oxygen sometimes diminishes fermentation aromas (esters and fatty acids) of dry white wines, this is caused by the resulting stimulation of alcoholic fermentation. The risk of slow or stuck fermentations, associated with strictly anaerobic conditions, is more serious than a minimal loss of transient fermentation aromas.

The juice should be aerated during pump-over of the tank—maintaining suitable air and liquid contact. Oxygen gas can also be directly injected into fermenting juice by an aerating device. This oxygen addition (2–4 mg/l) should occur in the first days of alcoholic fermentation during the yeast exponential growth phase. Oxygen permits the synthesis of sterols—essential cell membrane components and yeast survival factors during the stationary phase. Aeration becomes necessary for the completion of alcoholic fermentation when juice turbidity is low and sugar concentrations are high. Adjusting juice clarity, measuring and correcting (when necessary) yeast-assimilable nitrogen concentrations, and aerating at the right moment along with yeast inoculation are the principal factors governing successful fermentations in dry white winemaking.

13.7.4 Temperature Control

In traditional white winemaking, alcoholic fermentation is carried out in small containers (barrels or small tuns) located in cool cellars with temperatures between 12 and 16°C. Under these conditions, the fermentation temperature remains close to the cellar temperature at the start and end of fermentation. It rarely exceeds 22–25°C during the most active phase of fermentation. These barrel and tun fermentation conditions still exist in historically prominent white wine regions (Burgundy, Sauternes, Graves, Loire Valley, Alsace, etc.). In the last 15 years, many wineries, in a desire to improve white wine quality, have reverted back to the age-old technique of barrel fermenting. Temperature control problems are much less of an issue in these small containers.

In contrast, the temperature of high-capacity fermenting tanks must be controlled to avoid excessive temperatures during fermentation. Today, most wineries have tanks equipped with temperature control systems. A cooling system maintains water at a low temperature (4–6°C) in an insulated tank. A second system distributes the cool water through exchangers, placed inside the fermenting tank or set up as a jacket on the outside. These cooling systems were developed fairly recently. In the 1950s, the migration from small to large fermentors occurred without taking into account the heat exchange consequences of these changes.

As in red winemaking, excessively high temperatures (above 30°C) can be the cause of stuck fermentations, but this problem rarely occurs today, since winemakers have means of cooling juice before and during fermentation. They are also aware of the need to avoid excessive temperatures to limit aroma loss. Fermentation temperatures above 20°C diminish the amount of esters produced by yeasts and increase higher alcohol production (Bertrand, 1968). The effect of temperature on varietal wine aroma is much less clear. At very high temperatures (28–30°C), the rapid release of carbon dioxide entrains certain substances, causing aroma loss, but lower-temperature fermentations at 18°C, or which reach 23–24°C for a few hours, do not necessarily produce wines with a significant varietal aroma difference. Fermentations at 18°C or lower are therefore not a means of enhancing the fruitiness of aromatic varieties. Prefermentation operations and selection of yeast strains have a much greater influence. The temperature kinetics of high-capacity fermentors should simply be modeled after temperature evolution in small-volume fermentations, with a maximum temperature of around 22–23°C at mid-fermentation—progressively decreasing to the cellar temperature by the end of fermentation.

Untimely temperature drops should be avoided at all stages of fermentation in both barrels and tanks. For example, a tank temperature should not be lowered from 23 to 16°C in a few hours with the aim of avoiding the need for subsequent temperature control. The heat shocks that yeasts undergo under these conditions promote slow and even stuck fermentations (Section 3.7.1).

13.7.5 Completion of Alcoholic Fermentation

The duration of dry white wine fermentation depends on several parameters: juice extraction conditions, sugar and yeast-assimilable nitrogen concentrations, turbidity, yeast strain, aeration, and fermentation temperature. The winemaker can adjust and control all of them. A slow or stuck fermentation is most often the result of carelessness and always affects wine quality. If properly conducted, the alcoholic fermentation of a white wine should not exceed 10 days. Longer fermentations should be avoided, except in the case of exceptionally high sugar concentrations.

Juice density is measured daily to monitor alcoholic fermentation kinetics. When specific gravity drops to approximately 0.993–0.994, sugar concentrations are then measured daily to verify the completion of fermentation. Fermentation is considered complete when less than 0.1–0.3 g/l (sum of glucose and fructose) remains in the wine. The fermentors are then carefully topped off. Subsequent operations depend on whether malolactic fermentation is carried out.

If malolactic fermentation is not desired, the wine temperature is lowered to around 12°C. Before sulfiting, the lees are stirred daily by agitation or pumping, avoiding oxygen dissolution. This operation makes use of the reducing power of yeast lees to protect wine from oxidation. The formation of reduction odors in the lees is simultaneously avoided (Section 13.9).

After one to two weeks, the wine is sulfited at 4–5 g/hl. Until recently, lees aging in high-capacity tanks was not considered possible, due to the appearance of reduction odors. The lees were rapidly eliminated by racking shortly after sulfiting. Today, by taking certain precautions, white wines can be lees-aged even in tanks (Section 13.9).

13.7.6 Malolactic Fermentation

Malolactic fermentation is always sought with red wines but is used less often for white wines. Its use depends on the variety and wine region.

Chardonnay in Burgundy and Chasselas in Switzerland are two classic examples of lees-aged white wines that systematically undergo malolactic fermentation after alcoholic fermentation. The primary objective of this transformation is to deacidify the wine. This is especially true of premium white Burgundies. Before malolactic fermentation, their total acidity can be as high as 7g/l expressed as H_2SO_4 with a corresponding low pH. Malolactic fermentation also increases the biological stability of the wine. For example, to avoid an accidental malolactic fermentation in the bottle, Champagne wines undergo malolactic fermentation after alcoholic fermentation under controlled conditions. Malolactic fermentation also contributes to the aroma complexity of Chardonnay wines. A Chardonnay wine that has not undergone malolactic fermentation cannot be considered a great Chardonnay. Malolactic fermentation does not lessen the varietal aroma of Chardonnay; on the contrary, it develops and stabilizes certain aromatic and textural nuances, making the wine more complete. Unfortunately, since the varietal aroma of Chardonnay is not fully understood, enology still cannot provide an explanation of these phenomena at the molecular level.

Today, many Chardonnay wines made throughout the world according to the Burgundian model undergo malolactic fermentation more for the aroma consequences than for deacidification and stabilization. In many cases, the juices are even acidified to be suitable for malolactic fermentation. These practices may shock a European winemaker but are employed to produce a certain type of wine.

For most other varieties, such as Sauvignon Blanc, Semillon, and Chenin Blanc, and all Alsatian, German, and Austrian varieties, malolactic fermentation noticeably lowers the fruitiness of white wines. Other methods should be used to lower acidity when this operation is necessary, but the role of lactic acid bacteria on the aroma of wines made from these varieties should be studied, at least to justify the tradition of avoiding malolactic fermentation with these varieties.

When malolactic fermentation is desired, the wines are topped off and maintained on the lees after alcoholic fermentation, without sulfiting, at a temperature between 16 and 18°C. The containers must be fully topped off and the lees stirred weekly to avoid oxidation. With proper winemaking methods, in particular moderate sulfiting, malolactic fermentation spontaneously initiates after a latent phase of variable length that can be shortened with the use of commercially prepared malolactic inoculum (Section 12.7.5). Some wineries even keep non-sulfited wines, having completed malolactic fermentation, at low temperatures from one year to another to use as a malolactic starter culture. In regions where malolactic fermentation is systematically conducted, its initiation does not pose any particular problems, since the entire facility (notably the barrels) contains an abundant bacterial inoculum, and malolactic fermentation is difficult to avoid. If wine begins to oxidize while waiting for the onset of malolactic fermentation, it should be lightly sulfited (2g/hl). Malolactic fermentation is not compromised, and the aroma of the wine is preserved. Once the malic acid is broken down, the wines are sulfited at 4–5g/hl and maintained on the lees until bottling.

13.8 Making Dry White Wines in Barrels

13.8.1 Principles

Age-worthy dry white wines capable are traditionally fermented and matured in small containers. This practice was widespread at the beginning of the 20th century. In France, it has continued in certain prestigious Burgundy appellations. At the beginning of the 1980s, barrel fermenting and aging of white wines surged in popularity, affecting nearly all wine regions in the world. However, the use of barrels is not suitable for all wines; also, implementing a barrel program is difficult and very costly.

The yeasts play an essential role in the originality of the traditional Burgundian method of barrel-aging white wine. Contrary to red wine, which is barreled after the two fermentations, white juice is barrel-fermented and then aged on lees in the same barrel for several months without racking. During this aging process, there are interactions between the yeasts, the wood, and the wine. Unknown to enology for a long time, these phenomena are better understood today. They encompass several aspects: the role of exocellular and yeast cell wall colloids, oxidation–reduction phenomena linked to the presence of lees, the nature and transformation by yeasts of volatile substances released by the wood into the wine, and barrel fermenting and aging techniques.

13.8.2 The Role of Exocellular and Yeast Cell Wall Colloids

The yeast cell wall is composed of colloidal carbohydrates—essentially β-glucans and mannoproteins whose detailed molecular structure is now well understood (Section 1.2.2).

The macromolecular components of the yeast cell wall, particularly the mannoproteins, are partially released during alcoholic fermentation and especially during lees aging. In the laboratory with a model medium, contact time, temperature, and agitation of the yeast biomass promote the release of these substances (Volume 2, Section 3.7) (Llaubères et al., 1987). All of these conditions occur in traditional barrel aging on the lees. A wine that is barrel-fermented and barrel-aged on total lees with weekly stirring (bâtonnage) has a higher yeast colloidal carbohydrate concentration than a wine fermented and aged on the fine lees in a tank for the same time period (Figure 13.13). The difference in concentration can reach 150–200 mg/l.

The release of mannoproteins is the result of enzymatic autolysis of the lees. β-Glucanases present in the yeast cell wall (Section 1.2.2) maintain a residual activity for several months after cell death. They hydrolyze cell wall glucans. The latter are anchor points for mannoproteins, which are thus released into the wine.

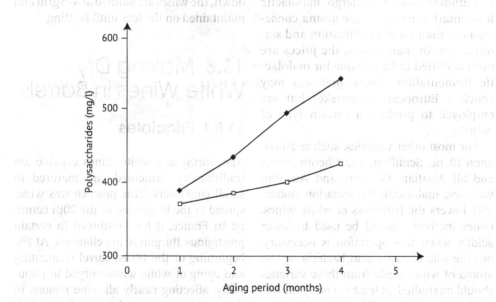

FIGURE 13.13 Evolution of the total polysaccharide concentration in white wine during tank aging on fine lees (□) or barrel aging on total lees (■).

The direct organoleptic influence of polysaccharides on the body or fullness of lees-aged white wines has never been clearly established, but the polysaccharides released during lees aging (from the yeast cell wall, for example) are capable of binding with phenolic compounds in white wines (Chatonnet *et al.*, 1992). The total polyphenol index and yellow color thus steadily decrease in the course of barrel aging on the lees. Moreover, after several months of aging, wines that are barrel-aged on the total lees are less yellow than the same wine aged on the fine lees in a tank (Figure 13.14). In particular, the lees limit the ellagitannin concentration originating from oak. Tannins given off by the wood are fixed on the yeast cell walls and also on the polysaccharides (mannoproteins) released by the lees. A wine conserved on the lees therefore has a lower overall tannin concentration as well as a much lower proportion of free (reactive) tannins (Figure 13.15).

In addition, lees aging lowers white wine sensitivity to oxidative pinking. This problem, characterized by a color evolution toward a grayish pink (Simpson, 1977), occurs when wine is slightly oxidized during stabilization or bottling.

Young white wines, in particular Sauvignon Blanc, whose musts have been carefully protected from oxidation, are especially sensitive to this color change. The compounds involved in these phenomena are not known. Contrary to anthocyanidins, this color cannot be removed by varying the pH and sulfiting, but the pink color does disappear upon exposure to light. Even if the pink color of these wines generally disappears after several months of bottle aging, this problem can lead to commercial disputes. The sensitivity of white wine to this oxidation can be evaluated by measuring the difference in absorbance of the wine at 500 nm, 24 hours after adding hydrogen peroxide (Figure 13.16). This value multiplied by 100 is the sensitivity index. If it is greater than 5, there is a definite risk of pinking. Wine sensitivity to pinking remains fairly constant during the course of aging racked wine on the fine lees, in barrel or in tank, but diminishes rapidly on the total lees (Table 13.17). The yeast lees probably adsorb the precursor molecules responsible for pinking, but neither casein fining nor PVPP treatment is capable of significantly decreasing wine sensitivity to pinking. The addition of ascorbic acid (10 g/hl) at bottling is the only effective preventive treatment.

The release of mannoproteins during lees aging also increases tartrate and protein stability in white wines (Volume 2, Sections 1.7.7, 5.6.3, and 5.6.4).

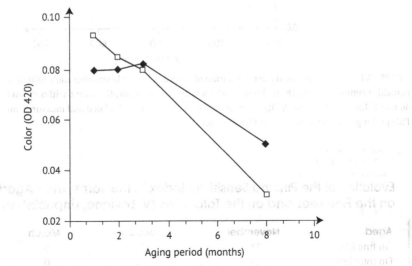

FIGURE 13.14 Evolution of the yellow color (OD 420) of a wine tank aged on the fine lees (♦) and aged in new barrels on the total lees (□) (Chatonnet *et al.*, 1992).

560 Chapter 13 White Winemaking

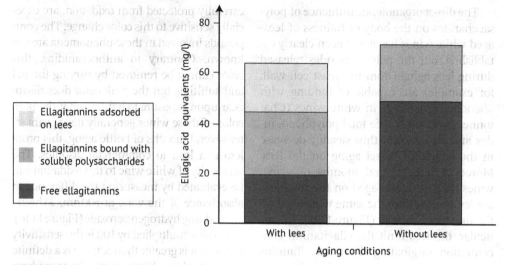

FIGURE 13.15 Proportions of ellagitannins of a barreled wine aged on lees or without lees (Chatonnet et al., 1992).

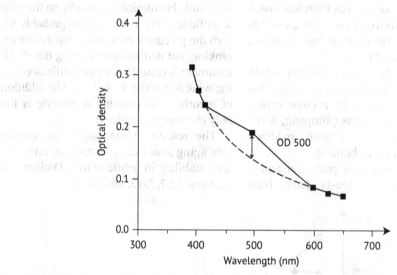

FIGURE 13.16 Pinking sensitivity determination of a white wine 24 hours after the addition of hydrogen peroxide (Simpson, 1977). Dashed line = interpolation of the absorption curve without pinking, using OD 400, 410, 420, 600, 625, 650. Δ OD 500 = difference between the calculated and measured values at OD 500. PSI (pinking sensitivity index) = Δ OD 500 × 100.

TABLE 13.17

Evolution of the Pinking Sensitivity Index in the Same Wine Aged on the Fine Lees and on the Total Lees (V. Lavigne, Unpublished Results)

Aged	November	January	March	April
On fine lees	17	13	8	8
On total lees	17	4	0	0

13.8.3 Oxidation–Reduction Phenomena Linked to the Presence of Lees

Maintaining white wines on the total lees in tanks after sulfiting is difficult without taking certain precautions (Section 13.9). Unpleasant sulfurous odors rapidly appear, making racking necessary. With proper must clarification and sulfiting (Section 13.5.2), barrel aging permits prolonged contact with total lees without the development of reduction off-odors. Conversely, when a dry white wine is separated from its lees and stored in new barrels, it more or less rapidly loses its fruitiness and develops oxidative odors. These resinous, waxy, and camphor-like odors intensify during bottle aging. The lees are thus indispensable to the proper evolution of dry white wine in barrels. They act as a reducing agent, in a manner similar to tannins in the aging of red wine.

White wines have a higher oxidation–reduction potential in barrels than in tanks (Dubourdieu, 1992). Inside the barrel, this potential decreases from the wine surface toward the lees (Figure 13.17). Over time, the barrels seem to lose some of their oxidative properties. Wood ellagitannins, released in lesser quantities as the barrel ages, contribute to its oxidizing power. A tendency toward reduction consequently occurs more often in used barrels than in new barrels. Stirring homogenizes the wine's oxidation–reduction potential (Figure 13.18). Lees reduction is blocked, as well as surface wine oxidation. The stirring of lees-aged wines is as indispensable in new barrels as in used barrels, but for different reasons. It protects wine in new oak from oxidation, whereas it prevents wine reduction in used barrels.

During aging, the lees release certain highly reductive substances into the wine, which limit oxidative phenomena occurring in the wood. These same compounds appear to slow premature aging of bottled white wines. The nature and formation mechanisms of these compounds are described in the next section.

13.8.4 Nature of Volatile Substances Released by Wood and Their Transformation by Yeasts

Among the many volatile substances released by oak into wine, volatile phenols, β-methyl-γ-octalactones, and phenol aldehydes are the main compounds responsible

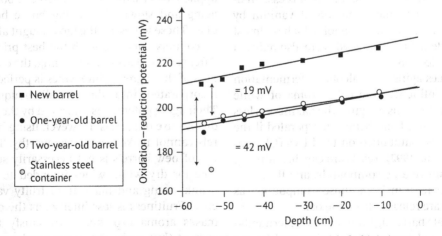

FIGURE 13.17 Influence of the aging method on the oxidation–reduction potential of white wine (Dubourdieu, 1992).

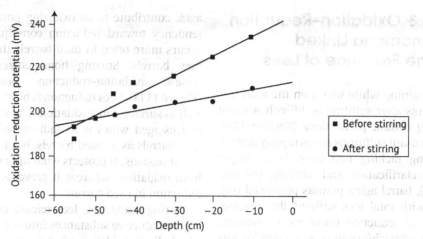

FIGURE 13.18 Influence of stirring on the oxidation–reduction potential of barrel-aged white wine (Dubourdieu, 1992).

for the oaky aroma of barrel-aged wines. Volatile phenols, in particular eugenol, give wine smoky and spicy aromas. The cis- and trans-methyl-γ-octalactones are responsible for the coconut aroma. Phenol aldehydes, essentially vanillin, produce vanilla notes. Furan aldehydes have grilled aromas, but their perception threshold is much higher than the concentrations found in wine. Their sensory impact is thus negligible.

Barrel-fermented wines overall are less oaky than the same wines barreled after alcoholic fermentation (Chatonnet et al., 1992). This phenomenon is essentially linked to the reduction of vanillin by yeasts into vanillyl alcohol, which is almost odorless. Furan aldehydes are also reduced into alcohols.

Lees aging after alcoholic fermentation also influences the oaky aroma of white wine. In terms of aging, the oakiness is less pronounced and better incorporated if the wine is maintained on total lees (Chatonnet et al., 1992). Yeasts are capable of fixing certain volatile compounds, and they continue to transform those compounds as they are released from the oak.

Late barreling (after alcoholic fermentation) and prematurely eliminating lees will produce white wines with excessive oakiness. These methods have unfortunately been used in the past in certain regions, where winemakers have applied red wine aging techniques to white wine.

13.8.5 White Wine Barrel-Aging Techniques

Due to its high cost and the care required, the use of new barrels is only economically feasible for the production of relatively expensive, premium wines produced in limited quantities. Barrel aging should be applied to dry white wines capable of bottle aging and slowly developing bottle bouquet. These wines are the most sought after by connoisseurs and fetch the best prices. After several years of bottle aging, the oakiness of these great white wines is perfectly incorporated into the overall bouquet. Their aging potential is improved by the judicious use of barrels. However, using barrels cannot make a wine age-worthy. The use of new barrels is not necessarily suitable for dry white wines intended to be drunk young and made from fruity varieties. Fruitiness is less intense, as the oak masks aroma expression. To satisfy the current (but perhaps temporary) demand for oaky wines in certain markets, it can be tempting to barrel-ferment and barrel-age ordinary wines. In this case, the barrel is simply a means of compensating for an aromatically deficient wine. The widespread use of these practices may actually lower the consumer appeal of oaky wines over time.

The most popular barrels in France for white winemaking are made from fine-grain oak from forests in the center of France, notably Allier. In these mature forested areas, sessile oak (*Quercus sessilis*) essentially constitutes the oak population. Fine-grain oak contains much higher concentrations of aroma compounds, in particular β-methyl-γ-octalactones, than coarse-grain oak. The latter—largely pedunculate oak (*Quercus pedunculata*)—comes from isolated trees or from coppice forests with standards (*taillis sous futaie*). The coarse-grain oak (Limousin) is less aromatic but contains much higher tannin concentrations. White wines made with Limousin oak therefore have a more pronounced yellow color and tannic character (Table 13.18). The technique is rarely used for wines, particularly whites, but spirits are generally aged in coarse-grain oak.

TABLE 13.18

Influence of Oak Origin (Allier, Limousin) and Toasting Intensity[a] on the Color and Oakiness of White Wines (Chatonnet, 1995)

Character	Control tank	Allier				Limousin			
		L	M	H	VH	L	M	H	VH
Total polyphenols (OD 280/PVPP)	3	4	3.9	3.9	3.8	5.2	4.3	4.7	4.4
Color (OD 420)	0.1	0.12	0.13	0.13	0.08	0.42	0.47	0.48	0.48
Furfural (mg/l)	0	0.9	3.6	4.9	3.5	1.8	2.55	4.8	4.3
5-Methylfurfural (mg/l)	0	0.8	1.1	0.75	0.5	0.9	0.95	0.8	0.4
Furfuryl alcohol (mg/l)	0	0.5	5.1	4.8	4.2	4	3.6	4.3	1.8
Total furans (mg/l)	0	2.2	9.8	10.45	8.2	6.7	7.1	9.9	6.5
trans-Methyl-octalactone (mg/l)	0	0.13	0.17	0.053	0.037	0.067	0.051	0.023	0.012
cis-Methyl-octalactone (mg/l)	0	0.29	0.14	0.089	0.114	0.095	0.095	0.055	0.058
Total methyl-octalactones (mg/l)	0	0.42	0.31	0.142	0.151	0.162	0.146	0.078	0.07
Guaiacol (µg/l)	2	10	18.5	38	65	6	12	21	33
4-Methyl guaiacol (µg/l)	0	10	14	24	29	10	11	14	18
4-Vinyl guaiacol (µg/l)	150	98	114	149	117	104	110	99	74
4-Ethyl guaiacol (µg/l)	0	9	9	14	15	4	4	4	13
Eugenol (µg/l)	0	27	29	38	28	13	13	19	23
Phenol + o-cresol (µg/l)	8	25	26	47	41	26	27	17	35
p-Cresol (µg/l)	—	1	1	2	—	0	1	1	0
m-Cresol (µg/l)	—	2	3	4	—	2	2	1	0
4-Vinylphenol (µg/l)	300	197	206	319	210	187	211	214	131
Vanillin (mg/l)	0	0.29	0.35	0.36	0.2	0.2	0.64	0.43	0.1
Syringaldehyde (mg/l)	0	0.49	0.69	1.4	1.8	0.27	0.4	—	—
Total phenol aldehydes (mg/l)	0	0.88	1.04	1.76	2	0.47	1.04	—	—

[a] L, light toast; M, medium toast; H, heavy toast; VH, very heavy toast.

Toasting, carried out during barrel production, considerably influences the aromatic impact of the oak on wine. Barrels are toasted to between medium and heavy so that the very fragrant fine-grain oak does not dominate the fragile aroma of white wines (Table 13.18).

Intermediate-grain wood from Burgundy may also be used for white winemaking. This wood is not very tannic; it is less fragrant than fine-grain oak and is best adapted to medium toasting.

Fine-grain oak from central and northern Europe, in particular Russia, also produces acceptable barrel wood. It has a similar composition to that of oak grown in central France. Under identical winemaking conditions, white wines made in fine-grain Russian barrels and French barrels are very similar in taste (Chatonnet et al., 1997).

American white oak (*Quercus alba*) is very fragrant. It is rarely used for premium white winemaking, as excessive concentrations of β-methyl-γ-octalactone are apt to be released, totally masking the wine's character. American oak is recommended for rapidly imparting oakiness on ordinary white wine.

Barrel preparation for white winemaking is relatively simple. New barrels, delivered unsulfited, are simply rinsed with cold water and drained for a few minutes before use. Used barrels, stored empty and regularly sulfured, are apt to release SO_2 into the juice during filling. As a result, abnormally high levels of H_2S are formed during alcoholic fermentation and are capable of generating strong reduction odors (Lavigne, 1996). This phenomenon is particularly pronounced if fermenting juice is barreled. Used barrels must consequently be filled with water 48 hours before use to eliminate the SO_2 likely to be released in the fermenting juice.

The barrels are placed in a cool cellar (16°C) and are filled either before or at the start of alcoholic fermentation. A 10% headspace should be left to avoid foam overflow during maximum fermentation intensity. The fine grape solids and/or the yeasts should be carefully put into suspension to homogenize the juice before barreling. After barreling, the tank dregs (grape solids and lees) should be scrupulously distributed in each barrel of the lot. If done at the start of fermentation, barreling takes the place of aeration during the yeast growth phase. If the barrels are filled with juice before fermentation, aeration (by introducing air or oxygen) is necessary when fermentation has started. If these precautions are not taken, fermentation will be irregular from one barrel to another in the same lot. In barrel as in tank, difficult fermentations are often the result of human error or negligence.

As soon as fermentation is nearly complete, the barrels are topped off with juice from the same lot. Sluggish fermentations can often be reactivated by topping off the barrel with a wine lot that has recently completed a successful fermentation. This technique is equivalent to using a starter (10%) composed of a population in the stationary phase—resistant to inhibition factors. At the end of alcoholic fermentation, the barrels are stirred daily until sulfiting (Section 13.7.6). Wines undergoing malolactic fermentation are not sulfited until its completion.

During barrel aging, stirring and topping-off should occur weekly, with free SO_2 concentrations maintained around 30 mg/l.

13.9 Controlling Reduction Off-Aromas During White Wine Aging

13.9.1 Evolution of Volatile Sulfur Compounds in Dry White Wine During Barrel or Tank Aging

Barrel-fermented and barrel-aged dry white wines are most often maintained on

13.9 | Controlling Reduction Off-Aromas During White Wine Aging

yeast lees throughout the entire aging process. In this type of winemaking, if reduction off-aromas do not appear during alcoholic fermentation, they rarely occur later. Stirring wine frequently to put the lees in suspension and limited oxidation through the barrel staves inhibit the formation of off-aromas from sulfur-containing compounds in the wine.

During the barrel aging of wine, volatile thiols, H_2S, and methanethiol—normally present at the end of fermentation—decrease progressively (Figure 13.19). This phenomenon occurs more rapidly in new barrels, probably due to greater oxygen dissolution and the oxidizing effect of new oak tannins (Lavigne, 1996).

Despite the relative ease of barrel aging of dry white wine on the total lees, the winemaker must still pay close attention to the winemaking factors (clarification and sulfiting) that influence the production of sulfur-containing compounds by yeasts. In fact, even if a wine is in a new or used barrel, racking and the definitive separation of the foul-smelling lees from the wine must be carried out, if a reduction flaw exists at the end of alcoholic fermentation. The quality of barrel aging is greatly compromised, particularly in new barrels; in the absence of lees, the dry white wine is not protected from oxidation.

Controlling reduction off-aromas in dry white wines during aging in high-capacity tanks is more difficult. The presence of lees inevitably leads to the development of reduction odors within the first month of aging, whatever the reduction state of the wine after alcoholic fermentation (Figure 13.20). In most cases, tank-aged dry white wines are systematically racked and separated from their lees. Under these conditions, if the gross lees are eliminated early enough, before reduction aroma defects occur, the wine can be aged on the fine lees without risk. Early racking helps to stabilize light sulfur-containing compound concentrations. This opinion was clearly stated in the previous version of the *Handbook of Enology* (Ribéreau-Gayon et al., 1976).

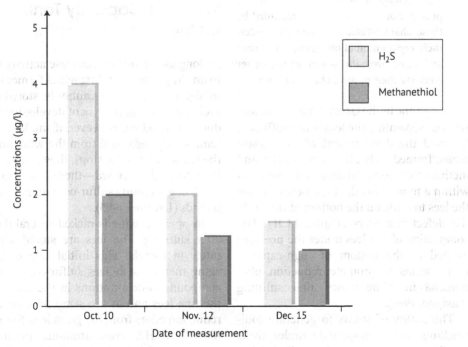

FIGURE 13.19 Evolution of light sulfur-containing compounds in barrel-fermented and barrel-aged white wine (without flaws) on total lees (Lavigne, 1996).

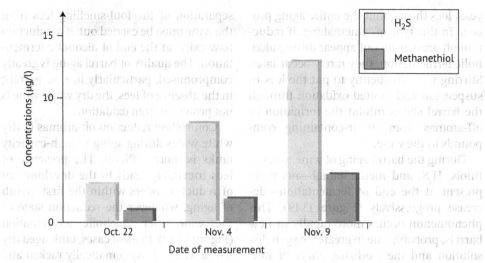

FIGURE 13.20 Evolution of light sulfur-containing compounds in white wine on total lees in tank (Lavigne, 1996).

It was based simply on observation, without any supporting analytical data:

> The principal danger of storing white wines on yeast lees in high-capacity tanks is the development of hydrogen sulfide and mercaptan odors. But even if the presence of yeasts is not accompanied by these characteristic off-odors and tastes, their rapid elimination results in fresher and more aromatic wines which better conserve their positive characteristics.

Racking wine in the tank with aeration and without separating the lees is not sufficient to avoid the development of unpleasant odors. Immediately after racking, H_2S and methanethiol concentrations diminish, but within a month, i.e. the time necessary for the lees to settle on the bottom of the tank, the defect reappears (Figure 13.21). The compacting of the lees under the pressure exerted at the bottom of high-capacity tanks seems to promote reduction phenomena in white wines after sulfiting (Lavigne, 1996).

The ability of yeasts to generate foul-smelling sulfur compounds under these conditions progressively diminishes during aging and totally disappears after a few weeks. The loss of the lees sulfite reductase activity, catalyzing the reduction of SO_2 into H_2S, explains this development.

13.9.2 Aging Dry White Wine in a High-Capacity Tank on the Lees

As long as the sulfite reductase activity remains in yeasts, dry white wines fermented in high-capacity tanks cannot be stored on their lees without the risk of developing reduction off-odors. However, if the lees are temporarily separated from the wine until the reductive activity stops, they can be reincorporated afterward—there is no longer a risk of generating sulfur-containing compounds (Lavigne, 1995).

In practice, wine is racked several days after sulfiting. The lees are stored separately in barrels. This initial step of the aging method stabilizes sulfur-containing compound concentrations in the wine on the fine lees and at the same time avoids reduction odors from the gross lees. Simultaneously, H_2S concentrations progressively diminish in the barreled lees. Only

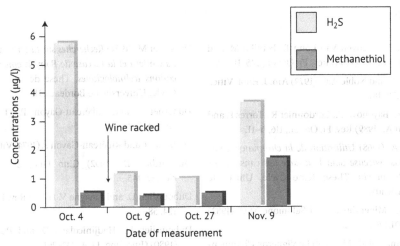

FIGURE 13.21 Evolution of light sulfur-containing compounds in a tank-aged white wine racked with its lees (Lavigne, 1996).

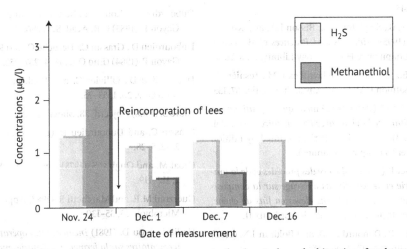

FIGURE 13.22 Evolution of H_2S and methanethiol concentration in a tank-aged white wine after the reincorporation of its lees (Lavigne, 1996).

one day after separation from the wine, these lees no longer contain methanethiol. After approximately one month, the lees are reincorporated into the wine. At this stage, not only do the lees no longer generate sulfur-containing compounds, but their addition also provokes an appreciable decrease in the concentration of methanethiol in the wine (Figure 13.22).

The use of fresh lees is an authorized enological practice, used to correct the color of prematurely oxidized white wine. The ability of lees to adsorb certain volatile thiols in wine has been discovered more recently (Lavigne and Dubourdieu, 1996).

Yeasts, taken at the end of fermentation and added to a model solution containing methanethiol and ethanethiol, are capable of adsorbing these volatile thiols. The latter are fixated by the yeast cell wall mannoproteins. During aeration, a disulfide bond is formed between the cysteine of the cell wall mannoproteins and the thiols from the wine.

References

Alexandre H., Nguyen Van Long T., Feuillat M. and Charpentier C. (1994) Rev. Fr. Oenol., 145, 11–20.

Arnold R.A. and Noble A.C. (1979) Am. J. Enol. Vitic., 30, 3, 179–181.

Baumes R., Bayonove C., Cordonnier R., Torres P. and Seguin A. (1989) Rev. Fr. Oenol., 116, 6–11.

Bertrand A. (1968) *Utilisation de la chromatographie en phase gazeuse pour le dosage des constituants volatils du vin*, Thèse 3ème Cycle, Université de Bordeaux.

Bisiach M., Minervini G. and Salomone M.C. (1982) Bull. OEPP, 12, 5–28.

Blank G. and Valade M. (1989) Le Vigneron Champennois, 6, 333–345.

Boidron J.N. (1978) Ann. Technol. Agric., 27, 1, 141–145.

Boulton R.B., Singleton V.L., Bisson L.F. and Kunkee R.E. (1995) *Principles and Practices of Winemaking*. Chapman & Hall Enology Library, New York.

Casalta E., Vernhet A., Sablayrolles J.-M., Tesnière C. and Salmon J.-M. (2016) Am. J. Enol. Vitic., 67, 133.

Chaptal J.A.C. (1801) *Traité théorique et pratique sur la culture de la vigne avec l'art de faire le vin*, Vol. 2, pp. 41–42. Delalain, Paris, reprinted by Edition d'art et technique (Morannes).

Chatonnet P. (1995) *Influence des procédés et de la tonnellerie et des conditions d'élevage sur la composition et la qualité des vins élevés en fûts de chêne*, Thèse Doctorat, Université de Bordeaux II.

Chatonnet P., Dubourdieu D. and Boidron J.N. (1992) Sci. Aliments, 12, 4, 666–685.

Chatonnet P., Sarishivilli N.G., Oganessyants L.A., Dubourdieu D. and Cordier B. (1997) Rev. Fr. Oenol., 167, 46–51.

Chauvet S., Sudraud P. and Jouan T. (1986) Bull. OIV, 661–668, 1021–1043.

Cheynier V., Trousdale E.K., Singleton V.L., Salgues M. and Wylde R. (1986) J. Agric. Food Chem., 34, 217–221.

Cheynier V., Rigaud J., Souquet J.M., Barillère J.M. and Moutounet M. (1989) Am. J. Enol. Vitic., 40, 1, 36–42.

Cheynier V., Rigaud J., Souquet J.M., Duprat F. and Moutounet M. (1990) Am. J. Enol. Vitic., 41, 346–349.

Crowell E.A. and Guymon J.F. (1963) Am. J. Enol. Vitic., 14, 214.

Dubernet M. (1974) *Recherches sur la tyrosinase de Vitis vinifera et la laccase de Botrytis cinerea. Applications technologiques*, Thèse de Doctorat 3ème Cycle, Université de Bordeaux II.

Dubernet J. and Ribéreau-Gayon P. (1973) Œno One, 4, 283.

Dubernet J. and Ribéreau-Gayon P. (1974) Vitis, 13, 233.

Dubourdieu D. (1992) Œno One, 26, Hors Série, 137–145.

Dubourdieu D. and Lavigne V. (1990) Rev. Fr. Oenol., 124, 58–61.

Dubourdieu D., Hadjinicolaou D. and Bertrand A. (1980) Œno One, 14, 4, 247–261.

Dubourdieu D., Hadjinicolaou D. and Ribéreau-Gayon P. (1981) Œno One, 18, 4, 237–252.

Dubourdieu D., Koh K.H., Bertrand A. and Ribéreau-Gayon P. (1983) C.R. Acad. Sci., 296, 1025–1028.

Dubourdieu D., Grassin C., Deruche C. and Ribereau-Gayon P. (1984) Œno One, 18, 4, 237–252.

Dubourdieu D., Ollivier C. and Boidron J.N. (1986) Œno One, 20, 1, 53–76.

Fabre S. (1988) Objectif, 28, March, 21–25.

Grassin C. and Dubourdieu D. (1986) Œno One, 20, 2, 12, 5–28.

Groat M. and Ough C.S. (1978) Am. J. Enol. Vitic., 29, 2, 112–119.

Guerzoni M.E. and Marchetti R. (1987) Appl. Environ. Microbiol., 53, 35–130.

Hadjinicolaou D. (1981) *Incidence des opérations préfermentaires sur la fermentescibilité des moûts et les caractères organoleptiques des vins blancs*, These de Doctorat, Université de Bordeaux II.

Haushoffer H. (1978) Ann. Technol. Agric., 27, 1, 221–230.

Hiéret J.P. (1986) *L'outillage traditionnel de la vigne et du vin en Bordelais*. Presses universitaires de Bordeaux, Universite de Bordeaux III, Bordeaux.

Lafon-Lafourcade S., Geneix C. and Ribéreau-Gayon P. (1979) Appl. Environ. Microbiol., 47, 1246–1249.

Larue F., Geneix C., Park M.-K., Murakami Y., Lafon-Lafourcade S. and Ribéreau-Gayon P. (1985) Œno One, 19, 41–52.

Lavigne V. (1996) Rev. Fr. Oenol., 155, 36–39.

Lavigne V. and Dubourdieu D. (1996) Œno One, 30, 4, 201–206.

Lavigne V. and Dubourdieu D. (1997) Rev. Oenol., 85, 23–30.

Lavigne-Cruège V. (1996) *Recherches sur les composés volatils soufrés formés par la levure au cours de la vinification et de l'élevage des vins blancs secs*, Thèse Doctorat, Université de Bordeaux II.

Llaubères R.M., Dubourdieu D. and Villetaz J.C. (1987) J. Sci. Food Agric., 41, 277–286.

Martinière P. and Sapis J.C. (1967) Œno One, 2, 64.

Maurer R. and Meidinger F. (1976) Deu. Weinbau, 31, 11, 372–377.

Milisavljevic D. (1963) *In 1er Symposium International d'Œnologie de Bordeaux*, p. 315.

Moutounet M., Rigaud J., Souquet J.M. and Cheynier V. (1990) Rev. Fr. Oenol., 124, 32–38.

Müller-Späth H. (1977) Die Weinwirtschaft, 6, 1–12.

Müller-Späth H. (1988) Objectif, 28, Mars, 21–25, 15–19.

Nikolantonaki M. (2010) *Incidence de l'oxydation des composés phénoliques sur la composante aromatique des vins blancs*, Thèse Doctorat, Université Victor Ségalen, Bordeaux II.

Nikolantonaki M. and Darriet P. (2011) J. Agric. Food Chem., 59, 3264.

Nikolantonaki M., Jourdes M., Shinoda K., Teissedre P.L., Quideau S. and Darriet, P. (2012) J. Agric. Food Chem., 60, 2647.

Ollivier C. (1987) *Recherches sur la vinification des vins blancs secs*, Thèse Diplôme d'Etudes et de Recherches de l'Université de Bordeaux II.

Ollivier C., Stonestreet T., Larue F. and Dubourdieu D. (1987) Œno One, 21, 1, 59–70.

Ough C.S. (1969) Am. J. Enol. Vitic., 20, 2, 93–100.

Ough C.S. and Berg H.W. (1971) Am. J. Enol. Vitic., 22, 3, 194–198.

Peynaud E. (1971) *Connaissance et travail du vin*. Dunod, Paris.

Peyrot des Gachons C. (2000) *Recherches sur le potentiel aromatique des raisins de Vitis vinifera L. Cv. Sauvignon blanc*. Thèse de doctorat, Université de Bordeaux 2.

Rapp A., Mandery H. and Niebergall H. (1986) Vitis, 25, 79–84.

Ribéreau-Gayon J. and Ribéreau-Gayon P. (1954) *Xème Congrès International des Industries Agricoles et Alimentaires*, Madrid.

Ribéreau-Gayon P., Lafon-Lafourcade S. and Bertrand A. (1975) Œno One, 9, 2, 117–139.

Ribéreau-Gayon J., Peynaud E., Ribéreau-Gayon P. and Sudraud P. (1976) *Traite d'Oenologie Sciences et Techniques du Vin*, 3, 363–364. Dunod, Paris.

Rigaud J., Cheynier V., Souquet J.M. and Moutounet M.(1990) Rev. Fr. Oenol., 124, 27–31.

Robertson G.L. (1979) Am. J. Enol. Vitic., 30, 2, 182–186.

Salgues M., Olivieri C., Chabas M. and Pineau J. (1984) Bull. OIV, 57, 638, 309–311.

Schanderl H. (1959) *Die Mikrobiologie des Mostes und Weines*. Eugen Ulmer, Stuttgart.

Schneider V. (1989) Die Weinwirtschaft, 1, 15–20.

Sergent D. (1996) *Adaptation du Sauvignon et du Sémillon à la diversité géologique et agro-pédologique de la région des Graves. Incidences du climat et des sols sur le comportement de la vigne, la maturation des raisins et les vins*, Thèse de Doctorat de l'Université Victor Ségalen Bordeaux 2.

Serrano M., Paetzold M. and Dubourdieu D. (1989) Œno One, 23, 4, 251–262.

Simpson R.F. (1977) Vitis, 16, 286–294.

Singleton V.L., Sieberhagen H.A., de Wet P. and Van Wyk C.J. (1975) Am. J. Enol. Vitic., 26, 2, 62–69.

Thibon C., Dubourdieu D., Darriet P. and Tominaga T. (2009) Food Chem., 114, 1359.

Thibon C., Cluzet S., Mérillon J.M., Darriet P. and Dubourdieu D. (2011) J. Agric. Food Chem., 59, 1344.

CHAPTER 14

Other Winemaking Methods

14.1 Rosé Wines
14.2 Botrytized Sweet Wines (Sauternes and Tokaji)
14.3 Champagne and Sparkling Wines
14.4 Fortified Wines
14.5 Flor Wines

14.1 Rosé Wines

14.1.1 Definition

In many countries, especially within the European Union (EU) and the International Organisation of Vine and Wine (OIV), laws have been established to define wine. However, red, white, and rosé wines are never specifically defined, although a classification system would be particularly useful for rosé wines. Rosé wines have some similarities to red wines; they are often made from red grape varieties and contain a small quantity of anthocyanins and tannins. They are also refreshing, like many white wines, and for this reason white winemaking techniques are also used in their production.

Various characteristics give rosé wines their charm. These fruity wines have a light structure and are served chilled; they can accompany an entire meal. Their consumption has risen sharply over the last 15 years. Some have obtained a good reputation (e.g. Côtes de Provence) for producing fresh, fruity rosé wines and are present as premium wines on the international market. This is linked to better control of winemaking conditions, as well as the selection of varieties and the specific research for maturity criteria adapted to subsequent winemaking practices. Moreover, these wines are less and less the byproduct of drawing juice out of tanks to enhance the concentration of the remaining red wine or even to attenuate certain defects in red wine grapes, e.g. insufficient ripeness, rot

---

Handbook of Enology, Volume 1: The Microbiology of Wine and Vinifications, Third Edition.
Pascal Ribéreau-Gayon, Denis Dubourdieu, Bernard B. Donèche and Aline A. Lonvaud.
© 2021 John Wiley & Sons Ltd. Published 2021 by John Wiley & Sons Ltd.

or off-flavors. In France, rosé wines are usually dry, while in other countries, notably the United States, many off-dry or sweet rosés may contain 10–20 g/l residual sugar.

Due to the diversity of grape crops and winemaking techniques used, it is practically impossible to establish a technological definition of rosé wines. Further complicating the situation, mixing red and white grapes is authorized in certain cases, but blending red and white wines is prohibited, with very few exceptions. Color is therefore the only criterion for defining rosé wines— falling somewhere between the color of red and white wines. Characteristic value ranges for the different color parameters should be established for the various kinds of rosé wines (see Table 14.1). In the last twenty years, the trend is towards lighter-colored rosé wines. In addition, rosé wines are usually dry, sometimes sparkling, while several examples of off-dry or sweet rosés contain 10–20 g/l residual sugar.

There is a wide range of rosé wines in terms of color intensity, and several studies have been conducted to characterize them by analytical parameters (Garcia-Jares et al., 1993a,b; Flanzy et al., 2009; Lambert et al., 2015). Clairets, at one end of the spectrum, are light red wines having some body; they require extended skin maceration and are softened by malolactic fermentation. But the general trend is towards more lightly colored wines, from rosés with raspberry and red currant hues to those with salmon hues. The range ends with barely colored, almost white, wines with shades of rose petal or onion peel. *Blanc de noir* wines resemble white wines but are made from red grapes or have been in accidental contact with red wines. The expression *blanc de blanc* was created to refer to white wines made from white grape varieties.

Although appearing slightly yellow in color, white wines made from unmacerated red grapes can contain a small amount of anthocyanins, in a colorless state due to the SO_2. A pink color after the addition of concentrated HCl indicates their existence. This reaction is very sensitive and clearly differentiates *blanc de noir* wines from *blanc de blanc* wines, which do not change color. Slight contamination in white wines can occur even in correctly washed and maintained oak containers that have contained red wine. Other complex polyphenol forms contributing to rosé color have been characterized in the past years (Cheynier et al., 2013). In addition to hydroxycinnamic acids, anthocyanins and flavan-3-ols, phenolic compounds associated with the reactivity of anthocyanins, have been identified in rosé wines, in particular depending on oxidative phenomena, such as pyranoanthocyanins, carboxypyranoanthocyanins, phenyl- and catechyl-pyranoanthocyanins.

Many red grape varieties are suitable for making rosé wines. The Côtes de Provence vineyards specialized in rosé wine production have developed a blend of grape varieties that provide well-balanced color, bouquet, and body (Flanzy, 1998). Cinsault,

TABLE 14.1

Comparison of the Composition of Rosé and Red Wines

	Color intensity/cm	Total polyphenol index	Anthocyanins (mg/l)
Rosé	0.3–1.2	3–15	5–40
Nouveau-type red wine	2.2–5.1	10–30	90–250
"Ageworthy" red wine	>6	>40	>350

Source: Adapted from Blouin and Peynaud (2001).
Color intensity: Σ OD 420/OD 520/OD 620 (for 1 mm thickness).

Syrah, and Mourvèdre add aromatic depth, finesse, and elegance to the basic varieties, Carignan and Grenache.

Among the variety of methods for making rosé wines (Sections 14.1.3 and 14.1.4), immediate pressing and *saignée* (or drawing-off of juice from the skins) are the most common. Carbonic maceration (Section 12.9) is not very widely used, but has been cited to produce interesting, complex aromas in full-bodied rosé wines (André *et al.*, 1980). However, it frequently results in wines that are too deep in color for a classic rosé. As a result, they must be blended with lighter-colored wines.

It is important to monitor grape ripeness. Grapes intended for fine, crisp rosé wines should not be overripe (potential alcohol not exceeding 12% by volume and relatively high acidity), while those intended for fuller-bodied, softer rosés need slightly higher potential alcohol levels and lower acidity. Only healthy grapes should be used, as wine quality is likely to suffer if over 15% of the raw material is affected by rot.

It is best to harvest the grapes when the temperature is cool, at night or in the early morning, to preserve their fruitiness. Most grapes for rosé wines are handpicked, but a properly adjusted harvesting machine also gives good results.

In addition, the protection of the harvest from oxidative phenomena, either by inerting with nitrogen gas or liquid carbon dioxide, frequently leads to an increased preservation of the aromatic component of freshness in these wines.

14.1.2 Importance of Color in Characterizing the Various Types of Rosé Wines

Characterizing rosé wines by analytical parameters has long been sought. Various means can be used. Initially, spectrophotometric measurements were used, with determination of color intensity (sum of the absorbance values at 420 and 520 nm, which has since been supplemented with the measurement of absorbance at 620 nm) and hue (ratio of absorbance values at 420 and 520 nm) (Sudraud *et al.*, 1968; Glories, 1984). These spectrophotometric analyses are accompanied by the measurement of the main phenolic compounds in wines (Table 14.2).

In addition, these measurements have been enriched with chromametric methods by means of tristimulus data (CIE L*a*b* method) (Blouin et al. 2002). These are the coordinates of L* related to clarity, between L* = 0 (black) and L* = 100 (white), then of the red-green chromatic component a*, ranging from green (−100) to red (+100); and finally of the yellow-blue chromatic component b*, ranging from blue (−100) to yellow (+100). In addition, a color chart has been developed that associates the L*a*b* references with the different color tones (Cayla *et al.*, 2010).

Thus, a rosé wine with a salmon pink color corresponds to a CIE L*a*b* value of L*96.9, a*3.3, and b*3.7 (color intensity

TABLE 14.2

Example of Hydroxycinnamic Acid, Anthocyanin and Tannin Concentrations in French Rosé Wines (Ducasse *et al.*, 2015)

Wine	Variety	Hydroxycinnamic acids (mg/l)	Anthocyanins (mg/l)	Tannins (mg/l)
Val de Loire rosé	Cabernet franc	179	8	22
Bordeaux rosé	Cabernet franc	50	24	40
Tarn rosé	Négrette	60	10	20
Provence rosé	Grenache-Shiraz	64	8	2

0.14, hue 1.26), while a rosé wine with a red currant color corresponds to a CIE L*a*b* value of L*88.6, a*14.7, and b*6.6 (color intensity 0.5, hue 0.9).

Rosé wine color is directly influenced by grape variety. It depends on the content of polyphenols in the skins and their dissolution speed, not only for anthocyanins but also complex polyphenol forms (Lambert et al., 2015). All things being equal, Syrah and Merlot produce more highly colored wines than Grenache or Mourvedre. Color is also highly related to the level of ripeness, to juice extraction conditions (skin contact duration, temperature) and to wine aging conditions (lees) as well as to any corrections that might be made by fining during the winemaking process. Color also varies according to the vintage. It should be remembered that limiting skin contact in order to obtain more lightly colored rosé wines also limits the extraction of the grape aroma potential and potentially the diversity of the aroma tones in the wines.

14.1.3 Rosé Winemaking by Direct Pressing

This widely used method consists in using white winemaking techniques on red grapes, with direct pressing of fresh grapes. A certain degree of maceration is necessary to obtain color. This is done directly in the press tank, while the crushed grapes are being drained.

Pressing conditions have a significant effect on wine quality. Increasing the pressure increases extraction of total phenolics. In addition, after each time that the press cake is broken up, extracted tannin concentrations increase more quickly than anthocyanin concentrations, and the yellow hue increases. The successive press juices should therefore be selected during extraction (with regular pH measurement on the juice and evaluation of astringency) and carefully blended with the free-run juice. The press juice from the last pressing cycle may be eliminated because, in addition to its vegetal taste, it supplies more tannins than anthocyanins.

During loading of the harvest, or during pressing, inerting under nitrogen or carbon dioxide can be beneficial in limiting oxidative phenomena. Then, the must is generally sulfited (3 to 5 g/hl). The care taken in settling the must is as important as in white winemaking. The choice of low turbidity (< 50 NTU) or close to 200 NTU associated with the type of yeast selected for fermentation will have an impact on the aromatic component of the wines.

Extreme clarification of the must should be avoided. As in the treatment of white must (Section 13.5), turbidity levels below 50 NTU may lead to difficult fermentations, while levels above 250 NTU may result in vegetal off-odors. Low doses (0.5–2 g/hl) of pectolytic enzymes may facilitate juice settling.

Fining with different products, carried out on the must at the time of settling or during fermentation, is a means of limiting color intensity and modulating the yellow hue. Complementing the traditional wine fining products (casein, gelatin, bentonite), pea and potato proteins are well-adapted alternatives, as is PVPP (polyvinyl polypyrrolidone) to correct color intensity or yellow tones (Gil et al. 2020). Many commercial preparations have been developed in recent years for these applications, with mixtures of different products. Residues from settling may be clarified in the same way as those from white must (Section 13.5.5). Fining can also take place during alcoholic fermentation, provided that the wine is not aged on total lees. Remember that it is not recommended to use bentonite in conjunction with pectolytic enzymes.

Commercial yeasts should be selected for their fermentation capacity and performance in revealing aromas (with a dominance of fermentative aromas on the one hand, and varietal aromas on the other). Temperature choices for alcoholic fermentation will influence the aromatic component of the wine, as in dry white winemaking. The poor fermentability of certain musts may lead to problems in completing fermentation, alleviated by adding nitrogen and especially oxygen.

In the past, malolactic fermentation was not customary—the freshness and fruitiness of these wines was considered indispensable. Today, this second fermentation is sometimes used to make these wines fuller. It is often difficult to carry out and requires a more moderate sulfiting of the grape crop.

Rosé wines should be kept at relatively low temperatures to preserve their aromas, with an adequate dose of free sulfur dioxide (20 mg/l). Immediately after sulfiting, wines are slightly discolored and appear more yellow; but in the long term the color is more stable, with a more pronounced pure red nuance. This type of wine is generally intended to be enjoyed young, so it is important to avoid loss of color. The role of anthocyanins in the preservation of aromatic markers associated with the fruity aroma component of rosés has been established (Murat *et al.*, 2003).

14.1.4 Making Rosé Wines by Skin Contact or *Saignée* Method

Deeper-colored, fuller-bodied rosé wines are produced by leaving the skins and seeds in contact with the juice for a short time, to extract more anthocyanins and tannins. However, excessive skin contact may result in too much color, accompanied by marked astringency and bitterness and limited capacity to maintain freshness in wines over time. On the other hand, doing maceration at a low temperature ($\leq 10°C$), for a controlled period of time, is a means to obtain juices not too high in color.

The juice may be kept in contact with the grape solids in the press for short periods of time (two to 20 hours): this technique is known as pre-fermentation skin contact. This process may also take place in a tank for a longer period (10–36 hours), then some of the juice is drawn off and fermented as a rosé wine. Skin contact is primarily aimed at making true rosé wine, while the *saignée* method is mainly intended to produce a more concentrated red wine from the remaining must in the tank, with the rosé wine made from the drawn-off juice as a byproduct. In the Côtes de Provence vineyards, which specialize in rosé wine production, 40% are made using skin contact, 10% by the *saignée* method, and the remaining 50% by direct pressing of the red grapes (Masson, 2001). Pre-fermentation skin contact is known to enhance softness and fruit, while reducing acidity.

In both cases, the crushed, stemmed, generally not sulfited and inerted grapes are transferred either directly to a pneumatic press with the drains closed, or to a maceration tank. After a variable period on the skins (2–24 hours), the juice is separated from the solids, either by pressing or by drawing off all or part of the liquid from the tank. The must is then fermented in the same way as juice from grapes that have been pressed immediately.

Contact time, temperature and sulfiting are factors that influence phenolic compound dissolution and color in rosé wines (Cayla *et al.*, 2010; Castino, 1988). Sulfur dioxide is known to have a certain solvent power (Section 8.7.5). It is not manifested during traditional red winemaking, due to the preponderant effects of other factors (duration, temperature and pumping-over). Yet when maceration is limited, the effect of sulfiting is obvious. Table 14.3 presents old results comparing the impact of the winemaking technique on the color intensity and phenolic compound concentrations of rosé wines. Sulfiting promotes anthocyanin dissolution and color enhancement. It is not easy to control the conditions that will produce the required color and phenolic structure, as they depend on the specific characteristics of the wine.

TABLE 14.3

Comparison of Different Rosé Wines Made from the Same Grapes (Sudraud et al., 1968)

Winemaking method	Total phenolics index	Anthocyanins (mg/l)	Tannins (mg/l)	Color intensity[a]	Tannin/ anthocyanin ratio
Direct pressing	6	7	100	0.41	14.3
Skin contact for 12 hours and *saignée*:	11	26	320	0.52	12.3
- Without SO$_2$					
- SO$_2$ at 10 g/hl	16	100	760	1.53	7.6

[a] See Table 14.2.

The success of such rosé winemaking is based above all on the use of healthy and perfectly ripe grapes from high-quality varieties that are not necessarily suited for making traditional red wines (grapes rich in assimilable nitrogen, moderate water stress in the vineyard).

Pre-fermentation skin contact trials for eight hours at 5°C, 10°C and 15°C were performed for rosé winemaking, and changes in color (CI: color intensity), phenolic compounds and volatile compounds were monitored in musts and wines until six months after bottling (Salinas et al., 2005). A skin contact temperature of 15°C led to wines with the highest CI and malvidin 3-glucoside values and the lowest alcohol and ethyl acetate contents. These wines were the only ones in which terpene alcohols were released after six months of bottle aging. The wines made at 5°C have the highest ester levels, which also stay stable during storage. When the skin contact temperature is used as a variable for differentiation in discriminant analysis, the volatile compounds seem to be important contributors. The color and phenolic compound parameters appear to be important when the sampling time is used as a differentiation variable. In sensory and ranking tests, the best scores are obtained for wines made with skin contact at 15°C, whereas the least preferred wines are those made with skin contact at 5°C. A study by Puertolas et al. (2011) examined the effect of skin contact time (zero to six hours) and skin contact temperature (4 to 20°C) on the anthocyanin content of rosé wines made from Cabernet Sauvignon grapes treated with pulsed electric fields (PEF) using parameters of 5 kV/cm and 50 pulses. Wines made from PEF-treated grapes always have higher anthocyanin contents than the control wines after two months of bottle aging. For a fixed skin contact time of two hours, the PEF treatment enables a reduction in the skin contact temperature needed to obtain an anthocyanin content of 50 mg/l, dropping it from 20°C to 4°C. This drop in temperature may decrease the oxidation of phenols and the loss of aroma compounds, which thus increases the quality of the resulting rosé wines. PEF could be a technique of the future, making it possible to control the extraction of anthocyanins in rosé wines.

Lastly, the aromas of rosé wines have been analyzed by gas chromatography coupled with sensory analyses (Murat et al., 2001; Masson et al., 2009). The study demonstrated how various volatile compounds of fermentation and varietal origin contribute to rosé aroma. Ethyl esters and acetates of higher alcohols are predominantly responsible for the fruity and amyl notes of rosé wines. In addition, research on specific varietal compounds known to be found in other types of wines has enabled identification of certain key compounds in the aroma of rosé wines from different varieties, notably two volatile thiols (3-mercaptohexanol, or 3-sulfanylhexanol (3-SH), and 3- mercaptohexyl acetate, or 3-sulfanylhexyl acetate), two furan compounds (furaneol and homofuraneol) and a C13-norisoprenoid (β-damascenone) (Murat et al., 2001; Masson et al., 2009).

14.2 Botrytized Sweet Wines (Sauternes and Tokaji)

14.2.1 Introduction

Due to their lack of tannins, conventional white wines tolerate a large diversity of sensory structures such as varying acidities and the presence of carbon dioxide (sparkling wines) or sugar (sweet wines). Most sweet wines are white wines. Only a few special red wines, or fortified red wines, such as port, contain sugar, but they also receive additional alcohol.

Sweet wines correspond to an incomplete fermentation, leaving a certain proportion of grape sugar that has not been transformed into alcohol. Wines are arbitrarily judged to be off-dry (*demi-sec*), sweet (*moelleux*), and very sweet (*liquoreux*) according to their sugar concentration: up to 20, 36, and above 36 g/l, respectively. The sugar concentration is sometimes expressed in potential alcohol: for example, 12 + 2 signifies a wine containing 12% alcohol by volume and 36 g of sugar per liter (2% volume of potential alcohol).

Off-dry and sweet winemaking is fairly similar to dry winemaking, but the grapes must have a sufficient sugar concentration, and the fermentation must be stopped before completion, either naturally or by a physical or chemical process.

Very sweet winemaking is different. The required high sugar concentration cannot be attained during ripening. Certain processes are needed to concentrate the juice, and certain winemaking steps are unique to these wines.

Drying, freezing, and noble rot are used to concentrate juice. Due to the importance of noble rot, it will be discussed in more detail later in this section.

Grapes can be dried naturally by the sun, when left on the vine, or by artificial heating (Section 11.2). This overripening process can be used to make different types of wine. The drying of the grapes results in a varying degree of concentration, but the enzymatic systems of the fruit permit a greater concentration of sugars than acids. Grapes can also be dried for up to several months in a closed room, which may or may not be heated. This method results in musts containing up to 400 g of sugar per liter, capable of producing very sweet wines. The fermentation of these juices is difficult, and the price of these wines is high, due to volume loss and production costs.

Freezing grapes on the vine produces ice wine (Eiswein), well known in Alsace and Germany. The grapes are left on the vine until the winter frosts. The temperatures, −6 to −7°C, lead to the partial freezing of the least ripe grapes. By pressing the grapes at low temperatures, only the juice from the ripest grapes, containing the most sugar, is extracted. Cryoextraction (section "Cold pressing (cryoextraction)") seeks to reproduce this natural process artificially. Making such wines is subject to weather conditions and not always possible every year. The method is difficult and expensive and should only be used to make premium wines.

14.2.2 Noble Rot

The biology of *Botrytis cinerea* and its development in the form of noble or gray rot have been described previously (Section 10.6). In the case of noble rot, this overripening process permits the production of great botrytized sweet wines. These exceptional wines can only be made under specific conditions. Their production is therefore limited.

The Sauternes-Barsac region is certainly one of the most highly esteemed areas for noble rot sweet wines, but other regions exist in France (Loupiac, Sainte-Croix-du-Mont, Monbazillac, Anjou), in Germany (Moselle), and in Hungary (Tokaj).

Noble rot presupposes fungus development on perfectly ripe grapes. Semillon and Sauvignon Blanc grapes must attain 12–13% vol. potential alcohol and have a pH of less than 3.2 before any fungal development. At this time, the berries are golden with slightly brown thick skins. This result can only be attained on certain *terroirs*, with low crop yields (40–45 hl/ha), before berry concentration by noble rot. Mycelial filaments penetrate through microfissures and decompose the grape skin. This decomposition is the result of an intense enzymatic maceration of the grape skin. The grape then attains the *pourri plein* (full rot) stage and has a brownish color. The berry does not burst: it maintains its shape, but the skin no longer maintains the role of a protective barrier against the external environment. The berry acts like a sponge and becomes concentrated as the water evaporates. The grapes are harvested when they attain the *rôti* stage. The concentration in the berry leads to an increase in internal osmotic pressure that causes the death of the fungus. The second phase of *B. cinerea* development must therefore occur soon after the full rot phase, before subsequent *Botrytis* development can result in gray rot. The distinction between noble rot and gray rot is not always obvious.

Late-season weather and harvesting conditions are essential for noble rot development. Maximum grape concentration should not always be sought, because gray rot may develop when conditions are not ideal. Attempts have been made to harvest at the full rot stage and then concentrate the grapes by artificial means.

Alternating humid and sunny periods are essential for reaching the perfect state of ripening. Gray rot occurs when *B. cinerea* develops under extremely humid conditions. The development of noble rot requires a particular climate, ideally with morning fogs to assure fungal growth, followed by warm afternoon sunshine to concentrate the grapes, for a relatively long period of two to four weeks. In Bordeaux vineyards, these meteorological conditions correspond with the establishment of a high-pressure front extending the anticyclone from the Azores to the northeast. Noble rot can also develop rapidly in the Gironde region after a short period of rain, caused by oceanic depressions, followed by a sunny and dry spell (low humidity, 60%) with winds from the north to northeast. This type of weather is generally associated with the presence of an anticyclone in northeastern Europe.

Noble rot develops progressively on different grape clusters and even on different grapes on the same grape cluster. The grapes must therefore only be picked when they attain their optimum ripeness. Selective harvesting ensures that grape-pickers only remove the noble-rotted grape clusters or grape cluster fractions during each picking. Climatic conditions dictate the number of selective pickings each year—up to three or four. Harvesting can continue until November in the Northern Hemisphere. Due to the evolution of grape crop quality from one day to another according to climatic conditions and the evolution of rot, juice should be vinified separately according to harvest date.

Grape growing for the purpose of making botrytized sweet wines requires more meticulous care than for making dry white wines. This is particularly the case in oceanic climates, favoring the early implantation of *Botrytis* on sensitive varieties. In the Sauternes region, Semillon and Sauvignon Blanc require shorter cane pruning and the rigorous control of vegetative growth, in particular early leaf removal (before *veraison*) in the cluster zone.

It is evident that the distinction between noble rot and gray rot is not always clear-cut. Late-season weather conditions and picking conditions are essential factors. It is a mistake to seek out maximum sugar concentration at all times, since this entails a risk of gray rot development. There is no point in trying to generalize the production of these exceptional wines, which can be made only under very special conditions.

Variable proportions of bunch rot can coexist with noble rot. The involvement of acetic acid bacteria results in the production of volatile acidity. Gray rot can also form odors and tastes resembling mushrooms, mold, iodine, and phenol. In addition, considerable amounts of carbonyl-based compounds may be produced, which bind sulfur dioxide and make these wines difficult to stabilize.

They are produced by acetic acid bacteria in the *Gluconobacter* genus present on the grapes. 5-Oxofructose is one of the main substances responsible for this phenomenon (Sections 8.4.3 and 8.4.6).

14.2.3 Composition of Musts Made from Grapes Affected by Noble Rot (Section 10.6.3) and the Resulting Wines

Noble rot can reduce crop volume by up to 50%. Grape quality depends on low crop yields of 15–25 hl/ha.

The fungus consumes a large quantity of grape sugar to ensure its growth. According to Ribéreau-Gayon *et al.* (1976),

> 50% more sugar and twice the juice volume is obtained from healthy grapes with respect to noble-rotted grapes, for the same acreage. Natural concentration and the resulting considerable crop yield losses are responsible for these richer and higher-quality wines.

This fungus also consumes a considerably higher proportion of acid than sugar. This phenomenon is beneficial to wine quality, since the acidity increases much more slowly than the sugar concentration. It is not a simple concentration by water evaporation, since the phenomenon is coupled with a biological deacidification. *Botrytis cinerea* has the rare property of breaking down tartaric acid: its concentration decreases more than the malic acid concentration, and the pH consequently increases by 0.2 units.

Table 10.13 takes the modifications caused by noble rot into account. The lower sugar and acidity concentration due to weight and volume loss in the 1,000 berries corresponds with a significant increase in the sugar concentration and a slight decrease in acidity of the juice.

Botrytis cinerea forms glycerol and gluconic acid from sugar. These two compounds play an important role in characterizing rot quality. Glycerol is produced at the start of *Botrytis* development. Its concentrations increase as the rot becomes more noble. Gluconic acid is formed much later and corresponds with poor rot evolution. Wines made from healthy grapes should contain less than 0.5 g of gluconic acid per liter, noble rot wines between 1 and 5 g and gray rot wines more than 5 g. The higher the glycerol/gluconic acid ratio, the better the rot quality is.

The development of *B. cinerea* also results in the production of two polysaccharides (Dubourdieu, 1982). One, with a complex structure, has antifungal properties and inhibits alcoholic fermentation. The other is a β-glucan with protective colloid properties, and it impedes clarification of new wine. It is produced in a viscous gel, located between the skin and the pulp. Its level of diffusion in the juice is related to grape handling and treatment conditions. All brutal mechanical operations, such as crushing, pumping, and pressing, diffuse glucan in the juice and make the wine more difficult to clarify.

Botrytis cinerea development corresponds with enzymatic changes in the grape. Laccase, an oxidation enzyme, replaces the grape's tyrosinase and is much more active on phenolic compounds than the latter. As a result, most phenolic substrates of the grape are already oxidized by the time of harvest. Phenolic composition is profoundly changed, owing to enzymes of *B. cinerea* and those of the grape (Section 10.6.4). In addition, *Botrytis* synthesizes a number of enzymes (cellulase, polygalacturonase), permitting its penetration into the grape. An esterase that hydrolyzes fermentation esters has been isolated, thus accounting for the differences in aroma between wines made from grapes affected by noble rot and other white wines.

Another characteristic of wines made from grapes affected by noble rot is their relatively high volatile acidity. Its origin can be accidental, due to the presence of lactic acid bacteria in juice and especially acetic acid bacteria on grapes. Yeasts also form some volatile acidity, due to the corresponding fermentation difficulties of these juices (Section 2.3.4).

Operations likely to reduce the production of volatile acidity include supplementing the must with ammonia nitrogen, which should be adjusted to 190 mg/l at the start of fermentation, combined with seeding with a suitable yeast and aeration (Section 14.2.5). The aim is to increase the cell population to a maximum, as this minimizes the formation of volatile acidity. For this reason, EU legislation has specified higher limits for volatile acidity in botrytized wines. Lafon-Lafourcade and Ribéreau-Gayon (1977) sought to specify the origin of this volatile acidity. Two lactic acid isomers and ethyl ester were analyzed in wines. Lactic acid bacteria produce large quantities of the former and acetic acid bacteria of the latter. The authors concluded the following:

- If D(−)-lactic concentrations are more than 200 mg/l, lactic spoilage may exist.
- If L(+)-lactic concentrations are more than 100 mg/l, malolactic fermentation is responsible.
- If ethyl ester concentrations are more than 160 mg/l, acetic acid bacteria are involved.

If these three parameters are less than the indicated values, the yeasts are entirely responsible for the volatile acidity. Volatile acidities between 0.9 and 1.3 g/l expressed as H_2SO_4 (1.1–1.6 g/l expressed as acetic acid) in many Sauternes wines were found to be produced exclusively by yeasts, with no bacterial involvement.

14.2.4 Noble Rot Juice Extraction

Pressing grapes

The general rules of white winemaking should be followed when transporting noble-rotted grape crops. The depth of grapes in crates and bins, in particular, should be kept to a minimum during transport to avoid spontaneous crushing.

Upon their arrival at the winery, the grapes should be handled with care. Rough handling provokes the excessive formation of suspended solids and vegetal flavors. It would also make the wines more difficult to clarify because of the diffusion of glucan from *B. cinerea* in the juice. Winemaking techniques should make use of gravity, as much as possible. Pumps on self-emptying gondolas are often too brutal.

Noble-rotted grapes are generally crushed and macerated in the liberated juice. This operation helps to extract the sugar. The grapes are not destemmed in order to facilitate the circulation of juice in the pressed skins, but manual destemming is sometimes done after the first pressing with the stems. In this manner, the tannic and vegetal substances of the stems are not transferred to the must during subsequent pressings at high pressure. The extraction of noble-rotted juice is difficult: a high pressing pressure must be used, and the press cake must be broken up between pressings.

Due to their high viscosity, these grape crops cannot be drained before pressing. They are directly transferred to the press cage by gravity or a conveyor belt. Pressing is certainly the most difficult and essential operation in botrytized sweet winemaking, and incorrect operation of the presses can sacrifice quality. These operations in particular must be carried out slowly and delicately.

Continuous presses should not be used. They are too brutal even when equipped with a large-diameter screw turning very slowly. They shred the grape crop and produce juices with a lot of suspended solids and glucan-rich wines that are difficult to clarify.

The older vertical cage hydraulic presses are effective and produce few suspended solids in the juice. The press cake is broken up manually with these presses and the pressing cycles are slow. Moving-head presses are therefore preferred because of their ease of operation and satisfactory yield. After two to three pressing cycles, the last cycle can be done by a hydraulic press, which enables higher pressing pressures. Pneumatic presses, used for other types of grape crops, are not well adapted for pressing noble-rotted grapes. Their pressing pressure is insufficient. Although manufacturers have created special models with up to 3 bar of pressure, these pneumatic presses do not properly extract the juices with the highest sugar concentrations, i.e. above 22 or 23% potential alcohol by volume.

Selecting the best juice from the various pressing cycles is a difficult problem. With healthy grapes, the drained juice and the juice from the first pressing cycles are the richest in sugar; the last pressing cycles are vinified separately. With noble-rotted grapes, the juice with the highest sugar concentration and the highest iron and tannin concentrations is the most difficult to extract: it is released in the last pressing cycles. The addition of this juice to the blend can therefore improve wine quality but should be done prudently.

Cold pressing (cryoextraction)

Cold pressing, or selective cryoextraction, enables the selection of the most sugar-rich and therefore ripest grapes. Chauvet et al. (1986) first proposed this technique (Section 11.3.4).

Cryoextraction is based on a law of physics. Raoult's law states that the freezing point of a solution lowers as the solute concentration increases. When the temperature of a white grape crop is lowered to 0°C, only the grapes containing the least amount of sugar are frozen. By pressing at this temperature, a selected juice is obtained, which represents only part of the total juice volume. The potential alcohol content of this juice, however, is higher. By further lowering the freezing point and pressing temperature of the grape crop, the number of frozen berries is increased. As a result, the selected must volume is further diminished, and the potential alcohol content increased. Table 14.4, which gives the results of an experiment, specifies the increase in potential alcohol content for grape crops pressed at different temperatures. The selected must volume represents between 60 and 80% of the total volume, depending on the temperature.

The selection of the most sugar-rich and thus ripest grapes by cryoextraction is the primary cause of improved wine quality. Freezing may also concentrate grapes diluted by rainy harvest seasons. A phenomenon called supraextraction is also involved. Freezing and thawing grapes results in tissue destruction and better sugar extraction.

In practice, the grape crop is collected in small containers and frozen in a walk-in freezer at varying temperatures as low as −16°C. The grapes are then pressed in a horizontal moving-head press or in a pneumatic press. The freezing point of the richest grapes is sufficiently low that their juice may be extracted without difficulty. Table 14.5 gives the results obtained for five Sauternes vineyards in 1985. Installing a cryoextraction system is expensive, but energy costs are reasonable. In any case, this technique can only be applied for limited grape crop volumes.

TABLE 14.4

Effect of Pressing Temperature (Cryoextraction) on the Volume and Potential Alcohol (% vol.) of Selected and Residual Musts from Noble Rot Grapes (Sauternes) (Chauvet et al., 1986)

Control musts: potential alcohol (% vol.)	Pressing temperature (°C)	Selected musts		Residual musts		Volume ratios selected must/ blended must
		Volume (hl)	Potential alcohol (% vol.)	Volume (hl)	Potential alcohol (% vol.)	
16.7	−5	116	19.4	19	11.3	86
16.0	−6	81.5	19.6	21.5	12.3	79
16.0	−8.5	74.5	20.7	33	9.8	69
16.2	−10	65	22.0	37	13.1	64
16.0	−11	67	23.2	33	10.7	67
17.0	−15	60	23.4	40	12.6	60

The improvement in must quality is not only limited to an increase in potential alcohol content. These musts have the same characteristics as their late-harvested counterparts: they are more difficult to ferment, the risk of volatile acidity production is greater, and the sulfur dioxide binding rate is higher. However, these cryoextraction wines are clearly preferred over the control wines due to their increased concentration, finesse, and typicity.

This technique is applicable to all types of white grape crops but is especially interesting for noble-rotted grape crops. Excessive humidity during overripening can prevent the grapes from obtaining the desired concentration levels. Prolonging the concentration phase would only increase grape deterioration and lower crop quality, but, by picking early, cryoextraction can be used to eliminate excess water and select the grapes capable of producing high-quality sweet wines. The quantity is diminished but quality is maintained. This method can complement manual sorting during the harvest but cannot replace it.

Sulfiting juice

Sulfiting has several effects during white winemaking (antioxidase, antioxidation, antimicrobial activity, etc.). In the case of juice intended for sweet wines, this practice has been reconsidered in the past 15 years or so, for the following reasons:

TABLE 14.5

Improvement of Sauternes Must Quality in 1985 by Using Cryoextraction (Chauvet et al., 1986)

Vineyard	Grape temperature at pressing (°C)	Potential alcohol (% vol.)	
		Control must	Selected must
A	−13	15.9	21.5
B	−9	13.2	19.8
C	−7	13.0	19.0
D	−7	13.6	20.3
E	−9	18.0	23.3

1. Sulfiting protects against oxidation by inhibiting laccase, produced by *B. cinerea*. Due to the considerable oxidation of grape phenolic substrates during overripening, however, the oxidation risks are less significant than one would think.
2. By blocking fermentation for several hours, it permits a coarse clarification by sedimentation. However, the level of clarification of juice made from noble-rotted grapes is not as high as for juice intended for dry white wines. In addition, clarification is above all increased by the storage of juice at low temperature to delay onset of fermentation.
3. It is also capable of inhibiting the development of acetic acid bacteria, often present in large quantities on botrytized grapes, as well as lactic acid bacteria, thus preventing early onset of malolactic fermentation. However, these points must be addressed right from the beginning by selection of grapes via proper sorting.

Sulfiting juice intended for botrytized sweet wine production has often been criticized. This operation leads to increased concentrations of bound sulfur dioxide, which remains definitively in the wine. Subsequent SO_2 additions must therefore be limited to remain within legal total SO_2 limits, thus compromising the microbiological stabilization of wine.

Lastly, the sulfiting of juice prior to fermentation is generally no longer necessary for sweet white winemaking with noble-rotted grapes, except if there is a risk of spoilage by lactic acid bacteria or acetic acid bacteria. In the latter case, light sulfiting at 3–5 g/hl is recommended.

Juice clarification, bentonite treatment, and juice corrections

Clarification is an indispensable operation in dry white winemaking. In most cases, clarification improves aroma finesse and sensory qualities of wines. It is not as necessary in the case of sweet wines made from noble-rotted grapes, in particular when the harvested crop is composed exclusively of *"pourri confit"* grapes (literally "candied rotted" or advanced stage of noble rot).

However, the natural settling of botrytized musts is difficult to effect: the suspended solids and the highly dense juice have similar specific weights. Must viscosity is also high due to the high sugar concentration and the presence of colloidal carbohydrates produced by *B. cinerea*. In practice, natural settling for 18–24 hours results in partial clarification, permitting large particles to be eliminated. The addition of pectolytic enzymes to noble-rotted juice presents no clarification advantages. Pectinases are secreted by *B. cinerea* in the grape and are found in abundant quantities in the juice. The average pectinase activity ratio between healthy grapes and noble-rotted grapes is estimated to be 1:10. Natural settling for three to four days at 10°C enables a more effective clarification. This kind of clarification is recommended in particular when the harvested crop has a wide range of rot quality.

As with dry white winemaking, over-clarification can lead to major fermentation problems and increased acetic acid production. Must turbidity should not be as low as in dry white winemaking (100–200 NTU); 500–600 NTU or even a slightly higher turbidity is perfectly acceptable. Moreover, botrytized sweet wines are not subject to the same problems related to insufficient clarification as dry white wines—the development of reduction odors and vegetal tastes, oxidability, etc.

Bentonite has been widely used as a settling aid for juices in large fermentors during dry white winemaking but is no longer used to make barrel-fermented wines that are aged on lees. The effectiveness of this treatment in the making of noble rot sweet wines has long been questioned. The high colloid concentration in these wines appears to diminish the adsorbing power of bentonite with respect to proteins. Moreover, adding bentonite before fermentation as opposed to after racking does not facilitate sedimentation. The presence of bentonite can also make racking, following the completion of alcoholic fermentation, more difficult. For these different reasons, wines are treated with bentonite a few months before bottling after a protein instability laboratory test.

Depending on the legislation, various adjustments such as the addition of sugar or modification of acidity can be made to juice before the initiation of fermentation. These adjustments should be limited to avoid an imbalance of the wine.

The addition of nitrogen often promotes the fermentation of these musts, depleted by the development of *B. cinerea*.

The addition of 50 mg of thiamine per hectoliter is often beneficial. It is not only a yeast growth factor but can also limit the binding rate of sulfur dioxide during wine storage by promoting the decarboxylation of ketonic acids (pyruvic acid and α-ketoglutaric acid) (Lafon-Lafourcade et al., 1967).

When dry white wines and noble rot sweet wines are made in the same winery, adding a small quantity of lees from healthy grape must, after clarification, can considerably improve the fermentability of noble-rotted musts and diminish the production of volatile acidity (Dubourdieu, unpublished results). Depending on the available amount of suspended solids from healthy grape must, 2-4l of solids should be added to each barrel (1-2%), resulting in a turbidity increase of 200-400 NTU. The sediment used should be sulfited, unfermented, and conserved at a low temperature.

14.2.5 Fermentation Process

Fermentation difficulties
Noble-rotted musts are known to be difficult to ferment. The high sugar concentration is the principal limiting factor, but nutrient deficiencies provoked by the growth of *B. cinerea* are also responsible. Among possible deficiencies in the must, nitrogen is a key factor. Masneuf et al. (1999) analyzed 32 musts intended for dry white winemaking and reported a mean available nitrogen content of 182 mg/l, measured by formol titration. The same analysis of 20 samples of botrytized must intended for sweet white wine production yielded a mean value of 84 mg/l only (Bely et al., 2003). Polysaccharidic fungal substances affecting fermentation kinetics also cause fermentation problems. For an identical sugar concentration, juice is more difficult to ferment when grapes are extensively attacked by *B. cinerea*.

It is advisable to adjust the ammonium sulfate content to 190 mg/l at the beginning of fermentation to facilitate the process and minimize the production of volatile acidity. Later addition of nitrogen supplements is likely to have a negative impact.

Inoculation of the juice with yeast is strongly recommended. The chosen yeast strain should be highly ethanol tolerant and should produce little volatile acidity under difficult fermentation conditions. Dry yeast should not be introduced directly into the juice to start fermentation for this type of winemaking. A yeast starter should be prepared in diluted must, supplemented with NH_4^+ and yeast hulls, then seeded with dried active yeast at a dose of 2.5 g/hl of the total volume to be inoculated. The starter is added to the must on the second day of fermentation, at a rate of 2% of the total volume. This increases the maximum yeast population, which controls fermentation rate and volatile acidity production (Section 2.3.4). In one experiment, adding yeast in this way reduced the final volatile acidity content by 20%.

Noble-rotted wines were traditionally fermented in small wooden barrels. This method creates favorable conditions for this kind of winemaking. The juice from each day of harvesting can be separated according to quality. Temperatures are also better controlled, since the fermentation occurs at near ambient temperatures, but the cellar may need to be heated if juice temperatures are too low or if the ambient temperature drops too low. Barrel fermentation also permits continuous microaeration, promoting yeast activity and a complete fermentation. Barrel-fermented musts with high sugar concentrations produce more alcohol than the same must in a large fermentor, because the fermentation stops earlier in a tank. Furthermore, the presence of carbon dioxide protects fermenting must from oxidation, and phenolic compounds, the main oxidation substrates, are destroyed in the grapes by *B. cinerea* as it develops.

Tank-fermented musts have an even greater need for aeration, since this fermentation occurs under stricter anaerobic conditions. In sweet winemaking, oxygen should be introduced during the stationary phase of the yeast population growth cycle, rather than during the growth phase, as is done when making other types of wine (Section 3.7.2).

Later aeration has been shown to prevent excessive increases in volatile acidity, which mainly occurs in the early stages of fermentation. Temperature control is also indispensable under these conditions to ensure that the fermentation temperature remains within reasonable limits (20–24°C).

Rapid and vigorous fermentations result in less aromatic wines and should be avoided, but exaggeratedly slow fermentations do not promote good quality.

Stopping fermentation (*mutage*)

Sugar concentration and alcohol content determine the taste equilibrium of this kind of wine. The sweetness of sugar must mask the burning characteristic of alcohol. Reciprocally, the latter must balance the heaviness of a high sugar concentration. For this type of wine, the alcohol content and potential alcohol should approximately satisfy the following relationship: $12 + 4$ to 5, $13 + 4$ to 5, $14 + 4$, or more rarely $15 + 5$. This equilibrium is generally obtained by blending several wine batches, some containing more sugar and others more alcohol.

Fermentation rarely stops spontaneously at the exact alcohol/sugar ratio desired. In some cases, it can go too far; in other cases, it becomes excessively slow, and the increase in volatile acidity is more significant than the decrease in the sugar concentration. At this point, the fermentation must be stopped. This operation generally consisted in adding sulfur dioxide to the wine.

Sulfur dioxide gas was traditionally added directly to warm wine, since the yeasts are more sensitive to SO_2 at higher temperatures. More recently it is considered preferable, before adding SO_2, to wait for the wines to cool and eliminate a portion of the yeasts by racking. For a given antiseptic dose, the lower the initial population, the smaller the residual population will be. The wine should be protected from air while racking, if possible by an inert gas atmosphere, to avoid the oxidation of ethanol and the formation of traces of acetaldehyde, which has a strong binding power with SO_2.

In any case, a massive dose of sulfur dioxide (20–30 g/hl) should be added to block all yeast activity instantaneously and avoid even limited acetaldehyde production by the yeasts. Several days after the addition, the free SO_2 concentration should be verified. A value of 60 mg/l is suitable for storage. Adjustments can be done at this time if necessary.

14.2.6 Aging and Stabilization

The sensory quality of botrytized sweet wines improves considerably after several months of barrel aging and several years of bottle aging. The bouquet takes on finesse and complexity—reminiscent of candied fruit and toasted almonds. The wine becomes harmonious on the palate. The sweetness is perfectly balanced by the alcohol, and a note of acidity gives a refreshing

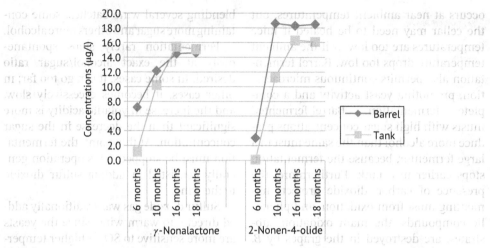

FIGURE 14.1 Changes in lactone contents (γ-nonalactone, 2-nonen-4-olide) associated with the initial *Botrytis* infection of grapes, during aging of botrytized sweet wines (Semillon, 2013; Stamatopoulos and Darriet, 2017).

finish. The wines are not heavy and syrupy, in spite of their high sugar concentration. These transformations remain poorly understood even today. They occur under conditions, particularly oxidation–reduction conditions, reminiscent of red wine aging.

It should be remembered that fermentation and aging help to reveal the aroma composition of sweet wines. Some volatile compounds from the oak enrich this range of aromas by promoting certain aroma pairings. Excessively oxidative aging can inversely affect the aroma composition, in particular in new oak.

Premium botrytized white wines are primarily barrel-aged, with the bung in the vertical position, for 10–18 months, sometimes even two years, though rarely more nowadays.

The conditions for fermentation and aging profoundly change the aroma composition of sweet wines. In particular, winemaking conditions affecting the oxygenation level and container type have a very significant effect. The analysis of aroma compounds, including those that have been recently identified, during the aging of botrytized sweet wines in tanks, new barrels, or used barrels, coupled with their sensory analysis (preference testing), has helped to specify the key parameters for evolution of the aroma of these wines. Firstly, lactones associated with *Botrytis* (2-nonen-4-olide, γ-octalactone, γ-nonalactone, δ-decalactone, massoia lactone) have been observed to increase steadily during the aging of sweet wines, particularly in oak barrels (Figure 14.1). This demonstrates the advisability of a sufficiently long period of barrel aging to ensure formation of certain volatile compounds (lactones) associated with the typicity of botrytized sweet wines (Volume 2, Section 7.7). This formation likely results from the transformation of aroma precursors that initially come from the *Botrytis* infection of grapes. Secondly, phenols decrease during this same period as the wines become oxygenated. The reactivity of these compounds with tannins (in particular oak tannins) can promote this decrease. However, it should be noted that these compounds have a different reactivity when compared with that observed in dry white wines from the same varietals (Volume 2, Section 7.5). Thus, the contents of phenylacetaldehyde, furanones, and fermentation esters decrease during aging, whether in barrels or in tanks (Stamatopoulos and Darriet, 2017).

Furthermore, during barrel aging, and possibly with oak alternatives (oak chips and tank staves), the release of volatile compounds from the oak helps to enrich the wines in compounds with vanilla and coconut aromas, in particular β-methyl-γ-octalactone and eugenol, which are involved via a perceptive interaction phenomenon in candied fruit notes (Volume 2, Section 7.7). In the end, barrel or tank aging significantly modifies the aroma composition of sweet wines, in particular with an increase in candied fruit notes in oak-aged wines with respect to the same wines aged in tanks.

From a practical point of view, topping-up is done weekly throughout aging. Barrels are hermetically sealed much earlier than in red wine aging (one to two weeks after *mutage*). Due to the risk of refermentation, all wine must be handled with the utmost cleanliness during the topping-off operation.

The first racking is generally done at the beginning of December, after the first cold spell. The objective is to separate the coarsest lees and conserve the finest. Rackings are then carried out every three months. At each racking, the barrels are carefully rinsed with cold water and then sterilized with hot water at 80°C or with steam. After being drained, they are sulfited before being refilled.

Microbiological stabilization is difficult with these wines. Refermentations are always possible, despite the low fermentability of these wines linked to the involvement of noble rot. First of all, the SO_2 binding rate can be high (Section 8.4). It is not always possible to obtain the necessary free SO_2 concentration while remaining within the total SO_2 limits imposed by legislation. Free SO_2 should be maintained at approximately 60 mg/l. This difficulty rarely occurs in the presence of noble rot and is generally linked to at least a partial involvement of gray rot. Due to their antiseptic properties, sorbic acid and fatty acids (C8 and C10) can be used as chemical adjuvants to sulfur dioxide (Sections 9.2 and 9.3) during the storage of these wines.

Various physical processes can complement the effect of SO_2 and help to stabilize these wines (Section 9.4). Wine conservation at low temperatures (around 0°C) hinders but does not definitively inhibit yeast growth, but heating wine to 50–55°C for several minutes can totally destroy the yeast population (Section 9.4). Heat-sterilized wines must be stored under sterile conditions to prevent subsequent contamination. Sterile storage, however, poses practical problems and is not possible in wooden casks (Volume 2, Section 12.2.3).

The difficulty of storing high-sugar wines, capable of refermenting, has incited the development of other techniques. A dry wine and a "sweetening reserve" can be prepared separately. The sweetening reserve is a partially fermented juice containing 2–2.5% vol. alcohol and 150–200 g of sugar per liter. Only the second wine fraction is difficult to store, but its volume is limited (15–20% of the total volume). The wine can therefore be sulfited at a high concentration (100 mg/l) and stored at a low temperature. The two wines are blended just before bottling in a sterile environment. Wine aroma conservation is maximized with this method, but the technique is only applicable to wines containing a maximum of 30 g of sugar per liter. Very sweet botrytized wines cannot under any circumstances be made by this method.

In addition, the presence of glucan makes botrytized sweet wines difficult to clarify by filtering or fining. This substance acts as a protective colloid (Volume 2, Section 11.5.2). For properly barrel-aged wines, bentonite is currently the only possible fining treatment, but it is sometimes not even necessary. In this case, the majority of clarification problems are due to poor rot quality and unsuitable working conditions.

14.2.7 Tokaji Wine

Tokaji, or Tokay, is a famous botrytized sweet wine produced in Hungary. Several types of wine exist under the same name. Tokaji wines are generally evocative of sweet wine, although there are several other types as varietal wines made from Furmint and Hárslevelű (cuvées), Tokaji Szamorodni (dry and sweet), Tokaji Fordítás and Máslás, Tokaji Eszencia and Tokaji Aszú. These wines are made from the six legally-permitted grape varieties under the Tokaji name: Furmint, Hárslevelű (Lindenleaf), Yellow Muscat (Goldmuskateller), Kövérszőlő, Zéta, and Kabar. Furmint is the most widely produced variety, yielding nearly 70% of the entire crop, and serves predominantly as a base wine for the production of Aszú.

These wines are made with grapes which are concentrated on the vine, by both drying and noble rot. The grapes are ground (mechanically today) to obtain a type of paste. A high-quality new wine in the final stage of fermentation, rich in alcohol, acidity, and extract, is then poured over this paste. The wine and paste are then macerated for 24–36 hours, permitting the sugar and different aromatic elements of the Aszú grapes to be diffused in the wine. This mixture is then pressed to separate the pomace and the wine. The wine is then aged in a cask. The amount of paste added to the wine corresponds to the different types of Tokaji Aszú. The wine is classified by the number of back-baskets (*puttonyos*) of paste (20–25 kg) per 136-l cask. The following types of Tokaji can thus be distinguished:

- Three *puttonyos* per cask of Tokaji Aszú: This contains at least 60 g of natural sugar per liter, and it must be aged for at least three years in barrel and bottle.
- Four *puttonyos* per cask of Tokaji Aszú: This contains at least 90 g of natural sugar per liter and is aged for at least four years.
- Five *puttonyos* per cask of Tokaji Aszú: This contains at least 120 g of natural sugar per liter and is aged for at least five years.
- Six *puttonyos* per cask of Tokaji Aszú: This contains at least 150 g of natural sugar per liter and is aged for at least six years.

Since 2013, it is no longer mandatory to mention the number of *puttonyos* on the label, and the three and four *puttonyos* categories have been eliminated. The quality of Azsú *eszencia* (essence of Azsú) is even higher. It is produced in certain specific vineyards with vine-dried and noble-rotted grapes, ideal for creating this wine. The corresponding must has a high sugar concentration and is difficult to ferment. The wine contains about 250 g of sugar per liter, and its alcohol content is between 6 and 8% vol.

Tokaji wines are improved by air-exposed aging in cool cellars. These cellars are galleries dug into limestone that maintain the wine at a constant 10°C temperatures. Due to their high sugar and alcohol concentration, these wines are not very sensitive to microbial spoilage and can be conserved in partially filled casks. It is possible that aging with greater protection from air, as is done with Sauternes wines, would also produce favorable results.

14.3 Champagne and Sparkling Wines

14.3.1 Introduction

There are many methods for making "sparkling wines," but this term refers exclusively to wines that have undergone alcoholic fermentation in a closed vessel. Artificial carbonation by saturation with carbon dioxide gas (CO_2) does not produce the same quality bead, and wines made effervescent by this method are known as "artificially carbonated wines."

Champagne is the most prestigious sparkling wine. It is produced according to *Appellation d'Origine Contrôlée* (AOC) regulations, in a delimited area, using strictly defined grape varieties and winemaking techniques.

Sparkling wine is generally made in a two-stage process. The first consists in making a base wine with particular characteristics. The wine is blended and cold-stabilized, then fermented for the second time. In the case of Champagne, this secondary fermentation (called *prise de mousse*) takes place in the bottle that will ultimately be delivered to the consumer, once the yeast deposit has been disgorged. The use of the expression *méthode champenoise* for sparkling wines produced, with the same method, outside the delimited Champagne appellation is now prohibited in the EU and has generally been replaced by *méthode traditionnelle* (traditional method). The advantages of secondary fermentation in bottle (Section 14.3.3) have also been combined with clarification by filtration, achieved by emptying the wine into a vat after its secondary fermentation, then filtering and treating it, if necessary, prior to bottling in the final bottles for shipment (Section 14.3.5).

In the case of sparkling wines made by the "Charmat" or "closed tank" method, the secondary fermentation takes place in airtight vats (Section 14.3.5). The wine is then filtered, bottling liqueur is added, and the blend is bottled under pressure, ready for delivery.

In the past, some wines became sparkling spontaneously, when they were bottled with some residual sugar before fermentation was completed. They were unreliable to make, as the secondary fermentation was uncontrolled and could produce insufficient pressure, or, on the contrary, result in excessive pressure that caused the bottle to explode. Several variations on this technique are still used today but are now much better controlled. Fermentation is stopped by refrigerating the wine once it has reached a specific residual sugar level. Once it has been bottled, it is allowed to warm up naturally, and fermentation is completed. This technique is relatively simple to use as the chilled must after stopped fermentation may be stored as long as required, then bottled at an appropriate time.

In the late 17th century, problems in controlling the secondary fermentation led Champagne winegrowers to devise their present system of separating the complete fermentation of a dry white base wine from the secondary fermentation in bottle, with the addition of a controlled amount of sugar corresponding exactly to the carbon dioxide pressure required.

14.3.2 Fermenting Base Wines

Principles The aim is to make a base wine with a moderate alcohol content, generally a maximum of 11% vol., as more ethanol will be formed during the secondary fermentation in bottle and the total overall alcohol content should remain below 13% vol. The base wine should also have a certain level of fresh acidity to ensure the right balance in the finished product. The grapes are thus harvested at an earlier stage of ripening than in other AOC vineyards. Consequently, the grapes must be pressed very carefully to avoid skin contact and the resulting bitterness and vegetal character. In the case of Champagne, this is especially vital when pressing the red varieties, Pinot Noir and Pinot Meunier, that are blended with the white Chardonnay. Picking and pressing conditions must, therefore, be carefully controlled.

Handpicking is followed by sorting to eliminate any defective, damaged, or rotten grapes, as a relatively low percentage of spoiled grapes has a negative impact on quality. The whole grapes are transported in 45–50 kg containers, with holes to drain off any juice, thus avoiding skin contact and accidental fermentation, as well as maintaining the grapes under aerobic conditions. These containers are carefully washed after each use.

Pressing and extracting the must

The grapes must be pressed soon after picking, without crushing, to avoid contact between skin and juice, and the various fractions of must are kept separately. Pressing takes approximately four hours. Two types of presses are used. The hydraulic presses traditionally used in Champagne are round or square, with a large surface area to ensure that the grape layer is no more than 80–90 cm thick. Pressure is kept low to avoid crushing the skins. Horizontal presses have also been used for a number of years now, but only plate presses without chains or, preferably, pneumatic presses (Valade and Blanck, 1989). The different fractions of the must reflect the uneven ripeness within the grapes and the varying composition of the cell sap in the berries. The *cuvée* corresponds to the juice from the middle part of the grape flesh, which is both the sweetest and the most acidic. The outer part of the flesh is sweet, but less acidic, due to the salification of the organic acids in the vicinity of the skin. The flesh closest to the seeds has the highest acidity and the least sugar.

Pressing methods are standardized, and the juice is collected in small vats, known as *belons*. Traditional Champagne presses used to hold 4,000 kg of grapes. Two or three pressings in quick succession, with loosening of the pomace after each one (Valade and Pernot, 1994), produced 2,050 l of top-quality must (enough to fill 10 of the 205-l barrels used in Champagne). This was known as the *cuvée*. The next two pressings produced 410 l (2 barrels) of *première taille*, and a third gave a further barrel of *deuxième taille*. The final pressing in a standard hydraulic press produced 200–300 l of press wine (*rebêche*), intended for distillation rather than for making Champagne. In presses with a large surface area, the edges of the pomace are subjected to less pressure than the center, so the pomace is brought from the edges toward the center between each pressing. These presses may be automated (Valade and Pernot, 1994).

The must is separated in a similar way in the case of horizontal presses, which are, however, easier to use.

A 1993 regulation specifies that 4,000 kg of grape must produce 25.50 hl of clarified must, allowing for 2–4% sediment. Only the *cuvée* (20.50 hl) is used to make prestigious Champagnes, while the *taille* (5 hl) produces fruitier, faster-maturing wines that are generally included in non-vintage *brut* blends.

Analysis of the different pressings, as they come out of the press, shows significant variations in composition (Table 14.6). As pressing continues, total acidity decreases, as do both the tartaric and malic acid levels. Mineral content and pH increase, as do the phenolic content and color intensity, while the sugar level remains relatively constant. The aroma intensity and finesse of wines made from successive pressings also diminishes, but we do not have the means to analyze these changes.

It is essential to take great care when harvesting, pressing, clarifying, and separating the must to maximize the quality of the finished Champagne (Moncomble et al., 1991). Not only do these precautions make it possible to produce practically colorless must with very little sediment from red grapes, but they are also vital to preserve the finesse and quality of the end product.

Must clarification and fermentation

For the reasons outlined above, the must should be perfectly clarified prior to fermentation (Pernot and Valade, 1994). Generally, the must is clarified immediately after pressing at the pressing room near the vineyards. Some Champagne producers clarify the must again on arrival in the fermentation cellar. Sulfite is added (5–8 g/hl), and the sediment is left to settle naturally. In some

TABLE 14.6
Physicochemical Characteristics of Champagne Must (Valade and Blanck, 1989)

	Cuvée			Second pressing			Third pressing		
	1982	1985	1986	1982	1985	1986	1982	1985	1986
Density	1,072	1,075	1,073	1,071	1,075	1,072	1,071	1,075	1,073
Sugar (g/l)	163	171	164	161	171	163	161	170	165
Total acidity (g/l H_2SO_4)	8.1	8.5	9.2	6.8	6.9	8	5.9	6	6.4
Tartaric acid (g/l)	7.4	8.2	8.6	6.5	6.9	7.8	5.9	6.5	7.1
Malic acid (g/l)	6.6	7.7	8.6	5.7	6.8	7.8	5.1	6.5	7.1
pH	3.05	3.1	3.05	3.18	3.27	3.17	3.33	3.43	3.43
Potassium (mg/l)	1,790	1,650	1,965	1,965	1,995	2,385	2,160	2,320	3,175
Total nitrogen (mgN/l)	575	650	790	570	675	850	585	760	955
Available nitrogen (mgN/l)	202	327	420	192	327	460	185	345	490

years, depending on the condition of the harvested grapes, pectinases may be added to facilitate flocculation and settling.

Champagne musts contain large concentrations of nitrogen compounds, especially proteins, which contribute to the quality of the finished Champagne, especially persistence of foam. However, proteins are also involved in instability problems, leading to turbidity. Some producers add tannins (5 g/hl) to flocculate any unstable proteins. Bentonite may also be used for this purpose, at doses not exceeding 30–50 g/hl.

Traditionally, the first alcoholic fermentation (*bouillage*) took place in oak barrels, in cellars where the temperature was never higher than 15–20°C. Some producers still barrel-ferment their wines to enhance aroma complexity and flavor. However, most base wines are now fermented in coated steel or, above all, stainless steel vats, as they are easy to maintain at temperatures below 20°C. The aim is for the fermentation to continue smoothly and uninterruptedly until the residual reducing sugar content is below 2 g/l. Although the must is usually chaptalized, this is not always necessary as the aim is to achieve an alcohol content no higher than 10–11% vol. Completing fermentation is not usually a problem, especially as the must is systematically inoculated with selected yeasts.

Winemaking methods in other sparkling wine-producing areas are generally based on those used in Champagne, but may be simplified to reduce costs. Although other grape varieties do not have the same quality as the Chardonnay, Pinot Noir, and Pinot Meunier used in Champagne, they make suitable base wines, provided they are picked with a sufficiently high acidity level, e.g. Chenin Blanc in the Loire Valley, Ugni Blanc in Bordeaux, Mauzac in Limoux, Maccabeo in Catalonia, etc.

Malolactic fermentation

Champagne is an elegant, fruity wine that requires a certain level of acidity (around 6 g/l H_2SO_4) to ensure freshness and ageworthiness. The drop in acidity that occurs during the secondary fermentation must be taken into account in making the base wine (Section 14.3.3).

When Emile Peynaud (cited by Ribéreau-Gayon et al., 1976) analyzed Champagnes in the 1950s, he found that malolactic fermentation was uncommon and there was a considerable difference in flavor between wines with and without malic acid. He added that there had been many changes since that time (increase in pH, decrease in sulfiting, etc.), resulting in a greater vulnerability to bacterial activity and increasingly frequent uncontrolled malolactic fermentations, unheard of in the past.

There is still some discussion today concerning the beneficial effect of malolactic fermentation on Champagne aromas. If it is properly controlled, it improves the quality of acidic wines, especially Chardonnay, as the bacterial activity enhances their aromas (Section 13.7.6). In other cases, it may result in wines lacking freshness that age too rapidly and may even necessitate the addition of tartaric acid to raise the acidity level (Volume 2, Section 1.4.3).

To ensure microbiological stability and avoid the serious consequences of malolactic fermentation during secondary fermentation (*prise de mousse*) or bottle aging (*conservation sur lattes*), malic acid must be eliminated from the base wine prior to bottling. This solution has been found most effective and is the most widely used, although other methods are under investigation.

The factor with the greatest impact on malolactic fermentation is the sulfur dioxide content. It is inhibited by even small doses of free SO_2, since the bound SO_2 also plays a role. Malolactic fermentation becomes very difficult, or even impossible, at total SO_2 levels above 80–100 mg/l. Careful observation (Ribéreau-Gayon et al., 1976) showed a correlation between the total SO_2 content and malolactic fermentation in Champagne during the secondary fermentation in bottle. Current filtration technology makes it possible to sterile-bottle the base wine and inoculate it with pure yeast, thus avoiding malolactic fermentation in bottle without excessive use of SO_2. Lysozyme may also assist in stabilizing the base wine (Gerbaux et al., 1997; Pilatte et al., 2000).

Inversely, it is not always easy to start malolactic fermentation at the right time in white wines with very high acidity (Section 13.7.6), and it is essential to adjust the SO_2 content and temperature with that in mind. It is also possible to inoculate a properly prepared starter, but this is a rather laborious operation. Seeding with reactivated bacterial biomass, initially developed for red wines (Section 12.7.5), has considerably improved malolactic fermentation conditions in Champagne base wines (Laurent and Valade, 1993), and suitable products are now commercially available.

14.3.3 Secondary Fermentation in Bottle: The Champagne Method

Preparing and bottling a *cuvée* The newly fermented base wines are clarified, racked, filtered, and fined in the usual way. Base wines may be fined using isinglass (1.5–2.5 g/hl) or gelatin (4–7 g/hl), with or without tannins (2–4 g/hl) (Marchal et al., 1993). Tannin may be required to deal with the instability resulting from these wines' relatively high protein content, while it is essential to maintain a sufficient level of protein to give a high-quality bead. Moderate doses of bentonite may also assist in protein stabilization.

A *cuvée* is prepared by blending wines of different origins (vineyards classified 100, 86%, etc.), qualities, and, possibly, vintages. This is indispensable to maintain the quality and character identified with the producer from year to year and is still mainly determined by tasting. There is a single AOC for all Champagnes, while the other two appellations in the region, Côteaux Champenois and Rosé de Riceys, are only applicable to still wines. The hierarchy among Champagnes is mainly dependent on the selection of base wines used to blend the *cuvée*.

Once the *cuvée* has been blended, the base wine is cold-stabilized to prevent tartrate precipitation. In some cases, it may be fined just before or after cold stabilization.

Preparation for bottling may also include filtration. Barrel-fermented wines are adequately clarified by simple settling of the lees, but wines fermented in tanks always require filtration, especially immediately after cold stabilization.

During bottling, the tirage liqueur (syrup containing 500 g/l sucrose) is added for the secondary fermentation, calculated to produce the carbon dioxide required for a pressure of 5–6 bar at 10–12°C. Theoretically, 20 g/l sucrose must be fermented to produce 5 bar pressure. Table 14.7 indicates the quantities of sugar required to produce the desired pressure in the bottle after fermentation.

An active yeast starter, consisting of selected strains of *Saccharomyces cerevisiae*, is added at the same time to ensure that the secondary fermentation (*prise de mousse*) will be successfully completed in the bottle. This starter may consist of fermenting must, but it is better to make a preparation with active dried yeasts. Dried yeasts develop well in must that does not contain alcohol, and may be inoculated directly, particularly in white winemaking (Section 13.7.2). However, in a medium containing alcohol, e.g. when required to restart a stuck fermentation (Section 3.8.3), they must be reactivated prior to use so that they are in a suitable physiological condition. Laurent and Valade (1994) recommended an effective method for preparing dried yeasts for use in secondary fermentation in bottle. It is advisable to inoculate the bottle with an initial population of 1.5×10^6 cells/ml. Below that amount, fermentation is slower, and some sugar may remain unfermented, while above that level (e.g. 2×10^6 cells/ml), fermentation is faster, but some yeast strains may produce yeasty off-odors.

The *tirage* liqueur and yeast starter can be added in the tanks prior to bottling, or to each bottle individually during the bottling process, e.g. via two injections with metering pumps.

Other substances may also be added at the time of bottling (e.g. 3 g/hl bentonite or

TABLE 14.7

Sugar Content of the *Tirage* Liqueur According to the Pressure Required (Valade and Laurent, 2001)

Pressure required (bar)			
At 10°C	At 20°C	Sugar (sucrose) (g/l)	Sugar (glucose + fructose) (g/l)
4.0	5.4	16.0	17.0
4.5	6.1	18.0	19.1
5.0	6.8	20.0	21.2
5.5	7.5	22.0	23.3
6.0	8.2	24.0	25.0
6.5	8.8	26.0	27.6

0.2–0.7 g/hl alginate) to facilitate elimination of the yeast sediment when the bottles are disgorged.

Secondary alcoholic fermentation and aging on the lees

A plastic bead, known as a bidule, is inserted in the bottle, which is then closed with a crown cap. The bottles are stored in horizontal stacks (*en lattes*), interspersed with laths of wood to steady the layers. It is important that the bottles be placed horizontally, firstly to ensure that they remain airtight during fermentation and secondly to provide a maximum interface for exchanges between the wine and its lees. Fermentation takes one month, or sometimes longer, at the constant temperature of 11–12°C in underground cellars in Epernay and Reims. This slow, even, low-temperature fermentation is another quality factor in producing fine Champagne, especially the finesse and persistence of the foam when the bottle is opened. Carbon dioxide pressure increases gradually (Valade and Laurent, 1999), inhibiting yeast growth and slowing the fermentation rate, especially at low pH and high alcohol levels. Attempts have been made to enhance the fermentation rate by adding nutrients, with mixed results. It is more effective to increase the initial yeast inoculation, as well as to adapt the yeast strain and starter preparation conditions to the type of base wine.

The wine is still not proper "Champagne," even when the secondary fermentation in bottle is completed, i.e. all the sugar has been fermented. The wine spends a long period aging on its yeast sediment, which gradually decreases in volume and becomes more compact. The bottle must remain in a horizontal position to provide a maximum wine-sediment contact area. The yeasts release substances (Section 14.3.4) into the wine, initially by excretion and then by diffusion from the dead yeast cells. These are mainly amino acids, either synthesized by the yeast or previously absorbed from the wine. Autolysis involving cell wall enzymes has also been observed. All these complex phenomena play a significant role in Champagne quality. The improvement in quality during this stage is correlated to the composition of the base wines, which explains why other sparkling wines benefit less from aging on the lees. Non-vintage Champagne is aged on the lees for a minimum of 15 months, while the minimum for vintage Champagne is three years, but they may stay on their lees for up to eight years, or even longer for some top *cuvées*.

As long as the sparkling wine remains in contact with the yeast sediment under anaerobic conditions, the lees act as a redox buffer, and the wine is perfectly preserved. Champagne from bottles several decades old tasted in Champagne cellars were found to be in perfect condition, as they had never been disgorged. Once the bottles have been disgorged, not only does the Champagne stop improving, but there is also a risk of defects developing due to redox phenomena.

The main risk during storage, especially if bottles are exposed to light, is the development of off-odors in the wine. These reduction odors are due to the formation of thiol groups by photodegradation of sulfur amino acids naturally present in Champagne. The reaction, photosensitized by riboflavin (vitamin B2), produces methanethiol and dimethyl disulfide, which are responsible for the sensory defect known as lightstrike (Volume 2, Section 8.6.5) (Maujean and Seguin, 1983). Maujean et al. (1978) showed that this drop in redox potential due to light exposure only occurred in disgorged Champagne. The formation of thiol groups and the resulting lightstrike also depend on the reduction conditions of the wine prior to light exposure. This defect may be prevented by using glass bottles with low transmission values at wavelengths below 450 nm. Adding ascorbic acid, together with SO_2, just before the bottles are finally corked (when the dosage is added), is an effective preventive measure.

Riddling and removing the yeast sediment

The next stage consists in gathering the yeast sediment on the inside of the bidule. This is traditionally done by riddling the bottles on special racks, which hold the bottles neck down, at a variable angle. Riddling consists in turning the bottles with a slightly jerky movement, bringing them gradually to a vertical position, completely upside down, over a period of a month or more.

This operation takes a variable amount of time, generally from three weeks to one month, depending on the type of wine and its colloidal structure, as well as the type of yeast and its capacity to form clumps. Riddling is an awkward stage in the Champagne production chain, due to the space required for the riddling racks, the labor-intensive process, and the fact that the bottles are immobilized for a relatively long period of time. A great deal of work has been done to simplify this operation. The first approach consists in adding various substances to the wine in tank, prior to bottling, intended to facilitate settling of the yeast sediment. While the results have not been negligible, this technique has not made any great improvement in the process.

More significant progress has been made by reproducing the intermittent movements of riddling on the scale of a box pallet (several hundred bottles). Each box pallet is installed on a movable base, which is tilted manually to change the angle of the bottles, gradually bringing them into the vertical neck-down position. This system may be mechanized and programmed (gyropalette) to riddle the bottles much more efficiently, completing the cycle in one week instead of the one month required for manual riddling. This system is now widely used, in spite of the high initial investment required.

Another approach to simplifying riddling consists in using yeast enclosed in tiny calcium alginate beads for the secondary fermentation in bottle (Duteurtre *et al.*, 1990; Valade and Rinville, 1991). The sediment settles in the bidule almost immediately when the bottle is turned upside down, and riddling is no longer necessary. Of course, this assumes that fermentation and aging on the lees continue normally with the enclosed yeasts. The second condition is that yeast cell growth does not burst the beads, producing a powdery deposit that is difficult to eliminate. This problem is now avoided by using a double coating of alginate on the beads. Several million bottles have now been processed with enclosed yeast, and work is continuing to monitor the aging and development of the wines. Once this technique has been demonstrated not to affect quality, it will be possible to envisage its use in large-scale production.

Disgorging and final corking

Once the sediment has settled in the bidule, the wine is disgorged. In the past, this operation was done manually by removing the cap quickly while raising the bottle slightly so that the few milliliters of wine containing the sediment would be expelled without emptying the bottle or losing too much carbon dioxide pressure.

Nowadays, the bottle necks are almost always frozen prior to disgorging, in an automated system that also adds the dosage liqueur, corks the bottles, and fits the wire closure. The bottles are held upside down, and the necks plunged into a low-temperature salt solution that freezes about 2 cm of wine above the cap, trapping the sediment. When the bottles are turned upright, the cap is removed, and the frozen plug with the bidule is expelled.

The bottle is then topped up with dosage liqueur (*liqueur d'expédition*), a syrup made of reserved wine containing approximately 600 g/l of sugar, used to adjust the final sugar level of the Champagne. "Brut" Champagne generally has 10–15 g/l (1–1.5% dosage), while "demi-sec" has 40 g/l (4% dosage).

The dosage liqueur can be acidified with citric acid, if necessary. It also contains the quantity of sulfur dioxide required to eliminate any dissolved oxygen and may be supplemented with ascorbic acid (50 mg/l). This offsets the sudden oxidative effect of disgorging: the redox potential may increase by 150 mV, or even more, depending on the redox buffer capacity of the wine.

According to Peynaud (cited by Ribéreau-Gayon et al., 1976), "dosage is not simply a matter of sweetening the wine, but of improving it. The quality of the dosage liqueur, the way it is aged, the types of wine used, the quality of the sugar, and the preparation formula all play a major role in the quality of the finished product." The dosage liqueur contributes to the overall flavor balance.

14.3.4 Composition of Champagne Wines

Analysis of Champagne wines

The analysis results in Table 14.8 show the effect of bottle fermentation on the wine's composition. The alcohol content increases by 1.3% vol. during the secondary fermentation and may drop by a few tenths during preparation for shipping, depending on the composition of the dosage liqueur. If the base wine is not properly cold-stabilized, total acidity may decrease slightly during the secondary fermentation due to the precipitation of potassium hydrogen tartrate and by the breakdown of small amounts of

TABLE 14.8

Comparison of Wine Composition Before and After Second Fermentation (1993 Vintage, Mean Analysis Values) (Tribaut-Sohier and Valade, 1994)

	Blend at the time of bottling	After secondary fermentation and addition of dosage for brut wine
Density at 20°C (g/dm³)	990.5	993.9
Alcohol at 20°C (% vol.)	11.0	12.2
Sugars (g/l)	1.3	12.7
pH	3.02	3.05
Total acidity (g/l H_2SO_4)	4.7	4.7
Volatile acidity (g/l H_2SO_4)	0.27	0.3
Free SO_2 (mg/l)	8	8
Total SO_2 (mg/l)	38	58
Tartaric acid (g/l)	3.5	3.2
Malic acid (g/l)	0.2	0.2
Potassium (mg/l)	330	325
Calcium (mg/l)	85	70
Copper (mg/l)	0.17	0.13
Iron (mg/l)	2.1	2.8
Sodium (mg/l)	8	12
Magnesium (mg/l)	60	70
Total nitrogen (mgN/l)	303	410
Ammonia nitrogen (mg/l)	13	20
OD 520 nm	0.038	0.028
OD 420 nm	0.087	0.106
Color intensity	0.125	0.134
Shade	2.59	3.89
Conductivity (mS/cm) at 20°C	1.32	1.32
Saturation temperature at 20°C	10°C	12.1°C

residual malic acid under the action of the yeasts. Otherwise, there is little variation in the acid content unless a small amount of citric acid is added to the dosage liqueur. This decrease in overall acidity results in an increase in pH.

One of the most significant characteristics of Champagne must and wine is their high nitrogen content, especially in the form of amino acids (Desportes *et al.*, 2000) (Table 14.9), which facilitates the initial and secondary fermentations. The amino acid content of Champagne is twice or thrice as high as that of Bordeaux wines (Ribéreau-Gayon *et al.*, 1976). The same authors gave the following analysis results for Champagne and Bordeaux: 462 and 184 mg/l of total nitrogen, 11.2 and 6.3 mg/l of ammonia nitrogen, and 216 and 100 mg/l of amino acid nitrogen, respectively.

The total nitrogen content of Champagne varies from 150 to 600 mg/l (Maujean *et al.*, 1990), and that of the must is considerably higher. Chardonnay and Pinot Noir/Meunier grape varieties have a high nitrogen content, and Champagne is

TABLE 14.9

Average Amino Acid Content of Base Wines Made from Different Champagne Grape Varieties (Assayed on Sulfonic Acid Resin and Detected Using Ninhydrin) (Results in Milligram per Liter) (Desportes *et al.*, 2000)

	Chardonnay	Pinot Meunier	Pinot Noir
Aspartic acid	6	28	11
Threonine	14	69	16
Serine	14	43	10
Asparagine	N.I.	27	24
Glutamic acid	38	55	30
Glutamine	67	82	36
Proline	777	165	222
Glycine	27	15	13
Alanine	118	209	95
Citrulline	38	15	11
Valine	32	18	9
Cysteine	24	N.I.	N.I.
Methionine	7	9	9
Isoleucine	4	12	5
Leucine	14	22	21
Tyrosine	41	28	20
β-Alanine	11	2	3
Phenylalanine	14	18	13
γ-N-butyric acid	67	} 78	} 74
Ethanolamine	4		
Ornithine	7	11	10
Lysine	8	11	16
Histidine	8	4	6
Arginine	20	321	105
Total	1,360	1,242	759
Proline/arginine	38.85	0.51	2.11

N.I., not integrated.

the winegrowing region where it reaches the highest levels. Table 14.9 compares the mean amino acid content of base wines made from different Champagne grape varieties.

According to the literature, Champagne must contains 25–100 mg/l of proteins (in BSA equivalent), while the level is considerably lower in base wine: 14–32 mg/l (Tusseau and Van Laer, 1993). In harvested grapes, 75% of total nitrogen is in amino acid form, while it accounts for 95% in new base wine (Tusseau et al., 1989).

Several proteins have molecular masses between 20 and 30 Kda, while one with a molecular mass of 62 Kda is probably combined with sugars (glycoproteins). They have isoelectric points between 2.5 and 6.5 (Brissonnet and Maujean, 1993).

Besides proteins and polypeptides from the must, the sparkling properties of Champagne also involve colloidal carbohydrates (polysaccharides and glycoproteins) (Marchal et al., 1996; Berthier et al., 1999) released from the yeast cell walls during aging on the lees (Feuillat et al., 1988; Tusseau and Van Laer, 1993). This yeast autolysis is certainly accompanied by more radical transformations. The amino acid content of some sparkling wines has been reported to increase depending on the contact time, and stirring the yeast back into suspension has been recommended to enhance this phenomenon.

Boidron et al. (1969) compared the volatile fermentation compounds involved in Champagne aromas with those found in other sparkling wines. Champagnes characteristically have lower concentrations of methanol and higher alcohols, propanol, ethyl butyrate, and isoamyl acetate, which have a negative effect on aroma. This is probably due to the winemaking conditions (e.g. temperature). Other more positive aroma compounds such as ethyl caprate and ethyl lactate (related to malolactic fermentation) are more abundant in Champagne.

In the past, it was relatively common to find large residual yeast populations in Champagne bottles. Yeast counts between 0.16 and 4.8×10^3 cells/ml have been found in Champagnes on the market (Ribéreau-Gayon et al., 1976). Current riddling and disgorging techniques, particularly sedimentation additives and fine-tuned riddling programs, have made substantial progress in eliminating residual yeasts.

Effervescence in Champagne wines

The high carbon dioxide pressure responsible for effervescence is an essential characteristic of Champagne.

When Champagne is poured into a glass, the foam, which is an important quality factor, appears even before the liquid. It is well known in wine tasting that a poor initial visual impression has a negative impact on the overall assessment, and this is certainly the case with the foam of a sparkling wine (Robillard, 2002). A good-quality bead consists of tiny bubbles that remain separate and spherical in shape. Large bubbles produce an unattractive, grayish bead that usually disappears very rapidly.

Effervescence also reveals the wine's aromas, as the bubbles contain odor compounds in addition to carbon dioxide (Maujean, 1996).

It is, therefore, important to consider the criteria for the formation and stability of foam in sparkling wines. The following analysis is based on a 1997 review by A. Maujean (Laboratory of Enology, Reims University) and B. Robillard (Moët et Chandon Research Laboratory), as well as several other publications (Maujean, 1989; Robillard et al., 1993; Liger-Belair and Jeandet, 2002).

The bubbles in sparkling wine are due to carbon dioxide, formed during the secondary fermentation and dissolved in the wine. A bubble of CO_2 must push the surrounding molecules apart before it can emerge. A great deal of energy is required to form a liquid/CO_2 interface, but this is minimized by nucleation phenomena.

Bubbles may be formed directly from dissolved gas (induced homogeneous nucleation). When Champagne is shaken up, e.g. during shipment, parent bubbles produce smaller bubbles, some of which are stabilized by contact with proteins and float to the surface. The drop in pressure when the bottle is opened causes them to explode, producing other smaller bubbles, which explode in turn, and so on. This chain reaction is responsible for a violent gush of wine, which may leave the bottle half-empty (Maujean, 1996).

Bubbles are more usually formed by adsorption of the gas on a solid particle (induced heterogeneous nucleation). It has been demonstrated that a minimum radius of 0.25 µm is required for the bubble's internal pressure to be sufficiently low in relation to that of the wine to enable the bubble to grow and rise through the liquid. Plastics have a higher surface energy than glass, creating a greater affinity for CO_2, so that bubbles coming off a plastic surface will be larger than those released in a glass vessel.

Several factors are involved in effervescence kinetics following degassing (Casey, 1987; Liger-Belair et al., 2000; Liger-Belair, 2002, 2003).

The first factor is the physical nature of the solid surface (particles in suspension or vessel wall), particularly the number and radius of microcavities on which the bubbles are formed, detaching themselves once they have reached a certain diameter. This produces a bead, or line of bubbles, which always rise from the same spot. Of course, the microcavities must be hydrophobic or they would be filled with wine.

Other factors, such as viscosity and chemical composition, are inherent to the liquid. It is quite probable that some carbon dioxide molecules are immobilized by binding with other substances. Electrostatic interactions may also lead to the adsorption of CO_2 on the surface of macromolecules, as shown by the significant changes in effervescence kinetics when proteins or polysaccharides are added to synthetic wines (Maujean et al., 1988).

The instability of the foam is defined by three parameters:

1. Swelling bubbles: The gas from small bubbles is absorbed into larger bubbles. This results in a coarse irregular foam with an unattractive appearance.
2. Draining: This refers to the liquid that drains out of the foam over time. It leads to a reduction in foam volume and a distortion in the shape of the bubbles. The foam gradually dries out (e.g. as on the head of a beer glass).
3. Coalescence: A break in the film between two smaller bubbles produces a larger one, resulting in a coarse foam that disappears quickly.

Several experimental processes have been proposed for measuring the spontaneous or forced degassing kinetics in sparkling wines (Maujean et al., 1988), as well as for assessing the persistence of foam (Maujean et al., 1990; Robillard et al., 1993).

The Mosalux apparatus (Maujean et al., 1990) is used to determine three characteristics of sparkling wine bubbles:

1. *Foamability*, or maximum foam depth, expresses the liquid's capacity to contain gas once it starts effervescing. It is visible in the foam formed when the wine is poured into a glass.
2. *Foam height* describes the constant height of the foam when the liquid is bubbling in the glass and corresponds to the height of the bead in the glass.
3. *Foam stability* measures the time required for the foam to disappear once the liquid stops effervescing. This parameter is only of theoretical interest in laboratory work.

Measurements show that foamability and foam stability are mutually independent—wines may produce a lot of foam, but it is not necessarily very stable. A close correlation has been observed between foamability and protein content. A decrease in protein content of a few mg/l can lead to a 50% drop in foamability (Malvy et al., 1994). However, Maujean et al. (1990) did not find any simple correlation between protein content and foam stability.

The solubility of proteins affects their impact on foaming in sparkling wine. Hydrophobic proteins may also be adsorbed at the liquid–gas interface, on the "bubble skin," stabilizing it by decreasing surface tension. Proteins with lower molecular weights are more rapidly adsorbed at the interface. Proteins that have an effect on effervescence have isoelectric points in the vicinity of wine pH (2.5–3.9). This characteristic does not promote solubility but makes the protein more hydrophobic. Thus, proteins affect foamability by changing the surface tension when they are adsorbed at the liquid–gas interface of the bubbles. Glycoproteins have an even greater impact on foaming, as the hydrophilic glycoside fraction increases the viscosity of the liquid film between the bubbles and reduces the draining of the liquid phase. Although yeast mannoproteins are less hydrophobic than plant glycoproteins, they are present in large quantities in Champagne wines and apparently contribute to their stability (Feuillat et al., 1988).

Foam depth decreases during aging on the lees but is largely compensated by the improvement in stability.

It is well known that the various stages in the winemaking process have an impact on foam quality. Robillard et al. (1993) examined the impact of filtering base wines. This operation removes solid or colloidal particles that provide a base for bubble formation (nucleation), considerably reducing the intensity of effervescence and thus the foam stability of the corresponding sparkling wine. The smaller the pores of the filter medium, the more marked the impact on foam stability.

Treatment with plant charcoal or bentonite also causes a considerable decrease in foamability, related to the reduction in protein content. In contrast, fining with gelatin, combined with silica gel or tannin, improves foaming qualities.

14.3.5 Other Secondary Fermentation Processes

Transfer method

The aim of this method is to benefit from the advantages of secondary fermentation in small volumes, with aging on the yeast lees, while avoiding the problems associated with riddling and disgorging. Once the secondary fermentation is completed, the wine is filtered and transferred to another bottle. This process is not permitted for Champagne, although there is a tolerance for quarter-bottles (splits) that are filled after filtration, following secondary fermentation in full-size bottles. This process is still occasionally used to prepare half-bottles, but its use is due to be prohibited in the near future.

After secondary fermentation and aging, the bottles are simply taken to the racking area. They are emptied automatically into a metal vat, under carbon dioxide pressure to prevent degassing. The tank itself is filled with CO_2 at a pressure equivalent to that created in the bottles by fermentation.

The wine in the vat is refrigerated to −5°C by circulating liquid coolant through a suitable heat exchanger. This makes the CO_2 more soluble. Dosage liqueur is also added to the vat, and the wine is left to rest for a few days. It is then plate-filtered to remove all the yeasts and bottled. As the wine is kept at low temperatures under pressurized carbon dioxide, it retains all the dissolved CO_2.

This system has a number of advantages. It eliminates the labor costs of riddling and disgorging, as well as the time the wine is immobilized on the riddling racks. Dosage liqueur is much more evenly distributed. It is also possible to blend several batches of wine after the secondary fermentation to obtain the desired quality.

Cold stabilization prevents tartrate precipitation, and filtration ensures that the yeasts are completely eliminated, leaving the wine perfectly clear.

If these operations are properly conducted, they give satisfactory results. However, wines made by the Champagne method were always preferred in comparative tastings, probably due to the fact that small amounts of oxygen are dissolved in the wine during transfer operations, however carefully they are controlled. It has also been demonstrated that exchanges occur between the carbon dioxide molecules resulting from the secondary fermentation and the industrial-grade gas used to protect the wine during the transfer process. Finally, filtration may modify the wine's foaming qualities.

The Charmat (closed tank) method
Secondary fermentation in the bottle is technically demanding and is, therefore, only justified for high-quality products made from fine base wines that are likely to benefit from aging on the yeast lees.

As long aging is not economically viable for cheaper products, a simpler, less expensive process (the Charmat method) has been developed to produce sparkling wine from lower quality grapes.

Figure 14.2 shows a simplified diagram of a system for secondary fermentation in a closed tank. The various base wines are blended and transferred to the secondary fermentation tank (C), and yeast starter (tank A) and syrup (tank B) are added to provide the quantity of sugar required for the secondary fermentation and the dosage of the finished product. The fermentation tank (C) is equipped with heating and cooling systems to maintain a temperature of 20–25°C. When pressure in the tank reaches 5 bar, fermentation is stopped by reducing the temperature and sulfiting slightly. The wine is then transferred to a refrigerated tank (D) and kept at −5°C for several days for cold stabilization.

The wine is filtered and then transferred to another tank (E), connected to the bottling line. The entire operation is carried

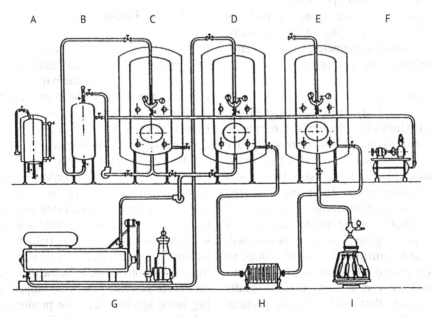

FIGURE 14.2 A closed tank facility for sparkling wine: A, yeast starter preparation tank; B, sugar addition tank equipped with a mixing system; C, pressure-rated secondary fermentation tank; D, refrigeration tank; E, racking tank; F, air compressor for isobaric racking; G, refrigeration unit; H, filter; I, bottler.

out under pressurized carbon dioxide (F) to prevent degassing.

These large-volume processes certainly cannot achieve the same quality as bottle fermentation. This is partly due to the fact that the wine is not aged in contact with the lees, as a sufficient level of interaction can only be achieved in a small container. The quality of the grapes used and the speed of the process also have an impact. In the Charmat method, the yeast is often eliminated after only a few days' fermentation to reduce costs. In view of the other factors involved, it is by no means obvious that aging on the lees would improve quality. Systems have, however, been developed for maintaining the wine in tank on its lees, stirring them into suspension to accelerate exchanges. The success of this operation depends on the quality of the base wine and keeping fermentation temperatures low to slow down the reactions.

The Charmat method may give better results than bottle fermentation in hot climates, as it preserves the base wine's freshness and fruit. Finally, the Charmat method is most appropriate for producing sparkling wines from aromatic varieties such as Muscat, as aging on the lees attenuates the Muscat character, without significantly improving quality.

Asti Spumante

This is probably the most famous sparkling Muscat. Unfortunately, when the must is fermented to produce a completely dry wine, it loses all the distinctive grape aromas and has an unpleasant, bitter taste. Long experience has led to the development of a low-temperature fermentation process that is interrupted every time it starts speeding up. The must is clarified, fined, and centrifuged as many times as necessary until the yeast and available nitrogen have run out. Analytical results show clearly that total nitrogen, particularly available nitrogen, decreases every time the fermenting must is filtered, probably due to fixation on yeast cells. The resulting wine is relatively stable, due to nitrogen deficiency, with 5–7% alcohol by volume and 80–120 g/l sugar. This wine used to be put into bottles for secondary fermentation, but it was irregular and uncontrollable. Secondary fermentation now takes place in a sealed tank (Charmat process), using a blend of wines from different vineyards, clarified by fining with gelatin/tannin and filtration. Fermentation starts at 18–20°C and is then slowed down by reducing the temperature. When the pressure reaches 5 bar, the wine is chilled to 0°C and clarified again. The temperature is then reduced to −4°C for 10–15 days to stabilize the wine. Following further filtration or centrifugation, bottling takes place in an environment pressurized with CO_2 to prevent degassing. Some producers use sterile filtration, and others pasteurize the wine to prevent it from fermenting again in bottle. The finished Asti Spumante contains 6–9% alcohol by volume and 60–100 g/l sugar. A number of other sparkling wines are produced using similar methods.

14.4 Fortified Wines

14.4.1 Introduction

Fortified wines are characterized by their high concentrations of alcohol and sugar. They are derived from the partial fermentation of fresh grapes or grape juice; the addition of alcohol prematurely stops the fermentation. This fortification can be done in one step or in several.

These wines were clearly created in the past in response to technical problems encountered in hot regions. Sugar-rich grapes and high temperatures resulted in explosive fermentations, easily leading to stuck fermentations. The partially fermented wine was unstable, especially since sulfiting was far from mastered at the time. Lactic acid bacteria subsequently developed, causing lactic spoilage and the production of volatile acidity (Section 3.8). The addition of alcohol during fermentation was a simple means of stabilizing the wine and produced an alcoholic and sweet product with an agreeable taste. As late as the 1960s, these wines represented a significant part of Californian and Australian production.

Today, other means can be used to produce standard types of wines under these climate conditions. Grapes are harvested at sugar concentrations compatible with complete fermentations, even in relatively hot climates. Fermentations are better controlled through sulfiting, aeration, and temperature control. They are also complete. Malolactic fermentation can now occur without bacterial spoilage.

Due to greater demand, traditional dry red and white wines have replaced fortified wines at many wineries. Today, only the most famous fortified wines remain. Specific natural factors and suitable technology permit these wines to develop their fine aromas and rich flavors. French *vins doux naturels* and Port wines are certainly among the most prestigious fortified wines, but other fortified wines from Greece, Italy, and other Mediterranean countries also exist.

The OIV defines fortified wines as "special wines having a total alcohol content (both potential and actual) above 17.5% vol. and an alcohol content between 15 and 22% vol." Two types of fortified wines exist:

1. Spirituous wines receive only brandy or rectified food-quality alcohol during fermentation.
2. Sweet wines can receive concentrated must or *mistelle* in addition to brandy or alcohol.

In both cases, the natural alcohol potential of the grape juice must be at least 12% vol. At least 4% vol. of the alcohol in the final product must come from fermentation.

Aging conditions, up to bottling, vary depending on the type of fortified wine. Due to their high alcohol content, these wines are very resistant to oxidative phenomena. Some actually develop their desired characteristic through a certain degree of oxidation. These wines undergo a true barrel aging. Other finer and more delicate fortified wines are protected from air and are bottle-aged.

14.4.2 French Fortified Wines (*Vins Doux Naturels*)

Definition These famous wines (VDN) are found in a dozen *appellations* across three regions in the south of France (Brugirard et al., 1991). Banyuls, Rivesaltes, Maury, and various Muscat *appellations* are among the best known. These wines fall under the OIV definition of fortified wines, but French legislation taxes the two types of fortified wines mentioned in Section 14.4.1 differently. With VDN wines, only the added alcohol is taxed. The other fortified wines are taxed on the total alcohol, including the alcohol from fermentation of the must.

Production conditions are more constraining in France than as specified by the OIV. Not only is the area covered by each appellation clearly defined, but the grape varieties are also specified. Non-Muscat varieties are Grenache, Macabeu (Maccabeo), and Malvoisie, while only Muscat of Alexandria and Muscat à Petits Grains are permitted in fortified Muscat wines.

Crop yield limits are set at 40 hl/ha, with only 30 hl/ha allowed to be used for making VDN. Grape juice must contain at least 252 g of sugar per liter (approximately 14.5% vol. potential alcohol). The proportion of alcohol added at the time of fortification must be between 5 and 10% of the must volume. The must is fortified when the fermentation has already transformed a little more than half of the natural sugar. The final product must contain between 15 and 18% vol. alcohol content and at least 21.5% vol. total alcohol. Total alcohol includes the alcohol and potential alcohol, which corresponds to the quantity of alcohol that the residual sugar could produce by fermentation:

$$\text{Residual sugar content}(g/l)/16.83 = \text{potential alcohol}(\%\text{vol.})$$

The minimum residual sugar content varies from 59 to 125 g/l, depending on the appellation.

The initial must concentration and the percentage of added alcohol are verified by the following relationship P/α, where P is residual sugar weight (g/l) and α is polarimetric deviation, which depends on the proportion of glucose and fructose, itself related to the quantity of sugar fermented.

In grape juice, the glucose/fructose (G/F) ratio is equal to 1. Glucose diminishes more rapidly than fructose during fermentation. A French fortified wine must have a P/α of between -2.00 and -3.00. Fraud is suspected below -3.5. A fortified wine artificially made from dry wine, alcohol, and sucrose or concentrated must (G/F = 1) would have a P/α of -5.23.

Table 14.10 provides supplemental information concerning the chemical composition of French fortified wines.

Winemaking

Several types of French fortified wines (VDN) exist. The white VDNs are made from white or gray Grenache or Macabeu grapes. They do not generally undergo skin contact, but occasionally have a short skin contact time. They are light, fruity non-oxidized wines made to be drunk young.

Red VDNs are made with skin contact. The juice and pomace are generally separated after several days in tank. Fortification most often occurs on the separated juice, but in certain cases the alcohol is added to the pomace, and skin contact is continued for 10–15 days. Richly colored fortified wines with high concentrations of dry extract are obtained by this alcoholic maceration. These wines are capable of being aged for a long time.

TABLE 14.10

Analytical Characteristics of French Fortified Wines (VDN) (Brugirard et al., 1991)

	Minimum values	Average values	Maximum values
Alcohol content (% vol.)	15.0	15.0–17.0	18.0
Total alcohol (% vol.)	21.5	21.5–22.5	23.0
Specific gravity at 20°C	1.010	1.015–1.030	1.035
Sugar (g/l)	45	70–125	150
P/α (15°C)	−1.5	−2.0 to −2.5	−3.5
Total dry extract (g/l)	90	110–140	170
Reduced dry extract (g/l)	18	20–26	40
Ashes (g/l)	1.4	1.8–2.5	3.5
Alkalinity of ash (g/l K_2CO_3)	1.3	1.2–2.2	3.1
Total acidity (g/l H_2SO_4)	2.0	0.30–3.5	5.0
Volatile acidity (g/l H_2SO_4)	0.15	0.30–0.60	Legal
pH (20°C)	2.90	3.60–3.80	4.20
Tannin (Folin index):			
White VDN	15	25–40	55
Red VDN	20	30–50	70
Aldehydes (mg/l)	25	60–120	150
Higher alcohols (mg/l)	50	70–90	150
Glycerol (g/l)	3.0	6.0–10.0	12.0
Butylene glycol (g/l)	0.30	0.50–0.80	1.20
Lactic acid (g/l)	0.19	0.30–0.4	0.63
Free SO_2 (mg/l)	0	0–15	20
Total SO_2 (mg/l)	Traces	100–150	Legal

After separating the must, Muscat wines are made similarly to white wines. However, skin contact increases aroma extraction; making these wines therefore requires a lot of care to respect the finesse of the aromas.

Grape maturity is regularly assessed to determine the harvest date according to the variety. During overripening, the maturity must be carefully monitored because sugar concentrations may increase sharply to reach 250–270 g/l (15–16% vol. potential alcohol). Acidity also diminishes considerably. The full aroma potential of Muscat wines is obtained at a sugar concentration of around 225 g/l. *Botrytis cinerea* negatively affects fortified wine quality, especially in the case of Muscat. Botrytized grapes should not have any skin contact.

The first step of winemaking with skin contact consists in moderately crushing and destemming the grapes. The grapes are then transferred to the tank and sulfited at 5–10 g/hl. The fermentation temperature is set at approximately 30°C to favor maceration. Skin contact times vary from two to eight days, if the fortification occurs after must separation. In this case, the fermentation speed should be reduced beforehand. Wines are macerated for 8–15 days when continuing the skin contact after fortification.

When there is no skin contact, the grapes are drained and pressed to extract the juice, according to traditional white winemaking methods (Section 13.3). Immediately after extraction, the juice is stabilized by sulfiting at 5–10 g/hl and preferably refrigeration. The must is then clarified by natural settling and racking or centrifugation. Yeast starter may be used, and fermentation temperatures are kept relatively low, 20–25°C (and even 18°C for Muscat), to avoid loss of aroma.

Fortification (Mutage)

The addition of alcohol to fermenting must stops yeast activity, increases the dissolution of phenolic compounds during maceration, and provokes the precipitation of insoluble substances. A near-neutral wine brandy is used. The alcohol addition may be done in several steps to slow and spread out the fermentation phenomena.

The moment of fortification is chosen according to the density, which decreases during fermentation. The density must not drop below a certain established limit, called the fortification point. Choosing the correct fortification point is essential to wine quality. The wine must have a sugar concentration corresponding to the type of product desired and in compliance with legislation.

Fortification tables are used to achieve the exact alcohol content required, using either wine spirits at a minimum of 96.0% alcohol by volume or a blend of spirits and must. The addition of wine spirits is done with either 90% vol. alcohol or with varied blends of alcohol and must. The second form of addition arose from the need to have a tax official present for tax purposes when using alcohol. In the past, wines were stabilized with high SO_2 doses to stop the fermentation while waiting for the authorization to use alcohol. Nowadays, the wine spirits are denatured by mixing with must that has just started fermenting, in the presence of a government inspector. The diluted spirits have an alcohol content between 40 and 50% by vol. and can be used freely for fortification.

It is recommended to stop fermentation before fortification (*mutage*) by refrigerating the must or eliminating the yeasts by centrifugation or filtration.

Sulfiting destined to neutralize the acetaldehyde formed and to block oxidations definitively stabilizes the wine. A free SO_2 concentration between 8 and 10 mg/l should be maintained. Approximately 10 g of SO_2 per hectoliter should be added, considering the high pH (3.5–4.0), sugar concentration, and alcohol content of these wines.

Conservation and aging

Due to their diversity, many storage and aging methods exist for these VDN wines. All are generally aged for a year in tank, undergoing repeated rackings to ensure clarification. Different methods specific to each type of wine are used after this period.

Muscat wines are stored in tanks until bottling. Precautions are taken to avoid oxidation and to protect aromas: 15–17°C temperature, sufficient humidity, use of inert gas, etc.

After a selection based on tasting, many red VDN wines, having undergone skin contact, are placed in 6 hl casks exposed to the sun. Oxidation phenomena cause these wines to take on an amber tint and characteristic aromas. The wines are often fined and cold-stabilized before being placed in casks. Carrying out these operations at the time of bottling could thin the wine. A simple filtration at this time is preferable. A once-traditional method to obtain the same oxidative transformations consists in leaving glass carboys with a slight headspace outside, exposed to natural climate variations, but this is now rarely used.

The finest and most delicate white and red VDN wines can be matured in 225 l oak barrels in cellars at moderate temperatures (15–17°C) without any particular oxidative phenomena, according to traditional fine winemaking methods. The wine is matured for approximately 30 months and bottled after fining with gelatin. Reduction phenomena after the wine is bottled are responsible for the actual aging process.

Rancio wines are made traditionally and locally. The production of these wines is not codified. The method consists in maintaining a 6 hl barrel partially filled. Each year, wine is removed from the barrel to be bottled and replaced with newer wine.

VDN wines are subject to the same clarification and stabilization problems as other wines. Iron haze, protein haze, tartrate deposits, and colored matter can cloud the wines. Standard preventive measures can help to avoid these problems. Enzymatic browning is another type of spoilage linked to grape rot.

The high alcohol content of these wines gives them a certain level of microbial stability, but spoilage is still possible due to their high sugar concentration and elevated pH. Some yeasts tolerate 16–17% vol. ethanol and are capable of causing refermentations. In addition, these yeasts resist high concentrations of free sulfur dioxide. Particular strains of *Lactobacillus hilgardii* have been identified in certain French VDNs. They are apt to develop, provoking lactic spoilage, which is responsible for abundant deposits and sensory defects.

Standard operations lower the risks of defects—hygiene, fining, filtration, sensible use of sulfur dioxide, etc.—but pasteurization is the only treatment that completely eliminates germs and stabilizes wine. The correct use of this method does not cause organoleptic modifications, even with Muscat. Sterile filtration can also be used.

14.4.3 Port Wines

Production conditions Port wines come from the steeply sloping Douro region in Portugal (Ribéreau-Gayon *et al.*, 1976; Barros, 1991). The schistous soil, the jagged relief, the large temperature variations between seasons, low rainfall, and intense sunlight characterize the Douro. These conditions lead to highly aromatic and highly pigmented grapes with high concentration of sugar and phenolic compounds. The *terroirs* are ranked on a decreasing scale from A to F according to soil nature, grape variety, vine age, altitude, exposition, etc. There is a great diversity of cultivated varieties in this region (15 red and 6 white). The grapes are picked very ripe but are not dried on the vine. They are sorted very carefully to eliminate bad grape clusters and spoiled grapes. The must, with a minimum of 11% potential alcohol by volume, but which usually contains 12–14% vol., is sulfited (9–10 g/hl) and may be acidified, if necessary.

A relatively slow partial fermentation is sought. Extraction of skin components occurs during a concurrent maceration. The wine was traditionally fermented in *lagares*,

80-cm-high granite vats containing 2.5–110 hl, an ideal shape for ensuring that all the grapes would be perfectly crushed. The grapes were trodden for several hours each day until the third day of skin contact. The fermentation occurred simultaneously. The pomace cap was immersed by mechanical means at regular intervals. When the desired density was attained, the *lagare* was opened, and the wine flowed into the casks. Brandy was added to the wine in casks to stop the fermentation and raise the alcohol content to 18–19% vol.

Today, most wines are made in modern wineries, and the crushing and skin contact operations are mechanized. The open or closed tanks are equipped with automatic pump-over and mechanical mixing systems. The manual work has all but disappeared. These perfectly controlled technical modifications have improved and regularized the quality of Port wine while increasing profitability.

Upon arrival at the winery, the grape crop is destemmed and carefully crushed to facilitate skin contact. The must is sulfited but generally not inoculated with yeasts to avoid explosive fermentations. The temperature is maintained at around 30°C during fermentation.

After reaching 4–5% vol. alcohol, the fermenting must drained from the tank is clarified, possibly by a rotating filter, before being fortified. Correctly choosing the fortification point is essential to the quality of Port wine and to obtaining the level of sweetness desired. The quality of the Port also depends on the quality of the brandy used for fortification. All brandies used are submitted to analytical and taste tests; they contain 77–78% vol. alcohol. Pneumatic and mechanical horizontal presses are currently replacing traditional vertical presses, since the former are easier to use. Pressing is moderate. During fortification, a fraction of the tannin- and color-rich press wine is added to the free-run wine.

Maturation and characteristics of Port wines

Figure 14.3 summarizes the maturation and aging process of the different types of Port wines. During the winter following the harvest, after the first racking, the wines are ranked according to taste. The best batches, during an exceptional year, may be reserved to be declared as Vintage Port, but most wines are blended.

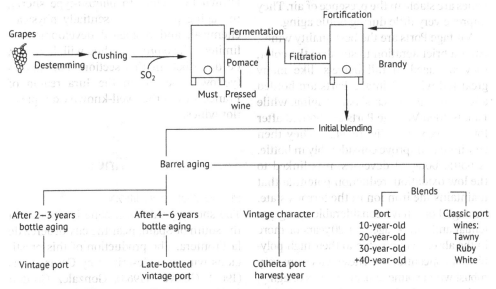

FIGURE 14.3 Flow diagram for the production of various styles of Port wine (Barros, 1991).

The blends are aged in 5–6 hl oak barrels (*pipas*) for several years under oxidative conditions that maintain a high oxidation–reduction potential. Metal ions, in particular copper and iron, play an essential role in polyphenol oxidation. Even in the bottle, these wines conserve a high oxidation–reduction potential. The iron remains in its ferric state, as if all of the reducing components had been destroyed by oxygen. Their prolonged oxidation and intense esterification give these wines a rich and complex bouquet.

During aging, the tannins become softer as they polymerize or combine with anthocyanins, while coloring matter precipitates and the color changes. The less oxidized Ruby Ports maintain the fruitiness and robustness of young wines. They have a more or less dark red color. The older, more oxidized Tawny Ports are golden red or golden.

White Ports undergo a certain level of maceration and are aged under the same oxidative conditions as blends. With certain exceptions, the wines are not oxidized to maintain their fruity aroma and pale color.

Superior-quality products (10-year-old, 30-year-old Ports, etc.) also undergo oxidative aging.

At the time of bottling, these oxidized wines are stable in the presence of air. They improve very little during bottle aging.

Vintage Ports are the best quality wines. After a brief aeration to stabilize the color, they are aged in full barrels, like many great red wines. Vintage Ports are bottled after two to three years' barrel aging, while Late Bottled Vintage Ports are bottled after four to six years (Figure 14.3). They then continue to improve considerably in bottle. A bottle bouquet develops: it is linked to the low oxidation–reduction potential that maintains the iron ion in the ferrous state. Vintage Ports have considerable aging potential and can be aged for 20 years or more in the absence of air, due to their high polyphenol concentration. These wines are very robust when young and, after years of aging, maintain a high extract concentration, a characteristic fruitiness, and relatively high color intensity, with red-mauve tones dominating. Once bottled, these wines are sensitive to oxygen: when the bottle is opened, the wine rapidly loses its qualities.

The year can still be mentioned on the bottle of non-vintage, quality Port wines. These vintage character wines are called Colheita Ports.

14.5 Flor Wines

14.5.1 Definition

The OIV defines "flor wines" as:

> wines whose principal characteristic is to be submitted to a biological aging period in contact with air, by the development of flor yeasts (film-forming yeasts) after alcoholic fermentation of the must. Brandy, rectified alcohol or agricultural spirits can be later added to the wine. In this case, the alcohol content of the finished product must be equal to or greater than 15% vol.

Sherry (in English and German; Jerez in Spanish; Xérès in French) is the best-known flor wine. In *oloroso*-type sherry, the aging process is essentially physical/chemical, and biological development is limited. The *oloroso* method will, however, be described in this section. *Vins jaunes* (yellow wines) from the Jura region of France are another well-known example of flor wines.

14.5.2 Sherry Wines

Production conditions

The sherry production zone is situated in the south of Spain, near the city of Jerez de la Frontera. The production of this prestigious wine was described by Casas Lucas (1967), Goswell (1968), Gonzalez Gordon (1990), and Jeffs (1992). This section is based on the work of E. Peynaud (in Ribéreau-Gayon et al., 1976), updated by J.F. Casas Lucas in 1994.

The Palomino cultivar constitutes nearly 95% of the grape production for this wine. The remaining 5% consists of the Pedro Ximénez variety. The vine is cultivated on different *terroirs*, creating a production hierarchy. The musts contain 12–14% vol. potential alcohol and an acidity of 2–3 g/l expressed as H_2SO_4 (or 3–4.5 g/l in tartaric acid).

The grapes are carefully picked and placed in 15 kg crates. In the past, the grapes were traditionally exposed to the sun for a day on a mud floor (*almijar*). This practice, known as *soleo*, results in a 10% loss in grape weight, an increase in sugar and tartaric acid concentrations, and a decrease in malic acid concentration. Although favorable to quality, the *soleo* practice has all but disappeared. Pedro Ximénez grapes may still undergo this practice, attaining a high sugar concentration within 15 days (Section 11.2.2). These grapes are used to prepare a sweet wine (cream sherry), which is used in variable proportions to sweeten dry wines.

The sherry winemaking method is based on white winemaking principles without skin contact. The juice extraction conditions are consequently of prime importance: moderate crushing, no contact with metal, slow and light pressing, and juice selection after pressing. The clarification and refrigeration of these juices are becoming general practices.

Plastering (adding calcium sulfate to must) is a traditional practice in this region. This operation permits the suspended solids to settle more rapidly, and the wines obtained are more limpid and their color is more brilliant. There is a decrease in pH, a diminution of ash alkalinity (due to the precipitation of acids in the form of salts), and an increase in total acidity and buffering power. Approximately 2 g of $CaSO_4(2H_2O)$ is added per liter, lowering the pH by 0.2 units. The wine may also be acidified with tartaric acid (1.5 g/l maximum).

Sulfur dioxide is used during winemaking and storage to disinfect the barrels, but concentrations must be limited so as not to hinder the development of flor yeast. Good hygiene practices prevent undesirable microbial contamination.

In the past, the fermentation occurred in 516 l oak barrels (*botas de extraccion*) filled with 450–467 l of must. Today, relatively low-capacity stainless steel containers are increasingly used to limit excessive temperature increases.

Biological aging principles for flor yeasts

The wines, still on their lees, are tasted during the months following the completion of fermentation. The best wines, considered the most apt for aging, are racked, fortified to 15–15.5% vol., and stored in a container filled to 5/6ths of its capacity.

The alcoholic content of these wines prevents microbial spoilage, but flor yeasts spontaneously develop on the surface of the wine. After a certain degree of development, another tasting results in a new classification, determining the appropriate type of aging (*crianza*) for each *bota*: biological or oxidative.

During biological aging, the flor develops, sometimes during several years. Certain yeasts are capable of developing on the surface of 15–16% vol. alcohol content wine in contact with air. This film is produced by yeasts specific to the region, coming from either grapes or previously used barrels. These yeasts develop under aerobic conditions by oxidizing ethanol. They belong to the *Saccharomyces* genus. Over the years, taxonomy has included these yeasts in the *S. cerevisiae* species or has identified them as *Saccharomyces oviformis*, *Saccharomyces bayanus*, etc. These are therefore not ordinary mycodermal yeasts, responsible for various wine diseases due to poor storage methods.

Martinez *et al.* (1997) identified the following yeast strains in Sherry *flor*:

74% S. cerevisiae beticus
14% S. cerevisiae montuliensis

8% *S. cerevisiae cheresiensis*
0.3% *S. cerevisiae rouxii*
3% *strains not typical of Sherry flor (Pichia, Hansenula, and Candida)*

The different strains develop and form *flor* at different rates, and with different metabolic effects; for example, e.g. *montuliensis* produces the highest concentrations of acetaldehyde.

The yeast film is called flor and the biological aging process has been known as *crianza de flor* for a long time in Jerez. The more or less rapid and intense flor formation and its appearance and color (white, cream, golden, burned) depend on many factors, especially the nature of the yeast and the chemical composition of the wine. The flor rarely forms above 16.5% vol. alcohol, and it is impossible above 17% vol. The presence of a little sugar is favorable; the presence of phenolic compounds is unfavorable and darkens the wine color. Sulfur dioxide, nitrogen compounds, and other substances are also involved. All of these factors exert an influence that is reflected in future wine aroma and quality.

During this type of aging, the wine does not remain permanently in the same container: it is periodically transferred to different *botas*. These transfers are fractional, following the solera system. *Botas* are piled in rows. The barrels (*botas*) in each horizontal row (*escola*) are full of wine from the same crianza (that is, same degree of aging). During aging, the wine is moved around, blended, and redistributed to obtain the most uniform wine possible at the time of bottling. This system also permits new wine to be added regularly, which helps to maintain the flor.

Figure 14.4 summarizes the *solera* system. This example contains three *escolas*: 720 l are taken from the six *botas* of the lowest row, or *solera*, for bottling. The *solera* barrels are filled with 720 l coming from the five *botas* of the preceding *criadera*. Finally, 720 l from the four *botas* of the highest row fill the last five *botas*. These four *botas* are filled with new wine.

The *bota de asienta* intended for sherry aging has a volume of 600 l. The fill volume is 5/6ths, or 500 l.

In practice, the *solera* systems are much larger—they usually contain several hundred barrels. Transfers are made in groups of 12–18 *botas*.

Wine is transferred three to four times per year in the *solera* system. The transfer volume depends on the type and age of the wine desired. The ratio of total system volume to annual volume removed determines average wine age.

Wine transformations during biological aging

The biochemical transformations provoked by the *crianza de flor* have been studied. As oxygen is consumed by the flor, its proportion decreases in the barrel

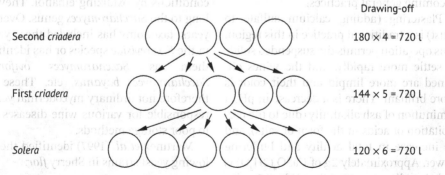

FIGURE 14.4 Solera system, showing the partial drawing-off and redistribution of wine into the next lowest row of barrels of an older *criadera* stage. This operation was begun by removing 120 l from each barrel on the lowest level (or solera) for bottling.

headspace and is replaced by carbon dioxide. The wine transfers, however, aerate the wine. The oxidation–reduction potential of the wine (250–300 mV) indicates a moderately reduced state. The flor acts as an isolating layer, protecting the wine from excessive oxidation. In this manner, the wine ages normally (fino type), acquiring a pale yellow color.

Volatile acidity diminishes to 0.10 g/l expressed as H_2SO_4 (0.12 g/l as acetic acid). Acetaldehyde formation is an essential characteristic of the *crianza de flor*, slowing during aging to produce a total of 220–380 mg/l. Due to its chemical reactivity, acetaldehyde is a precursor of many chemical substances that contribute to the bouquet of sherry wine (diethyl acetal, 50–60 mg/l). Sotolon (Section 10.6.4) is a characteristic element of the aroma of *fino* wines.

The glycerol concentration attains 7–9 g/l immediately after alcoholic fermentation. During the first *crianza* phases, it is significantly depleted. After three years, its concentration can fall to a few tenths of gram per liter.

Lactic acid is also formed, reaching 22 mEq/l. This production cannot be explained by alcoholic (7 mEq/l) and malolactic (6 mEq/l) fermentation alone. Malolactic fermentation is nevertheless complete and contributes to wine quality.

Free amino acid concentrations diminish during flor aging, but the evolution of each amino acid varies, depending on the situation. After seven years of *flor* aging, the concentration of proline, the most abundant amino acid in *fino*-type sherries, representing 70% of the initial amount of total amino acids, decreases to only 31% (Botella *et al.*, 1990).

The *crianza de flor* in Sherry wineries produces *fino* wines. *Manzanilla* is produced according to the same principle in the Sanlucar de Barrameda region. This style of wine is aged for at least three years. At the end of this long aging process, the *crianza de flor* disappears. The aging process can, however, be continued chemically (*oloroso* wines) for six years or more. These products have the following names according to their age: *fino, amontillado, amontillado viejo,* and *amontillado muy viejo*.

Oxidative aging of oloroso wines

A flor film develops on nearly all sherry wines. This flor develops several months after the completion of alcoholic fermentation. An initial fortification at 15–15.5% vol. alcohol is conducted for wines most suited to biological aging. After a few months of being aged under the film-like growth, the wines are tasted and classified, confirming those that are destined to be aged biologically and deciding which wines must undergo oxidative aging. The latter are selected according to film growth conditions. If the film-like growth is not established under suitable conditions, the growth is completely stopped by an additional fortification to 17.5–18% vol.

From this moment on, the wines receiving the additional fortification will age in the absence of a yeast film, without flor yeast activity. Only physicochemical phenomena occur. During the aging process, some of the substances responsible for the fruity character in the wine are oxidized by oxygen. The barrel wood also plays a role in the oxidation process. Its texture acts as a sort of semipermeable membrane. The reactions involved in these phenomena are slow and poorly understood. Basic wood substances are extracted and oxidized.

Oloroso wines can be aged according to the *solera* system or in a more static manner without blending. This static method produces vintage wines (*anadas*). *Oloroso* wines are generally richer in color and more robust than *fino* wines. *Raya olorosa* wines correspond to a lower class of wine than *oloroso*.

Before bottling, the different types of sherry are clarified by fining with albumin or powdered blood. They are also sometimes stabilized by a bentonite addition.

There is a risk of bacterial spoilage during both aging processes (biological and chemical). For this reason, each barrel is regularly tasted during the process. At the slightest doubt about quality, the wine is transferred to sherry vinegar production.

14.5.3 Vins Jaunes (Yellow Wines) from Jura

Although they have their own character (Chevennement *et al.*, 2001), there are a certain number of parallels with sherry. Like sherry, they undergo an aging process with flor development, but no alcohol is added to them.

These wines are made from the Savagnin grape. The base wine contains approximately 12% vol. alcohol. The wine is placed in small barrels; the barrels are topped off and sealed. They are aged in a cellar for six years, without topping off, producing a headspace in the barrels. A film-like growth progressively forms on the surface of the wine. This flor is composed of aerobic film-forming yeasts that develop by respiration and cause various transformations—in particular the oxidation of ethanol into acetaldehyde. The yeast most often encountered in yellow wines belongs to *S. cerevisiae*. The inoculation is spontaneous with these wines, causing the film growth to be irregular and sometimes resulting in spoilage. The risks of increased volatile acidity are greater than in sherry wines and increase as the wine alcohol content decreases.

To improve production conditions, film growth can be accelerated by inoculation with a film-producing yeast culture and by leaving headspace when filling the barrel, instead of waiting for it to occur spontaneously by evaporation. Maintaining a low temperature (12–13°C) also limits bacterial spoilage.

In practical terms, *vin jaune* is aged for over six years and is subjected to alternating cold winter (5–10°C) and hot summer (25–30°C) temperatures. These variations cause the development and elimination of a series of *flor* blooms over the years, resulting in the coexistence of live yeast in the surface *flor* and dead yeast cells that are deposited in the lees and autolyzed. The yeasts have an intense metabolic activity at 25°C, but are much less active at 10°C (Charpentier *et al.*, 2002).

The yellow wines from Jura are characterized by their high acetaldehyde concentration (600–700 mg/l), their deep color, and their particular organoleptic characteristics, such as hints of nuts, in which furanones, such as sotolon, play a role (Martin et al., 1992).

References

André P., Benard P., Bourgeois M. and Flanzy C. (1980) Ann. Technol. Agric., 29, 497.

Barros P. (1991) *La Technologie des Vins de Liqueur*. Office International de la Vigne et du Vin, Paris.

Bely M., Rinaldi A. and Dubourdieu D. (2003) Biosci. Bioeng., 96, 6, 507.

Berthier L., Marchal R., Debray M., Bruet E., Jeandet P. and Maujean A. (1999) J. Agric. Food Chem., 47, 2193.

Blouin J., Pierre-Loti-Viaud D., Crachereau N. and Richard J.P. (2002) Œno One, 36, 31.

Blouin J. and Peynaud E. (2001) *Connaissance et Travail du Vin*. Dunod, Paris.

Boidron J.N., Avakiants S.P. and Bertrand A. (1969) Œno One, 3, 43.

Botella M., Pérez-Rodriguez L., Domecq B. and Valpuesta V. (1990) Am. J. Enol. Vitic., 41, 12.

Brissonnet F. and Maujean A. (1993) Am. J. Enol. Vitic., 44, 3, 297.

Brugirard A., Fanet J., Seguin A. and Torres P. (1991) *La dégustation et le service des vins doux naturels à Appellation d'Origine Contrôlée*, Université des vins du Roussillon, 66300 Tressere (France).

Casas Lucas J.F. (1967) *Fermentation et vinification*. Institut National de la recherche Agronomique, Paris.

Casey J.A. (1987) Aust. Grapegrow. Winemak., 280, 55.

Castino M. (1988) Vigne Vini, 12, 31.

Cayla L., Pouzalgues N. and Masson G. (2010) 33rd OIV World Congress of Vine and Wine. 8th General Assembly of the OIV, June 20-25, 2010, Tbilisi, Georgia 8 pp., 2010.

Charpentier C., Dos Santos A.M. and Feuillat M. (2002) Rev. Fr. Oenol., 195, 3.

Chauvet S., Sudraud P. and Jouan T. (1986) Rev. Oenol., 39, 17.

Chevennement R., Cibey R., Grispout P., Levaux J. and Sintot D. (2001) Rev. Fr. Oenol., 127, 26.

Cheynier V., Wirth J., Morel-Salmi C., Souquet J.M., Mazauric J.P., Meudec E., Fulcrand H., Lambert M., Verbaere A., Mazerolles G., Sommerer N., Dieval J.B., Ducasse M.A. and Masson G.(2013) Rev. Fr. Oenol., 260, 30.

Desportes C., Charpentier M., Duteurtre B., Maujean A. and Duchiron F. (2000) J. Chromatogr. A, 9893, 281.

Dubourdieu D. (1982) *Recherches sur les polysaccharides sécrétés par Botrytis cinerea dans la baie de raisin*, Thèse Doctorat, Universite de Bordeaux II.

Ducasse M.A., Cayla L., Masson G., Anneraud C., Chretien P., Davaux F., Boulet J.C., Meudec F., Verbaere A., Sommerer N. and Cheynier V. (2015) in Œno2015 Proceedings 10th Symposium Enology Bordeaux, Vigne Vin Publications Internationales ed. 723 p.

Duteurtre B., Ors P., Charpentier M. and Hennequin D. (1990) Le Vigneron Champenois, 7-8, 21.

Feuillat M., Charpentier C., Picca G. and Bernard G. (1988) Rev. Fr. Oenol., *Cahier Scientifiques*, 13, 463.

Flanzy C. (1998) *Œnologie. Fondements Scientifiques et Technologiques*. Tec & Doc, Lavoisier, Paris.

Flanzy C., Masson G. and Millo F. (2009) *Le vin Rosé*. Féret ed, Bordeaux.

Garcia-Jares C.M., Rozès N. and Medina B. (1993a) Œno One, 27, 1, 35.

Garcia-Jares C.M., Médina B. and Sudraud P. (1993b) Rev. Fr. Oenol., 140, 19.

Gerbaux V., Villa A., Monamy C. and Bertrand A. (1997) Am. J. Enol. Vitic., 48, 49–54.

Gil M., Louazil P., Iturmendi N., Moine V., Cheynier V. and Saucier C. (2019) Food Chem., 295, 493.

Glories Y., (1984). Œno one, 18, p. 253.

Gonzalez Gordon M.M. (1990) *Sherry, the Noble Wine*. Quiller Press, London.

Goswell R.W. (1968) Process Biochem., 3, 47.

Jeffs J. (1992) *Sherry*, 4th edition. Faber and Faber, London.

Lafon-Lafourcade S. and Ribereau-Gayon P. (1977) C.R. Acad. Agric., 63, 551.

Lafon-Lafourcade S., Blouin J., Sudraud P. and Peynaud E. (1967) C.R. Acad. Agric., 53, 1046–1051.

Lambert M., Meudec E., Verbaere A., Mazerolles G., Wirth J., Masson G., Cheynier V. and Sommerer N. (2015) Molecules, 20, 7890.

Laurent M. and Valade M. (1993) Le Vigneron Champenois, 6, 5.

Laurent M. and Valade M. (1994) Le Vigneron Champenois, 1, 7.

Liger-Belair G. (2002) Ann. Phys. Fr., 27, 4, 101.

Liger-Belair G. (2003) Sci. Am., 280, 1, 68.

Liger-Belair G. and Jeandet P. (2002) Rev. Fr. Oenol., 193, 45.

Liger-Belair G., Marchal R., Robillard B., Dambrouch T., Maujean A., Vignes-Adler M. and Jeandet P. (2000) Langmuir, 16, 1881.

Malvy J., Robillard B. and Duteurtre B. (1994) Sci. Aliments, 14, 87.

Marchal R., Sinet C. and Maujean A. (1993) Bull. OIV, 751–752, 691.

Marchal R., Bouquelet S. and Maujean A. (1996) J. Agric. Food Chem., 44, 1716.

Martin B., Etiévant P., Le Quéré J. L. and Schlich P. (1992) J. Agric. Food Chem., 40, 475.

Martinez P., Pérez Rodriguez L. and Benitez T. (1997) Am. J. Enol. Vitic., 48, 160.

Masneuf S., Muzet M.L., Choné X. and Dubourdieu D. (1999) Vitis, 249, 20.

Masson G. and Schneider R. (2009) Am. J. Enol. Vitic., 60, 1.

Maujean A. (1989) Rev. Fr. Oenol., 120, 11.

Maujean A. (1996) *L'Amateur de Bordeaux*, December, 32.

Maujean A. and Seguin N. (1983) Sci. Aliments, 3, 589 and 603.

Maujean A., Haye M. and Feuillat M. (1978) Œno One, 12, 277.

Maujean A., Gomérieux T. and Garnier J.M. (1988) Bull. OIV, 61, 683.

Maujean A., Pinsaut P., Dantan H., Brissonnet F. and Cossiez E. (1990) Bull. OIV, 63, 405.

Moncomble D., Valade M. and Pernot N. (1991) Le Vigneron Champenois, 5, 14.

Murat M.L., Tominaga T. and Dubourdieu D. (2001) Œno one, 35, p.99.

Murat M. L., Tominaga T., Saucier C., Glories Y. and Dubourdieu D. (2003) Am. J. Enol. Vit., 54, 135.

Pernot N. and Valade M. (1994) Le Vigneron Champenois, 6, 6.

Pilatte E., Nygaard M., Gaigao Y., Krentz S., Poweer J. and Lagarde G. (2000) Rev. Fr. Oenol., 185, 26.

Puértolas E., Saldaña G., Alvarez I. and Raso J. (2011) Food Chemistry, 126, 1482.

Ribéreau-Gayon J., Peynaud E., Ribéreau-Gayon P. and Sudraud P. (1976) *Sciences et Techniques du Vin, Vol. III: Vinifications—Transformations du vin*. Dunod, Paris.

Robillard B. (2002) Rev. Fr. Oenol., 193, 49.

Robillard B., Delpuech E., Viaux L., Malvy J., Vignes-Adler M. and Duteurtre B. (1993) Am. J. Enol. Vitic., 44, 387.

Salinas M.R., Garijo J., Pardo F., Zalacain A. and Alonso G.L. (2005) J. Sci. Food Agric., 85, 1527.

Stamatopoulos P. and Darriet P. (2017) *in Actes de 13e Journée Technique CIVB*, Bordeaux, February 14, 2017, 94–103.

Sudraud P., Bar M. and Martiniere P. (1968) Œno one, 2, 349.

Tribaut-Sohier I. and Valade M. (1994) Le Vigneron Champenois, 9, 16.

Tusseau D. and Van Laer S. (1993) Sci. Aliments, 13, 463.

Tusseau D., Benoit C. and Valade M. (1989) *Actualites Oenologiques*, Vol. 89. Dunod, Paris.

Valade M. and Blanck G. (1989) Rev. Fr. Oenol., 118, 23.

Valade M. and Laurent M. (1999) Le Vigneron Champenois, 6, 67.

Valade M. and Laurent M. (2001) Le Vigneron Champenois, 3, 40.

Valade M. and Pernot M. (1994) Le Vigneron Champenois, 113, 6.

Valade M. and Rinville C. (1991) Le Vigneron Champenois, 3, 22.

INDEX

acetic acid, 196
acetic acid bacteria, 238, 442
  classification, 230–231
  evolution of, 238–240
  grape must development, 237–238
  isolation and identification, 231–232
  metabolism of
    acetoin formation, 236–237
    ethanol, 233–236
    lactic acid and glycerol, 236
    sugars, 233–234
  principal characteristics and cytology, 229–230
  principal physiological characteristics, 232–233
acetic acid bacteria cells, 229
Acetobacter, 230, 231–232, 237, 238
acetoin formation, 91
  acetic acid bacteria, 236–237
acetoinic molecules, 190
acidification, 408–410
acidity, 440
active dry yeasts (ADYs), 23, 117, 440, 553
  strain, 553
active yeasts, 139–141
acute toxicity, 246, 247
additive techniques, 416–418
adenosine triphosphate (ATP), 73, 74
  ATPase, 10
aeration, 447–449
  of wine, 295
alcoholic fermentation, 3, 77–78, 513, 555, 556
amidation, 93, 94
amino acids, 94, 95, 203
anaerobic bacteria, 150

anaerobic respiration, 176
anamorphic ascomycetes, 28
anamorphic basidiomycetes, 28
anthocyanins, 332
  content, 356
  in rosé wines, 573
van Leeuwenhoek, Antonie, 3
Araldite®, 443
arginine deiminase, 193
arginine, nitrogen metabolism of grapes, 329
aromas
  composition, 370–372
  freeing of, 429
  substances, 384
    evolution, 332–343
    monitoring, 345
artificial drying, 401
artificially carbonated wines, 588
ascomycetes, 4
ascorbic acid, 280
  antioxidant properties, 294
  organoleptic protection of aerated wines, 297
  oxidation reaction, 293, 294
  properties and mode of action, 293–294
  protection against enzymatic oxidations, 294–295
  protection against iron casse, 295–296
asporogenous yeasts, 4
Asti Spumante, 602
automated red winemaking methods
  continuous winemaking, 494–497
  thermovinification, heating the harvest, 497–499
automatic pump-over systems, 445

bacteriophages, 224–226
barrel-fermented musts, 585
barrel/tank aging, 239
basidiomycetes, 4
beer yeast, 4
bentonite treatment, noble rot, 583–584
berry, 310–315
  abscisic acid accumulation, 314
  cluster structure, 310, 311
  defective pollination, 314–315
  environmental conditions, 310
  flowering (bloom), 311
  formation, 310–315
  fruit development, 310
  genetic factors, 310
  hormone balance during development, 316, 317
  inflorescence formation processes, 311
  at maturity, 310, 311
  pollination, 314
biochemical transformations of grapes during ripening, 399
biosynthesis pathways of phenolic compounds, 331
biotin, 115, 116–117
bipolar electrodialysis, 300–301
1,3-bisphosphoglycerate (1,3-BPG), 75, 179
Blanc de noir wines, 572
BLAST algorithm, 168
Bordeaux wines, 23
  classification, 27
Bordeaux-style winemaking, 451
Bota de asienta, 610
Botrytis cinerea, 347, 518–521
  chemical composition of, 380–386
    acids, 382–383

Botrytis cinerea (cont'd)
   aroma substances, 384–386
      glucose and fructose, 380–382
      nitrogen substances, 383
      phenolic compounds, 383–384
   development of, 389, 579
   laccase, 472
   glycerol and gluconic acid, 579
   grape sensitivity, 376–379
   infection process, 379
   synthesis, 580
botrytized grapes, 435
botrytized sweet wines, 577–588
   aging and stabilization, 585–588
   noble rot (*see* Noble rot)
   Tokaji wine, 588
bound SO_2, 206
bunch stem necrosis, 375
Burgundian method, 558
2,3-butanediol formation, 91

Cabernet Sauvignon, 459, 460
   grapes at harvest, vintage in
      Bordeaux, 362, 364
cap punching (pigeage), 442
carbamate kinase, 193
carbohydrates, 103
carbonic maceration, 573
   anaerobic metabolism, 501–504
   characteristics of wines made
      by, 509–511
   fermentor undergoing, 507–508
   free-run and press juice
      composition, 508–509
   gaseous exchanges, 501
   grape transformations, 504, 505
   microbiology of, 504, 506
   principles, 499–501
carbon 13 isotope
   discrimination, 354
catabolite repression by
   glucose, 17
cell multiplication and
   enlargement, grape
   development and
   maturation, 330
cellular oxidations, 229–230
chalcone synthase, 332
Champagne
   malolactic fermentation, 591–592
   musts
      clarification and
         fermentation, 590–591
      physicochemical
         characteristics of, 591

secondary fermentation in
   bottle
      alcoholic fermentation and
         aging on lees, 594
      disgorging and final
         corking, 595–596
      preparing and bottling a
         cuvée, 592–594
      yeast sediment, riddling and
         removing, 595
   transfer method, 600–601
   wine composition
      analysis of, 596–598
      effervescence in, 598–600
chaptalization, 416–417
Chardonnays, white wine, 516
chemolithotrophs, 73
chemoorganotrophs, 156, 175
chitin, 8
   chitin synthase, 15
chlorophyllous plants, 73
chrome–nickel–molybdenum
   steel, 443
CIE L*a*b* method, 573
cis-vaccenic acid (C18), 150
citramalic acid, 92
citric acid metabolism, 185–187
clarification, noble rot, 583
cluster microclimate, 340
coarse-grain oak (Limousin), 563
coenzyme A (CoA), 78, 79
cold pressing, noble rot, 581–582
color extraction and
   stabilization, 428–429
colorimetric method, 520, 521
commercial enzymes in
   winemaking, 427–429
concentrated must, 418
concrete tank, 443
"Contains Sulfites", labeling 248
continuous winemaking,
   494–497
conventional red
   winemaking, 431
cozymase, 74
Crabtree effect, 83
crop thinning, 374
crushing, 435–436
cryoextraction, noble rot, 581–582
crystallized lysozyme, 287
cyclic adenosine monophosphate
   (cAMP), 15
cysteine, 196
cytoplasmic granulations, 152
cytosol, 15

dairy industry bacteria, 185
damaged grapes, spoilage
   pathway, 389
deacidification, 410–412
24(28)-dehydroergosterol, 10, 13
denaturing gradient gel
   electrophoresis
   (DGGE), 169
deoxy-D-xylulose 5-phosphate
   synthase (DXS) gene, 335
destalking, 436
destemming, 436–440
deuteromycetes, 4
diacetyl, 91
   in wine, 196
diammonium phosphate, 112
diammonium sulfate, 112
diatomaceous earth rotary
   vaccum filters, 551
α-dicarbonyl compounds in
   wine, 196
diethyl dicarbonate (DEDC),
   285–286
dimethyl dicarbonate
   (DMDC), 280
dimethylglyceric acid, 92
diphosphatidylglycerol, 149, 150
diphosphatidylglycerol or
   cardiolipin (PG), 10, 11
direct osmosis, 414
disease and adverse weather,
   effects of, 374–375

early co-inoculation, 491
early-ripening varieties, 367
electrophoresis, in
   agarose gel, 167
Embden–Meyerhof–Parnas
   pathway, 74
Embden–Meyerhof pathway, 177
endoplasmic reticulum (ER),
   15–16
enzymatic formation mechanism
   of aldehydes and C6
   alcohols, 423
enzymatic hydrolysis mechanism
   of terpene glycosides, 422
enzymatic transformations, grape
   after harvest, 418–427
ergosterol, 10, 13, 117, 118
ethanol, 103
ethyl pyrocarbonate (Baycovin)
   in wine, 280
European Food Safety Authority
   (EFSA), 289

European viticultural climate zones, 361
evolution of oxidation–reduction potential of reduced wines, 295, 296
extended maceration, 470

facultative anaerobes, 156
facultative heterofermenters, 155
fatty acids, lactic acid bacteria, 150, 151
fermentation activators
fermentative metabolism, 156
fermentor capacity, 444
Firmicutes, 155
Flash Detente treatment, 465
flavin adenine dinucleotide (FAD), 81
flor wines
  OIV defines, 608
  sherry wines
    biological aging for flor yeasts, 609–610
    production conditions, 608–609
    wine transformations during biological aging, 610–611
  Vins Jaunes (Yellow Wines) from Jura, 612
flow cytometry, 104
flower and fruit evolution during bloom and fruit set, 314, 315
foamability, 599
  height, 599
  instability of, 599
  stability, 599
fortification (mutage), 605
fortified wines
  aging, 603
  characteristics, 602–603
  French (Vins Doux Naturels)
    analytical characteristics of, 604
    conservation and aging, 605–606
    definition, 603
    fortification (mutage), 605
    winemaking, 604–605
  OIV definition, 603
  port wines
    maturation and characteristics of, 607–608
    production, 606–607

Fourier transform infrared spectrometry, 347
free amino acid, 611
free amino nitrogen (FAN), 112
free and bound terpenes, 336
freeing of aromas, 429
free-run wine, 477, 478–480
French wines, 246
fructose 1,6-bisphosphate, 179
fuel, 73
fungal development under grape skin, 381

general amino acid permease (GAP), 12, 13, 96
glucan, 6
β-glucanases, 558
β-1,3-glucan synthase, 15
Gluconacetobacter, 230, 231
gluconic acid, 382
Gluconobacter, 230, 231–232, 237
  Gluconobacter oxydans, 237
glutamate dehydrogenase (GDH) enzyme, 329
glutamine synthetase (GS), 93 enzyme, 329
glutathione, 301
glyceraldehyde 3-phosphate, 75, 179
glycerol accumulation, 83–84
glycerol concentration of contaminated grapes, 381
glyceropyruvic fermentation pathway, 78, 79
glycogen, 15
glycolipids, 150
glycolysis, 74–77, 177, 178
  glycolytic pathway, 328
glycosidases, 421–422, 441
glycosylation, 6, 7
gram-negative bacteria, 146
gram-positive bacteria, 146, 155
gram stain, 230
grape
  berry tissues at maturity, 317, 318
  cell division, 315
  cluster composition at ripeness, 319–322
  description and composition, 310–322
    berry, 310–315
  developmental stages, 315–317
  development process of mold, 310

environmental parameters and growing practices, 310
gibberellic acid on seedless grapes, 315
maturation (*see* maturation (ripening) process, grape
maturation, growth phase, 317
maturation (ripening) process, 309
maturity level, 309
morphology, 317–319
physiological and biochemical phenomena, 309
veraison (berry color change), growth phase, 316–317
grape crop, cleaning and sorting, 407
grape harvest, 402–404
grape maturation *see* maturation (ripening) process, grape
grape reaction product (GRP), 424
grape ripening *see* ripeness
grape selection criterion, 389
and selective must extraction, 407–408
grape solid maceration, 418
grape transformation technologies, 301
grape tyrosinase on hydroxycinnamic acids, mode of action of, 424
grapevine fructification, 309
grapevine reproductive cycle, 310, 312
grapevine water deficit, 358
grapevine water status
  on grape aroma potential, 356–358
grapevine water supply on grape ripening, 352–360
  heavy rain, 353
  leaf water potential, 354
  monitoring vine water levels, 353–355
  water deficit, 352
gray rot grapes (*Botrytis cinerea*), 375–376, 518–521
  chemical composition of, 380–386
  infection process, 379, 386
  malic and tartaric acid degradation, 387
  oxidative phenomena, 388
green grape skin, phytoalexins, 377

harvested grapes, 389–391
  acidity adjustments, 408–412
harvest transport, 405–407
heat concentration, 413
heat resistance of wine yeasts, 290–291
heterosis effect, 49
heterothallic yeast strain (HO), 21, 22
heterothallism, 22
hexokinases, 75
hexoses, 75
high-speed rotary crushers, 436
homofermentative bacteria, 155, 158
horizontal shell-and-tube heat exchangers, 413
hot post-fermentation maceration, 461, 462
hydraulically controlled pistons, 445
hydrolytic enzymes, 419–422
hydroxycinnamic acid in rosé wines, 573
8-hydroxygeraniol, 336
β-hydroxypropionaldehyde, 189, 190

immunological method, B. cinerea, 390f
inert gases, 297
  carbon dioxide concentration, 299–301
  wine storage, 297–299
inflorescence formation processes, 311
  induction and initiation phases, controlling factors, 311, 313
integral proteins, 10, 11
International Organisation of Vine and Wine (OIV), 571
iron casse, ascorbic acid, 295–296
3-isobutyl-2-hydroxypyrazine (IBHP), 339
  biosynthesis pathway, 338–340

juice extraction, 514, 526–537
  closed tank membrane press, 530–531
  crushing and destemming, 533–534
  general principles, 526–527
  immediate continuous extraction, 527–528

maceration/skin contact, 534–536
moving-head press, 529–531
pectolytic enzymes, 427
pneumatic pressing, 532–533
vertical screw press, 528–529
whole-cluster pressing, 528–530, 532

killer phenomenon
  factor, 24
  physiology and genetics of, 24–25
  in winemaking, 25–27
Komagataeibacter, 230, 231

laccase activity, 400
  and other oxidases, 383
lactic acid, 611
  L-lactic acid, 183, 184
lactic acid bacteria, 452, 471, 482, 484
  components of
    cell wall, 146–148
    cytoplasm, 146, 152–153
    genetic material, 153–154
    nucleoid, 146
    plasma membrane, 148–152
    reproduction of bacteria, 154
  general principle, 156–158
  genotypic analysis methods
    determining G + C percentage, 163
    DNA–DNA hybridization, species identification, 163
    DNA–DNA hybridization, using probes, 163–165
    genome sequencing, 170
    PCR (see polymerase chain reaction)
    quantitative PCR, 168
    REA-PFGE, identification of strains, 165–166
  metabolism of
    catabolic reactions, 175
    chemoorganotrophic, 175
    hydrogen peroxide, 176
    organic acids in wine (see organic acids in wine)
    sugars (see sugars)
    wine composition and quality, 195–197
    in winemaking, 189–195
Oenococcus oeni, 170–171

phenotypic analysis
  analysis of metabolism, 158–162
  fatty acid and protein composition, 162–163
  microscopic observation, 158
taxonomy of
  bacterial species, 154–155
  wine classification of, 155–157
wine development (see wine development)
lactic spoilage, 196, 217
lactobacillic acid (C19), 150
Lactobacillus, 155, 156, 158
  L. hilgardii growth, 206
lactone concentrations, 384, 385
lateral/horizontal strike harvesting techniques, 402–404
leaf thinning, 374
leaf water potential, 354
Leuconostoc, 155, 156, 158, 170
light, on floral induction, 348
low-temperature pressing, 407–408
lysophosphatidylglycerol, 149
lysozyme
  in dairy and cheese industries, 287
  difficult alcoholic fermentation, 288
  from egg white, sulfur dioxide, antimicrobial properties, 279
  lactic acid bacteria and malolactic fermentation development delay, 288
  malolactic fermentation inhibition in white wines, 287–288
  microbiological stabilization after malolactic fermentation, 288
  nature and properties, 286–287
  use and labeling of wines, 289
  in winemaking, applications, 287–289

maceration, 431
  grape quality, 466–468
  grape sulfiting and alcohol production, fermentation, 462–463

pomace (flash detente)
and pulsed electric
fields, 464–466
principles of, 455–457
pump-overs and punchdowns,
459–460
role of, 452–454
skin, 534–536
temperature influence, 460–462
time influence, 457–459
types of, 455
wine tannin concentrations,
468–469
Malassez counting chamber, 104
malate dehydrogenase
(MDH), 183
malic acid, 326–328, 382
L-malic acid, 183
transformation, 182–185
by yeast, 92, 93
malolactic enzyme, 183
Champagne, 591–592
malolactic fermentation (MLF),
207, 432, 556–557
automated red winemaking
methods
continuous winemaking,
494–497
thermovinification, heating
the harvest, 497–499
conditions required for
acidity influence, 488
aeration influence, 489
bacterial growth, 487
optimum growth, 488
sulfiting influence, 489–490
temperature influence,
488–489
development delay, lactic acid
bacteria, 288
history, 480–484
inhibition in white wines,
287–288
inoculation
alcoholic fermentation, 491
with commercial *Oenococcus
oeni* preparations after
reactivation, 492–493
with commercial *Oenococcus
oeni* preparations not
requiring a reactivation
phase, 493–494
with non-proliferating
bacteria, 491–492

microbiological stabilization
after, 288
monitoring, 485–487
wine transformations, 484–485
yeast, 184
mannoproteins, 6, 9, 15, 558
manual harvesting, 404
maturation (ripening) process,
grape, 309
aroma composition, 370–372
aroma substances
evolution, 332–343
biochemical processes, 322
carotenoid degradation, 338
cell wall changes, 330–331
characteristics, 322
definition of, 399
evolution of respiration, 325
growth phase, 317
leaf treatments, 370–372
level, 309
metabolic pathway
modifications, 325
minerals accumulation, 328–329
nitrogen compounds evolution,
329–330
nutrient supply, 322
organic acids evolution, 325–328
peripheral network, 322
phenolic compounds
production, 331–332
phytosanitary protection,
370–372
soil composition and
fertilization, 369–370
sugar accumulation, 323–325
variety and rootstock, 366–368
weather conditions during the
year, 361–366
maturity index, 345–348
mechanical harvesting, 402
meiosis in yeasts, 21
Merlot wines, 459, 460
mesoinositol, 115, 116
methionine, 196
methoxypyrazines, 338–341
methylerythritol phosphate
(MEP) pathway, 334
mevalonic acid (MVA), 334
micro-imagery by nuclear
magnetic resonance, 347
millerandage (shot berries/hens
and chicks), 314
minerals accumulation, 328–329

mitochondria, 17–18
mousiness, 197
mucic acid, 382
multiple loci VNTR analysis
(MLVA), 169
"multiplex" PCR protocol, 167
musts
clarification, 427–428
composition, 579–580

N-acetylglucosamine, 146
N-acetylmuramic acid, 146
NADP glutamate dehydrogenase
($NADP^+$-GDH), 93
native yeasts, 117
must, 440
natural grape drying
(passerillage), 401
neutral white wines, 516
nicotinamide, 115, 116
nicotinamide adenine
dinucleotide, 75
5-nitrofurylacrylic acid, 280
alcohol intoxication, 280
carcinogenic, 280
fungicide, 280
nitrogen and sulfur leaf
spraying, 370–372
nitrogen catabolite repression
(NCR), 96, 111
nitrogen compounds, 103
evolution, 329–330
nitrogen fertilization, 369
nitrogen spraying, 370–372
nitrogen substances, 383
noble rot grapes, 375–376
chemical composition
of, 380–386
development of, 578
fermentation process, 584–585
vs. gray rot, 578–579
harvesting conditions, 578
humid and sunny periods, 578
infection process, 379
juice extraction
bentonite treatment, 583–584
clarification, 583
cold pressing
(cryoextraction), 581–582
juice correction, 583–584
pressing grapes, 580–581
sulfiting juice, 582–583
Semillon and Sauvignon Blanc
grapes, 578

nondestructive techniques, 348
N-*o*phthaldialdehyde (NOPA) method, 554
norisoprenoid derivatives, 336–338
nouveau-type wines, 432, 459
nucleoid, 153
nutrition in wine
  energy sources, 202
  nutrients, vitamins, and trace elements, 202–204

octanoic and decanoic acids, 283–285
  sweet wine sterilization, alcoholic fermentation, 284
Oenococcus alcoholitolerans, 170
Oenococcus oeni, 138, 148, 150, 151–152, 170–171, 194, 196, 202, 218, 225, 226
off-vine grape drying, 401
oleanolic acid, 117, 118
oleic acid (C18, unsaturated), 12, 13, 117, 118
Oloroso wines, 611–612
on-vine grape drying, 400
on-vine overripening methods, 399
optimal enological maturity, 402
organic acids, 103
  evolution, 325–328
  in wine
    citric acid metabolism, 185–187
    malic acid transformation, 182–185
    tartaric acid metabolism, 187–189
organoleptic protection of aerated wines, 297
ornithine transcarbamylase, 193
osmotic pressure, 414
overripening, grape quality, 400–401
oxaloacetate, 183, 185
oxidation mechanisms
  of botrytized grape must by laccase, 425
  enzymes, 422–427
  of healthy grape must by tyrosinase, 425
  quinones trapping by glutathione, 425
oxidative fermentation, 230

oxidative phosphorylation, 17, 74, 81, 82
oxidative yeast and acetic acid bacteria, 389
oxidoreductases, 424–427
5-oxofructose, 255
oxygenases, 422–424

pantothenic acid, 115, 116
Pasteur effect, 82
pasteurization, 289–293
Pasteur, Louis, 4
pectins solubilization, 330, 331
pectolytic enzymes, 420–421
Pediococcus, 153, 155, 156
  P. damnosus, 206
peptidoglycan, 146, 148
permeases, 12
peroxidases, 426–427
"petite" mutants, 18
phenolic compounds, 383–384
  enzymatic oxidation of, 538, 539
  extraction of, 535
  pre-fermentation maceration, 537
  production, 331–332
phenological observations on red grape development and maturation, 362
phenylalanine ammonia-lyase (PAL), 332
phosphatidylcholine (PC), 10, 11
phosphatidylethanolamine (PE), 10, 11
phosphatidylglycerol, 149, 150
phosphatidylinositol (PI), 8, 10, 11
phosphatidylserine (PS), 10, 11
phosphoanhydride bonds, 73
phosphoenolpyruvate, 179
phototrophs, 73
phytosanitary protection, 370–372
pimaricin, fungistatic effect, 280
plasma membrane, of lactic acid bacteria, 148
plasma technology, 300–301
plate-and-frame filters, 552
polarographic method, 520
polymerase chain reaction (PCR)
  advantage of, 166
  DGGE, 169–170
  electrophoresis, 167
  quantitative, 168
  RAPD and VNTR, 169

  and sequencing of the 16S RNA gene, 168–169
  species or strains detection, 167–168
  template DNA fragment, 166–167
polyphenols, 208
polyvinylpolypyrrolidone (PVPP), 526
pomace, 441–442
porins, 230
port wines
  maturation and characteristics of, 607–608
  production, 606–607
potassium, 328–329, 369, 370
predawn leaf water potential, 354
pre-fermentation practices, 400
press wines, 477–480
  basket, 529–530, 531
  Bucher, 531, 532
  closed tank membrane, 530–531
  horizontal, 590
  moving-head, 529–531
  pneumatic, 532–533
  in rosé wines, direct, 574–575
  vertical screw, 528–529
  whole-cluster, 528–530, 532
procyanidins, 332
proteases, 419–420
protein synthesis, 330
pulp, berry, 320–322
pulsed-field electrophoresis, 18
pump-over process, 449, 450
pyridoxal phosphate (PLP), 94, 98
pyridoxamine phosphate (PMP), 94, 97
pyridoxine, 115, 116
pyruvate decarboxylase (PDC), 77

quality cultivars, 367
quantitative PCR, 168

randomly amplified polymorphic DNA (RAPD), 169
rectified concentrated must (RCM), 417–418
red winemaking, 269
  carbonic maceration
    anaerobic metabolism, 501–504
    characteristics of wines made by, 509–511

fermentor
  undergoing, 507–508
  free-run and press juice
    composition, 508–509
  gaseous exchanges, 501
  grape transformations,
    504, 505
  microbiology of, 504, 506
  principles, 499–501
classic steps in, 433
controlling alcoholic
    fermentation
  ambient conditions effect
    of, 446–448
  monitoring temperature and
    fermentation, 451–452
  pump-overs and must
    aeration, 447–451
conventional, 431
draining off the skins
  free-run wine into tanks or
    barrels, 473, 475
  moment for, 469–471
  premature fermentor
    draining due to external
    factors, 471–474
grape composition, 432
grape quality, 432
maceration, 431
  grape quality, 466–468
  grape sulfiting and
    alcohol production,
    fermentation, 462–463
  pomace (flash detente)
    and pulsed electric
    fields, 464–466
  principles of, 455–457
  pump-overs and
    punchdowns, 459–460
  role of, 452–454
  temperature influence,
    460–462
  time influence, 457–459
  types of, 455
  wine tannin
    concentrations, 468–469
malolactic fermentation,
    432–433
  conditions required for,
    487–490
  history, 480–484
  inoculation, 490–494
  monitoring, 485–487
  wine transformations, 484–485

mechanical processing,
    harvested grapes
  crushing, 435–437
  destemming, 436–440
  harvest reception, 433–435
pressing
  composition and use of press
    wines, 477–480
  pomace, 475–477
tank filling
  fermentor construction,
    443–444
  fermentor equipment,
    444–446
  and operations, 440–441
  principal tank systems,
    441–443
restriction enzyme analysis
    by pulsed field gel
    electrophoresis (REA-
    PFGE), 165–166
restriction fragment length
    polymorphism (RFLP), 44
resveratrol production, parasite
    infection, 378
reverse osmosis, 413–416
riboflavin, 115, 116
ripeness, grape, 343–345
  biochemical maturation process
    light, effects of, 348–349
    temperature, 349–352
  evaluation of, 345–348
  grapevine water supply, 352–360
  optimal sugar/acid ratio, 344
  sampling and study of
    maturation, 345
  skin, 320
Rohmer pathway, 334
roller crushers, 436
rootstocks, 366–367
ropy *Pediococcus* strains, 194
rosé wines
  analytical parameters, 573
  blanc de blanc, 572
  characteristics, 571
  color characterization in,
    573–574
  Cotes de Provence
    vineyards, 572
  definition, 571–573
  making process
    direct pressing, 574–575
    skin contact/*saignée*
      method, 575–577

noble rot (*see* Noble rot)
  vs. red wine composition, 572

Saccharomyces cerevisiae
  See also yeasts
  ADY strain, 553
  ecology of, 62–66
  electrophoretic (pulsed field)
    profile of, 51, 52
  fermentation mechanism
    of, 542
  karyotype analysis, 18
  killer activity in, 25
  PCR associated with δ
    sequences, 53
  RAD-seq method, 56
  strains, 550, 553–554
S-adenosyl-L-methionine
    (SAM), 339
Saignée method, 575
salty soils, 370
saturated short-chain fatty
    acids, 283–285
Sauvignon Blanc grapes,
    357, 516–517
Schizosaccharomyces, 92, 93
S-conjugates
  of cysteine and glutathione in
    Semillon grapes, 385
  precursors of volatile thiols,
    341–344
secondary fermentation
  in bottle
    alcoholic fermentation and
      aging on lees, 594
    disgorging and final
      corking, 595–596
    preparing and bottling a
      cuvée, 592–594
    yeast sediment, riddling and
      removing, 595
  Charmat (closed tank) method,
    601–602
seeds, berry, 319
self-emptying tanks, 445, 446
sherry wines
  biological aging for flor yeasts,
    609–610
  oloroso wines, oxidative aging
    of, 611–612
  production conditions, 608–609
  wine transformations during
    biological aging, 610–611
skin, berry, 319

skin maceration, 534–536
soil on grape composition and wine quality, 369
soleo, 401
Solera system, 610
sorbic acid, 279
   antibacterial properties, 282
   antimicrobial properties, 281–282
   doses, sweet wine conservation, 281
   formula for, 280
   fungistatic concentration, 281
   physical and chemical properties, 280–281
   precautions, 280
   stability and sensory impact, 282–283
   sulfur dioxide, antimicrobial properties, 279
   trans-trans isomer, 280
   uses of, 283
sour rot and acid rot (*pourriture aigre*), 388–389
sparging of wine by carbon dioxide, 300
sparkling wines
   "Charmat"/"closed tank" method, 589
   closed tank facility for, 601
   definition, 588
   fermenting base wines
      malolactic fermentation, 591–592
      must clarification and fermentation, 590–591
      pressing and extracting must, 590
      principles, 589–590
   méthode champenoise for, 589
   méthode traditionnelle, 589
   physicochemical characteristics, 591
   processing stage, 589
spindle pole bodies (SPB), 18
spraying nitrogen and sulfur on leaves, 370
stainless steel tanks, 443
stalk/rachis, berry, 319
stearic acid (C18, saturated), 12, 13
steel tank, 443
stem water potential, 354
stilbene phytoalexins in the grapevine, 377, 378

"stress proteins", 209
strict heterofermenters, 155
strict homofermenters, 155
striker mechanisms, 402, 404
stuck fermentations, 104
substrate-level phosphorylation, 74
subtractive techniques, 412–416, 415
sugar accumulation, 323–325, 365
sugar addition, 416, 419
sugar assimilation and secondary product formation, noble rot attack, 382
sugar concentrations, 412–418
sugar degradation pathways
   alcoholic fermentation, 77–78
   glyceropyruvic fermentation, 78
   glycolysis, 74–77
   respiration, 78–82
sugars, metabolism of
   acetic acid bacteria, 233–234
   lactic acid fermentation
      heterofermentative of hexoses, 179–181
      homofermentative of hexoses, 177–179
      pentoses, 180–182
sulfiting, 268, 288, 489–490
   noble rot, 582–583
   in white winemaking, 542, 546
sulfur dioxide
   ascorbic acid, 293–297
   diethyl dicarbonate (DEDC), 285–286
   gas, 585
   inert gases, 297–301
   lysozyme, 286–289
   octanoic and decanoic acids, 283–285
   sorbic acid, 279–283
   yeasts destruction by heat, 289–293
sulfur dioxide (SO_2) in must and wine treatment
   antimicrobial properties of
      antibacterial activities, 264–265
      antifungal activities, 262–264
      wine conservation properties, different forms, 262, 263
   antioxidase, 244
   antiseptic, 243–244
   chemistry of

      bound sulfur dioxide, 250–252
      free sulfur dioxide, 248–250
   empirical laws of binding, 262
   equilibrium reactions, 261
   molecules binding
      acetaldehyde, 252–253
      balance in wines made from botrytized grapes, 258–260
      dicarbonyl group molecules, 256–257
      keto acids, 253–255
      phenolic compounds., 258
      sugars and sugar derivatives, 255–256
   physiological effects, 246–248
   slow reaction, 244
   temperature influence, 261–262
   in winemaking
      advantages and disadvantages, 265
      dissolving power and general effects on taste, 269–270
      inhibition, activation, and selection of yeasts, 267–268
      protection against oxidation, 265–267
      selection between yeasts and bacteria, 268–269
   in winery
      diminution of sulfur dioxide by oxidation during storage, 272–274
      forms of sulfur dioxide used, 274–275
      storage and bottling concentrations, 272, 273
      sulfiting wines by sulfuring barrels, 275–276
      winemaking concentrations, 270–272
   wine types of, 245, 246
sweet wine sterilization, alcoholic fermentation, 284
synthetic pesticide residues, 369
syringaldazine, 390

tangential microfiltration, 300–301
tannin, in rosé wines, 573
tartaric acid, 326, 347
   metabolism, 187–189
teleomorphic ascomycetes, 28
teleomorphic basidiomycetes, 28
template DNA, 166–167

terpene alcohols by enzymatic
    hydrolysis, 429
terpene compounds, 334–336
terpenoids, 335
terroirs, 370
thermal denaturation, 426
thermovinification, 497–499
thiamine, 115, 116–117
thiamine pyrophosphate (TPP),
    77, 78
total phenols index (TPI), 526
transposon yeast (Ty) elements, 19
tricarboxylic acid cycle/Krebs
    cycle, 78, 80
trimming/hedging, 374

vacuum concentration, 413
variable number of tandem
    repeats (VNTR), 169
viable but non-culturable (VBNC)
    bacteria, 212–214, 239
vine age, 367–368
vinegar taint, 238
vine growth management, 372
vine–soil–climate equilibrium, 368
vineyard practices for vigor
    control, 373–374
Vins Jaunes (Yellow Wines)
    from Jura, 612
vintage quality and climate
    conditions, 362, 363
virus-like particles (VLP), 24
vitamin C, 280
    antioxidant properties, 294
    organoleptic protection of
        aerated wines, 297
    oxidation reaction, 293, 294
    properties and mode of
        action, 293–294
    protection against enzymatic
        oxidations, 294–295
    protection against iron casse,
        295–296
viticulture, 355
    terroirs, 370
volatile thiol precursor
    concentrations, 384

water balance, grapevine water
    supply, 353
water deficit on early ripening,
    358–359
water status
    on grapevine growth, 355–356
    and vintage effect, 359–360

water supply, 369
    modifications in vineyard, 360
water transfer, 414
white winemaking, 269
    aging
        barrel, 562–564
        controlling reduction
            off-aromas, 564–567
    in barrel
        aging techniques, 562–564
        fermentation, 555–556
        principles, 557–558
        volatile compounds
            from oak, 561–562
        yeast cell wall colloids and
            exocellar role, 558–560
    bentonite treatments, 552
    clarification
        effects on fermentation
            kinetics, 547–550
        effects on wine
            composition, 544–547
        for grape solids
            deposits, 551–552
        methods, 548, 550–551
        suspended solids and
            lees, formation and
            composition, 543–544
    crushing, 533–534
    cryoselection, 536–537
    destemming, 534
    diversity of styles, 515
        aromatic white wines, 517
        Chardonnays, 516
        neutral white wines, 516
        Sauvignon Blanc, 516–517
    exocellular role, 558–560
    fermentation
        alcoholic, 513, 555, 556
        aroma components, 517
        barrel, 555–556
        malolactic, 556–557
    gray rot *(Botrytis
        cinerea)*, 518–521
    harvesting for, 525–526
    juice extraction, 514, 526–537
        closed tank membrane
            press, 530–531
        crushing and destemming,
            533–534
        general principles, 526–527
        immediate continuous
            extraction, 527–528
        maceration/skin contact,
            534–536

moving-head press, 529–531
pneumatic pressing, 532–533
vertical screw press, 528–529
whole-cluster pressing,
    528–530, 532
juice oxidation
    complementary techniques,
        542–543
    cooling grapes and juices, 542
    mechanisms, 539–542
    protecting from, 538–543
    sulfiting, 542
    traditional and current
        techniques, 538–539
lees
    formation, 543–544
    high-capacity tank, 566–567
    oxidation–reduction
        phenomena and, 561
maceration/skin contact,
    514, 534–537
maturity assessment, 521–524
nitrogen sources and juice
    aeration, 554–555
pre-fermentation operations,
    essential role of, 513–515
quality and picking criteria,
    517–526
sulfur compounds, volatile, 564–566
supraextraction, 536–537
tank filling, 552–553
temperature control, 555–556
volatile compounds from
    oak, 561–562
yeast cell wall colloids, 558–560
yeast inoculation, 553–554
wine development
    of bacteriophages, 224–226
    microbial interactions during
        winemaking
        interactions between lactic
            acid bacteria, 223–224
        interactions between
            yeasts and lactic acid
            bacteria, 219–223
    microflora evolution of
        bacterial species, 214–215
        different phases of bacterial
            development, 216–218
        population, 210–212
        VBNC bacteria, 212–214
    nutrition
        energy sources, 202
        nutrients, vitamins, and trace
            elements, 202–204

wine development (cont'd)
  physicochemical factors of
    bacterial growth
      adaptation phenomena,
        209–210
      ethanol influence, 206–207
      oxygen effect, 208
      phenolic compounds
        effect, 207–208
      pH influence, 204–205
      sulfur dioxide effect, 205–206
      temperature effect, 207
      winemaking rules, 204
wine lactic acid bacteria, 153
winemaking
  arginine formation, 192–194
  biogenic amines formation,
    190–192
  exocellular polysaccharides
    synthesis, 194–195
  glycerol degradation, 189–190
  SO_2 in must and wine treatment
    advantages and
      disadvantages, 265
    dissolving power and general
      effects on taste, 269–270
    inhibition, activation, and
      selection of yeasts, 267–268
    protection against
      oxidation, 265–267
    selection between yeasts and
      bacteria, 268–269
  white wines (see
    winemaking, white)
"winemaking rules", 204
winemaking, white (see white
    winemaking)
winery
  diminution of sulfur dioxide
    by oxidation during
    storage, 272–274
  forms of sulfur dioxide used,
    274–275
  storage and bottling
    concentrations, 272, 273
  sulfiting wines by sulfuring
    barrels, 275–276
  winemaking concentrations,
    270–272
wine storage using inert gases,
  297–299
winter pruning, 373–374
wooden tanks, 443

yeast-assimilable nitrogen
  concentration,
  370, 371, 554
yeast cell, 4, 5
  wall colloids, 558–560
yeast development
  fermentation activators
    adding yeast starter, 120–122
    growth factors, 115–117
    hydrolyzed yeast
      extracts, 119–120
    survival factors, 117–119
  inhibition of fermentation
    ethanol, 122
    use of yeast hulls, 123–125
    vine treatment spray
      residues, 125
  monitoring and controlling
    fermentations
    avoiding foam formation, 108
    counting yeasts, 104
    fermentation control
      systems, 105–108
    fermentation
      kinetics, 104–105
    measuring temperature, 105
  nutrition requirements
    carbon supply, 110
    mineral requirements, 115
    nitrogen supply, 110–115
  physicochemical factors,
    yeast growth and
    fermentation kinetics
    clarification on white
      grapes, 134–135
    influence of oxygen, 131–133
    temperature effect, 126–131
  stuck fermentations
    active yeasts, 139–141
    causes of, 135–138
    consequences of, 138–139
  yeast growth cycle
    and fermentation
    kinetics, 108–110
yeast inoculation, 553–554
yeast metabolism
  of nitrogen compounds
    amino acid synthesis
      pathways, 93–95
    assimilation mechanisms of
      ammonium, 96
    catabolism of amino
      acids, 96–99

    formation of higher alcohols
      and esters, 98–101
  sugar degradation pathways
    alcoholic fermentation,
      77–78
    glyceropyruvic
      fermentation, 78
    glycolysis, 74–77
    respiration, 78–82
  sugar-utilizing metabolic
    pathways, regulation of
    alcoholic fermentation
      and glyceropyruvic
      fermentation, 83–84
    degradation of malic acid by
      yeast, 92–93
    fermentation and
      respiration, 82–83
    formation and accumulation
      of acetic acid by
      yeasts, 85–92
    secondary products
      formed from pyruvate
      by glyceropyruvic
      fermentation, 84–85
    secondary products of
      the fermentation of
      sugars, 91–92
yeasts
  in alcoholic fermentation, 3, 4
  cell wall
    cellular organization of, 9
    chemical structure and
      function of, 6–9
    general role of, 5
  classification of species
    genus *Saccharomyces* and
      the position of wine
      yeasts, 37–47
    grape and wine yeast
      genera, 28
    interspecific hybrids, 47–49
    taxonomy and species
      delimitation, 28–37
  cytoplasm and its organelles
    cytosol, 15
    endoplasmic reticulum
      (ER), 15–17
    Golgi apparatus, 16, 17
    mitochondria, 17–18
    vacuole, 16–17
  definition, 4
  destruction by heat, 289–293

ecology of grape and wine
  succession of, 57–62
eukaryotes, 4
killer phenomenon (*see* killer
  phenomenon)
living organism, 4
nucleus, 18–19
plasma membrane
  chemical composition
    and organization,
    9–13
  functions of the, 13–15

reproduction and
  biological cycle
  sexual reproduction, 20–23
  vegetative
    reproduction, 19–20
secondary products, 4
wine yeast strains,
  identification of
  general principles, 49
  genetic fingerprinting, 52
  genome DNA sequences, 52
  genome sequencing, 56

karyotype analysis, 50–52
mitochondrial DNA
  analysis, 49–50
PCR associated with δ
  sequences, 53–54
PCR with microsatellites,
  54–56
yeast cell, 4, 5

zymase, 74
zymosterol, 10, 13, 118

Printed in the USA/Agawam, MA
September 8, 2022